In Case of a Transportation Emergency Involving a Compressed Gas

In the United States, ask for advice through CHEMTREC, the Chemical Transportation Emergency Center at the Chemical Manufacturers Association in Washington, DC

48 contiguous states, Puerto Rico, Virgin Islands, Alaska, Hawaii, and if transporting Canadian products in the United States (toll free)	(800) 424-9300
District of Columbia and foreign locations (exclusive of Canada)	(202) 483-7616
For non-emergency information only, call The Chemical Referral Center	(800) 262-8200
([illegible] f[illegible]	(202) 887-1315

In Canada, ask for advice through CANUTEC, Transport of Dangerous Goods Branch, Transport Canada, Ottawa, Ontario.

In an emergency, from all points within Canada, call collect 24 hours a day	(613) 996-6666
For non-emergency information only, call	(613) 992-4624

HANDBOOK OF

Compressed Gases

Notice

HANDBOOK OF

Compressed Gases

Third Edition

COMPRESSED GAS ASSOCIATION
Arlington, Virginia

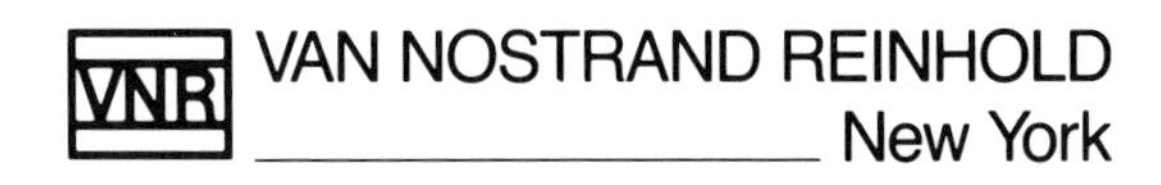

Library of Congress Catalog Card Number 89-36187
ISBN 0-442-21881-8

Printed in the United States of America

Van Nostrand Reinhold
115 Fifth Avenue
New York, New York 10003

Van Nostrand Reinhold International Company Limited
11 New Fetter Lane
London EC4P 4EE, England

Van Nostrand Reinhold
480 La Trobe Street
Melbourne, Victoria 3000, Australia

Nelson Canada
1120 Birchmount Road
Scarborough, Ontario M1K 5G4, Canada

16 15 14 13 12 11 10 9 8 7 6 5 4 3 2

Library of Congress Cataloging-in-Publication Data

Handbook of compressed gases / Compressed Gas Association.—3rd ed.
p. cm.
ISBN 0-442-21881-8
1. Gases, Compressed. I. Compressed Gas Association.
TP761.C65H36 1989
665.7—dc20

89-36187
CIP

Foreword

In the field of compressed gases and related equipment, there is an expanding core of essential knowledge that people handling and using these materials should be familiar with or should know where to find when necessary.

The focus of this book concerns the properties and the accepted means of transportation, storage, and handling of compressed gases. This *Handbook* is simultaneously intended as an overview of the subject and a source of supplementary information. It is also intended to serve as a guide to pertinent federal regulatory requirements and published standards of the Compressed Gas Association and other standards-writing bodies.

Readers are advised that the CGA technical pamphlets remain the official statement of policy by the Association on a particular matter. Reference is made throughout this text to the numerous technical pamphlets published by the Compressed Gas Association. Some of these publications have been incorporated by reference into federal, state, provincial, and local regulations. Since these pamphlets are reviewed on a periodic basis, wherever the text of this *Handbook* may be found in conflict with corresponding information in the CGA technical pamphlets, the latter shall take precedence.

Preface

Most people in our society never stop to consider just how widespread the use of industrial and medical gases has become. Recall the familiar oxy-acetylene torch used to replace a muffler or weld the steel frame of a building, the portable oxygen pack giving mobility to those with respiratory problems and similar equipment for breathing used by firefighters and underwater divers, the carbon dioxide cylinder under every restaurant counter putting the bubbles into our soft drinks, and the less obvious refrigerant gases that ensure the preservation of food in our refrigerators. These are but a few of the more common and innumerable ways in which society has employed gaseous substances for the betterment of mankind. More uses will be mentioned in Chapter 1 of this book.

This third edition of the *Handbook of Compressed Gases* comes at a time when the uses of compressed gases continue to grow as fast as new technologies emerge. For example, gases known only as scientific curiosities a few decades ago are now routinely used in semiconductor and other high-technology manufacturing. An increasing awareness of possible applications and the necessary understanding required to implement these applications continues to emerge. New concerns result in changing regulations covering the transportation, handling, and use of these materials. This third edition of the *Handbook* is an attempt to keep pace with these changes.

In 1988, the Compressed Gas Association observed its 75th Anniversary. Throughout the years since its inception, those who have led the Association have reaffirmed and steadfastly adhered to its foremost objective of enhancing the safety of the industry. In general, this goal is pursued through activities aimed at fulfilling the Association's purposes as stated in its Constitution:

> To develop, promote and coordinate technical, educational and standards activities in the compressed gas industry; to promote public and occupational safety; to safeguard public and private property and the environment; and to cooperate with other trade and technical organizations in safety and technical matters, standards activities and public programs relating to the compressed gas industry.

The publication of this edition of the *Handbook of Compressed Gases* is once again derivative of this intent.

The pervasiveness of compressed gas utilization means that thousands of people come into contact with such gases during the course of their daily lives. Many of these people may have no particular technical or chemical expertise and yet they are dealing with potentially very hazardous materials: hazardous by virtue of pressure, flammability, oxygen deficiency, corrosivity, toxicity, and other physical and chemical properties. The compressed gas industry is extremely proud of the fact that, despite this wide range of hazards, the safety record of its products is among the best in general industry.

This edition of the *Handbook* represents a thorough revision and comprehensive expansion of the text. It aims to provide both trained engineers or scientists, and those without detailed technical training or experience, a basic understanding of the properties, safety considerations, and regulatory framework concerning compressed gases and compressed gas equipment. For the reader who requires a more in-depth knowledge on specific topics, this *Handbook* will provide a useful guide to the standards and recommended practices published by the Compressed Gas Association and other organizations.

The Association is indebted to Handbook Committee Chairman Frank J. Heller (Bartlesville, Oklahoma), whose perseverance and leadership have been marked by distinction, not only during the preparation of this edition, but through several decades of contribution to the compressed gas industry. Also Handbook Committee Vice Chairmen Allen L. Mossman and A. J. Nathanielsz are to be commended for the substantial effort they expended to review and suggest improvements to the many pages of draft manuscript. Many valuable contributions were made by numerous other technical experts within the membership of the Association. Among these, Robert J. Engstrom, William S. Kalaskie, David N. Simon, Felix Smist, Edward Van Schoick, William West, and Lionel Wolpert deserve special thanks.

Harrison T. Pannella, CGA Staff Manager, should be acknowledged for his coordination of the Handbook Committee, and for combining his technical and editorial skills to produce this edition.

The Compressed Gas Association gratefully acknowledges E. I. du Pont de Nemours & Co. for its help in preparing vapor pressure curves for many of the gases covered in this book. Likewise, the Association is grateful to AGA Gas Inc. for permission to reprint the text in Chapter 3.

The contributors to this book have endeavored to provide the most accurate information available. These efforts notwithstanding, we recognize that such works are generally not without error, whether due to the difficulties of compiling complex technical data into a finished publication, or merely from the passage of time and changes in the body of technical information itself. We therefore welcome being advised of any errors detected by readers and appreciate the citation of relevant reference sources where applicable.

Carl T. Johnson, President
Compressed Gas Association, Inc.

Introduction

The *Handbook of Compressed Gases* is intended to serve the needs of a wide audience. Among others, this group will include laymen, teachers and students, workers and supervisors in many industrial sectors, scientists and engineers, and managerial personnel. The broad range of people who may have occasion to become familiar with the subject of compressed gases and compressed gas equipment is indicative of the pervasive role these materials play throughout our society. Much diversity in the amount of formal technical training will also be found within these groups. There will also be differences in the interests of readers, especially between those within the industry, who often have acquired considerable familiarity with the subject over the years, and on the other hand, readers who may be approaching the subject for the first time.

In an attempt to satisfy the needs of all such readers, this *Handbook* has been divided into four parts. The contents of each part will be briefly described in the Part Introductions. By structuring the book in this way, the contributors hope to enable readers to better focus on a particular area and level of interest. References to regulations, standards, and other citations are indicated by bracketed numbers within the text. These references are listed in accordance with their order of appearance at the end of each chapter or gas monograph.

Contents

PART I

Compressed Gases and Related Equipment

Part I presents basic information concerning compressed gases and cryogenic liquids. Introductory presentations give an overview of the uses to which these materials are put, the regulatory framework involved in their shipping and handling, the scientific basis for understanding the behavior of gases, the types of containers used to contain these materials, and basic guidelines for their safe handling. Chapter titles indicate the areas covered by the chapters.

The Compressed Gas Association, Inc. (CGA) issues general industry standards and recommendations for compressed gases along with a number of publications which are concerned with highly specialized aspects of the industry. Reference to these standards and recommendations, as well as those of other organizations, is made throughout the chapters of Parts I and II and in the individual gas monographs in Part III. A full descriptive listing of CGA publications and audiovisuals appears as an appendix in Part IV.

A brief explanation of the International System of Units (SI) and some of the conversion factors used in this book can be found in the Units of Measure section at the end of Chapter 1. A Glossary of Terms and a List of Abbreviations are given as appendices in Part IV.

CHAPTER 1

Compressed and Liquefied Gases Today

INTRODUCTION

This chapter introduces the substances designated as compressed and liquefied gases in descriptive terms, and mentions some uses and other general considerations which may provide useful background information, especially to anyone unfamiliar with the field. Included are descriptions of the various groupings and families into which gases may be categorized, as well as information concerning standard units of measurement.

As mentioned in the introductory material, the aim of this *Handbook* is to provide accurate and authoritative information about the compressed and liquefied gases having current commercial importance, and to provide such information in a manner that is easily accessible to both general readers and professionals alike. For this reason, published standards for procedures, equipment design, and so forth, have not been reproduced in this book, although much of the essential content of certain standards is presented in summary form. Clear references are provided to applicable standards and regulations so that readers who need more detailed information will know where it can be found.

Modern Uses of Compressed and Liquefied Gases

The myriad uses of compressed and liquefied gases have become so pervasive in modern society that we often take them for granted. Yet without them, civilized life as we know it would simply not be possible.

Compressed and liquefied gases have enabled humans to venture deep below the seas and to speed far into space. Gases provide the tremendous power to launch spacecraft, as well as the micro-thrusts needed to control and maneuver space vehicles (see Fig. 1-1).

High-technology uses of compressed and liquefied gases have been instrumental in producing many new benefits, including computers, whose integrated semiconductor circuits can be produced only in high-purity inert gas atmospheres and with special gases which are used in the manufacture of components of such small dimensions.

Medical techniques which involve special gas mixtures such as those required in hyperbaric chambers and other oxygen therapies have achieved remarkable results. Magnetic resonance imaging (MRI), which is used for diagnostic purposes, and special low-temperature surgical procedures, known as cryosurgery, also require the use of compressed gases (see Fig. 1-2).

The fields of refrigeration, air conditioning, metalworking, medicine, food processing, plastics production, vehicular travel, and aerosol packaging are but a few of the other technologies that depend to some extent on compressed and liquefied gases. In short, the use of compressed and liquefied gases is an essential part of modern life.

A comprehensive description of the uses to which compressed and liquefied gases are put could in itself fill the pages of a book.

Fig. 1-1. The NASA Space Shuttle Program represents one of the more dramatic applications for compressed gases. (*Photo courtesy of NASA*)

Fig. 1-2. Compressed gases in medical service include cryogenic liquids for operating superconducting magnets during magnetic resonance imaging (MRI).

Some further discussion of the uses of specific gases is given in the remainder of this chapter and in the gas monographs of Part III. The physical world inhabited by humankind largely consists of three states of matter: solid, liquid, and gas. Suffice it to say that the history of gas utilization is the history of how humans have used one of these states for the improvement of life.

What Are Gases?

We designate as gases any substances that boil at atmospheric pressure and any temperature between absolute zero, −459.67°F (−273.15°C), and temperatures up to about 80°F (26.7°C). Eleven of the 92 chemical elements, not including transuranium elements or radioisotopes, have boiling points within this range. These elements are: hydrogen, nitrogen, oxygen, fluorine, chlorine, helium, neon, argon, krypton, xenon, and radon. In addition, there are an apparently unlimited number of chemical compounds that fit this definition of gases, as well as numerous mixtures, of which air is the most common.

What Is a "Compressed Gas"?

Currently, there are nearly 200 different substances commonly shipped in compressed gas containers that can be considered compressed gases. Compressed gases have traditionally been defined as:

> any material or mixture having in the container an absolute pressure exceeding 40 psi (pounds per square inch) at 70°F (275.8 kPa at 21.1°C) or, regardless of pressure at 70°F (21.1°C), having an absolute pressure exceeding 104 psi at 130°F (717 kPa at 54.4°C), or any liquid material having a vapor pressure exceeding 40 psi absolute at 100°F (275.8 kPa at 37.8°C) as determined by ASTM (American Society for Testing and Materials) Test D-323.

Note: Absolute pressure in a container is the reading of a pressure gauge plus the local atmospheric pressure which is affected by the local atmosphere and altitude. The generally accepted standard for atmospheric pressure is 14.696 psi (101.325 kPa) at sea level. For approximate purposes, the pressure addition to gauge pressure may be rounded to 15 psi.

More specific definitions based upon hazard class may be found in current regulations of the U.S. Department of Transportation (DOT) and Transport Canada.

Nonliquefied and Liquefied Gases

Compressed gases may be divided into two major groups, depending on their physical state in containers under certain pressures and temperatures and their range of boiling points. These groups are denoted as nonliquefied gases and liquefied gases.

Nonliquefied gases are those which do not liquefy at ordinary terrestrial temperatures and under pressures which range up to 2000–2500 psig (13 789–17 237 kPa). Nonliquefied gases are elements or compounds that have relatively low boiling points, ranging from approximately −150°F (−101°C) on down (see Table 1-1). Of course, these gases do become liquids if cooled to temperatures below their boiling points. When these gases become liquefied at these very low temperatures, they are generally referred to as cryogenic liquids.

TABLE 1-1. BOILING POINTS OF SOME TYPICAL NONLIQUEFIED GASES.[a]

Oxygen	Helium	Nitrogen
−297°F	−452°F	−320°F
182.8°C	−268.9°C	−195.5°C

[a]At one atmosphere.

TABLE 1-2. BOILING POINT AND PRESSURES OF SOME TYPICAL LIQUEFIED GASES AT 70°F.

Ammonia	Chlorine	Propane	Carbon Dioxide
−28°F	−29°F	−44°F	−109°F (sub-limes to gas)
−33.3°C	−33.9°C	−42.2°C	−78.3°C

- The pressure of ammonia at 70°F (21.1°C) is 114.1 psig (786.7 kPa, gauge)
- The pressure of chlorine at 70°F (21.1°C) is 85.5 psig (589.5 kPa, gauge)
- The pressure of propane at 70°F (21.1°C) is 109.7 psig (756.4 kPa, gauge)
- The pressure of carbon dioxide at 70°F (21.1°C) is 838 psig (5777.8 kPa, gauge)

Oxygen, helium, and nitrogen are examples of nonliquefied gases in wide use as compressed gases and cryogenic liquids. Note that the lower limit for cryogenic temperatures is fixed by absolute zero −459.67°F (−273.15°C). The upper limit is somewhat arbitrary and is usually taken to be about −130°F (−90°C).

Container charging pressures for these gases are more than 2000 psig (13 789 kPa) at 70°F (21.1°C). When cooled to cryogenic temperatures near their boiling points, charging pressures can be very low. However, some storage vessels will have design pressures up to about 250 psig (1724 kPa) to allow for a rise in pressure with ambient heating.

Liquefied gases are those which become liquids to a very large extent in containers at ordinary temperatures and at pressures from 25 to 2500 psig (172.4–17 237 kPa) (see Table 1-2). Liquefied gases are elements or compounds that have boiling points relatively near atmospheric temperatures. These range from about −130°F to 25°F or 30°F (−90°C to −3.9 or 1.1°C). Such liquefied gases solidify at cryogenic temperatures. Of these, only carbon dioxide has come into commercial use in solid form as "dry ice."

Liquefied gases are shipped under rules that limit the maximum amount that can be put into a container to allow space for liquid expansion with rising ambient temperatures.

Still another, although minor, grouping is that of dissolved gas. This category actually involves only one widely used gas, acetylene. The acetylene gas is dissolved in acetone which is absorbed into a porous cellular material that fills the inside of an acetylene cylinder. By storing the acetylene dissolved in acetone in the small cells of the porous material, it is possible to use shipping pressures of 250 psig (1724 kPa) at 70°F (21.1°C). Acetylene contained free in ordinary containers or piping is unstable at pressures above 15 psig (103 kPa) and can decompose with explosive violence. The presence of the porous filler material in acetylene cylinders renders them unique in design and unsuitable for use with other gases.

Besides its wide use as a fuel gas, acetylene is used in a number of chemical processes. Some consumers of acetylene in bulk and most industrial suppliers of acetylene produce it directly by the reaction of calcium carbide and water.

MAJOR FAMILIES OF COMPRESSED GASES

Compressed and liquefied gases are often described according to loosely knit families. Such designations can be somewhat arbitrary and are usually based on either a common source, a similar use, or a related chemical structure.

Part III of this book is comprised of monographs on individual gases and indicates predominant chemical characteristics

Fig. 1-3. A modern air separation plant liquefies gases from the atmosphere and extracts the constituents through fractional distillation.

for each gas. However, it is important to remember that any particular gas may exhibit several such characteristics to varying extents. The following paragraphs describe some typical families, or groupings, by which compressed gases may be categorized.

Atmospheric Gases

Atmospheric gases comprise one of the leading families. The most plentiful member of this family is nitrogen, constituting 78 percent of the atmospheric air by volume. The second most abundant member is oxygen, which constitutes 21 percent of the atmosphere. Nitrogen is a basic constituent of the amino acids which serve as the building blocks of proteins in all living organisms. Oxygen is essential to animal life by virtue of its role in metabolic respiration.

But these two gases have a great many other uses as well. Oxygen is used to produce stronger steels less expensively. It is also used together with acetylene and other fuel gases such as propane to weld or cut steel in fabricating structures and machinery. It is used in medical treatment and respiratory therapy, to name but a few of the most prominent applications. Liquid nitrogen is used as a refrigerant for the transportation of perishable goods and in the quick-freezing of fruits and vegetables.

Most of the remaining one percent of the atmosphere consists of gases sharing the property of chemical inertness. These inert gases are chiefly argon, with minute amounts of helium, neon, krypton, xenon, and radon. The last four are frequently called the "rare gases" due to their scarcity. Argon, helium, krypton, and neon are used to obtain the glowing colored messages we see in the so-called "neon signs." Inert gas atmospheres provided by argon or helium are also essential to the welding of stainless steels and special metals like titanium and zirconium.

Hydrogen also occurs minutely in the atmosphere, as do a large variety of trace constituents, small amounts of carbon dioxide, and large amounts of water vapor.

Carbon dioxide fire extinguishers are commonly used for protection against chemical and electrical fires, while carbon dioxide dissolved under pressure makes the sparkle in "carbonated" beverages. And solid carbon dioxide, called dry ice, has largely replaced water ice in the handling of some perishable foods where mechanical refrigeration is not available.

Nitrogen, oxygen, argon, and the rare gases are commercially produced by cooling the air down to liquid form and then distilling off "fractions" having different boiling points, much the same as petroleum fractions are distilled at higher temperatures to produce fuel, gasoline, and various petrochemicals. This process is called fractionation (see Fig. 1-3).

Not all gases come from the atmosphere. Some high-purity helium is obtained through air liquefaction and fractionation, but most helium obtained today comes from wells of natural gas in which it occurs in concentrations of a few percent.

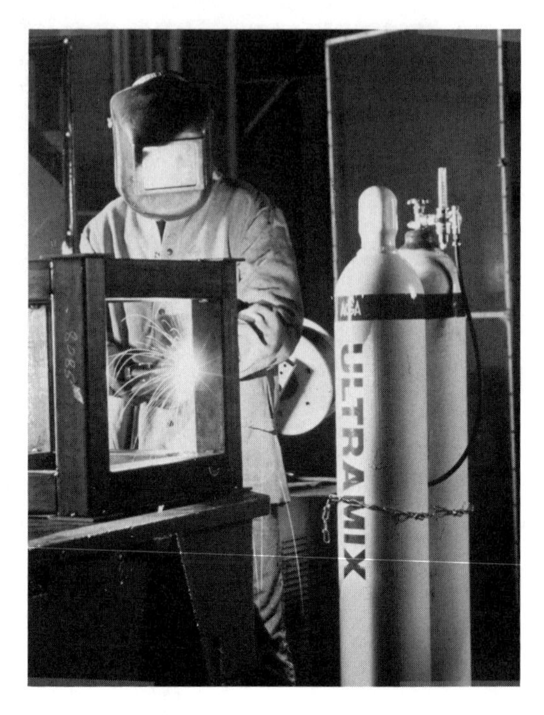

Fig. 1-4. Modern welding techniques employ compressed gases in numerous ways.

Fuel Gases

Fuel gases burned using air or oxygen to produce heat make up a large family of gases related through a major use. Its members are notably the hydrocarbon gases and liquefied petroleum gases, or LP-gases, such as propane and butane. Methane, the largest component of natural gas, is another leading representative of the family, and welding gases such as acetylene and hydrogen are somewhat special representatives (see Fig. 1-4). Liquefied petroleum gases are widely used as portable fuel where pipeline natural gas is not available, and as propellants for aerosol products. Some newer gases in the fuel gas family are the inhibited methylacetylene-propadiene mixtures, commonly referred to as MAPP gas, which resemble LP-gas in a number of ways.

Fuel gases such as hydrogen are required for welding certain metals. Liquefied methane, also referred to as LNG, or liquefied natural gas, finds use in "peak shaving." That is, electric power plants liquefy the gas from pipelines over a long period and use the gas when electric usage peaks. LNG is also finding increasing use in heating and vehicle propulsion.

Refrigerant Gases

Another extensive family is that of refrigerant gases. A refrigerant gas is one that liquefies easily under pressure, for it works by being mechanically compressed into a liquid. In this state, it absorbs large amounts of heat as it circulates in cooling coils, eventually vaporizing back into a gas.

Among the refrigerant gases, the halogenated hydrocarbons, or halocarbons, are the most popular. This family is very large, since there are many combinations by which volatile hydrocarbons can be halogenated. Halogenated hydrocarbons serve well as refrigerant gases because most of them are chemically inert to a large extent, and they can be selected, mixed, and compounded to provide physical properties desired in particular refrigerant applications. Due to the recently recognized deleterious effects of certain halogenated hydrocarbons on the ozone layer of the earth's stratosphere, a search is under way for substitute materials.

Dry or anhydrous ammonia liquefies under low pressure and was the earliest widely used refrigerant gas. While it is still used to some extent for this purpose, usually in larger refrigeration operations, most anhydrous ammonia is consumed as an agricultural fertilizer.

Any liquefied gas is a strong candidate for membership in the refrigerant family, and even some gases more commonly used in other ways, such as methane, have been classified as refrigerant gases by the American Society of Heating, Refrigerating, and Air Conditioning Engineers. While not refrigerant gases in the mechanical sense, due to their extremely low temperatures, cryogenic liquids, such as liquid nitrogen, also find use in food preservation and processing (see Fig. 1-5).

More About "Families"

Classification of gases by "families" can sometimes be misleading, because individual gases may often be used in a number of very different ways. Nitrous oxide, for example, belongs to several families, for it is a prominent member of the "medical gases" family, a widely used propellant gas, and also a reliable refrigerant gas.

Similarly, oxygen enjoys a large and growing application in medicine, where it is used alone as well as mixed with carbon dioxide, nitrogen, or helium for many kinds of respi-

Fig. 1-5. Compressed gases find wide utilization in the processing, preservation, and preparation of foods. Cryofreeze process utilizes liquid nitrogen to quick freeze hamburger patties.

ration therapy. At the same time, it is also used in welding and steel making.

Poison Gases

Gases considered in North America to be members of the poison gas family are generally those that the U.S. Department of Transportation and Transport Canada have classified as poison gases to ensure public safety in transportation. Some examples of these gases that have commercial importance are phosgene, which serves as an intermediate in the production of barbiturate drugs, and arsine, which is used as a doping agent in the production of semiconductor integrated circuits. Arsine is also the raw material used in the production of gallium arsenide, an advanced semiconductor material. See Fig. 1-6.

Both of these gases are listed as Class A poisons and "extremely dangerous" by the U.S. DOT. In Canada, they are identified for special treatment within the poison gas classification. Specific information on the safe handling of gases of this type is provided in the individual gas monographs.

Gases with No Family Ties

Many other gases that are important commercially have no marked family ties. For example, the methylamines serve as chemical intermediate sources of reactive organic nitrogen, and methyl mercaptan helps in the synthesis of insecticides. The mercaptans are also used in odorizing natural gas to allow easy leak detection.

Sulfur dioxide is a liquefied gas employed widely as a preservative and bleach in processing foods, as well as a bleach in manufacturing sulfite papers and artificial silk. It is also used as an additive to irrigating water to improve alkaline soils in the American Southwest.

Carbon monoxide is used in making other chemicals, including the urethane plastics. It is also used in some metallurgical processes, including one for refining high-purity nickel, and in another for removing some gases from molten aluminum in conjunction with chlorine and nitrogen.

Fluorine is used to make sulfur hexafluoride, a gas with high dielectric strength used in electrical equipment. As previously noted, anhydrous ammonia is used to fertilize crop-

Fig. 1-6. The use of compressed gases in the manufacture of semiconductors for the electronics industry has become a major application. Here, a portable "clean room" and advanced pipe welding techniques are used to ensure highest possible gas purity.

lands, resulting in higher yields at lower cost. Nitrous oxide and other anesthetic gases are needed in modern surgery. Acetylene, butadiene, carbon monoxide, chlorine, the methylamines, and vinyl chloride make possible the production of plastics, synthetic rubber, and modern drugs. Nitrous oxide and carbon dioxide are used as propellants for edible products such as whipped dessert toppings. Other products such as perfumes, shampoos, shaving lathers, paints, and insecticides are pressurized with gases such as butane, a liquefied petroleum or LP-gas. Chlorine is used in swimming pools and in the treatment of drinking water.

Specialty gases are another important and growing segment of the compressed gas industry. The semiconductor-manufacturing industry uses special gases such as silane and other gases in very pure form to manufacture the microchips that are the "brains" of computers of all sizes.

From the sampling of the uses of compressed and liquefied gases in modern society given in this chapter, the thousands of uses to which compressed and liquefied gases may be put in modern society become readily apparent. This in turn means that people from many different backgrounds will have occasion to use these materials in one way or another. In publishing this book, it is the aim of the Compressed Gas Association and the compressed gas industry to make a significant contribution that this be done knowledgeably and safely.

SAFETY PROGRAMS

Due to the obvious potential hazards resulting from the inherent stored energy, chemical reactivity, and possible biological effects of compressed and liquefied gases, the necessity for sound safety practices in their storage, transportation, handling, and use has been a primary concern of the industry.

In fact, it was as a result of this recognized need that the Compressed Gas Association was founded in 1913 by a handful of companies engaged in the production of compressed gases and related products. Today, approximately 250 member companies of the Compressed Gas Association continue to support the development of improved and up-to-date industry standards pertaining to safe practices and equipment design and maintenance. This work often is done in cooperation with other trade associations and professional societies as well as with the participation of regulatory authorities.

Emergency Response

In more recent years, work by CHEMTREC, the emergency response agency of the Chemical Manufacturers Association (CMA), in assisting local authorities in handling transportation emergency problems with chemicals, including compressed and liquefied gases, has had a significant impact in lessening the potential damage that might result from such incidents. The Compressed Gas Association has cooperated in this work and has endeavored to expand its effectiveness by instituting its own Compressed Gas Emergency Action Plan (COMPGEAP) to work in conjunction with CHEMTREC. In Canada, CGA Canada has worked with CANUTEC, the emergency response information agency of Transport Canada, to assure that information and emergency response contacts are readily available. More discussion of emergency response plans will be found in Chapter 5.

A Glance at Safety Hazards

Practically all gases can act as simple asphyxiants by displacing the natural oxygen in the air. The chief precaution against this potential hazard is adequate ventilation of all enclosed areas in which unsafe concentrations may build up. It is imperative to avoid entering unventilated areas that may contain high concentrations of gas without first using a self-contained or hose-line air supply.

Some gases can also have a toxic effect on the human system, either through being inhaled or through having high vapor concentrations or liquefied gas come in contact with the skin or eyes. Precautions against liquefied gases and cryogenic liquids that are toxic or very cold, or both, include thorough knowledge and training for all personnel handling such gases, and the development of foolproof procedures and equipment for handling them in both normal and emergency situations.

With flammable gases, it is necessary to guard against the possibility of fire or explosion. Ventilation again represents a prime precaution against these hazards, together with safe procedures and equipment to detect leaks. Potential sources of ignition are to be avoided. Should a fire break out, suitable fire-extinguishing apparatus and preparations can help limit damage.

Oxygen poses a combustion hazard of a special kind. Although it does not itself ignite, it lowers the ignition point of flammable substances and greatly accelerates combustion. It is especially worthwhile to note that grease and oil are commonplace materials which must not come in contact with oxygen and oxygen equipment.

Hazards pertaining to the possible rupture of a cylinder or other vessel containing gas at high pressure can be avoided by careful handling at all times. This includes securing cylinders to keep them from falling over, avoiding situations whereby a cylinder valve may be broken off, and never using compressed gas cylinders as rollers or for any other purpose other than to contain gas.

Chapter 5 will provide a more thorough discussion on safety guidelines, as will other sections of this book. The Compressed Gas Association publishes numerous industry

TABLE 1-3. CONVERSION FACTORS PERTINENT TO COMPRESSED GASES. U.S. Units to SI Units

Category	U.S. Unit	Multiplied by	SI Unit
Pressure[a]	lb/in^2 (psi)	6.894757	kPa
Pressure	kg/cm^2	97.06650	kPa
Pressure	atm	101.325	kPa
Temperature	°F	(°F − 32)/1.8	°C[b]
Density	lb/cu ft	16.01846	kg/m^3
Volume[c]	cu ft	0.02831685	m^3
Specific volume	cu ft/lb	0.06242796	m^3/kg
Heat	Btu/lb	2.326	kJ/kg
Heat	Btu/cu ft	37.25895	kJ/m^3
Heat	Btu/gal	278.7163	kJ/m^3
Specific heat	Btu/(lb)(°F)	4.1868	kJ/(kg)(°C)
Mass	lb	0.4535924	kg
Length	inch	0.0254	m

[a]In this book, unless otherwise stated, gauge pressure is indicated by the abbreviations *psig* and *kPa,* whereas absolute pressure is indicated by the abbreviations *psia* or *kPa abs.*

[b]The recommended SI unit of temperature is the degree Kelvin (K). However, degree Celsius (°C) values are acceptable for commonly used temperature measurements. A one-degree difference on the Celsius scale is the same as a degree difference on the Kelvin scale; 0 K equals −273.15°C.

[c]In the United States, a standard cubic foot for industrial gas use is defined at 70°F (21.1°C) and 14.696 psia (101.325 kPa abs). In Canada, a standard cubic meter for industrial gas use is defined at 15°C (59°F) and 101.325 kPa abs (14.696 psia). To convert cu ft at 70°F to m^3 at 15°C, multiply by 0.02772878.

standards and recommendations which are referred to throughout this text and are listed with a brief description in Part IV.

UNITS OF MEASURE

The specific types of metric units used in this book to designate metric equivalents of quantitative values given in U.S. units are SI units (International System of Units; see Table 1-3). The SI units have been adopted as an American National Standard by the American National Standards Institute (ANSI) in a booklet designated ANSI-IEEE Std. 268-1982, *Metric Practice.* [1]

The SI system represents the modern version of the metric system and is an attempt to standardize systems of measurement used by different countries throughout the world. The use of SI units for metric applications has also been approved by the National Institute of Standards and Technology (NIST), the American Society for Testing and Materials (ASTM), the American Society of Mechanical Engineers (ASME), the Canadian Standards Association (CSA), the International Organization for Standardization (ISO), and other organizations.

Among SI units used for different kinds of properties important in the compressed gas field are:

Pressure	kilopascal	kPa
Temperature	degrees Celsius	°C
Density	kilograms per cubic meter	kg/m^3
Volume	cubic meter	m^3
Specific volume	cubic meter per kilogram	m^3/kg

REFERENCE

[1] ANSI-IEEE Std. 268-1982, *Metric Practice,* American National Standards Institute, 1430 Broadway, New York, NY 10018.

CHAPTER 2

Regulations Pertaining to Compressed Gases

INTRODUCTION

Those who produce, supply, or use compressed gases, as well as carriers involved in the transportation of compressed gases, must comply with a variety of government safety regulations in the United States, Canada, and other countries throughout the world. These regulations are promulgated and enforced by regulatory agencies at the federal, state, provincial, and local levels of government in many countries. International shipments of compressed gases by water and air are regulated by the International Maritime Organization (IMO) and the International Civil Aviation Organization (ICAO), respectively.

Compressed gases may be flammable, nonflammable, corrosive, poisonous, oxidizing, or inert. But whichever of these categories may best describe their respective chemical properties, all compressed gases present potential hazards unless packaged, transported, and used under safe conditions and in accordance with effective regulations. The ultimate purpose of meaningful government regulations in this area is to protect the public from the potential problems of hazardous materials which may be encountered during the manufacture, transportation, and use of these materials.

This chapter primarily will discuss regulations in the United States and Canada, recognizing that other countries may have similar regulations. Ultimately, it is the responsibility of the manufacturer, transporter, and user of compressed gases to become knowledgeable in the applicable government regulations affecting his or her operations. This chapter cannot deal in detail with all applicable regulations, but can only give the reader guidance in searching out the regulations relating to his or her particular interest. Additional guidance to regulatory requirements on specific topics will be found throughout the chapters of this *Handbook*.

Industry Standards

Technical associations develop safety standards for the guidance of members and the general public. The latter often include organizations or individuals who make use of the products of a particular industry. While these industry safety standards are not regulations and do not have the effect of law, in many cases such standards have been adopted by federal, state, provincial, and local authorities. In such cases, industry safety standards do become regulations. Nevertheless, where an industry safety standard exists, it behooves the manufacturer or user of a product or process to comply with the standard.

The Compressed Gas Association is among the technical associations publishing standards for the safe transportation, handling, and use of compressed gases. Another is the National Fire Protection Association, located at Batterymarch Park, Quincy, Massachusetts. Brief descriptions of the publications of the Compressed Gas Association appear as an appendix in Part IV of this *Handbook*.

Transportation of Compressed Gases

Compressed gases are packaged and transported in cylinders, portable tanks, over-the-road vehicles known as cargo tanks or tank trucks, and railroad tank cars, as well as other approved specialized containers. Containers of compressed gases are shipped by highway, rail, water, and air. Certain constraints on the allowable modes of transportation will sometimes be encountered depending upon the specific gas and the hazard classification of the gas.

Government transportation regulations require that hazardous materials which meet the definition of a compressed gas be shipped in containers which comply with certain design specifications and performance requirements, and that, in most cases, the containers be equipped with prescribed pressure relief devices. Transportation regulations are divided into several broad and distinct areas: (1) design specifications for each container type; (2) qualifications, maintenance, material compatibility, and filling requirements for containers; and (3) marking, labeling, documentation, and placarding requirements. They may also, as in Canada, cover certain areas of required training and certification as well as emergency response and incident reporting. A summary overview of the basic considerations pertaining to the third area is given at the end of the next section. The reader will find additional information on areas (1), (2), and (3) in subsequent chapters, particularly Chapters 4 and 9.

TRANSPORTATION REGULATIONS

Federal and International Regulatory Authorities—DOT in the United States; Transport Canada in Canada

The primary agencies regulating compressed gas shipments in North America are the U.S. Department of Transportation (DOT) and the Transportation of Dangerous Goods Directorate, Transport Canada (TC), of the Canadian government. The DOT transportation regulations prior to 1967 were administered by the Interstate Commerce Commission (ICC). Prior to 1980, Transport Canada confined its jurisdiction to the regulation of dangerous goods by water and air, with the Canadian Transport Commission bearing the responsibility for movement by rail. Movement by road was essentially unregulated except by some provinces in some areas. Transport Canada now administers regulatory jurisdiction over all modes of transport for shipments of compressed gases.

U.S. Department of Transportation

In the United States, the Department of Transportation promulgates regulations and issues exemptions for the domestic transportation of hazardous materials, including compressed gases, by highway, rail, water, and air. These regulations are contained in Title 49 of the *Code of Federal Regulations* (49 CFR), Parts 100 to 199. [1] With respect to compressed gases, the most frequently used parts of the DOT Hazardous Materials Regulations are as follows:

- 49 CFR Part 171, General Information, Regulations, and Definitions
- 49 CFR Part 172, Hazardous Materials Tables and Hazardous Materials Communications Regulations
- 49 CFR Part 173, Shippers: General Requirements for Shipments and Packagings
- 49 CFR Part 174, Carriage by Rail
- 49 CFR Part 175, Carriage by Aircraft
- 49 CFR Part 176, Carriage by Vessel
- 49 CFR Part 177, Carriage by Public Highway
- 49 CFR Part 178, Shipping Container Specifications
- 49 CFR Part 179, Specifications for Tank Cars

In addition to the above regulations governing the transportation of compressed gases by highway, rail, water, and air, the DOT also regulates the driver and the motor vehicle carrying compressed gases by highway under the Motor Carrier Safety Regulations prescribed by the DOT Federal Highway Administration. The following parts are found in Title 49 of the *Code of Federal Regulations* [2]:

- 49 CFR Part 390, General (including definitions)
- 49 CFR Part 391, Qualification of Drivers
- 49 CFR Part 392, Driving of Motor Vehicles
- 49 CFR Part 393, Parts and Accessories Necessary for Safe Operation
- 49 CFR Part 394, Recording and Reporting of Accidents
- 49 CFR Part 395, Hours of Service of Drivers
- 49 CFR Part 396, Inspection and Maintenance
- 49 CFR Part 397, Transportation of Hazardous Materials by Motor Vehicle

Copies of the above DOT regulations are available from the Superintendent of Documents, U.S. Government Printing Office, Washington, DC 20402. *Tariff BOE 6000,* the Title 49 CFR DOT regulations as republished by the Bureau of Explosives, may be obtained through the Bureau of Explosives, Association of American Railroads, 50 F Street, N.W., Washington, DC 20001. The Motor Carrier Safety Regulations are included in the *Federal Regulations Manual for Private Truck Operators,* which is available on a subscription basis from the National Private Truck Council, 1320 Braddock Place, Alexandria, VA 22314. (See Fig. 2-1.)

Transport Canada

Early in the 1970s, the Canadian government became interested in adopting appropriate regulations covering the transportation of dangerous goods (hazardous materials) by road. At the same time, an increasing desire on the part of the Canadian government to expand its international trade outside North America resulted in a decision to look at the regulatory standards for the transportation of dangerous materials established by the United Nations.

In 1980, the Canadian government passed the Transportation of Dangerous Goods Act, which permitted the government to publish regulations for the transportation of dangerous goods in all modes. The intent was to consolidate the existing regulations pertaining to transportation by air, rail, and water, as well as to introduce a new set of regulations covering the movement of these materials by highway. Because of the requirements of the Canadian Constitution, it was necessary for this Act to give powers to the Governor General in Council to enter into agreement with the various provincial governments for the implementation of the Act and the regulations with respect to road transportation. Since that time, the various provincial governments have enacted appropriate laws which permit them to adopt regulations for the transportation of dangerous goods by road. In January 1985, the Canadian government promulgated the *Transportation of Dangerous Goods Regulations,* and since that time all Canadian provinces have indicated their intention to adopt these federal regulations directly or adopt similar regulations of their own. [3]

One major difference between the U.S. Department of Transportation and Transport Canada regulations is that the Canadian government has introduced a new classification for certain types of gases under the hazard class of "corrosive gases." In addition, the Canadian regulations call for pictorial placards and labels, and the use of words is virtually eliminated.

The *Transportation of Dangerous Goods Regulations* can be obtained from the Canadian Government Publishing Centre, Supply and Service Canada, Ottawa, Ontario, Canada K1A 0S9. [3]

International Maritime Organization

International shipment of compressed gases by water is governed by the International Maritime Organization (IMO), which

§172.101 Hazardous Materials Table

(2)	(3)	(3A)	(4)	(5) Packaging		(6) Maximum net quantity in one package		(7) Water shipments		
Hazardous materials descriptions and proper shipping names	Hazard class	Identi-fication number	Label(s) required (if not excepted)	(a) Exceptions	(b) Specific require-ments	(a) Passenger carrying aircraft or railcar	(b) Cargo only aircraft	(a) Cargo vessel	(b) Pas-senger vessel	(c) Other requirements
6-Nitro-4-diazotoluene-3-sulfonic acid (dry)	Forbidden									
p-Nitroaniline. *See* Nitroaniline										
N-Nitroaniline	Forbidden									
+ Nitroaniline	Poison B	UN1661	Poison	173.364	173.373	50 pounds	200 pounds	1,2	1,2	
m-Nitrobenzene diazonium perchlorate	Forbidden									
Nitrobenzene, liquid *or* Nitrobenzol, liquid (*oil of mirbane*)	Poison B	UN1662	Poison	173.345	173.346	1 quart	55 gallons	1,2	1,2	
Nitro carbonitrate. See Blasting agent, n.o.s.										
Nitrocellulose, colloided, granular *or* flake, wet with not less than 20% alcohol *or* solvent, *or* block, wet with not less than 25% alcohol	Flammable liquid	NA2059	Flammable liquid	173.118	173.127	1 quart	25 pounds	1,3	1	
Nitrocellulose, colloided, granular *or* flake, wet with not less than 20% water	Flammable solid	NA2555	Flammable solid	173.153	173.184	25 pounds	100 pounds	1,3	1	
Nitrocellulose, dry. See High explosive										
Nitrocellulose, wet with not less than 30% alcohol *or* solvent	Flammable liquid	NA2556	Flammable liquid	173.118	173.127	1 quart	25 pounds	1,3	1	
Nitrocellulose, wet with not less than 20% water	Flammable solid	NA2555	Flammable solid	173.153	173.184	25 pounds	100 pounds	1,3	1	
+ Nitrochlorobenzene, meta *or* para, solid	Poison B	UN1578	Poison	173.364	173.374	50 pounds	200 pounds	1,2	1,2	
+ Nitrochlorobenzene, ortho, liquid	Poison B	UN1578	Poison	173.345	173.346	1 quart	55 gallons	1,2	1,2	
Nitroethane	Flammable liquid	UN2842	Flammable liquid	173.118	173.119	15 gallon	55 gallons	1,2	1	
Nitroethylene polymer	Forbidden									
Nitroethyl nitrate	Forbidden									
Nitrogen *or* Nitrogen, compressed	Nonflammable gas	UN1066	Nonflammable gas	173.306	173.302 173.314	150 pounds	300 pounds	1,2	1,2	
Nitrogen, refrigerated liquid (*cryogenic liquid*)	Nonflammable gas	UN1977	Nonflammable gas	173.320	173.316 173.318	100 pounds	1,100 pounds	1,3	1,3	
Nitrogen dioxide, liquid	Poison A	UN1067	Poison gas and Oxidizer	None	173.336	Forbidden	Forbidden	1	5	Segregation same as for nonflammable gases. Stow away from organic materials
Nitrogen peroxide, liquid	Poison A	NA1067	Poison gas and oxidizer	None	173.336	Forbidden	Forbidden	1	5	Segregation same as for nonflammable gases. Stow away from organic materials
Nitrogen tetroxide, liquid	Poison A	NA1067	Poison gas and oxidizer	None	173.336	Forbidden	Forbidden	1	5	Segregation same as for nonflammable gases. Stow away from organic materials
Nitrogen trichloride	Forbidden									
Nitrogen trifluoride	Nonflammable gas	UN2451	Nonflammable gas	None	173.302	Forbidden	300 pounds	1	5	Stow away from living quarters and organic materials
Nitrogen triiodide	Forbidden									
Nitrogen triiodide monoamine	Forbidden									
Nitroglycerin, liquid, desensitized. See High explosive, liquid										
Nitroglycerin, liquid, not desensitized. See 173.51	Forbidden									
Nitroglycerin, spirits of. See Spirits of nitroglycerin										
Nitroguanidine, dry. See High explosive										
Nitroguanidine nitrate	Forbidden									
Nitroguanidine, wet with not less than 20% water	Flammable solid	UN1336	Flammable solid	173.153	173.184	25 pounds	100 pounds	1,2	4	
1-Nitro hydantoin	Forbidden									
Nitrohydrochloric acid	Corrosive material	UN1798	Corrosive	None	173.278	Forbidden	5 pints	1	5	
Nitrohydrochloric acid, diluted	Corrosive material	UN1798	Corrosive	None	173.278	Forbidden	5 pints	1	5	
Nitro isobutane triol trinitrate	Forbidden									

Fig. 2-1. Typical page from the Hazardous Materials Table, U.S. *Code of Federal Regulations*, 49 CFR 172.101.

follows the recommendations of the United Nations Committee of Experts on the Transport of Dangerous Goods under the United Nations Economic and Social Council. The *International Maritime Dangerous Goods Code* is designed to assist compliance with general requirements of the International Convention for the Safety of Life at Sea (SOLAS) regarding the carriage of hazardous materials by sea. [4] The *International Maritime Dangerous Goods Code* is published in five volumes, with Volume II covering compressed gases.

The IMO Code groups all compressed gases into Class 2, and further divides this class into Division 2.1 for flammable gases, Division 2.2 for nonflammable gases, and Division 2.3 for poisonous gases. While there are many similarities between the DOT and IMO requirements, there are some significant differences in shipping names, hazard classifications, marking, and labeling of compressed gases. The *International Maritime Dangerous Goods Code* is available from the International Maritime Organization, London, United Kingdom. [4]

International Civil Aviation Organization

International shipment of compressed gases by air is governed by the International

Civil Aviation Organization (ICAO), which follows the recommendations of the United Nations Committee of Experts on the Transport of Dangerous Goods. These air transportation regulations are published for ICAO in the *Technical Instructions for the Safe Transport of Dangerous Goods by Air,* available from Intereg Group Inc., 5724 N. Pulaski Road, Chicago, Illinois 60646. [5] ICAO is located at 1000 Sherbrooke Street West, Suite 400, Montreal, Quebec, Canada H3A 2R2.

In addition to the ICAO requirements, domestic and international air carriers have published regulations governing conditions under which they will carry hazardous materials, including compressed gases. Domestic and international regulations, including ICAO's Technical Instructions, also are given in *Dangerous Goods Regulations,* published by the International Air Transport Association (IATA), 2000 Peel Street, Montreal, Quebec, Canada H3A 2R4. [6] Although the ICAO regulations and air carrier tariffs will authorize the air transportation of many compressed gases, individual airlines may embargo the carriage of certain classes of compressed gases.

Basic Considerations Concerning Transportation of Compressed Gases

The following subsections will summarize the basic transportation requirements that are found in the regulations discussed earlier in this chapter from the perspective of the type of compressed gas being packaged. It must be emphasized that there are many differences in the various regulations and all regulations are subject to change. In the United States and Canada, amendments to the regulations are published in the *Federal Register* and *Canada Gazette,* respectively. The international regulatory authorities, IMO and ICAO, amend their regulations by issuing revised pages or revised editions which are published from time to time. It is, therefore, imperative that shippers and carriers consult the latest edition of the regulations applicable to the country, or countries, involved and the specific mode, or modes, to be utilized.

This summary is divided into the following considerations:

- Classification
- Marking and labeling
- Documentation
- Compatibility
- Placarding

Classification

All regulations require the shipper, who is in many cases the manufacturer, to determine if the material to be offered for transportation is a hazardous material (or dangerous good) under the definitions which appear as a part of every regulation. The shipper must examine all properties of the material and determine which hazard class, or classes, is or are applicable. In the case of multiple hazards, the shipper must determine which is the primary hazard and then also determine the subsidiary hazards.

All regulations contain a section, list, or table which contains approximately 3000 materials (including many gases) by technical name. For specifically listed materials the appropriate proper shipping name, hazard class or classes, labels, and packaging requirements are given. Many of the standard compressed gases are listed by technical name, and therefore, proper shipping names, hazard class(es), and labels are readily determinable. However, many compressed gas mixtures or blends classified as compressed gases are not listed, so the shipper must make the decisions on primary and subsidiary hazards (if any) and labeling requirements. For materials which are not listed by name, the shipper must select the most appropriate "N.O.S." (not otherwise specified) shipping name and that shipping name determines the proper packaging, marking, labeling, and placarding.

Under several regulatory systems, compressed gases (nonliquefied and liquefied) are in United Nations (UN) Class 2, which is frequently divided into the following divisions: 2.1, flammable gases; 2.2, nonflammable gases; and 2.3, poisonous gases. The *Transportation of Dangerous Goods Regulations* contain an additional division, 2.4, which is comprised of several corrosive gases listed by name in the regulations. [3] The primary hazard of any compressed gas is always one of these divisions, but a subsidiary hazard may be another division of Class 2 or other hazard classes.

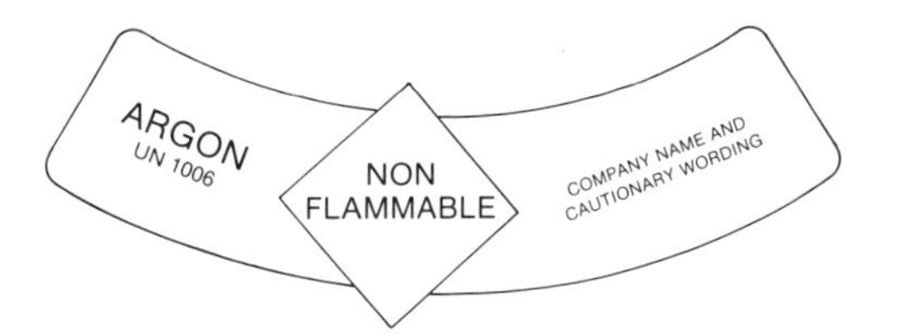

Fig. 2-2. Schematic example of CGA shoulder decal used to identify cylinder contents in accordance with Appendix A of CGA C-7, *Guide to the Preparation of Precautionary Labeling and Marking of Compressed Gas Containers.*

Marking and Labeling

Every compressed gas, when shipped, must be marked and labeled in accordance with the applicable regulations. Hazard labels are designed in accordance with United Nations standards and take the form of a square on point (i.e., diamond-shaped). They utilize colors, symbols, and frequently warning words to communicate the potential hazard of the compressed gas. The UN class number, 2 for compressed gases, is frequently shown in the lower point of the label reflecting the primary hazard. The hazard class number is omitted on labels identifying subsidiary hazards.

Under certain limited conditions found in some regulations, CGA container shoulder decals designed in accordance with CGA C-7, *Guide to the Preparation of Precautionary Labeling and Marking of Compressed Gas Containers,* may be substituted for the conventional hazard labels. [7] This labeling exception never applies to air or water shipments and only applies to flammable and nonflammable compressed gases when shipped by private or contract motor carrier. The shipper must be thoroughly familiar with the applicable national regulations when using this labeling exception (see Fig. 2-2).

Compressed gas cylinders must always be clearly and durably marked with the proper shipping name and the identification number assigned to that shipping name by the appropriate regulations. Under some regulations, the components of compressed gas mixtures must be shown as part of the marking requirement. Bulk containers, such as cargo tanks, portable tanks, tank cars, and multi-unit tank car tanks must also be marked with the appropriate identification number. Freight containers of one commodity must under some regulations display the appropriate identification number.

Documentation

All national and modal regulations require the shipper to complete some sort of shipping document which completely and accurately describes each hazardous material (dangerous good) in the shipment. Shipping papers must invariably include the proper shipping name, the hazard class (either written out or by UN class number) and the identification number assigned to the shipping name. In most cases, the number and type of packages are also mandatory. Weights or volumes are required. In many cases, the ingredients of mixtures or blends must be shown. Most regulations require that the shipper sign a certification statement.

The actual format of the shipping documents varies with the country and/or mode involved. Most regulations do not have a mandatory form. Some carrier organizations, such as the International Air Transport Association, have agreed upon and published forms which the shipper must use. Again, it is necessary for the shipper to consult the applicable regulations to determine the exact wording and format requirements. There also may be record retention requirements in some regulatory systems.

Compatibility

When shipping more than one hazard class, the shipper must always check the applicable regulations to determine if the combination of hazards is permissible in the same vehicle, rail car, or freight container. Also, compatibility must be considered when placing more than one hazard class in an overpack, which can be briefly described as a protective outer packaging for multiple compressed gas containers. Regardless of the hazard class, the shipper is obligated not to combine materials which may react dangerously.

Placarding

Motor vehicles, rail cars, and freight containers used to transport compressed gases must be placarded if the gross weight of the shipment exceeds 1000 lb (454 kg). Very hazardous gases, such as poison gases, require placarding for any quantity; other gases require placards if the gross weight, cylinder plus contents, exceeds 1000 lb (454 kg). Placards must be displayed on both ends and both sides of transport units.

Placards are designed in accordance with UN standards and, like hazard labels, take the form of a square on point (diamond-shaped) utilizing colors, symbols, and frequently warning words to communicate the potential hazards of the compressed gas. The UN class number, 2 for compressed gases, is usually shown in the lower point of the placard. Some regulations require that subsidiary hazards also be identified by placards, and in this case the UN class number is omitted.

For mixed loads of two or more hazards, some regulations allow for use of a *DANGEROUS* or *DANGER* placard in lieu of display of a placard for each hazard class. However, there are very strict limitations on when the simplified placarding rules can be applied.

Bulk transport units, such as tank cars, cargo tanks (tank wagons), portable tanks, tube trailers, and multi-unit tank car tanks (frequently known as ton containers), must also display the appropriate identification numbers. The regulations generally allow two options: (1) use of a combination of the appropriate placard as described above plus the "UN orange panel" showing the appropriate number or (2) modification of the placard to allow for the display of the number in a white rectangle in the center of the placard. However, the second alternative may not be authorized for certain gases. The appropriate regulations must be consulted. Regulations also contain detailed specifications on size of placards and numbers and specifications on the "orange panel" placards.

MEDICAL GAS AND MEDICAL DEVICE REGULATIONS

Food and Drug Administration

In the United States, the Food and Drug Administration (FDA) regulates medical gases as drugs, and other certain compressed gases and gas apparatus as medical devices. Manufacturers of medical gases and medical devices are required to be registered with the FDA and to produce these gases and devices in compliance with FDA regulations.

Regulations for labeling, registration, and good manufacturing practice requirements for drugs are contained in Title 21 of the *Code of Federal Regulations* (21 CFR), Parts 201, 207, 210, and 211. [8] Regulations for labeling, registration, and good manufacturing practice requirements for medical devices are published in Title 21 of the U.S. *Code of Federal Regulations* (21 CFR), Parts 801, 807, and 820. [9]

Typical compressed medical gases classified as drugs are: oxygen USP; nitrous oxide USP; carbon dioxide USP; helium USP; nitrogen NF; and mixtures of these gases. The abbreviations USP and NF indicate that the product conforms to the requirements of the *United States Pharmacopeia/National Formulary.* [10] Typical medical devices include lung diffusion gases, blood gases, and concentrators and similar equipment.

The USP and NF standards give the basic measures required for medical gas strength, quality, and purity. The *United States Pharmacopeia/National Formulary* is available from the United States Pharmacopeial Convention, Inc., 12601 Twinbrook Parkway, Rockville, Maryland 20852. The USP and NF standards include requirements for packaging, labeling, identification, impurity levels, and assay procedures for certain medical gases.

Certain compressed gases, when used as food ingredients, are regulated by the FDA under Title 21 of the U.S. *Code of Federal Regulations* (21 CFR) Parts 182 and 184. [11] Such gases include carbon dioxide, nitrogen, helium, propane, normal butane, isobutane, and nitrous oxide. These gases have been granted GRAS status (Generally Recognized As Safe) by the FDA for use as direct human food ingredients.

Compressed medical gases are also regulated by some state and local agencies. In Canada, medical gases are regulated by Health and Welfare Canada. Additionally, many practices in regard to handling and storing of gases for medical purposes are regulated by Provincial Ministries of Health. In general, specifications for medical gases in Canada are in line with USP and NF requirements. [10]

EMPLOYEE SAFETY AND HEALTH REGULATIONS

Occupational Safety and Health Administration

The Occupational Safety and Health Administration (OSHA) under the United States Department of Labor promulgates regulations "to assure safe and healthful working conditions for working men and women." These OSHA regulations are published in Title 29 of the *Code of Federal Regulations* under Part 1910 relating to General Industry Standards [12], Part 1915 relating to Shipyard Employment [13], and Part 1926 relating to Construction Industry Standards. [14]

Under the General Industry Standards of 29 CFR Part 1910, the following sections will be of particular interest to users of compressed gases:

- 29 CFR 1910.94, Ventilation
- 29 CFR 1910.95, Occupational Noise Exposure
- 29 CFR 1910.101, Compressed Gases (general requirements)
- 29 CFR 1910.102, Acetylene
- 29 CFR 1910.103, Hydrogen
- 29 CFR 1910.104, Oxygen
- 29 CFR 1910.105, Nitrous Oxide
- 29 CFR 1910.110, Storage and Handling of Liquefied Petroleum Gases
- 29 CFR 1910.111, Storage and Handling of Anhydrous Ammonia
- 29 CFR 1910.251–.254, Welding, Cutting, and Brazing
- 29 CFR 1910.307, Hazardous (classified) Locations
- 29 CFR 1910.1000, Air Contaminants
- 29 CFR 1910.1047, Ethylene Oxide
- 29 CFR 1910.1200, Hazards Communication

Compressed gases exhibit flammable, corrosive, poisonous, asphyxiating, or oxidizing properties, as well as being under pressure. Therefore, it is important that employees become familiar with the safety and health hazards of compressed gases prior to handling and use.

Under 29 CFR Part 1910.1200, manufacturers, importers, and distributors of compressed gases are required to label their containers and to provide Material Safety Data Sheets. [12] Employers are required to provide employees with information and training on hazardous chemicals in their work area. It is the responsibility of the employer to be familiar with the applicable OSHA regulations to assure the safety and health of the employees.

Compressed gas producers are always willing to provide safety and health information to customers for the education and training of their employees. The Compressed Gas Association has available numerous safety pamphlets and a number of audiovisual programs for use in employee training. A brief description of these is provided as an appendix in Part IV of this *Handbook.* CGA publications are periodically revised by committees of technical experts, and users are urged to maintain the most recent editions of those publications which meet their needs.

Cylinder Labeling Regulations

Under the regulations of the Occupational Safety and Health Administration in Title 29 of the U.S. *Code of Federal Regulations,* Part 1910.1200, section (f), the manufacturer, importer, or distributor of compressed gases must ensure that each container is labeled, tagged, or marked with appropriate hazard warnings, in addition to certain other information. The Compressed Gas Association, through its Labeling and Placarding Committee, has published CGA C-7, *Guide to the Preparation of Precautionary Labeling and Marking of Compressed Gas Containers.* [7] The labeling provisions in this pamphlet were patterned after ANSI Z129.1, *Precautionary Labeling for Hazardous Industrial Chemicals.* [15] CGA C-4, *American National Standard Method of Marking Portable Compressed Gas Containers to Identify the Material Contained,* establishes requirements for marking the product name on gas cylinders. [16] Compliance with these labeling standards will fulfill the requirements as specified in the OSHA labeling regulations.

STERILANT/FUMIGANT GAS REGULATIONS

Under the Federal Insecticide, Fungicide and Rodenticide Act (FIFRA), the United States Environmental Protection Agency (EPA) regulates gases used as pesticides and fumigants. These regulations are published in Title 40 of the U.S. *Code of Federal Regulations.* [17] Some gases used as pesticides and/or fumigants include ethylene oxide, propylene oxide, and methyl bromide, and mixtures in inert gases such as carbon dioxide and dichlorodifluoromethane.

Manufacturers of pesticides/fumigants are required to register each pesticide and each producing facility with the EPA and to report each year to the EPA all types and amounts produced during each year. Production and filling plants for pesticide/fumigant gases are subject to inspection by EPA inspectors.

REFERENCES

[1] *Code of Federal Regulations,* Title 49 CFR Parts 100–199 (Transportation), Superintendent of Documents, U.S. Government Printing Office, Washington, DC 20402.

[2] *Code of Federal Regulations,* Title 49 CFR Parts 200–399 (Transportation), Superintendent of Documents, U.S. Government Printing Office, Washington, DC 20402.

[3] *Transportation of Dangerous Goods Regulations,* Canadian Government Publishing Centre, Supply and Services Canada, Ottawa, Ontario, Canada K1A 0S9.

[4] *International Maritime Dangerous Goods Code,* International Maritime Organization, 4 Albert Embankment, London, United Kingdom SE1 7SR.

[5] *Technical Instructions for the Safe Transport of Dangerous Goods by Air,* Intereg Group Inc., 5724 N. Pulaski Road, Chicago, IL 60646.

[6] *Dangerous Goods Regulations,* International Air Transport Association (IATA), 2000 Peel Street, Montreal, Quebec, Canada H3A 2R4.

[7] CGA C-7, *Guide to the Preparation of Precautionary Labeling and Marking of Compressed Gas Containers,* Compressed Gas Association, Inc., 1235 Jefferson Davis Highway, Arlington, VA 22202.

[8] *Code of Federal Regulations,* Title 21 CFR Parts 200–299 (Food and Drugs), Superintendent of Documents, U.S. Government Printing Office, Washington, DC 20402.

[9] *Code of Federal Regulations,* Title 21 CFR Parts 800–1299 (Food and Drugs), Superintendent of Documents, U.S. Government Printing Office, Washington, DC 20402.

[10] *United States Pharmacopeia/National Formulary,* United States Pharmacopeial Convention, Inc., 12601 Twinbrook Parkway, Rockville, MD 20852.

[11] *Code of Federal Regulations,* Title 21 CFR Parts 170–199 (Food and Drugs), Superintendent of Documents, U.S. Government Printing Office, Washington, DC 20402.

[12] *Code of Federal Regulations,* Title 29 CFR Parts 1900–1910 (Labor), Superintendent of Documents,

U.S. Government Printing Office, Washington, DC 20402.

[13] *Code of Federal Regulations,* Title 29 CFR Parts 1911–1925 (Labor), Superintendent of Documents, U.S. Government Printing Office, Washington, DC 20402.

[14] *Code of Federal Regulations,* Title 29 CFR Part 26 (Labor), Superintendent of Documents, U.S. Government Printing Office, Washington, DC 20402.

[15] ANSI Z129.1, *Precautionary Labeling for Hazardous Industrial Chemicals,* American National Standards Institute, 1430 Broadway, New York, NY 10018.

[16] CGA C-4, *American National Standard Method of Marking Portable Compressed Gas Containers to Identify the Material Contained* (ANSI/CGA C-4), Compressed Gas Association, Inc., 1235 Jefferson Davis Highway, Arlington, VA 22202.

[17] *Code of Federal Regulations,* Title 40 CFR Parts 150–189 (Protection of the Environment), Superintendent of Documents, U.S. Government Printing Office, Washington, DC 20402.

CHAPTER 3

General Properties Of Gases[1]

EQUATIONS OF STATE

The term *pressure* is used throughout this chapter to mean the absolute pressure, not the gauge pressure, which is the pressure above atmospheric pressure. Absolute pressure is gauge pressure plus atmospheric pressure. Absolute temperature and temperature differences are measured in Kelvin (K). $T = t(°C) + 273.15$. (See Table 3-1.)

Ideal Gases

When we work with gases, we frequently need to determine the relative quantities of a gas in its different states. An ideal gas consists of particles of negligible volume. The forces of repulsion and attraction between the particles do not vary with the distance between the particles.

Ideal Equation of State

$$pv = R_i T \tag{3-1}$$

or

$$pV = \frac{m}{M} RT \tag{3-2}$$

Real Gases

Real gases deviate to various degrees from this ideal equation of state. The deviations become considerable at high pressures and at temperatures near a gas's condensation point.

Under the conditions at which real gases are often used, vapor molecules constitute a significant portion of the total volume. All known intermolecular forces vary with distance. If this distance effect extends over more than a small portion of the mean distance between the molecules, this can lead to deviations from the ideal gas law. Intermolecular forces also include an attraction component that increases as the temperature decreases and the pressure increases. At the point of condensation, the forces of attraction have become predominant.

Graph of State. Changes in the state of real gases can be illustrated in a graph of the relationships between p, v, and T. Usually p and v are the coordinate axes, T is a constant, and the resulting curves are called isotherms. Figure 3-1 shows isotherms for a real gas in transition between gas and liquid phases. For an ideal gas, $p \cdot v$ is a constant, producing a hyperbola with the coordinate axes as asymptotes.

This graph of state is adequate for a general study of a gas near its condensation point, but is of little use for studying thermodynamic functions or the small differences between real and ideal gases at low densities.

In many industrial processes one of the thermodynamic functions is constant or nearly constant. Examples include compression in a high-speed compressor (near-con-

[1]Adapted and reprinted with permission from *AGA Gas Handbook,* 1985.

TABLE 3-1. SYMBOLS USED IN EQUATIONS OF STATE.

Symbols		SI Units	U.S. Units
m	mass	kg	lb
M	mass of one kmol	kg/kmol	lb/mol
V	volume	m^3	ft^3
v	specific volume	m^3/kg	ft^3/lb
p	pressure	Pa or N/m^2	psi
T	temperature	K	K
t	temperature	°C	°F
R	molar gas constant $R = 8\,314$ J/(kmol·K)	J/(kmol·K)	J/(kmol·K)
R_i	individual gas constant $R_i = \frac{R}{M}$	J/(kg·K)	J/(kg·K)
z	compressibility factor		

stant entropy) and a gas flowing through an orifice (constant enthalpy). These concepts are discussed below. This type of process typically is described in thermodynamic graphs, such as pressure-enthalpy, temperature-entropy, and so forth, which depict real processes simply and accurately.

Compressibility Factor

To calculate changes in the states of real gases more accurately, one can use the ideal gas law and then correct for the deviations of real gases. A correction factor, z, the compressibility factor, is defined as follows:

$$z = \frac{pv}{R_i T} \tag{3-3}$$

Deviations from the ideal equation of state occur when z diverges from the value of 1. The compressibility factor is generally dependent on both temperature and pressure. It has been measured and tabulated for most gases.

THERMODYNAMIC FUNCTIONS

The study of the relationships between various forms of energy is called thermodynamics. Its importance cannot be overestimated: Almost all processes involve the conversion of energy from one form to another. (See Table 3–2.)

Within chemistry, the most important purposes of thermodynamics are to determine the equilibrium point of a chemical reaction and to predict whether a reaction is spontaneous under defined conditions. Thermodynamics cannot supply any information on the rate at which the reaction takes place.

First Law of Thermodynamics

The first law of thermodynamics is most simply described as the indestructibility of

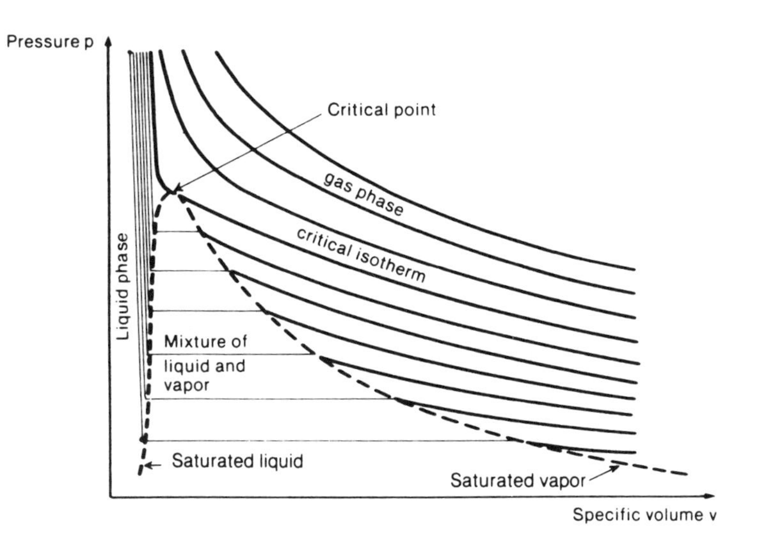

Fig. 3–1. Graph of state for a real gas.

TABLE 3-2. SYMBOLS USED IN THERMODYNAMIC FUNCTIONS.

Symbols		SI Units	U.S. Units
m	mass	kg	lb
q	specific heat quantity	J/kg	Btu/lb
Q	heat quantity ($Q = mq$)	J	Btu
u	specific internal energy	J/kg	Btu/lb
U	internal energy ($U = mu$)	J	Btu
w	specific external work	J/kg	Btu/lb
W	external work ($W = mw$)	J	Btu
p	pressure	Pa or N/m^2	psi
v	specific volume	m^3/kg	ft^3/lb
V	volume ($V = mv$)	m^3	ft^3
T	temperature	K	K
c_p	isobaric specific heat capacity (constant pressure)	J/(kg·K)	Btu/(lb·K)
c_v	isochoric specific heat capacity (constant volume)	J/(kg·K)	Btu/(lb·K)
h	specific enthalpy	J/kg	Btu/lb
H	enthalpy ($H = mh$)	J	Btu
s	specific entropy	J/(kg·K)	Btu/(lb·K)
S	entropy ($S = ms$)	J/K	Btu/K
g	specific free enthalpy	J/kg	Btu/lb
G	free enthalpy ($G = mg$)	J	Btu
M	mass of one kmol	kg/kmol	lb/mol
R	molar gas constant	J/(kmol·K)	Btu/(mol·K)
R_i	individual gas constant	J/(kg·K)	Btu/(mol·K)

energy. A change in the total energy within a closed system is always equal to the exchange of energy between a closed system and its surroundings.

An outside heat input (ΔQ) passes to an arbitrary gas bulk (m). This causes a volume increase (ΔV) and a temperature increase (ΔT). The temperature increase entails an increase in the internal energy; the volume increase means that external work is done. The first law of thermodynamics means that we can set.

$$\Delta Q = \Delta U + \Delta W \tag{3-4}$$

(Other forms of energy are assumed to remain unchanged.)

We now switch to specific quantities:

$$\Delta Q = m\Delta q \qquad \Delta U = m\Delta u \qquad \Delta W = m\Delta w$$

In differential form, this becomes:

$$dq = du + dw \tag{3-5}$$

the first law of thermodynamics. The dq is positive when heat is transferred to the system; dw is positive when the volume of the system increases.

Figure 3-2 shows that the external work is the sum of the movement of ΔF along the distance Δr, that is, $\Delta W = \Delta F \Delta r$.

But $\Delta F = p\Delta A$, and thus, $\Delta W = p\Delta A\Delta r$. Here, $\Delta A\Delta r = \Delta V$ and, therefore, $\Delta W = p\Delta V$. In differential form, for 1 kg of gas, $dw = pdv$.

The first law of thermodynamics is then:

$$dq = du + pdv \tag{3-6}$$

Specific Heat Capacity

In general form, $dq = cdT$, where c = specific heat capacity. Specific heat capacity is the amount of heat required to raise the temperature of 1 kg of a substance by 1 K. Specific heat capacity is dependent on temperature. It is useful to express it as a polynomial with respect to temperature.

$$c = (a + a_1)\cdot(T + a_2)\cdot(T^2 + \ldots$$

Often, the first two terms are adequate.

Let us study a couple of special cases here.

1. *Heat input takes place at constant volume* (marked with a subscript v), a so-called isochoric process.

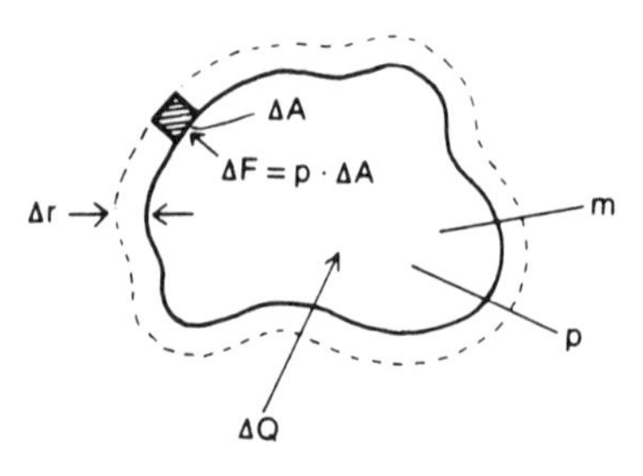

Fig. 3-2. Providing heat to a gas bulk.

$$dv = 0 \text{ and thus } dq_v = du$$

We then get $dq = c_v dT = du$, and the original equation can be written:

$$dq = c_v dT + p dv \qquad (3\text{-}7)$$

2. *Heat input takes place at constant pressure* (marked with a subscript p), a so-called isobaric process. According to the definition, $dq_p = c_p dT$. After substitution we obtain:

$$dq_p = c_v dT + p dv = c_p dT$$

The ideal gas law $pv = R_i T$ in differential form is $p dv + v dp = R_i dT$. With p constant, $dp = 0$ and the expression is $p dv = R_i dT$. If this is substituted in the expression $dq_p = c_v dT + p dv$, we get:

$$c_p dT = c_v dT + R_i dT \qquad (3\text{-}8)$$

or

$$c_p - c_v = R_i \qquad (3\text{-}9)$$

or

$$Mc_p - Mc_v = R \qquad (3\text{-}10)$$

With the so-called kinetic gas theory, it can be shown that, for monoatomic, ideal gases, $c_p/c_v = 5/3 = 1.667$. This ratio decreases as the number of atoms in the molecule increases.

For diatomic gases, $c_p/c_v = 1.40$. For triatomic gases, $c_p/c_v = 1.30$ etc., approaching unity as the number of atoms increases.

Pressure-Volume Work

In the previous cases the external work is often called pressure-volume work. In a graph of the process, with volume as the abscissa and pressure as the ordinate, the external work is the area between the process line and the v axis. See Fig. 3-3.

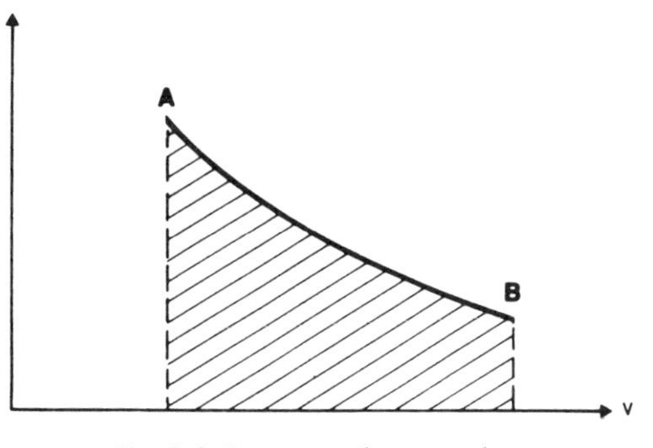

Fig. 3-3. Pressure-volume graph.

Sample Problem

A container with a volume of 2.00 liters (0.0706 ft^3) contains hydrogen sulfide at 0°C (32°F) and 760 mm Hg (14.7 psia). The gas is heated to 100°C (212°F), with the container volume remaining constant. How much heat is absorbed by the gas?

We can obtain the following information on hydrogen sulfide: molecular weight = 34.08; specific heat capacity (c_p) = 1.02 kJ (kg·K) (at 25°C).

The process takes place at constant volume.

$$dq = dq_v = c_v dT \quad \textit{and} \quad Q_v = mc_v(T_2 - T_1)$$

The ideal gas law (Eq. 3-2) gives $m = pVM/RT$.

$$p = 101.3 \times 10^3 \text{ Pa}$$

$$\therefore m = \frac{(101.3 \times 10^3)(2 \times 10^{-3})(34.08)}{(8314)(273)} = 3.042 \times 10^{-3} \text{ kg}$$

From the expression $M_{c_p} - M_{c_v} = R$, we get:

$$c_v = c_p - \frac{R}{M} = 1.02 \times 10^3 - \frac{8314}{34.08} = 776 \text{ J/(kg·K)}$$

and

$$Q_v = mc_v(T_2 - T_1) = 3.042 \times 10^{-3}(776)(100) = 236 \text{ J}$$

Note: The specific heat capacity varies with the temperature; however, when extreme accuracy is required and high temperature variables exist, you must first calculate the values between the temperature limits.

Enthalpy

Because many technical processes are carried out at constant pressure, it has been found convenient to introduce the terms *en-*

thalpy, H, and *specific enthalpy, h,* for the heat energy in this special case. Earlier, the phrase *heat content at constant pressure* was often used, frequently abbreviated to *heat content.*

The change in enthalpy can be defined as $dH = mc_p dT$ and the change in specific enthalpy as $dH = c_p dT$. We then get:

$$dh = c_p dT = du + p dv \tag{3-11}$$

Integration from 0 to h, from 0 to u, and from 0 to v, gives (p is constant):

$$h = u + pv \tag{3-12}$$

and

$$H = U + pV$$

This means that the lower integration limit is absolute zero, $T=0$.

For numerical calculations, a temperature scale with a more relevant zero point is often chosen.

Kirchhoff's Law

Assume that we have determined the change in enthalpy for a chemical process at a mean temperature T by means of a calorimeter test. If A represents all reactants, and B all products, the change in enthalpy for the process A to B can be written:

$$\Delta H = H_B - H_A$$

But the mass before and after the reaction is unchanged. Thus, we can write the change in specific enthalpy as follows:

$$\Delta h = h_B - h_A$$

This expression, derived with respect to the temperature, is:

$$\frac{d(\Delta h)}{dT} = \frac{d}{dT}(h_B - h_A) = \frac{dh_B}{dT} - \frac{dh_A}{dT}$$

Here

$$\frac{dh_\Delta}{dT} = c_{pA} \quad \text{and} \quad \frac{dh_B}{dT} = c_{pB} \tag{3-13}$$

where c_{pA} and c_{pB} are the specific heat capacities for the reactant mixture and the product mixture, respectively. Kirchhoff's law states that:

$$\frac{d(\Delta h)}{dT} = \Delta c_p \tag{3-14}$$

where Δc_p is the difference between the specific heat capacities of the products and the reactants.

Kirchhoff's law is used to determine the temperature dependence of a change in the specific enthalpy of a chemical process. In most cases, Δc_p is small at moderate temperature changes, which means that the change in specific enthalpy is relatively independent of the temperature at which the process is carried out. The relationship assumes that no changes occur in the state of aggregation of reactants or reaction products within the relevant temperature range.

Specific Latent Heat of Vaporization

Figure 3–4 shows a temperature-specific entropy ($T - s$) graph for a "pure" substance. The vaporization range is limited by the lower boundary curve, A–B–K and the upper, K–C. To the left of the lower bound-

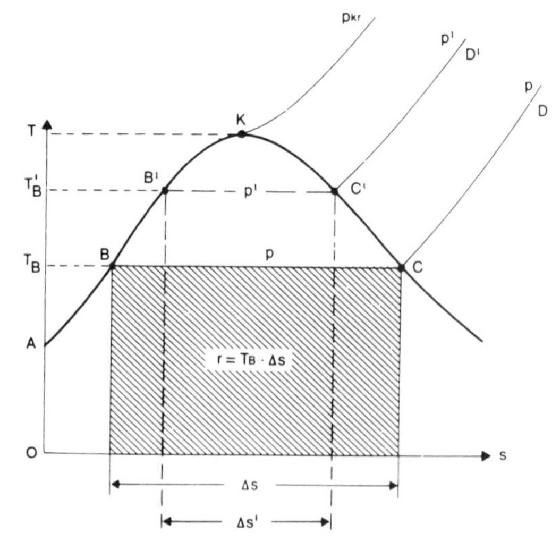

Fig. 3-4. T-s graph for a pure substance.

ary curve, the medium is completely liquid; to the right of the upper boundary curve, it is completely gaseous.

For a pure substance, the vaporization temperature is constant at constant pressure. An isobar, then, follows the curve, A–B–C–D. Vaporization begins at point B and is complete at point C. During this process, the temperature, T_B, is constant. The amount of heat required for vaporization, *r,* is represented by the area between line B–C and the *s* axis ($T = 0$). The figure shows that the latent heat of vaporization decreases at high pressures, reaching zero at the so-called *critical point,* K. Above this point, a diffuse transition takes place between liquid and gas phase.

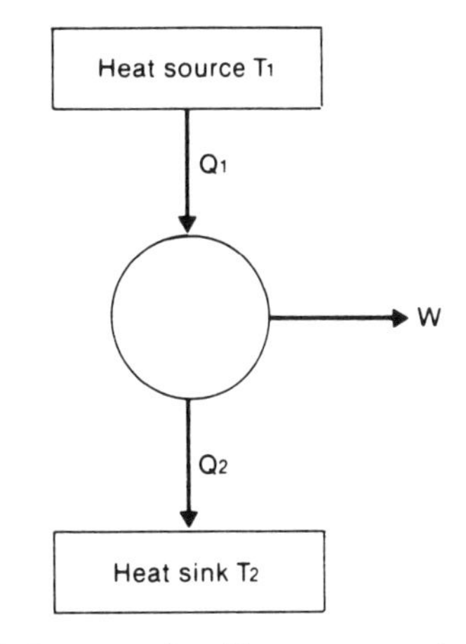

Fig. 3-5. Carnot engine. The process can be reversed, whereby mechanical work is supplied and heat is raised from T_2 to T_1.

Entropy

At the beginning of the nineteenth century, Carnot found, while studying thermal engines, that heat could not be converted completely to mechanical work, which should be possible according to the first law of thermodynamics. If the concept of absolute temperature, *T,* is introduced, the maximum amount of work that can be extracted can be written:

$$W = Q_1 \cdot \frac{T_1 - T_2}{T_1} \tag{3-15}$$

The quantity of heat Q_1 is from a heat source with a constant temperature T_1, and $Q_2 = Q_1 - W$, and is delivered to a heat sink with a constant temperature T_2. See Fig. 3–5. This imaginary, ideal process is called a Carnot process. It is an example of a reversible process, because it can also proceed in the opposite direction and both the system and its surroundings can be returned to their original state. Thus, using a Carnot process, we can determine the zero point of the absolute temperature scale in accordance with Eq. 3-15.

In a reversible process, all heat transfer takes place without a temperature drop. The prerequisite for a heat flow into or out of a system is always a temperature drop in the direction of transfer. The greater the temperature difference, the greater the flow of heat energy per unit of time. The logical consequence is that isothermal heat transfer requires an infinitely long time. The same must then apply to reversible processes, which exchange heat with the surroundings.

In a pressure-volume graph, we have previously seen that the area between the abscissa and the line representing a process is proportional to the external work ($W = \int p\,dV$). See Fig. 3–3. If we wish to represent heat energy as a similar area, with temperature T as the ordinate, the abscissa will be $\int dQ/T$. See Fig. 3–6.

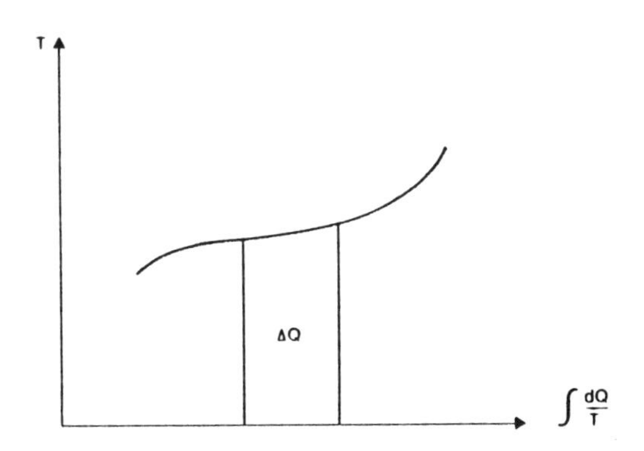

Fig. 3–6. Graphic description of the entropy concept.

The quantity obtained is called entropy (or, for the unit mass, specific entropy) and is designated *S(s)*. It is of utmost importance for an understanding and treatment of many thermodynamic relationships.

The mathematical definition of entropy is thus:

$$dS = \frac{dQ}{T} \quad \text{or} \quad ds = \frac{dq}{T} \tag{3-16}$$

Let us look at the entropy change in the surroundings of a Carnot process. The heat quantity Q_1 enters the process at temperature T_1. Since T_1 is constant,

$$\Delta S_1 = -\frac{Q_1}{T_1}$$

The heat quantity Q_2 leaves the process at T_2, which is also constant, and we obtain

$$\Delta S_2 = +\frac{Q_2}{T_2}$$

After substituting Eq. 3-15, we obtain

$$Q_2 = Q_1 \frac{T_2}{T_1}$$

and thus

$$\Delta S_2 = \frac{Q_1 T_2}{T_1} \cdot \frac{1}{T_2} = \frac{Q_1}{T_1}$$

We find that the entropy change in the surroundings is $\Sigma \Delta S = 0$, which characterizes all reversible processes. (The Carnot engine itself has not changed, and has therefore not undergone any entropy change.) See Fig. 3-7.

In an actual technical process, there are always temperature drops associated with heat transfers. Let us assume that a heat source must have a temperature of $T_1 + \Delta T_1$ and a heat sink, a temperature of $T_2 - \Delta T_2$, for the process to proceed at the desired rate. See Fig. 3-7. The entropy extraction from the heat source is now only

$$\Delta S_1 = -\frac{Q_1}{T_1 + \Delta T_1}$$

and the entropy input to the heat sink increases to

$$\Delta S_2 = \frac{Q_2}{T_2 - \Delta T_2}$$

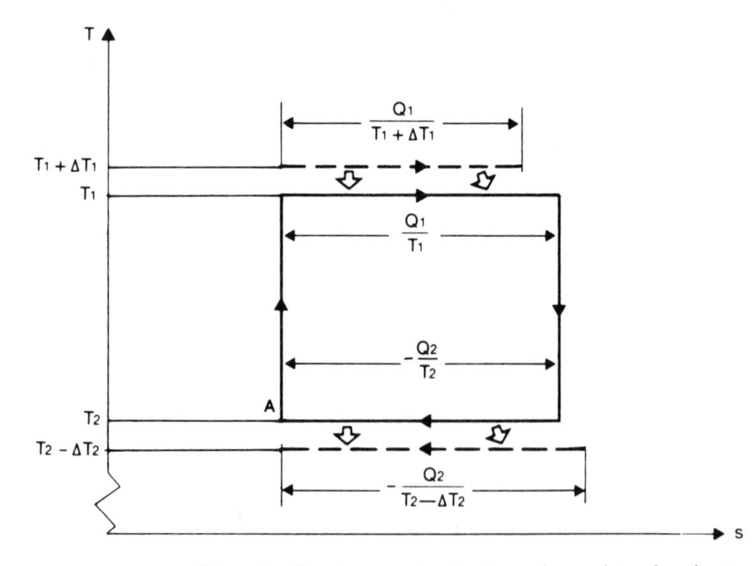

Fig. 3-7. Carnot process or reversible cycle. Starting at point A, the cycle consists of an isentropic compression, an isothermal expansion, an inentropic expansion and an isothermal compression. All steps are reversible. The open arrows symbolize the heat flow in a real technical process where there are temperature drops in the heat transfer to and from the process medium. The entropy flows from the heat source to the heat sink respectively, correspond to the hatched lines in that case.

The overall entropy change is now greater than zero and the process is no longer reversible.

If we introduce the specific entropy to the general heat equation, Eq. 3-7, ($dq = c_v dT + p dv$), we get:

$$ds = c_v \frac{dT}{T} + \frac{P}{T} dv \qquad (3\text{-}17)$$

For an ideal gas, $pv = R_i T$. Then

$$\frac{p}{T} = \frac{R_i}{v}$$

and

$$ds = c_v \frac{dT}{T} + R_i \frac{dv}{v}$$

This expression is always integratable, term by term:

$$s = c_v \ln T + R_i \ln v + \text{constant}$$

and

$$\Delta s = c_v \ln \frac{T_2}{T_1} + R_i \ln \frac{v_2}{v_1}$$

(applicable to an ideal gas passing from state 1 to 2).

This shows that entropy is a state function, that is, entropy for a given state is not dependent on the manner in which this state was reached. This can also be shown to apply to a real gas.

For all real processes, entropy within an isolated system (the process and its surroundings) will increase. Thus, the following applies to an isolated system:

Condition for possible (spontaneous) reaction: $\Delta S > 0$
Condition for equilibrium: $\Delta S = 0$

Second Law of Thermodynamics

The second law of thermodynamics can be formulated in several ways. One formulation states, as Carnot found, that heat energy cannot be completely converted to mechanical work. Another formulation states that heat energy cannot be transferred spontaneously from a body at a lower temperature to another at a higher temperature.

Historically, only exothermic reactions were thought to exist, but research soon proved the existence of endothermic reactions. An evaluation of many processes without subsequent energy exchange verified the existence of endothermic processes within isolated systems. See Fig. 3-8.

Such systems pass spontaneously from state A to B as soon as the two containers come into contact with each other. Since no energy exchange takes place with the surroundings, the internal energy of the systems remains constant and $\Delta U = 0$. The reverse processes (B $\rightarrow$ A) also have $\Delta U = 0$. But they never take place spontaneously. The change A $\rightarrow$ B is irreversible and proceeds with an increase in entropy. All processes that take place spontaneously are irreversible.

Free Enthalpy (Gibbs Free Energy)

Chemical reactions are possible at simultaneously constant pressure and temperature. Reaction feasibility cannot be determined by the entropy condition.

Gibbs introduced a new state function:

$$G = H - TS \qquad (3\text{-}19)$$

where G represents free enthalpy. If we convert Eq. 3-19 to specific free enthalpy according to $G = mg$, the expression is:

$$g = h - Ts$$

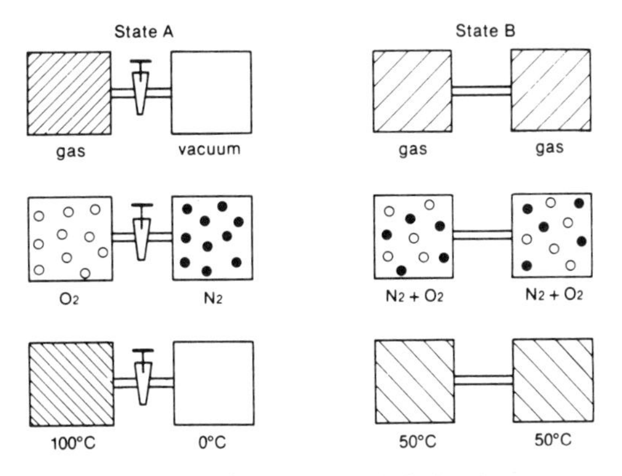

Fig. 3-8. Irreversible processes in isolated systems.

If we differentiate this, we get:

$$dg = dh - Tds - sdT$$

For a process at constant temperature:

$$dg = dh - Tds$$

(*Note:* T being constant does not necessarily mean that dh is zero.) Compare this with the expression for enthalpy. Phase transformations or chemical reactions can involve a change in internal energy and perform pressure-volume work without a change in temperature.

Thus, for a chemical, thermodynamic process at both constant pressure and constant temperature:

$$\Delta g = \Delta h - T\Delta s \qquad (3\text{-}20)$$

Δh and Δs can be determined by calorimetric tests. (*Note:* Δh and Δs must be related to the relevant pressure and temperature.) Δg can be calculated in this manner. If $\Delta g > 0$, the reaction in question is not possible; if $\Delta g < 0$, the reaction is possible (spontaneous). The change in free enthalpy indicates nothing about the rate at which the reaction takes place. A slow process can be accelerated by the addition of a catalyst.

CRITICAL PROPERTIES

TABLE 3-3. SYMBOLS USED FOR CRITICAL PROPERTIES.

$E(r)$	attraction energy between molecules
r	distance between molecules
ϵ	maximum attraction energy between two molecules
σ	collision diameter, E(r) = 0
P	pressure
P^c	critical pressure
P^*	reduced pressure
T	temperature
T^c	critical temperature
T^*	reduced temperature
V	volume
V^c	critical volume
V^*	reduced volume

Molecular Motion

Gas molecules are acted on by various forces. Molecular motion strives to keep the gas uniformly distributed in the available volume. The molecules are acted on by forces of attraction and repulsion. Forces of repulsion act at very short distances and derive from the mutual repulsion of the electron shells. A widely used empirical formula that describes this interplay of forces is the Lennard-Jones potential. See Eq. 3-21 and Fig 3-9. (See also Table 3-3.)

$$E(r) = 4\epsilon \left[\left(\frac{\sigma}{r}\right)^{12} - \left(\frac{\sigma}{r}\right)^{6}\right] \qquad (3\text{-}21)$$

Liquid State. According to Eq. 3-21, the forces of attraction and repulsion vary with the distance between the molecules, by 10^6 and 10^{12}, respectively. When the attraction energy exceeds the kinetic energy (i.e., energy of motion), the molecules adopt the liquid state of aggregation.

Unlike gas, the liquid molecules do not disperse uniformly in a given volume. The molecules in the liquid are still in motion.

Vapor Pressure. Individual molecules in the liquid have different amounts of kinetic energy, distributed roughly along a normal curve. Some molecules will have energy exceeding the intermolecular forces of attraction. These molecules will vaporize. The number of molecules, per unit time and unit area, that vaporize dictates the vapor pressure of the substance. Since kinetic energy is directly proportional to temperature, vapor pressure is dependent solely on temperature.

Gaseous State—Boiling Point. If vapor pressure exceeds the pressure exerted by the surrounding atmosphere, the substance eventually will vaporize completely. The temperature at which a substance vaporizes is its boiling point. If the surrounding pressure is 0.1013 MPa (1.013 bar, 14.7 psig), this temperature is described as the normal boiling point.

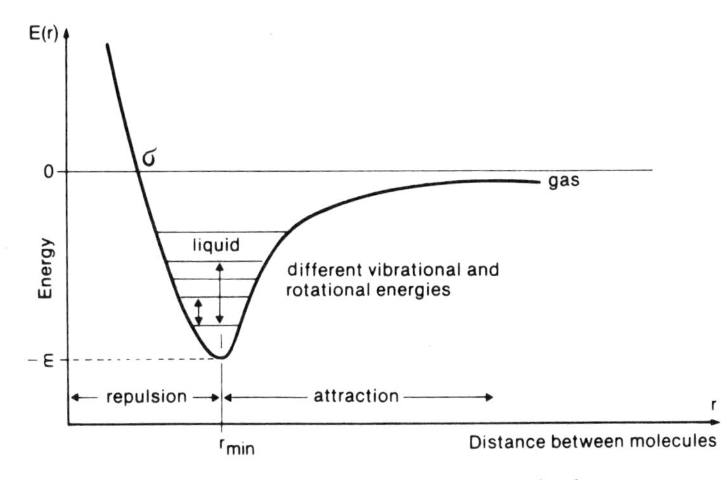

Fig. 3-9. Lennard-Jones potential for a molecule.

Critical Temperature, Pressure, and Volume

At sufficiently high temperatures, the kinetic energy will exceed the maximum attraction energy between molecules, no matter how high the pressure. The substance will then never become liquid. This limit temperature is called the critical temperature. It represents the upper limit for condensation, beyond which there is no distinct transition between gas and liquid. Other properties at this temperature are also referred to as critical, such as critical volume and critical pressure. Above its critical temperature, a gas cannot liquefy and a liquid cannot continue to exist. See Fig. 3-10.

Reduced Quantities. Some agreement in the quantities of state of the real gases can be obtained if pressure, volume, and temperature are expressed as ratios with respect to the corresponding critical quantities. These ratios are usually called reduced quantities

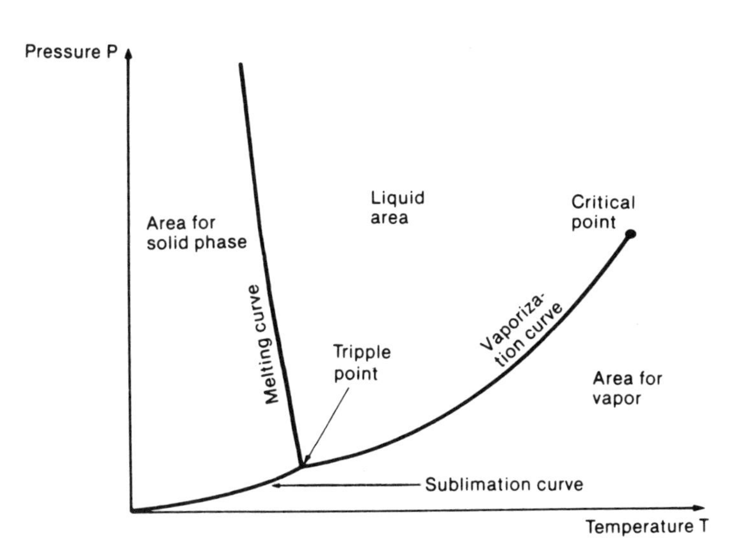

Fig. 3-10. Pressure-temperature graph.

and are marked with an asterisk. (*Note:* these reduced quantities are dimensionless.)

$$p^* = \frac{P}{p^c} \qquad T^* = \frac{T}{T^c} \qquad V^* = \frac{V}{V^c} \quad (3\text{-}22)$$

TRANSPORT PROPERTIES

Thermal conductivity, viscosity (internal friction), and diffusion are referred to collectively as transport properties. For gases, they are all molecular properties. (See Table 3–4).

Heat is conducted through a gas by the transfer of energy during collisions from molecules with higher energy to those with lower energy. The fraction of energy transferred is determined by the specific nature of the intermolecular potential. This applies particularly to rotational and vibrational energy. In addition, in a flowing gas, the amount of movement, or impulse, is transferred between gas layers of different velocities, creating viscosity.

In the case of diffusion, different concentrations of a gas in a given volume are equalized as the molecules strive to disperse uniformly. This dispersion is slowed by intermolecular collisions.

Because the molecular speed and collision frequency increase with temperature, the intensity of the transport processes generally increases with temperature.

When the pressure changes, the corresponding changes in both the mean free path between collisions and the collision frequency nullify each other. Thus, the transport properties are, in principle, independent of pressure.

TABLE 3–4. SYMBOLS USED FOR TRANSPORT PROPERTIES.

Symbols	SI Units	U.S. Units
Φ materials flow density	kmol/(m²·s)	mol/(ft²·s)
D diffusion coefficient	m²/s	ft²/s
c concentration	kmol/m³	mol/ft³
τ shear stress	Pa or N/m²	psi
η dynamic viscosity	(Pa·s) or (N·s)/m²	cP
ρ density	kg/m³	lb/ft³

Thermal Conductivity

In a stationary gas, all heat transfer is conductive. An example of this is the boundary layer immediately adjacent to a wall surface, such as a heat exchanger surface. The thermal conductivity of a gas is, in principle, independent of pressure but increases with rising temperature. Gases of low molecular weight have a higher thermal conductivity than heavier gases (cf. hydrogen and air: the thermal conductivity of hydrogen expressed as W/(m·K) is nearly eight times greater than that of air).

Intermolecular potential is directly related to the interaction of molecules of different gases, which affects the thermal conductivity of a given gas mixture. Caution should be observed when calculating the thermal conductivity of gas mixtures because a simple proportioning (''weighting'') of the components' thermal conductivities can lead to considerable error if those conductivities are substantially different.

Diffusion

If two gases are in contact but there is no convection, they will eventually mix by diffusion. This process is analogous to heat transfer through conduction. The counterpart of thermal conductivity is the diffusion coefficient, *D*. This coefficient is generally not constant but is dependent upon the concentrations of the two substances. The variation over the entire concentration range, however, is generally less than 10 percent. Fick's first diffusion law is:

$$\Phi = D \frac{dc}{dx} \qquad (3\text{-}23)$$

For diffusion around a boundary layer between two gases as shown in Fig. 3–11, the following equation applies:

$$c_1(x, t) = c_{1b} \cdot \frac{c_{1a} - c_{1b}}{2} \left[1 + \mathrm{erf} \frac{x}{2\sqrt{D_{1,2}\, t}} \right] \qquad (3\text{-}24)$$

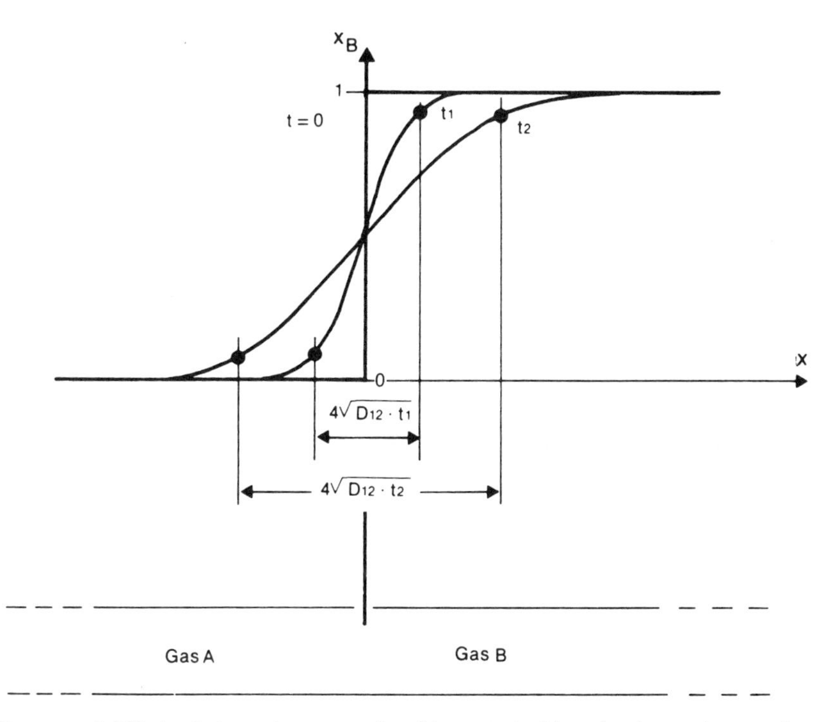

Fig. 3-11. Progress of diffusion between two gases placed in contact with each other at time equals zero (t = 0).

where

c_{1a} = the concentration of substance 1 in gas a at $t < 0$

c_{1b} = the concentration of substance 1 in gas b at $t < 0$

$c_1(x,t)$ = the concentration of substance 1 at point x at time t

$D_{1,2}$ = the binary diffusion coefficient for substances 1 and 2

erf = the error integral

$$\left(\text{erfw} = \frac{2}{\sqrt{\pi}} \int_0^w e^{-s^2}\, ds \right)$$

Equation 3-24 assumes that the binary diffusion coefficient is independent of the concentration and that the diffusion process involves only a small fraction of the two gas volumes.

To get a practical grip on the extent of the diffusion, we can set

$$\frac{x}{2\sqrt{D_{1,2}\, t}}$$

equal to -1 or $+1$. We then find that this corresponds to a concentration of 92 percent and 8 percent, respectively, of the gas that had been on the left-hand side. The distance between these two points is a measure of the extent of the diffusion and is calculated as:

$$2x = 4\sqrt{D_{1,2}\, t}$$

For the gases hydrogen and argon, for instance, $D_{\text{H-Ar}} = 70 \times 10^{-6}$ m²/s at 0°C, 100 kPa (753 $\times$ 10^{-6} ft²/s at 32°F and 14.5 psi) and we find that after one hour, the width of the diffusion zone is $2x$, approximately 2 m (6.5 ft).

As a rule, the diffusion coefficient increases sharply with temperature and decreases with pressure. The following relationship is generally accurate:

$$D = D_0 \left(\frac{T}{T_0} \right)^{1.5} \cdot \frac{P_0}{P} \tag{3-25}$$

where D_0 is the diffusion coefficient at the reference temperature and pressure T_0 and P_0, respectively.

As a rough approximation, the diffusion coefficient for a two-gas system is equal to the weighted mean of the self-diffusion coefficients of the two gases. When greater accuracy is required, a reference such as Landolt-Börnstein should be used.

When gas is forced through a fine porous membrane, the flow rate is inversely proportional to the square root of the molecular weight. This process is called effusion and can sometimes be used for isotope separation.

Viscosity

The viscosity of gases is purely Newtonian, that is, dependent solely on pressure and temperature. Absolute or dynamic viscosity is defined as the shear stress between adjacent layers of different velocities in a flowing medium.

If we designate the shear stress at τ, and the velocity gradient,

$$\frac{(dv_x)}{(dy)}$$

where x is the flow direction, we obtain:

$$\frac{(dv_x)}{(dy)} = \frac{\tau}{\eta}$$

where the proportionality constant η is called dynamic viscosity.

Kinematic viscosity, v, is used in fluid mechanics, and is defined as:

$$v = \frac{\eta}{\rho}$$

The dynamic viscosity of a gas increases with temperature but is fairly independent of pressure. Calculating the viscosity of gas mixtures is subject to substantial error similar to that described above for thermal conductivity.

PHYSICAL AND CHEMICAL PROPERTIES

TABLE 3-5. SYMBOLS USED FOR PHYSICAL AND CHEMICAL PROPERTIES.

Symbols	SI Units	U.S. Units
M molecular weight		
A solubility coefficient	$(Pa \cdot m^3)/$ kmol	$(psi \cdot ft^3)/$ mol
B bunsen coefficient	atm	psi
c concentration	$kmol/m^3$	mol/ft^3
ρ density	kg/m^3	lb/ft^3
Subscripts:		
g gas		
s solvent		

Solubility

The solubility of gases in liquids varies within wide limits. It decreases rapidly with rising temperature, reaching zero when the liquid reaches its boiling point. Very high gas solubilities may be due to the influence of chemical processes. Examples are aqueous solutions of ammonia or carbon dioxide.

The solubility of gases in liquids generally follows Henry's law, which states that, at equilibrium, the partial pressure of a gas lying above a solution is proportional to the concentration of the gas in the solution (see Table 3-5):

$$P_g = A_g - c_g$$

The proportionality constant A_g, for a given gas and a given solvent, is dependent solely on the temperature. In general, Henry's law applies only to highly diluted solutions and is therefore not applicable to systems such as ammonia-water, carbon dioxide-water, and acetylene-acetone.

If concentration is expressed in cubic meters of gas (0°C, 101.3 kPa) (32°F, 1 atm) per cubic meter of solvent, and the partial pressure is measured in atmospheres, the proportionality constant is termed the bunsen coefficient. The relationship between

A_g and B_g is expressed by the following formula:

$$A_g = \frac{(101.3)(10^5)(22.41)(\delta_s)}{(B_g)(M_s)}$$

Under high pressure, many gases, including the noble gases, form aqueous solutions, in relatively fixed concentrations, that can take on solid form. Such solutions are called clathrates. Example: argon clathrate < 7.99 $Ar{\cdot}46H_2O$.

Gas Mixtures

Gas mixtures often exhibit complex behavior at the transition between gas and liquid phases. The temperature at which a liquid mixture will start to boil, the bubble point, is dependent on the pressure. See Fig. 3–12.

When a gas mixture is cooled, the temperature at which condensation begins, the dew point, is also pressure dependent. The bubble and dew points do not generally coincide.

There is usually a given pressure and temperature for every gas mixture at which the proportional composition of its liquid and gas phases is identical. This point, called the plait point, varies with the contents of the mixture. For pure gases, the plait point does not correspond to the critical point. For example, a mixture can occur in liquid form at a temperature above the plait point. See Figs. 3–12 through 3–14.

When the proportions of a mixture are varied, the plait point changes; for a two-substance mixture, a curve plotted through the plait points at different proportions will terminate, at either end, in the critical points of the pure substances. At other points in the graph of state, the compositions of the two phases are not identical, producing two curves: a bubble point curve and a dew point curve. See Fig. 3–13.

These relationships can be illustrated in a three-dimensional diagram, Fig. 3–14, of which Figs. 3–12 and 3–13 constitute plane projections.

Under certain conditions, gas mixtures exhibit retrograde boiling and condensation, that is, boiling takes place during a temperature decrease or a pressure increase. Similarly, retrograde condensation is the reverse of normal condensation. Follow the vertical broken line in Fig. 3–12 and notice what happens when the pressure changes.

Gases generally possess unlimited capacity to mix together. Such mixtures do not separate as long as both components are in the gas phase. There are, however, some gas mixtures that have limited intermiscibility, that is, a "mixing gap." A two-component example is the mixture of ammonia and nitrogen, whose solubility diagram is shown in Fig. 3–15. Note, however, the extreme conditions under which this phenomenon occurs.

Reactivity

The reactivity of gases varies within very wide limits and is often the basis of their use for a given purpose. Among the most inert substances known are the so-called noble

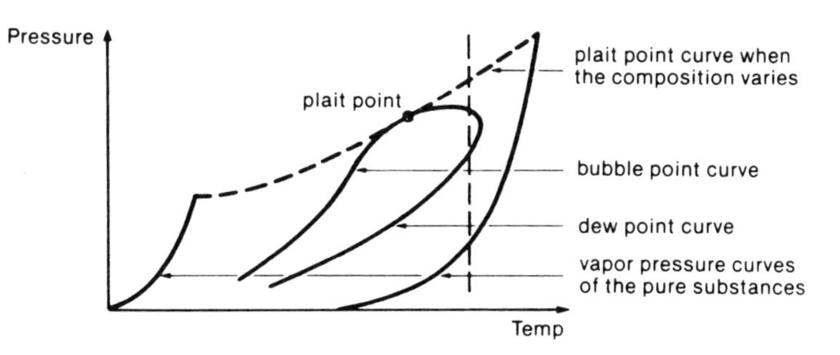

Fig. 3–12. Graph of state for a binary mixture.

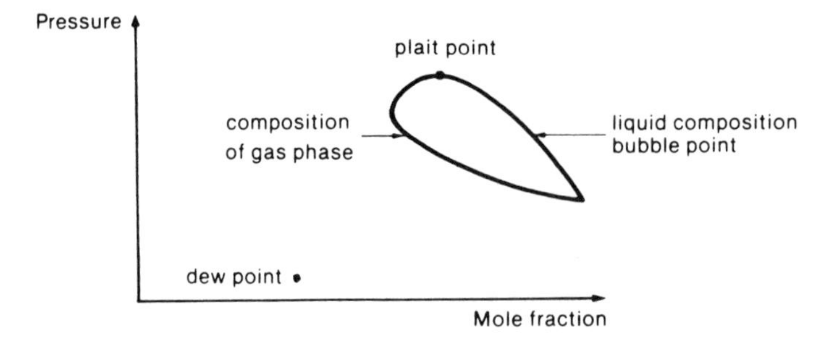

Fig. 3-13. Graph of state for a binary system at constant temperature.

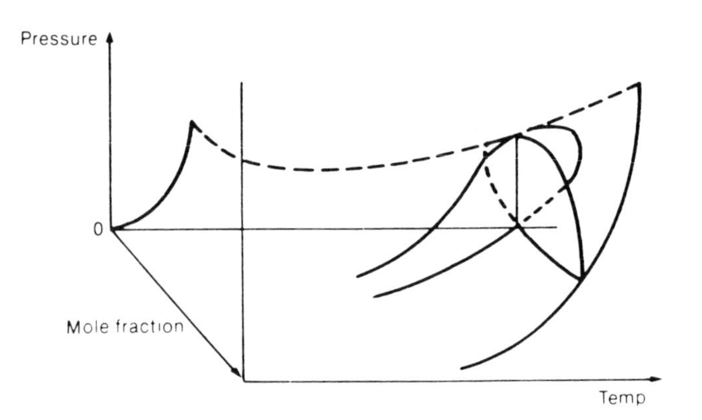

Fig. 3-14. Graph of state for a binary system.

gases (also known as rare gases or inert gases). Of these, only krypton and xenon are capable, if extremely "reluctantly," of forming (impermanent) chemical compounds. Nitrogen is often sufficiently inert for many purposes.

At the other end of the scale we find the halogens, which are highly reactive and potent oxidants. Together with reactive and flammable gases, the halogens form systems that are capable of liberating large amounts of chemical energy in the form of heat.

Biological Effects

Many of the industrial gases are of fundamental importance for biological processes. Most life forms obtain their energy by metabolizing nutrients with oxygen, producing the end products water and carbon dioxide. In photosynthesis, carbon dioxide and water are converted to oxygen and biological matter with the aid of the energy in sunlight.

Nitrogen is incorporated in protein structures and is taken up by plants in the form of nitrates or ammonium compounds.

Many gases are highly toxic to animals or plants. Examples are carbon monoxide, sulfur dioxide, hydrogen sulfide, ozone, and ethylene oxide.

Acetylene and nitrous oxide have a powerful anesthetic effect. As discovered during deep-sea diving work under high pressure, the noble gases, possibly with the exception of helium, also have a mildly anesthetic or narcotic effect at high pressures.

Some gases have an interesting, specific biological action. Ethylene, which is sometimes called the ripening hormone, is regularly used to stimulate the ripening of various kinds of fruit. Ripening apples give off small quantities of ethylene. This biological

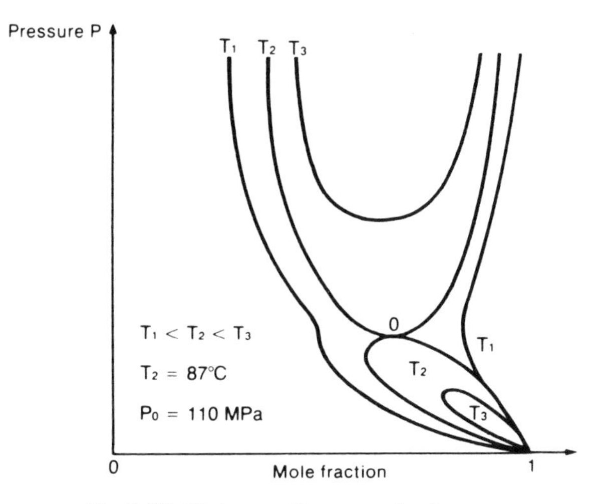

Fig. 3-15. Mixing gap for ammonia-nitrogen.

action can have a less favorable effect on cucumbers, however, which in most cases are preferred unripe.

DETERMINATION OF MOISTURE CONTENT

TABLE 3-6. SYMBOLS USED FOR DETERMINATION OF MOISTURE CONTENT.

Symbols	SI Units	U.S. Units
c concentration	kmol/m^3	mol/ft^3
M molecular weight		

Determining the moisture content of industrial gases is an important task. Even small amounts of moisture can cause problems due to the special properties of water (condensation, freezing, or clathrate formation). When a moist gas is compressed, the partial pressure of the water vapor can easily rise to such a level that the water condenses during subsequent chilling. This is particularly troublesome at the high charging pressures that are used in modern gas cylinders. For example, moisture can cause problems in gas regulators.

Because of its dipolar moment, water has a strong tendency to be adsorptively bound to various materials. This, together with the small diameter of the water molecule, is the reason for the high diffusion capacity of water through most organic materials. Correct determination of moisture content is important and requires the right temperature and measurement procedure. (See Table 3-6.)

Methods of Measurement

Different methods of dew point measurement can be used to determine low moisture contents. For measurements in the parts per million range, the condensation temperature is very low and the precipitated quantities of liquid are small. The actual measurement can be difficult to determine, especially as the condensation points of other components of a gas are approached. Dew points of −18 to −34°C (0° to −30°F) can be measured readily on a conventional dew-pointer, whose operational principle is the condensation of moisture on a mirror.

An indirect method for moisture content determination uses a capacitor with an aluminium oxide dielectric, designed in such a manner that moisture can easily migrate into and out of the dielectric. The amount of water absorbed by the dielectric is compared with the moisture content of the surrounding gas. The capacitor is calibrated to indicate the moisture content of the gas. The most difficult part of this process is generating an H_2O calibration standard.

A third measurement process uses an electrolytic cell in which the moisture is absorbed and electrolyzed. The electrolytic current constitutes a measure of the moisture content of the gas.

In all moisture content measurement, it is important that the entire measuring system, including the detector, be made of metallic materials. Plastics and rubber may not be used.

Pressure

Because water can be adsorbed so readily by most surfaces—especially if they are rough, contaminated, or corroded—long purging times are necessary, and low-dew-point gas (≤ 3 ppm) must be used. The moisture content of gas from a container will vary, therefore, depending on the container pressure. As the container pressure decreases, more and more adsorbed water is released from the walls and the moisture content of the gas increases. This is particularly noticeable as the pressure in the container approaches atmospheric pressure.

Figure 3–16 shows the theoretical curve for the variation in moisture content with the container pressure. To obtain gas with a low moisture content, the cylinder should be left with approximately 200–300 psig residual pressure. The moisture problem can be greatly reduced by special surface treatment of the inside of the container and proper handling. Do not leave the valve open.

Units

Moisture content is usually expressed in kilograms per cubic meter or parts per million. In the latter case, the unit may be parts by weight, ppm_{weight}, or by volume, ppm_{vol}, sometimes designated vpm. The moisture concentration is thus expressed independently of pressure, temperature, and type of gas.

Conversion between these units is as follows:

$$c[ppm_{vol}] = \frac{Mg}{M_{H_2O}} \cdot c[ppm_{weight}] \qquad (3\text{-}26)$$

$$c[ppm_{vol}] = \frac{22.41}{1000\ M_{H_2O}} \cdot c[g/m^3] \qquad (3\text{-}27)$$

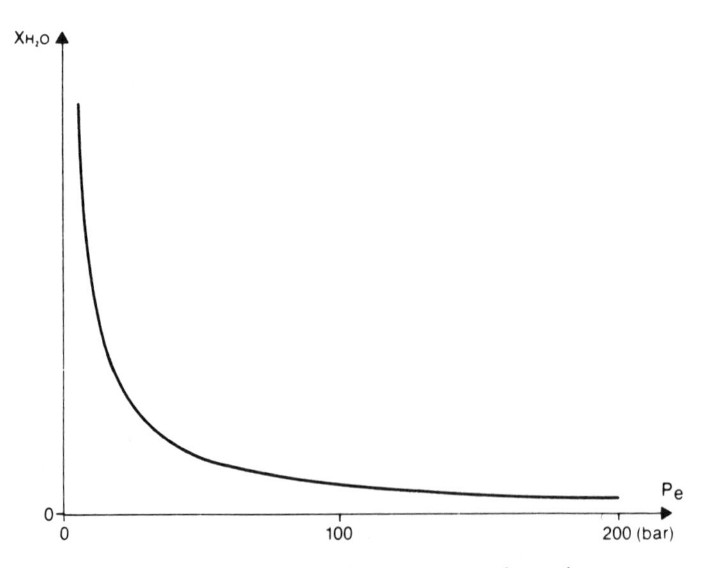

Fig. 3–16. Relationship between moisture content and container pressure.

REFERENCES

[1] Dymond, J. H., and Smidt, E. B., *The Virial Coefficients of Pure Gases and Mixtures, A Critical Compilation,* Oxford University Press, Oxford, 1980.

[2] Flügge, S., *Handbuch der Physik, Band XII, Thermodynamik der Gase,* Springer-Verlag, Berlin, 1958.

[3] Franke, H., *Lexikon der Physik,* Franckh'sche, Stuttgart.

[4] Grenthe, I., Norén, B., Ots, H., and Sellers, P., *Termodynamik, Elektrokemi.* Lunds Universitet (*Thermodynamics, Electrochemistry,* The University of Lund), Sweden, 1975.

[5] Landolt-Börnstein, *Zahlenwerte und Funktionen, Band IV, Teil 4, Wärmetechnik,* Springer-Verlag, Berlin, 1967.

[6] Mogensen, P., *Kompendium i Gasteknologi (Compendium in Gas Technology),* AGA AB, Innovation, Lidingö, Sweden, 1981.

[7] Sears, F. W., *Thermodynamics, The Kinetic Theory of Gases and Statistical Mechanics,* Addison-Wesley, Cambridge, MA, 1955.

[8] Ulvås, C., *Energi (Energy),* Esselte Studium, Lund, Sweden, 1978.

[9] Zemansky, M. W., *Heat and Thermodynamics,* 4th ed., McGraw-Hill, New York, 1957.

CHAPTER 4

Compressed Gas Containers and Appurtenances

INTRODUCTION

This chapter describes and illustrates the major types of containers used to transport and store compressed and liquefied gases, including those liquefied at very low or cryogenic temperatures. To keep pace with developments in the industry, the Compressed Gas Association, through its technical committees, continually reviews and recommends changes to existing container specifications promulgated by the U.S. Department of Transportation (DOT) or Transport Canada (TC). Moreover, new specifications for compressed gas containers are also proposed from time to time. This chapter provides an overview of the following types of compressed gas containers and their associated appurtenances:

- Cylinders and small containers
- Regulators and control valves for cylinders and other containers
- Containers for shipping by rail in bulk
- Cargo tanks and tube trailers
- Portable tanks and ISO containers
- Containers for shipping by water in bulk
- Cryogenic containers
- Stationary storage containers
- Pipelines

Readers should see Chapter 5 for general information on the safe handling of cylinders and other containers. Specific topics related to compressed gas containers are covered in Chapters 6 through 10. Information about the types of containers used with specific gases can be found in the gas monographs of Part III.

DOT, TC, and ASME Regulations for Compressed Gas Containers

In North America, most of the shipping and storage containers made for compressed gases are built to comply with one major source of detailed specifications. These are the generally identical regulations adopted by the U.S. Department of Transportation and Transport Canada, and the *ASME Boiler and Pressure Vessel Code,* particularly Section VIII on Unfired Pressure Vessels, issued by the American Society of Mechanical Engineers. [1], [2], and [3] The DOT and TC regulations are obligatory for containers shipped in interstate or interprovincial commerce and apply mainly to shipping containers. The ASME Code applies mainly to stationary storage vessels, although truck cargo tanks and portable tanks must conform to the ASME Code under the respective DOT and TC specifications.

In the United States, regulatory authority for containers used to transport compressed gases formerly resided with the Interstate Commerce Commission (ICC). Similarly, such authority in Canada was formerly the jurisdiction of the Canadian Transport

Commission (CTC). The specification markings on some containers still in service may reflect this.

Inspectors examine compressed gas containers built in accordance with the applicable regulations as they are made and approve them for container markings (usually stamped) which indicate that the marked container fulfills the regulatory specifications identified by the marks. Independent inspectors authorized by the U.S. Department of Transportation are required for high pressure cylinders. Inspectors are required to file a report on each individual container that is made under the specifications of these regulations before the container goes into service, and each container made is identified and thus registered by a serial number and manufacturer's symbol. Compliance with one or more of these regulations is often stipulated in state and insurance company requirements as well as in federal regulations. (See Fig. 4-1.)

CYLINDERS AND SMALL CONTAINERS

Cylinders for compressed gases are generally defined in the DOT and TC specifications as containers having a maximum water capacity of 1000 lb (453.6 kg) or less. This is approximately the equivalent of 120 gallons (454.2 L).

Specifications 3AX, 3AAX, and 3T permit the use of larger cylinders. A popular size for mounting on a vehicle chassis has a water capacity of approximately 5000 lb (2268.0 kg). Cylinders are made in a wide variety of lengths and diameters and range in capacities from the authorized maximum down to a cubic foot or less. Specification 3AX, 3AAX, and 3T cylinders have a minimum water capacity of 1000 lb (453.6 kg) with no specified upper volumetric limit.

Those cylinders which have relatively broad and squat proportions are generally made for low pressure liquefied gas service;

Fig. 4-1. Typical air separation facility showing cold boxes, storage tanks, and bulk cryogenic transport vehicles.

cylinders which have relatively tall and thin proportions are primarily used for high pressure service of the permanent gases. Cylinders are most often made with flat bottoms or are fitted with footrings so that they may stand on end and be securely fastened in an upright position while their contents are being withdrawn. Some cylinders, though, have hemispherically-rounded sealed ends and are designed for withdrawal of contents while securely fastened in a horizontal or slanted position. One end of the cylinder (or both ends, in some special cases) is tapered into a neck which is tapped with screw threads for attachment of a cylinder valve.

Cylinders are the type of compressed gas containers most widely authorized for different means of shipment. They are the most generally accepted type of container for air shipment in the case of most gases. (See Fig. 4-2.)

Cylinder Manufacture

Cylinders are made from seamless tubing, brazed or welded tubing, billets in the billet-piercing process, or flat sheets drawn to cylindrical shapes in large punch-press dies. Sealed ends are made either by spinning in a lathe while being heated by flame at forging temperatures, by forging, or by die-drawing. The ends of some cylinders are closed and sealed by spinning or forging. In some instances, the sealed end of such cylinders is drilled, threaded, and then "plugged" with an additional metal piece.

Fig. 4-2. A variety of compressed gas cylinder types.

See Title 49 of the U.S. *Code of Federal Regulations* (49 CFR) Part 178, or equivalent Canadian regulations, for detailed specification requirements for cylinder manufacture. [1] and [2]

Provision for Interchangeable Use of DOT and TC Specification Cylinders

Under the DOT and TC regulations, a cylinder meeting any one specification may be authorized for shipping a number of different gases, or any one gas may be authorized for shipment in a number of different specification cylinders. This differs from the common European practice, in which a cylinder is marked for dedicated service for a single designated gas. The one exception in North American practice involves acetylene, which is authorized for shipment as a "dissolved" gas only in cylinders filled with a porous monolithic mass and meeting Specification DOT-8, DOT-8AL, TC-8, or TC-8AL. Such cylinders must not be charged with any gas except acetylene.

Required Cylinder Markings

Under the DOT and TC regulations, cylinders must be marked to indicate (1) the specification under which they were made, (2) the service pressure for which they were designed, (3) a serial number of the manufacturer, (4) the symbol of the inspector (if required), and (5) a symbol indicating the manufacturer. The markings are usually stamped into the shoulder of the cylinder (the part sloping up to the neck) or into the valve guard ring welded to the cylinder. The complete detailed marking requirements for any given cylinder will be found under the appropriate specification in Title 49 of the U.S. *Code of Federal Regulations* (49 CFR) Part 178, or equivalent Canadian regulations. [1] and [2]

The markings of a typical cylinder might be arranged as follows on one side of the shoulder:

DOT-3A2015

8642
XYZ
JCN

In this example, the DOT specification is 3A, the service pressure is 2015 psig at 70°F (13 893 kPa at 21.1°C), the serial number is 8642, the manufacturer is XYZ, and in this instance the owner's symbol is JCN. These same markings could be arranged in a horizontal line around the shoulder and might appear as follows:

DOT-3A2015   8462   XYZ   JCN

Certain other markings are required. Cylinders are pressure tested as part of the manufacturing procedures, and the month and year of this initial qualifying test are required as permanent markings on the cylinder. The manufacturing date stamp will appear as 1-89 if manufactured in January 1989, and it should be placed so that the dates of subsequent retests required for requalification can be added at that time immediately below the original test date. Also, the DOT and TC specifications require that the word *SPUN* be stamped near the specification mark on cylinders which have the bottom end closure produced by spinning, or the word *PLUG* when the bottom end closure has been effected by spinning, drilling, and plugging. The symbol of the inspector (if required) is commonly placed between the month and year of the manufacturing test date. Because of space limitations, the test date and other markings are sometimes on the opposite side of the shoulder from the markings previously described. The complete markings on a cylinder might be as follows:

DOT-3A2015-SPUN

8642   1-AB-89
XYZ
JCN

AB represents the symbol of the inspector. (See Fig. 4-3.)

If a plus (+) sign appears immediately after the test date of a Specification 3A, 3AX, 3AA, 3AAX, or 3T cylinder, it means that the cylinder is authorized for charging up to 10 percent in excess of the marked service pressure. The requirements for this authorization will be found in 49 CFR 173.302. [1]

For Specification DOT-3A or 3AA cylinders, if a five-pointed star is stamped after

Fig. 4-3. Permanent markings shown stamped into the shoulder of a cylinder include: the DOT (of TC) specification (3AA); the service pressure in psig (2265); the cylinder manufacturer's symbol, the manufacturer's serial number (K161110), and the cylinder owner's symbol (BTWECO). The month and date of the initial qualification test along with the identifying mark of the inspector stamped on the opposite side of the shoulder are not shown.

the most recent test date (or following the plus mark, if applied), it indicates that the cylinder may be retested every ten years instead of every five years. The details of the requirements for qualifying a cylinder for the ten-year retest interval will be found within 49 CFR 173.34, which covers requirements for the qualification, maintenance, and use of cylinders. [1]

For cylinders stamped DOT-3 or DOT-3E without any pressure marking, the service pressure is 1800 psig (12 411 kPa). Such cylinders manufactured in recent years will be stamped DOT-3E1800.

Required markings on cylinders must not be changed except for the service pressure, which may be lowered after approval by DOT in conformance with requirements to be found within 49 CFR 173.34. [1]

Common Cylinder Types

Among the cylinder specification types manufactured in the largest quantities are the following.

DOT-3A. These are seamless carbon-steel cylinders and are produced usually for high pressure service in excess of 1800 psig (12 411 kPa) and ranging as high as 10 000 psig (68 948 kPa).

DOT-3AA. These are seamless alloy-steel cylinders used generally for the same purposes as Specification 3A cylinders. The alloy steels are heat treated by quenching and tempering to produce a higher strength so that a higher working stress can be used. This permits a lighter weight cylinder for a given size and service pressure than can be produced under Specification 3A.

DOT-3AL. These are seamless cylinders made of definitely prescribed aluminum alloys. These cylinders require a minimum service pressure of 150 psig (1034 kPa) and a maximum water capacity of 1000 lb (453.6 kg).

DOT-3E. These are small seamless carbon-steel cylinders made for a service pressure of 1800 psig (12 411 kPa) and limited in size to 2 inches maximum (5.08 cm) diameter and a length of 2 ft (69.96 cm) maximum.

DOT-3HT. These are lightweight seamless alloy-steel cylinders designed for service inside an aircraft. They are limited to a water capacity of 150 lb (68.04 kg) and are usually made for a service pressure of 1800 psig (12 411 kPa) or higher.

DOT-8 or 8 AL. These are welded steel cylinders for acetylene service only and are made to the single authorized service pressure of 250 psig (1724 kPa). Such specification cylinders may have a longitudinal welded seam and circumferential seams. These cylinders must also contain a permanent monolithic porous inner filling material which absorbs a specified amount of a solvent, usually acetone, which absorbs the acetylene.

DOT-3AAX. These are alloy-steel cylinders (trailer tubes) similar to DOT-3AA cylinders, except that the minimum volumetric capacity is 1000 lb (453.6 kg) of water. Common dimensions are 22 inches diameter by 34 ft long (55.9 cm by 10.4 m).

DOT-3T. These are higher strength alloy-steel cylinders (trailer tubes) with a minimum volumetric capacity of 1000 lb (453.6 kg) of water and a minimum service pressure of 1800 psig (12 411 kPa). Common dimensions are 22 inches diameter by 34 ft long (55.9 cm by 10.4 m).

DOT-4B, 4BA, and 4BW. These are welded or brazed low-carbon steel cylinders with a volumetric capacity of not over 1000 lb (453.6 kg) of water, normally made for a service pressure of 240 psig (1655 kPa).

DOT-39. These are nonrefillable (single-trip) cylinders made of seamless, welded or brazed steel or aluminum with a maximum volumetric capacity of 55 lb (24.9 kg) of water and a maximum service pressure of 500 psig (3447 kPa). A maximum volumetric capacity of 10 lb (4.54 kg) of water can be produced for service pressures over 500 psig (3447 kPa).

Small Containers Exempt from Cylinder Requirements

Under the DOT and TC regulations, limited quantities of compressed gases are exempted from the requirements under certain conditions. The details of these exceptions will be found in 49 CFR 172.101 and 173.306. [1]

REGULATORS AND CONTROL VALVES FOR CYLINDERS AND OTHER CONTAINERS

To insure the proper discharge of compressed gases from cylinders and other containers, the correct regulators and control valves must be used for any nonliquefied or liquefied compressed gas. This section primarily discusses regulators and control valves used for cylinders.

Nonliquefied Gases

To insure the safe withdrawal of a nonliquefied gas from a cylinder, the pressure must be reduced to a safe value. This is most commonly done with an automatic pressure regulator like that shown in Fig. 4-4.

Cylinders are delivered to users with cylinder valves complying with national standards. Refer to Chapter 8 and CGA V-1, *American National, Canadian, and Compressed Gas Association Standard for Compressed Gas Cylinder Valve Outlet and Inlet Connections.* [4] The regulator is attached to the cylinder valve outlet with the regulator connection. Basically, the regulator consists

Fig. 4-4. Two stage pressure regulator.

of a spring-loaded (or gas-loaded) diaphragm that controls the opening or closing of the delivery orifice. The diaphragm control can be set by hand to maintain a constant delivery pressure at any value within the range for which it is designed. Once the delivery pressure is set, the diaphragm acts to open or close the delivery orifice to keep the delivery pressure constant. Gauges can be attached to the regulator to show the delivery pressure chosen and the cylinder pressure. A flow-control valve that is also part of the regulator controls the volume of gas delivered at the chosen delivery pressure.

Essential Factors in Choosing Regulators

The choice of a regulator for use in a specific gas service depends on four factors:

(1) The design and materials for safe and trouble-free operations with the specific gas and pressures involved.
(2) The range of delivery pressures required.
(3) The degree of accuracy of delivery pressure to be maintained.
(4) The flow rate required.

There are two basic types of automatic pressure regulators: the single-stage, and the double or two-stage. Generally, a two-stage regulator will deliver a more constant pressure under more widely varying operating conditions than will a single-stage regulator. A single-stage regulator will show a slight variation in delivery pressure as the cylinder pressure drops, as well as a drop in delivery pressure as the flow rate is increased. The two-stage regulator is less subject to such delivery pressure variations. The single-stage regulator will also have a higher "lockup" pressure (the pressure increase above delivery set point necessary to stop flow) than the two-stage regulator at relatively large flow rates.

Liquefied Gases

Liquefied gases in a cylinder exist in liquid and gaseous form at a pressure equal to the vapor pressure of the particular gas. Usually the gas phase of the liquefied gas is drawn for use. The cylinder pressure will remain constant at the vapor pressure of the material as long as there is any liquid remaining in the cylinder. When the contents of the cylinder are withdrawn to the point that no liquid remains, the pressure in the cylinder will begin to diminish. Therefore, a single-stage automatic pressure regulator may be used for a constant delivery pressure while there is still liquid in the cylinder. For example, 80 percent of the contents of a carbon dioxide cylinder can be drawn off at room temperature before there is no liquid phase of the gas left and the cylinder pressure drops below the vapor pressure. For withdrawing all of the liquefied gas in a cylinder at a constant delivery pressure, a two-stage regulator should be used.

Certain problems can develop when removing the gas phase of liquefied gas. Rapid removal of the gas may cause the liquid to cool too rapidly, causing the pressure and flow to drop below the required level. To prevent this, cylinders may be heated in a water bath with the temperature no higher than 120°F (48.9°C).

For the controlled removal of the liquid phase of a liquefied gas, a manual flow-control valve is used. Special liquid flow regulators are also available. Removal of the liquid must be done at the vapor pressure of the material. Care must be taken to prevent blockage of the gas line downstream from a user's heat exchanger, which would cause excessive pressure buildup in the heat exchanger and the cylinder. Pressure relief devices should be installed in all transfer lines to relieve any sudden and dangerous hydrostatic or vapor pressure buildups.

Handling and Use of Automatic Regulators

When an automatic pressure regulator is attached to a cylinder, the threads must not be forced. If the regulator does not fit, it must in no way be forced. The poor fit of a regulator may indicate that it is the wrong regulator for that specific type of gas. However, users should also make sure in advance that the correct regulator for the gas service intended is chosen by acting on the advice of a responsible source, such as the supplier or producer of the gas or regulators.

Use the following procedure to obtain the proper delivery pressure from an automatic regulator:

(1) After the regulator has been attached to the cylinder valve outlet, rotate the delivery pressure adjusting screw counterclockwise until it turns freely.

(2) Open the cylinder valve slowly until the tank gauge on the regulator registers the cylinder pressure. At this point, the cylinder pressure should be checked to see if it is at the expected value. A large error may indicate leakage in the cylinder valve.

(3) With the flow-control valve at the regulator outlet closed, turn the delivery pressure adjusting screw clockwise until the required delivery pressure is reached. Control of flow can be regulated by means of the flow-control valve installed in the regulator outlet or by a supplementary valve placed by the user in a pipeline downstream from the regulator. The regulator itself should not be used as a flow control by adjusting the pressure to obtain different flow rates. This defeats the purpose of the pressure regulator. In some cases, where higher flows are obtained in this manner, the pressure setting may be in excess of the design pressure of the user's system of piping and devices for employing the gas.

Manual Flow Controls

Manual flow controls may be used when an intermittent flow is needed and an operator will be present at all times. A manual flow control, illustrated in Fig. 4-5, is simply a valve operated manually to deliver the proper amount of gas. A very fine control of the flow of gas can be obtained. However, it is important to remember that dangerous pressures can build up in a closed system or in one that becomes plugged, since under these circumstances there is no means for the automatic prevention or release of excessive pressures.

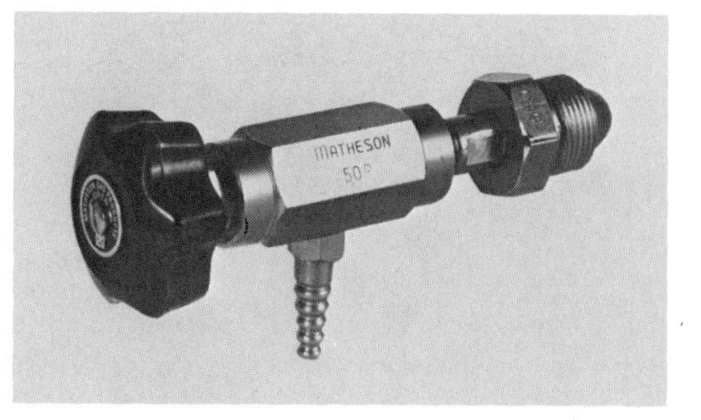

Fig. 4-5. Manual needle valve.

Cylinder Valves

There are a number of different types of cylinder valves affixed to the cylinders that are deliverd to users. Regulators are attached directly to these cylinder valves by the user. Four commonly used types of cylinder valves are illustrated in Figs. 4-6 through 4-9. Many valve outlet and inlet connections have been standardized by the Compressed Gas Association for the different families of gases to prevent the interchange of regulator equipment between gases which are not compatible. These standardized connections have been adopted by the American National Standards Institute and the Canadian

Fig. 4-6. Typical cylinder valve.

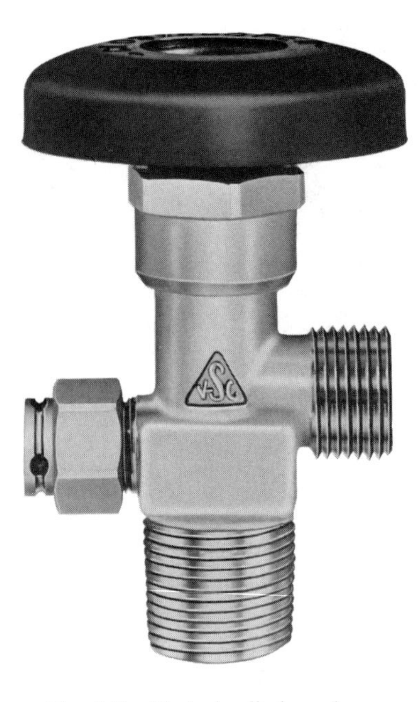

Fig. 4-7. Typical cylinder valve.

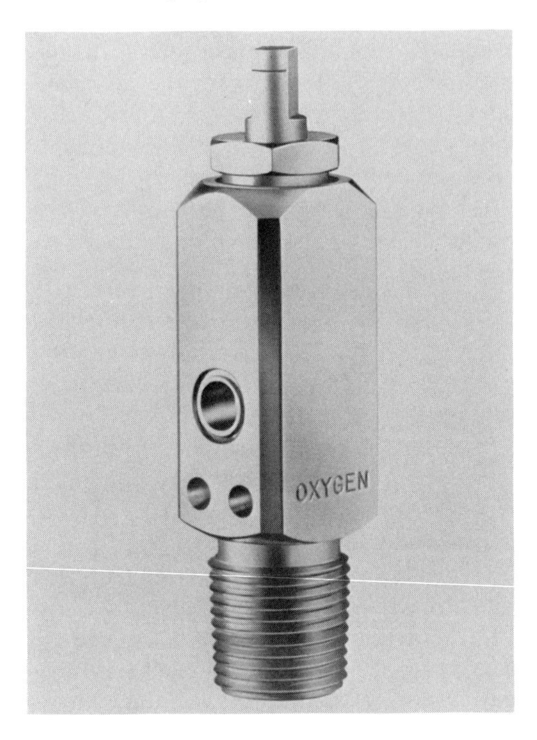

Fig. 4-9. Typical cylinder valve.

Standards Association. Refer to Chapter 8 and CGA V-1, *American National, Canadian, and Compressed Gas Association Standard for Compressed Gas Cylinder Valve Outlet and Inlet Connections.* [4]

The use of adapters to change the outlet size of a cylinder valve defeats the whole purpose of standardizing the valve outlets. Adapters should be used with care only on gases definitely known to be compatible. Equipment for certain gases, such as oxygen, should never be used with other gases. Gases which are oil-pumped can cause an oil film to coat internal parts of gas delivery systems. The introduction of oxygen into such an oil-coated system can cause a fire or explosion.

Fig. 4-8. Typical cylinder valve.

CONTAINERS FOR SHIPPING BY RAIL IN BULK

Bulk rail shipment of compressed gases is authorized under DOT and TC regulations in containers of three major kinds: single-unit tank cars; multi-unit tank cars; and Specification 107A tank cars. Adaptations of existing tank car specifications, and proposed specifications for new types of cars, must be approved by the Tank Car Committee of the Association of American Railroads (AAR).

In the United States, federal regulations covering the interstate rail transportation of compressed gases are published in Title 49 of the U.S. *Code of Federal Regulations* (49

CFR) Parts 100–199. [1] DOT regulations require that compressed gases be shipped in containers manufactured to DOT specifications and maintained in accordance with DOT regulations. In Canada, containers in which compressed gases are shipped must comply with specifications and regulations of Transport Canada. [5]

For selection of the proper tank car specification for a particular product, refer to 49 CFR 173.314 of the DOT regulations or equivalent Canadian regulations. Qualification, maintenance, and use of tank cars must comply with 49 CFR 173.31. In the United States, the placarding and handling of tank cars must comply with 49 CFR Part 174. Tank car specification requirements are found in 49 CFR Part 179. [1] In Canada, placarding is covered by Section 5 of the *Transportation of Dangerous Goods Regulations,* while tank specifications and handling are covered by other TC regulations. [2] and [5]

Shippers are strongly recommended to refer to the most current federal regulations for information. Sections of the regulations referenced in this *Handbook* may have changed since the date of publication. The most current issue of the *AAR Specifications for Tank Cars* should also be reviewed. [6]

Tank Car Specifications

A typical compressed gas tank car specification is TC/DOT-105A300W. Each part of the specification number has some significance: TC is the designation of the Transport Canada; DOT is the U.S. Department of Transportation designation. These jurisdictions permit the use of their designations individually or together in the manner shown. The first number, 105 in the example, is the DOT class of the tank car. A Class 105 tank car is an insulated, pressure car. For Class 105, 112, and 114 tank cars, the letter following the car class will indicate whether or not the tank car has thermal (fire) protection or tank head puncture protection. The letter *A,* shown in the example, indicates the tank car has neither. The letter *J* indicates the tank car has jacketed thermal protection and head puncture protection. The letter *T* indicates the tank car has a coating for thermal protection and has head shields for puncture protection. The letter *S* indicates a tank car with head shields but without thermal protection.

The second set of numbers is the marked test pressure of the tank, 300 psig (2068 kPa) in the example TC/DOT-105A300W. The last letter, or set of letters, gives information about the tank material and the method of fabrication. The letter *W,* in the example, indicates a steel fusion-welded tank. If the letters were *ALW,* it would indicate an aluminum fusion-welded tank. The letter *X* would signify a tank with a fusion-welded longitudinal seam and forge-welded head seams. If the suffix letter is *F,* it denotes a forge-welded tank.

A Class TC/DOT-113 tank car is vacuum-insulated, and the insulation system is designed to meet the holding time requirements of the specification. These cars are designed for specific loading and shipping temperatures and have certain material and fittings requirements. A typical specification is TC/DOT-113C120W. The first letter following the class number in this example is *C.* This designates a tank suitable for temperatures down to −260°F (−162°C). The letter *A* would designate a tank suitable for temperatures down to −423°F (−253°C), and the letter *D* would signify a tank suitable for −155°F (−104°C). The second set of numbers, 120 in the example, is the test pressure in psig. The last letter, *W* in this example, indicates a fusion-welded tank.

Association of American Railroad tank cars are for nonregulated commodity services. Most AAR tank cars have TC and DOT counterparts.

Single-Unit Tank Cars

Single-unit tank cars consist of a single large pressure tank which is permanently mounted to the car underframe. The tanks can be compartmented or noncompartmented. Both insulated and noninsulated tank cars are used in compressed gas service. Cryogenic liquids and refrigerated liquids must be shipped in insulated tank cars.

Single-unit tank cars built to TC/DOT specifications after November 30, 1970, must not exceed 34 500 gallons (130.6 m^3) capacity and 263 000 lb (119 294 kg) gross weight on rail. Since the early 1960s, most tank cars for liquefied gases that are shipped in large quantities, such as propane and ammonia, have nominal water capacities in the range of 30 000 gallons (113.6 m^3) to 33 500 gallons (126.8 m^3).

Tank car specifications authorized by the United States and Canada for compressed gases are the Class TC/DOT-105, 112, and 114 tank cars. (See Fig. 4-10.) The DOT-105 tank car is insulated, but the other two specification cars are not.

Single-unit tank cars for compressed gas typically have a manway located at the top center of the tank. The venting, loading, and unloading valves are directly bolted to seatings on the manway cover plate and are protected with a housing. When used, the gauging device, sampling valve, and thermometer well are located on the cover plate within the protective housing.

Class TC/DOT-114 tank cars may have the manway located other than at the top of the tank, and the valves and fittings need not be mounted on the manway cover plate. Purge nozzles may be provided in each head. TC/DOT-114 cars may have bottom outlets, but cars commonly used for compressed gases have top fittings.

Single-unit tank cars are protected from excessive internal pressure by a spring-loaded pressure relief valve located on the manway cover plate. Excess flow check valves are installed on the liquid eduction and vapor lines of many tank cars in compressed gas service. These excess flow valves are intended to contain the tank contents in transit, or in a wreck environment in the event that the valve is sheared off or other incidents where uncontrolled release of product may occur.

The typical compressed gas tank car has two valves in line with the track centerline. The two valves are connected to liquid eduction pipes that extend to the bottom of the cars. These are referred to as liquid valves or eduction valves. A third valve is mounted toward the side of the car, terminating in the tank's vapor space. Occasionally, a fourth valve is located on the opposite side in contact with the vapor space. These valves are referred to as vapor valves.

Class TC/DOT-113 tank cars are vacuum-

Fig. 4-10. TC/DOT-112J340W tank car used to transport LP-gas or anhydrous ammonia.

insulated cars having an inner container and carbon-steel outer shell. (See Fig. 4-11.) They are used to transport cryogenic liquids such as ethylene and hydrogen. Class AAR-204 tank cars are similar to the Class TC/DOT-113 tank car but can only be used to transport products not regulated by TC or DOT. For example, atmospheric gases and helium, when shipped as cryogenic liquids so that the pressure in the container will not exceed 25.3 psig (174 kPa), are not regulated in the U.S. except for incident reports, marking, and placarding requirements.

Class 107A Tank Cars

The Class TC/DOT-107A tank car specification is for seamless steel tanks to be mounted on or forming part of a car. TC/DOT-107A tank cars are uninsulated high pressure service cars having several permanently mounted seamless forged-and-drawn steel tanks designed to a maximum stress level in the shell. (See Fig. 4-12.)

The stenciled specification is TC/DOT-107A****, where the marked test pressure is indicated by the figures substituted for the **** in the specification. The letter *A* in the specification has no significance. The gas pressure at 130°F (54.4°C) in the tank must not exceed 70 percent of the marked test pressure of the tank.

The 107A tank car is typically a car carrying clustered sets of long tubular tanks and is used to transport bulk quantities of nonliquefied gases such as argon, helium, hydrogen, nitrogen, and oxygen at high pressure. A common tank car configuration consists of about 30 tubes connected or manifolded to a common header at one end of the car. Test pressures for the 107A tank cars range up to 3500 psig (24 132 kPa).

Multi-unit or TMU Tank Cars

Multi-unit tank cars consist of a railroad flatcar that carries up to 15 large cylindrical, uninsulated, pressure tanks crosswise on the car. (See Fig. 4-13.) Class TC/DOT-106A and 110A specification tanks are used in multi-unit tank service. The tanks are designed to be removed from the car structure for filling and emptying. The filled tanks are lifted onto or off the car by crane or hoist.

Each multi-unit tank has a water capacity of at least 1500 lb (680 kg) and not more than 2600 lb (1179 kg). All openings are located in the heads, and the loading and unloading valves are protected by a detachable protective housing.

Fig. 4-11. A typical AAR 204W tank car used for transporting cryogenic liquids.

Fig. 4-12. TC/DOT Specification 107A tank car.

The listed DOT specifications include 106A500X, 106A800X, 110A500W, 110A-800W, and 110A1000W. Class TC/DOT-110A tanks have fusion-welded heads formed concave to pressure; Class TC/DOT-106A tanks have forge-welded heads formed convex to pressure.

The popular name ton containers, or ton multi-unit tanks (TMU), is derived from their first use, which was to transport a ton of liquefied chlorine in each tank. TMU tanks are also authorized for the highway shipment of some compressed gases.

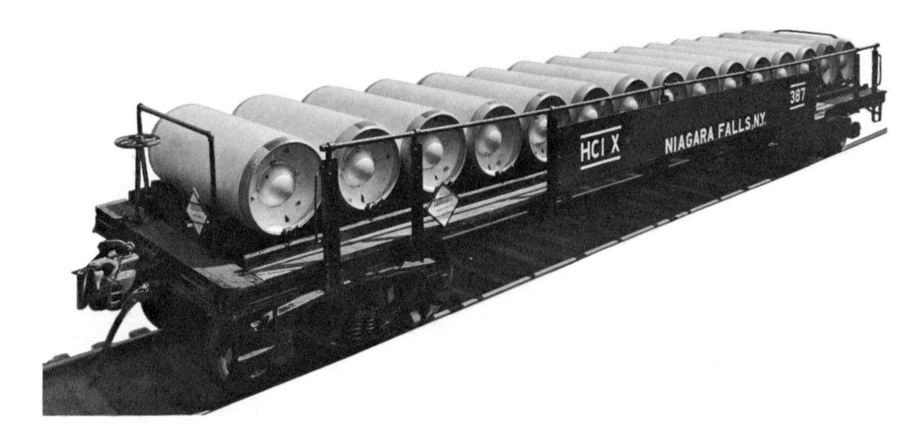

Fig. 4-13. Multi-unit tank car (TMU) for rail transport of compressed gases.

CARGO TANKS AND TUBE TRAILERS

Cargo Tanks

A cargo tank is defined as any tank permanently attached to or forming a part of any motor vehicle or any bulk liquid or compressed gas packaging not permanently attached to any motor vehicle, which by reason of its size, construction, or attachment to a motor vehicle, is loaded or unloaded without being removed from the motor vehicle.

Cargo tank capacities range up to 11 300 gallons (42.77 m^3). Although no minimum capacity is given for them in the specifications, cargo tanks are normally large-capacity tanks permanently mounted on truck bodies, or they form a semitrailer body. In some instances, truck tractors draw double tank trailers containing compressed gas.

In the United States, the cargo tank selected for transporting a specific compressed gas must be a container authorized for that product and the container must be qualified in accordance with DOT regulations. Refer to 49 CFR 173.33 for a complete listing of DOT tank specifications authorized for use. [1] Requirements for qualification, maintenance, and use of cargo tanks are found in this section of the federal regulations. Refer to 49 CFR 173.315 and 173.318 for detailed shipping requirements for compressed gases in cargo tanks. [1]

Canadian requirements are outlined in CSA B620, *Highway Tanks and Portable Tanks for the Transportation of Dangerous Goods,* and CSA B622, *CSA Standard for the Selection and Use of Tank Trucks, Tank Trailers and Portable Tanks for the Transportation of Dangerous Goods for Class 2 By Road,* which are published by the Canadian Standards Association. [7] and [8] The specifications most often authorized by DOT for compressed gas cargo tanks are MC-330, TC/MC-331 and TC/MC-338, and TC-341. Specification 330 is obsolete for new construction but may continue in use. In Canada, Specification TC-341 applies to inert cryogenic liquids at pressures less than 40 psig (276 kPa). (See Fig. 4-14.)

Cargo tanks complying with Specifications MC-330 or 331 must be of ASME Pressure Vessel Code construction, and must have a design pressure of not less than 100 psig (689 kPa) nor more than 500 psig (3447 kPa). [3] Tanks of either specification may be insulated to meet product requirements. Refer to the individual gas monographs in Part III of this *Handbook* for further product information.

Cargo tanks complying with Specification 338 or 341 are insulated tanks used for the transportation of refrigerated liquids (cryogenic liquids) such as ethylene. (See Fig. 4-15). Specification 338 cargo tanks must comply with the ASME Code and must have a design pressure of at least 25.3 psig (174 kPa) but not more than 500 psig (3447 kPa). Refer to 49 CFR 173.318 for detailed shipping requirements in the United States. Specification design requirements for cargo tanks transporting compressed gases must comply with 49 CFR Part 178, Subpart J. Shipping papers, marking, and placarding of cargo tanks shipping compressed gas must comply with 49 CFR Part 172. [1] In Canada, the *Transport of Dangerous Goods Regulations* will apply. [2]

Each cargo tank must be provided with one or more pressure relief devices which, unless otherwise specified for a particular product, must be pressure relief valves of the spring-loaded type. Refer to Chapter 7 for detailed information on pressure relief devices.

Tube Trailers

High pressure nonliquefied gases, such as hydrogen and helium at ambient temperatures, are often shipped by highway in tube trailers. These are truck semitrailers on which a number of very long gas cylinders have been mounted and manifolded into a common header. (See Figs. 4-16 and 4-17.) Tube trailer service pressures are as high as 2600 psig (17 926 kPa) or more. The tubular cylinders of the trailers are often made according to the cylinder Specifications 3A or 3AA, or to cylinder Specification 3AX,

Fig. 4-14. MC-331 cargo tanks. (top) bobtail (bottom) transport vehicle

3AAX, or 3T. Specifications 3AX, 3AAX, and 3T are commonly used in new production for containers approximately 22 inches (55.88 cm) in diameter instead of the older $9\frac{5}{8}$-inch (24.45-cm) diameter tubes which were made to Specifications 3A and 3AA. Specifications 3AX, 3AAX, and 3T cylinders have a minimum size of 1000 lb (454 kg) water capacity under the TC and DOT regulations. Tube trailers in the United States have been built to carry as much as 180 000 standard ft^3 (5097 m^3) of helium. In some Canadian provinces, capacities up to 200 000 ft^3 (5663 m^3) of hydrogen are possible.

PORTABLE TANKS AND ISO CONTAINERS

Portable Tanks

A portable tank is defined by TC and DOT as any packaging, except a cylinder having a 1000-lb (454-kg) or less water capacity, over 110 U.S. gallons (380 L) capacity and designed primarily to be loaded into or on, or temporarily attached to, a transport vehicle or ship. A portable tank is equipped with skids, mounting, or accessories to facilitate handling of the tank by mechanical means. As defined, a portable tank is not a cargo

Fig. 4-15. An insulated trailer for bulk transport of cryogenic liquids (Specification CGA-341) in liquid nitrogen service.

tank, tank car tank, tank of the TC/DOT-106A or 110A type, or trailers carrying 3AX, 3AAX, or 3T cylinders. If a portable tank is used as a cargo tank, it must comply with all the requirements prescribed for cargo tanks.

Portable tanks complying with TC/DOT Specification 51 are authorized for shipping many compressed gases. Specification 51 provides for steel tanks of seamless or welded construction which must be in excess of 1000 lb water capacity (454 kg). The tank service pressure cannot be less than 100 psig (689 kPa) nor more than 500 psig (3447 kg). (See Fig. 4-18.)

There are other portable tank specifications which may be continued in use. For complete listings, refer to Title 49 of the U.S. *Code of Federal Regulations,* specifically 49

Fig. 4-16. Semitrailer using large-diameter high-pressure tubes, commonly known as jumbo tube trailer.

Fig. 4-17. Semitrailer using small-diameter high-pressure tubes, commonly known as a small tube trailer.

CFR 173.32 or Canadian Standards Association Standard B620. [1] and [7] Requirements for qualification, maintenance, and use of portable tanks are also found in 49 CFR 173.32 and in the *Transportation of Dangerous Goods Regulations.* In the United States, refer to 49 CFR 173.315 for detailed shipping requirements for compressed gases in portable tanks; in Canada, refer to the *Transportation of Dangerous Goods Regulations.* [1] and [2]

In the United States, specification design requirements for cargo tanks transporting compressed gases must comply with 49 CFR Part 178, Subpart H. [1] Shipping papers, marking, and placarding of portable tanks shipping compressed gases must comply with 49 CFR Part 172. [1] In Canada, specifications are provided in CSA B620, *Highway Tanks and Portable Tanks for the Transportation of Dangerous Goods.* [7] Shipping papers, marking, and placarding are described

Fig. 4-18. Specification 51 portable tank.

in the *Transportation of Dangerous Goods Regulations.* [2]

Each portable tank must be provided with one or more pressure relief devices which, unless otherwise specified for a particular product, must be pressure relief valves of the spring-loaded type. Refer to Chapter 7 for detailed information on pressure relief devices.

ISO Containers

ISO containers for compressed gases are gas tanks permanently mounted in a frame conforming to the standards of the International Organization for Standardization (ISO). A catalog of ISO standards is available from International Organization for Standardization, Central Secretariat, 1 rue de Varembe, Case postale 56, CH-1211 Geneve 20, Switzerland/Suisse.

The frame of an ISO container and its corner castings are specially designed and dimensioned to be used in multimodal transportation service on container ships, special highway chassis, and container-on-flatcar (COFC) railroad equipment. (See Fig. 4-19.) The gas tank in an ISO frame is designed to the requirements of TC/DOT Specification 51 or International Maritime Organization (IMO) Type 5 for the particular gas being transported. See the previous section for details on Specification 51 portable tanks.

Tank containers which meet the requirements of AAR.600, "Specifications for Acceptability of Tank Containers in COFC Service," may be accepted for rail transportation on container cars having end-of-car cushioning. Positive lock corner casting restraint or another secure tie-down method is required. AAR.600 is part of *AAR Specifications for Tank Cars* (M-1002), published by the Association of American Railroads. [6]

ISO tanks are becoming an important factor in international trade.

CONTAINERS FOR SHIPPING BY WATER IN BULK

Practically all types of containers authorized for shipping compressed gases on land

Fig. 4-19. ISO container for international shipment.

are also authorized under some conditions for water shipment; they include single-unit tank cars which are approved for cargo vessels or railroad car ferry vessels, and tank trailers which are approved for cargo vessels and trailerships in the shipment of certain gases.

Tankships and Tank Barges

Some tankships and many tank barges are built with fixed pressure tanks primarily for bulk water transport of compressed and liquefied gases. Regulations of the United States Coast Guard and Transport Canada set forth detailed requirements for fixed tanks or barges used for shipping ammonia and chlorine. Special permission of these regulatory agencies has authorized tank vessels for other compressed gases. The first of any importance was a ship equipped to transport LP-gases in intercoastal services and, more recently, oceangoing tankers that carry liquefied methane in insulated and refrigerated tanks from North America to Europe and from North Africa to North America. (See Figs. 4-20 and 4-21.)

Among other gases that are shipped in substantial quantities by tank barge are inhibited butadiene, anhydrous dimethylamine, liquefied hydrogen, methyl chloride, and vinyl chloride.

In Canada, movement of compressed gases by barge is covered in part by the *Transport of Dangerous Goods Regulations,* and by regulations under the Canadian Shipping Act. [2] The latter regulations are entitled *Dangerous Goods Shipping Regulations* and *Dangerous Bulk Materials Regulations.* [9] and [10]

CRYOGENIC CONTAINERS

A wide variety of containers have been developed for shipping gases liquefied at very low or cryogenic temperatures which range from about −130°F (−90°C) down to the neighborhood of absolute zero, or −459.67°F (−273.15°C). The containers usually have evacuated, high-efficiency insulation, and most of them dissipate heat absorbed by the contained cryogenic fluid by venting small amounts of vapor. (See Figs. 4-22 and 4-23.)

Fig. 4-20. Barge for bulk transport of cryogenic liquid.

Fig. 4-21. Tankship for transporting liquefied natural gas.

In the United States, the transportation of most cryogenic fluids is regulated by the U.S. Department of Transportation, and in Canada, by Transport Canada. Regulations covering the design, fabrication, and use of cryogenic liquid containers are published in Title 49 of the U.S. *Code of Federal Regulations* in Parts 171 through 179, and by the *Transportation of Dangerous Goods Regulations* in Canada. [1] and [2] Regulated containers for cryogenic fluids include TC/DOT Specification 4L cylinders, TC/MC-338 cargo tanks or semitrailers, and Class 113 railroad tank cars. TC/DOT-4L cylinders are generally used for shipping liquid argon, helium, nitrogen, and oxygen at pressures up to 625 psig (4309 kPa), liquid neon up to 295 psig (2034 kPa), and liquid hydrogen up to 17 psig (117 kPa). (See Figs. 4-24 and 4-25.)

TC/MC-338 or TC-341 cargo tanks are used to ship a variety of flammable and nonflammable cryogenic liquids at pressures up to 500 psig (3447 kPa). Liquid hydrogen and

Fig. 4-22. Cryogenic containers ranging from 3 liters to 47 liters capacity.

Fig. 4-23. Cryogenic containers ranging from 160 liters (3 on right) to 210 liters capacity.

liquid ethylene are authorized for rail shipment in specification DOT-113A60W and 113C120W tank cars, respectively. Nonspecification tank cars such as AAR240W are used to transport nonflammable liquefied atmospheric gases at pressures below 25.3 psig (174 kPa).

Cryogenic fluids are generally subject to DOT regulations when they are transported at pressures above 25.3 psig (174 kPa), or are flammable regardless of pressure. The *Transportation of Dangerous Goods Regulations* of Transport Canada regulates all material encompassed by its criteria for compressed gases. [2] Situations involving the shipment of a cryogenic fluid in a nonspecification container above 25.3 psig (174 kPa) can be authorized by DOT exemption. Both authorities retain the right to issue exemptions or permits of exception from their regulations.

A number of international organizations publish design and operating requirements for containers used to transport cryogenic liquids outside North America. Included among these organizations are:

- ADR/RID—European Agreement Concerning the International Carriage of Dangerous Goods by Road (ADR) and Rail (RID)
- IATA—International Air Transport Association
- ICAO—International Civil Aviation Organization
- IMO—International Maritime Organization

See Chapter 2 for further information on these international organizations.

STATIONARY STORAGE CONTAINERS

While compressed and liquefied gases are often stored by users of smaller quantities in the shipping containers in which they are received, that is, in banks or storerooms of cylinders, in TMU tanks and portable tanks, in high pressure tube trailers or Specification 107A tank cars, some users and manufacturers require a more efficient means for handling larger quantities on-site. (See Fig. 4-26.)

Stationary storage containers are pressure vessels conforming to the *ASME Boiler and Pressure Vessel Code.* [3] Compressed gases in stationary storage tanks are often trans-

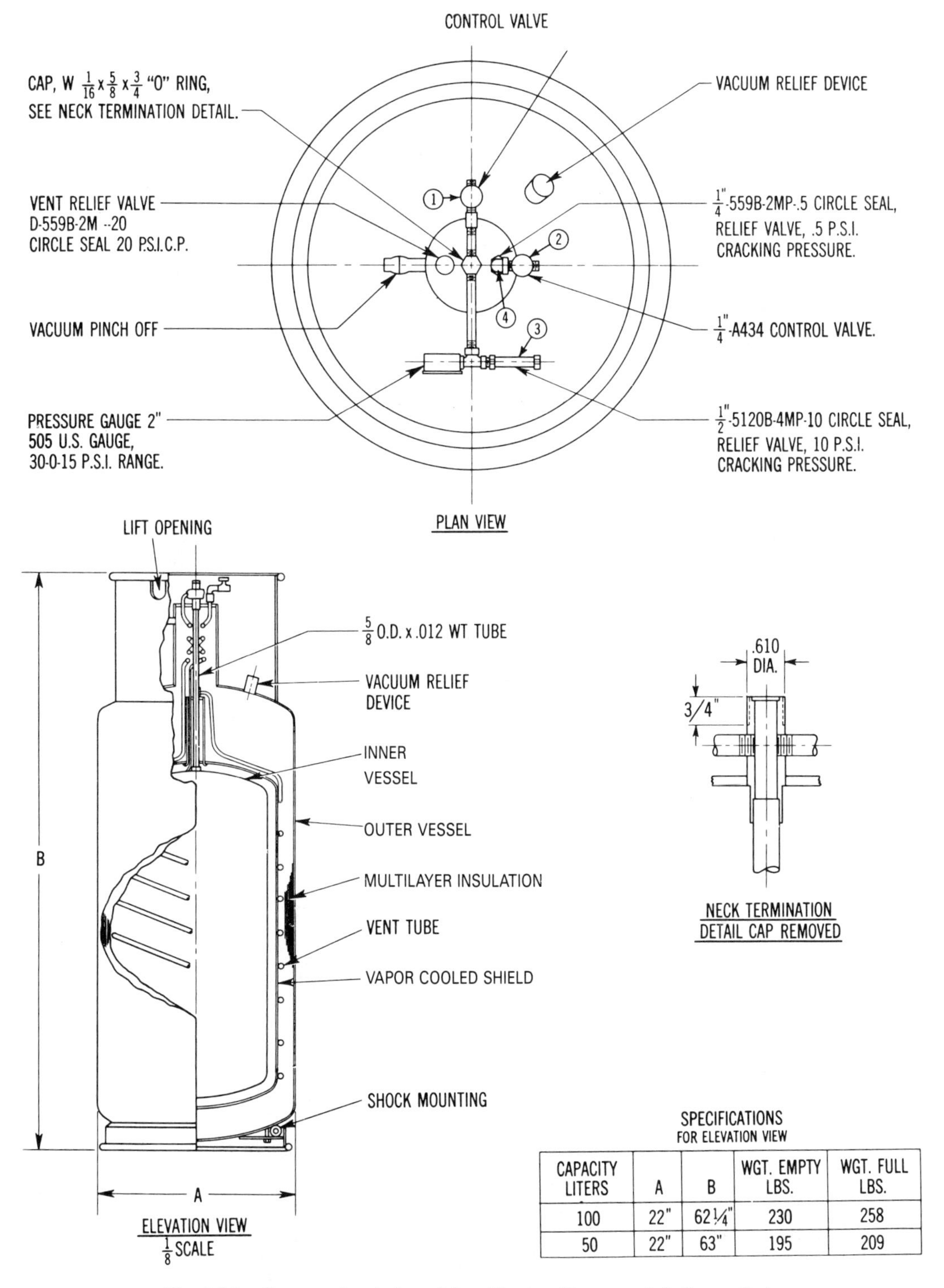

CAPACITY LITERS	A	B	WGT. EMPTY LBS.	WGT. FULL LBS.
100	22"	62¼"	230	258
50	22"	63"	195	209

Fig. 4-24. Cross-sectional view of Specification 4L cryogenic helium cylinder.

Fig. 4–25. Typical customer liquid oxygen storage tank.

Fig. 4–26. Aerial view of tanks for typical storage installation for propane or ammonia in open farm country. Siding and unloading gear for receiving shipments by rail appear at bottom left; truck-loading pumps for local deliveries are at center right.

ferred by manufacturers into shipping containers as compressed or liquefied gases, or cryogenic liquids.

Gases such as argon, oxygen, and nitrogen are often stored in a more compact manner as cryogenic liquids, with vaporizing units sometimes provided for converting the liquid to vapor by utilizing the surrounding ambient air as the heating medium. The liquid storage container is constructed of stainless steel or an equivalent metal suitable for low temperature service in accordance with the ASME Code. [3] The container is usually ASME Code stamped for a design working pressure of about 250 psig (1724 kPa) with a gastight carbon-steel jacket enclosing the container and the space between filled with insulation. A high vacuum is usually maintained in this annular space. These containers usually range in size from several hundred gallons up to about 10 000 gallons (37 854 L).

Bulk liquefied petroleum gases are normally stored in uninsulated carbon-steel vessels that are generally ASME Code designed for a working pressure of 250 psig (1724 kPa). [3] These individual containers range in size from a water capacity of 500 gallons to 90 000 gallons (1893 L to 340 687 L), and are the stationary containers used at commercial, industrial, or LP-gas bulk plant installations. For storage of very large quantities of LP-gas, for example, capacities of 2 000 000 gallons (7 570 824 L) or more such as are found at marine terminals, an insulated, refrigerated, and consequently low pressure storage vessel is used. Cavern storage is also used for LP-gas storage along the national pipeline network. (See Fig. 4–27.)

Anhydrous ammonia and most refrigerant fluorocarbons are normally stored at commercial and industrial locations in uninsulated ASME-designed vessels of carbon-steel construction with a design pressure of about 250 psig (1724 kPa). Compressed gases such as carbon dioxide having higher vapor pressures are usually stored in similar carbon-steel ASME-fabricated vessels, but

Fig. 4–27. Low, earth-covered dome caps 160,000-barrel frozen pit for storing LP-gas liquefied at moderately low temperature and pressure for which the surrounding earth serves as insulation.

are insulated and equipped with mechanical refrigeration to control the liquid carbon dioxide within required temperature and pressure limits. These vessel capacities are normally in the range of 6000 gallon to 45 000 gallon (22 712 L to 170 345 L).

PIPELINES

Transportation of compressed gases in large quantities is accomplished most efficiently by the use of pipelines. Cross-country pipelines are generally run 2 to 3 ft (61 to 92 cm) underground, although special conditions may require either significantly greater depth or aboveground construction.

Fluids in the hydrocarbon group represent the bulk of pipeline utilization, with natural gas being the predominant user within that family. The increased use of carbon dioxide for oil field stimulation has created a dramatic growth in pipeline construction for that commodity. Other compressed gases distributed by pipeline include ammonia, chlorine, nitrogen, and oxygen.

Pipelines operate over a wide range of working pressures from the 100 psig (690 kPa) range up to 7000 psig (48 263 kPa) in nitrogen service. The current upper extreme of design pressure for permanently installed pipelines does not represent a limit, and there are presently uses for compressed nitrogen, primarily in the oil field, at pressures up to 15 000 psig (103 421 kPa).

The U.S. Department of Transportation is cited predominantly throughout this text as the authority governing numerous aspects of the compressed gas industry in the United States. As the name implies, DOT is concerned with "transportation" in the vehicular sense and thus primarily attempts to safeguard the means by which compressed gas containers and their contents are transported over land, water, or air.

Pipelines, on the other hand, represent transportation systems in which the container remains stationary and only the fluid is transported. This important distinction gives rise to a different family of considerations in overseeing the design, construction, and operation of pipelines. No longer is the container readily accessible for periodic inspections. Periodic pressure tests become impractical given that pipelines are normally in continuous operation. The likelihood of finding an unanticipated product being conveyed by the line is far lower than the possibility of cross-filling a multi-use portable container.

Due to the preceding considerations, the responsibility for guiding pipeline design, construction, and operation rests not with the DOT, but with the American Society of Mechanical Engineers, which, working through Committee B31 of the American National Standards Institute, has generated the governing document ANSI B31.8, entitled, *American National Standard Code for Pressure Piping—Gas Transmission and Distribution Piping Systems.* [11]

It is interesting to note that the word *gas* as used in ANSI B31.8 has a different meaning than that normally used. ANSI B31.8 was originally written to deal with the transmission of gases suitable for use as domestic or industrial fuel (such as natural gas, manufactured gas, and liquefied petroleum gas). This is a more restrictive definition than that used in this *Handbook*. As a practical matter, however, common usage has extended the use of applicable portions of ANSI B31.8 to other gases such as the atmospheric gas group.

The intent of ANSI B31.8 is to provide for the safety of the general public and to foster employee safety insofar as this safety is affected by basic design, quality of materials and workmanship, and requirements for testing, operation, and maintenance of gas transmission and distribution facilities. The user of ANSI B31.8 is cautioned, however, that government agencies having jurisdiction may have issued regulations at variance with provisions of the code. Where such conflict exists, the code does not apply. The scope of the code is most readily illustrated by the diagram in Fig. 4-28. The contents of the code are briefly outlined below.

(1) *Materials and Equipment:* This section deals with procedures to be followed in

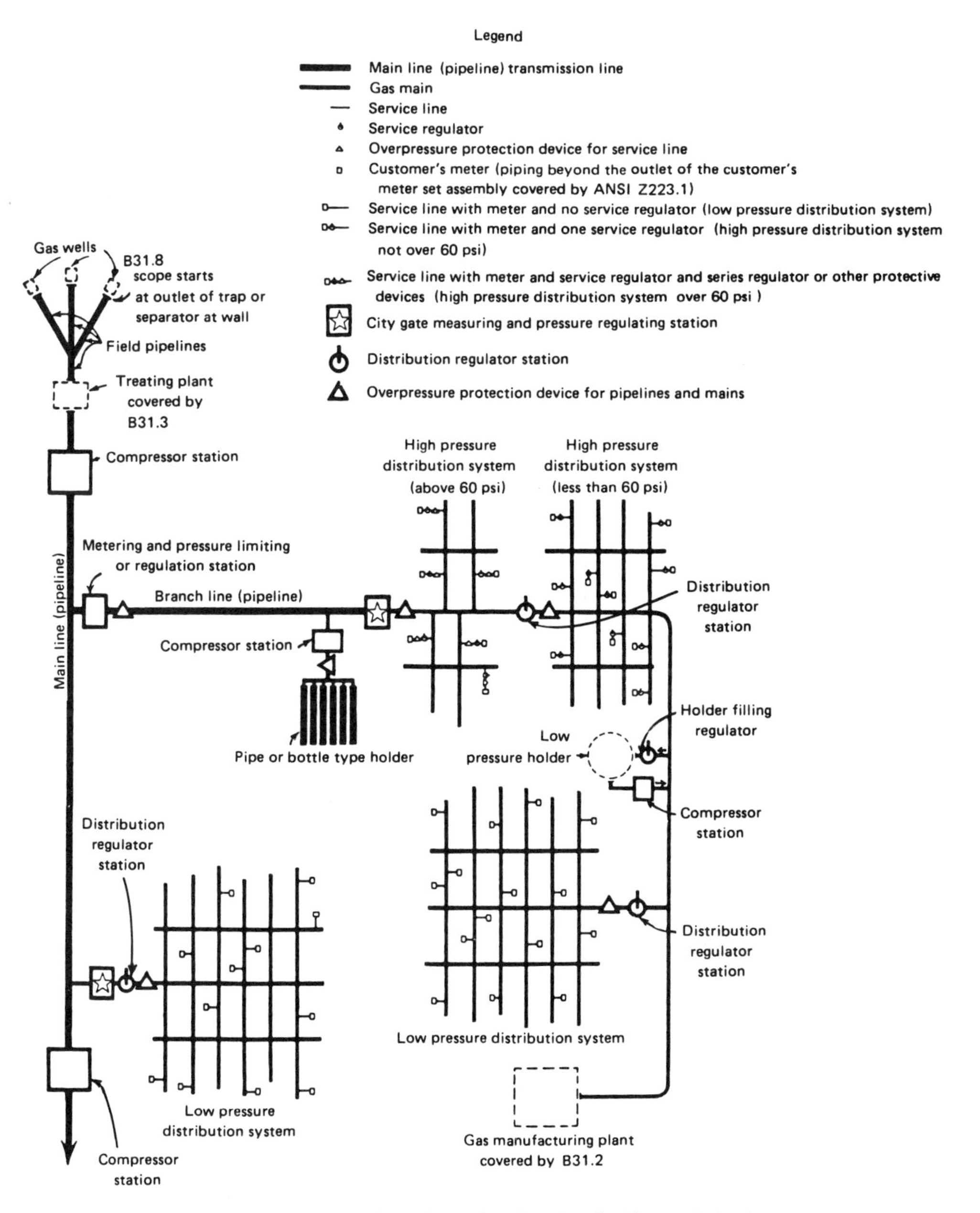

Fig. 4-28. Pipeline schematic. (*Reprinted with permission.*)

qualifying new or used pipe for use under ANSI B31.8. [11]

(2) *Welding:* Detailed in this section are many aspects of welding, including joint preparation, procedure and welder qualification, preheat, stress relieving, welding and inspection tests, and repair procedures.

This section of ANSI B31.8 introduces another document finding widespread use in the pipeline industry, API Standard 1104, *Standard for Welding Pipe Lines and Related Facilities,* issued by the American Petroleum Institute. [12] The purpose of this standard is to present the methods for the production of high quality welds through the use of qualified welders using approved welding procedures, materials, and equipment, and methods for the production of high quality radiographs to insure the proper analysis of welding quality.

(3) *Piping System Components and Fabrication Details:* Valves, flanges, bolting, gaskets, and other fittings are discussed in this section. It also embraces considerations on branch connections and extruded outlets. Requirements for provisions for expansion, flexibility, supports, and anchorage for aboveground piping are discussed, as well as anchorage for underground piping.

(4) *Design, Installation, and Testing:* Cross-country pipelines travel through all types of terrain and through desolate and populous areas. In keeping with the purpose of ANSI B31.8 to protect the safety of the public, this section concerns itself with limiting design stress levels in the pipe to successively lower levels as the population density along the pipeline route increases. In addition, this section establishes the guidelines for compressor stations, pressure-relieving devices, mainline block valve spacing, and gas holders. Inspection during construction and testing of the system are also covered in this section.

(5) *Operating and Maintenance Procedures:* ANSI B31.8 recognizes that due to the many variables involved, no detailed procedures can be formulated that will encompass all cases. This section does, however, outline prudent operating and maintenance strategies which must be followed by operating companies.

(6) *Corrosion Control:* External pipe corrosion is normally controlled by coating the exterior surface of the pipe and by installing cathodic protection equipment on the line. In addition, pipelines must be electrically insulated at all interconnections with foreign systems unless provisions are made for mutual cathodic protection of the systems.

The discussion in this section on internal corrosion control will remind the reader of the definition of *gas* as used in ANSI B31.8 [11]. Corrosive gas, therefore, is given as "gas having a water dew point normally exceeding gas temperature." The reader of this text is cautioned that while using ANSI B31.8 to design a pipeline within safe mechanical limits, this code does not address the question of compatibility between other gases and pipe materials.

(7) *Miscellaneous:* This section deals with odorization of fuel gases as well as special precautions for pipelines paralleling overhead electric transmission lines on the same right-of-way. The use of extreme care in grounding and the prevention of damaging stray fault currents in such parallel construction cases cannot be overemphasized.

In summary, ANSI B31.8 should be used to govern pipeline design, installation, and operation for the purpose of insuring the mechanical safety of the line. [11] The designer and owner must look to qualified engineering expertise when dealing with other than fuel gases in order to ensure the chemical compatibility of the materials used with the gas being conveyed. They must also look to other guides for specific cleaning procedures. One such guide is CGA G-4.4, *Industrial Practices for Gaseous Oxygen Transmission and Distribution Piping Systems.* [13]

REFERENCES

[1] *Code of Federal Regulations,* Title 49 CFR Parts 100–199 (Transportation), Superintendent of Documents, U.S. Government Printing Office, Washington, DC 20402.

[2] *Transportation of Dangerous Goods Regulations,* Canadian Government Publishing Centre, Supply and Services Canada, Ottawa, Ontario, Canada K1A 0S9.

[3] *ASME Boiler and Pressure Vessel Code* (Section VIII), American Society of Mechanical Engineers, 345 E. 47th St., New York, NY, 10017.

[4] CGA V-1, *American National, Canadian, and Compressed Gas Association Standard for Compressed Gas Cylinder Valve Outlet and Inlet Connections* (ANSI/CSA/CGA V-1), Compressed Gas Association, Inc., 1235 Jefferson Davis Highway, Arlington, VA 22202.

[5] *Transportation of Dangerous Commodities by Rail,* Canadian Government Publishing Centre, Supply and Services Canada, Ottawa, Ontario, Canada K1A 0S9.

[6] *AAR Specifications for Tank Cars* (M-1002), Association of American Railroads, 50 F Street, N.W., Washington, DC 20001.

[7] CSA B620, *Highway Tanks and Portable Tanks for the Transportation of Dangerous Goods,* Canadian Standards Association, 178 Rexdale Blvd., Rexdale (Toronto), Ontario, Canada M9W 1R3.

[8] CSA B622, *CSA Standard for the Selection and Use of Tank Trucks, Tank Trailers and Portable Tanks for the Transportation of Dangerous Goods for Class 2 By Road,* Canadian Standards Association, 178 Rexdale Blvd., Rexdale (Toronto), Ontario, Canada M9W 1R3.

[9] *Dangerous Goods Shipping Regulations,* Canadian Government Publishing Centre, Supply and Services Canada, Ottawa, Ontario, Canada K1A 0S9.

[10] *Dangerous Bulk Materials Regulations,* Canadian Government Publishing Centre, Supply and Services Canada, Ottawa, Ontario, Canada K1A 0S9.

[11] ANSI B31.8, *American National Standard Code for Pressure Piping—Gas Transmission and Distribution Piping Systems,* American National Standards Institute, 1430 Broadway, New York, NY 10018.

[12] API 1104, *Standard for Welding Pipe Lines and Related Facilities,* American Petroleum Institute, 1220 L Street, N.W., Washington, DC 20005.

[13] CGA G-4.4, *Industrial Practices for Gaseous Oxygen Transmission and Distribution Piping Systems,* Compressed Gas Association, Inc., 1235 Jefferson Davis Highway, Arlington, VA 22202.

CHAPTER 5

Safety Guidelines for Compressed Gases and Cryogenic Liquids

INTRODUCTION

The very properties that make compressed gases, liquefied compressed gases, and cryogenic liquids useful in almost every area of modern life can also make them dangerous when mishandled. Some of these gases are stored in containers at high pressure, others at very low "cryogenic temperatures" (e.g., temperatures below −130°F or −90°C). Some are flammable, some are inert (which can pose hazards of asphyxiation due to displacement of air), others are oxidizers, while still others are corrosive or toxic.

Years of experience with these products have resulted in safe practices and equipment which, if properly employed, result in complete safety. However, the safe handling of compressed gases, liquefied compressed gases, and cryogenic liquids requires a thorough knowledge of the physical and chemical properties of these products, the specialized equipment used in their storage, handling, and use, and the proper procedures to be followed in normal use and handling. Personnel handling these products must also be trained in the correct procedures to be followed in the event of an emergency. Due to the potential hazards involved, it is imperative that all personnel handling these products receive thorough training in these subjects.

This chapter describes in general some of the potential hazards associated with compressed gases, liquefied compressed gases, and cryogenic liquids, and presents some basic guidelines for safe handling of these products. However, the reader is advised that the general guidelines presented in this chapter are not complete with respect to all types of gases or containers, and that some applications require additional safety precautions. More detailed information may be found in:

- CGA P-1, *Safe Handling of Compressed Gases in Containers* [1]
- CGA P-2, *Characteristics and Safe Handling of Medical Gases* [2]
- CGA P-2.1, *Recommendations for Medical-Surgical Vacuum Systems in Health Care Facilities* [3]
- CGA P-2.6, *Transfilling of Liquid Oxygen to be Used for Respiration* [4]
- CGA P-5, *Suggestions for the Care of High Pressure Air Cylinders for Underwater Breathing* [5]
- CGA P-6, *Standard Density Data, Atmospheric Gases and Hydrogen* [6]
- CGA P-7, *Standard for Requalification of Cargo Tank Hose Used in the Transfer of Compressed Gases* [7]
- CGA P-8, *Safe Practices for Air Separation Plants* [8]
- CGA P-9, *The Inert Gases—Argon, Nitrogen and Helium* [9]
- CGA P-12, *Safe Handling of Cryogenic Liquids* [10]

CGA P-13, *Safe Handling of Liquid Carbon Monoxide* [11]
CGA P-14, *Accident Prevention in Oxygen-Rich and Oxygen-Deficient Atmospheres* [12]
CGA P-15, *Filling of Industrial and Medical Nonflammable Compressed Gas Cylinders* [13]

A complete list of the publications, standards, and audiovisual training programs of the Compressed Gas Association, Inc., can be found with a brief description of the coverage of each as an appendix in Part IV of this *Handbook*.

Detailed information regarding the physical and chemical properties of individual gases and gas mixtures, specific handling and safety precautions to be followed, physiological effects on humans, and emergency procedures, including special firefighting considerations, are covered in the individual gas monographs.

GENERAL SAFETY CONSIDERATIONS FOR COMPRESSED GASES

Compressed gases are normally stored in containers at high pressures and must always be handled with care. While compressed gas containers are designed to meet rigid government specifications and are safe for their intended use, they can be dangerous if misused or abused. Therefore, care should be taken to ensure that compressed gas containers are stored, handled, and used in accordance with safe practices.

U.S. Department of Transportation (DOT) regulations published in Title 49 of the U.S. *Code of Federal Regulations,* and similar regulations published by Transport Canada (TC) in the *Transportation of Dangerous Goods Regulations,* require that compressed gas containers must not be filled, except with the owner's consent, that such cylinders must be periodically inspected and tested to ensure that they are in safe condition for filling, and that filling shall be undertaken only by properly qualified personnel. [14] and [15] Each cylinder must bear the proper DOT or TC label to identify the product contained.

Requirements for marking, labeling, inspection, testing, filling, and disposition of cylinders are covered in Chapter 9. For further details, refer to CGA P-1, *Safe Handling of Compressed Gases in Containers* [1], CGA P-15, *Filling of Industrial and Medical Nonflammable Compressed Gas Cylinders* [13], CGA C-4, *American National Standard Method of Marking Portable Compressed Gas Containers to Identify the Material Contained* [16], CGA C-7, *Guide to the Preparation of Precautionary Labeling and Marking of Compressed Gas Containers* [17], and CGA C-2, *Recommendations for the Disposition of Unserviceable Compressed Gas Cylinders with Known Contents.* [18]

Caution: Compressed gases must never be transferred from one container to another except by the gas manufacturer or distributor. Compressed gas containers must not contain gases capable of combining chemically with each other or with the container material so as to endanger its integrity.

Where the user is responsible for the handling of the container and connecting it for use, such containers must bear a legible label or marking identifying the contents and giving precautionary warnings. Containers not bearing such a legible label must not be used, but must be returned to the gas supplier. See Fig. 5-1.

Caution: Color should never be used as the sole means for identifying the contents of a compressed gas container.

Container Storage and Handling

Compressed gas containers must be handled with care and should be stored and secured in an upright position in a safe place where they will not be knocked over. Storage rooms should be well ventilated, and containers must not be subjected to ambient temperatures above 125°F (51.7°C). Container storage areas at user facilities must be prominently posted with the names of the

Fig. 5-1. Use the label on the cylinder to identify the cylinder's contents. Valve protection caps should remain in place during transportation.

gases to be stored, and gases of different types should be grouped taking into account the properties of the gases contained. Flammable gases must be stored separately from oxidizing gases.

Oxygen containers in storage must be separated from flammable gas containers or combustible materials (especially oil or grease) by a minimum distance of 20 ft (6.1 m), or by a noncombustible barrier at least 5 ft (1.5 m) high having a fire resistance rating of at least one-half hour.

Where removable caps are provided for valve protection, the user should keep such caps on the containers at all times, except when the containers are connected to dispensing equipment.

Compressed gas containers must never be dropped, slid, rolled, or allowed to come in contact with sharp objects. They must not be exposed to fire or flames from welding torches. Flammable gases must be stored in properly labeled secured areas away from possible ignition sources, and kept separate from oxidizing gases. Toxic or poisonous gases must be stored in a properly labeled secured area and protected from unauthorized access by personnel not properly equipped and trained in their use.

Pressure Regulation

Caution: The release of high pressure gas from cylinders can be hazardous unless adequate means are provided for reducing the gas pressure to usable levels and for controlling the gas flow. Accordingly, pressure-reducing regulators should always be used when withdrawing the contents of a cylinder, as such devices deliver a constant safe working pressure.

A suitable pressure-regulating device must be used where gas is admitted to a system of lower pressure rating than the supply pressure, and where, due to the gas capacity of the supply source, the system pressure rating may be exceeded. This is a requirement regardless of the presence of a pressure relief device protecting the lower pressure system.

Likewise, a suitable pressure relief device must be used to protect a system utilizing a compressed gas where the system has a pressure rating less than the compressed gas supply source, and where, due to the gas capacity of the supply source, the system pressure rating may be exceeded.

Pressure-regulating devices, pressure relief devices, valves, cylinder connections, and hoseline (if used) should be inspected at frequent intervals to ensure that they are undamaged and in safe working condition. Compressed gas containers with devices and appurtenances that appear to be damaged in any way should not be used, and the supplier should be contacted for instructions on the disposition of the container.

Container valves must always be opened slowly, and valve outlets must always be pointed away from the user and other persons. Valves without handwheels must be opened using only the special wrenches provided for this purpose by gas suppliers. Valves must never be hammered in attempting to open or close them.

Before disconnecting an empty container for return to a supplier, the cylinder valve must be closed, the pressure regulator removed, and the valve protection cap (if used) replaced.

For more detailed information on pressure relief and safety devices, refer to Chapter 7.

Valve Outlet Connections

Cylinder valve outlet and inlet connections, cylinder ancillary equipment connections, transfer connections, and regulator connections have been standardized by the industry for the handling of specific types of compressed and liquefied gases and to prevent the misapplication of these potentially hazardous materials. Users of compressed gases are cautioned not to tamper with or replace connections supplied by an authorized gas supplier.

For more detailed information about cylinder valve outlet and inlet connections, cylinder ancillary equipment connections, bulk transfer connections, and regulators, refer to Chapter 8.

The threads on a cylinder valve outlet, as well as on regulators and other auxiliary equipment, should be examined at frequent intervals to ensure they are undamaged. Threads on regulators and other ancillary equipment must match those on the container valve outlet.

Caution: Do not attempt to force connections that do not fit.

Detailed information regarding compressed gas containers, pressure regulators, pressure relief devices, and valve outlet connections is contained in Chapters 6 through 9. Additional safe handling procedures for compressed gases can be found in CGA P-1, *Safe Handling of Compressed Gases in Containers* [1], and in CGA AV-1, an audiovisual safety training program of the same title.

Special Considerations Concerning Medical Gases

Medical gases are prepared under carefully controlled conditions. Purity specifications for certain gases are prescribed by the *United States Pharmacopeia/National Formulary* (USP/NF). [19] Medical gases are normally shipped under pressure in portable containers in accordance with the regulations of the U.S. Department of Transportation (DOT), and in Canada under the regulations of Transport Canada (TC). [14] and [15]

Medical gases in the United States must comply with the Federal Food, Drug, and Cosmetic Act and its implementing regulations. The registration requirements of these regulations will be found in Title 21 of the U.S. *Code of Federal Regulations.* [20] Under the provisions of this Act, certain medical gases must conform to the standards of the *United States Pharmacopeia/National Formulary.* [19] All medical gases must be appropriately packaged and labeled. In Canada, medical gases must comply with the regulations of Health and Welfare Canada. [21]

The *United States Pharmacopeia/National Formulary* contains a section or monograph on each of the following gases: air, carbon dioxide, cyclopropane, helium, nitrogen, nitrous oxide, and oxygen. Definite standards and methods of testing are prescribed in each USP/NF monograph to assure a product of appropriate quality and purity.

While all of the general guidelines for safe handling, storage, and use of compressed gases described above apply to containers used in the storage and use of medical gases, some additional guidelines are necessary due to the need to protect the purity of medical gases and to protect against the possible improper administration of these gases. Among these are the following:

(1) Never attempt to mix gases in containers, or to transfer gas from one container to another. Medical gases and gas mixtures should be used only as prepared by a qualified medical gas supplier.
(2) Never interchange regulators or other appliances used with one gas with similar equipment intended for use with other gases.
(3) Identify the gas content by the label on the cylinder before using. If the cylinder is not identified by a label showing the gas contained, return the

container to the supplier without using.

Note: While the medical gas industry has adopted a color code to aid in the identification of medical gas cylinders as described in CGA C-9, *Standard for Color Marking of Compressed Gas Containers Intended for Medical Use,* the user must not rely on such color coding for identification of the gas contained. [22]

(4) Do not deface or remove any markings or labels which are used to identify the contents of a container.
(5) The user shall not change, modify, tamper with, or obstruct the discharge ports of pressure relief devices.
(6) Repairs of, or alterations to, containers or pressure relief devices should be performed only by the supplier.
(7) Never use containers for any purpose other than to supply the contained gas as received from the supplier.
(8) Never store cylinders in the operating room.
(9) Never use medical gases where the cylinder is liable to become contaminated by the feedback of other gases or foreign material unless protected by suitable traps or check valves.
(10) Make sure that the threads on regulator-to-cylinder valve connections or the pin-indexing devices on yoke-to-cylinder valve connections are properly mated. Never force connections that do not fit.
(11) Never permit gas to enter the regulating device suddenly. Always open the container valve slowly.
(12) Piping and manifold systems for medical gases should be constructed only in accordance with NFPA 99, *Standard for Health Care Facilities,* and CGA P-2.1, *Standard for Medical-Surgical Vacuum Systems in Health Care Facilities.* [23] and [3]

For further details concerning safe handling of medical gases, refer to CGA P-2, *Characteristics and Safe Handling of Medical Gases,* and the audiovisual training program CGA AV-4 of the same title. [2]

LIQUEFIED COMPRESSED GASES

Liquefied compressed gases are generally defined as elements or compounds that have boiling points from about −130°F to 30°F (−90°C to −1.1°C) at atmospheric pressure.

Examples of liquefied compressed gases include carbon dioxide, nitrous oxide, anhydrous ammonia, butadiene, chlorine, methyl chloride, methyl mercaptan, sulfur dioxide, vinyl chloride, a number of refrigerant gases including the fluorocarbon gases, and the liquefied petroleum gases (LP-gases) such as butane, isobutane, propane, and propylene.

Liquefied compressed gases are usually stored at pressures less than those for nonliquefied gases, but storage pressures are usually still well above final-use pressures, and therefore pressure regulators and pressure relief devices must be used.

Liquefied compressed gas containers are filled by weight to no more than the legal filling density as established by DOT or TC regulations, except that in some cases, particularly for LP-gas service, fill gauges (e.g. "dip tubes") are permitted in lieu of weighing.

Safe handling of liquefied compressed gases requires a thorough knowledge of the physical and chemical properties of these materials, the use of equipment designed specifically for the handling of these products, and strict compliance with regulations and standards governing their storage, transport, and handling. For information regarding containers used in the transport and storage of liquefied compressed gases, refer to Chapter 4.

Special precautions are necessary in the handling of certain hazardous liquefied compressed gases, particularly the following: butadiene (inhibited), vinyl chloride, anhydrous ammonia, LP-gases, chlorine, methyl chloride, sulfur dioxide, and the fluorinated hydrocarbons. For specific information on the physical and chemical properties of liquefied compressed gases, physiological effects, safe handling procedures, specific precautions, and emergency procedures to be followed, refer to the individual gas monographs in Part III.

The DOT regulations covering the handling, storage, and transfer of liquefied compressed gases can be found in Title 49 of the U.S. *Code of Federal Regulations,* Parts 107, 171, 172, 173, 174, 177, 178, and 179. [14]

CRYOGENIC LIQUIDS

Cryogenic liquids are gases that have been transformed into extremely cold refrigerated liquids which are stored at temperatures below −130°F (−90°C). They are normally stored at low pressures in specially constructed, multi-walled, vacuum-insulated containers.

Examples of gases commonly handled as cryogenic liquids include oxygen, nitrogen, argon, neon, krypton, xenon, hydrogen, and helium. Liquefied natural gas (LNG) and/or liquid methane and carbon monoxide are also handled as cryogenic liquids.

The potential hazards that accompany these products may result from:

(1) Extreme cold which can freeze human tissue on contact, and which can also cause embrittlement of carbon steel, plastics, and rubber
(2) Extreme pressure which can result from rapid vaporization of the refrigerated liquid due to rising temperature from leakage of heat into the cryogenic container or system
(3) Asphyxiation due to displacement of air by escaping liquid and the resultant rapidly expanding gas (in the case of inert gases)
(4) Fire or explosion in the case of escaping flammable gases such as hydrogen, carbon monoxide, or methane, or from escaping liquid oxygen, which while not itself a flammable gas, can combine with organic materials with explosive violence

In order to handle cryogenic liquids safely, it is important to know the physical and chemical properties of these products. Table 5–1 is a useful guide to the physical properties of commonly used cryogenic liquids. Table 5–2 shows the basic fire-extinguishing agents used on cryogenic liquids. Table 5–3 is a guide to the explosive and fire hazards of common cryogenic liquids. For more detailed information, refer to the individual gas monographs in Part III for the particular cryogenic liquid in question.

Personnel Safety

Because of the potential hazards resulting from the extremely low temperatures of cryogenic liquids, all personnel handling them must be properly trained in the use of specialized equipment designed for the storage, transfer, and handling of these products.

Heavy leather protective gloves, safety shoes, aprons, and eye protection must be worn to prevent possible contact with the extremely cold surfaces of uninsulated piping, transfer connections, valves, and other equipment, or from the cold liquid or boil-off vapors which may result from spilled or splashed liquid.

Any transfer operations involving open containers such as dewars must be conducted slowly to minimize boiling and splashing of the cryogenic liquid, and such operations must be conducted only in well-ventilated areas to prevent the possible accumulation of inert gas which can replace the oxygen in the atmosphere and cause asphyxiation.

Cryogenic liquids must be handled and stored only in containers and systems specifically designed for these products and in accordance with proven safe practices as described in CGA P-12, *Safe Handling of Cryogenic Liquids.* [10] Cryogenic liquid containers are described in Chapter 4.

Equipment and systems designed for the storage, transfer, and dispensing of cryogenic liquids must be constructed of materials compatible with the products handled and the temperatures encountered. All such

TABLE 5-1. PHYSICAL PROPERTIES OF CRYOGENIC LIQUIDS

(Also see Table 5-3)

GAS Chemical Symbol	Xenon Xe	Krypton Kr	Methane CH_4	Oxygen O_2	Argon Ar	Carbon Monoxide CO	Nitrogen N_2	Neon Ne	Hydrogen H_2	Helium He
Boiling Point, 1 atm										
°F	-163	-244	-259	-297	-303	-313	-320	-411	-423	-452
°C	-108	-153	-161	-183	-186	-192	-196	-246	-253	-268
Melting Point, 1 atm										
°F	-169	-251	-296	-362	-309	-341	-346	-416	-435	—[1]
°C	-112	-157	-182	-219	-189	-207	-210	-249	-259	—
Density at boiling point and 1 atm										
lb/cu ft	191	151	26	71	87	49	50	75	4.4	7.8
(kg/m^3)	(3059.5)	(2418.8)	(416.5)	(1137.3)	(1393.6)	(784.9)	(800.9)	(1201.4)	(70.48)	(124.94)
Heat of vaporization at boiling point										
Btu/lb	41	46	219	92	70	93	86	37	193	9
(Joule/kg)	(95 366)	(106 996)	(509 394)	(213 992)	(162 820)	(216 318)	(200 036)	(86 062)	(448 918)	(20 934)
Volume expansion ratio, liquid at 1 atm and boiling point to gas at 70°F (21.1°C) and 1 atm	559	693	625	860	842	680	696	1445	850	745
Flammable	No	No	Yes	No[2]	No	Yes	No	No	Yes	No

[1]Helium does not solidify at 1 atm pressure.

[2]Oxygen does not burn, but supports and accelerates combustion. However, high concentration oxygen atmospheres substantially increase combustion rates of other materials, and may form explosive mixtures with other combustibles. Flame temperatures in oxygen are higher than those in air.

systems must be equipped with pressure relief devices to prevent excessive pressure buildup due to the vaporization of the cryogenic liquid as heat leaks into the system. Piping must be equipped with pressure relief devices in a manner that prevents the buildup of excessive pressure due to vaporization of liquid between valves. Only transfer lines designed for cryogenic service should be used.

Caution: Do not smoke or permit smoking or open flames in any area where flammable

TABLE 5-2. FIRE-EXTINGUISHING AGENTS FOR CRYOGENIC LIQUID IN FIRE.

Extinguishing Agent	Oxygen	Hydrogen, Methane, Carbon Monoxide
Water	Preferred	Used to protect adjacent equipment or property and to spray personnel. Not to be applied directly onto burning vapor or cryogenic liquid, since the water will evaporate additional flammable material.
Soda Acid	No Effect	Unacceptable.
CO_2	No Effect	Fair. Apply at base of flame.
Dry Powder	No Effect	Good. Apply at base of flame.
Methyl Bromide	Unacceptable	Not normally used unless authorized and supplied for individual premises or equipment.

TABLE 5-3. EXPLOSIVE AND FIRE HAZARDS OF COMMON CRYOGENIC LIQUIDS

	Oxygen	Nitrogen	Argon	Helium	Krypton	Xenon	Neon	Methane	Hydrogen	Carbon Monoxide
Explosive hazard with combustible materials	Yes	No	No	No	No	No	No	No	No	No
Explosive hazard with oxygen or air	—	No	No	No	No	No	No	Yes (within flammable limits)	Yes (within flammable limits)	Yes (within flammable limits)
Pressure rupture if liquid or cold vapor is trapped	Yes	Yes	Yes	Yes	Yes	Yes	Yes	Yes	Yes	Yes
Fire hazard type										
Combustible	Nil	Nil	Nil	Nil	Nil	Nil	Nil	Yes	Yes	Yes
Promotes ignition	Yes	No	No	No	No	No	No	Yes	Yes	Yes
Condenses air and expands flammable range	No	Yes	No	Yes	No	No	Yes	No	Yes	No
Flammable limits in air, Percent by volume	—	—	—	—	—	—	—	5-15	4-75	13-74
Spontaneous ignition temperature in air at 1 atm (101.325 kPa).										
°F	—	—	—	—	—	—	—	1000	1085	1204
°C	—	—	—	—	—	—	—	538	585	651
Minimum ignition energy, mJ	—	—	—	—	—	—	—	0.30[1] 0.45	0.02	
Flame temperature in air										
°F	—	—	—	—	—	—	—	3407	3722	3812[2]
°C	—	—	—	—	—	—	—	1875	2050	2100
Flame velocity in air										
ft/sec	—	—	—	—	—	—	—	1.28	8.9	1.08
cm/sec	—	—	—	—	—	—	—	39	271	33
Limiting oxygen index volume[3], Percent	—	—	—	—	—	—	—	12	5	5.7

[1]Some components of LNG, such as ethylene, have ignition energies as low as 0.08 mJ.
[2]The maximum flame temperature for carbon monoxide occurs at about the stoichiometric mixture (66.6% carbon monoxide, 33.33% oxygen). The range of flame temperatures is 1560°C to 2100°C.
[3]Minimum oxygen concentration to support flame propagation when stoichiometric fuel-air mixture is diluted with nitrogen.

liquids or gases or liquid oxygen are stored, handled, or used, or where they are loaded or unloaded.

Special Liquid Oxygen Precautions

Liquid oxygen containers, piping, and equipment must be properly cleaned for oxygen service and must be kept clean and free of grease, oil, or other hydrocarbon materials which can combine with oxygen with explosive violence. Detailed information on the means of cleaning is provided in CGA G-4.1, *Cleaning Equipment for Oxygen Service* [24], and in Chapter 10 of this *Handbook*. Liquid oxygen systems at consumer sites must comply with NFPA 50, *Standard for Bulk Oxygen Systems at Consumer Sites.* [25] See Fig. 5-2.

Caution: Keep all combustible materials,

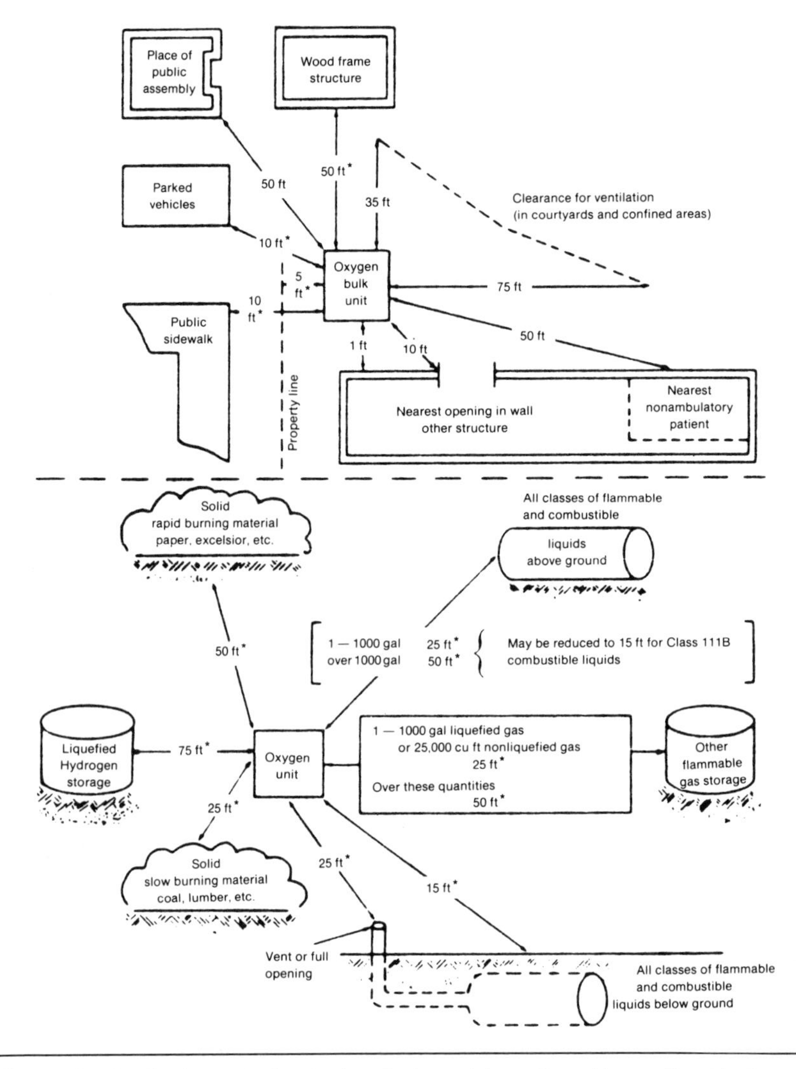

*NOTE: These distances do not apply where protective structures having a minimum fire resistance rating of two hours interrupt the line-of-sight between uninsulated portions of the bulk oxygen storage installation and the exposure. The protective structures protect uninsulated oxygen storage containers or supports, control equipment and system piping (or parts thereof) from external fire exposure. Liquid oxygen storage containers are insulated. Such containers may provide line-of-sight protection for uninsulated system components.

NOTE: Changes to NFPA 50 may cause these distances to change. The text of NFPA 50 shall govern. [25] Reprinted by permission from NFPA 50, (1985)

Fig. 5-2. Distances between bulk oxygen systems and exposures.

especially oil or grease, away from oxygen. While oxygen is nonflammable, it vigorously accelerates and supports combustion. Do not permit liquid oxygen or oxygen-rich air atmospheres to come in contact with organic materials or flammable substances of any kind.

Among the organic materials that can react violently with oxygen or oxygen-rich atmospheres when ignited are: oil, grease, asphalt, kerosene, cloth, tar, and dirt that may contain oil or grease. A single hot spark can be sufficient to trigger ignition of such materials when in contact with an oxygen-rich atmosphere. If liquid oxygen spills on asphalt or other combustible substances (i.e., oil-soaked concrete or gravel), do not walk on or roll equipment over the area of the spill, and keep all sources of ignition away from the area for at least 30 minutes after all frost or fog has disappeared.

Note: Neither liquid nor gaseous oxygen can be effectively blanketed by such fire-extinguishing agents as carbon dioxide, dry chemical, or foam. If a fire should occur where liquid or gaseous oxygen is present, it is necessary to cool combustible materials below their ignition temperatures to stop the fire. One method of doing this is to use large quantities of water in spray form.

Caution: Liquid oxygen, if spilled on clothing or another combustible substance, can pose a serious hazard of fire or explosion due to the rapid chemical reaction between the two substances. If liquid oxygen is spilled on clothing, the clothing must be removed at once and aired before reuse. Areas where liquid oxygen may be spilled during transfer operations must be free of hycrocarbons or other combustible materials.

For additional information on the handling of oxygen, refer to the section entitled Oxygen and Oxidizing Gases later in this chapter, and to the monograph on oxygen in Part III.

Special Liquefied Natural Gas (LNG) Precautions

Any area where liquefied natural gas (LNG) is stored must be posted *NO SMOKING—FLAMMABLE GAS.* Open flames and general-purpose electrical equipment must be prohibited. Storage and transfer must be under positive pressure to prevent infiltration of air and other gases. Liquefied natural gas systems at utility plants and consumer sites must comply with NFPA 59A, *Standard for the Production, Storage, and Handling of Liquefied Natural Gas (LNG).* [26] For additional information on safe handling of LNG, refer to the monograph on methane in Part III.

Special Liquid Hydrogen Precautions

Liquid hydrogen is highly flammable and should be stored, transferred, and handled out-of-doors, unless a specific review for safe handling indoors has been made. Where liquid hydrogen is stored or handled, smoking, open flames, and general-purpose electrical equipment must be prohibited, and the area must be marked *NO SMOKING—FLAMMABLE GAS.* Liquid hydrogen must be stored and transferred only in equipment specially designed for hydrogen service and under positive pressure to prevent the infiltration and solidification of air or other gases. Liquid hydrogen systems at consumer sites must comply with NFPA 50B, *Liquefied Hydrogen Systems at Consumer Sites.* [27] For additional information on safe handling of liquid hydrogen, refer to the monograph on hydrogen in Part III.

Special Liquid Helium Precautions

Liquid helium must be stored and transferred only under positive pressure to prevent the infiltration and solidification of air and other gases. The problem of backflow diffusion of air into liquid helium (and liquid hydrogen) vessels and equipment must be emphasized. It is a major safety hazard, and if it occurs, blockage of openings may result leading to rupture of the container. For additional information on liquid helium and other inert cryogenic liquids, refer to the individual gas monographs in Part III. Detailed information concerning the safe handling of the various inert, oxidizing, and flammable cryogenic liquids can be found in CGA P-12, *Safe Handling of Cryogenic Liquids.* [10] See Fig. 5-3.

FLAMMABLE GASES

Flammable gases are defined by the U.S. Department of Transportation as those that when mixed with air are flammable in concentrations of 13 percent or less by volume in air; or, if the gas has a flammable range wider than 12 percent in air, regardless of its lower flammable limit. Examples of flammable gases include acetylene, butadiene, carbon monoxide, ethane, ethylene, hydrogen, hydrogen sulfide, the liquefied petroleum

Fig. 5-3. Cryogenic containers such as these must be kept upright and should only be moved in handcarts or equipment designed for this purpose.

gases (i.e., butane, isobutane, propane, and propylene), methane, methylacetylene-propadiene (MAPP—liquid), methyl chloride, silane, and vinyl chloride. The amount of flammable material is respectively much greater per unit volume when a flammable gas is stored in liquefied form.

Storage Facilities

When flammable gases are stored in separate buildings or separate rooms without occupancy, provisions should be made to avoid hazardous exposure from and to adjoining buildings, equipment, property, and concentrations of people. The walls, partitions, and ceilings of such buildings or storage areas must be constructed of noncombustible materials having a fire resistance rating of at least one hour, and where applicable, in accordance with the rating requirements of the National Fire Protection Association (NFPA) standard for the gas involved. Fire doors and windows must be in accordance with NFPA 80, *Standard for Fire Doors and Windows.* [28] Such buildings must be heated by steam, hot water, or other indirect means, not by flame. Electrical equipment must be in accordance with NFPA 70, *National Electrical Code,* Article 501 for Class I, Division 2 locations. [29]

Flammable gases must not be stored or used near open flames, near hot surfaces, adjacent to oxidizers, near electrical power lines, or near underground electrical equipment. Adequate portable fire extinguishers of carbon dioxide or dry chemical types must be available for fire emergencies at storage installations. *NO SMOKING—FLAMMABLE GAS* signs must be posted conspicuously in storage areas of buildings or at the entrances to special storage rooms.

Flammable gas containers stored inside of buildings or rooms with other occupancies must be kept at least 20 ft (6.1 m) from flammable liquids or oxygen containers, and from highly combustible materials and similar substances, and not near arcing electrical equipment, open flames, or other sources of ignition.

Flammable gas containers stored inside of industrial buildings at consumer sites, except those in use or attached for use, must be limited to a total gas capacity of 2500 ft^3 (70.8 m^3) of acetylene or nonliquefied flammable gas, or a total water capacity of 735 lb (333 kg) for liquefied petroleum gas or stabilized methylacetylene-propadiene. [Note: 735 lb (333 kg) water capacity is equivalent to approximately 309 lb (140 kg) of propane, 368 lb (167 kg) of stabilized methylacetylene-propadiene, or 375 lb (170 kg) of butane.]

Caution: A flame must never be used for detection of flammable gas leaks. A flammable gas leak detector, soapy water, or other suitable solution must be used.

Special Carbon Monoxide Precautions

Carbon monoxide is odorless, colorless, toxic, and flammable. Because of these characteristics, all leaks must be eliminated before a system is put into operation by testing with nitrogen. At least two persons should be assigned to any operation involving carbon monoxide, and a suitable supply of oxygen should be available to administer to personnel who may become exposed to excessive carbon monoxide concentrations. Exposure to concentrations of about 4000 ppm may be fatal in less than one hour. Exposure to very high concentrations can result in death almost immediately. Therefore, all personnel handling or transferring carbon monoxide should be protected with self-contained breathing apparatus.

Additional information is provided in the monograph for carbon monoxide in Part III. Also refer to CGA P-13, *Safe Handling of Liquid Carbon Monoxide,* for information concerning the cryogenic liquid form of this product. [11]

Special Precautions Regarding Acetylene

Acetylene is a highly reactive flammable gas and can be stored safely only in cylinders specially designed for acetylene service. For details on the chemical and physical properties of acetylene, the special design of acetylene cylinders, and safe handling and emergency procedures to be followed, refer to the monograph on acetylene in Part III.

Fire Prevention and Fire Fighting

Fire is best prevented by elimination of all leakage of flammable gases, whether in gaseous or liquefied form, and by storage of such gases in accordance with safe practices as described previously. Experience has shown that flammable cryogenic liquids that leak and vaporize are easily ignited. Therefore, a primary effort should be focused on proper containment, detection of any leaks, and ventilation of the area.

Should a fire occur in or near any area where flammable gases are stored, call the fire department immediately and make every reasonable effort to keep the fire from reaching the stored flammable gases without endangering the safety of personnel.

A Fire-Fighting Plan

Any facility where flammable gases are stored or handled should have a thorough training program for all employees as to what to do in a fire emergency. Such a program should include frequent fire drills, proper location and upkeep of fire-fighting apparatus, and liaison between the facility's fire-fighting personnel and local fire department personnel.

A fire plan should be developed with sketch maps and diagrams showing the locations of stored flammable materials, fire alarms, fire brigade call points, emergency exits, escape routes, rescue and safety equipment, hydrants, fire-fighting equipment, and plant controls to be activated in the event of a fire emergency. All personnel should be aware of special or unusual risks associated with stored flammable materials.

While it is not possible to outline specific fire-fighting techniques that will cover all types of fires involving flammable gases or cryogenic liquids, some general rules should be observed.

(1) Everyone not actively engaged in fighting the fire should be evacuated from the area.

(2) If a flammable cryogenic liquid is involved, the flammable mixture zone may extend beyond the normal fog cloud produced by condensing water vapor in the air. Personnel should be evacuated well outside the fog area.

(3) The single best fire-fighting technique is to shut off the flow of flammable liquid or gas.

(4) If electrical equipment is involved, be sure that the power supply is disconnected before using water to fight the fire, or use carbon dioxide or dry chemical extinguishers to fight the fire. *Note:* Fires involving fuel supported by liquid or gaseous oxygen cannot be effectively blanketed by such agents as carbon dioxide, dry chemical, or foam. It is necessary to cool combustible materials below their ignition temperatures to stop the fire. In such cases, use large quantities of water in spray form.

(5) Use the spray to cool any burning material below its ignition temperature. If possible, do not spray cold areas of equipment or direct water into a cryogenic fluid. Firehoses with stream-to-spray nozzles should be available where large quantities of flammable liquefied gases or cryogenic liquids are handled.

(6) Depending on the circumstances, it is usually not advisable to extinguish a flammable cryogen in a confined area. If the flammable gas supply cannot be shut off, the continued escape of unburned gas can create an explosive mixture. This mixture may be reignited by other burning material or hot surfaces. In such cases, it is usually better to allow the gas to burn itself out in a confined area and keep adjacent objects cool with water, rather than risk exposing personnel at the site to a potential explosion.

(7) If an inert cryogenic liquid is involved, judgment should be used in deciding whether to allow the gas to escape, with the possible risk of asphyxiation of firefighters if in a confined area, or to attempt to cut off the gas flow by means of the associated valving. In no case should pressure relief devices be blocked or prevented from operating.

(8) Observe the special precautions mentioned earlier in this chapter when fighting fires where oxygen, carbon monoxide, LNG, or hydrogen are present.

Further specific information for handling flammable gases, including special fire-fighting considerations, can be found in the individual gas monographs provided in Part III.

INERT GASES

Inert Gas Hazards

Inert gases such as argon, carbon dioxide, helium, krypton, neon, nitrogen, and xenon are simple asphyxiants which can displace oxygen in the air, which is necessary to sustain life, and can cause rapid suffocation due to oxygen deficiency. A portable self-contained breathing apparatus or air-line mask must be worn by any personnel entering areas containing an oxygen-deficient atmosphere, or in any case where the oxygen concentration in the atmosphere is suspected to be less than 19 percent by volume.

Inert gases handled as cryogenic liquids can, if released, generate extremely large volumes of inert gas which can rapidly displace the oxygen in the atmosphere. For example, one volume of liquid nitrogen at its boiling temperature at 1 atm vaporizes to 696.5 volumes of nitrogen gas when warmed to room temperature (70°F or 21.1°C at 1 atm). The gas volume expansion ratio of liquid neon is 1445 to 1. Therefore, any time that there is a leak or spill of inert gas, there is a serious hazard of oxygen deficiency in the atmosphere. The cold boil-off vapors of these inert gases can generally be seen as a fog due to condensing moisture in the air. However, these vapors quickly transform into invisible inert gas at room temperature.

It is advisable for all personnel working with inert gases in potentially oxygen-deficient atmospheres to use the buddy system to ensure that an individual who may be working in an oxygen-deficient atmosphere will have someone unexposed to that atmosphere to provide assistance if needed. In such cases, both worker and "buddy" should be equipped with a portable self-contained breathing apparatus.

Oxygen-Deficient Atmospheres

The normal oxygen content of air is approximately 21 percent by volume. Depletion of the normal oxygen content in air, either by combustion or by displacement with inert gas, is a serous potential hazard to personnel. When the oxygen content of air is reduced to 15 or 16 percent, an individual breathing the air is mentally incapable of diagnosing the situation, as the symptoms of sleepiness, fatigue, lassitude, loss of coordination, errors in judgment, and confusion will be masked by a state of euphoria, giving the victim a false sense of security and well-being.

Caution: Human exposure to atmospheres containing 12 percent or less oxygen will bring about unconsciousness without warning, and so quickly that the individual cannot help or protect himself or herself. This is true whether the condition is reached by an immediate change of environment or by gradual depletion of oxygen in the atmosphere.

Whenever personnel are working in potentially oxygen-deficient atmospheres (as with inert gases), they must be equipped with a portable self-contained breathing apparatus or air-line breathing equipment.

More detailed information is available in CGA P-14, *Accident Prevention in Oxygen-Rich and Oxygen-Deficient Atmospheres* [12], and CGA P-12, *Safe Handling of Cryogenic Liquids.* [10] The Compressed Gas Association also publishes a brief safety bulletin, CGA SB-2, "*Oxygen Deficient Atmospheres,*" which can serve as a means to emphasize to personnel the potential seriousness of this invisible hazard.

Nonflammable Liquefied Gas Hazards

Nonflammable liquefied gases such as carbon dioxide, sulfur hexafluoride, and nitrous oxide also pose the hazard of asphyxiation by reducing the oxygen content in air if large quantities are spilled or released. Carbon dioxide is much heavier than air and can accumulate in low areas, replacing the oxygen necessary to support life, as can sulfur hexafluoride. Nitrous oxide, an anesthetic gas, is both heavier than air and an oxidizer.

OXYGEN AND OXIDIZING GASES

Oxygen and Oxidizing Gas Hazards

Oxygen and gas mixtures containing large quantities of oxygen react chemically with organic materials to produce heat. ***This reaction can take place with explosive violence.*** Therefore, keep all combustible materials and potential sources of ignition away from oxygen or gas mixtures containing high concentrations of oxygen.

All equipment to be used for oxygen service, whether the oxygen is in a gaseous or liquid state, must be specifically cleaned for oxygen service to remove any traces of oils, greases, or other hydrocarbon materials. Procedures for cleaning of components and systems to be used for oxygen service are contained in Chapter 10 and in CGA G-4.1, *Cleaning Equipment for Oxygen Service.* [24]

Caution: Never permit oil, grease, or other combustible substances to come in contact with cylinders, valves, regulators, gases, hose, and fittings used for oxidizing gases such as oxygen and nitrous oxide, which may combine with these substances with explosive violence. Never lubricate valves, regulators, gauges, or fittings with oil or any other combustible substance. Do not handle cylinders with oily hands or gloves. And never store cylinders where oil, grease, or other readily combustible substances may come in contact with them.

Oxygen and gas mixtures containing oxygen also pose the potential for oxidation of steel cylinders, which can reduce the strength of these containers. Such cylinders must be regularly inspected to ensure that such oxidation has not reduced the strength of the cylinder to the point where it is unsafe for use. Proper inspection procedures for steel cylinders are described in CGA C-6, *Standards for Visual Inspection of Steel Compressed Gas Cylinders.* [30]

Caution: Clothing on which liquid oxygen is spilled or exposed to an oxygen-rich atmosphere can pose a serious hazard of fire or explosion with consequent serious personal injury due to the rapid chemical reaction between the two substances. If liquid oxygen is spilled on clothing, the clothing must be removed at once and aired out before reuse. See CGA P-14, *Accident Prevention in Oxygen-Rich and Oxygen-Deficient Atmospheres,* for more detailed recommendations regarding clothing which may become exposed to oxygen-rich atmospheres. [12] Areas where liquid oxygen may be spilled during transfer operations must be free of hydrocarbons or other combustible materials. Never transfer liquid oxygen where it may spill onto gravel or another porous material that may contain oil drippings.

Oxidizing gases such as fluorine, chlorine, nitrogen trifluoride, and nitrous oxide are covered in individual gas monographs in Part III.

POISON (TOXIC) GASES

Poison (Toxic) Gas Hazards

Caution: Poison gases such as arsine, diborane, methyl bromide, nitric oxide, nitrogen dioxide, phosgene, and phosphine pose serious potential hazards to personnel and therefore require special handling. These products must never be handled except by specially trained personnel who are fully aware of the potential hazards involved and who are equipped with such special personal safety apparatus as is necessary in the handling of these products.

Personnel handling and using highly toxic or poison gases must have available for immediate use, in emergencies, gas masks (with cartridge-type adsorbents) or self-contained breathing apparatus of a design approved by the U.S. Bureau of Mines, the National Institute of Occupational Safety and Health (NIOSH), or other approving authority for the particular gas service involved. Gas masks may be used only under conditions where the concentrations of the toxic or poison gas involved will not exceed the rating of the particular gas mask used and where the oxygen content of the atmosphere is not less than 19 percent by volume. Such equipment must be located convenient to the place of work, but kept out of the area most likely to become contaminated.

Storage of highly toxic or poison gases must be outdoors, or in a separate noncombustible building without other occupancy, or in a separate room without other occupancy and of noncombustible construction with a fire resistance rating of at least one hour. Storage locations must be clearly marked and protected against tampering or entry by unauthorized persons. The total quantity of highly toxic or poison gases stored at a user's site should be limited to foreseeable requirements.

Highly toxic or poison gases can be filled and utilized only in forced ventilated areas, or, preferably, in hoods with forced ventilation, or outdoors. Such gases, when emitted from equipment, must be discharged into appropriate scrubbing equipment which will remove the toxic gases from the effluent gas stream.

Before using a highly toxic or poison gas, the user must read all the information on the container label and the Material Safety Data Sheet (MSDS) associated with the product. All personnel working in the immediate area must be instructed as to the toxicity of the gas or gases being used, the appropriate methods of protection against harmful exposure, and first aid treatment in case of exposure.

Exposure limits for employees handling highly toxic or poisonous gases are expressed in parts per million (ppm) of mixtures of the gas in air or milligrams per cubic meter over a period of eight hours. Permissible exposure limits (PELs) are listed in Subpart Z of Part 1910 of the U.S. Occupational Safety and Health Administration (OSHA) regulations in Title 29 of the U.S. *Code of Federal Regulations.* [31] Information regarding threshold limit values (TLVs) is available in the annually updated *Threshold Limit Values and Biological Exposure Indices* published by the American Conference of Governmental Industrial Hygienists (ACGIH). [32]

Because of the hazardous nature of highly toxic and poisonous gases, persons handling such gases are advised to contact the supplier for more complete information with regard to usage and first aid. More detailed information regarding the chemical and physical properties of poison gases, their effects on humans, and safety precautions to be followed in handling these products may also be found in the specific gas monographs in Part III.

CORROSIVE GASES

Corrosive (Acid and Alkali) Gas Hazards

Caution: Corrosive gases attack human tissue and other materials, and special protective clothing and self-contained breathing apparatus must be used by personnel handling these substances.

Precautions must be taken to avoid contacting the skin or eyes with acid or alkaline gases. Goggles or face shields and rubber (or other suitably chemically resistant material) gloves and aprons must be worn when handling these products. Open shoes or sneakers must not be worn when handling these products. Personnel handling and using corrosive (acid or alkali) gases must have available for immediate use in emergencies, gas masks or self-contained breathing apparatus of a design approved by the U.S. Bureau of Mines, the National Institute of Occupational Safety and Health (NIOSH), or other approving authority for the particular gas to be used.

Areas in which acid or alkaline gases are filled or used should be equipped with an emergency shower and eyewash fountain. Drenching with copious amounts of water is the accepted first aid procedure in the event of exposure of the skin or eyes to corrosive gases. Persons accidentally exposed to such gases should receive prompt attention by a physician.

Acid or alkaline gases should be utilized only in a well-ventilated area, and quantities stored at a user's site should be limited to those required for foreseeable use.

Examples of acid gases include boron trichloride, boron trifluoride, chlorine, dichlorosilane, hydrogen bromide, hydrogen chloride (anhydrous), hydrogen fluoride, and sulfur dioxide. Examples of alkali gases include ammonia, monomethylamine, dimethylamine, and trimethylamine. The physical and chemical properties of these gases, their physiological effects on humans, special precautions to be followed, and first aid measures to be taken in the event of an emergency are covered in individual gas monographs in Part III.

OSHA, FDA, EPA, AND CANADIAN DEPARTMENT OF HEALTH AND WELFARE REGULATIONS

In addition to the regulations related to transportation, users of compressed gases, liquefied compressed gases, and cryogenic liquids must be aware of regulations of the Occupational Safety and Health Administration (OSHA), the Food and Drug Adminstration (FDA), and the Environmental Protection Agency (EPA) in the United States. In Canada, provincial Occupational Health and Safety regulations and regulations of Health and Welfare Canada must be met.

OSHA is concerned with employee safety and health and has specific rules for several gases, including oxygen, hydrogen, acetylene, and nitrous oxide under Title 29 of the U.S. *Code of Federal Regulations.* See 29 CFR Part 1910, Subpart H, "Hazardous Materials" (29 CFR) 1910.102 through 1910.105. [31] LP-gases are covered under 29 CFR 1910.110. Additional gases covered under specific OSHA rules are ammonia, chlorine, carbon dioxide, ethylene oxide, and some other gases with particular hazards such as the materials listed under 29 CFR Part 1910, Subpart Z, "Toxic and Hazardous Substances." [31] OSHA also requires the preparation and distribution of Material Safety Data Sheets (MSDS) in accordance with 29 CFR 1910.1200, "Hazard Communication." [31] More detailed information regarding OSHA regulations can be found in Chapter 2.

The FDA in the United States, and Health and Welfare Canada, are the regulatory agencies concerned with maintaining quality standards for compressed gases used in food and medical applications. Packagers, repackagers, and relabelers of medical gases must be registered with the appropriate federal agency.

These federal agencies inspect registered facilities on a routine basis to assure compliance with the regulations. Failure to comply can result in possible halting of production, seizure of product, and penalties or fines and/or imprisonment. For more details on medical gas regulations, refer to Chapter 2.

EPA regulations affect cylinder fillers because of the need to properly dispose of some gases upon return for refilling. Certain toxic gases must be released only through scrubbers or other safe disposal means, or can be recycled only when suitable quality can be assured. More detailed information regarding EPA regulations can be found in Chapter 2.

EMERGENCY RESPONSE

Due to the potentially hazardous nature of many compressed gases, liquefied compressed gases, and cryogenic liquids during manufacture, storage, transport, and use, there is an essential need for effective, informed, and timely emergency response:

- To prevent or minimize injury to persons or property or damage to the environment in the event of an incident
- To facilitate the expeditious handling of any cleanup
- To analyze the causes of the incident and the effectiveness of the emergency response so that future corrective or preventative measures can be taken

In recent years, the Compressed Gas Association has been engaged in the process of implementing a plan for the industry, entitled the Compressed Gas Emergency Action Plan (COMPGEAP), which will help to coordinate emergency response activities of member companies during physical distribution activities. The COMPGEAP plan is activated through the CHEMTREC emergency response network established by the Chemical Manufacturers Association, which is designed to speed emergency response information on hazardous materials to sites anywhere within the United States. In Canada, the process is implemented through CANUTEC. However, in both countries the primary responsibility for emergency response lies with the manufacturer, transporter, or user of the hazardous material(s).

Developing an Emergency Response Plan

For any company manufacturing, storing, handling, or transporting compressed gases, liquefied compressed gases, or cryogenic liquids, the development and implementation of an emergency response plan must be considered a first priority. Such a plan must take into consideration the specific hazards associated with the materials being manufactured, stored, or transported, and a thorough analysis of all possible emergency situations that could occur. These events should be analyzed in terms of their probability of occurrence and their potential outcomes. Techniques for such analyses can vary from very sophisticated fault-tree analysis to a simple checklist, depending on the level of risk involved.

The types of events that may require an emergency response include such situations as:

- Storms, floods, hurricanes, tornadoes, and earthquakes
- Accidental release of cryogenic fluids
- Accidental release of flammable, toxic, or highly reactive gases
- Fires, power outages, explosions, or bomb threats
- Highway, rail, or other transport incidents

and any other possible events which could trigger the release of hazardous material into the environment.

When analyzing the probability of emergency events, it is important to recognize that in many cases one type of emergency may often trigger subsequent events which may have even greater consequential implications than the initiating event.

Although most written emergency response plans contain the same basic elements, two distinct types of plans emerge for consideration: (1) plans applicable to a facility or site location and (2) plans associated with transportation incidents.

On-Site Emergency Response

When developing a facility or site emergency response plan, it is not enough to focus on the effects of possible accidents within the site itself. Also to be considered are the effects on the surrounding community and the need to coordinate emergency response planning with local officials, including police, fire, and medical emergency rescue personnel.[1]

As a starting point, a plot plan of the site should be prepared showing the locations of existing structures, property lines, entrances, exits, fire hydrants, sprinkler systems, emergency equipment, plant controls and shutoffs, and areas where specific hazardous materials are handled or stored.

To determine which materials may be hazardous, copies of Material Safety Data Sheets (MSDSs) should be obtained and evaluated. Information should be gathered and analyzed regarding the areas immediately adjacent to the site, including information on local topography, area classification (i.e., residential, industrial, commercial), transportation routes, and available community and emergency services such as police, fire, emergency rescue, and medical services.

Once this has been completed, an analysis should be made of the potential scenarios that might trigger an emergency involving either the site alone or the site and surrounding community.

A written emergency response plan for a site location should include, at a minimum, the following:

(1) A list of emergency numbers to be used to alert plant personnel, local police, fire, emergency rescue, and medical personnel
(2) Evacuation procedures for the site and (if appropriate) the community adjacent to the site
(3) A list of emergency equipment available and the locations of such equipment
(4) A detailed plot plan designating areas where hazardous materials are handled or stored
(5) Material Safety Data Sheets (MSDSs) detailing the chemical and physical

[1]An excellent example of industry's voluntary efforts at meeting this need is the Community Awareness and Emergency Response (CAER) program established by the Chemical Manufacturers Association in 1985. The program consists of two distinct parts: first, the integration of industry and community emergency response plans, and second, the development of a program to inform the community at large of the chemicals in their locality and potential exposure during an emergency.

The concept of CAER, originally a voluntary, industry-created action program, became law when the federal government enacted the Superfund Amendments and Reauthorization Act (SARA) Title III regulation in October 1986 (Public Law 99-499). This law requires formation of regional response districts and comprehensive emergency response plans. Also incorporated into this law is the Emergency Planning and Community Right-To-Know Act (Sections 311 and 312 of Title III) affording community members the power to determine what chemicals are located at a given site.

properties of hazardous materials being stored or handled, and specific emergency response measures to be implemented to contain these materials in the event of an emergency

(6) A schedule for periodic discussions of the plan with local officials and police, fire, rescue, medical, and other emergency response personnel

(7) A plan for employee training to ensure that all employees understand the emergency response plan and their respective responsibilities in the event of an emergency

(8) A schedule for implementing periodic fire drills and for testing other emergency response procedures, including evacuation of the site.

Transportation Emergencies and Response

The chief differences between transportation emergencies involving hazardous materials and those occurring at a site involve the greater unpredictability of transportation incidents, the substantial risk of public exposure, and the difficulty of locating, organizing, and coordinating the efforts of the resources necessary to bring the incident under control.

The primary responsibility for responding to a transportation emergency lies with the product carrier, although the product shipper, especially in situations where a hazardous material is being transported, usually has the greatest expertise in handling an emergency involving the product. Generally the shipper's role is advisory in nature, providing information on product hazards, handling, and cleanup to those at the scene who are responsible for dealing with the emergency.

The first line of defense in handling a transportation emergency is to make sure that the carrier has the basic hazard and other MSDS-type information on the product(s) being transported and knows whom to notify in the event of an emergency. The U.S. Department of Transportation and Transport Canada may require that such information be indicated on the driver's hazardous material shipping papers. All vehicles carrying hazardous materials must be placarded to indicate the hazardous nature of the materials being transported.

CHEMTREC Services (United States)

Emergency response assistance in any transportation emergency within the United States involving compressed gases (as well as other chemicals) can be obtained through the Chemical Manufacturers Association's Chemical Transportation Emergency Center (CHEMTREC) by calling toll free 800-424-9300. CHEMTREC provides 24-hour-a-day response, seven days a week, and offers immediate advice for those at the scene of emergencies, then promptly notifies the shipper for more detailed assistance and the appropriate follow-up. Shippers of compressed gases are encouraged to register with CHEMTREC.

For further information about CHEMTREC services, write to: Manager, CHEMTREC/CHEMNET, Chemical Transportation Emergency Center, Chemical Manufacturers Association, 2501 M Street, N.W., Washington, DC 20037; or phone: (202) 887-1255.

CANUTEC (Canadian Transport Emergency Centre or Centre Canadien D'Urgence Transport)

Transport Canada also operates a 24-hour emergency respose information service called CANUTEC, which can be reached in emergencies by calling (613) 996-6666. For general information about this service, call (613) 992-4624.

Canadian regulations covering the transportation of dangerous goods require shippers of certain dangerous goods in excess of specific quantities to register an Emergency Response Plan (ERP) with Transport Canada. The registered number must be shown on any shipment of this material over the specified quantity. In all cases, shipping documents for dangerous goods requiring an ERP must also provide a phone number through which the plan can be activated at any time during the day.

In Canada, the Canadian Chemical Producers Association (CCPA) operates a mutual aid program for its members, the Transportation Emergency Assistance Plan (TEAP). In addition, the Propane Gas Association of Canada (PGAC) operates a mutual aid response plan for its members. TEAP is designed to handle emergencies involving any member's products, while the PGAC plan deals with incidents involving liquefied petroleum gas.

Further information on these plans can be obtained from the CCPA at 350 Sparks Street, Suite 850, Ottawa, Ontario K1R 7S8, and the PGAC at 500 Fourth Avenue S.W., Suite 1202, Calgary, Alberta T2P 2V6.

Emergency Response by Producers

In addition to the emergency response information available from CHEMTREC and CANUTEC, many shippers of hazardous materials have the capability of dispatching an emergency response team to the site of an incident involving hazardous materials. Such teams are trained and equipped to provide product information and guidance to local emergency personnel and in many cases can provide special equipment such as acid suits, portable breathing apparatus, recovery drums, or the capability of making on-site cargo transfers, if necessary. See Fig. 5-4.

Implementing the Emergency Response Plan

Once a written emergency response plan has been developed, it must be implemented at all levels throughout the organization through communications and training programs for employees, as well as communications and coordination with local emergency response agencies.

In order for an emergency response plan to be effective, it is essential that a rapport be established with local emergency response agencies such as fire and police departments

Fig. 5-4. Emergency response team trains for containment of a leaking cylinder using a cylinder encapsulation device.

and local medical facilities. Meetings should be held with members of these agencies to educate them to the specific hazards defined in the plan and the planned responses to be taken in the event of an emergency. Outside agency representatives should be invited to the site so that they can become familiar with the site layout, the processes and materials involved, the number of employees at the site, the locations and types of emergency equipment available at the site, and the specific properties and hazards associated with any hazardous materials being produced, stored, or used at the site.

Testing the Emergency Response Plan

Because of infrequent emergency response events relative to the large number of shipments in transportation or hours of production at facilities, the emergency response plan must be tested periodically to ensure that it will be effective when an emergency arises and that all employees and local emergency response personnel understand their responsibilities in carrying out the plan. Periodic tests of the emergency response plan must take into consideration the frequency of changes in personnel both on-site and among outside agencies.

When such tests are carried out, the occurrence of a particular event should be simulated and all responsible personnel should be required to respond in accordance with the plan. Such tests give those in charge of the emergency response program an opportunity to evaluate the effectiveness of the plan, to identify any flaws in the system, and to test the understanding and readiness of personnel assigned to critical responsibilities.

Post-Emergency Activities

After an emergency situation has been brought under control, the next objective should be to implement necessary cleanup operations and get the operation back to normal. At the same time, an investigation should be started to determine the cause(s) of the incident, to identify corrective measures to be taken to prevent future emergencies, and to evaluate the effectiveness of the emergency response effort.

Before initiating these activities, the area where the incident occurred should be secured to prevent the disruption, displacement, or destruction of physical evidence, to ensure that salvageable equipment and materials are not removed, and to prevent unauthorized persons from entering the site and incurring possible injury. In certain instances, photographs taken of the site and specific details of equipment can be useful in reconstructing events and assessing damage. Once this has been done, the physical evidence should be cataloged and removed for further examination.

After the post-emergency investigation has been completed, the results of this investigation, together with any resulting changes or improvements in procedures, equipment, or the emergency response plan, should be communicated to concerned employees and to local emergency response agencies.

Conclusion

Effective emergency response requires a commitment at all levels within an organization starting from the top down. Such response can be effective only when the necessary resources are made available for planning, training, coordination, testing, and implementation.

Often, the only positive result that comes from an actual emergency incident is reinforcement of the fact that such incidents can and do happen, and that resources expended on emergency preparedness are a sound investment.

SAFETY TRAINING

Thorough safety training is a must for anyone handling compressed gases, liquefied compressed gases, or cryogenic liquids, as these materials, if mishandled, can pose serious threats to safety and health. Start with careful selection of personnel charged with the responsibilities for handling any of these potentially hazardous materials. Choose reliable, intelligent people with a strong sense of responsibility and arrange to train them thoroughly. Because employees may be promoted or transferred, or may leave the company, and replacements and any new employees must be adequately trained before assuming their duties, training programs should be an ongoing consideration of management. If possible, the safety training program should include instruction by a medical officer, safety supervisor, and fire protection engineer.

The compressed gas manufacturer or supplier is a logical source of technical data describing the physical and chemical properties of the material in question. In addition, the manufacturer or supplier may have useful information on other characteristics of the products, the protective equipment necessary for handling these products, proper procedures for storage and handling, procedures to minimize the likelihood of accidents, and the corrective measures and emergency procedures to be followed in the event of an incident. Some gas manufacturers and suppliers will furnish qualified technical representatives or training aids to train personnel in every step required in handling these hazardous materials. Take advantage of these services where they are available.

Information on the physical and chemical properties of individual gases, safe handling procedures, and emergency procedures, including fire fighting recommendations, can also be found in the individual gas monographs in Part III of this *Handbook,* as well as through the Material Safety Data Sheets (MSDS) available from gas manufacturers and suppliers.

Before allowing any employee to handle hazardous materials, make sure the employee has studied the MSDS data and other pertinent information and is completely versed in the contents. In addition to training employees regarding the characteristics of the gases being handled, the employer is responsible for training employees in the applicable rules and regulations regarding the handling of these materials.

CGA Safety Training Aids

The Compressed Gas Association offers a wide variety of technical publications and standards, safety bulletins, safety posters, and audiovisual training aids which can be helpful in implementing a sound training program, and new training aids are being developed on a continuing basis. Among the audiovisual safety training programs produced by CGA are the following:

CGA AV-1, *Safe Handling and Storage of Compressed Gases in Containers*
CGA AV-2, *Pre-Trip Inspection of Compressed Gas Tank Cars*
CGA AV-3, *Filling of Industrial and Medical Nonflammable Compressed Gas Cylinders*
CGA AV-4, *Characteristics and Safe Handling of Medical Gases*
CGA AV-5, *Safe Handling of Liquefied Nitrogen and Argon*
CGA AV-6, *Hazardous Material Shipping Papers and Placards for Cylinder Truck Operations*
CGA AV-7, *Characteristics and Safe Handling of Carbon Dioxide*
CGA AV-8, *Characteristics and Safe Handling of Cryogenic Liquid and Gaseous Oxygen*
CGA AV-9, *Handling Acetylene Cylinders in Fire Situations*

A complete list of available CGA publications and audiovisuals can be found in Part IV of this *Handbook.*

REFERENCES

[1] CGA P-1, *Safe Handling of Compressed Gases in Containers,* Compressed Gas Association, Inc., 1235 Jefferson Davis Highway, Arlington, VA 22202.

[2] CGA P-2, *Characteristics and Safe Handling of Medical Gases,* Compressed Gas Association, Inc., 1235 Jefferson Davis Highway, Arlington, VA 22202.

[3] CGA P-2.1, *Recommendations for Medical-Surgical Vacuum Systems in Health Care Facilities,* Compressed Gas Association, Inc., 1235 Jefferson Davis Highway, Arlington, VA 22202.

[4] CGA P-2.6, *Transfilling of Liquid Oxygen to be*

Used for Respiration, Compressed Gas Association, Inc., 1235 Jefferson Davis Highway, Arlington, VA 22202.

[5] CGA P-5, *Suggestions for the Care of High Pressure Air Cylinders for Underwater Breathing,* Compressed Gas Association, Inc., 1235 Jefferson Davis Highway, Arlington, VA 22202.

[6] CGA P-6, *Standard Density Data, Atmospheric Gases and Hydrogen,* Compressed Gas Association, Inc., 1235 Jefferson Davis Highway, Arlington, VA 22202.

[7] CGA P-7, *Standard for Requalification of Cargo Tank Hose Used in the Transfer of Compressed Gases,* Compressed Gas Association, Inc., 1235 Jefferson Davis Highway, Arlington, VA 22202.

[8] CGA P-8, *Safe Practices for Air Separation Plants,* Compressed Gas Association, Inc., 1235 Jefferson Davis Highway, Arlington, VA 22202.

[9] CGA P-9, *The Inert Gases—Argon, Nitrogen and Helium,* Compressed Gas Association, Inc., 1235 Jefferson Davis Highway, Arlington, VA 22202.

[10] CGA P-12, *Safe Handling of Cryogenic Liquids,* Compressed Gas Association, Inc., 1235 Jefferson Davis Highway, Arlington, VA 22202.

[11] CGA P-13, *Safe Handling of Carbon Monoxide,* Compressed Gas Association, Inc., 1235 Jefferson Davis Highway, Arlington, VA 22202.

[12] CGA P-14, *Accident Prevention in Oxygen-Rich and Oxygen-Deficient Atmospheres,* Compressed Gas Association, Inc., 1235 Jefferson Davis Highway, Arlington, VA 22202.

[13] CGA P-15, *Filling of Industrial and Medical Nonflammable Compressed Gas Cylinders,* Compressed Gas Association, Inc., 1235 Jefferson Davis Highway, Arlington, VA 22202.

[14] *Code of Federal Regulations,* Title 49 CFR Parts 100–199 (Transportation), Superintendent of Documents, U.S. Government Printing Office, Washington, DC 20402.

[15] *Transportation of Dangerous Goods Regulations,* Canadian Government Publishing Centre, Supply and Services Canada, Ottawa, Ontario, Canada K1A 0S9.

[16] CGA C-4, *American National Standard Method of Marking Portable Compressed Gas Containers to Identify the Material Contained,* Compressed Gas Association, Inc., 1235 Jefferson Davis Highway, Arlington, VA 22202.

[17] CGA C-7, *Guide to the Preparation of Precautionary Labeling and Marking of Compressed Gas Cylinders,* Compressed Gas Association, Inc., 1235 Jefferson Davis Highway, Arlington, VA 22202.

[18] CGA C-2, *Recommendations for the Disposition of Unserviceable Compressed Gas Cylinders with Known Contents,* Compressed Gas Association, Inc., 1235 Jefferson Davis Highway, Arlington, VA 22202.

[19] *United States Pharmacopeia/National Formulary,* United States Pharmacopeial Convention, Inc., 12601 Twinbrook Parkway, Rockville, MD 20852.

[20] *Code of Federal Regulations,* Title 21 CFR Parts 200–299 (Food and Drugs), Superintendent of Documents, U.S. Government Printing Office, Washington, DC 20402.

[21] *Food and Drugs Act and Regulations,* Canadian Government Publishing Centre, Supply and Services Canada, Ottawa, Ontario, Canada K1A 0S9.

[22] CGA C-9, *Standard for Color Marking of Compressed Gas Containers Intended for Medical Use,* Compressed Gas Association, Inc., 1235 Jefferson Davis Highway, Arlington, VA 22202.

[23] NFPA 99, *Standard for Health Care Facilities,* National Fire Protection Association, Batterymarch Park, Quincy, MA 02269.

[24] CGA G-4.1, *Cleaning Equipment for Oxygen Service,* Compressed Gas Association, Inc., 1235 Jefferson Davis Highway, Arlington, VA 22202.

[25] NFPA 50, *Standard for Bulk Oxygen Systems at Consumer Sites,* National Fire Protection Association, Batterymarch Park, Quincy, MA 02269.

[26] NFPA 59A, *Standard for the Production, Storage, and Handling of Liquefied Natural Gas (LNG),* National Fire Protection Association, Batterymarch Park, Quincy, MA 02269.

[27] NFPA 50B, *Liquefied Hydrogen Systems at Consumer Sites,* National Fire Protection Association, Batterymarch Park, Quincy, MA 02269.

[28] NFPA 80, *Standard for Fire Doors and Windows,* National Fire Protection Association, Batterymarch Park, Quincy, MA 02269.

[29] NFPA 70, *National Electrical Code,* National Fire Protection Association, Batterymarch Park, Quincy, MA 02269.

[30] CGA C-6, *Standards for Visual Inspection of Steel Compressed Gas Cylinders,* Compressed Gas Association, Inc., 1235 Jefferson Davis Highway, Arlington, VA 22202.

[31] *Code of Federal Regulations,* Title 29 CFR Parts 1900–1910 (Labor), Superintendent of Documents, U.S. Government Printing Office, Washington, DC 20402.

[32] *Threshold Limit Values and Biological Exposure Indices,* American Conference of Governmental and Industrial Hygienists, 6500 Glenway Avenue, Bldg. D-7, Cincinnati, OH 45211-4438.

PART II

Specific Technical Information for Compressed Gas Equipment

Part II addresses several specific areas of significant importance concerning equipment used in conjunction with compressed gases. The bulk shipment of liquefied compressed gases, such as in tank cars and cargo tanks, is an important means of enabling economical distribution of those gases produced in relatively high commercial volumes. Chapter 6 will be of interest to those who are engaged in such shipments as suppliers or as users.

Pressure relief and safety devices have uniquely contributed to the safe utilization of compressed gases. Chapter 7 will provide insights into the evolution and systematic application of the various types of pressure relief devices now in use. Likewise, the systematic use of cylinder valve connections, connections to ancillary equipment, and bulk transfer connections has evolved into a methodology that is intended to help ensure that the proper gas is actually connected for the intended application. This methodology, as well as some interesting background information, will be found in Chapter 8.

Chapter 9 covers specific essential information for anyone who has day-to-day involvement with compressed gas cylinders. The topics of marking, labeling, inspection, testing, filling, and disposition all contribute to ensuring the continued safety of the many millions of compressed gas cylinders currently in use. In the United States and Canada, there are approximately one-half as many compressed gas cylinders of various types as there are people.

The last chapter in Part II, Chapter 10, provides specific coverage on the requirements for preparing compressed gas equipment for use with oxygen. In pure form and at high pressure, the chemical reactivity of oxygen, that is, its oxidizing property, becomes highly pronounced. Careful procedures must be employed to handle it safely.

CHAPTER 6

Handling Bulk Shipments of Liquefied Compressed Gases

INTRODUCTION

The liquefied compressed gases commonly shipped to users in bulk by rail or highway are: anhydrous ammonia, liquefied petroleum gas (LP-gas), butadiene, chlorine, methyl chloride, methyl mercaptan, sulfur dioxide, vinyl chloride, and broadly, liquefied fluorinated hydrocarbons. Carbon dioxide, hydrogen chloride, nitrous oxide, and vinyl fluoride are shipped as liquefied compressed gases at low temperatures. However, due to special considerations, these gases are not covered in this chapter. Similarly, this chapter does not address bulk shipments of cryogenic liquids such as oxygen, nitrogen, argon, hydrogen, and helium.

Typical shipping containers for bulk shipments of liquefied compressed gases are single-unit tank cars, multi-unit tank cars (TMUs), and cargo tanks. Safely unloading single-unit tank cars or cargo tank trucks, or safely removing and unloading containers from multi-unit tank cars, involves applying known safety procedures which will be discussed in this chapter.

Factors contributing to the safe handling and unloading of liquefied compressed gases include:

(1) Understanding their physical and chemical properties
(2) Using equipment specifically designed to be compatible with these physical and chemical properties
(3) Fully complying with the regulations and standards governing their handling
(4) Implementing appropriate procedures designed to minimize accidents which result from human failure

Proper safety precautions must be observed at all times during any stage of loading or unloading operations. Safe procedures are provided in regulations, in standards, in the literature of technical associations, and in the instructions of compressed gas manufacturers.

Check Regulations and Industry Standards

U.S. Department of Transportation (DOT) or Transport Canada (TC) regulations, and other state, provincial, and local regulations as may be applicable in a particular jurisdiction as well as industry standards, must be followed when handling compressed gases. Statements made in this chapter, and throughout this book, are intended only to provide complementary information and explanation of these regulations and standards and not to supplant them.

Obviously, within the scope of this chapter, it is not possible to set forth all the precautions to be taken when handling a particular gas. It is possible only to mention some of the more important considerations. In all cases, the compressed gas manufacturer must be consulted for thorough advice concerning specific handling and transfer problems.

Definitions

In addition to single-unit railroad tank cars, a number of other vehicles are used to transport liquefied compressed gases. The following terms are commonly used when referring to such vehicles and may be encountered in regulations or other literature on this subject.

Cargo tank: A cargo tank, as defined by the U.S. Department of Transportation and Transport Canada, is any tank permanently attached to or forming a part of any motor vehicle; or any compressed gas packaging not permanently attached but which by reason of its size, construction, or attachment to a motor vehicle, is loaded or unloaded without being removed from the motor vehicle. The capacity of cargo tanks ranges up to approximately 11 300 gal (42.77 m^3).

Motor vehicle: A motor vehicle includes a vehicle, machine, tractor, trailer, semitrailer, or any combination thereof, propelled or drawn by mechanical power and used on the highway.

Transport vehicle: A transport vehicle means a motor vehicle or rail car used for the transportation of cargo by any mode.

PERSONNEL AND TRAINING

Careful selection of the personnel charged with responsibility for loading or unloading bulk shipments of liquefied compressed gases is an important consideration. Reliable, intelligent people with a high sense of responsibility should be chosen and trained thoroughly. Because employees may be promoted or transferred, or may leave the company, training programs need to be ongoing and continuous. Replacement personnel must be thoroughly trained before assuming duties. It is recommended that instructors in the training program include a medical officer, safety supervisor, and fire protection engineer.

When training personnel, compressed gas manufacturers are a logical source for technical information describing the physical and chemical properties of the compressed gas in question. Manufacturers can provide information concerning physical and chemical characteristics of the product, personnel protective equipment, and storage and handling procedures aimed at preventing accidents. The manufacturer can also communicate corrective and first aid steps to be taken in the event of an accident. Trainees must study this information until they are completely versed with the contents. Some compressed gas manufacturers furnish qualified technical representatives or training aids to train personnel in every step of the loading and/or unloading procedure.

Training programs must impart a knowledge of all applicable rules and regulations. For personnel operating in the United States, this includes in particular those found in Title 49 of the U.S. *Code of Federal Regulations,* (49 CFR) Parts 107, 171, 172, 173, 174, 177, 178, and 179. [1]

Personnel should be made aware of the civil penalties for which they may become liable. In the United States, these can be found in 49 CFR Part 107. [1] There may also be local, state, or provincial regulations which must be complied with. In Canada, the *Transportation of Dangerous Goods Regulations* includes specific requirements for training and certification of all personnel involved in handling dangerous goods. [2] In this context, the term *handling* includes shipping and receiving as well as loading and unloading operations. Pertinent Hazardous Materials Regulations of the U.S. Department of Transportation from Title 49 of the U.S. *Code of Federal Regulations* are republished in a modified format in Tariff BOE-6000, *Hazardous Materials Regulations of the Department of Transportation by Air, Rail, Highway, Water, and Military Explosives by Water, including Specifications for Shipping Containers,* which is available from the Association of American Railroads in Washington, DC. [3]

GENERAL PRECAUTIONS

If any one of the following conditions is encountered, either company personnel responsible for maintaining the transport vehicle fleet, the compressed gas supplier or manufacturer, and/or the vehicle owner, should be immediately contacted for assistance and instructions as appropriate:

(1) A tank car is received in bad order. Do not attempt to load it. A loaded car returned with a bad order card should be unloaded if it can be done safely. Check with safety personnel in the absence of an established company procedure. Empty bad order cars should be moved to repair facilities.

(2) A failure of a fitting is apparent or a leak is present that cannot be readily repaired by a simple adjustment or tightening of the fitting. Isolate the transport vehicle and permit only properly instructed and protected personnel to enter the area.

(3) The transport vehicle cannot be unloaded after following all instructions.

(4) An accident of any kind has occurred. In the United States, a transportation accident must be reported to DOT by the carrier as required by 49 CFR 171.15–16. [1] Applicable Canadian regulations define requirements for immediate and/or 30-day reporting of incidents involving dangerous goods and packaging or containers.

No attempt should be made to correct leaks at unions or other fittings in a transfer line with a wrench while the line is under pressure. Never break a hose coupling which is under pressure.

Personnel should wear protective gloves to prevent contact with liquids during loading and unloading. Liquefied gases vaporize rapidly upon release at normal temperatures and atmospheric pressure. Contact with such liquids by skin surfaces and subsequent evaporation causes severe burns (frostbite). When large quantities of liquid are involved, deep and severe freezing of the body area in contact with the liquid results.

Large-lensed safety spectacles or goggles should be worn to prevent liquid from contacting the eyes. Gas masks, protective clothing, fire extinguishers, and related equipment of approved design and in good condition should be stored where they are readily accessible in case of a leak. This equipment should be maintained in good condition. Personnel should be thoroughly instructed in the use of this equipment and continuously trained by occasional "dry runs."

Frequent inspection of loading and unloading hoses, piping, and fittings for wear, deterioration, or abuse is recommended.

Slip Tube Gauging Device

When gauging a tank car equipped with a slip tube gauging device, one should never stand over the gauging device. After removal of the cover, a hand is placed on top of the control valve on top of the slip tube rod while releasing the top lock so that the rod will not rise in the event the brake is loose. While holding the top of the control valve, the spring-held handbrake is released. The rod is not supposed to move freely. If it does move freely, the brake is put on and the packing gland nut is tightened until the rod does not move freely. During this entire operation, one must keep one hand on top of the gauge rod. Proper operation requires that some hand lifting force should be used to operate the gauge rod.

During loading or unloading, the gauge rod should never be pulled all the way out until it hits the stop ring. The rod should only be pulled out far enough to be above the required or expected liquid level of the product in the tank. The function of the stop ring on the rod is to permit changing of packing rings while the tank is under pressure and when the procedure for making this change is followed. The stop ring should not be allowed to frequently hit the seal under the packing rings because the seal may become damaged and the packing rings could no longer be safely changed while the tank is under pressure.

For safety reasons, the following must be emphasized:

- Never stand on the gauge rod.
- Restrain movement of the rod with the hand on top of the control valve on top of the gauge rod until the rod is secured in its desired position below the seal.
- Never stand on the brake lever or use a hammer on any part of the gauging device.

TANK CARS

Tank Car Specifications

General information describing compressed gas shipping containers is presented in Chapter 4. Single-unit tank cars for transporting flammable compressed gases may be of the Class 105, 112, or 114 type and must be equipped with thermal fire protection and tank head puncture resistance systems such as head shields. See 49 CFR 179.105. [1] *Note:* An exception to this is that Class 105 type tank cars of not greater than 18 500 gallon (70 m^3) capacity and built before September 1, 1981, do not require thermal and head protection.

Single-unit tank cars for transporting anhydrous ammonia may be of the Class 105, 112, or 114 type and must have head shields. An exception to this is that Class 105 type tank cars of not greater than 18 500 gallon (70 m^3) capacity and built before September 1, 1981, do not require head shields.

Head shields can be items separate from the tank, located outside each head, or they can be integral with the jacket head in the case of a jacketed tank. The head shield is made part of the jacket by using $\frac{1}{2}$-inch (1.27-cm) thick material for at least the lower half of the jacket head.

The U.S. Department of Transportation, Transport Canada, and the Tank Car Committee of the Association of American Railroads review tank car specifications on an ongoing basis for any needed improvement. Therefore, those engaged in shipping and receiving compressed gases by tank car need to keep abreast of current DOT and/or TC regulations for information on proper shipping containers. [1] and [2] Also, the most current version of the *AAR Specifications for Tank Cars* should be reviewed because sometimes the AAR requirements for tank cars are more stringent than the DOT regulations. [4]

Tank cars are never filled completely with liquid but have a vapor space, termed outage, the needed volume of which is calculated for each specific gas and temperature condition.

The arrangement of most compressed gas tank car valves and fittings is similar. (See Fig. 6–1.) The valves are located on a manhole cover plate, usually on top of the car and within a protective housing. These are sometimes incorrectly referred to as dome fittings, although the compressed gas design tank has no dome as such. A few tank cars built to the Class 114 specification have been equipped with bottom outlets. Since these cars are few in number and are found in special service, this discussion is confined to tank cars with top fittings.

Usually, there are two valves in line with the railroad track. These are connected to liquid eduction pipes that extend to the bottom of the tank cars. They are referred to as liquid valves or eduction valves. Another valve is mounted toward the side of the car, terminating in the tank's vapor space. Sometimes an additional valve is located on the opposite side. Such valves are referred to as vapor valves. A pressure relief valve is always present, usually in the center of the cover plate.

Additional cover plate fittings may include a liquid level gauge, a thermometer well, and a sample line. Excess flow check valves are installed on the liquid eduction lines of most single-unit tank cars handling compressed gases. For liquefied flammable gases, such valves are required by DOT regulations on both the eduction and vapor lines. These excess flow valves are intended to contain the tank contents during transit, or in a wreck environment, in the event of valve wipeoff. They are neither intended nor designed for containment of the tank car contents in the event of a loading or unloading line break.

- A Liquid eduction valve
- B Vapor valve
- C Pressure relief valve
- D Gaging device
  1. Gaging pointer
  2. Gage rod lock
  3. Gage rod valve
  4. Gage rod
  5. Gage rod brake
  6. Packing gland nut
  7. Protective housing
  8. Gasket
  9. Gaging rod shield vent holes
  10. Lubricator assembly
- E Sample valve
- F Thermometer well
- G Excess flow valves
- H Liquid eduction pipe
- I Sample line
- J Screen
- K 4-in. insulation
- L Liquid level

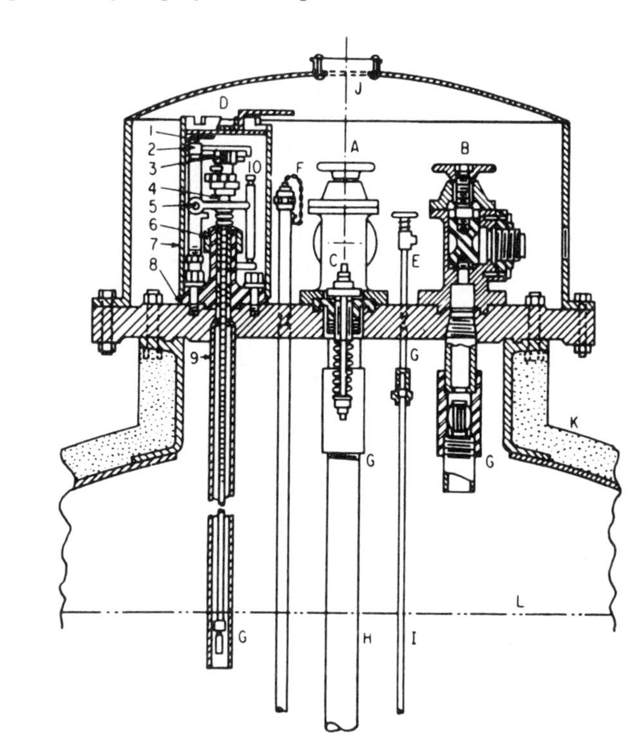

Fig. 6-1. Tank cars for liquefied compressed gases have similar valve arrangement.

Piping Compatibility at Unloading Sites

Because many unloading systems have hose or piping smaller than the tank car piping system, they may not be capable of the flow rates required to close the tank car excess flow valves should a break occur during unloading. To prevent loss of product in the event of a hose rupture in the unloading system, an additional, smaller, excess flow valve should be a part of the line connected to the tank car and located on the tank car side of the hose. However, those responsible for such operations are cautioned to check state, provincial, and local regulations to be certain an excess flow valve is considered proper protection. The "authority having jurisdiction," such as a fire marshal, is a good source to contact for information on local restrictions. A listing of some state authorities is provided as an appendix in Part IV.

The designed closing flow for such an excess flow valve should be about 50 percent greater than the line flow during unloading. The unloading line should never be of a pipe size smaller than that of the excess flow valve installed in the line.

In a loading line, an excess flow valve is not practical because it would close upon excess flow into the tank car but would not protect against loss of vapor out of the tank car. Protection can be better obtained by installing a backflow check valve in the loading line at the tank car end of any hose or swivel-type piping.

Another method of preventing loss of contents from the tank car is to install an emergency shutoff valve in the loading or unloading line near the tank car valve. In fact, NFPA 58, *Standard for the Storage and Handling of Liquefied Petroleum Gases,* paragraph 4-2.3.6 states: "When a hose or swivel-type piping is used for loading or unloading railroad tank cars, an emergency shut-off valve complying with 2-4.5.4 shall be used at the tank car end of the hose or swivel-type piping." [5] Paragraph 2-4.5.4 requires the valve to have automatic shutoff through thermal actuation, manual shutoff from a remote location, and manual shutoff at the installed location. NFPA 58 does not apply to marine terminals, pipeline terminals, natural gas processing plants, refineries, utility gas plants, chemical plants, or tank farms. Tank farm storage at industrial locations is covered by NFPA 58. [5]

Occasionally, an operator will open a tank car liquid valve too quickly and the excess flow valve will slam shut. Never hammer tank car fittings in an attempt to open a closed excess flow valve. To open a closed excess flow valve, close the tank car liquid angle valves. The pressure will slowly equalize and the excess flow valve will reopen with an audible click. The tank car liquid angle valves can then be opened slowly without further premature closing of the excess flow valve.

Single-unit tank cars are protected from excessive internal pressure by a spring-loaded pressure relief valve located on the manhole cover plate. The valve setting depends upon the specification of the tank car. Loading and unloading procedures must be such that the pressure developed does not cause the pressure relief valve to open.

INITIAL PRECAUTIONS AT RAIL LOADING AND UNLOADING RACKS

Rail track at loading and unloading spots should be essentially level so cars can be properly gauged and unloaded completely. Brakes must be set and wheels blocked on the car. Caution signs as prescribed in 49 CFR Part 174 of the DOT regulations, or similar TC regulations, must be so placed on the track or cars to give necessary warning to persons approaching the cars from the open end(s) of the siding and must be left up until after the liquid transfer is complete and the cars' loading or unloading lines are disconnected. [1] and [2] Once caution signs are posted, tank cars so protected must not be coupled or moved. Other cars must not be permitted on the same track except after notifying the person who placed the signs. It is also recommended that the unloading spot be isolated from other rail car movement by locked-out switches or derailing devices.

Training aids for instructing personnel about pre-trip inspection of the tank car are available from sources such as the Compressed Gas Association. Before connecting the loading lines, the inspection procedure should include the following steps:

(1) Make certain the tank car is authorized for the product to be shipped.
(2) If the car has a product label or marking different from the product to be loaded, make certain the product residue in the tank is compatible with the product to be loaded. Also, remove the improper marking or change markings to correspond with the new product. Check applicable regulations to determine which products require a product marking or label on the container.
(3) Check the condition of the undercarriage.
(4) Check safety appliances (handrails, grab irons, etc).
(5) Check the condition of the tank paint and the stenciling, marking, and labeling.
(6) Check the condition of the manway bolts and gaskets. Gasket material should be compatible with products, and gaskets should be maintained or replaced regularly.
(7) Note the test due date for both the tank and the pressure relief valve. Do not load if either test is overdue or if the pressure relief valve is leaking or inoperable.
(8) Check the condition of internal and external valves and fittings.
(9) Check the condition of the tank.

After disconnecting the loading lines, the inspection should continue and include the following:

(1) If a product marking is not required on the tank, product markings which do

not refer to the loaded product should be removed.

(2) If a product marking is required, the tank should be marked with the name of the loaded product in accordance with Appendix C of the *AAR Specifications for Tank Cars* and DOT regulations at 49 CFR Parts 172 and 173, or the *Transportation of Dangerous Goods Regulations,* as appropriate. [4], [1], and [2]

(3) All valves should be closed and tightened and not leaking.

(4) Caps or plugs on sample valves, loading and unloading valves, and gauging device valves must be installed and wrench tightened. Gauging devices must be securely in place and not leaking.

(5) The protective housing cover must be secure, pinned, and with proper seals in place.

(6) Tank cars should be placarded in accordance with DOT regulations at 49 CFR Part 172 or the *Transportation of Dangerous Goods Regulations,* as appropriate. [1] and [2]

Remember that it is illegal to ship a defective or leaking tank car. If a car cannot be put into suitable condition for shipping, notify the proper company personnel and request instructions. In some cases, the tank car owner will be the proper person to contact for assistance.

UNLOADING TANK CARS

With single-unit tank cars and multi-unit tank cars, full responsibility for safe unloading rests with the consignee. Proper equipment and well-trained personnel are a must at unloading locations.

Transfer lines and flexible hoses must be of materials suitable for the product being unloaded, as recommended by the compressed gas manufacturer or industry standards. Plugging or capping the ends of the lines when not in use will prevent the accumulation of moisture and dirt. In the vicinity of racks where flammable gas is being unloaded, NFPA Class 1, Group D (explosion-proof) lights, switches, motors, and other electrical appliances as required by Articles 500 and 501 of NFPA 70, *National Electric Code,* must be used. [6] In the case of flammable gases, a firm "no smoking" rule for all personnel in the unloading rack area must be enforced. Check the area for all open flames and extinguish them before starting unloading operations.

In preparing the tank car for unloading, the seal on the tank car protective housing cover is broken, as are other seals as required. The car is made ready for unloading by carefully and cautiously removing the plugs from the car's liquid and vapor valves to be used in unloading. If gauging the tank car is necessary and a slip tube gauge is provided, personnel must take care not to place their heads or bodies directly over the tank car gauging device when releasing the hold-down latch. Tank pressure may force the slip tube up rapidly and with considerable force. Likewise, personnel should never place their heads over the pressure relief valve opening.

After the unloading lines have been connected and all connections are tight, the system may be pressurized and checked for leaks. Throughout the entire period of unloading, and while the car is connected to the unloading device, the tank car must be monitored. Tank cars should not be allowed to stand with unloading connections attached after unloading has been completed.

The different methods of unloading may require different valve types and related equipment. In all cases, the liquid unloading line should be equipped with a pressure gauge and shutoff valve. As mentioned earlier, the unloading lines should be equipped with properly sized excess flow valves or emergency shutoff valves to protect against loss of product in the event of a hose failure.

Liquefied compressed gases are usually transferred from single-unit tank cars by means of pressure differential. All equipment must be designed for the particular gas being transferred. The following is a listing of methods for obtaining the pressure differential required and the gases to which the method may be applied:

- By compressor: anhydrous ammonia, LP-gas, butadiene, methyl chloride, vinyl chloride, sulfur dioxide, and liquefied fluorinated hydrocarbons
- By vaporizer: butane
- By gas repressuring: LP-gas, butadiene, and vinyl chloride
- By direct-acting liquid pump: LP-gas, anhydrous ammonia, and vinyl chloride
- By air padding: chlorine and sulfur dioxide

Unloading with a Compressor

A typical compressor unloading setup is shown in Fig. 6–2. The suction side of the compressor is connected to the vapor line of the storage tank and the discharge side is connected to the vapor valve on the tank car. Note that oil-lubricated compressors should be equipped with an oil mist extractor on the compressor discharge to prevent contamination. One or both liquid eduction lines on the tank car are connected to the storage tank with appropriate piping. The compressor withdraws and compresses vapor from the storage tank and transfers it to the tank car's vapor space. In this manner, the desired pressure differential is created to transfer liquid to the storage tank from the tank car.

To use the compressor unloading system after the piping hookup has been completed, both liquid eduction valves on the tank car are opened slowly and completely. Then all other valves in the liquid line are opened, working from the tank car to the storage tank. The storage tank filling valves are opened slowly. If the tank car pressure is higher than that of the storage tank, the valves in the vapor line should remain closed and the compressor is not operated. When the rate of liquid flow drops to an unsatisfactory level, the vapor valves between the tank car and the storage tank are opened and the compressor is started.

The sight glass or flowmeter (when one is installed) in the liquid discharge line, the pressure gauges in the transfer lines, or the sample line on the tank car are typical methods used to determine when all the liquid is removed from the tank car. When this has been established, the car is prepared for disconnecting (with some exceptions such as in the case of chlorine) by closing down liquid transfer lines first. This is done by starting at the storage tank and closing all valves up to and including the tank car's liquid valve.

Withdrawal of vapors from the tank car after it has been emptied of liquid is accomplished by shutting down the compressor, closing the liquid valves starting at the storage tank and working to the tank car, and then turning the compressor's reversing

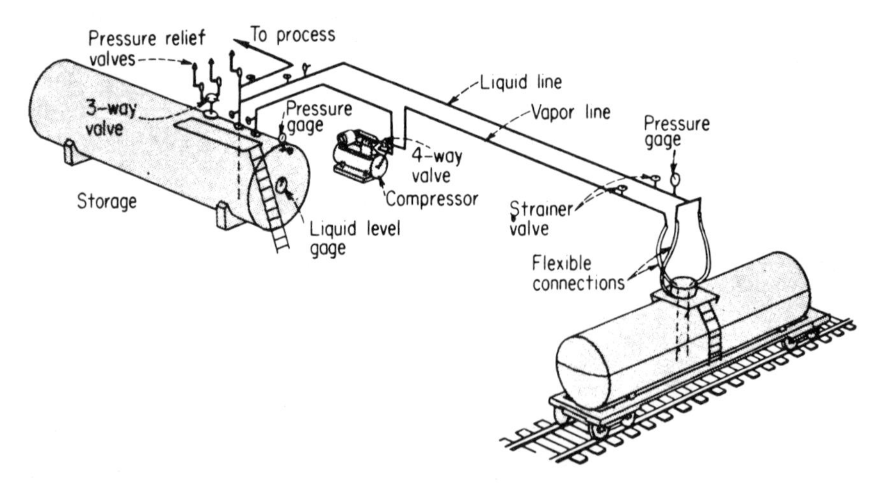

Fig. 6–2. Typical compressor unloading setup.

valve. When the compressor is started, most of the vapors in the tank car will be withdrawn to storage. The amount of vapors withdrawn is largely based on the operating costs of the compressor versus the value of the vapors or safety considerations.

When vapor transfer is to be discontinued, the tank car's vapor valve is closed first, and then the other valves are closed successively by working from the tank car to the source of pressure. Once all transfer lines have been bled to atmospheric pressure, the tank car is disconnected.

Unloading with a Vaporizer

Another method of unloading involves using a vaporizer. This piece of equipment is sometimes called an evaporator. Liquid from storage is charged into the vaporizer and the vaporizer's vapor section is connected to the vapor valve on the tank car. When heat is exchanged at the vaporizer, the vapor pressure generated exerts pressure on top of the liquid in the tank car, causing it to flow to storage. By this method, all the liquid can be transferred to storage but the vapor remains in the car. A steam vaporizer method of tank car unloading is illustrated in Fig. 6-3.

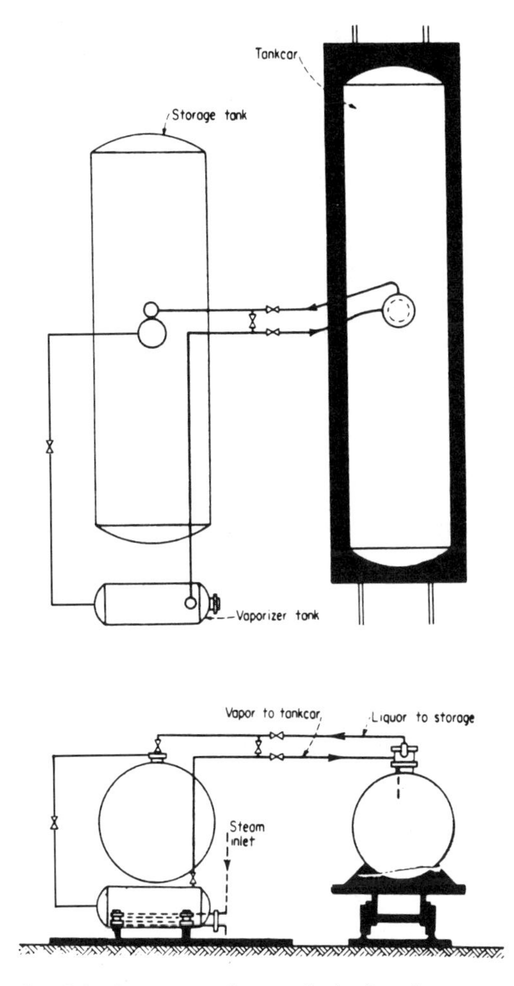

Fig. 6-3 Steam vaporizer method of tank car unloading.

Unloading with a Pump

A tank car can be unloaded with a liquid pump, but it is impossible to remove any of the vapors in the tank car with the liquid pump. For this reason, a compressor is almost always used. The basic procedure when using a liquid pump is the same as that followed when using the compressor, except that the pump is connected in the liquid line from the tank car to the storage tank and a vapor-equalizing line is connected to each tank. The vapor line between the storage tank and the tank car must be large enough so that the pressure in the tank car will not drop appreciably below the pressure in the storage tank. Otherwise, the liquid pump may vapor lock and unloading will be difficult if not impossible.

If at the start of unloading the tank car pressure is above the storage tank pressure, the valves in the vapor line should not be opened. The higher tank car pressure will help the pump and unloading will be faster. Open the vapor valves only when the storage tank pressure is equal to, or higher than, the tank car pressure.

Unloading by Gas Pressure

Unloading by gas pressure (or any method of unloading that leaves the car filled with noncondensible gas) should not be used without the permission of the tank car owner or lessee. The noncondensible gas contaminates the next load, makes loading the car very difficult, and may cause premature operation of the pressure relief valve. Therefore, all noncondensible gases will have to be vented before the tank car is reloaded. The loading rack must be equipped to handle this situation. If gas pressuring must be used to unload a tank car, the gas used must be clean, dry, and noncorrosive. For some products, an inert gas is required. A dryer may be needed to remove moisture from the gas. The pressure supplied must not be high enough to actuate the tank car pressure relief valve.

To unload, the liquid hoses are connected to the tank car liquid eduction valves and storage tank manifold lines. The pressuring hose is connected to the tank car vapor valve and to the gas supply line. This line must have an approved check valve in it to prevent the flow of gas from the tank car into the gas supply system. Both tank car liquid eduction valves are opened slowly but completely. Then all the other valves in the liquid line are opened, working from the tank car to the storage tank. The storage tank filling valve is slowly opened, with care taken not to open this valve too far if the tank car pressure is above the storage tank pressure. Otherwise, the tank car excess flow valves may close. The tank car vapor valve or the valves in the pressuring gas supply line are not opened if the tank car pressure is higher than the storage tank pressure. When the liquid flow drops to an unsatisfactory rate with the storage tank fill valve wide open, the pressuring gas supply to the tank car is opened. The tank car pressure should be maintained 5 to 10 psi (34 to 69 kPa) above the storage tank pressure by controlling the pressurizing gas supply to the tank car.

A flow of gas instead of liquid through the sight flow glass in the unloading line indicates that the tank car is empty. This should be checked by opening the tank car sample valve, if so equipped. When the tank car is empty, the repressuring gas supply is shut off, then all valves in the liquid line are closed, working from the storage tank to the tank car. All of the valves on the tank car are closed next. After the pressure is bled from all the hoses, the hoses are disconnected and the plugs replaced in the tank car valves.

Unloading with Air Padding

For some products, unloading with air padding may be prohibited by regulations. When air padding is allowed and is used for unloading, it is imperative that clean, oil-free, cooled, dry compressed air be introduced into the tank car's vapor space through its vapor valve to transfer the car's liquid. The setup consists of an air compressor and its tank equipped with a pressure regulator, a pressure relief filter, and an air dryer with appropriate valves and gauges similar to those shown in Fig. 6-4. A dependable check valve must be incorporated in the air-padding line as also shown. Never use a plant air system for air padding since vapors may be drawn into the plant air system.

HANDLING MULTI-UNIT TANK CARS

Multi-unit tank car tanks are commonly known as TMU tanks, ton tanks, or ton containers. This name originated from their capacity in terms of chlorine. Currently, these tanks are most often transported in highway van trailers properly blocked and braced against movement. Occasionally, they are still shipped on a rail car underframe. This was the original transportation mode for these tanks many years ago (with rail shipments usually consisting of 15 containers per car).

Before removing containers from the rail car frame, it is necessary to observe the same DOT or TC regulations concerning personnel, setting brakes, blocking wheels of the car, and posting caution placards as in unloading single-unit cars. Full containers are removed from the underframe by a lifting hook in combination with a hoist on a trolley or a jib boom. A lifting magnet, rope, or chain sling should never be used to unload

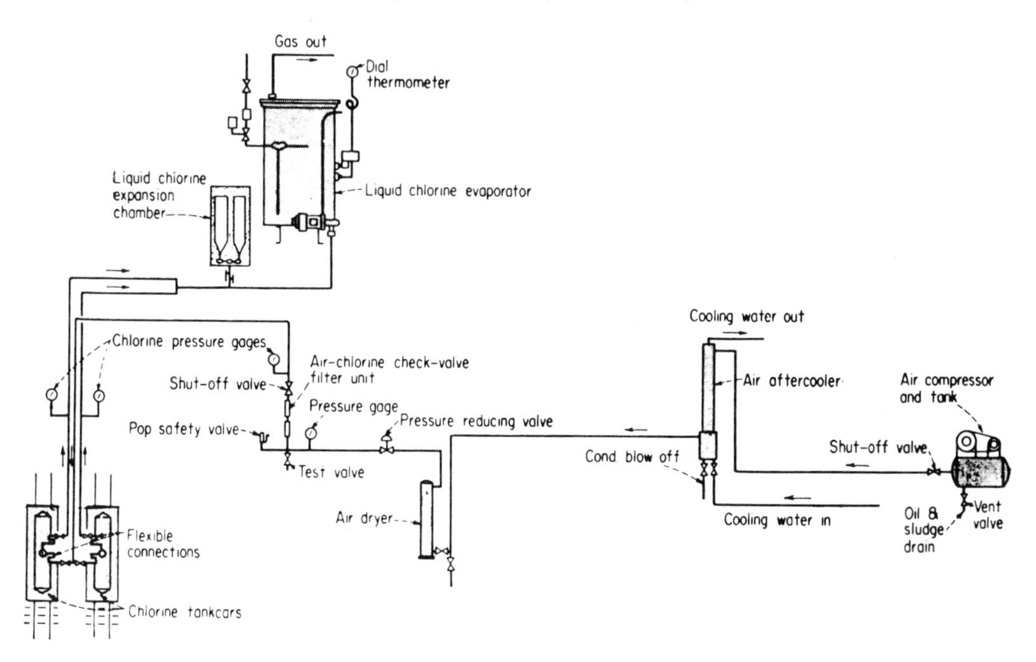

Fig. 6–4. Unloading with air padding.

or otherwise handle these containers. Containers must be protected from shock which might damage valves, fusible plugs, or the container itself. Keep valve-protecting hoods in place at all times when containers are being moved. When containers are trucked from the rail siding into the storage area, they must be placed on saddles on the truck. It is preferable to clamp them down to prevent shifting and rolling.

Each container has two valves, each equipped with a $\frac{1}{2}$-inch (12.7-mm) eduction pipe. When the container is positioned horizontally with a slight downward pitch and one valve is directly above the other, the top eduction pipe ends in the vapor phase and the eduction pipe of the bottom valve in the liquid phase. Either liquid or vapor can be withdrawn by connecting to the appropriate valve. These two valves are protected by hoods. These hoods should be kept in place at all times except when the container is connected for withdrawing its contents.

Some containers may be protected by fusible plugs in each head. In some cases, the fusible metal is designed to soften or melt at temperatures as low as 157°F (69.4°C). There should be no tampering with the fusible plugs under any circumstances.

In both the United States and Canada, regulations provide that multi-unit containers may be transported under certain conditions on trucks or semitrailers. They must be chocked or clamped on the truck to prevent shifting. Adequate facilities must be available when transfer in transit is necessary. TMU containers may be removed from the rail car underframe when the car is spotted on carrier tracks provided that pertinent regulations are complied with. The carrier may give permission for the unloading only if a private siding is not available within a reasonable trucking distance of the final destination, and the consignee must furnish an adequately strong mechanical hoist to lift containers from the car and deposit them directly on vehicles furnished by the consignee. The containers are then transported by truck to the user's plant, where similar handling equipment must be available. Cautions concerning the handling of these containers as described earlier must be strictly observed.

TMU containers should be stored on the user's premises in a cool, dry place and be protected against heat sources. A convenient storage rack can be made by supporting the containers at each end on a railroad rail or an I-beam. When containers are not being used, the valve-protective hoods should be kept in place at all times. Storing full and empty containers in different places will avoid confusion in handling. It is good practice to tag empty containers.

Containers should not be stored near elevators or gangways, or in locations where heavy objects may fall and strike them. Likewise, containers should never be stored near combustible or flammable materials. When containers hold flammable gases, they must be kept away from all sources of ignition. Storage rooms should be well ventilated and so arranged that any container can be removed with a minimum of handling of other containers. When practical, the storage room should be fireproof. Storage in subsurface locations is to be avoided. Containers stored outdoors must be kept free of debris and tall grass, and away from public access. They should be kept clean and inspected regularly.

Containers storing flammable gas or gas that affects the respiratory system should not be placed where fumes can enter a ventilating system or where wind can carry fumes to populated areas.

Except when recommended by the compressed gas manufacturer, TMU containers should not be manifolded to withdraw contents from two or more containers simultaneously. To discharge the contents of a container, place it so that it is in a nearly horizontal position with a slight downward pitch (about 1 inch or 2.54 cm overall) toward the valves. The assembly for withdrawing a container's contents usually consists of a transfer line equipped with a pressure-reducing valve, rotometer, control valve, and diffuser. In chlorine service, the system may include only pressure and control valves.

Test the TMU container connections and transfer piping for leaks as described later in this chapter under the handling information for each specific gas. Sometimes it may be necessary to use heat to help the flow of either gas or liquid from sulfur dioxide or a liquefied fluorinated hydrocarbon ton container. When heat is used, the method approved by the compressed gas manufacturer must be followed. For handling sulfur dioxide, refer to CGA G-3, *Sulfur Dioxide.* [7] It is important to exercise great care as fusible plugs in ton containers may melt. Fusible plugs must be vapor tight at not less than 130°F (54.4°C). To minimize the potential for cold flow of the fusible plug material, it is recommended that containers never be allowed to reach a temperature above 125°F (51.7°C). Blow torches, steam hoses, or use of an open flame from any source must never be used to heat containers.

CARGO TANKS

DOT requirements for carriage of compressed gases by public highway are found in Title 49 of the U.S. *Code of Federal Regulations,* primarily in 49 CFR Part 177. [1] In Canada, the *Transportation of Dangerous Goods Regulations* address this area. [2] In part, these regulations provide that while loading or unloading a flammable gas there must be no smoking, lighting of matches, or carrying of any flame or lighted cigar, pipe, or cigarette. A hazardous material must never be loaded or unloaded unless the handbrake is securely set and other reasonable precautions are taken to prevent motion of the vehicle. The truck should be parked so that the cargo tank is level for loading or gauging. For unloading, park the cargo tank truck so that the outlet opening is in the lowest part of the tank.

Flammable gas must not be loaded or unloaded with the motor vehicle engine running unless the engine is used for the operation of the transfer pump of the vehicle. Unless the delivery hose is equipped with a shutoff valve at its discharge end, the engine of the motor vehicle must be stopped at the finish of the unloading operation while the discharge connection is disconnected.

Tank motor vehicles must not be moved, coupled, or uncoupled when the loading connections are attached to the vehicle. Semitrailers or trailers must not be left without the power unit unless chocked or prevented from moving by equivalent means.

Most liquefied compressed gases are, or may be, shipped in cargo tanks conforming to Specifications MC-330 or TC/MC-331, TC/MC-338, and TC-341. See 49 CFR 173.315, [1], or in Canada, see the Candian Standards Association standards CSA B620, *Highway Tanks and Portable Tanks for the Transportation of Dangerous Goods,* and CSA B622, *CSA Standard for the Selection and Use of Tank Trucks, Tank Trailers and Portable Tanks for the Transportation of Dangerous Goods for Class 2 By Road.* [8] and [9] The transport vehicle must be marked and placarded in accordance with the appropriate regulations. See 49 CFR Part 172 and equivalent regulations of Transport Canada. [1] and [2] Chlorine cargo tanks may be shipped only if the contents are to be unloaded at one point. Specification MC-330 and TC/MC-331 cargo tanks constructed of quenched and tempered steel can be used to transport anhydrous ammonia if it has a minimum water content of 0.2 percent by weight. Such tanks can be used to transport LP-gases determined to be noncorrosive.

For a product being loaded, the vapor pressure (psig) at 115°F (46.1°C) must not exceed the design pressure of the cargo tank. Also, cargo tanks must not be filled liquid full as space must be provided for the liquid to expand as the temperature rises. The permissible filling density for each product is found in 49 CFR Part 173 for the United States and CSA B622 for Canada. [1] and [9] Cargo tanks may be filled by weight or volume unless otherwise specified for a particular product. Highway load limits must also be observed.

The actual loading operation will vary depending on the design of the cargo tank and the method of measurement used at that particular loading point. To load properly, the loader must usually know the specific gravity at the temperature of the liquid when loaded. Unless the tank has been cleaned, the product last shipped must be compatible with the product to be shipped. For example, do not load LP-gas into a tank which previously contained anhydrous ammonia.

During the unloading of a cargo tank, the motor carrier who transports hazardous materials must ensure that a qualified person is in attendance at all times. However, the carrier's obligation to monitor unloading ceases when:

(1) The carrier's obligation for transporting the materials is fulfilled.
(2) The cargo tank has been placed on the consignee's premises.
(3) The motive power has been removed from the cargo tank and removed from the premises.

When the carrier is not responsible, the consignee is responsible for the safe transfer of the product and must have a qualified person in attendance at all times. Definitions of *attends* and *qualified person* can be found in 49 CFR 177.834(i). [1] In Canada, requirements for training and certification are contained in Part IX of the *Transportation of Dangerous Goods Regulations.* [2]

Where responsibility for unloading cargo tank trucks rest with the common carrier or transport owner, his or her operator should be trained in all phases of the operation. The operator should be thoroughly familiar with all federal, state, provincial, and local regulations.

Anhydrous ammonia is unloaded from cargo tanks by a compressor or a liquid pump, LP-gas by a compressor or liquid pump, and methyl chloride and sulfur dioxide by a compressor. Liquefied fluorinated hydrocarbons are usually unloaded from cargo tank trucks by a turbine pump, although a compressor may be used. DOT and TC regulations for unloading cargo tank trucks, as well as precautions previously noted in this discussion, are to be observed. Procedures are generally similar to those described for unloading single-unit tank cars. Even where responsibility for unloading cargo tank trucks rests with the carrier or transport owner, the consignee should be as familiar with the entire unloading procedure as is the transport's operator in charge of transfer.

RETURN OF EMPTY TRANSPORT EQUIPMENT

Consignee Becomes Shipper

After the unloading operation has been completed, the consignee must prepare the transport vehicle for return to the supplier. Thus, the consignee now becomes the shipper.

After unloading a single-unit tank car, closing the valves, and putting the plugs or caps in place, the protective housing cover is latched and closed in place. Valves on TMU containers should be closed and the protecting hoods in place when readied for return. For multi-unit tank cars, load the containers and secure them in place. The placards must be removed from the empty tank car and replaced or reversed to show a *RESIDUE* placard in accordance with TC or DOT regulations. See 49 Part 172. [1] The return billing paper must show *Residue* or *Residue: Last contained,* followed by the name of the material last contained in the tank and the hazard class of the material. The word *Placarded* must be included in the United States. In Canada, it is necessary to indicate the number and class of placards provided (e.g. 4 × 2.1 placards provided).

After a cargo tank has been emptied of a hazardous material, it must remain placarded for the return shipment unless it is (1) reloaded with a material not subject to DOT regulation or (2) sufficiently cleaned and purged of vapors to remove any potential hazard. The return shipping paper for a cargo tank containing the residue of a hazardous material may contain the words *Residue* or *Residue: Last contained* followed by the name of the material.

SPECIAL PRECAUTIONS FOR SPECIFIC GASES

Butadiene, Inhibited

Butadiene is a flammable gas which must be inhibited when offered for transportation because of hazards arising from the potential for polymerization. In high concentrations, it is an anesthetic that can cause respiratory paralysis and death after prolonged breathing. Lower concentrations may produce slight irritation of eyes, nose, and throat. Butadiene is mildly aromatic, and blurring of vision and nausea are characteristic symptoms of exposure. The liquid will cause a freezing burn of the skin.

Large butadiene fires are difficult to extinguish. Shutting off the source of the fuel and keeping all adajcent occupancies wetted with water spray are means to help control a fire. Small fires can be extinguished with CO_2 or dry-chemical first aid extinguishers.

Pressurized air should never be used for unloading. Butadiene should be unloaded through a closed system using a vapor return line and compressor, or by use of pressurized inert gas (such as nitrogen). When all the liquid is removed from the transport vehicle, valves should be closed and no air permitted to enter the tank. Butadiene transport vehicles should not be used for any other product unless the tank has been cleaned, freed of gas, and purged of air. See the monograph on butadiene in Part III for additional information.

Vinyl Chloride

Vinyl chloride monomer is a flammable gas which is easily ignited and which produces hazardous combustion gases largely composed of hydrogen chloride and carbon monoxide. The odor of vinyl chloride vapors is pleasant, and when inhaled, vinyl chloride acts as an anesthetic.

An explosion hazard can exist when drawing samples or venting to the atmosphere. Fire involving large quantities of liquid are difficult to extinguish since vinyl chloride is not miscible with water and is lighter than water. Most small fires can be extinguished with carbon dioxide or dry-chemical agents if properly applied. Vinyl chloride can be unloaded by pump, compressor, or inert gas pressure. Air must not be permitted to enter the tank. Additional information concerning vinyl chloride is provided in the monograph in Part III.

Anhydrous Ammonia

Anhydrous ammonia is classified by the DOT as a nonflammable gas. In Canada, it is classified as a corrosive gas, class 2.4.

In transferring ammonia from containers, including single-unit tank cars, TMU containers, and cargo tank trucks, never use compressed air as it will contaminate the ammonia.

The continuous presence of the sharp, irritating odor of ammonia is evidence of a leak. Leaks of ammonia can be located by allowing fumes from an open bottle of hydrochloric acid (from a squeeze bottle of sulfuric acid or from a sulfur dioxide aerosol container) to come in contact with leaking ammonia vapor. This produces a dense fog. Leaks may also be detected with moist phenolphthalein or litmus paper. Sulfur tapers for detecting ammonia leaks are not recommended. When there is a leak around an ammonia container valve stem, it usually can be corrected by tightening the packing gland nut which has a left-hand thread.

When a leak occurs in a congested area where atmospheric dissipation is not feasible, the ammonia can be absorbed in water. Its high solubility in water may be utilized to control the escape of ammonia vapor. Applying a large volume of water from a fog or spray nozzle lessens vaporization, as the vapor pressure of ammonia in water is much less than that of liquid ammonia. Do not neutralize liquid ammonia with acid; the heat generated by the reaction may increase the fumes.

Only an authorized person should attempt to stop a leak. If there is any question as to the seriousness of the leak, a gas mask of the type approved by the U.S. Bureau of Mines or the National Institute for Occupational Safety and Health (NIOSH) for use with ammonia must be worn. Have all persons not equipped with such masks leave the affected area until the leak is stopped.

Also, provide personnel subject to exposure to ammonia with a hat, gloves, suit, and boots made of rubber or other suitable material impervious to ammonia. Garments worn beneath rubber outer clothing should be of cotton. Some protection to the skin may be obtained by applying protecting oils prior to possible exposure to ammonia. Approved eye goggles must be worn if the eyes are not protected by a full face mask.

Although ammonia is flammable in air only within the narrow range of 16 to 25 percent by volume, the mixture of oil with ammonia broadens this range. Therefore, precautions to keep sources of flame or sparks from areas that involve ammonia storage or use are necessary.

In the event a fire does break out in an area containing ammonia, every effort should be made to remove portable containers from the premises without endangering the safety of personnel. If the containers cannot be removed, inform the firefighters of their location.

For data concerning the physiological effects of ammonia, protective equipment, and first aid measures, see CGA G-2, *Anhydrous Ammonia,* and the monograph on ammonia in Part III. [10]

LP-Gases

All LP-gases are classified by the DOT and TC as flammable gases and must be treated as such. Reference should be made to NFPA 58, *Standard for the Storage and Handling of Liquefied Petroleum Gases.* [5]

Liquid LP-gas leaks in transfer piping are indicated by frost at the point of leakage due to the low boiling point of the material. Extremely small leaks and leaks in vapor transfer piping can be detected by applying soap suds or a similar material to the suspected area. Under no circumstances should a flame be used a detect a leak.

The most important safety considerations in unloading LP-gas are: (1) avoiding unnecessary releases of the product, (2) keeping open flames and other sources of ignition away from the unloading area, and (3) making sure suitable first aid type fire extinguishers are available (dry-chemical or carbon dioxide types are suitable for LP-gas fires).

It is important for personnel to be familiar with the characteristics of LP-gas. For instance, the importance of not extinguishing a fire unless the source of the leakage (as by closing a valve) is stopped must be emphasized. If the fire is extinguished and leakage is allowed to continue, unburned vapor could accumulate and possibly present a more serious hazard than if the escaping gas were allowed to burn.

If a fire is in progress and an LP-gas storage tank is exposed, it is of prime importance to keep the container cool (especially the vapor space) by applying hose streams of water to the tank until the fire is properly extinguished. Likewise, if fire threatens a single-unit tank car, a TMU car, or a cargo transport truck, immediately remove these from the area. If this is not possible, apply hose streams of water, as in protecting storage tanks.

Monographs providing additional information on the various LP-gases can be found in Part III.

Chlorine

In both its liquid and gaseous form, chlorine is neither flammable nor explosive. It currently is classified as a nonflammable gas by the U.S. Department of Transportation. In Canada, it is classified as a corrosive gas, class 2.4. The United Nations' classification is poison gas. Its principal hazard arises from inhalation. Data describing chlorine's physiological effects, handling, employee training and protection, and chemical characteristics and physical properties is available in the Chlorine Institute's *Chlorine Manual.* [11] Additional information on chlorine can also be found in the monograph in Part III.

Usually, chlorine tank cars are filled at low temperature and pressure. Normally, the inherent pressure of the vapor in the tank car is sufficient to accomplish withdrawal of the liquid chlorine to the process, but sometimes, especially during the winter, the car is air padded to accomplish liquid withdrawal. When air padding is required, introduce only clean, oil-free, cooled, dry compressed air into the tank car through its vapor valve.

In Fig. 6-4, a typical arrangement for unloading liquid chlorine from tank cars is shown. Liquid flow is through the tank car's liquid eduction valve to a liquid evaporator within battery limits, then direct to process. Whenever liquid chlorine can be trapped between two valves, the line must be protected by an expansion chamber.

Absorbing Chlorine in a Liquid. If chlorine is to be absorbed in a liquid, there is a tendency for the liquid to suck back into the container when the container becomes empty due to the creation of a partial vacuum. This has resulted in numerous accidents, and measures should be taken to avoid this occurring.

As soon as the container is empty and its pressure has dropped to zero, the container valve should be shut. This is followed by venting air into the line leading from the container after the valve has been shut off to prevent liquid from "sucking back" into the line. For this purpose, a "vacuum break" valve or loop on the chlorinator well line should be installed.

Unloading. In general, guidelines for unloading all chlorine containers apply to the tank car operation. These include:

(1) Never tampering with pressure relief devices.
(2) Opening container valves slowly.
(3) Making sure that threaded connections are the same as those on the container valve outlets. Never force connections that do not fit. The outlet threads on

valves of TMU containers are not tapered pipe threads.

Containers or valves should never be altered or repaired by unauthorized personnel. Use only reducing valves and gaskets designed for chlorine. Consult the chlorine manufacturer for details. The Chlorine Institute has compiled industry recommendations for chlorine tank cars in Pamphlet 66, *Chlorine Tank Car Loading, Unloading, Air Padding, Hydrostatic Testing.* [12]

Leaks. If ammonia vapor is directed at a leak, a white cloud will form indicating the source of the leak. A plastic squeeze bottle containing aqua ammonia can be used. If such a wash bottle is used, the dip tube should be cut off so that squeezing the bottle directs vapor, not liquid, out of the nozzle. Contact of aqua ammonia with brass or copper is to be avoided. Commercial 26° Baumé aqua ammonia should be used. If a leak cocurs in equipment or piping, the chlorine supply should be shut off, the pressure relieved, and necessary repairs made. Gas leaks around a valve stem may usually be contained by tightening the packing nut.

Employee Safety and Training. Safety in handling chlorine depends upon the effectiveness of employee training, proper safety instructions, intelligent supervision, and the use of suitable equipment. Severe exposure to chlorine can occur whenever chlorine is handled or used. Suitable protective equipment for emergency use should be available outside of chlorine rooms near the entrance, away from areas of likely contamination. For further details, consult the reference material available from the Chlorine Institute.

Emergency Measures. An emergency plan for chlorine is essential. This should include a procedure for the training of personnel who could become involved and periodic drills to review the response. Assistance is available to help in emergencies, but the first action must be taken by the people on scene.

The best source of assistance for the majority of chlorine consumers is the supplier. A first step in any emergency plan should be the conspicuous posting of the supplier's telephone number and the instruction to call that number for assistance. Help also is available in the United States from CHLOREP through CHEMTREC. In Canada, CANUTEC provides information and assistance. The CHEMTREC or CANUTEC numbers also should be posted.

In case of fire, chlorine containers should be removed from the fire zone immediately. As soon as there is any indication of the presence of chlorine in the air, take steps to correct the condition. Chlorine leaks always get worse if not corrected promptly. Keep on the windward side of the leak and higher than the leak. Since gaseous chlorine is approximately $2\frac{1}{2}$ times as heavy as air, it tends to lie close to the ground.

If a chlorine leak occurs, authorized, trained personnel equipped with suitable respiratory protection should investigate. All other persons should be kept away from the affected area until the cause of the leak is discovered and corrected. If the leak is extensive, warn all persons in the path of the fumes.

Do not spray water on a chlorine leak. To do so makes the leak worse because of the corrosive action of wet chlorine. When a leak occurs in equipment in which chlorine is being used, immediately close the chlorine container valve.

If a chlorine container is leaking in such a position that chlorine is escaping as a liquid, the container should be turned if possible so that chlorine gas escapes instead. The quantity of chlorine escaping from a gas leak is about $\frac{1}{15}$ the amount that escapes from a liquid leak through the same size hole. Leaks at valve stems are often stopped by tightening the valve packing nuts or closing the valve.

If a chlorine leak occurs in transit in a populated area, it is recommended that the vehicle keep moving, if possible, until it reaches an open area where the escaping gas will be less hazardous. If a chlorine leak occurs in transit and the conveying vehicle is wrecked, the container or containers should be shifted so that gaseous chlorine, rather than liquid, is escaping. If possible, the container should be transferred to a suitable vehicle and taken to open country.

The severity of a chlorine leak can be lessened by reducing the pressure on the leaking container. This may be done by absorbing chlorine gas from the container into caustic soda solution. Evaporation of some of the liquid chlorine cools the remaining liquid, reducing its pressure.

At regular points of storage and use, emergency preparations for disposing of chlorine from leaking cylinders or ton containers should be made. Chlorine may be absorbed into caustic soda or soda ash solution. Caustic soda solution is preferred as it absorbs chlorine most readily.

The following publications are among those published by the Chlorine Institute:

Pamphlet 1, *Chlorine Manual* [11]
Pamphlet 6, *Piping Systems for Dry Chlorine* [13]
Pamphlet 17, *Cylinder & Ton Container Procedure for Chlorine Packaging* [14]
Pamphlet 49, *Handling Chlorine Tank Motor Vehicles* [15]
Pamphlet 66, *Chlorine Tank Car Loading, Unloading, Air Padding, Hydrostatic Testing* [12]

Methyl Chloride

Methyl chloride is classified as a flammable gas in both the United States and Canada. In Canada, a subsidiary poison risk also applies. It burns feebly but forms explosive mixtures with air. The end product of high-temperature decomposition may be toxic. For additional data concerning physiological effects, protective equipment, chemical and physical properties, and so forth, see the monograph on methyl chloride in Part III.

Throughout the entire single-unit tank car unloading operation, particularly while connecting and disconnecting, great caution must be taken to make sure the working area is free of heated surfaces, flames, static electricity, railroad locomotives, gasoline tractors, and all other sources of ignition. Tank cars and TMU containers should be grounded electrically.

Under no circumstances should water or other materials be introduced into tank cars which contain, or have contained, methyl chloride. Soapy water should be used in testing for leaks; in freezing weather or around very cold pipes or equipment, use glycerine. An open flame must never be used to test for leaks. This is prohibited, and it applies not only to tank cars but to any containers holding methyl chloride.

When single-unit tank cars are unloaded, after all the liquid has been transferred, the greater part (but not all) of the methyl chloride vapors may be recovered by creating a slight pressure differential. Observe caution in this operation, as a slight residual pressure must always remain on the tank car so that no air is drawn into it to form explosive mixtures when the pipes are disconnected.

When unloading tank cars, if a leak occurs which cannot be readily repaired by simple adjustment or tightening of the fittings, telephone the methyl chloride manufacturer at once for instruction, and evacuate the area around the car immediately. Permit only properly protected and instructed personnel to enter the contaminated area.

In case of TMU container leaks, all sources of ignition must be removed from the area at once, and if the leak cannot be stopped, transfer the methyl chloride to another container. As to the procedure, obtain and follow the detailed instructions of the methyl chloride manufacturer or supplier.

Fire and Explosion Hazards. Avoid all sources of ignition, of whatever nature, when unloading single-unit tank cars and TMU containers holding methyl chloride.

Methyl chloride fires are gas fires. The most effective method of extinguishing them is to shut off the flow of vapor by closing the valves. Carbon dioxide or dry chemical may be used to extinguish the flame to permit access to shutoff valves. If the valve is in the area of the fire, attack it if possible, close it, and then attack and extinguish the secondary fire which consists of other burning material ignited by the gas fire.

Circumstances may make it impossible to attack the valve, and in such a case, the flame may be allowed to continue burning while the surrounding area and objects are cooled with water spray. Provide employees engaged in extinguishing fires with gas masks to protect them from methyl chloride vapors and the toxic combustion products formed.

Sulfur Dioxide

Sulfur dioxide is classified by DOT as a nonflammable gas. In both its gaseous and liquid form it is neither flammable nor explosive. It is a respiratory and a skin and eye irritant. Consult CGA G-3, *Sulfur Dioxide,* for information concerning chemical and physical properties, physiological effects, personal protective equipment, and emergency action procedures. [7] CGA G-3 also provides numerous recommendations for handling and unloading all types of sulfur dioxide shipping containers.

Leaks. Only personnel trained for and designated to handle emergencies should attempt to stop a leak. Respiratory equipment of a type suitable for sulfur dioxide must be worn. All persons not so equipped must leave the affected area until the leak has been stopped.

If sulfur dioxide vapor is released, the irritating effect of the vapor will force personnel to leave the area long before they have been exposed to dangerous concentrations. To facilitate their rapid evacuation, there should be sufficient well-marked and easily accessible exits. If, despite all precautions, a person should become trapped in a sulfur dioxide atmosphere, he or she should breathe as little as possible and open his or her eyes only when necessary. Partial protection may be gained by holding a wet cloth over the nose and mouth. Since sulfur dioxide vapor is heavier than air, the upper floors of buildings will normally have lower concentrations, but personnel may also get trapped if assistance is not quickly forthcoming.

Sulfur dioxide is fairly soluble in cool water (less than 100°F or 37.8°C), and therefore the vapor concentration can be reduced by the use of spray or fog nozzles. If water is added to liquid sulfur dioxide, it will cause an increase in evaporation rate (unless the water is very cold). The evaporation will slow after sufficient water is added.

When possible, leaking containers or vessels can be vented to a lime or caustic soda solution. The reduction in pressure that results from venting will slow the leak, and may permit stopping it. Water should never be sprayed at or into a tank or system which is leaking sulfur dioxide. The presence of water causes sulfur dioxide to be very corrosive, and water directed into a tank would also increase the venting rate.

When sulfur dioxide is released into the environment, the appropriate regulatory agency should be notified. Sulfur dioxide does not have a reportable quantity listed in 49 CFR 172.101. [1] In the event of a release however, all state, provincial, municipal, and/or local reporting regulations must be complied with. It is most important that the response groups in the area affected be notified as quickly as possible. If the producer or supplier cannot be reached, in the United States contact CHEMTREC, or in Canada contact CANUTEC. See the front of this *Handbook* for telephone numbers, etc.

A sulfur dioxide container exposed to a fire should be removed. If for any reason it cannot be removed, the container should be kept cool with a water spray until well after the fire is out. Fire fighting personnel should be equipped with protective clothing and respiratory equipment. For further information regarding handling leaks and emergencies, see DOT P5800, *Emergency Response Guidebook.* [16]

Fluorocarbons

The U.S. Department of Transportation and Transport Canada classify most liquefied fluorinated hydrocarbons as nonflammable gases. Some of those presently shipped in single-unit tank cars, TMU containers, and cargo tank trucks are: dichlorodifluoromethane (R12), trichloromonofluoromethane (R11), chlorotrifluoromethane (R13),chlorodifluoromethane (R22), dichlorotetrafluoroethane (R114), trichlorotrifluoroethane (R113), and chloropentafluoroethane (R115). R11, R113, and R114 are not currently regulated as compressed gases by DOT.

These gases are odorless, and leaks cannot be detected by sense of smell. Frosting is evidence of a large leak, while smaller leaks may be located by means of a halide torch.

Avoid contact with the liquid and excessive inhalation of vapor. In case of a severe leak, persons entering an area of dense concentration of vapor should wear an air gas mask of the type approved by the U.S. Bureau of Mines or the National Institute of Occupational Safety and Health (NIOSH) for liquefied fluorinated hydrocarbon service.

REFERENCES

[1] *Code of Federal Regulations,* Title 49 CFR Parts 100–199, (Transportation), Superintendent of Documents, U.S. Government Printing Office, Washington, DC 20402.

[2] *Transportation of Dangerous Goods Regulations,* Canadian Government Publishing Centre, Supply and Services Canada, Ottawa, Ontario, Canada K1A 0S9.

[3] Tariff B0E-6000, *Hazardous Materials Regulations of the Department of Transportation by Air, Rail, Highway, Water, and Military Explosives by Water, including Specifications for Shipping Containers,* Association of American Railroads, 50 F Street NW, Washington, DC 20001.

[4] *AAR Specifications for Tank Cars,* Association of American Railroads, 50 F Street NW, Washington, DC 20001.

[5] NFPA 58, *Standard for the Storage and Handling of Liquefied Petroleum Gases,* National Fire Protection Association, Batterymarch Park, Quincy, MA 02269.

[6] NFPA 70, *National Electric Code,* National Fire Protection Association, Batterymarch Park, Quincy, MA 02269.

[7] CGA G-3, Sulfur Dioxide, Compressed Gas Association, Inc., 1235 Jefferson Davis Highway, Arlington, VA 22202.

[8] CSA B630, *Highway Tanks and Portable Tanks for the Transportation of Dangerous Goods,* Canadian Standards Association, 178 Rexdale Blvd., Rexdale (Toronto), Ontario, Canada M9W 1R3.

[9] CSA B622, *CSA Standard for the Selection and Use of Tank Trucks, Tank Trailers and Portable Tanks for the Transportation of Dangerous Goods for Class 2 By Road,* Canadian Standards Association, 178 Rexdale Blvd., Rexdale (Toronto), Ontario, Canada M9W 1R3.

[10] CGA G-2, *Anhydrous Ammonia,* Compressed Gas Association, Inc., 1235 Jefferson Davis Highway, Arlington, VA 22202.

[11] Pamphlet 1, *Chlorine Manual,* Chlorine Institute, Inc., 2001 L Street, N.W., Suite 506, Washington, DC 20036.

[12] Pamphlet 66, *Chlorine Tank Car Loading, Unloading, Air Padding, Hydrostatic Testing,* Chlorine Institute, Inc., 2001 L Street, N.W., Suite 506, Washington, DC 20036.

[13] Pamphlet 6, *Piping Systems for Dry Chlorine,* Chlorine Institute, Inc., 2001 L Street, N.W., Suite 506, Washington, DC 20036.

[14] Pamphlet 17, *Cylinder & Ton Container Procedure for Chlorine Packaging,* Chlorine Institute, Inc., 2001 L Street, N.W., Suite 506, Washington, DC 20036.

[15] Pamphlet 49, *Handling Chlorine Tank Motor Vehicles,* Chlorine Institute, Inc., 2001 L Street, N.W., Suite 506, Washington, DC 20036.

[16] DOT P5800, *Emergency Response Guidebook,* Superintendent of Documents, U.S. Government Printing Office, Washington, DC 20402.

ADDITIONAL REFERENCES

Standard 1202 (Feb. 1960), *Measuring, Sampling and Calculating Tank Car Quantities and Calibrating Tank Car Tanks (Pressure Type Tank Cars),* American Petroleum Institute, 1220 L Street N.W., Washington, DC 20005.

CHAPTER 7

Pressure Relief and Safety Devices

INTRODUCTION

Almost all compressed gas containers are fitted with pressure relief devices. A pressure relief device is a pressure- and/or temperature-activated device used to prevent the pressure from rising above a predetermined maximum, and thereby prevent rupture of a normally charged cylinder when subjected to a standard fire test as required by Title 49 of the U.S. *Code of Federal Regulations* (49 CFR 173.34(d)), or equivalent regulations of Transport Canada.[1] [1], [2], and [3]

Objectives

This chapter deals with the various types of pressure relief devices used in the compressed gas industry and the standards developed to describe their design criteria, performance characteristics, and applications. The sections covered are:

- Pressure relief devices for compressed gas cylinders
- Pressure relief devices for cargo and portable tanks
- Pressure relief devices for compressed gas storage containers
- Other safety devices

PRESSURE RELIEF DEVICES FOR COMPRESSED GAS CYLINDERS

This section summarizes the information in CGA S-1.1, *Pressure Relief Device Standards—Part 1—Cylinders for Compressed Gases.* [4] Included are descriptions of pressure relief devices for use on cylinders having capacities of 1000 lb (453.6 kg) of water or less (see 49 CFR 173.34(d) for exceptions), TC/DOT-3AX, 3AAX, and 3T cylinders having capacities over 1000 lb, and TC/DOT-4L insulated cylinders containing cryogenic liquids.

Pressure Relief and Safety Devices

The Compressed Gas Association has classified pressure relief devices according to type using the letter designation CG followed by a numeral. Each of these types is described in the following subsections.

Type CG-1 (Pressure Relief Rupture Disk). A rupture disk (synonymous with the name *burst disk* within the industry) is a pressure-operated device which affords protection against development of excessive pressure in cylinders. This device is designed to sense ex-

[1]Prior to April 16, 1981, pressure relief devices required approval by the Bureau of Explosives of the Association of American Railroads. Subsequent to April 16, 1981, the U.S. Department of Transportation promulgated new regulations amending 49 CFR 173.34 to eliminate the need for pressure relief device approval by the Bureau of Explosives. It is now the responsibility of the individual manufacturer or shipper to conduct flow and/or fire tests on new pressure relief device combinations to show compliance with CGA S-1.1, *Pressure Relief Device Standards—Part 1—Cylinders for Compressed Gases* [4]; CGA C-12, *Qualification Procedure for Acetylene Cylinder Design* [5]; and CGA C-14, *Procedures for Fire Testing of DOT Cylinder Safety Relief Device Systems* [6], as applicable, and to retain test records of this compliance.

cess pressure in a cylinder and will function when the cylinder pressure is of sufficient magnitude by the rupture or bursting of the rupture disk element, thereby venting the contents of the cylinder. The rupturing of the rupture disk element results in a nonreclosing orifice.

Rupture disk devices installed on compressed gas cylinders may be either an integral part of the cylinder valve assembly or may be installed on the cylinder as an independent attachment. The materials of construction selected must be compatible with the fluid in the cylinder as well as the cylinder valve materials with which the rupture disk device comes in contact in order to minimize corrosion.

One of the most common types of rupture disk devices consists of (1) a gasket, (2) a rupture disk, and (3) a rupture disk holder. These components may be supplied as independent parts (see Fig. 7–1) or may be in the form of factory-assembled devices designed to be replaced as a unit (see Fig. 7–2).

The gasket is the part which provides the proper seal to prevent leakage of the cylinder contents past the rupture disk assembly and

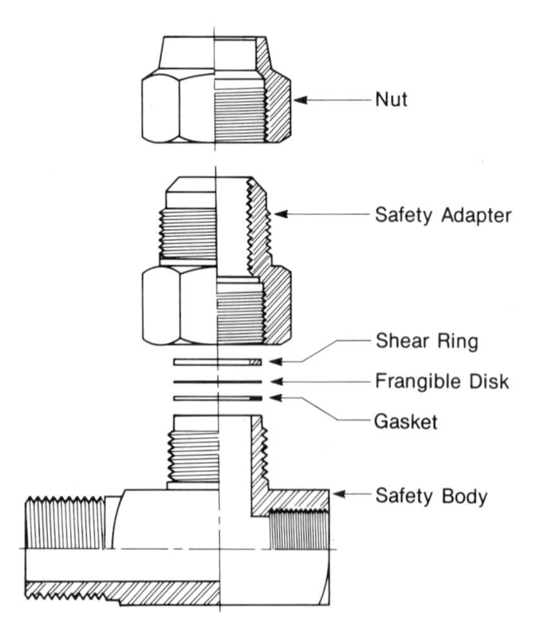

Fig. 7–1. Rupture disk device (CG-1) made up of independent parts.

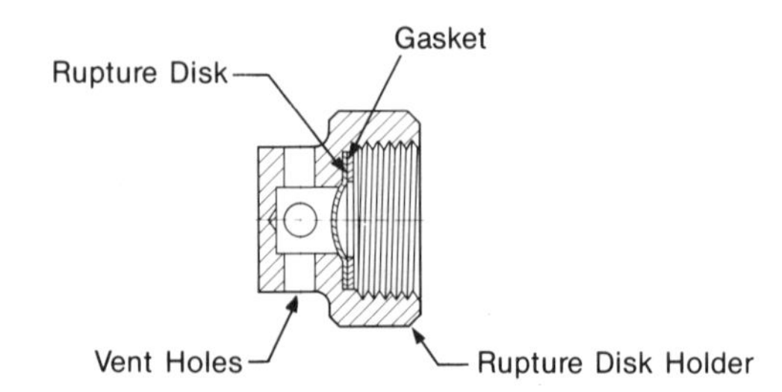

Fig. 7–2. Rupture disk device (Type CG-1) factory-assembled as a complete unit.

may be constructed of metallic or nonmetallic materials.

The rupture disk is the operating part of the pressure relief device and, when installed in a proper rupture disk holder, is designed to burst at a predetermined pressure to permit discharge of the cylinder contents. Such disks are usually made of metallic materials and may be of flat, preformed, reinforced, grooved, or scored construction. Nonmetallic materials are also used for specific applications.

The rupture disk holder is the part of the pressure relief device which contains the opening against which the rupture disk mates. The rupture disk holder usually also contains the discharge porting or passages, beyond the operating parts of the device, through which fluid must pass to reach the atmosphere. In many cases, the discharge holder is provided with radial vent holes through which the fluid in the cylinder vents to the atmosphere. This radial discharge design provides an anti-recoil feature which minimizes rocketing of compressed gas cylinders during discharge of the contents through the pressure relief device. Other types of discharge ports may also be provided in rupture disk holders to suit specific application requirements.

Most rupture disk devices are designed with holders having either sharp-edged or radius-edged orifices to which the rupture disk mates. The sharp-edged orifice produces a shear-type actuation mode whereby the disk ruptures in shear, producing a characteristic leaf-type configuration after functioning.

Since the actuation modes of each type of

holder described above are completely different, it is important that only original manufacturer's parts or assemblies be used in the repair or replacement of rupture disk devices, unless the interchangeability of parts has been proven by suitable test.

Cautions: The pressure relief rupture disk device is a primary safety component and hence the following precautions should be noted and adhered to:

(a) Only trained personnel should be permitted to service pressure relief devices.

(b) Tightening of the rupture disk assembly to the cylinder valve or to the cylinder itself should be in accordance with the manufacturer's instructions. Tightening to a torque less than the manufacturer's recommendations may result in a leaking device or a device that may rupture at a lower pressure than specified. Conversely, overtightening can also result in disk actuation at a lower pressure than specified due to an excessive twisting action which may create wrinkles or distortions in the disk.

Limitations: A rupture disk is a pressure-operated device which affords protection against excessive pressure. It protects against excessive pressure when the properties of the gas, cylinder design, and percentage of charge in the cylinder are such that exposure to excessively high temperatures will cause an increase in internal pressure sufficient to actuate the rupture disk before the cylinder walls are seriously weakened. The rupture disk also protects against excessive pressure due to improper charging practices such as overfilling.

A rupture disk is a nonreclosing device. Once the disk has ruptured, there is no way to prevent the complete release of the contents of the cylinder.

This device does not provide good protection against pressures caused by exposure to excessively high temperatures when the cylinder is only partially charged. The pressure rise may not be sufficient to actuate the rupture disk before the cylinder walls are seriously weakened.

Consideration should be given to environmental conditions to which the cylinder may be exposed. Severely corrosive atmospheres may contribute to premature rupture of the disk. To prevent corrosion of the rupture disk, care must be taken to select materials of construction that do not interact with either the contents of the cylinder or the anticipated environmental conditions.

Type CG-2 and CG-3 (Fusible Plugs). A fusible plug is a thermally-operated pressure relief device which affords protection against excessive pressure developed by exposure to excessive heat. Once sufficient heat melts the fusible metal, the contents of the cylinder will be vented. The CG-2 fusible metal has a nominal melt temperature of 165°F (73.9°C); the CG-3 fusible metal has a nominal melt temperature of 212°F (100°C).

Fusible plugs can be installed on the cylinder as independent devices, or fusible metal can be cast directly into a suitable orifice in the cylinder valve body. In some cases, a fusible plug may be installed as a separate device into the cylinder valve body.

The plug body is usually provided with a suitable method for wrenching to facilitate installation of the fusible plug. This may be accomplished by means of a hexagonal head, a recessed head, or other suitable design.

The plug body is designed to assist in retaining the fusible alloy in place. Designs used to accomplish this include a taper bore decreasing in diameter from the pressurized end, a step bore resulting from a series of bore diameters decreasing progressively in size from the pressurized end, or a straight bore or taper bore provided with threads. When a threaded bore is used, it is recommended that a thread form with rounded roots and crests be used to aid in filling with fusible metal.

In addition to the above recommendations to assist in retaining the fusible alloy in place, consideration should also be given to providing sufficient overall length in relation to the diameter of the bore.

Figure 7-3 illustrates typical fusible plug bodies with recommended design considerations. Figure 7-4 illustrates fusible metal

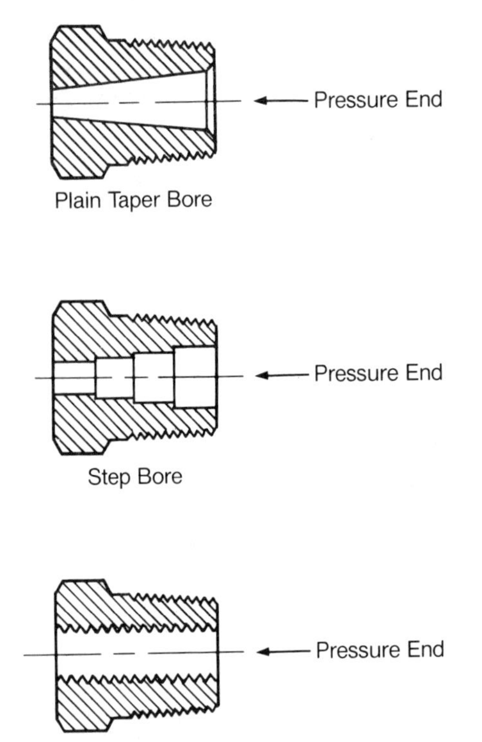

Fig. 7-3. Illustration of typical fusible plug bodies (for Type CG-2 or CG-3) with recommended design considerations.

cast directly into a cylinder valve body. Figure 7-5 illustrates a separate device threaded into a cylinder valve body.

Cautions: No attempt should be made to repair fusible plug devices that have been returned from service by melting out the alloy and recasting. Used fusible plugs that are not leaking may be cleaned externally, reinstalled, and subjected to the prescribed leak tests.

Limitations: Since the fusible plug is a thermally operated device, it is designed to function only when the fusible metal melts out. Hence, it does not protect against overpressure from improper charging practices.

Sufficient heat to melt the fusible metal is necessary for proper functioning of this type of device. Therefore, the location of such devices is an important consideration.

Industry practice limits the application of fusible plugs to cylinders with 500 psig (3447 kPa) service pressure or less to minimize the possibility of cold flow or extrusion of the fusible metal.

A fusible plug device is a nonreclosing device and hence, when it functions, it releases the entire contents of the cylinder.

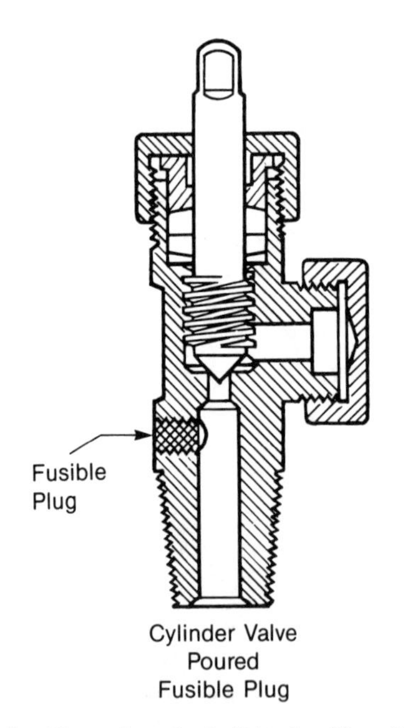

Fig. 7-4. Illustration of a fusible plug (Type CG-2 or CG-3) cast directly into a cylinder valve body.

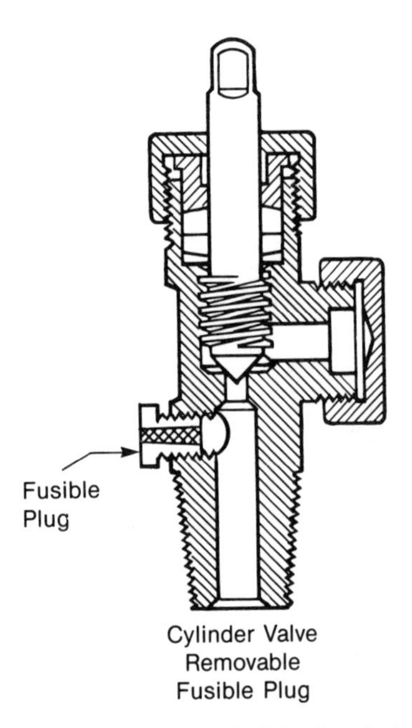

Fig. 7-5. Illustration of a fusible plug device (Type CG-2 or CG-3) threaded into a cylinder valve body.

Type CG-4 and CG-5 (Combination Rupture Disk/Fusible Plug). A combination rupture disk/fusible plug pressure relief device requires both excessive pressure and excessive temperature to cause it to operate. Sufficient heat is required to first melt out the fusible metal, after which the device will afford the same protection as the CG-1 rupture disk device.

The CG-4 combination device has fusible metal with a nominal melt temperature of 165°F (73.9°C). The CG-5 combination device has fusible metal with a nominal melt temperature of 212°F (100°C).

In this type of device, the rupture disk portion (CG-1) is directly exposed to the internal cylinder pressure, and so is directly upstream of the fusible metal. In general, the same components that make up the CG-1 device are used and the vent portion or downstream side of the rupture disk holder is suitably filled with fusible metal (see Figs. 7-6 and 7-7). The rupture disk is thus reinforced against rupturing by the fusible metal, and the fusible metal is reinforced against extrusion by the rupture disk.

Cautions: The same precautions noted for CG-1 devices should be adhered to for CG-4 and CG-5 devices.

Limitations: CG-4 and CG-5 combination devices function only in the presence of both excessive heat and excessive pressure, and sufficient heat must be present first to melt the fusible metal. Therefore, this device does not offer protection against overpressure from improper charging practices.

Type CG-7 (Pressure Relief Valves). A pressure relief valve is a spring-loaded pressure-operated device designed to relieve excessive cylinder pressure, reclose, and reseal to prevent further release of product from the cylinder after the cause of excessive pressure is removed and valve reseating pressure has been achieved.

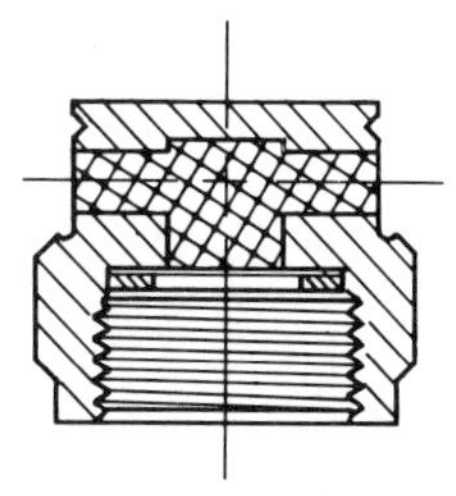

Fig. 7-6. Combination rupture disk/fusible plug device (Type CG-4 or CG-5) factory-assembled as a complete unit.

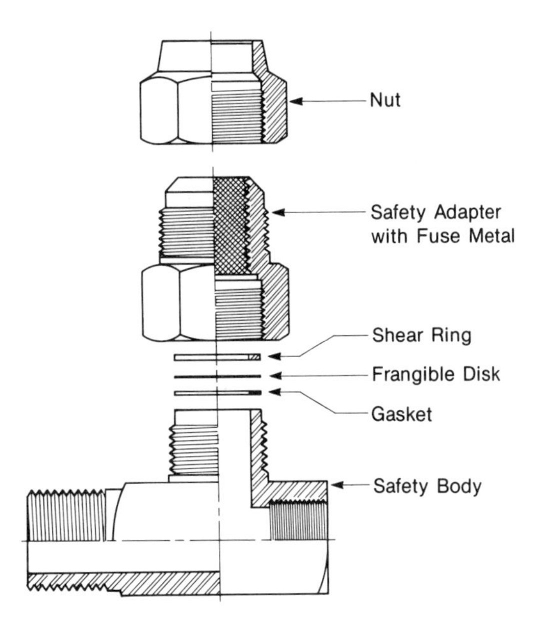

Fig. 7-7. Combination rupture disk/fusible plug device (Type CG-4 or CG-5) made up of independent parts.

The primary advantage of using the pressure relief valve is that functioning of this type of device will not release all of the contents of the cylinder, but is designed to reseal after reseating pressure has been achieved. This characteristic, in fire conditions, will minimize feeding the fire in the case of a flammable lading.

Typical pressure relief valves are illustrated in Figs. 7-8, 7-9, 7-10, 7-11, and 7-12. The relief valves shown here are independent devices designed to be threaded directly into suitable openings in the cylinder.

The pressure relief valve shown in Fig. 7-9 is housed as an integral part of the cylinder valve and hence does not require a separate opening into the cylinder.

Figure 7-12 illustrates the typical components in cross section. In this example, a gas-tight seal against cylinder pressure is achieved by the elastomeric seat against a

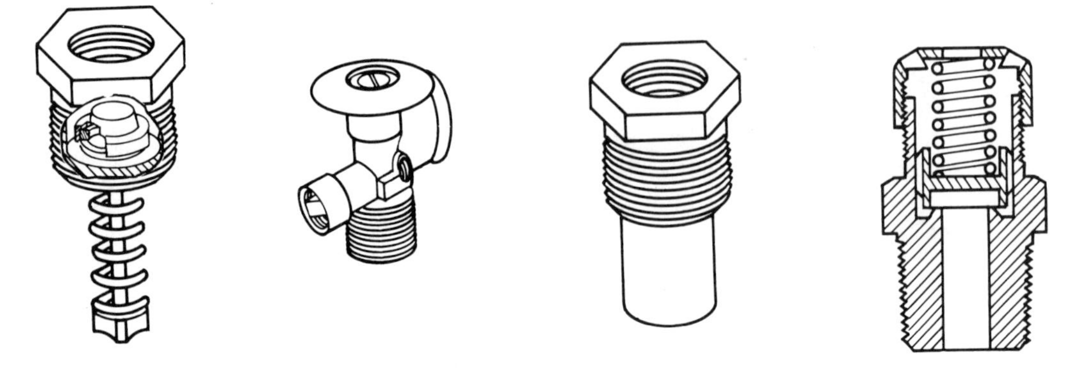

Fig. 7-8.-7-11. Typical pressure relief valves (Type CG-7).

suitable ridge in the valve body. In some designs a metal-to-metal seal is used. The necessary force required to achieve this seal is provided by the spring. The spring adjusting cap permits variation in the magnitude of the spring force.

The device shown in Fig. 7-8 uses essentially the same components as that shown in Fig. 7-12, except that the orientation is reversed to permit the spring and adjusting mechanism to be completely within the cylinder after installation.

Limitations: Pressure relief valves are designed to maintain the pressure in the cylinder at a limit as determined by the spring force. Therefore, such devices do not protect the cylinder against possible rupture when continued application of external heat or direct flame impingement weakens the cylinder wall to the point where its rupture pressure is less than the operating pressure of the relief valve.

Pressure relief valves are generally more susceptible to leakage around the seal formed by the seat than rupture disks or fusible plugs. These devices may be affected by the environment and/or lading, which may cause freezing, sticking, or otherwise improper operation.

Pressure Relief Devices for TC/DOT 4L Cylinders. TC/DOT 4L cryogenic containers (see Fig. 7-13) are double-walled, vacuum and multi-layer insulated cylinders designed for product withdrawal in either gaseous or liquid form and are generally used for liquefied argon, helium, hydrogen, neon, nitrogen, and oxygen.

TC/DOT 4L cylinders are typically comprised of a stainless-steel inner container encased within an outer steel vacuum shell. The typical insulation system between the inner and outer containers consists of multi-layer insulation and high vacuum. To protect the inner container from overpressurization, the unit includes a CG-7 pressure relief valve.

As a secondary pressure relief device, such containers are further protected from overpressurization by a CG-1 rupture disk. The cylinder may also contain an internal vaporizer, which converts cold liquid to warm gas for gas withdrawal.

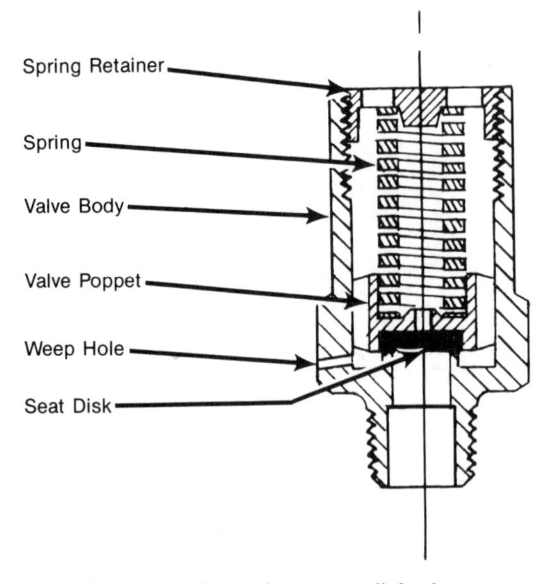

Fig. 7-12. External pressure relief valve.

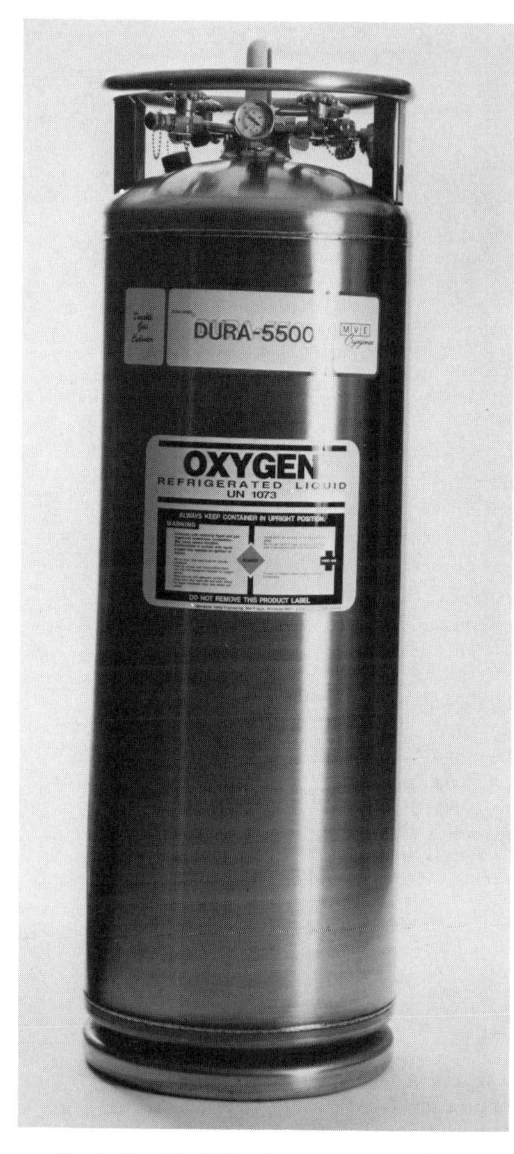

Fig. 7-13. TC/DOT 4L cryogenic container.

Tables 7-1 and 7-2

Table 7-1 shows the standard numerical code for gas classification (abbreviated as the FTSC code). In accordance with this code, each gas is assigned a four-digit classification number in which the first digit is a function of the gas fire potential (F), the second digit a function of the gas toxicity (T), the third digit a function of the state of the gas (S), and the fourth digit a function of the gas corrosiveness (C). For a more complete discussion of the FTSC rating system, consult CGA V-7, *Standard Method of Determining Cylinder Valve Outlet Connections for Industrial Gas Mixtures.* [7] The FTSC code is a very useful tool in grouping gases with similar classification number.

Table 7-2 is an alphabetical listing of gases and the pressure relief devices that are acceptable, as indicated by letter symbols. When more than one type of device is listed in Table 7-2 for a particular gas, only one type is required. The letter symbols used in Table 7-2 are defined at the end of the table. Interpretation of these symbols is necessary to determine the type of relief device to be used with a specific gas. While listed materials preceded by an asterisk are not compressed gases, and hence are not covered by CGA S-1.1, [4], recommended pressure relief device information has been included here to inform the reader of prevailing industry practice.

For certain gases, use of pressure relief devices is not permitted. For such gases, the pressure relief device column is marked "Prohibited."

PRESSURE RELIEF DEVICES FOR CARGO AND PORTABLE TANKS

This section summarizes the information on pressure relief devices contained in CGA S-1.2, *Pressure Relief Device Standards—Part 2—Cargo and Portable Tanks for Compressed Gases.* [8] Included are descriptions of pressure relief devices designed for use on cargo tanks and portable tanks for compressed gases having water capacities in excess of 1000 lb (453.6 kg) and built in compliance with DOT requirements or equivalent Canadian requirements. These devices are larger in size than those used on compressed gas cylinders due to the greater relieving requirements necessitated by the larger container capacities.

Types of Pressure Relief Devices

Pressure Relief Valves. Pressure relief devices used on cargo tanks and portable tanks are similar to the CG-7 devices described for compressed gas cylinders earlier in this chap-

TABLE 7-1. FTSC NUMERICAL CODE FOR GAS CLASSIFICATION.

1st Digit — Fire Potential

Code	Meaning
0	= inert
1	= supports combustion (oxidizing)
2	= flammable: lower limit of flammability less than 13% or flammable range greater than 12%
3	= pyrophoric
4	= highly oxidizing
5	= may decompose or polymerize and is flammable

2nd Digit — Toxicity

Code	Meaning
1	= nontoxic: over 500 ppm permitted for 8 hours exposure
2	= toxic: 50 to 500 ppm permitted (or very toxic with good warning properties) for 8 hours exposure.
3	= very toxic: less than 50 ppm permitted for 8 hours exposure
4	= DOT poison A or others of similar toxicity
5	= DOT poison A or others of similar toxicity used in the electronic industry

3rd Digit — State of Gas (in the cylinder at 70°F) (21°C)[a]

Code	Meaning
0	= noncryogenic liquefied gas (less than 500 psig) (3450 kPa)[b]—gas withdrawal
1	= noncryogenic liquefied gas (over 500 psig) (3450 kPa)—gas withdrawal
2	= liquefied gas (liquid withdrawal)[c]
3	= dissolved gas
4	= nonliquefied gas—or cryogenic gas withdrawal (less than 500 psig), (3450 kPa)
5	= Europe only
6	= nonliquefied gas between 500 and 3000 psig (3450 and 20 680 kPa)
7	= nonliquefied gas above 3000 and below 10 000 psig (20 680 and 68 900 kPa)
8	= cryogenic gas (liquid withdrawal) above −400°F (−240°C)
9	= cryogenic gas (liquid withdrawal) below −400°F (−240°C)

4th Digit — Corrosiveness

Code	Meaning
0	= noncorrosive
1	= non-halogen acid forming
2	= basic
3	= halogen acid forming

[a]The temperatures of the refrigerated (cryogenic) liquids are always below −130°F (−90°C).
[b]If pressure at 130°F (54.4°C) is over 600 psig (4140 kPa), use digit 1.
[c]When a separate outlet for liquid withdrawal is specified.

ter, but are much larger in size. They are of the spring-loaded type as illustrated in Fig. 7-14. However, these devices are larger in size than those used on compressed gas cylinders due to the greater relieving requirements necessitated by the larger container capacities.

These pressure relief devices meet the applicable DOT or TC requirements for design, materials, installation, set pressure tolerance, markings, and certification of capacity as given in current editions of one of the following standards:

- *Safety Relief Valves for Anhydrous Ammonia and LP-Gas* (UL 132) [9]
- *ASME Boiler and Pressure Vessel Code,* Section VIII, Division 1 [10]
- *Specifications for Tank Cars,* Appendix A, (AAR M-1002) [11]

Type CG-2 (Fusible Plug Device). This type of device is described earlier in this chapter under compressed gas cylinder pressure relief devices. Only the CG-2 fusible plug with a nominal melt temperature of 165°F

TABLE 7-2. ALPHABETICAL LIST OF GASES AND DEVICES ASSIGNED (see notes).

Note 1: When more than one type of device is listed in Table 7-2 for a particular gas, only one type is required.
Note 2: The symbols used in Table 7-2 are defined at the end of the table. Interpretation of these symbols is necessary to determine the type of relief device to be used with the specific lading.
Note 3: Type CG-4 and CG-5 devices are not acceptable for 110% fill; see 49 CFR 173.302(c).
Note 4: For certain gases, use of pressure relief devices is not permitted. For such gases the pressure relief device column is marked "Prohibited."

Cryogenic Liquids

FTSC Code	Name of Gas	CG-1 Disk	CG-2 165°F	CG-3 212°F	CG-4 165°F w/Disk	CG-5 212°F w/Disk	CG-7 RV
	Argon	G					
	Helium	G					
	Hydrogen	G					
	Neon	G					
	Nitrogen	G					
	Oxygen	G					

Gases

FTSC Code	Name of Gas	CG-1 Disk	CG-2 165°F	CG-3 212°F	CG-4 165°F w/Disk	CG-5 212°F w/Disk	CG-7 RV
5130	Acetylene			F			
1160	Air	A		KB	B	B	K
2100	Allene		M				A
	Allylene (see Methylacetylene)						
0202	Ammonia, Anhydrous (over 165 lb) (none required if under 165 lb)		E				
0303	Antimony Pentafluoride	Prohibited					
0160	Argon	A			B	B	K
2500	Arsine	Prohibited					
0503	Arsenic Pentafluoride	Prohibited					
	Boron Chloride (see Boron Trichloride)						
	Boron Fluoride (see Boron Trifluoride)						
0203	*Boron Trichloride		L				
0263	Boron Trifluoride				B	B	
4303	*Bromine Pentafluoride	Prohibited					
4303	*Bromine Trifluoride	Prohibited					
0403	*Bromoacetone	Prohibited					
0100	*Bromochlorodifluoromethane (R12B1) or (Halon 1211)	L					L
0100	*Bromochloromethane (Halon 1011)	None Required					
	Bromoethylene (see Vinyl Bromide)						
	Bromomethane (see Methyl Bromide)						
3100	Bromotrifluoroethylene (R113B1)	C					A
0100	Bromotrifluoromethane (R13B1 or Halon 1301)	A					A
5100	1, 3 Butadiene (Inhibited)						A
2100	Butane, Normal			M			A
2100	1-Butene						A
2100	2-Butene						A
0110	Carbon Dioxide	A					K
	Carbon Dioxide/Nitrous Oxide Mixture (liquid)	A					
	Carbon Dioxide/Oxygen Mixture (Gas)	A			B	B	K

*Not a compressed gas.

TABLE 7-2. (*Continued*)

FTSC Code	Name of Gas	CG-1 Disk	CG-2 165°F	CG-3 212°F	CG-4 165°F w/Disk	CG-5 212°F w/Disk	CG-7 RV
	Carbonic Acid (see Carbon Dioxide)						
2260	Carbon Monoxide				J	J	
	Carbon Oxysulfide (see Carbonyl Sulfide)						
	Carbon Tetrafluoride (see Tetrafluoromethane)						
	Carbonyl Chloride (see Phosgene)						
0413	Carbonyl Fluoride	Prohibited					
2301	Carbonyl Sulfide		B		BC		
4203	Chlorine		H				
4303	Chlorine Pentafluoride	Prohibited					
4303	Chlorine Trifluoride	Prohibited					
2100	Chlorodifluoroethane (R142b)		M	M			A
0100	Chlorodifluoromethane (R22)	A	M	M			A
0100	Chlorodifluoromethane/ Chloropentafluoroethane (Mixture) (R502)	A	M	M			A
	Chloroethane (see Ethyl Chloride)						
	Chloroethylene (see Vinyl Chloride)						
2100	Chlorofluoromethane (R31)						A
0100	Chloroheptafluorocyclobutane (RC317)	A					A
	Chloromethane (see Methyl Chloride)						
0100	Chloropentafluorethane (R115)	A					A
0100	1-Chloro-1,2,2,2-Tetrafluoroethane (R124)	A					A
0100	1-Chloro-2,2,2-Trifluoroethane (R133a)	A					A
5200	Chlorotrifluoroethylene (R1113)	C					A
0100	Chlorotrifluoromethane (R13)	A			P		
2400	Cyanogen	Prohibited					
0403	Cyanogen Chloride	Prohibited					
2100	Cyclobutane		M				A
2100	Cyclopropane	A	M				A
2160	Deuterium	N			J	J	
0213	Deuterium Chloride				B		
0203	*Deuterium Fluoride	None Required					
2500	Deuterium Selenide	Prohibited					
2301	Deuterium Sulfide		B		BC		
5360	Diborane				B	B	
1200	*Dibromodifluoroethane	None Required					
0200	*Dibromodifluoromethane (R12B2) (Halon 1202)	None Required					
	Dibromomethane (see Methylene Bromide)						
0100	*1,2 Dibromotetrafluoroethane (R114B2) (Halon 2402)	L					L
0100	*1,2 Dichlorodifluoroethylene	None Required					
0100	Dichlorodifluoromethane (R12)	A	M	M			A
0100	Dichlorodifluoromethane/Difluoroethane Mixture (R500)	A	M	M			A
0200	*1,2 Dichloroethylene (R1130)	None Required					
0100	*Dichlorofluoromethane (R21)	L					L
0100	*1,2 Dichlorohexafluorocyclobutane (RC316)	None Required					
2403	*Dichlorosilane	Prohibited					
0100	*1,1 Dichlorotetrafluoroethane (R114a)	L	M	M			L
0100	*Dichlorotetrafluoroethane (R114)	L	M	M			L
0100	*2,2 Dichloro-1,1,1-Trifluoroethane (R123)	None Required					
	Dicyan (see Cyanogen)						

*Not a compressed gas.

TABLE 7-2. (*Continued*)

FTSC Code	Name of Gas	CG-1 Disk	CG-2 165°F	CG-3 212°F	CG-4 165°F w/Disk	CG-5 212°F w/Disk	CG-7 RV
3300	*Diethylzinc	Prohibited					
2100	1,1 Difluoroethane (R152a)		M	M			A
2110	1,1 Difluoroethylene (R1132a)	A			B		
	Difluoromethane (see Methylene Fluoride)						
2202	*Dimethylamine, Anhydrous	None Required					
2100	Dimethyl Ether						A
3200	*Dimethylsilane	None Required					
2100	*2,2 Dimethylpropane						L
0403	Diphosgene	Prohibited					
2110	Ethane	J					
2100	*Ethylacetylene		L				L
2100	*Ethyl Chloride		L				L
0403	Ethyldichloroarsine	Prohibited					
2160	Ethylene	J					
5320	*Ethylene Oxide	(See 49 CFR 173.124)					
2100	*Ethyl Ether						L
2400	Ethyl Fluoride	Prohibited					
4343	Fluorine	Prohibited					
	Fluoroform (R23) (see Trifluoromethane)						
2400	Germane	Prohibited					
0160	Helium	A			B	B	K
	Helium/Oxygen Mixture	A			B	B	K
2400	Heptafluorobutyronitrile	Prohibited					
0203	Hexafluoroacetone		B		B		
2400	Hexafluorocyclobutene	Prohibited					
0100	Hexafluoroethane (R116)	A			B		
0100	Hexafluoropropylene (R1216)	A					A
2160	Hydrogen	N			J	J	K
0203	Hydrogen Bromide				E		
0313	Hydrogen Chloride				B		
5301	Hydrogen Cyanide	Prohibited					
0203	*Hydrogen Fluoride	None Required					
0203	Hydrogen Iodide				B		
2500	Hydrogen Selenide	Prohibited					
2301	Hydrogen Sulfide		B		BC		
4303	*Iodine Pentafluoride	Prohibited					
2100	Isobutane						A
2100	Isobutylene						A
0160	Krypton	A			B	B	K
0403	Lewisite	Prohibited					
2160	Methane	N			J	J	K
2100	Methylacetylene		M				A
0300	*Methyl Bromide	None Required					
2100	*3-Methyl-1-Butene						L
2200	Methyl Chloride						A
0403	Methyldichloroarsine	Prohibited					
2203	*Methylene Bromide	None Required					
2203	Methyl Fluoride				B		
0110	Methylene Fluoride (R32)	A					A
2200	*Methyl Formate	None Required					
0303	*Methyl Iodide	None Required					
2201	Methyl Mercaptan	None Required					

*Not a compressed gas.

TABLE 7-2. (*Continued*)

FTSC Code	Name of Gas	CG-1 Disk	CG-2 165°F	CG-3 212°F	CG-4 165°F w/Disk	CG-5 212°F w/Disk	CG-7 RV
3200	*Methylsilane	None Required					
2202	*Monoethylamine	None Required					
2202	Monomethylamine, Anhydrous	None Required					
0403	Mustard Gas	Prohibited					
2160	Natural Gas	N			J	J	K
0160	Neon	A			B	B	K
2400	*Nickel Carbonyl	Prohibited					
4461	Nitric Oxide	Prohibited					
0160	Nitrogen	A		KB	B	B	K
4401	*Nitrogen Dioxide	Prohibited					
4401	*Nitrogen Tetroxide	Prohibited					
4343	Nitrogen Trifluoride			B	B	B	
4301	Nitrogen Trioxide	Prohibited					
0203	Nitrosyl Chloride	None Required—10 lb weight and under					
0303	Nitrosyl Fluoride	Prohibited					
4110	Nitrous Oxide	A					
0303	Nitryl Fluoride	Prohibited					
0100	Octafluorocyclobutane (RC318)						A
0100	Octafluoropropane (R218)	A					A
4160	Oxygen	A			B	B	K
4343	Oxygen Difluoride	Prohibited					
4330	Ozone (Dissolved in R13)	Prohibited					
3300	*Pentaborane	Prohibited					
2400	Pentafluoropropionitrile	Prohibited					
4303	Perchloryl Fluoride	Prohibited					
0100	*Perfluorobutane		L				L
0200	*Perfluoro-2-Butene						L
0303	Phenylcarbylamine Chloride	Prohibited					
0403	Phosgene	Prohibited					
3510	Phosphine	Prohibited					
0403	Phosphorous Pentafluoride	Prohibited					
0203	Phosphorous Trifluoride				B		
2100	Propane			M			A
2100	Propylene						A
3360	Silane				B		
0203	*Silicon Tetrachloride	None Required					
0263	Silicon Tetrafluoride				B		
5300	Stibine	Prohibited					
0201	Sulfur Dioxide		B				
0100	Sulfur Hexafluoride	A				B	A
0203	Sulfur Tetrafluoride				B		
0300	Sulfuryl Fluoride		B				
5110	Tetrafluoroethylene-Inhibited (R1114)	A			B		
4343	Tetrafluorohydrazine	Prohibited					
0160	Tetrafluoromethane (R14)	A			B	B	K
2400	Tetramethyllead	Prohibited					
0100	*Trichlorofluoromethane (R11)	L					L
2203	*Trichlorosilane	None Required					
0100	*1,1,1 Trichlorotrifluoroethane (R113a)	None Required					
0100	*1,1,2 Trichlorotrifluoroethane (R113)	None Required					
3300	Triethylaluminum	Prohibited					
3300	Triethylborane	Prohibited					
2400	Trifluoroacetonitrile	Prohibited					

*Not a compressed gas.

TABLE 7-2. (*Continued*)

FTSC Code	Name of Gas	CG-1 Disk	CG-2 165°F	CG-3 212°F	CG-4 165°F w/Disk	CG-5 212°F w/Disk	CG-7 RV
0303	Trifluoroacetyl Chloride	Prohibited					
2100	1,1,1 Trifluoroethane (R143a)		M				A
0100	Trifluoromethane (R23)	A			E		
4363	Trifluoromethyl Hypofluorite	Prohibited					
0200	Trifluoromethyl Iodide				B		
2202	*Trimethylamine	None Required					
3200	*Trimethylsilane	None Required					
3300	Trimethylstibine	Prohibited					
0303	*Tungsten Hexafluoride	Prohibited					
0303	*Uranium Hexafluoride	Prohibited					
5200	*Vinyl Bromide		L				L
5200	Vinyl Chloride		E				A
2100	Vinyl Fluoride				B		
5200	Vinyl Methyl Ether		E				A
0160	Xenon	A			B		K

*Not a compressed gas.

DEFINITIONS OF SYMBOLS USED IN TABLE 7-2

A This device is required in one end of the cylinder only, regardless of length, with the exception of trailer tubes in which this device is required in both ends.

B When cylinders are over 65 inches (1651 mm) long, exclusive of the neck, this device is required at both ends. For shorter cylinders, the device is required in one end only.

C This device is permitted only in cylinders having a minimum required test pressure of 3000 psig (20 680 kPa) or higher, and is required in one end only. The bursting pressure of the disk shall be at least 75% of the minimum required test pressure of the cylinder.

D [Reserved]

E When cylinders are over 30 inches (762 mm) long, exclusive of the neck, this device is required at both ends. For shorter cylinders, the device is required in one end only.

F The number and location of pressure relief devices for cylinders of any particular size shall be proved adequate as a result of the fire test. Any change in style of cylinder, a filler, or quantity of devices can only be approved if found adequate upon reapplication of the fire test. The fire test shall be conducted in accordance with CGA C-12, *Qualification Procedure for Acetylene Cylinder Design.* [5]

G This device is required in one end of the cylinder only, regardless of length. A pressure-controlling valve as required in 49 CFR 173.316(b) of DOT regulations must also be used. [1] This valve must be both sized and set so as to limit the pressure in the cylinder to $1\frac{1}{4}$ times its marked service pressure less 15 psi (103 kPa) if vacuum insulation is used. The insulation jacket shall be provided with a pressure-actuated device which will function at a pressure of not more than 25 psig (172 kPa) and provide a minimum discharge area of 0.00012 in.^{2}/lb (0.171 mm^2/kg) water capacity of cylinder.

An alternate pressure relief valve, with a marked set pressure not to exceed 150% of the DOT service pressure, may be used in lieu of the rupture disk device if the flow capacity required for relief devices on TC/DOT Specification 4L insulated cylinders is provided at 120% of marked set pressure. See CGA S-1.1, *Pressure Relief Device Standards—Part 1—Cylinders for Compressed Gases.* [4] Installation must provide for (1) prevention of moisture accumulation at the seat by drainage away from that area, (2) periodic drainage of the vent piping, and (3) avoidance of foreign material in the vent piping.

H When cylinders are over 55 inches (1397 mm) long, exclusive of the neck, this device is required in both ends, except for cylinders purchased after October 1, 1944, which must contain no aperture other than that provided in the neck of the cylinder for attachment of a valve equipped with an approved pressure relief device. (Chlorine cylinders do not generally exceed 55 inches (1397 mm) in length, since 49 CFR 173.304(a)(2) Note 2 of the DOT regulations requires that cylinders purchased after November 1, 1935, must not contain over 150 lb (68 kg) of chlorine). [1]

TABLE 7–2. (*Continued*)

J This device is required in only one end of cylinders having a length not exceeding 65 in. (1651 mm), exclusive of the neck. For cylinders over 65 in. long this device is required in both ends, and each device shall be arranged to discharge upwards and unobstructed to the open air in such a manner as to prevent any impingement of escaping gas upon the containers.
K This device can be used up to 500 psig (3450 kPa) charging pressure.
L This device is recommended, but no pressure relief device is required by Title 49 of the U.S. *Code of Federal Regulations.* [1]
M May be used in addition to CG-7.
N For use only on cylinders over 65 inches (1651 mm) long. This device is required on both ends, and each device shall be arranged to discharge upwards and unobstructed to the open air in such a manner as to prevent any impingement of escaping gas upon the containers.
P For use only on cylinders over 65 inches (1651 mm) long. This device is required on both ends.

(73.9°C) is authorized for use on cargo tanks and portable tanks.

Rupture Disk Device. Conventional domed rupture disks are domed in the direction of the subsequently applied rupturing pressures as shown in Fig. 7–15. Such disks are domed by a means sufficient to cause a permanent set such that no further plastic flow will occur when the disk is subjected to its intended operating conditions.

Reverse-domed rupture disks are domed against the direction of applied rupturing pressure as shown in Fig. 7–16 and are designed to rupture by buckling under pressure.

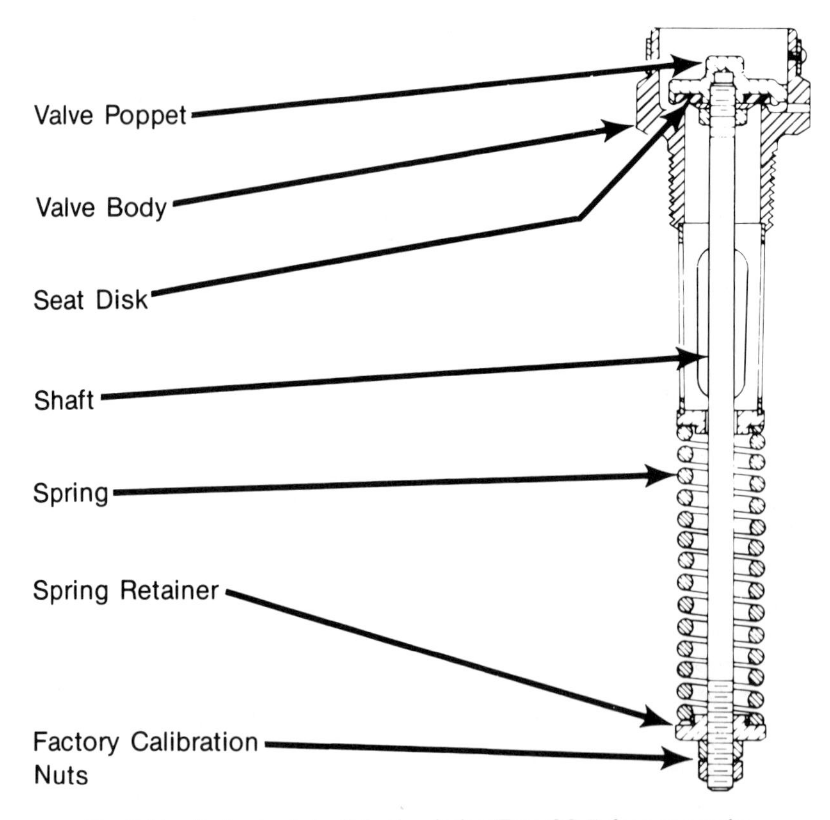

Fig. 7–14. Spring-loaded relief valve device (Type CG-7) for cargo tanks.

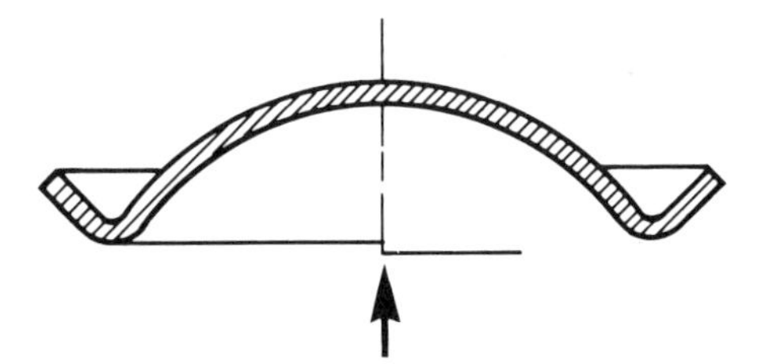

Fig. 7-15. Conventional domed rupture disk (Type CG-1) for cargo tanks.

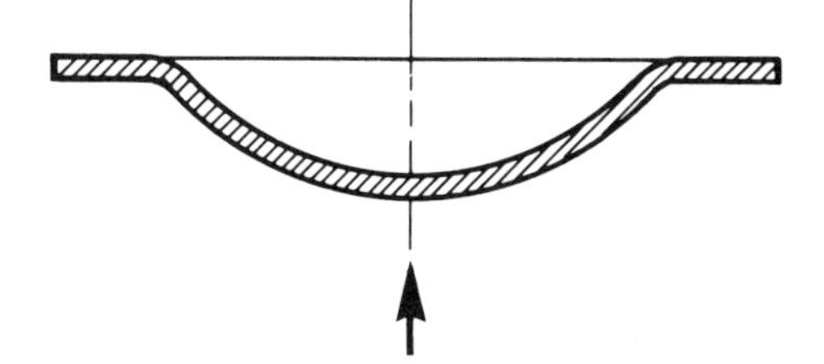

Fig. 7-16. Reverse-domed rupture disk (Type CG-1) for cargo tanks.

Rupture Disk Device in Combination with Pressure Relief Valve. Combination rupture disk/pressure relief devices, as illustrated in Fig. 7-17, consist of a conventional reclosing pressure relief valve in series with a rupture disk. The rupture disk is located between the pressure relief valve and the container. The relief valve is thus sealed from toxic and corrosive ladings and the rupture disk protected against corrosion from the environment.

Should the rupture disk rupture, the relief valve instantly pops fully open, and after tank pressure is relieved and returns to a safe working pressure, the relief valve will reseat and function as a conventional spring-loaded relief valve until it is convenient to replace the rupture disk portion of the device.

When this combination-type pressure relief

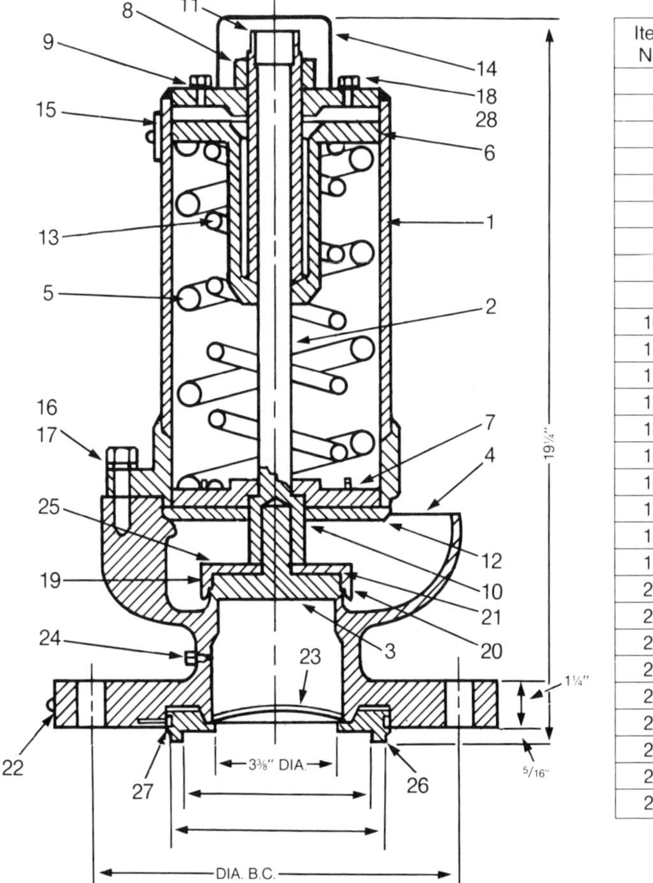

Item No.	No. Req.	Part Name
1	1	Top Guide
2	1	Stem
3	1	Plug
4	1	Body
5	1	Outer Spring
6	1	Follower
7	1	Guide
8	1	Top Nut
9	1	Top Gasket
10	1	Shaft Seal
11	1	Adjusting Screw
12	1	Seal Retainer
13	1	Inner Spring
14	1	Cap
15	1	Instruction Plate
16	4	Bolt
17	4	Lock Washer
18	1	Wire Seal
19	1	Retainer
20	1	Seat "O" Ring
21	1	Plug "O" Ring
22	1	Name Plate
23	1	Rupture Disc
24	1	⅛" Pipe Plug w/54 Hole
25	1	Bumper
26	1	Rupture Disc Flange
27	1	Set Screw
28	4	Bolt

Fig. 7-17. Rupture disk device (Type CG-1) in combination with pressure relief valve (Type CG-7) for cargo tanks.

device is installed, some regulations require that the space between the rupture disk and relief valve be vented to the atmosphere or monitored to detect any pressure buildup on the downstream side of the rupture disk which could prevent proper functioning of the rupture disk.

Breaking Pin Device in Combination with Pressure Relief Valve. This style of pressure relief device, as illustrated in Fig. 7-18, is very similar to the rupture disk relief valve combination, except that a breaking pin is used in place of the rupture disk. The breaking pin device is a non-reclosing pressure relief device actuated by inlet static pressure and is designed to function by the breaking of the load-carrying section of the pin which supports a pressure-containing member.

Should the breaking pin fracture, the relief valve instantly pops fully open and, after tank pressure is relieved and returns to a safe working pressure, the relief valve will reseat and function as a conventional spring-loaded relief valve until it is convenient to replace the breaking pin portion of the device.

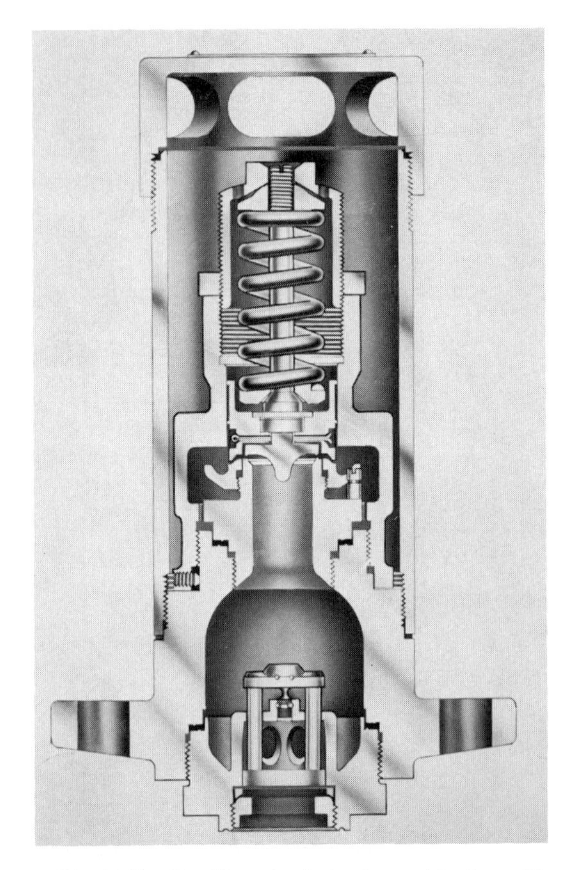

Fig. 7-18. Breaking pin device in combination with pressure relief valve (Type CG-7) for cargo tanks.

Where required by regulations, the space between the breaking pin and the relief valve must be vented to the atmosphere or monitored to detect any pressure buildup on the downstream side of the breaking pin which could prevent proper functioning of the breaking pin.

PRESSURE RELIEF DEVICES FOR COMPRESSED GAS STORAGE CONTAINERS

This section summarizes the description of pressure relief devices contained in CGA S-1.3, *Pressure Relief Device Standards—Part 3—Compressed Gas Storage Containers.* [12] Included is information on pressure relief devices for use on compressed gas storage containers constructed in accordance with the American Society of Mechanical Engineers Code or equivalent. [10]

The pressure relief devices in this section must meet the applicable portions of the following standards:

- Section VIII, Division 1 of Part AR of Section VIII, Division 2 of the *ASME Boiler and Pressure Vessel Code* [10]
- American National Standard K61.1, *Safety Requirements for the Storage and Handling of Anhydrous Ammonia* (CGA G-2.1) [13]
- NFPA 58, *Standard for the Storage and Handling of Liquefied Petroleum Gases* [14]

Types of Pressure Relief Devices

All four of the following types of pressure relief devices for compressed gas storage containers are similar to those described and illustrated in the preceding section concerning devices for cargo and portable tanks:

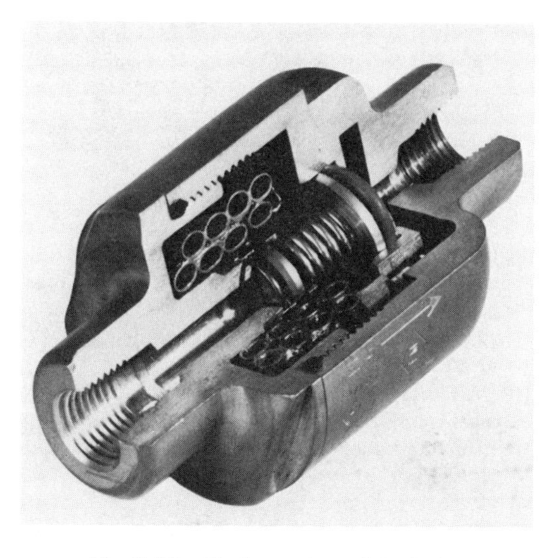

Fig. 7-19. Flash arrestor safety device.

(1) Pressure relief valve
(2) Rupture disk device
(3) Rupture disk device in combination with pressure relief valve
(4) Breaking pin device in combination with pressure relief valve

OTHER SAFETY DEVICES

This section describes and illustrates other safety devices.

Flash Arrestors

The flash arrestor, illustrated in Fig. 7-19, is widely used in the welding industry to help prevent an accidental mixture of gases such as oxygen and acetylene, which if ignited can be hazardous.

This device is generally installed in the hose lines carrying the gas and is normally in the fully open mode. If reverse flow should start, the device closes instantly. If a flash-back occurs, the device also closes instantly and the flame is extinguished within the device.

Excess Flow Check Valves

Excess flow check valves are installed in hoses and piping systems that carry pressurized fluids. In the normal open position of this device, fluid flow is permitted in either direction, as shown in Fig. 7-20(a).

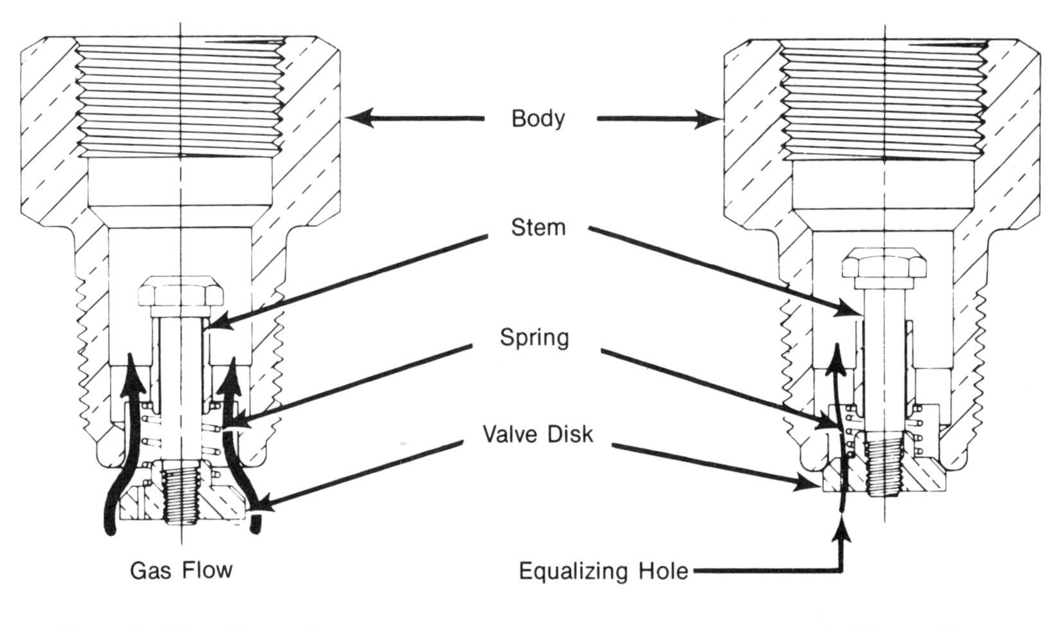

Fig. 7-20. Excess flow check valve.

In the event of a complete break in the downstream hose or pipe, the force of the fluid flow, or the high difference in pressure between the fluid container and downstream piping, will cause the excess flow valve to close, as illustrated in Fig. 7-20(b). This action not only saves the loss of fluid but is a definite safety factor if the fluid is flammable.

Figure 7-20(b) also shows an equalizing hole in the valve disk. If the excess flow check valve closes suddenly or "slugs" due to a sudden opening of a shutoff valve in the system causing a high fluid flow, the equalizing hole allows pressure on both sides of the disk to gradually equalize, thereby causing the excess flow valve to automatically reopen.

REFERENCES

[1] *Code of Federal Regulations,* Title 49 CFR Parts 100-199 (Transportation), U.S. Department of Transportation, Superintendent of Documents, U.S. Government Printing Office, Washington, DC 20402.

[2] *Transportation of Dangerous Goods Regulations,* Transport Canada, Canadian Government Publishing Centre, Supply and Services Canada, Ottawa, Ontario, Canada K1A 0S9.

[3] *Regulations for the Transportation of Dangerous Commodities by Rail,* Canadian Government Publishing Centre, Supply and Services Canada, Ottawa, Ontario, Canada K1A 0S9.

[4] CGA S-1.1, *Pressure Relief Device Standards—Part 1—Cylinders for Compressed Gases,* Compressed Gas Association, Inc., 1235 Jefferson Davis Highway, Arlington, VA 22202.

[5] CGA C-12, *Qualification Procedure for Acetylene Cylinder Design,* Compressed Gas Association, Inc., 1235 Jefferson Davis Highway, Arlington, VA 22202.

[6] CGA C-14, *Procedures for Fire Testing of DOT Cylinder Safety Relief Device Systems,* Compressed Gas Association, Inc., 1235 Jefferson Davis Highway, Arlington, VA 22202.

[7] CGA V-7, *Standard Method of Determining Cylinder Valve Outlet Connections for Industrial Gas Mixtures,* Compressed Gas Association, Inc., 1235 Jefferson Davis Highway, Arlington, VA 22202.

[8] CGA S-1.2, *Pressure Relief Device Standards—Part 2—Cargo and Portable Tanks for Compressed Gases,* Compressed Gas Association, Inc., 1235 Jefferson Davis Highway, Arlington, VA 22202.

[9] UL 132, *Safety Relief Valves for Anhydrous Ammonia and LP-Gas,* Underwriter's Laboratories Inc. Publications Stock, 333 Pfingsten Road, Northbrook, IL 60062.

[10] *ASME Boiler and Pressure Vessel Code,* Section VIII, Division 1, American Society of Mechanical Engineers, 345 East 47th Street, New York, NY 10017.

[11] *Specifications for Tank Cars,* Appendix A, (AAR M-1002), Association of American Railroads, 50 F Street, N.W., Washington, DC 20001.

[12] CGA S-1.3, *Pressure Relief Device Standards—Part 3—Compressed Gas Storage Containers,* Compressed Gas Association, Inc., 1235 Jefferson Davis Highway, Arlington, VA 22202.

[13] American National Standard K61.1, *Safety Requirements for the Storage and Handling of Anhydrous Ammonia* (CGA G-2.1), Compressed Gas Association, Inc., 1235 Jefferson Davis Highway, Arlington, VA 22202.

[14] NFPA 58, *Standard for the Storage and Handling of Liquefied Petroelum Gases,* National Fire Protection Association, Batterymarch Park, Quincy, MA 02269.

ADDITIONAL REFERENCES

CGA S-7, *Method for Selecting Pressure Relief Devices for Compressed Gas Mixtures in Cylinders,* Compressed Gas Association, Inc. 1235 Jefferson Davis Highway, Arlington, VA 22202.

CHAPTER 8

Cylinder Valve, Cylinder Ancillary Equipment, and Bulk Transfer Connections

INTRODUCTION

Safe and proper performance of compressed gas systems depends on the integrity of the many mechanical joints between system components. The proper design and use of mechanical connections is of utmost importance to the safety of personnel, equipment, and plant operations.

These connections must be rugged, durable, capable of resisting abuse, and easy to connect and disconnect. They must also be designed so that they do not permit unintended connections which may cause a hazardous condition. With nearly 200 pure gases and an infinite number of gas mixtures, this is a formidable task.

Objectives

This chapter deals with the various types of connections used in compressed gas systems and the industry standards developed to control their use. Topics covered are:

- Cylinder valve outlet connections
- Cylinder valve inlet connections
- Cylinder ancillary equipment connections
- Bulk transfer connections

CYLINDER VALVE OUTLET CONNECTIONS

This section summarizes the uniform American and Canadian standards for compressed gas cylinder valve outlet connections.

Connection Assignments Given in Table 8–1

Table 8–1 which appears at the end of this chapter, lists nearly 200 different pure chemicals and gases in alphabetical order which are commonly shipped in compressed gas cylinders. Listed to the right of each gas in Table 8–1 are the cylinder volumes and fill pressures for each gas when that is pertinent to the selection of the valve connection. Information concerning whether the connection is for liquid withdrawal of product from the cylinder and whether the connection is threaded or of the yoke type is also provided.

Listed to the right of each of these categories, as applicable, are the "Standard," "Limited Standard," and "Alternate Standard" connections for each gas. CGA V-1, *American National, Canadian, and Compressed Gas Association Standard for Compressed Gas Cylinder Valve Outlet and Inlet*

Connections, defines small cylinders as those with a capacity less than or equal to 110 in.^{3} (1.80 L; 1.81 kg water). [1] For a discussion of the valve connections that may be used on such cylinders, which include those termed lecture bottles, readers are referred to CGA V-1 (paragraph 2.8 of the 1987 sixth edition).

A "Standard" connection is the recommended connection for a particular gas or gases. A "Limited Standard" connection is one recommended for a particular gas or gases application where some limitation is imposed on its use. An "Alternate Standard" connection is one scheduled for obsolescence over a phaseout period, usually five years from the date the Alternate Standard is incorporated into CGA V-1. [1]

The use of a standard connection is always recommended over an alternate standard to encourage a complete implementation of the changeover during the five-year phaseout period.

Valve Outlet Connections for Gas Mixtures

The selection of the appropriate cylinder valve outlet connections on cylinders to be used for industrial gas mixtures is more complex and is covered by selection procedures described in CGA V-7, *Standard Method of Determining Cylinder Valve Outlet Connections for Industrial Gas Mixtures.* [2] In accordance with CGA V-7, each component of a gas mixture is assigned a mixture rating based on its FTSC number, a four-digit number each digit of which expresses the fire potential (F), toxicity (T), state of the gas (S), and corrosiveness (C). The higher the mixture rating number of a particuar gas, the more influential that gas will be in determining the correct connection for a mixture. Mixture rating numbers are integers ranging between one and six. A series of principles, described in CGA V-7, defines the procedure for selecting appropriate connections for mixtures. [2] Special instructions are given when two or more gases in a mixture have the same mixture rating.

CGA V-7 is not applicable to mixtures of medical gases, that is, gaseous drugs or gaseous medical devices. CGA V-1, *American National, Canadian, and Compressed Gas Association Standard for Compressed Gas Cylinder Valve Outlet and Inlet Connections,* covers connections for medical gas mixtures. [1]

Types of Valve Connections

Cylinder valve outlet connections fall into three basic designs: (1) threaded, (2) nonthreaded nonindexed yoke, and (3) nonthreaded pin-indexed yoke. Connections are identified by a three-digit number. Figure 8–1 illustrates a typical specification page for a threaded connection from CGA V-1, in this case, Connection CGA 540. Given at the top of the specification page for each connection is a full designation of the valve outlet thread, the rated or maximum cylinder pressure at which the connection is to be used, and the gas or gases assigned to the connection. Also included are cross-sectional views of each part of the connection, as well as a view of the connection assembly.

For each connection, the necessary dimensions, including tolerances necessary to manufacture the connection, are also provided. This serves to ensure the interchangeability of parts from different manufacturers, as well as noninterchangeability with other cylinder valve connections that would result in a hazardous cross-connect.

A typical nonthreaded non-pin-indexed yoke connection is shown in Fig. 8–2. The same format regarding connection number, rated pressure, gas assignment, and dimensional description that is provided for threaded connections is also used for these connections. Its primary use is in the self-contained underwater breathing apparatus (SCUBA) industry.

The third basic valve outlet design, shown in Fig. 8–3, is Connection CGA 870. This is a nonthreaded pin-indexed yoke connection.

COMPRESSED GAS ASSOCIATION, INC.

CONNECTION CGA 540

.903-14NGO-RH-EXT

STANDARD CYLINDER VALVE OUTLET CONNECTION FOR
PRESSURES UP TO 3,000 psig (20 680 kPa) FOR
Oxygen

WARNING — Do not use this thread for any other gas or for any gas mixture

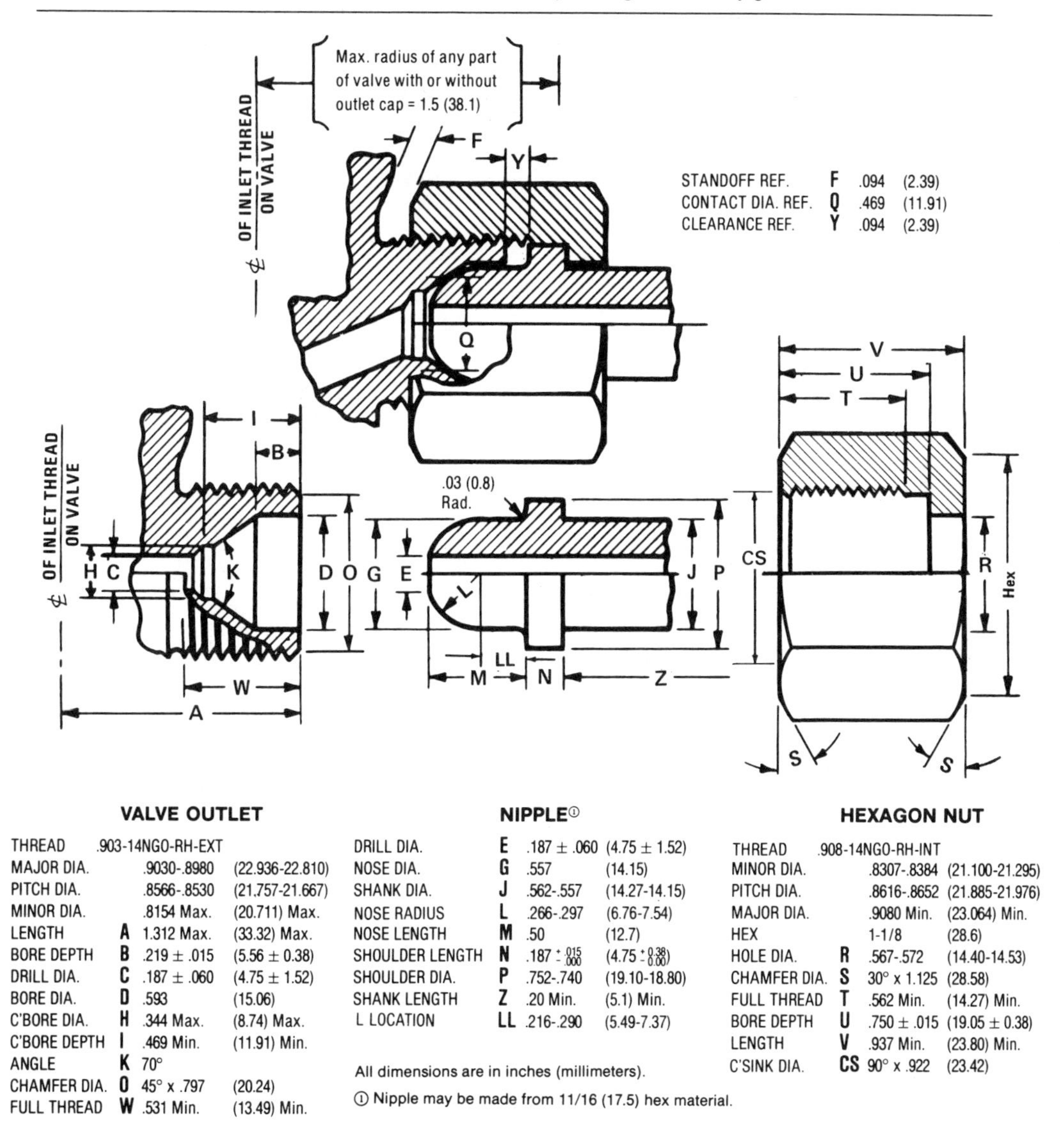

VALVE OUTLET

THREAD		.903-14NGO-RH-EXT	
MAJOR DIA.		.9030-.8980	(22.936-22.810)
PITCH DIA.		.8566-.8530	(21.757-21.667)
MINOR DIA.		.8154 Max.	(20.711) Max.
LENGTH	**A**	1.312 Max.	(33.32) Max.
BORE DEPTH	**B**	.219 ± .015	(5.56 ± 0.38)
DRILL DIA.	**C**	.187 ± .060	(4.75 ± 1.52)
BORE DIA.	**D**	.593	(15.06)
C'BORE DIA.	**H**	.344 Max.	(8.74) Max.
C'BORE DEPTH	**I**	.469 Min.	(11.91) Min.
ANGLE	**K**	70°	
CHAMFER DIA.	**O**	45° x .797	(20.24)
FULL THREAD	**W**	.531 Min.	(13.49) Min.

NIPPLE①

DRILL DIA.	**E**	.187 ± .060	(4.75 ± 1.52)
NOSE DIA.	**G**	.557	(14.15)
SHANK DIA.	**J**	.562-.557	(14.27-14.15)
NOSE RADIUS	**L**	.266-.297	(6.76-7.54)
NOSE LENGTH	**M**	.50	(12.7)
SHOULDER LENGTH	**N**	$.187^{+.015}_{-.000}$	($4.75^{+0.38}_{-0.00}$)
SHOULDER DIA.	**P**	.752-.740	(19.10-18.80)
SHANK LENGTH	**Z**	.20 Min.	(5.1) Min.
L LOCATION	**LL**	.216-.290	(5.49-7.37)

All dimensions are in inches (millimeters).

① Nipple may be made from 11/16 (17.5) hex material.

HEXAGON NUT

THREAD		.908-14NGO-RH-INT	
MINOR DIA.		.8307-.8384	(21.100-21.295)
PITCH DIA.		.8616-.8652	(21.885-21.976)
MAJOR DIA.		.9080 Min.	(23.064) Min.
HEX		1-1/8	(28.6)
HOLE DIA.	**R**	.567-.572	(14.40-14.53)
CHAMFER DIA.	**S**	30° x 1.125	(28.58)
FULL THREAD	**T**	.562 Min.	(14.27) Min.
BORE DEPTH	**U**	.750 ± .015	(19.05 ± 0.38)
LENGTH	**V**	.937 Min.	(23.80) Min.
C'SINK DIA.	**CS**	90° x .922	(23.42)

Fig. 8–1. Dimensional drawing from CGA V-1 of Connection CGA 540, a standard cylinder valve outlet connection.

COMPRESSED GAS ASSOCIATION, INC.

CONNECTION CGA 850

LIMITED STANDARD[1,2] CYLINDER VALVE YOKE CONNECTION FOR PRESSURES UP TO 3,000 psig (20 680 kPa) FOR Scuba[1] Air (R-729)

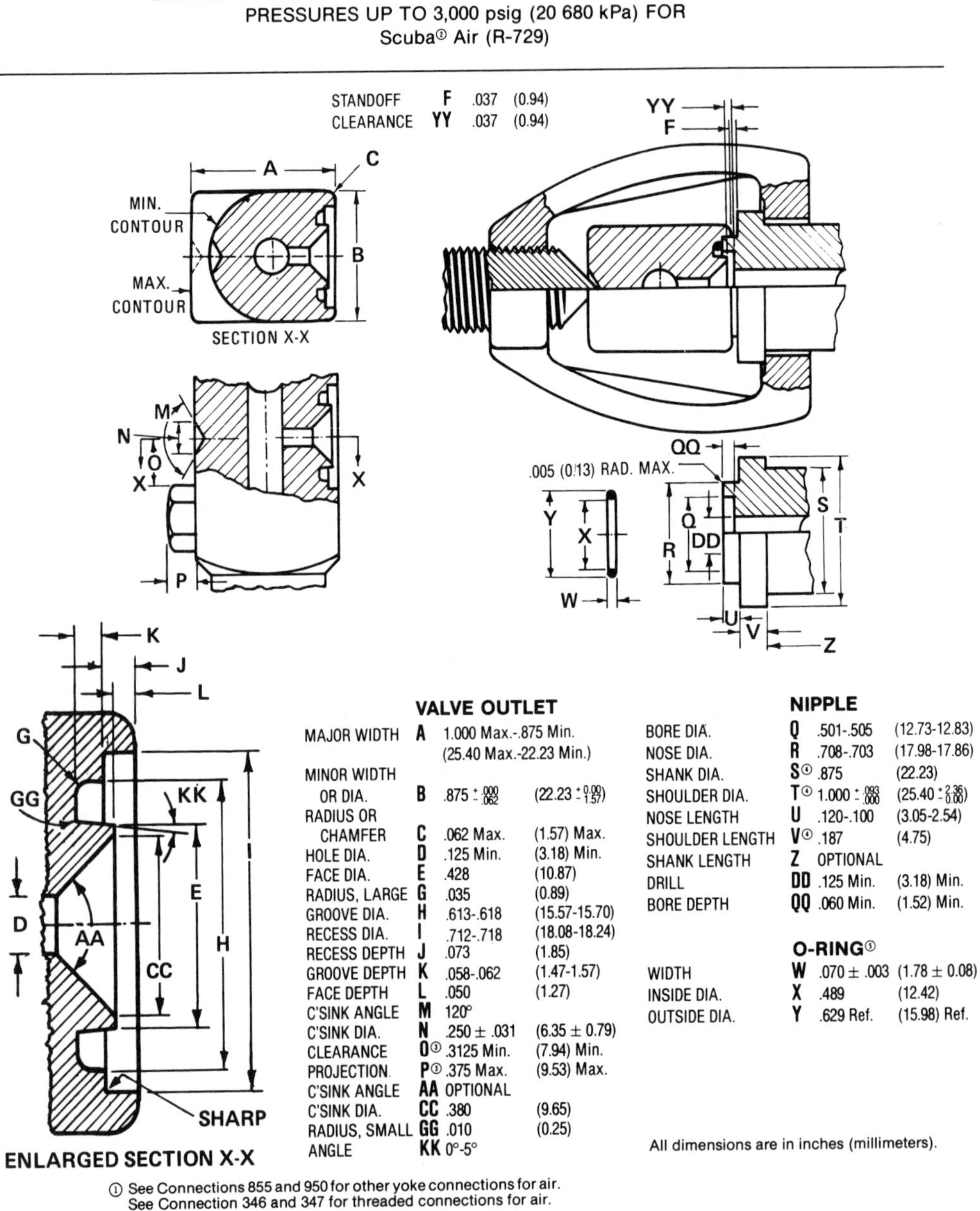

VALVE OUTLET

MAJOR WIDTH	**A**	1.000 Max.-.875 Min. (25.40 Max.-22.23 Min.)	
MINOR WIDTH OR DIA.	**B**	$.875^{+.000}_{-.062}$	$(22.23^{+0.00}_{-1.57})$
RADIUS OR CHAMFER	**C**	.062 Max.	(1.57) Max.
HOLE DIA.	**D**	.125 Min.	(3.18) Min.
FACE DIA.	**E**	.428	(10.87)
RADIUS, LARGE	**G**	.035	(0.89)
GROOVE DIA.	**H**	.613-.618	(15.57-15.70)
RECESS DIA.	**I**	.712-.718	(18.08-18.24)
RECESS DEPTH	**J**	.073	(1.85)
GROOVE DEPTH	**K**	.058-.062	(1.47-1.57)
FACE DEPTH	**L**	.050	(1.27)
C'SINK ANGLE	**M**	120°	
C'SINK DIA.	**N**	.250 ± .031	(6.35 ± 0.79)
CLEARANCE	**O**[3]	.3125 Min.	(7.94) Min.
PROJECTION	**P**[3]	.375 Max.	(9.53) Max.
C'SINK ANGLE	**AA**	OPTIONAL	
C'SINK DIA.	**CC**	.380	(9.65)
RADIUS, SMALL	**GG**	.010	(0.25)
ANGLE	**KK**	0°-5°	

NIPPLE

BORE DIA.	**Q**	.501-.505	(12.73-12.83)
NOSE DIA.	**R**	.708-.703	(17.98-17.86)
SHANK DIA.	**S**[4]	.875	(22.23)
SHOULDER DIA.	**T**[4]	$1.000^{+.093}_{-.000}$	$(25.40^{+2.36}_{-0.00})$
NOSE LENGTH	**U**	.120-.100	(3.05-2.54)
SHOULDER LENGTH	**V**[4]	.187	(4.75)
SHANK LENGTH	**Z**	OPTIONAL	
DRILL	**DD**	.125 Min.	(3.18) Min.
BORE DEPTH	**QQ**	.060 Min.	(1.52) Min.

O-RING[5]

WIDTH	**W**	.070 ± .003	(1.78 ± 0.08)
INSIDE DIA.	**X**	.489	(12.42)
OUTSIDE DIA.	**Y**	.629 Ref.	(15.98) Ref.

All dimensions are in inches (millimeters).

① See Connections 855 and 950 for other yoke connections for air. See Connection 346 and 347 for threaded connections for air.

② Limited to self-contained underwater breathing apparatus. See Paragraph 3.1.2.1 of *Introduction*.

③ Dimensions O and P are applicable only if projecting type pressure relief device is used.

④ Not required for integral yoke design.

⑤ Although O-Ring sizes are not identical, Connections 850 and 855 are interchangeable. Means of retaining O-Ring may be used, provided interchangeability with 855 is maintained.

Fig. 8-2. Dimensional drawing of a typical non-pin-indexed yoke connection, Connection CGA 850.

COMPRESSED GAS ASSOCIATION, INC. **CONNECTION CGA 870**

PIN-INDEXED YOKE, PINS 2-5

STANDARD MEDICAL CYLINDER VALVE YOKE CONNECTION FOR
PRESSURES UP TO 3,000 psig (20 680 kPa) FOR
Oxygen

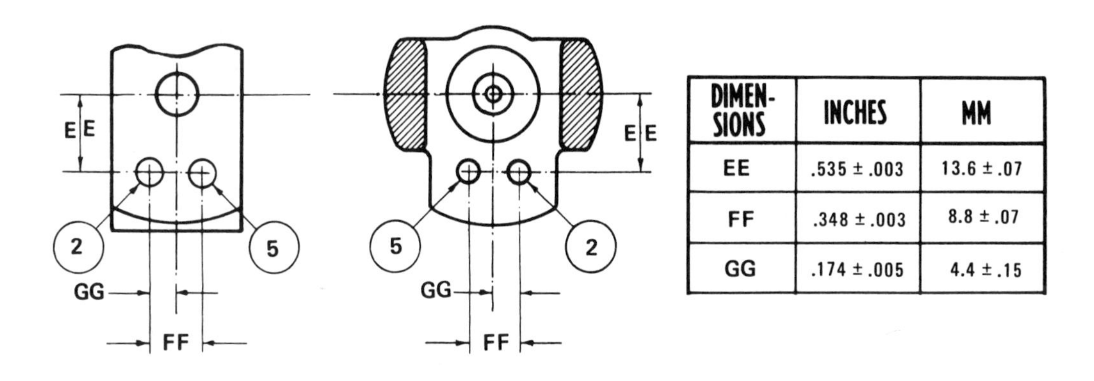

DIMEN-SIONS	INCHES	MM
EE	.535 ± .003	13.6 ± .07
FF	.348 ± .003	8.8 ± .07
GG	.174 ± .005	4.4 ± .15

(FOR OTHER DIMENSIONS, SEE DRAWING NO. 860)

Fig. 8-3. Connection CGA 870 pin-indexed basic valve outlet design.

The specifications page in CGA V-1 for each connection of this type again carries the connection number, rated pressure, and gas assignment. [1] The necessary dimensions for manufacture are also given.

Drawing No. 860, shown in two parts as Fig. 8-4 (basic dimensional drawing) and Fig. 8-5 (basic dimensions), is a necessary reference for each pin-indexed yoke connection since it contains all the additional information that is common to each yoke connection.

The nonthreaded pin-indexed yoke connection design assures noninterchangeability between connections by the use of a series of holes and matching pins. The holes are drilled in the valve body at specified locations in relation to the valve outlet, while matching locations for pins are provided in the yoke.

One series of outlet connections (CGA 870 through 960) has the hole and matching pin locations along the arc of a single radius. Each hole and matching pin location is assigned a single digit, numbered one through six, for identification. The various combinations of pins and holes that make up this series are shown in CGA V-1 on the specification page for each connection. Connection CGA 965 uses only a single hole and matching pin at Location No. 7.

A second series of outlets (at the time of publication only Connection CGA 973 was assigned) has the holes and matching pin locations along the arcs of two radii. Each hole and matching pin on the inner radius is assigned a double digit, with the first digit being the digit 1 (i.e., 11, 12, 13, 14, 15). Each hole and matching pin on the outer radius is assigned a double digit, with the first digit being the digit 2 (i.e., 21, 22, 23, 24). Each connection in this series is made up of a combination of one hole and matching pin from the inner radius and one hole and matching pin from the outer radius, for a total of 20 possible connections.

The dimensions of the holes and matching pin locations are arranged so that a connection from the first series will not cross-connect with one from the second series, thus maintaining the desired separation for the gas assignment(s) of each connection.

COMPRESSED GAS ASSOCIATION, INC. **DRAWING NO. 860**

PIN-INDEXED YOKE CONNECTIONS FOR MEDICAL GASES

BASIC DIMENSIONAL DRAWING FOR
CONNECTION NOS. 870 THRU 973

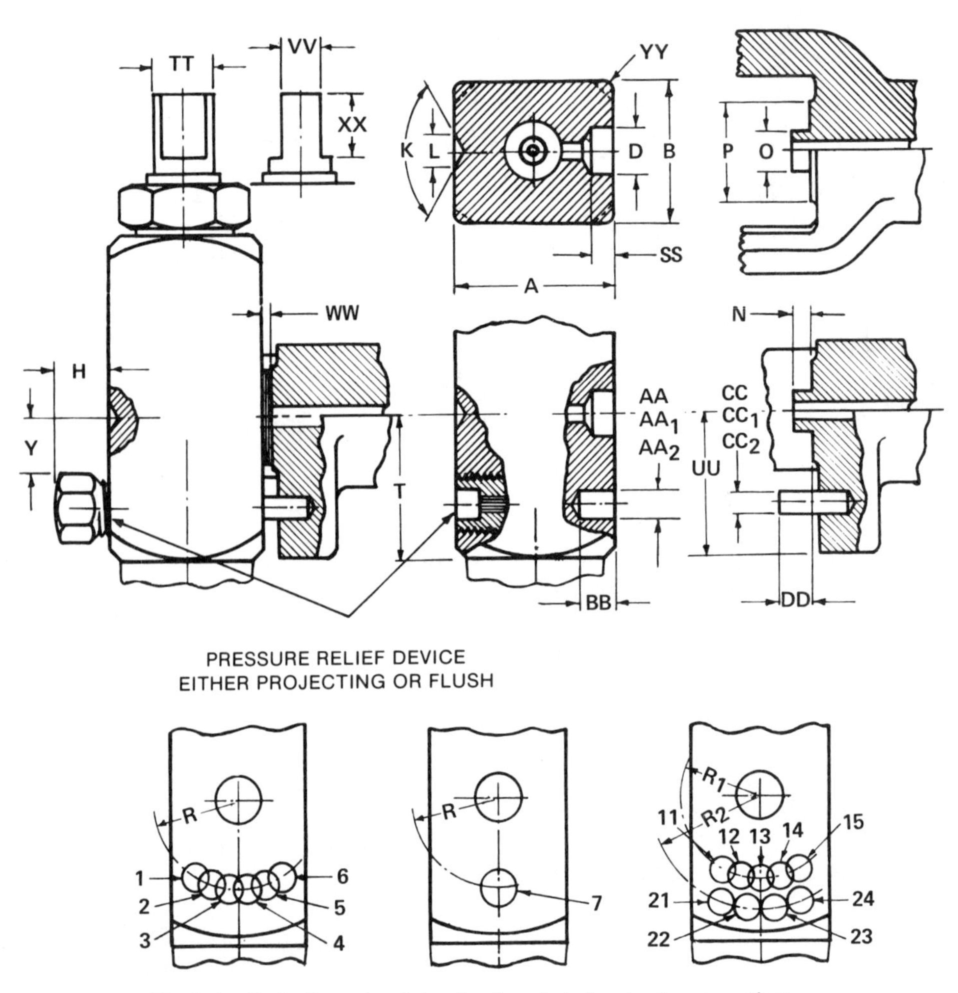

Fig. 8-4. Basic dimensional drawing for pin-indexed yoke connections.

For a detailed history of the development of standards for compressed gas cylinder valve outlet connections, refer to the Foreword of CGA V-1. [1] The Introduction to CGA V-1 describes the design guidelines used in the development of these standard connections. Both of these sections of CGA V-1 should be carefully reviewed by anyone wishing to understand the background and basis for the various outlet connections.

Since new connections and changes affecting existing connections may occur periodically, it is strongly recommended that the reader obtain the latest editions of CGA V-1 and V-7. [1] and [2] Whenever in doubt, contact the Compressed Gas Association for

COMPRESSED GAS ASSOCIATION, INC.

DRAWING NO. 860

PIN-INDEXED YOKE CONNECTIONS FOR MEDICAL GASES

BASIC DIMENSIONS FOR
CONNECTION NOS. 870 THRU 973

VALVE

DESCRIPTION		INCHES	MM
MAJOR WIDTH	**A**	1 ± .015	25 ± 0.2
MINOR WIDTH	**B**	.875 ± .015	$22.2^{+0.4}_{-0}$
BORE DIA.	**D**	.275 Min.	7 Min.
PROJECTION③	**H**	.375 Max.	9.6 Max.
COUNTERSINK ANGLE	**K**	100°-120°	100°-120°
COUNTERSINK DIA.	**L**	.203-.250	$6^{+0}_{-0.5}$
DISTANCE	**T**	.875 Min.	22 Min.
CLEARANCE③	**Y**	.312 Min.	8 Min.
HOLE DIA.			
SINGLE ROW	**AA**	.187-.191	$4.75^{+0.1}_{-0}$
SINGLE HOLE DIA.	**AA_1**	.228-.232	5.8-5.9
HOLE DIA.			
DOUBLE ROW	**AA_2**	.187-.191	$4.75^{+0.1}_{-0}$
HOLE DEPTH	**BB**	.219-.250	$5.5^{+0.5}_{-0}$
BORE DEPTH	**SS**	.140 Min.	3.6 Min.
STEM DIA.	**TT**	.373-.356	9.5-9.1
WRENCH FLATS	**VV**	.240-.225	6.1-5.7
FLAT LENGTH	**XX**	.394 Min.	10 Min.
RADIUS OR CHAMFER⑫	**YY**	.062 ± .010	1.6 ± .25

YOKE

DESCRIPTION		INCHES	MM
NOSE LENGTH	**N**	.140-.120	3.6-3.0
NOSE DIAMETER	**O**	.255-.235	$6.5^{+0}_{-0.2}$
SPOTFACE DIA.	**P**	.625 Min.	16 Min
RADIUS, 6 PINS @ 12°	**R**	.562 Nom.	14.3 Nom.
RADIUS, 5 PINS @ 15°	**R_1**	.472 Nom.	12 Nom.
RADIUS, 4 PINS @ 15°	**R_2**	.689 Nom.	17.5 Nom.
PIN DIA. SINGLE ROW	**CC**	.157-.155	4 ± 0.1
SINGLE PIN DIAMETER	**CC_1**	.213-.209	5.4-5.3
PIN DIA. DOUBLE ROW	**CC_2**	.157-.155	4 ± 0.1
PIN PROJECTION	**DD**	.219-.188	$5.5^{+0}_{-0.5}$
DISTANCE	**UU**	.866 Max.	22 Max.

WASHER

DESCRIPTION		INCHES	MM
PROTRUSION OF WASHER BEFORE COMPRESSION	**WW**	.094 Max.	2.4 Max.

NOTES

① Each connection (except no. 965) includes two pins in the yoke and two mating holes in the valve, properly indexed for safety.

② For precise positions of the pins and holes, see the respective connections that follow.

③ Dimensions **H** and **Y** are applicable only if projecting type pressure relief device is used.

④ The rotary movement of the yoke on the valve must be limited to ± 6 degrees prior to pin engagement.

⑤ Bore **D** and countersink **L** must align within .020" (0.5 mm) T I R of sides and of each other.

⑥ Pins in yoke must be made of corrosion resisting material having a minimum tensile strength of 60,000 PSI (414 mPa).

⑦ Design must be such that dimension **DD** cannot be reduced below minimum by force.

⑧ Face of valve outlet may be grooved or counterbored and face of yoke washer seating surface may be machined with concentric circles for effective washer seal.

⑨ A single washer shall be used on the valve outlet or yoke connection to insure a gastight seal. The washer shall be of such thickness that pin engagement of not less than .094" (3/32") (2.4 mm) is accomplished before washer is compressed.

⑩ Method of tightening yoke on valve optional except that no pointed object sharper than 100° included angle shall be used as a means to apply tightening pressure at the back of the valve.

⑪ For non pin-indexed yoke connections, see 2.1.5 of the introduction.

⑫ Larger chamfer permitted provided face width is 5/8" (16 mm) minimum.

⑬ These dimensions apply to valves manufactured after January 1, 1977. Valves made prior to this date complying with earlier editions of this standard are acceptable for continuing use.

⑭ Letter symbols (A, B, C, etc.) and dimensions in millimeters are as shown in ISO Recommendation R407, "Yoke Type Connections for Small Medical Gas Cylinders Used for Anaesthetic and Resuscitation Purposes," December 1964, except for distance T which was increased to accommodate the double row of pins. The American National Standard for Connections 870 through 973 was written in inches and when adopted as ISO R407, the translation of inches to mm was handled by ISO. Little attention was paid to possible discrepancies between the inch and mm dimensions until recently. The metric dimensions are now being reviewed by CGA with a view to recommending changes to make inch and mm dimensions more compatible. It is intended to confer with ISO countries to determine the most suitable manner of presenting dimensions and symbols for international use. In the meantime, it should be understood that the widest possible tolerances shown for each dimension—whether in metric or inches—is considered in compliance with this standard.

Fig. 8-5. Basic dimensions (pin-indexed yoke connections) for medical gases (see Fig. 8-4).

assistance in selecting the proper cylinder valve outlet connections.

CYLINDER VALVE INLET CONNECTIONS

This section summarizes the uniform American and Canadian standards for cylinder valve inlet connections in which a threaded joint is used to attach cylinder valves to cylinders and containers. A more complete discussion may be found in CGA V-1. [1]

Two basic types of thread are used: (1) straight thread and (2) tapered thread. The straight thread requires use of an O-ring or gasket to accomplish a gastight seal. See Fig. 8-6. This type of thread requires less installation torque than a tapered thread and is mostly used with aluminum cylinders. Common sizes currently in use are 0.750-16UNF and 1.125-12UNF. In most common use are the national gas tapered (NGT) threads which taper at a rate of $\frac{1}{16}$ inch per inch ($\frac{3}{4}$ inch per foot). Most NGT cylinder valve inlet connections are $\frac{3}{4}$-14NGT, where $\frac{3}{4}$ is the nominal size of the thread in inches and 14 is the number of threads per inch, that is, $\frac{1}{14}$ thread pitch. Other popular sizes are $\frac{3}{8}$-18NGT, $\frac{1}{2}$-14 NGT, and 1-11$\frac{1}{2}$NGT. The size of the thread is influenced by, among other things, the size and type of cylinder and customer preference.

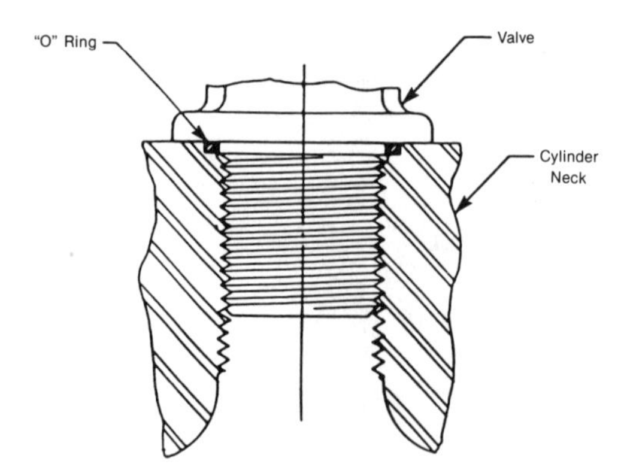

Fig. 8-6. Typical straight thread valve inlet connection.

Oversize Threads

Oversize NGT threads are also available; four and seven threads oversize are the most common. However, NGT threads can be provided for other oversize connections. The availability of oversize threads on valves prolongs the life of cylinders by compensating for the wear that occurs to cylinder threads from repeated valving operations. Oversize threads are also used by certain gas producers to differentiate their cylinders from others. Table 8-2 (Table 3 from CGA V-1) lists dimensions for features listed in Fig. 8-7 for the various tapered inlet thread sizes.

Uniquely, chlorine valves have their own oversize thread series: 0 threads oversize, 4 threads oversize, 8$\frac{1}{2}$ threads oversize, 14 threads oversize, and 28 threads oversize. Respectively, these are known as: $\frac{3}{4}$-14NGT (Cl)-1, $\frac{3}{4}$-14NGT (Cl)-2, $\frac{3}{4}$-14 (Cl)-3, $\frac{3}{4}$-14NGT (Cl)-4, and $\frac{3}{4}$-14NGT (Cl)-5 threads. The (Cl) designation signifies chlorine.

The Cl-5 type has been added recently. It is applicable to chlorine fuse plugs and cylinder valves. It is not used in chlorine ton container valves because it would interfere with the eductor tube attachment.

In the chlorine application, oversize thread specifications apply only to the external valve thread. There are no oversize internal chlorine thread specifications for cylinders because the oversize valves are only used in conjunction with worn, out of specification, oversize cylinder threads.

Establishing a Leak-Tight Seal

As listed in Column 2 of Table 8-2, hand-tight engagement (L_1) occurs when the pitch diameter of the first thread on the valve engages the thread in the cylinder neck with the same pitch diameter. Thus, for $\frac{3}{4}$-14NGT threads, this occurs nominally after an engagement of 0.339 inch or, 0.339 inch $\times$ 14 threads/inch = approximately 4$\frac{3}{4}$ threads. A manufacturing tolerance of plus or minus one turn is permitted on both the valve and the cylinder threads. Hand-tight engagement

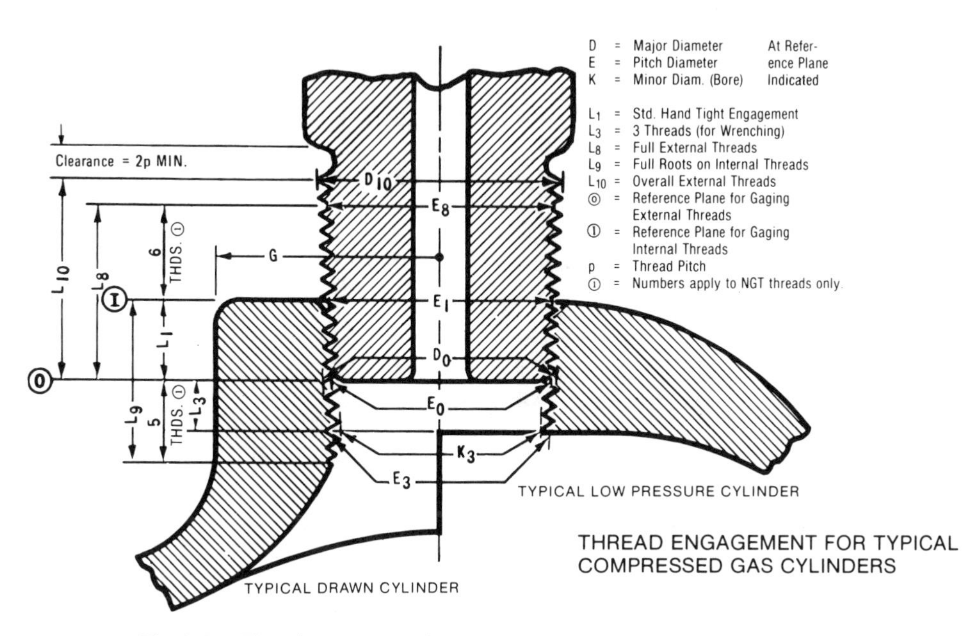

Fig. 8–7. Thread engagement for typical compressed gas cylinders.

can then be approximately $2\frac{3}{4}$ to $6\frac{3}{4}$ threads. Three turns are allowed for wrenching to establish a leak-tight seal. Total engaged threads can range from $5\frac{3}{4}$ to $9\frac{3}{4}$, neglecting chamfers on the first threads of the valve and cylinder neck.

Since the length of full threads on the valve (L_8) is 0.7676 inch, as shown in Column 7 of Table 8–2, the total number of full threads is 0.7676 inch × 14 threads/inch, or approximately $10\frac{3}{4}$ threads. If the thread tolerances are such that the valve threads are as large as they can be and the cylinder opening is as small as it can be, the valve enters only $2\frac{3}{4}$ turns for hand-tight engagement. That is, if the largest valve is installed in the smallest cylinder, there will be approximately five full threads showing after the three turns for wrenching.

Conversely, if the thread tolerances are such that the valve threads are as small as they can be and the cylinder opening is as large as it can be, the valve will enter $6\frac{3}{4}$ turns for hand-tight engagement. Thus, if the smallest valve is installed in the largest cylinder, there will be approximately one full thread showing after the three turns for wrenching. While it is highly improbable that these extremes will be experienced with new parts, this illustrates why the counting of exposed threads is a poor way of ascertaining a sufficiently engaged joint.

Using a predetermined amount of torque to establish a leak-tight seal also has some drawbacks. Variations in coefficients of friction, thread damage, type of sealant used, and so on, can influence the amount of applied torque that is necessary to create a seal. For example, if the first thread on the cylinder valve is severely damaged, much of the torque may be used just to overcome the resistance of the damaged thread.

The "handtight plus 3 turns for wrenching" identified in both CGA V-1 and NBS Handbook H28, *Federal Screw Thread Standards,* provides a method for engaging the valve to a steel cylinder that is not affected by manufacturing tolerances on the valve and cylinder threads. [3] This engagement is illustrated by the dimension (L_3) in Fig. 8–7. For aluminum cylinders and/or straight threads, consult the manufacturer.

TABLE 8–2. NATIONAL GAS TAPER (NGT) THREADS[6]

		EXTERNAL								INTERNAL					
		SMALL END			FULL THREADS		LARGE END					FULL THREADS			
SYMBOL (Designation of Thread) (1)	HAND-TIGHT ENGAGE-MENT L_1 (3)	MAJOR DIAM. D_0	PITCH DIAM. E_0	CHAMFER 45° x MIN. DIAM. GG	PITCH DIAM. E_8	LENGTH L_8 (4)	MAJOR DIAM. APPROX. D_{10}	OVERALL LENGTH APPROX. L_{10}	NECK RADIUS MIN. G	PITCH DIAM. AT FACE E_1	C'SINK 90° x MAX. DIAM. KK	BORE MAX. K_3	PITCH DIAM. E_3	LENGTH L_1+L_3	LENGTH OF FULL ROOT MIN. L_9 (5)
1	2	3	4	5	6	7	8	9	10	11	12	13	14	15	16
1/8—27NGT	0.1800 (4.572)	0.3931 (9.984)	0.3635 (9.233)	0.3281 (8.334)	0.3886 (9.870)	0.4022 (10.216)	0.4204 (10.678)	0.4375 (11.113)	0.2813 (7.145)	0.3748 (9.520)	0.4063 (10.320)	0.3269 (8.303)	0.3566 (9.058)	0.2911 (7.394)	0.3652 (9.276)
1/4—18NGT	0.2000 (5.080)	0.5218 (13.253)	0.4774 (12.126)	0.4219 (10.716)	0.5107 (12.972)	0.5333 (13.546)	0.5530 (14.046)	0.6250 (15.875)	0.3750 (9.525)	0.4899 (12.443)	0.5625 (14.288)	0.4225 (10.732)	0.4670 (11.862)	0.3667 (9.314)	0.4778 (12.136)
3/8—18NGT	0.2400 (6.096)	0.6564 (16.672)	0.6120 (15.545)	0.5625 (14.288)	0.6479 (16.457)	0.5733 (14.562)	0.6915 (17.564)	0.6875 (17.463)	0.4375 (11.113)	0.6270 (15.926)	0.6875 (17.463)	0.5572 (14.153)	0.6016 (15.281)	0.4067 (10.330)	0.5178 (13.152)
1/2—14NGT	0.3200 (8.128)	0.8156 (20.716)	0.7584 (19.263)	0.6875 (17.463)	0.8052 (20.452)	0.7486 (19.014)	0.8625 (21.908)	0.8125 (20.638)	0.5625 (14.288)	0.7784 (19.771)	0.8750 (22.225)	0.6879 (17.473)	0.7450 (18.923)	0.5343 (13.571)	0.6771 (17.198)
3/4—14NGT	0.3390 (8.611)	1.0248 (26.029)	0.9677 (24.580)	0.9063 (23.020)	1.0157 (25.799)	0.7676 (19.497)	1.0795 (27.419)	0.8750 (22.225)	0.6875 (17.463)	0.9889 (25.118)	1.0625 (26.988)	0.8972 (22.789)	0.9543 (24.239)	0.5533 (14.054)	0.6961 (17.681)
3/4—14NGT(CI)-1	0.3390 (8.611)	1.0248 (26.029)	0.9677 (24.580)	0.9063 (23.020)	1.0268 (26.081)	0.9461 (24.031)	1.0951 (27.816)	1.1250 (28.575)	0.6875 (17.463)	0.9889 (25.118)	1.0625 (26.988)	0.8972 (22.789)	0.9543 (24.239)	0.5533 (14.054)	0.9461 (24.031)
3/4—14NGT(CI)-2	0.3390 (8.611)	1.0427 (26.484)	0.9856 (25.034)	0.9219 (23.416)	1.0447 (26.535)	0.9461 (24.031)	1.1130 (28.270)	1.1250 (28.575)							
3/4—14NGT(CI)-3	0.3390 (8.611)	1.0628 (26.995)	1.0057 (25.545)	0.9375 (23.813)	1.0648 (27.046)	0.9461 (24.031)	1.1331 (28.781)	1.1250 (28.575)		SEE	NOTE (7)				
3/4—14NGT(CI)-4	0.3390 (8.611)	1.0873 (27.617)	1.0302 (26.167)	0.9688 (24.608)	1.0893 (27.668)	0.9461 (24.031)	1.1576 (29.403)	1.1250 (28.575)							
3/4—14NGT(CI)-5	0.3390 (8.611)	1.1498 (29.204)	1.0927 (27.755)	1.0313 (26.195)	1.1518 (29.256)	0.9461 (24.031)	1.2201 (30.991)	1.1250 (28.575)							
1-11½—NGT	0.4000 (10.160)	1.2832 (32.593)	1.2136 (30.825)	1.1250 (28.575)	1.2712 (32.288)	0.9217 (23.411)	1.3457 (34.181)	1.0000 (25.400)	0.8125 (20.638)	1.2386 (31.460)	1.3125 (33.338)	1.1278 (28.646)	1.1973 (30.411)	0.6609 (16.787)	0.8348 (21.204)
1-1/4-11½—NGT	0.4200 (10.668)	1.6267 (41.318)	1.5571 (39.550)	1.4688 (37.308)	1.6160 (41.046)	0.9417 (23.919)	1.6931 (43.005)	1.0625 (26.988)	1.0000 (25.400)	1.5834 (40.218)	1.6719 (42.466)	1.4713 (31.371)	1.5408 (39.136)	0.6809 (17.295)	0.8548 (21.712)
1-1/2-11½NGT	0.4200 (10.668)	1.8657 (47.388)	1.7961 (45.621)	1.7031 (43.259)	1.8550 (47.117)	0.9417 (23.919)	1.9360 (49.174)	1.1250 (28.575)	1.1563 (29.370)	1.8223 (46.286)	1.9063 (48.420)	1.7102 (43.439)	1.7798 (45.207)	0.6809 (17.295)	0.8548 (21.712)
3/4-14SGT(2)	0.4008 (10.180)	1.0470 (26.594)	0.9852 (25.024)	0.9219 (23.416)	1.0731 (27.257)	0.7030 (17.856)	1.1564 (29.373)	0.8750 (22.225)	0.6875 (17.463)	1.0353 (26.297)	1.1094 (28.179)	0.8556 (21.732)	0.9474 (24.064)	0.5714 (14.514)	0.7030 (17.856)

However, the number of turns required to establish a hand-tight engagement will vary depending on whether the threaded joint is bare metal, whether Teflon tape is applied to the valve, or whether a suitable luting compound is used.

One way to compensate for the above mentioned variables is to first tighten the joint without luting compound or Teflon tape as tight as possible with gloved hands, and count the turns needed to accomplish this. Next, apply the luting compound or Teflon tape that is going to be used in actual valve installations and repeat the above hand-tightening procedure, again counting turns. The difference between the number of turns to accomplish a hand-tight joint with and without luting compound (or Teflon tape) should then be added to the "3 turns for wrenching."

For example, consider the instance where 5 turns are needed to make a hand-tight engagement with bare metal. When luting compound or Teflon tape is used for this same valve, $4\frac{1}{2}$ turns are required to arrive at a hand-tight engagement. To establish a leak-tight seal, the valve would then be wrenched $3\frac{1}{2}$ turns rather than 3, or engaged a total of $8\frac{1}{2}$ turns, that is, 5 turns (bare metal hand-tight) plus $3\frac{1}{2}$ turns (plus 3 for wrenching plus $\frac{1}{2}$ from tape). In this manner, the effect of the tape or luting compound is adequately taken into account.

Other valve installation methods are successfully used, but all of them, when properly done, end up with approximately the same number of wrenched turns.

Effect of Tolerances on Taper and Bore Size

The NGT thread provides for a plus 0, minus 1 turn tolerance on the taper of the valve external thread so that any deviation from

TABLE 8–2. *(continued)*

All dimensions are basic and are given in inches (millimeters). All NGT threads are right hand.

(1) Symbol (Designation of Thread)

Oversize valves — For uses other than chlorine, oversize threads for revalving are generally but not always at 4 or 7 turns oversize. For chlorine, the ¾—14NGT(Cl)—1 is not oversize; the —2 is 4 turns overs·ze; the —3 is 8½ turns oversize, the —4 is 14 turns oversize and the —5 is 28 turns oversize.

(2) ¾ - 14 SGT

The ¾—14SGT (Special Gas Taper Thread) is a standard having a taper of 1½" per foot equivalent to (12.50 mm per 100 mm) on diameter with a 60° thread normal to the axis and 0.0618" (1.570 mm) deep. For this thread Col. 13, 14 and 15 are based on gages 0.7030" (17.856 mm) long. Cylinders are held to final inspection limits from basic to 1½ turns small, and valves to plus or minus 1 turn.

(3) Handtight Engagement

The basic condition of fit is that the External Thread with a pitch diameter of E_0 at the end (reference plane for gaging External Thread) shall enter by hand engagement to a distance L_1 into the Internal Thread with a pitch diameter of E_1 at the opening (reference plane for gaging Internal Thread.)

(4) Length

External Threads shall be threaded the approximate length L_{10} but gaged up to L_8. Dimension L_8 is equal to L_1 plus six (6) threads for all NGT threads and L_1 plus eight and a half (8½) threads for the NGT (Cl) threads. Dimension E_8 is measured at distance L_8 from E_0, and dimension D_{10} is measured at distance L_{10} from E_0. These longer External Threads are desirable if further tightening should be necessary. To facilitate gaging, provision should be made to allow the L_8 ring gage to advance a distance of 2 full threads beyond the L_8 length (one turn for allowable variation in pitch diameter and one turn for allowable variation in taper).

(5) Length of Full Root Min.

Full Internal Threads at the crests and roots shall extend throughout lengths L_1 plus L_3 (L_3 = 3 threads). This dimension determines the minimum metal on the inside of the neck to produce maximum bore K_3. Any metal below L_3 shall have tapped threads with full roots to a minimum length L_9 (L_1 + 5 threads for all NGT threads and L_1 + 8½ threads for the NGT (Cl) threads).

(6) Gaging NGT Threads

Because of their length and more rigid requirements for sealing compressed gases against leaks, NGT threads require special gages. They have been developed and are used by Federal Services for Inspecting NGT threads. They are fully described in FED-STD-H28/9. Care and operation of these gages are described in Military Supply Procedures Manual 8310 IGM-5-5008 of 12 January 1954. Any other method of gaging which will give the required results can be used.

(7) Internal thread dimensions for (Cl)-2 through (Cl)-5 threads, not applicable.

normal is in the direction of the valve thread falling away from the cylinder thread, as indicated in Fig. 8–8. Similarly, a plus 1, minus 0 turn tolerance on the taper of the cylinder internal thread results in any deviation from the nominal causing the cylinder thread to fall away from the valve thread. Sealing is forced to occur where the lower end of the valve thread engages the cylinder thread.

If the bore in the cylinder exceeds the maximum specified as (K_3) in Table 8–2, it can result in flat crests at the point where sealing is supposed to occur. It is therefore important that the bore not be oversize and that a good bore gauge that registers off the cylinder thread crests at a depth of (L_9) be used, as shown in Fig. 8–9. The first thread on a 4-thread oversize valve is the same size as the fourth thread on a standard valve, i.e., zero threads oversize.

To reduce propulsion effects in the event the cylinder valve becomes accidentally severed at the top of the cylinder neck, limits have been established on the size of the inlet opening in the valve. This diameter (d_o), shown in Fig. 8–10, should be kept to a minimum within the constraints of other design requirements.

Except for valves used in liquefied gas service, the inlet hole diameter should be restricted to 0.300 inch (7.6 mm) maximum beyond the length (L_1) + (L_3). For straight threads, this depth should not exceed the minimum full thread length on the cylinder valve. This recommendation applies to cylinders having a service pressure over 500 psig (3447 kPa) and approximately 125 lb water capacity.

Special design requirements, such as pressure relief device requirements, provisions for eductor tube installations, or compressed gas valves used in fire control systems which have high blowdown flow requirements, may dictate greater diameters.

For further information on cylinder valve inlet connections and other thread types presently in use, that is, Special Gas Taper Threads (SGT) and National Gas Straight Threads (NGS), refer to CGA V-1, *American National, Canadian, and Compressed Gas Association Standard for Compressed Gas Cylinder Valve Outlet and Inlet Connections.* [1]

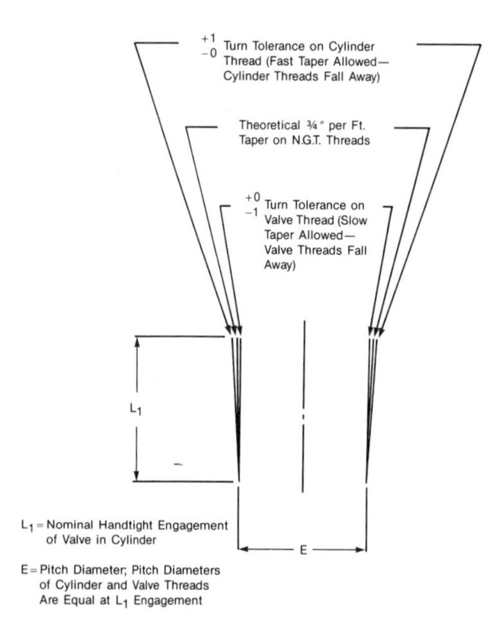

Fig. 8–8. Allowable tolerances on NGT tapered threads.

CYLINDER ANCILLARY EQUIPMENT CONNECTIONS

This section summarizes the requirements for gas cylinder ancillary equipment connections.

Pressurized gas ancillary equipment downstream from the cylinder valve outlet is joined to the system piping by various types of threaded connections. Such equipment includes pressure-reducing regulators, gauges, flowmeters, and station valves (i.e., manifold isolation valves, etc.).

Industry is standardizing such connections as the need arises. Connections from the regulator inlet stream have been standardized in medical applications to avoid the potentially deleterious effects to life and health that could occur if leaks were to occur at connections or if misconnections were to be made.

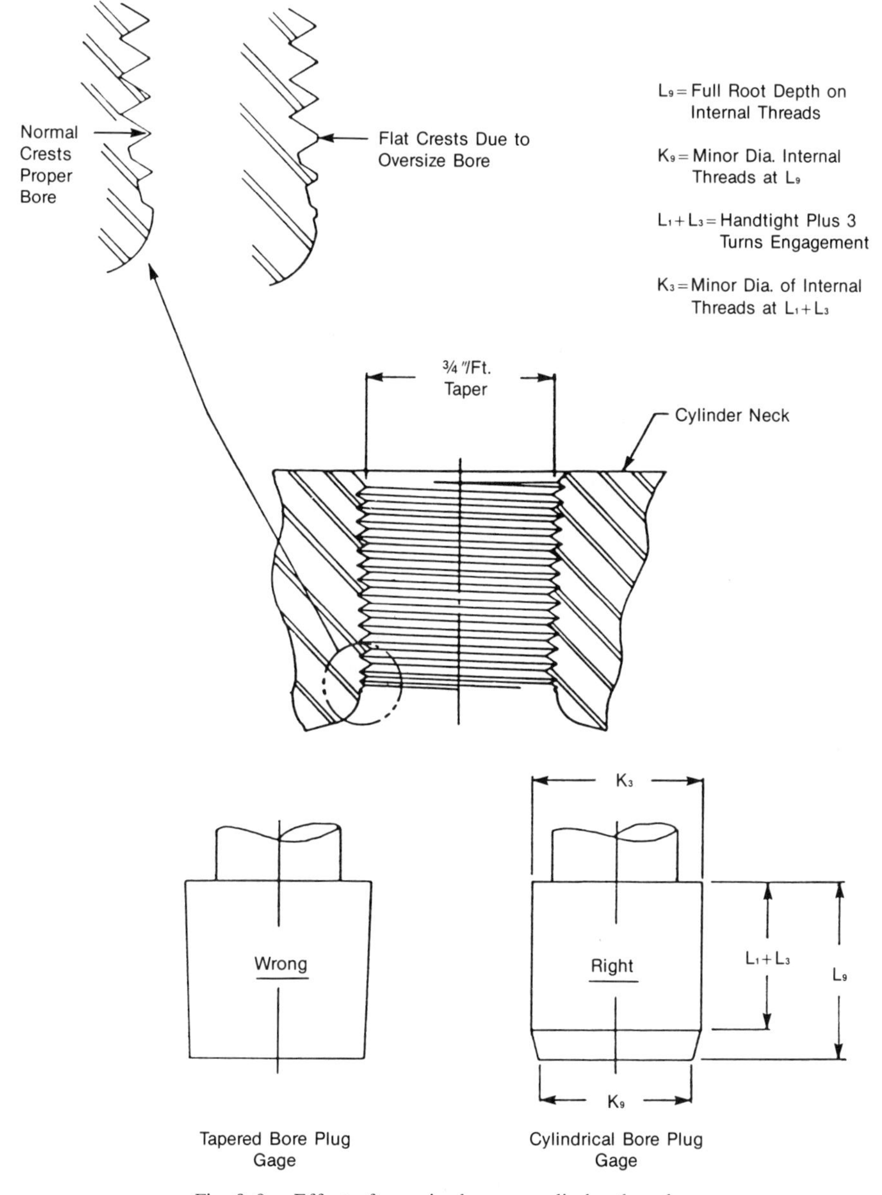

Fig. 8-9. Effect of oversize bore on cylinder threads.

Diameter Index Safety System (DISS)

CGA V-5, *Diameter Index Safety System,* describes standards for noninterchangeable connections for pressures of 200 psig (1379 kPa) and less for medical regulator outlets and connections for anesthesia, resuscitation, and therapy apparatus. [4] This system of connections, referred to as the Diameter Index Safety System (DISS), is based on a concept of noninterchangeable indexing achieved by a series of increasing and decreasing diameters in the component parts of the connections.

A typical connection design, Connection CGA 1000A for medical gases, is shown in

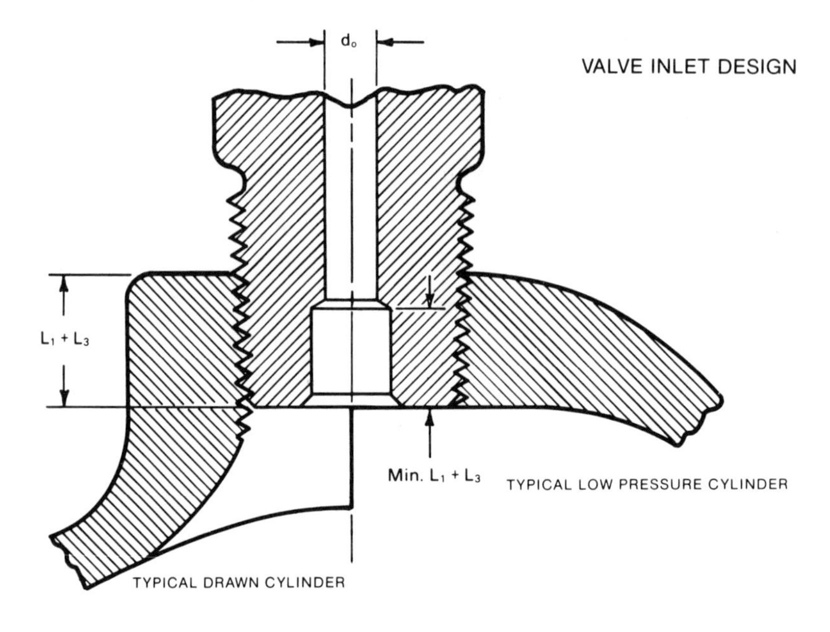

The diameter d_0 is the valve inlet diameter. This diameter should be kept to a minimum within the constraints of other design requirements, to minimize propulsion effects should the valve be accidentally severed at the top of the cylinder neck. Except for valves used in liquefied gas service, this inlet hole *diameter* should be restricted to .300 inch (7.62 mm) maximum beyond the length minimum $L_1 + L_3$. *For straight threads this depth should not exceed the minimum full thread length on the cylinder valve.* This *recommendation* is for cylinders having both a service pressure over 500 psig (3 450 kPa) and less than 125 pounds (56.7 kg) water capacity. Special design requirements such as pressure relief device requirements, provisions for eductor tube installations or compressed gas valves used in fire control systems, which have *high blow down flow* requirements, may dictate greater diameters.

Fig. 8–10. Valve inlet opening restrictions.

Fig. 8–11. The concept of increasing/decreasing part diameters is illustrated in Fig. 8–12.

The adequacy of these connections has been demonstrated through their extensive safe use since 1959, when they were first approved by the CGA. CGA E-7, *Standard for Medical Gas Regulators and Flowmeters,* was established in 1983. [5] This CGA standard specifies that, where applicable, connections to these components be in accordance with CGA V-5, *Diameter Index Safety System.* [4]

Special Application Connections

While no CGA standard exists for high pressure station (manifold) connections, cylinder valves with $\frac{3}{8}$-inch, $\frac{1}{2}$-inch, or $\frac{3}{4}$-inch NGT inlet connections are commonly used with the appropriate CGA outlet connection for the particular gas.

In response to the recognized potential hazards inherent in homeowner use of outdoor grill propane cylinders, the CGA has developed a standard connection for this service which cannot be cross-connected to any other connection. Similarly, the CGA has developed standards for connections used in the welding industry. CGA E-1, *Standard Connections for Regulator Outlets, Torches and Fitted Hose for Welding and Cutting Equipment,* has been adopted throughout the welding industry. [6]

Figure 8–13 illustrates a typical connection for oxygen and fuel gases which consists of a male connector, a tailpiece, and a connector nut. Oxy-fuel cutting and welding torch-to-hose connections are covered by CGA E-5, *Torch Standard for Welding and Cutting.* [7] Thread and connection configurations must comply with CGA E-1. [6]

COMPRESSED GAS ASSOCIATION, INC.

DRAWING NO. 1000-A

CO-STANDARD 1000-A SERIES LOW-PRESSURE CONNECTIONS FOR

Medical Gases

BASIC DIMENSIONS FOR CONNECTION NUMBERS 1020-A THRU 1200-A
(Dimensions in Inches)

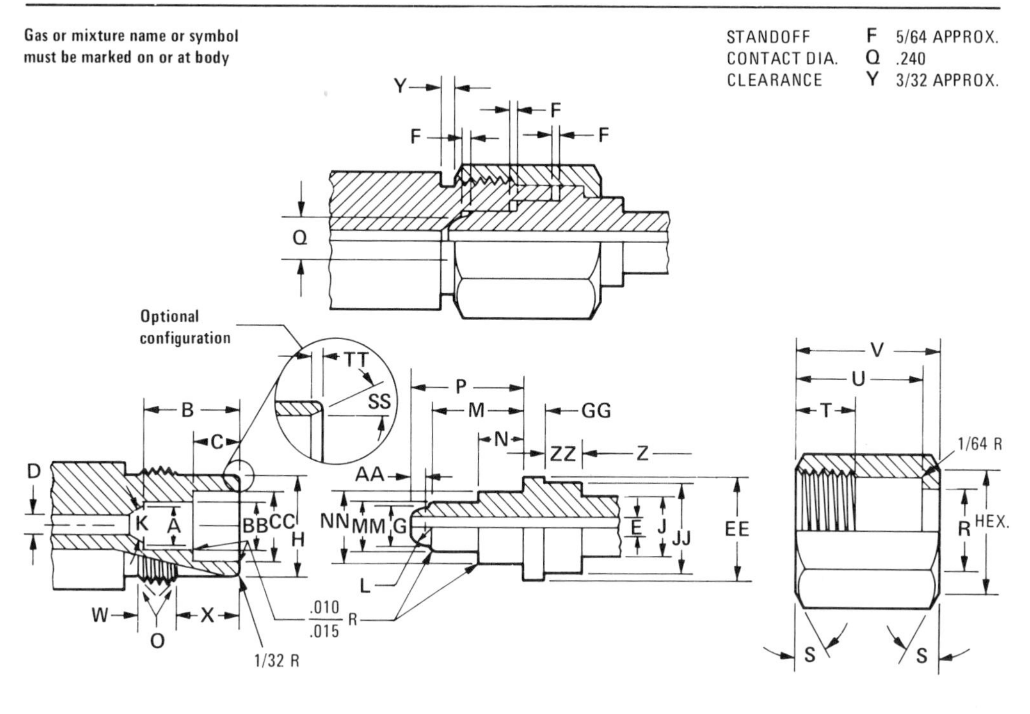

BODY			NIPPLE			HEXAGON NUT		
THREAD		.750-16UNF-2A-RH	DRILL	E	.161 MAX.	THREAD		.750-16UNF-2B-RH (MOD)
MAJOR DIA.		.7485-.7391	SHOULDER DIA.	EE	.672 ± .005	MINOR DIA.		.690-.696 (MOD)
PITCH DIA.		.7079-.7029	*NOSE DIA.	G	.296-.293	PITCH DIA.		.7094-.7159
MINOR DIA.		.6718 REF.	SHOULDER LGTH	GG	.125 ± .005	MAJOR DIA.		.7500 MIN.
*SEAT DIA.	A	.299-.302	NOSE RADIUS	L	.1480-.1465	COUNTERSINK		90° x 49/64
BORE DEPTH	B	.625 ± .005	*SM INDEX DIA.	MM	†	HEXAGON		7/8
*BORE DIA.	BB	†	LENGTH	M	.625 ± .005	HOLE	R	.546-.551
C'BORE DEPTH	C	.312 ± .005	LG INDEX DIA.	NN	†	CHAMFER	S	30° x 7/8 DIA.
*C'BORE DIA.	CC	†	LENGTH	N	.312 ± .005	FULL THREAD	T	3/8 MIN.
DRILL	D	.161 MAX.	HEAD LENGTH	P	.781 ± .005	BORE DEPTH	U	.870-.880
BODY DIA.	H	.656-.650	STEP DIA.	JJ	.536-.531	LENGTH	V	1.000 MIN.
ANGLE	K	70°	SHANK DIA.	J	OPTIONAL			
CHAMFER	O	45° x DIA. H	STEP LENGTH	ZZ	3/16 to 3/8			
COUNTERSINK	SS	25°	SHANK LENGTH	Z	OPTIONAL			
COUNTERSINK	TT	.031	RADIUS DIST.	AA	.120 ± .003			
FULL THREAD	W	1/4 MIN.						
LENGTH	X	.437 ± .005						

*Body diameters A, BB and CC as well as nipple diameters G, MM and NN should be concentric within .002 Full Indicator Movement (FIM). These are critical dimensions for safety that must be adhered to on final product whether plated or not.

†See page 11 for these dimensions.

Fig. 8-11. Co-standard series low-pressure connections for medical gases; dimensional drawing from CGA V-1.

COMPRESSED GAS ASSOCIATION, INC.

CONNECTION NOS. 1020 THRU 1200-A

STANDARD 1000 AND 1000-A SERIES DIAMETER-INDEX DIMENSIONS FOR

Medical Gases

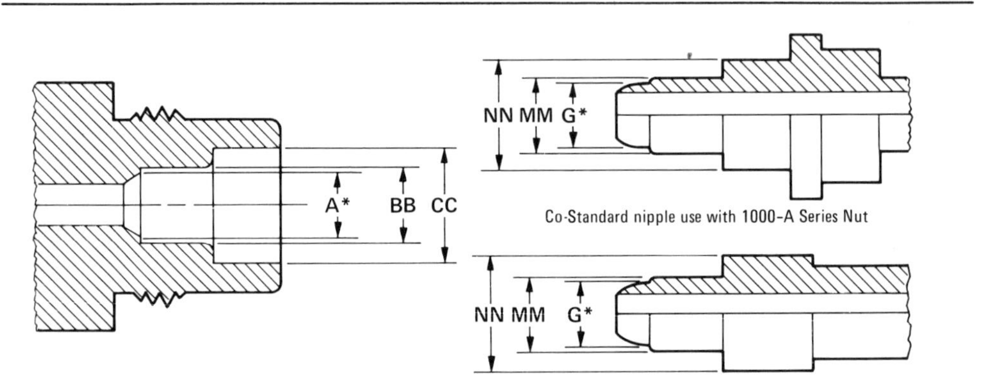

GAS NAME	GAS SYMBOL	CONN. NO.	DIAMETERS (INCHES) *BB	*CC	*MM	*NN
Special Mixtures. For Limited Experimental Applications. The word "SPECIAL" is permissible.		1020 1020-A	.299-.302	.539-.542	.296-.293	.536-.533
Nitrous Oxide	N_2O	1040 1040-A	.311-.314	.527-.530	.308-.305	.524-.521
Helium, Helium-Oxygen Mixtures (Helium over 80.5%)	He, He-O_2 Mixture	1060 1060-A	.323-.326	.515-.518	.320-.317	.512-.509
Carbon Dioxide, Carbon Dioxide-Oxygen Mixtures (CO_2 over 7.5%)	CO_2, CO_2-O_2 Mixture	1080 1080-A	.335-.338	.503-.506	.332-.329	.500-.497
Cyclopropane	C_3H_6	1100 1100-A	.347-.350	.491-.494	.344-.341	.488-.485
Nitrogen	N_2	1120 1120-A	.359-.362	.479-.482	.356-.353	.476-.473
Ethylene	C_2H_4	1140 1140-A	.371-.374	.467-.470	.368-.365	.464-.461
Air		1160 1160-A	.383-.386	.455-.458	.380-.377	.452-.449
Oxygen-Helium Mixtures (Helium not over 80.5%)	O_2-He Mixture	1180 1180-A	.395-.398	.443-.446	.392-.389	.440-.437
Oxygen-Carbon Dioxide Mixture (CO_2 not over 7.5%)	O_2-CO_2 Mixture	1200 1200-A	.407-.410	.431-.434	.404-.401	.428-.425

* Body diameters A, BB and CC as well as nipple diameters G, MM and NN should be concentric within .002 Full Indicator Movement (FIM). These are critical dimensions for safety that must be adhered to on final product whether plated or not.

FOR BASIC DIMENSIONS SEE DRAWINGS NOS. 1000 OR 1000-A

Fig. 8-12. Standard series diameter-index dimensions for medical gases connections.

COMPRESSED GAS ASSOCIATION, INC.

CONNECTION NOS. 020 & 021

Formerly IAA Class A

.375-24UNF-2A-RH/LH-EXT

STANDARD GAS WELDING AND CUTTING HOSE CONNECTIONS
RH Connection No. 020 for Oxygen
LH Connection No. 021 for Fuel Gases

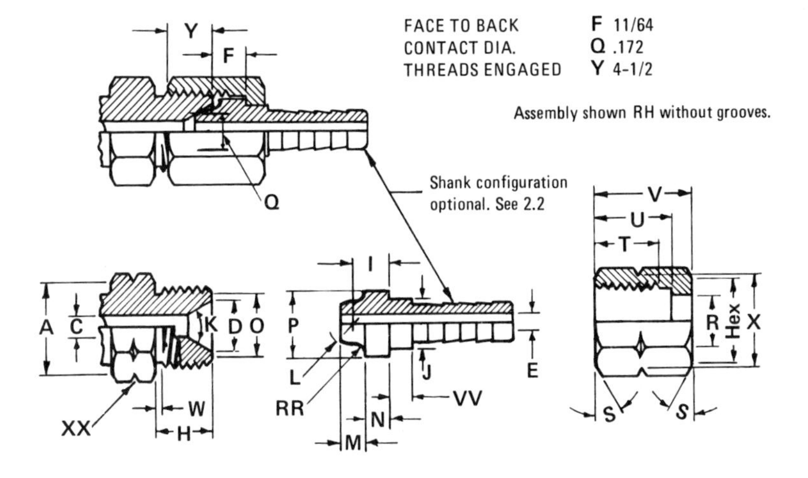

RH CONNECTOR NO. 020 LH CONNECTOR NO. 021	
THREAD	.375–24UNF–2A–RH/LH–EXT
MAJOR DIA.	.3739–.3667
PITCH DIA.	.3468–.3430
MINOR DIA.	.3228 REF.
CHAMFER	O 45° x 5/16 DIA.
SHANK LENGTH	H 9/32
NECK	W 1/16 x 5/16 DIA.
DRILL	C 3/32 MIN.
SEAT DIA.	D .245–.255
ANGLE	K 60°
BODY DIA. OR HEX	A 7/16 MIN.
GROOVE (LH ONLY)	*XX

TAILPIECE NO. 020–1/8 020–3/16	
DRILL	E See Table I
SHANK DIA.	J .248–.243
" LENGTH	VV 7/64 MIN.
NOSE RADIUS	L .099
RADIUS DISTANCE	I .187–.177
NOSE LENGTH	M 1/8
SHOULDER LENGTH	N 1/8
SHOULDER DIA.	P .328–.324
BLEND RADIUS	RR 3/64

RH NUT NO. 020 LH NUT NO. 021	
THREAD	.375–24UNF–2B–RH/LH–INT
MINOR DIA.	.330–.340
PITCH DIA.	.3479–.3528
MAJOR DIA.	.3750 MIN.
COUNTERSINK	90° x 25/64 DIA.
HEXAGON	7/16
HOLE	R .257–.262
HEX. CHAMFER	S 30° x 7/16 DIA.
FULL THREAD	T 1/4 MIN.
BORE DEPTH	U 3/8
LENGTH	V 15/32
GROOVE (LH ONLY)	X 60° x 15/32 DIA.

*Groove to identify LH thread only. Depth and location optional, but must not adversely affect connector strength.

Fig. 8-13. Standard gas welding and cutting hose connections; dimensional drawing from CGA E-1.

COMPRESSED GAS ASSOCIATION, INC.

CONNECTION NO. OX-150

1-1/2" OXYGEN CONNECTION

STANDARD FOR CRYOGENIC FLUID TRANSFER

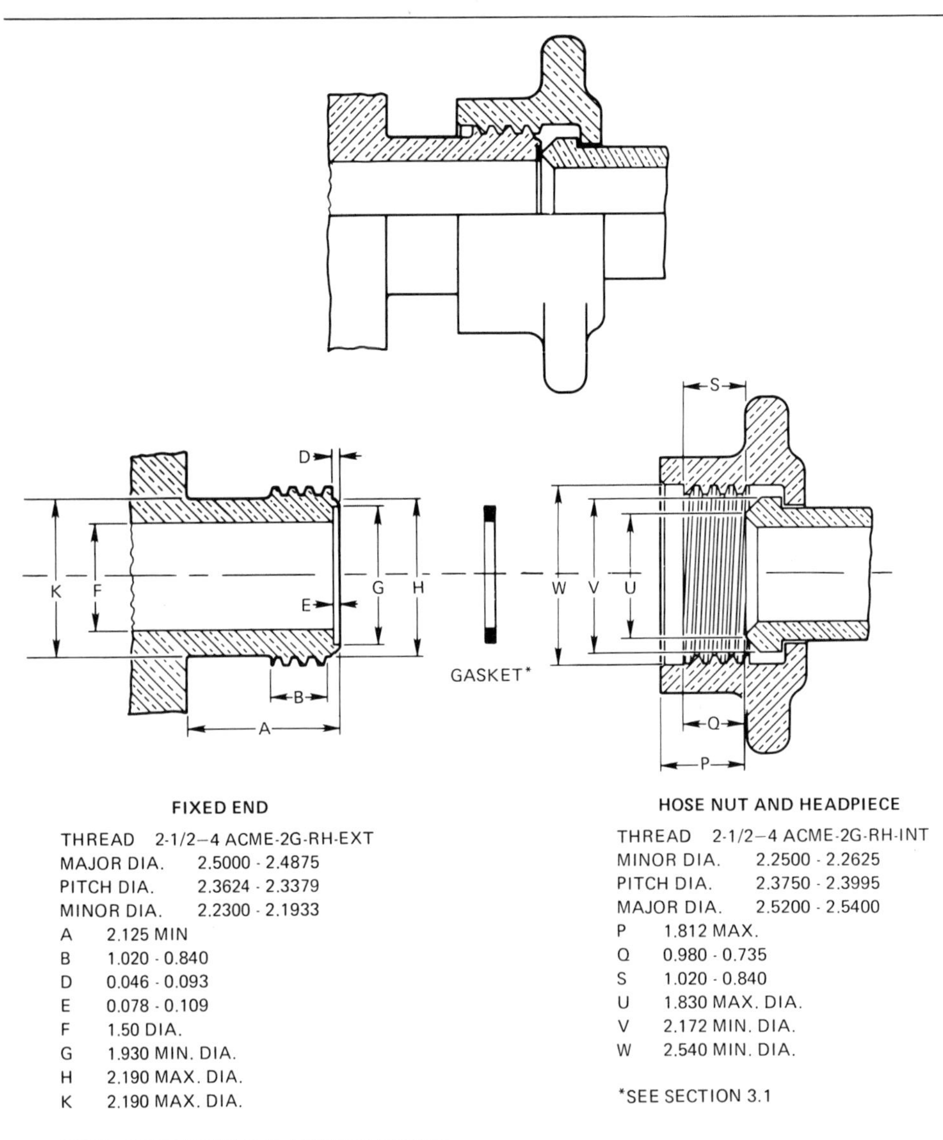

FIXED END

THREAD 2-1/2–4 ACME-2G-RH-EXT
MAJOR DIA. 2.5000 - 2.4875
PITCH DIA. 2.3624 - 2.3379
MINOR DIA. 2.2300 - 2.1933

A	2.125 MIN
B	1.020 - 0.840
D	0.046 - 0.093
E	0.078 - 0.109
F	1.50 DIA.
G	1.930 MIN. DIA.
H	2.190 MAX. DIA.
K	2.190 MAX. DIA.

HOSE NUT AND HEADPIECE

THREAD 2-1/2–4 ACME-2G-RH-INT
MINOR DIA. 2.2500 - 2.2625
PITCH DIA. 2.3750 - 2.3995
MAJOR DIA. 2.5200 - 2.5400

P	1.812 MAX.
Q	0.980 - 0.735
S	1.020 - 0.840
U	1.830 MAX. DIA.
V	2.172 MIN. DIA.
W	2.540 MIN. DIA.

*SEE SECTION 3.1

NOTE: DIMENSIONS ARE SHOWN IN INCHES

Fig. 8-14. Dimensional drawing for $1\frac{1}{2}$-inch oxygen connection for cryogenic fluid transfer from CGA V-6.

COMPRESSED GAS ASSOCIATION, INC.

CONNECTION NO. CO_2-100

1" CARBON DIOXIDE CONNECTION

STANDARD FOR CO_2 LIQUID AND VAPOR TRANSFER

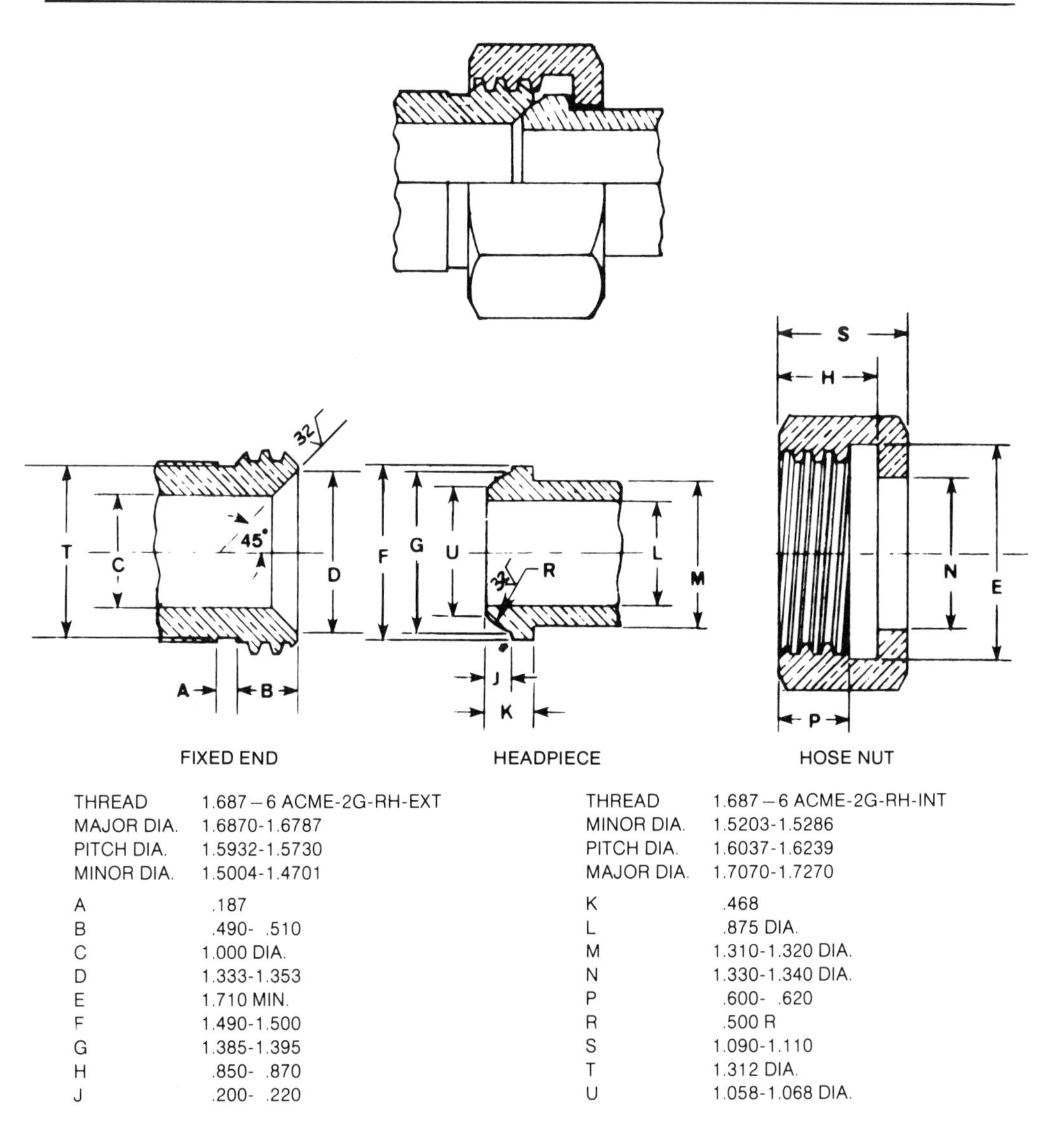

THREAD	1.687 – 6 ACME-2G-RH-EXT
MAJOR DIA.	1.6870-1.6787
PITCH DIA.	1.5932-1.5730
MINOR DIA.	1.5004-1.4701
A	.187
B	.490- .510
C	1.000 DIA.
D	1.333-1.353
E	1.710 MIN.
F	1.490-1.500
G	1.385-1.395
H	.850- .870
J	.200- .220

THREAD	1.687 – 6 ACME-2G-RH-INT
MINOR DIA.	1.5203-1.5286
PITCH DIA.	1.6037-1.6239
MAJOR DIA.	1.7070-1.7270
K	.468
L	.875 DIA.
M	1.310-1.320 DIA.
N	1.330-1.340 DIA.
P	.600- .620
R	.500 R
S	1.090-1.110
T	1.312 DIA.
U	1.058-1.068 DIA.

NOTE: DIMENSIONS ARE SHOWN IN INCHES. UNSPECIFIED TOLERANCES ARE ±.015

Fig. 8-15. Dimensional drawing of 1-inch carbon dioxide connection for liquid and vapor transfer from CGA V-6.1.

CGA E-4, *Standard for Gas Regulators for Welding and Cutting,* covers both cylinder inlet connections, which must conform with CGA V-1, and pipeline regulator inlet connections, which must comply with the requirements of CGA E-3, *Pipeline Regulator Inlet Connection Standards.* [8] and [9] CGA E-3 became fully effective on January 1, 1985.

BULK TRANSFER CONNECTIONS

This section covers transfer connections between bulk transport equipment for liquefied gases or cryogenic liquids and termination points such as station storage containers. These connections do not apply to ambient-temperature vessels such as cylinders.

As with cylinder valve connection philosophy, the intent of standardizing transfer connections is to minimize the proliferation of connection types for the same application and to avoid potentially hazardous cross connections. Standard transfer connections for the applications described above consist of: (1) a fixed end connector at the termination point, (2) a hose nut, (3) a headpiece, and, in the case of cryogenic liquids, (4) a gasket.

The fixed end connector is typically an integral part of the stationary container. The headpiece is the end connection of the transfer hose. The hose nut is a swivel-mounted nut captured on the hose by the headpiece. When the nut is tightened, a leak-tight seal is attained.

Drawings and specifications for standard transfer connections for cryogenic liquids are contained in CGA V-6, *Standard Cryogenic Liquid Transfer Connections.* [10] Drawings and specifications for carbon dioxide liquid and vapor transfer connections are contained in CGA V-6.1, *Standard Carbon Dioxide Transfer Connections.* [11]

Figure 8-14 shows one of the standard connections for cryogenic oxygen fluid transfer from CGA V-6, and Figure 8-15 shows one of the liquid and vapor standard connections for carbon dioxide from CGA V-6.1.

REFERENCES

[1] CGA V-1, *American National, Canadian, and Compressed Gas Association Standard for Compressed Gas Cylinder Valve Outlet and Inlet Connections* (ANSI/CSA/CGA V-1, Compressed Gas Association, Inc., 1235 Jefferson Davis Highway, Arlington, VA 22202.

[2] CGA V-7, *Standard Method of Determining Cylinder Valve Outlet Connections for Industrial Gas Mixtures,* Compressed Gas Association, Inc., 1235 Jefferson Davis Highway, Arlington, VA 22202.

[3] NBS Handbook H28, *Federal Screw Thread Standards,* National Institute of Standards and Technology, Superintendent of Documents, U.S. Government Printing Office, Washington, DC 20402.

[4] CGA V-5, *Diameter Index Safety System,* Compressed Gas Association, Inc., 1235 Jefferson Davis Highway, Arlington, VA 22202.

[5] CGA E-7, *Standard for Medical Gas Regulators and Flowmeters,* Compressed Gas Association, Inc., 1235 Jefferson Davis Highway, Arlington, VA 22202.

[6] CGA E-1, *Standard Connections for Regulator Outlets, Torches and Fitted Hose for Welding and Cutting Equipment,* Compressed Gas Association, Inc., 1235 Jefferson Davis Highway, Arlington, VA 22202.

[7] CGA E-5, *Torch Standard for Welding and Cutting,* Compressed Gas Association, Inc., 1235 Jefferson Davis Highway, Arlington, VA 22202.

[8] CGA E-4, *Standard for Gas Regulators for Welding and Cutting,* Compressed Gas Association, Inc., 1235 Jefferson Davis Highway, Arlington, VA 22202.

[9] CGA E-3, *Pipeline Regulator Inlet Connection Standards,* Compressed Gas Association, Inc., 1235 Jefferson Davis Highway, Arlington, VA 22202.

[10] CGA V-6, *Standard Cryogenic Liquid Transfer Connections,* Compressed Gas Association, Inc., 1235 Jefferson Davis Highway, Arlington, VA 22202.

[11] CGA V-6.1, *Standard Carbon Dioxide Transfer Connections,* Compressed Gas Association, Inc., 1235 Jefferson Davis Highway, Arlington, VA 22202.

TABLE 8–1.

TABLE OF VALVE CONNECTION ASSIGNMENTS FOR COMPRESSED GASES IN CYLINDERS (V-1—1987).

GAS		STANDARD		LIMITED STANDARD		ALTERNATE STANDARD
		EXISTING	ADDED IN 1987	EXISTING	ADDED IN 1987	OBSOLETE 1/1/92
Acetylene	Over 50 cu. ft. (1.39 m³)	510			300, 415①	410①
	Between 35 (970 L) and 75 cu. ft. (2.08 m³)			520		
	Approx. 10 cu. ft. (280 L)			200		
Air (R729)	Up to 3000 psig (20 680 kPa) Threaded	346			590	
	Up to 3000 psig (20 680 kPa) Yoke	950		850②	855②③	
	3001-5500 psig (20 690-37 900 kPa)		347			
	5501-7500 psig (38 000-51 700 kPa)		702			④
	Cryogenic Liquid Withdrawal	440				
Allene		510				
Allylene: *See Methylacetylene*						
Ammonia (R717)	Threaded	240, 705			660	
	Yoke	800, 845				
Antimony Pentafluoride		330				
Argon	Up to 3000 psig (20 680 kPa)	580				
	3001-5500 psig (20 690-37 900 kPa)		680			677
	5501-7500 psig (38 000-51 700 kPa)		677			
	Cryogenic Liquid Withdrawal	295				
Arsine		350			660	
Bis (trifluoromethyl) Peroxide: *See Hexafluorodimethyl Peroxide*						
Boron Chloride: *See Boron Trichloride*						
Boron Fluoride: *See Boron Trifluoride*						
Boron Trichloride			660			330
Boron Trifluoride		330				
Bromine Pentafluoride		670				
Bromine Trifluoride		670				
Bromoacetone			660			330
Bromochlorodifluoromethane (R12B1)			660	165, 182		668
Bromochloromethane			660	165, 182		668
Bromoethylene: *See Vinyl Bromide*						
Bromomethane: *See Methyl Bromide*						
Bromotrifluoroethylene (R113B1)		510				660
Bromotrifluoromethane (R13B1)			660	165, 182		668
1, 3-Butadiene		510				
Butane (R600)	Gas Withdrawal	510				
	Liquid Withdrawal	555				
1-Butene		510				
2-Butene		510				

① Limited Standard for Canada only. ② Limited to SCUBA (Self-Contained Underwater Breathing Apparatus) use. ③ Was formerly CGA 1310.
④ Connection 677 which had been assigned to include high pressure air in V-1 (1977) became obsolete for air with the publication of V-1 (1987).

TABLE 8-1. (*continued*)

TABLE OF VALVE CONNECTION ASSIGNMENTS FOR COMPRESSED GASES IN CYLINDERS (V-1—1987).

GAS		STANDARD		LIMITED STANDARD		ALTERNATE STANDARD
		EXISTING	ADDED IN 1987	EXISTING	ADDED IN 1987	OBSOLETE 1/1/92
α-Butylene: *See 1-Butene*						
β-Butylene: *See 2-Butene*						
1-Butyne: *See Ethylacetylene*						
Carbon Dioxide (R744)	Threaded	320				
	Yoke	940				
Carbonic Acid: *See Carbon Dioxide*						
Carbon Monoxide		350				
Carbon Oxysulfide: *See Carbonyl Sulfide*						
Carbon Tetrafluoride: *See Tetrafluoromethane*						
Carbonyl Chloride: See Phosgene						
Carbonyl Fluoride			660			750
Carbonyl Sulfide		330				
Chlorine	Threaded				660①	
	Yoke	820				
Chlorine Pentafluoride		670				
Chlorine Trifluoride		670				
1-Chloro-1, 1-difluoroethane (R142b)		510				660
Chlorodifluoromethane (R22)			660	165, 182		668
Chloroethane: *See Ethyl Chloride*						
Chloroethylene: *See Vinyl Chloride*						
Chlorofluoromethane (R31)		510				
Chloroheptafluorocyclobutane (RC317)			660	165, 182		668
Chloromethane: *See Methyl Chloride*						
Chloropentafluoroethane (R115)			660	165, 182		668
1-Chloro-1, 2, 2, 2-tetrafluoroethane (R124)			660	165, 182		668
1-Chloro-2, 2, 2-trifluoroethane (R133a)			660	165, 182		668
Chlorotrifluoroethylene (R1113)		510				660
Chlorotrifluoromethane (R13)			660	165, 182	320	668
Cyanogen			660			750
Cyanogen Chloride			660			750
Cyclobutane		510				
Cyclopropane	Threaded	510				
	Yoke	920				
Deuterium		350				
Deuterium Chloride		330				
Deuterium Fluoride			670		660	330
Deuterium Selenide		350			660	

① For use in the Specialty Gas industry only.

TABLE 8-1. *(continued)*

TABLE OF VALVE CONNECTION ASSIGNMENTS FOR COMPRESSED GASES IN CYLINDERS (V-1—1987).

GAS	STANDARD		LIMITED STANDARD		ALTERNATE STANDARD
	EXISTING	ADDED IN 1987	EXISTING	ADDED IN 1987	OBSOLETE 1/1/92
Deuterium Sulfide	330				
Diborane	350				
Dibromodifluoroethane		660	165, 182		668
Dibromodifluoromethane (R12B2)		660	165, 182		668
1, 2-Dibromotetrafluoroethane (R114B2)		660	165, 182		668
1, 2-Dichlorodifluoroethylene		660	165, 182		668
Dichlorodifluoromethane (R12)		660	165, 182		668
1, 2-Dichloroethylene (R1130)		660	165, 182		668
Dichlorofluoromethane (R21)		660	165, 182		668
1, 2-Dichlorohexafluorocyclobutane (RC316)		660	165, 182		668
Dichlorosilane		678			330
1, 1-Dichlorotetrafluoroethane (R114a)		660	165, 182		668
1, 2-Dichlorotetrafluoroethane (R114)		660	165, 182		668
2, 2-Dichloro-1, 1, 1-trifluoroethane (R123)		660	165, 182		668
Dicyan: *See Cyanogen*					
Diethylzinc		510			750
Difluorodibromoethane: *See Dibromodifluoroethane*					
Difluorodibromomethane *See Dibromodifluoromethane*					
1, 1-Difluoroethane (R152a)	510				660
1, 1-Difluoroethylene (R1132a)	350				
Difluoromethane: *See Methylene Fluoride*					
Difluoromonochloroethane: *See Chlorodifluoroethane*					
Dimethylamine	705				240
Dimethyl Ether	510				
Dimethylhexafluoroperoxide: *See Hexafluorodimethyl Peroxide*					
2, 2-Dimethylpropane	510				
Dinitrogen Oxide: *See Nitrous Oxide*					
Dinitrogen Tetroxide: *See Nitrogen Dioxide*					
Dinitrogen Trioxide: *See Nitrogen Trioxide*					
Diphosgene		660			750
Epoxyethane: *See Ethylene Oxide*					
Ethane (R170)	350				
Ethene: *See Ethylene*					

TABLE 8-1. (*continued*)

TABLE OF VALVE CONNECTION ASSIGNMENTS FOR COMPRESSED GASES IN CYLINDERS (V-1—1987).

GAS			STANDARD		LIMITED STANDARD		ALTERNATE STANDARD
			EXISTING	ADDED IN 1987	EXISTING	ADDED IN 1987	OBSOLETE 1/1/92
Ethylacetylene			510				
Ethylamine: *See Monoethylamine*							
Ethyl Chloride (R160)				300			510
Ethyldichloroarsine				660			750
Ethylene (R1150)		Threaded	350				
		Yoke	900				
Ethylene dichloride: *See Dichloroethylene*							
Ethylene Oxide			510				
Ethyl Ether			510				
Ethyl Fluoride				660			750
Ethylidene Fluoride: *See 1, 1-Difluoroethane*							
Ethyl Methyl Ether: *See Methyl Ethyl Ether*							
Ethyne: *See Acetylene*							
Fluorine			679				
Fluoroethylene: *See Vinyl Fluoride*							
Fluoroform (R23)				660	165, 182	320	668
Fluoromethane: *See Methyl Fluoride*							
Gases in Small Cylinders: *See "Introduction" Par. 2.8*							
Germane				350		660	750
Helium	Up to 3000 psig (20 680 kPa)	Threaded	580				
		Yoke	930				
	3001-5500 psig (20 690-37 900 kPa)			680			677
	5501-7500 psig (38 000-51 700 kPa)			677			
	Cryogenic Liquid Withdrawal		792				
Heptafluorobutyronitrile				660			750
Hexafluoroacetone			330				660
Hexafluorocyclobutene				660			750
Hexafluorodimethyl Peroxide				660			755
Hexafluoroethane (R116)				660	165, 182	320	668
Hexafluoro-2-propanone: *See Hexafluoroacetone*							
Hexafluoropropylene				660	165, 182		668
Hydriodic Acid, Anhydrous: *See Hydrogen Iodide*							
Hydrobromic Acid, Anhydrous: *See Hydrogen Bromide*							
Hydrochloric Acid, Anhydrous: *See Hydrogen Chloride*							
Hydrocyanic Acid, Anhydrous: *See Hydrogen Cyanide*							

TABLE 8–1. (*continued*)

TABLE OF VALVE CONNECTION ASSIGNMENTS FOR COMPRESSED GASES IN CYLINDERS (V-1—1987).

GAS		STANDARD		LIMITED STANDARD		ALTERNATE STANDARD
		EXISTING	ADDED IN 1987	EXISTING	ADDED IN 1987	OBSOLETE 1/1/92
Hydrofluoric Acid, Anhydrous: *See Hydrogen Fluoride*						
Hydrogen	Up to 3000 psig (20 680 kPa)	350				
	3001-5500 psig (20 690-37 900 kPa)		695			677
	5501-7500 psig (38 000-51 700 kPa)		703			677
	Cryogenic Liquid Withdrawal	795				
Hydrogen Bromide		330				
Hydrogen Chloride		330				
Hydrogen Cyanide			660			750
Hydrogen Fluoride			670		660	330
Hydrogen Iodide		330				
Hydrogen Selenide		350			660	
Hydrogen Sulfide		330				
Industrial Gas Mixtures: *See CGA Pamphlet V-7*						
Iodine Pentafluoride		670				
Isoamylene: *See 3-Methyl-1-butene*						
Isobutane (R601)		510				
Isobutene: *See Isobutylene*						
Isobutylene		510				
Isopropylethylene: *See 3-Methyl-1-butene*						
Krypton	Up to 3000 psig (20 680 kPa)	580				
	3001-5500 psig (20 690-37 900 kPa)		680			677
	5501-7500 psig (38 000-51 700 kPa)		677			
Laughing Gas: *See Nitrous Oxide*						
Lewisite [Dichloro (2-chlorovinyl) arsine]			660			750
Liquid Dioxide: *See Nitrogen Dioxide*						
Marsh Gas: *See Methane*						
Medical Gas Mixtures: *See Table 2, page 23*						
Methane (R50)	Up to 3000 psig (20 680 kPa)	350				
	3001-5500 psig (20 690-37 900 kPa)		695			677
	5501-7500 psig (38 000-51 700 kPa)		703			677
	Cryogenic Liquid Withdrawal	450				
Methanethiol: *See Methyl Mercaptan*						
Methoxyethylene: *See Vinyl Methyl Ether*						
Methylacetylene		510				
Methylamine: *See Monomethylamine*						
Methyl Bromide		330			320	
3-Methyl-1-butene		510				
Methyl Chloride (R40)		510			660	

TABLE 8-1. (*continued*)

TABLE OF VALVE CONNECTION ASSIGNMENTS FOR COMPRESSED GASES IN CYLINDERS (V-1—1987).

GAS	STANDARD		LIMITED STANDARD		ALTERNATE STANDARD
	EXISTING	ADDED IN 1987	EXISTING	ADDED IN 1987	OBSOLETE 1/1/92
Methyldichloroarsine		660			750
Methylene Fluoride (R32)	320				
Methyl Ether: *See Dimethyl Ether*					
Methyl Ethyl Ether	510				
Methyl Fluoride (R41)	350				
Methyl Iodide		660			
Methyl Mercaptan	330				750
2-Methylpropene: *See Isobutylene*					
Methyl Vinyl Ether: *See Vinyl Methyl Ether*					
Monochlorodifluoromethane: *See Chlorodifluoromethane*					
Monochloropentafluoroethane: *See Chloropentafluoroethane*					
Monochlorotetrafluoroethane: *See Chlorotetrafluoroethane*					
Monochlorotrifluoromethane: *See Chlorotrifluoromethane*					
Monoethylamine (R631)	705				240
Monomethylamine (R630)	705				240
Mustard Gas [Bis (2-chloroethyl) Sulfide]		660			750
Natural Gas Up to 3000 psig (20 680 kPa)	350				
3001-5500 psig (20 690-37 900 kPa)		695			677
5501-7500 psig(38 000-51 700 kPa)		703			677
Cryogenic Liquid Withdrawal	450				
Neon Up to 3000 psig (20 680 kPa)	580				
3001-5500 psig (20 690-37 900 kPa)		680			677
5501-7500 psig(38 000-51 700 kPa)		677			
Cryogenic Liquid Withdrawal	792				
Neopentane: *See 2, 2-Dimethylpropane*					
Nickel Carbonyl		660			750
Nickel Tetracarbonyl: *See Nickel Carbonyl*					
Nitric Oxide		660			755
Nitrogen Up to 3000 psig Threaded	580			555, 590	
(20 680 kPa) Yoke	960				
3001-5500 psig (20 690-37 900 kPa)		680			677
5501-7500 psig (38 000-51 700 kPa)		677			
Cryogenic Liquid Withdrawal	295				
Nitrogen Dioxide		660			160, 755
Nitrogen Peroxide: *See Nitrogen Dioxide*					
Nitrogen Sesquioxide: *See Nitrogen Trioxide*					

TABLE 8-1. (*continued*)

TABLE OF VALVE CONNECTION ASSIGNMENTS FOR COMPRESSED GASES IN CYLINDERS (V-1—1987).

GAS	STANDARD		LIMITED STANDARD		ALTERNATE STANDARD
	EXISTING	ADDED IN 1987	EXISTING	ADDED IN 1987	OBSOLETE 1/1/92
Nitrogen Tetroxide: *See Nitrogen Dioxide*					
Nitrogen Trifluoride		330			679
Nitrogen Trioxide		660			755
Nitrosyl Chloride	330			660	
Nitrosyl Fluoride	330				
Nitrous Oxide (R744a) Threaded	326				
Nitrous Oxide (R744a) Yoke	910				
Nitryl Fluoride	330				
Octafluorocyclobutane (RC318)		660	165, 182		668
Octafluoropropane (R218)		660	165, 182		668
Oxirane: *See Ethylene Oxide*					
Oxygen Up to 3000 psig (20 680 kPa) Threaded	540				
Oxygen Up to 3000 psig (20 680 kPa) Yoke	870				
Oxygen 3001-4000 psig (20 690-27 580 kPa)		577			
Oxygen 4001-5500 psig (27 590-37 900 kPa)		701			
Oxygen Cryogenic Liquid Withdrawal	440				
Oxygen Difluoride	679				
Ozone					755
Pentaborane		350			660, 750
Pentachlorofluoroethane		660	165, 182		668
Pentafluoroethane (R125)		660	165, 182		668
Pentafluoroethyl Iodide		660	165, 182		668
Pentafluoropropionitrile		660			750
Perchloryl Fluoride	670				
Perfluoroacetone: *See Hexafluoroacetone*					
Perfluorobutane		660	165, 182		668
Perfluoro-2-butene		660	165, 182		668
Perfluorocyclobutane: *See Octafluorocyclobutane*					
Perfluorodimethyl Peroxide: *See Hexafluorodimethyl Peroxide*					
Perfluoroethane: *See Hexafluoroethane*					
Perfluoropropane: *See Octafluoropropane*					
Phenylcarbylamine Chloride	330				
Phosgene		660		160	750
Phosphine	350			660	
Phosphorous Pentafluoride	330			660	
Phosphorous Trifluoride	330			660	

TABLE 8-1. (*continued*)

TABLE OF VALVE CONNECTION ASSIGNMENTS FOR COMPRESSED GASES IN CYLINDERS (V-1—1987).

GAS		STANDARD		LIMITED STANDARD		ALTERNATE STANDARD
		EXISTING	ADDED IN 1987	EXISTING	ADDED IN 1987	OBSOLETE 1/1/92
Propadiene: *See Allene*						
Propane (R290)	*Gas Withdrawal*	510			600	
	Liquid Withdrawal	555				
Propene: See Propylene						
Propylene (R1270)		510			600	
Propyne: See Methylacetylene						
"REFRIGERANTS" — Numerical Listing						
R11: *See Trichlorofluoromethane*						
R12: *See Dichlorodifluoromethane*						
R12B1: *See Bromochlorodifluoromethane*						
R12B2: *See Dibromodifluoromethane*						
R13: *See Chlorotrifluoromethane*						
R13B1: *See Bromotrifluoromethane*						
R14: *See Tetrafluoromethane*						
R21: *See Dichlorofluoromethane*						
R22: *See Chlorodifluoromethane*						
R23: *See Fluoroform*						
R31: *See Chlorofluoromethane*						
R32: *See Methylene Fluoride*						
R40: *See Methyl Chloride*						
R41: *See Methyl Fluoride*						
R50: *See Methane*						
R112: *See 1,1,2,2-Tetrachlorodifluoroethane*						
R112a: *See 1,1,1,2-Tetrachlorodifluoroethane*						
R113: *See 1,1,2-Trichlorotrifluoroethane*						
R113B1: *See Bromotrifluoroethylene*						
R114: *See 1,2-Dichlorotetrafluoroethane*						
R114a: *See 1,1-Dichlorotetrafluoroethane*						
R114B2: *See 1,2-Dibromotetrafluoroethane*						
R115: *See Chloropentafluoroethane*						
R116: *See Hexafluoroethane*						
R123: *See 2,2-Dichloro-1,1,1-trifluoroethane*						
R124: *See 1,-Chloro-1,2,2,2-tetrafluoroethane*						
R125: *See Pentafluoroethane*						
R133a: *See 1-Chloro-2,2,2-trifluoroethane*						
R142b: *See 1-Chloro-1,1-difluoroethane*						
R143a: *See 1,1,1-Trifluoroethane*						
R152a: *See 1,1-Difluoroethane*						

TABLE 8-1. (*continued*)

TABLE OF VALVE CONNECTION ASSIGNMENTS FOR COMPRESSED GASES IN CYLINDERS (V-1—1987).

GAS		STANDARD		LIMITED STANDARD		ALTERNATE STANDARD
		EXISTING	ADDED IN 1987	EXISTING	ADDED IN 1987	OBSOLETE 1/1/92
"REFRIGERANTS"—Numerical Listing (continued)						
R160: *See Ethyl Chloride*						
R170: *See Ethane*						
R218: *See Octafluoropropane*						
R290: *See Propane*						
RC316: *See Dichlorohexafluorocyclobutane*						
RC317: *See Chloroheptafluorocyclobutane*						
RC318: *See Octafluorocyclobutane*						
R600: *See Butane*						
R601: *See Isobutane*						
R630: *See Monomethylamine*						
R631: *See Monoethylamine*						
R717: *See Ammonia*						
R729: *See Air*						
R744: *See Carbon Dioxide*						
R744a: *See Nitrous Oxide*						
R764: *See Sulfur Dioxide*						
R1113: *See Chlorotrifluoroethylene*						
R1114: *See Tetrafluoroethylene*						
R1130: *See 1,2-Dichloroethylene*						
R1132a: *See 1,1-Difluoroethylene*						
R1140: *See Vinyl Chloride*						
R1141: *See Vinyl Fluoride*						
R1150: *See Ethylene*						
R1270: *See Propylene*						
Silane	Up to 500 psig (3 450 kPa)		510			
	Up to 3000 psig (20 680 kPa)	350				
Silicon Tetrafluoride		330				
Silicon Tetrahydride: *See Silane*						
Stibine		350				
Sulfur Dioxide (R764)			660			668
Sulfur Hexafluoride			590			668
Sulfur Tetrafluoride		330				
Sulfuryl Fluoride			660			330
1,1,1,2-Tetrachlorodifluoroethane (R112a)			660	165, 182		668
1,1,2,2-Tetrachlorodifluoroethane (R112)			660	165, 182		668

TABLE 8-1. (*continued*)

TABLE OF VALVE CONNECTION ASSIGNMENTS FOR COMPRESSED GASES IN CYLINDERS (V-1—1987).

GAS	STANDARD		LIMITED STANDARD		ALTERNATE STANDARD
	EXISTING	ADDED IN 1987	EXISTING	ADDED IN 1987	OBSOLETE 1/1/92
1,1,2,2-Tetrafluoro-1-chloroethane		660	165, 182		668
Tetrafluoroethylene (R1114)	350		165, 182		
Tetrafluorohydrazine	679				
Tetrafluoromethane (R14)	580			320	
Tetrafluorosilane: *See Silicon Tetrafluoride*					
Tetramethyllead					750
Tetramethylmethane: *See 2,2-Dimethylpropane*					
Trichlorofluoromethane (R11)		660			668
Trichloromonofluoromethane: *See Trichlorofluoromethane*					
1,1,1-Trichlorotrifluoroethane		660	165, 182		668
1,1,2-Trichlorotrifluoroethane (R113)		660			668
Triethylaluminum		510			750
Triethylborane		660			750
Trifluoroacetonitrile		660			750
Trifluoroacetyl Chloride	330				
Trifluorobromomethane: *See Bromotrifluoromethane*					
Trifluorochloroethylene: *See Chlorotrifluoroethylene*					
1,1,1-Trifluoroethane (R143a)	510				
Trifluoroethylene	510				
Trifluoromethane: *See Fluoroform*					
Trifluoromethyl Chloride: *See Chlorotrifluoromethane*					
Trifluoromethyl Hypofluorite	679				
Trifluoromethyl Iodide		660	165, 182		668
Trifluorovinyl Bromide: *See Bromotrifluoroethylene*					
Trimethylamine	705				240
Trimethylene: *See Cyclopropane*					
Trimethylmethane: *See Isobutane*					
Trimethylstibine					750
Tungsten Hexafluoride		670			330
Uranium Hexafluoride	330				
Vinyl Bromide	510			290	
Vinyl Chloride (R1140)	510			290	
Vinyl Fluoride (R1141)	350				

TABLE 8-1. *(continued)*

TABLE OF VALVE CONNECTION ASSIGNMENTS FOR COMPRESSED GASES IN CYLINDERS (V-1—1987).

GAS		STANDARD		LIMITED STANDARD		ALTERNATE STANDARD
		EXISTING	ADDED IN 1987	EXISTING	ADDED IN 1987	OBSOLETE 1/1/92
Vinylidene Fluoride: *See 1,1-Difluoroethylene*						
Vinyl Methyl Ether		510			290	
Xenon	Up to 3000 psig (20 680 kPa)	580				
	3001-5500 psig (20 690-37 900 kPa)		680			677
	5501-7500 psig (38 000-51 700 kPa)		677			

TABLE 8-1. (*continued*)

TABLE OF VALVE CONNECTION ASSIGNMENTS FOR COMPRESSED GASES IN CYLINDERS (V-1—1987).

GAS		STANDARD		LIMITED STANDARD		ALTERNATE STANDARD
		EXISTING	ADDED IN 1987	EXISTING	ADDED IN 1987	OBSOLETE 1/1/92
Medical Gas Mixtures for pressures up to 3000 psig:						
Carbon Dioxide & Oxygen (CO_2 not over 7%)	Threaded	280				
	Yoke	880				
Carbon Dioxide & Oxygen (CO_2 over 7%)	Threaded	500				
	Yoke	940				
Carbon Dioxide, Oxygen, Nitrogen	Threaded		500			
	Yoke		973			
Clinical Blood Gas Mixtures	Threaded		500			
	Yoke		973			
Gas Mixtures, Medical[1][3] Nonflammable, Noncorrosive	Threaded	500				
	Yoke		973			
Helium & Oxygen (He not over 80%)	Threaded	280				
	Yoke	890				
Helium & Oxygen (He over 80%)	Threaded	500				
	Yoke	930				
Lung Diffusion Mixtures	Threaded		500			
	Yoke		973			
Nitrous Oxide & Oxygen (N_2O 47.5 to 52.5%)	Threaded	280				
	Yoke	965				
Nitrogen & Oxygen (O_2 over 23.5%)	Threaded	280				
	Yoke		890			
Xenon & Oxygen (O_2 over 20%)	Threaded		280			
	Yoke		890			

① For a definition of the term Medical Gas see paragraph 6 page 11 of the Introduction.
② Nominal mixture concentration; normal mixture tolerances are allowable.
③ Gas mixtures labeled as drugs or medical devices and not having another connection assignment.

CHAPTER 9

Compressed Gas Cylinders: Marking, Labeling, Inspection, Testing, Filling, Disposition

INTRODUCTION

Portable compressed gas containers are subject to strict regulation throughout North America. In the United States, the Hazardous Materials Regulations of the Department of Transportation (DOT) apply. These are to be found in Title 49 of the U.S. *Code of Federal Regulations.* [1] In Canada, during the second half of the 1980s, the regulatory jurisdiction began a period of transition whereby authority for most of the areas of regulation concerning compressed gases was transferred from the Canadian Transport Commission (CTC) to Transport Canada (TC), the latter of which publishes the *Transportation of Dangerous Goods Regulations.* [2] As part of this transition, Canada has adopted the Canadian Standards Association specification standard B339, *Cylinders, Spheres and Tubes for the Transportation of Dangerous Goods,* and CSA B340, *Selection and Use of Cylinders, Spheres, Tubes and Other Containers for the Transportation of Dangerous Goods.* [3] and [4]

Regulations in the United States and Canada are essentially identical with regard to safety requirements. However, some significant differences in the Canadian regulations include the elimination of the 10 percent overfilling allowance above the specified service pressure for seamless steel cylinders and the replacement of the test pressure ratio of 5/3 service pressure in the United States by a ratio of 1.5 times service pressure in Canada. The Canadian standard also includes metricated values for dimensions, stresses, pressures, and so on.

Requirements for marking a portable compressed gas container are among those given in the Hazardous Material Regulations of the U.S. Department of Transportation and the regulations of Transport Canada. [1] and [2] Such containers are constructed in accordance with a particular specification of the recognized authority. Depending on the particular type of container under consideration and the country, the recognized authorities may include the U.S. Department of Transportation, Transport Canada, the Canadian Standards Association, or the American Society of Mechanical Engineers, which publishes the *ASME Boiler and Pressure Vessel Code.* [5]

Compressed gas containers are legibly and durably marked with the DOT or TC proper shipping name to identify the material contained therein. They are usually marked by means of stenciling or labeling. Generally, the marking is located at the valve end on the cylinder shoulder or sidewall. Lettering should be of a contrasting color to the background and have a minimum height of $\frac{3}{16}$ inch (4.76 mm).

Detailed information on proper marking can be found in CGA C-4, *American National Standard Method of Marking Portable Compressed Gas Containers to Identify the Material Contained,* and CGA C-7, *Guide to the Preparation of Precautionary Labeling and Marking of Compressed Gas Containers.* [6] and [7] Compressed gas containers for use in international trade must be marked in accordance with the International Organization for Standardization (ISO) publication ISO 448, *Gas Cylinders for Industrial Use—Marking for Identification of Content.* [8] Note that none of the requirements outlined in these publications are intended to supersede applicable federal, state, or provincial regulations such as follow:

(1) *Code of Federal Regulations,* Title 49 CFR, Parts 100–199, (Transportation) [1]
(2) *Code of Federal Regulations,* Title 21 CFR, Parts 200, 201, 207, 210, and 211 (Food and Drugs) [9]
(3) *Code of Federal Regulations,* Title 29 CFR, Parts 1900–1910 (Labor) [10]
(4) *Transportation of Dangerous Goods Regulations* [2]
(5) *Regulations for the Transportation of Dangerous Commodities by Rail* [11]
(6) In Canada, regulations under the Transportation of Dangerous Goods Act, 1980, respecting the handling, offering for transport, and transportation of dangerous goods.

PRECAUTIONARY LABELING AND MARKING OF COMPRESSED GAS CONTAINERS

The compressed gas industry has for many years recognized the value of precautionary labels on portable containers of compressed gases and cryogenic liquids. These labels identify the contents of the container and warn of the principal hazards. In preparing labels, the regulatory requirements set forth by the U.S. Department of Transportation (DOT), the Food and Drug Administration (FDA), the Environmental Protection Agency (EPA), the Occupational Safety and Health Administration (OSHA), and other federal and state regulating bodies must be considered. In Canada, the appropriate Canadian authorities should be consulted. These include the Transport of Dangerous Goods Branch of Transport Canada, and Health and Welfare Canada.

The Compressed Gas Association has prepared the publication CGA C-7, *Guide to the Preparation of Precautionary Labeling and Marking of Compressed Gas Containers,* which provides details on marking and labeling requirements for portable containers not exceeding 1000 lb (454 kg) water capacity. [7] Methods of preparing label information as established by the American National Standards Institute and published in ANSI Z129.1, *American National Standard for Precautionary Labeling of Hazardous Industrial Chemicals* [12], have been followed in CGA C-7, but have been modified where necessary to meet the specific labeling needs of the compressed gas industry.

The following items represent minimum labeling compliance requirements:

1. DOT Marking
   a. DOT Proper Shipping Name. As given in DOT's Hazardous Materials Table in Title 49 Part 172 of the U.S. *Code of Federal Regulations.* [1]
   b. Hazardous Materials Product Identification Number. As shown in the DOT Hazardous Materials Table. [1]
2. DOT Labeling

   A color-coded DOT 4-inch diamond-shaped label must be applied to the container to satisfy the labeling requirements set forth in 49 CFR Part 172 Subpart E. [1]

   For flammable and nonflammable gases, a color-coded $1\frac{1}{4}$-inch diamond label, as shown in Fig. 9–1, may be used as an alternative to satisfy labeling and marking requirements when transported by private or contract motor car-

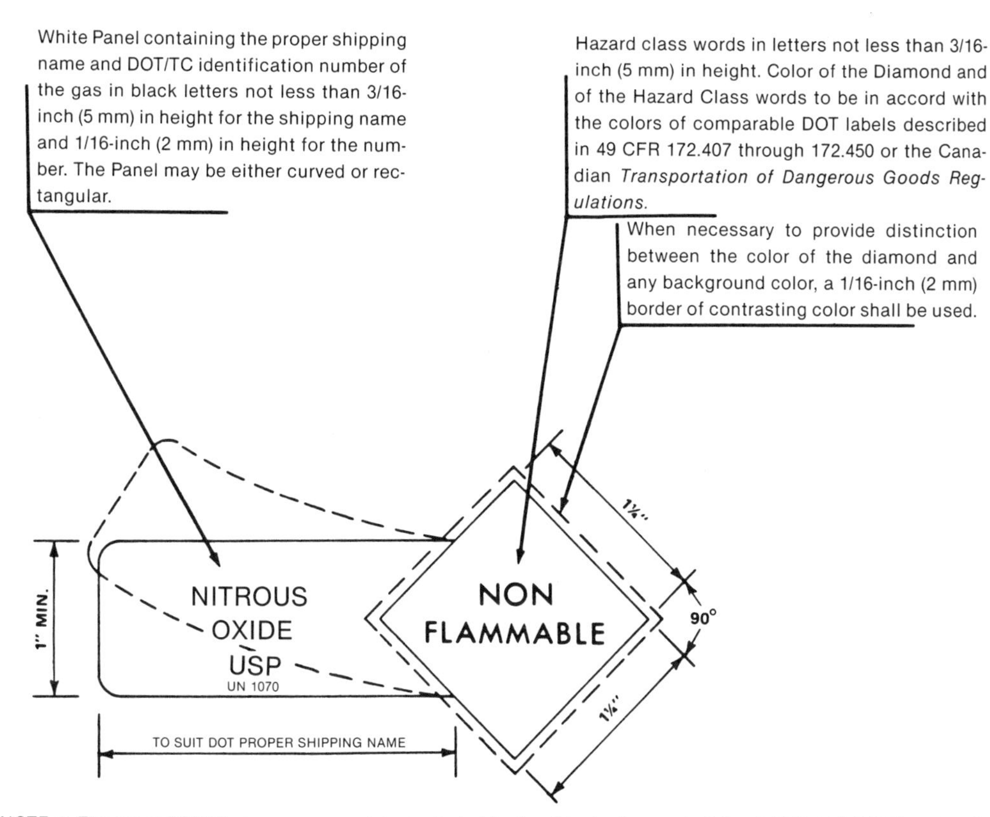

NOTE 1: The word "GAS" may or may not be included in the 1$\frac{1}{4}$-inch diamond of the BASIC MARKING or the CGA MARKING SYSTEM.
NOTE 2: Except in the case of a few specifically named poison gases, the Canadian *Transportation of Dangerous Goods Regulations* require a pictorial label without words. The class number "2" must be displaced in the bottom corner of all primary hazard labels.

Fig. 9-1. Schematic diagram for labeling of a nonflammable gas.

rier. For details on this, see Appendix A of CGA C-7, which describes the CGA Marking System. [7]

3. TC Marking and Labeling
   In Canada, labeling requirements are described in Paragraph 5.7(1) of Transport Canada's *Transportation of Dangerous Goods Regulations.* [2] A square on point (diamond-shaped) label, designated shipping name, and UN number must be used. Additional trade names or markings are permitted.
4. Hazard Warning and Precautionary Information
   An appropriate hazard warning must be shown on the label to comply with the OSHA Hazard Communication standard contained in Title 29 CFR 1910.1200. [10] Examples of hazard warnings and precautionary statements are given in the text of CGA C-7. [7] Appendix B of ANSI Z129.1 provides labeling guidelines for gases with serious chronic health hazards. [12]
5. Food and Drugs
   Compressed medical gases classified as drugs or devices require additional labeling pursuant to Title 21 CFR Part 201, labeling for drugs, and Part 801, labeling for devices. [9] Consult CGA C-7, Appendix B and C, for examples. [7] In Canada, the regulations of Health and Welfare Canada should be consulted.
6. Additional Requirements
   Some gases may be subject to specific labeling requirements set forth by OSHA, EPA, and the U.S. Department

of Agriculture or other agencies. These agencies should be consulted for their specific labeling requirements.

7. General
   Labels on compressed gases should be legible, durable, and displayed in a conspicuous place on the container so as to be easily read by the handler or user.

VISUAL INSPECTION OF COMPRESSED GAS CYLINDERS

Introduction

The Hazardous Materials Regulations of the U.S. Department of Transportation and similar regulations of Transport Canada require that cylinders be visually inspected to determine their suitability for ongoing service at a prescribed pressure. Note that for the purpose of this discussion, a tube trailer tube may be thought of as a large high pressure cylinder.

A visual inspection is used under the following circumstances:

(1) Under certain conditions of use, a formal visual inspection has been authorized in lieu of the periodic hydrostatic retest for certain low pressure cylinders dedicated for noncorrosive gas service. Specific details for this allowance can be found in DOT regulations under 49 CFR 173.34, [1], or in Canada, CSA B339 and B340. [3] and [4]

(2) As a supplement to a hydrostatic test and normal filling procedures to ensure the integrity of the cylinder or tube.

(3) Special conditions to evaluate damage from fire, misuse, etc.

This section examines inspection requirements and can serve as general guidance to cylinder owners, fillers, shippers, and users concerning the considerations involved in setting up visual inspection standard operating procedures to ensure that cylinders which do not meet the criteria for continued service will not be inadvertently returned to service.

This discussion is general in nature and is not intended to cover all conditions, sizes, types of defects, or methods of inspection. More complete and specific coverage of the subject can be found in CGA C-6, *Standards for Visual Inspection of Steel Compressed Gas Cylinders,* and 49 CFR 173.34 and 173.301. [13] and [1] For cylinders made from materials other than steel, see CGA C-6.1, *Standards for Visual Inspection of High Pressure Aluminum Compressed Gas Cylinders,* CGA C-6.2, *Guidelines for Visual Inspection and Requalification of Composite High Pressure Cylinders,* and CGA C-6.3, *Reinspection of Low Pressure Aluminum Compressed Gas Cylinders.* [14], [15], and [16]

Experience in inspecting cylinders, including the recognition of defects and the parameters of continued use, is of the utmost importance. Despite the availability of specific defect ranges, the subjective evaluation by experienced industry personnel and/or authorized inspection agencies must be an integral part of the acceptance, rejection, or downgrading cycle for continued cylinder service. Some cylinders may also be downgraded in accordance with CGA C-5, *Cylinder Service Life—Seamless, Steel, High Pressure Cylinders,* and CSA B339. [17] and [3] For more information, see CGA C-1, *Methods for Hydrostatic Testing of Compressed Gas Cylinders.* [18]

The inspection procedures are intended to identify cylinders that should be condemned due to:

(1) Leaking
(2) Internal corrosion
(3) External corrosion
(4) Bulging
(5) Metal fatigue[1]
(6) Manufacturing defects[1]
(7) Dents
(8) Internal deposits
(9) Other observable defects such as overheating in fire, arc burns, etc.
(10) Unauthorized or improper repair

[1]Certain defects are detectable only with the use of sophisticated inspection equipment.

Basic Elements of the Inspection Procedure

The basic elements involved in performing a visual inspection of a compressed gas cylinder include:

(1) Preparing the surface. This includes removal of extensive dirt, grease, etc., and may require paint removal by stripping or blasting, depending upon the nature of the inspection.
(2) Performing an external and internal visual inspection, utilizing the necessary measurement devices as required.
(3) Completing formal reporting and record-keeping procedures. See Fig. 9–2 for an example of a sample inspection report form.

Inspection Report Form

In 49 CFR 173.34 of the U.S. *Code of Federal Regulations* and equivalent Canadian regulations, DOT and TC state that results of required five-year (or when applicable, ten-year) periodic inspections of a cylinder be recorded and that a record be kept by the owner or the owner's authorized agent either until the expiration of the retest period or until the cylinder is again reinspected or retested, whichever occurs first. [1] and [2] Also, a cylinder which passes the inspection prescribed must have the date recorded on the cylinder in the manner currently prescribed for recording the retest date. Note that an *E* is to follow the date (month and year) when requalification by the visual external inspection method is allowed in lieu of hydrostatic retesting.

Definitions

Definitions for terms commonly used with respect to the visual inspection of compressed gas cylinders are as follows:

High Pressure Cylinders—High pressure cylinders are those with a marked service pressure of 900 psi (6205 kPa) or greater.

Low Pressure Cylinders—Low pressure cylinders are those with a marked service pressure below 900 psi (6205 kPa).

Condemned—No longer fit for service.

Reject—Not fit for service in present condition; however, may be requalified for service at a lower pressure by additional testing or by heat treatment, repair, or rebuilding as allowed by 49 CFR 173.34 of the DOT regulations and CSA B339, *Cylinders, Spheres, and Tubes for Transportation of Dangerous Goods* [1] and [3].

Minimum Allowable Wall Thickness—Minimum allowable wall thickness is the minimum wall thickness required by the specification under which the cylinder was manufactured.

Dents—Dents in cylinders are deformations caused by contact with a blunt object in such a way that an indentation is made without materially impairing the thickness of the metal.

Cuts—Cuts are deformations caused by contact with a sharp object in such a way as to cut into or upset the metal of the cylinder, decreasing the wall thickness at that point.

Corrosion or pitting—Corrosion or pitting in cylinders involves the loss of thickness by corrosive action. There are several kinds of corrosion or pitting, as described in the definitions that follow.

Isolated pitting—Pits of various depth and diameter but which stand alone as opposed to appearing in a group, as in line corrosion or general corrosion. Isolated pits of small diameter and shallow depth do not effectively weaken the cylinder.

Line corrosion—When pits are connected to others in a narrow band or line, such a pattern is termed *line corrosion*. This condition is more serious than isolated pitting.

Crevice corrosion—Corrosion which occurs in the area of the intersection of the footring or headring and the cylinder.

General corrosion—General corrosion is that which covers considerable surface

SAMPLE VISUAL INSPECTION REPORT FORM
Reference: DOT Regulations 173.34(e)(10)

COMPANY ____________________ DATE: __________ (Month) ______ (Year)

PLANT ____________________ RESPONSIBLE MANAGER ____________________ (Signature)

CYLINDER IDENTIFICATION					PROTECTIVE COATING		CYLINDERS INSPECTED FOR								DISPOSITION		
Serial No.	Identifying Symbol	ICC DOT Spec.	Mfg.	Date of Mfg.	Type	Condition	Corrosion and Pitting	Dents	Cuts, Digs and Gouges	Leaks	Fire Damage	Bulges	Neck Defects	Attachments	Disposition	Date Inspected	Inspector's Initials
90615	GAS INC	4BA240	ABCyl.	6-58	Paint	Good	✓	✓	✓	✓	✓	✓	✓	✓	OK	1-6-72	JHD
220196	ABC CYL	4BW240	ABCyl.	10-70	Galvanized	FAIR	✓	SC	✓	✓	✓	✓	✓	✓	SC	1-6-72	JHD
109640	OXYING	4AA280	ABCyl.	5-60	Paint	Good	✓	✓	✓	✓	✓	✓	✓	R	R	1-6-72	JHD
180015	XYZ CO	4BA290	ABCyl.	4-68	Paint	BURNED	✓	✓	✓	✓	SC	SC	✓	✓	SC	1-6-72	JHD
11225	ABCC.	4BA290	ABCyl.	9-55	Paint	FAIR	✓	✓	✓	SC*	✓	✓	✓	✓	SC	1-6-72	JHD

Disposition Code: OK – Return to Service
SC – Scrap (Condemned)
R – Hold for Authorized Repair

*REMARKS: CYL. No. 11225 - Scrapped Due to Bullet Holes

Fig. 9-2. Sample visual inspection report form.

area of the cylinders. It reduces the structural strength and is often accompanied by pitting.

Figures 9-3 through 9-7 show examples of the various types of corrosion.

Types of Inspection Methods and Equipment

Depth Gauges, Scales, and Straightedge Measurement. Exterior corrosion, denting, bulging, gouges, or digs are normally measured by simple direct measurement with scales or depth gauges. In brief, a rigid straightedge of sufficient length is placed over the defect and a scale is used to measure the distance from the bottom of the straightedge to the bottom of the defect. See Fig. 9-8. Also available are commercial depth gauges which are especially suitable for measuring the depth of small cuts or pits. It is important when measuring such depths to use a scale which spans the entire affected area. When measuring cuts, the upright metal should be removed or compensated for so that only the actual depth of metal removed from the cylinder wall is measured.

Ultrasonic Wall Thickness Measurement. There are a variety of ultrasonic, electronic wall measurement devices which will measure wall thickness and thus allow judgments to be made regarding the suitability of a cylinder for continued service.

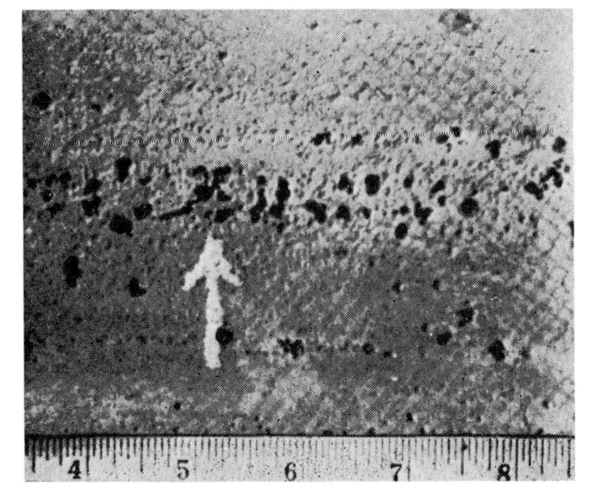

Fig. 9-4. Line corrosion.

Ultrasonic Flaw Detecting (Shearwave). This method and equipment can be utilized to detect surface and subsurface defects in cylinders.

Magnetic Particle Inspection. This method uses magnetism to find surface and subsurface discontinuities in steel and as such is adaptable to cylinder inspection to quickly locate surface and subsurface faults not readily visible to the naked eye.

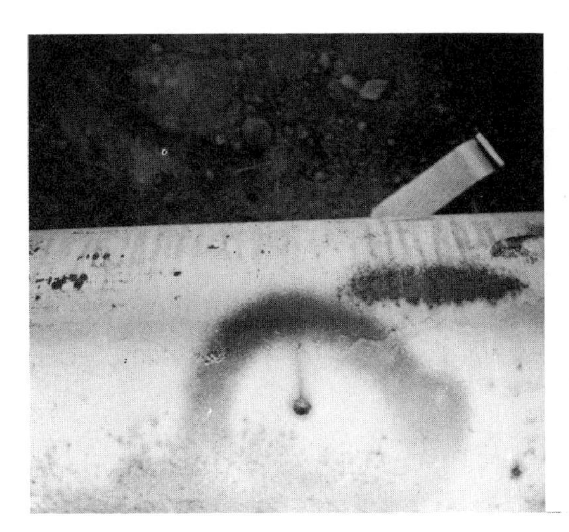

Fig. 9-3. Isolated pitting corrosion.

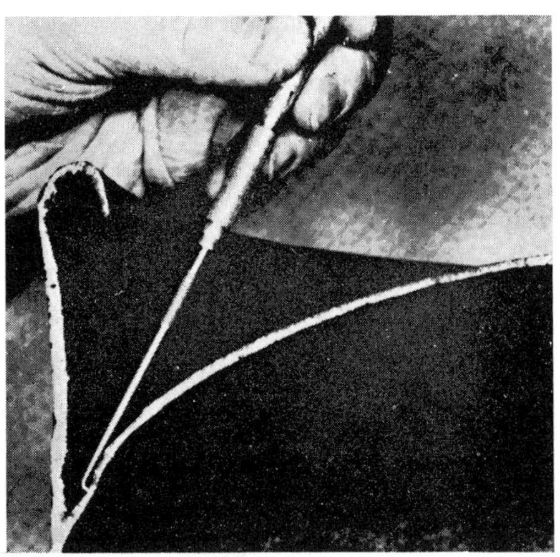

Fig. 9-5. Crevice corrosion near cylinder footring.

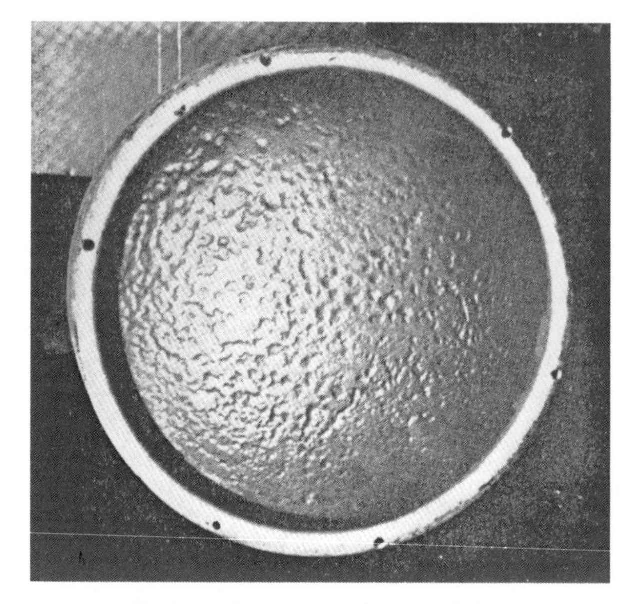

Fig. 9-6. General corrosion with pitting.

Penetrant Testing. Dye penetrant materials are available which show surface faults not readily visible to the naked eye.

Acoustic Emission Testing. Acoustic emission (AE) testing uses electronic sensors as receivers to register and locate defects in the steel detected as they grow under pressure applied to the vessel.

Leak Testing. Leak testing is the use of a solution, such as a soap solution, to observe a leak under pressure by the formation of bubbles as gas escapes from the leak.

Boroscope. The boroscope enables detailed visual inspection of internal cylinder and tube surfaces, and allows for close-up inspection of affected areas.

Nondestructive inspection methods vary from the simple straightedge to the sophisticated acoustic emission, and the method selected will depend upon the nature of the in-

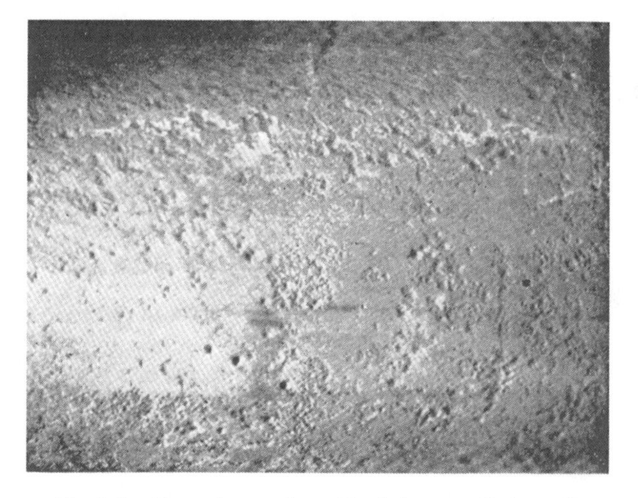

Fig. 9-7. General corrosion with pitting on cylinder wall.

Fig. 9-8. Measuring the length of a typical dent.

spection, the availability of equipment, and the training of the personnel.

Inspection Categories

Three major categories can be defined for cylinders subject to visual inspection:

- Low pressure cylinders exempt from hydrostatic testing
- Low pressure cylinders subject to hydrostatic testing
- High pressure cylinders

The first category covers cylinders exempt from hydrostatic retesting requirements of the DOT and TC by virtue of their exclusive use in certain noncorrosive gas service. They are not subject to internal corrosion and do not require internal shell inspection. If, due to unusual circumstances, internal corrosion is suspected, cylinders of these types should be hydrostatically tested.

The second category is low pressure cylinders not exempted from hydrostatic retest requirements by the applicable regulations. See 49 CFR 173.34. [1] Cylinders not exempted require a periodic hydrostatic retest, which includes an external and internal examination. Defect limits for the external examination are the same as for exempted cylinders. For nonexempted cylinders, there are additional procedures for internal inspection.

High pressure cylinders are those cylinders with a marked service pressure of 900 psi (6205 kPa) or greater. They are seamless; that is, no welding is permitted. High pressure cylinders must be inspected internally at least every time the cylinder is periodically retested.

Requirements concerning the external and internal inspection of low pressure and high pressure cylinders are discussed in the following sections. This material summarizes the inspection procedures provided in CGA C-6, *Standard for Visual Inspection of Steel Compressed Gas Cylinders.* [13] Except where noted, the focus of this presentation therefore pertains to low pressure and high pressure steel cylinders. While many of the same considerations are also pertinent to the inspection of aluminum cylinders and other types of cylinders made of materials other than steel, the respective CGA technical publications should be consulted for specific procedures and defect limitations concerning such cylinders. See CGA C-6.1, C-6.2, and C-6.3. [14], [15], and [16] Likewise, those concerned with the inspection of steel cylinders should refer to the most current edition of CGA C-6. [13]

Furthermore, there are other cylinders which are unique due to their service or construction, such as acetylene cylinders. These kinds of cylinders require special inspection procedures. See for example, CGA C-13, *Guidelines for Periodic Visual Inspection and Requalification of Acetylene Cylinders,* or CGA C-8, *Standard for Requalification of DOT-3HT Seamless Steel Cylinders.* [19] and [20] A complete listing of the technical standards published by the Compressed Gas Association pertaining to cylinders appears in Part IV of this *Handbook.*

EXTERNAL INSPECTION

A cylinder must be checked as outlined in the following paragraphs for corrosion, general distortion, or any other defect that might indicate a weakness which would render it unfit for service. Equipment should be available to allow for inspection of the cylinder bottom, as this area is especially susceptible to corrosion.

Preparation for Inspection

Cylinder sidewalls and bottoms must be in a condition that allows adequate visual inspection. If paint or other material has accumulated on the cylinder thick enough to prohibit full view of possible sidewall and bottom defects, such accumulation must be removed so that the surface can be adequately inspected. Removal of rust, scale, caked paint, or other foreign substances from the exterior surface may be accomplished by shot blasting, sand blasting, chemical stripping, sanding, or whatever approved method is available.

Corrosion Limits

It is beyond the scope of this presentation to fix corrosion limits for all types, designs, and sizes of cylinders and include them here. General criteria for low pressure and high pressure steel cylinders is given in the following paragraphs.

Low Pressure Steel Cylinders. For low pressure steel cylinders, failure to meet any of the following four general rules is cause for condemning the cylinder:

(1) A cylinder must be condemned when the tare weight is less than 90 percent of the original stamped tare weight. A cylinder must be rejected when the tare weight is less than 95 percent of the original stamped tare weight. A rejected cylinder may be requalified in accordance with 49 CFR 173.34 or CSA B339. [1] and [3] When determining tare weight, be sure that the cylinder is empty.

(2) A cylinder must be condemned when the remaining wall thickness in an area having isolated pitting is less than one-third of the original minimum allowable wall thickness.

(3) A cylinder must be condemned when line or crevice corrosion on the cylinder is 3 inches (7.6 cm) in length or over and the remaining wall is less than three-fourths of the original minimum allowable wall thickness, or when the line or crevice corrosion is less than 3 inches (7.6 cm) in length and the remaining wall thickness is less than one-half the original minimum allowable wall thickness.

(4) A cylinder must be condemned when the remaining wall in an area of general corrosion is less than one-half the original minimum allowable wall thickness.

Representative Cylinder Wall Thickness. To use the above criteria, it is necessary to know the minimum allowable wall thickness at manufacture. Table 9-1 provides the minimum allowable wall thickness for a number of common size low pressure steel cylinders.

High Pressure Steel Cylinders. For high pressure steel cylinders, the following general rules should be applied concerning inspection.

(1) A cylinder must be rejected when the remaining wall thickness in an area having crevice corrosion, line corrosion, or general corrosion is less than the minimum allowable wall thickness obtained using the maximum wall stress limitation of 49 CFR 173.302(c)(3) or equivalent Canadian regulations. [1] and [2]

(2) A cylinder must be rejected when the remaining wall thickness in an area having isolated pitting (of small cross sections only) is less than two-thirds of the average wall thickness at manufacture.

For specific examples of how to apply these corrosion limits for low pressure and high pressure steel cylinders, refer to CGA C-6, *Standards for Visual Inspection of Steel Compressed Gas Cylinders.* [13]

Dents

Considerations of appearance play a major factor in the evaluation of dents. Dents can be tolerated when the cylinder wall is not deformed excessively or abruptly. Dents are of concern where the metal deformation is sharp and confined, or, in the case of low pressure steel cylinders, where it is near a weld. Where metal deformation is not sharp, dents of larger magnitude can be tolerated.

Low Pressure Steel Cylinders. Where denting occurs so that any part of the deformation includes a weld, the maximum allowable

TABLE 9-1. ORIGINAL MINIMUM ALLOWABLE WALL THICKNESS FOR CERTAIN COMMON SIZE LOW PRESSURE CYLINDERS.

Nominal Cylinder Diameter inches	(mm)	DOT Specification Marking	Original Minimum Allowable Wall Thickness (*) inches	(mm)
4	(101.6)	4B500	.085	(2.159)
5	(127.0)	4B400	.109	(2.769)
5	(127.0)	4B240	.090	(2.286)
6	(152.4)	4B400	.090	(2.286)
6	(152.4)	4BA500	.080	(2.032)
6	(152.4)	4B500	.111	(2.819)
6	(152.4)	4B240	.090	(2.286)
6.75	(171.45)	4B300	.090	(2.286)
7	(177.8)	4BA300	.087	(2.210)
8	(203.2)	4B400	.125	(3.175)
8	(203.2)	4B240	.090	(2.286)
8	(203.2)	4B300	.105	(2.667)
8	(203.2)	4BA300	.087	(2.210)
9	(228.6)	4B240	.090	(2.286)
9	(228.6)	4BA240, 4BW240, 4BA300, 4BW300	.078	(1.981)
10	(254.0)	4BA240, 4BW240, 4BA300, 4BW300	.078	(1.981)
12	(304.8)	4B240	.105	(2.667)
12	(304.8)	4B240, 4BW240	.078	(1.981)
14.5	(368.3)	4B240	.125	(3.175)
14.5	(368.3)	4BA240, 4BW240	.087	(2.210)
14.5	(368.3)	4AA480	.185	(4.699)
14.5	(368.3)	3A480	.212	(5.384)
22	(558.8)	4B240	.191	(4.851)
22	(558.8)	4BA240, 4BW240	.130	(3.302)
24	(609.6)	4B240	.208	(5.283)
24	(609.6)	4BA240, 4BW240	.142	(3.607)
30	(762.0)	4B240	.251	(6.375)
30	(762.0)	4BA240, 4BW240	.172	(4.369)

(*) Some cylinders have thicker walls due to differences in manufacturing methods and inspection procedures. Values shown are for absolute minimums allowed by specifications. Higher values may be used if information showing thicker walls than those listed is obtained from the manufacturer of the cylinder. Use of ultrasonic testing is suggested for determining wall thickness.

dent depth is $\frac{1}{4}$ inch (0.64 cm). When denting occurs so that no part of the deformation includes a weld, the cylinder must be rejected if the depth of the dent is greater than one-tenth of the greatest dimension of the dent.

High Pressure Steel Cylinders. In general, industry practice for a 9-inch (228.6-mm) diameter × 52-inches (1295.4-mm) long high pressure steel cylinder is to accept dents up to $\frac{1}{16}$ inch (0.062 inch; 1.575 mm) depth when the major diameter of the dent is 2 inches (50.8 mm) or greater.

Cuts, Gouges, and Digs

Cuts, gouges, or digs reduce the wall thickness of the cylinder and in addition are considered to increase stress. Cuts, digs, or gouges may be measured with suitable depth gauges. Any upset metal must be smoothed off to allow true measurements without causing further damage to parent metal.

Low Pressure Steel Cylinders. For low pressure steel cylinders, depth limits are such that:

(1) Cylinders must be condemned at one-half of the limit set in (2) and (3) whenever the length of the defect is 3 inches (7.6 cm) or more.

(2) When the original wall thickness at manufacture is not known, and the actual wall thickness cannot be measured, a cylinder must be condemned if the cut, gouge, or dig exceeds one-half the minimum allowable wall thickness at manufacture.

(3) When the original wall thickness at manufacture is known, or the actual wall

thickness is measured, a cylinder must be condemned if the original wall thickness minus the depth of the defect is less than one-half the minimum allowable wall thickness.

High Pressure Steel Cylinders. Limits for cuts, gouges, or digs for high pressure steel cylinders are established by stress considerations. That is, a cylinder must be rejected when the remaining wall thickness is less than the minimum allowable wall thickness obtained using the maximum wall stress limitation defined in 49 CFR 173.302(c)(3) or equivalent Canadian regulations. [1] and [2]

Any defect of appreciable depth having a sharp bottom increases stress, and even though a cylinder may be acceptable from a stress standpoint, it is common practice to remove such defects. After any such conditioning operation, verification of the cylinder strength must be made by wall thickness measurement followed by hydrostatic testing.

Leaks

Leaks can originate from a number of sources such as defects in a welded or brazed seam, defects at the threaded opening, or defects from sharp dents, digs, gouges, or pits. Any leakage is cause for rejection. To check for leaks, the cylinder must be charged and carefully examined. All seams, pressure openings, sharp dents, digs, gouges, and pits must be coated with a soap or other suitable solution to detect the escape of gas. Any leakage is cause for rejection and possibly condemnation.

Evaluation of Fire Damage

Cylinders must be carefully inspected for evidence of exposure to fire. When inspecting for fire damage, common evidence of exposure to fire includes (a) charring or burning of paint or other protective coatings, (b) burning or scarfing of the metal, (c) distortion of the cylinder, (d) melted-out fuse plugs, or (e) burning or melting of the valve.

U.S. Department of Transportation regulations state that "A cylinder which has been subjected to the action of fire must not again be placed in service until it has been properly reconditioned," in accordance with 49 CFR 173.34. [1] Similar wording appears in TC regulations. [2] The general intent of this requirement is to remove from service cylinders which have been subject to the action of fire which has changed the metallurgical structure of the strength properties of the steel (or in the case of acetylene cylinders, has caused breakdown of the porous filler).

The evaluation of fire damage is normally determined by visual examination, with particular emphasis given to the condition of the protective coating. If there is evidence that the protective coating has been burned off any portion of the cylinder surface, or if the cylinder body is burnt, warped, or distorted, it is assumed that the cylinder has been overheated, and the requirements regarding fire damage of 49 CFR 173.34 must be complied with. If, however, the protective coating is only smudged, discolored, or blistered and is found by examination to be intact underneath, the cylinder should not be considered affected within the scope of this requirement. Hydrostatic testing ***must not*** be used to determine the extent of fire damage.

Arc and Torch Burns

Cylinders with arc or torch burns must be rejected. They may be evaluated for possible repair and requalification in accordance with 49 CFR 173.34 or CSA B339. [1] and [3] Defects of this nature may be recognized by one or more of the following conditions:

(1) Removal of metal by scarfing or cratering
(2) A scarfing or burning of the base metal
(3) A hardened heat-affected zone
(4) A deposit of weld metal or displacement of base metal

Bulges

Cylinders have a symmetrical shape. Cylinders which have definite visible bulges must be removed from service and evaluated.

Low Pressure Steel Cylinders. Bulges in low pressure steel cylinders can be measured as follows:

(1) Bulges on the cylinder sidewall can be measured by comparing a series of circumferential measurements.

(2) Bulges in the head, and also in some cases on the sidewall, can be measured by comparing a series of measurements of the peripheral distance between the valve spud and the center seam (if any) or an equivalent fixed location on the cylinder sidewall.

(3) Variations from normal cylinder contour can be measured directly by (a) measuring the height of a bulge with a scale or (b) comparing templates of bulged areas with similar areas not bulged.

Cylinders must be condemned when a variation of 1 percent or more is found in the measured circumferences or in peripheral distances measured from the valve spud to the center seam (or equivalent fixed point). An example for a 15-inch outside diameter cylinder follows:

Normal cylinder outside diameter	15 inches	(38.1 cm)
Cylinder circumference	47.12 inches	(119.68 cm)
Maximum circumference 47.12 + .01(47.12) =	47.59 inches	(120.88 cm)
Variation in circumference	0.47 inches	(11.94 mm)
Equivalent variation in diameter $0.47/\pi$ =	00.15 inches	(3.81 mm)

If the bulge is uniform around the cylinder, the limiting height of the bulge would be 0.15 inch/2 = 0.075 inch (1.90 mm) in this example.

High Pressure Steel Cylinders. High pressure steel cylinders with definite visible bulges such as bulges caused by fire damage must be condemned. Because high pressure cylinders are made of hardened steel, any evidence of bulging indicates a weakening of the strength of the steel such that the cylinder is unfit for continued service.

Neck Defects

Cylinder necks must be examined for cracks, folds, flaws, and distortion. Neck cracks are normally detected by testing the neck during charging operations by applying a soap solution and watching for bubble formation.

Cylinder neck threads must be examined whenever the valve is removed from the cylinder. When manufactured, cylinders have a specified number of full threads of proper form as required in applicable thread standards. More information on this can be found in CGA V-1, *American National, Canadian, and Compressed Gas Association Standard for Compressed Gas Cylinder Valve Outlet and Inlet Connections.* [21] When inspected, cylinders must be rejected if the required number of effective threads has been reduced so that a gas-tight seal cannot be obtained by reasonable valving methods. Common thread defects are worn or corroded crests and broken or nicked threads.

Attachments

Attachments on cylinders may lose their intended function through service abuse. These attachments and the associated portion of the cylinder must receive careful inspection. Welding is not permitted on high pressure cylinders.

The footring and headring and/or neckring on cylinders may lose their intended functions, which are, respectively: (1) to cause the cylinder to remain stable and upright and (2) to protect the valve. Rings must be examined for distortion, for looseness, and for condition of threads and, in the case of low pressure steel cylinders, for failure of welds. Appearances may often warrant removal of the cylinder from service.

When the cylinder bears a permanent attachment which covers a portion of the cylinder surface proper such as a footring, headring and/or neckring, or marking plate, it must receive particular attention during inspection to ascertain that it is in the same relation to the cylinder as at the time of its attachment. The entire region of attachment to the cylinder must be checked for possible entry of moisture to cylinder surface which cannot be seen. In the case of adhesive attachments, any evidence of a break in the seal is cause for removal of the attachment. The use of a dull probing tool is recommended. Plastic materials must be checked carefully for gouges or splits, which, if present, would also require their removal. When the cylinder bears a removable attachment, such as a removable boot, the attachment must be removed for visual inspection.

In the case of a marking plate that is not completely sealed, any evidence of corrosion between it and the wall necessitates removal of the plate and visual inspection of the cylinder wall. However, removal of the plate must be undertaken only by authorized repair facilities or original cylinder manufacturers, as stated in 49 CFR 173.34. [1]

Repair rules for a number of specification cylinders are established by DOT in 49 CFR 173.34 and parallel sections of TC regulations by use of CSA B339. [1], [2], and [3]

Inspection of Aluminum Cylinders

As previously mentioned, many of the considerations for visual inspection of steel cylinders are also applicable to aluminum cylinders. However, some differences exist with respect to acceptance criteria. For example, aluminum cylinders must be condemned when impairment to the surface (corrosion or mechanical defect) exceeds a depth where the remaining wall is less than three-fourths of the minimum allowable wall thickness required by the specification under which the cylinder was manufactured.

Also, aluminum cylinders that have been subjected to the action of fire or excessive heat must be condemned as required by 49 CFR 173.34 and equivalent Canadian regulations. [1] and [2]

CGA C-6.1, *Standards for Visual Inspection of High Pressure Aluminum Compressed Gas Cylinders,* and CGA C-6.3, *Reinspection of Low Pressure Aluminum Compressed Gas Cylinders,* provide details on the requirements for inspection of aluminum cylinders. [14] and [16] Other publications, such as CGA C-6.2, *Guidelines for Visual Inspection and Requalification of Composite High Pressure Cylinders,* provide similar information for cylinders made of other materials. [15]

INTERNAL INSPECTION

In addition to external inspection, low pressure steel cylinders not exempted from hydrostatic retesting and high pressure cylinders must be inspected internally at least every time the cylinder is periodically retested.

Hammer Test

For high pressure steel cylinders, the hammer test is a valuable indicator of internal corrosion and is a convenient test that can be made without removing the valve prior to each charging of the cylinder. The hammer test should be performed on empty unpressurized cylinders.

The hammer test consists of tapping the cylinder sidewall with a light blow using a half-pound (0.23 kg) ball-peen hammer or equivalent. A cylinder will normally have a clear ring. A dull ring would indicate internal corrosion, liquid, or accumulation of foreign material in the cylinder. Such cylinders must be inspected internally.

Preparation for Inspection

Cylinder inspection must only be undertaken if the possible dangers associated with the contents and pressure of the cylinders are recognized and proper precautions are taken.

Warning: Position the cylinder to prevent the safety nut from causing injury to personnel or damage to equipment if it becomes totally disconnected from the valve. Remove the valve from the cylinder only after making certain that the cylinder is empty. Cylinders containing flammable or toxic gas must be properly purged in a safe area using inert gas or water.

The interior of the cylinder is prepared for inspection by the removal of internal scale, dirt, or other condition as necessary to permit the inspection of the internal surface. Cylinders with interior coating must be examined for defects in the coating. If the coating is defective, it must be removed.

Note: The Chlorine Institute, 2001 L St., NW, Ste. 506, Washington, DC 20036, Pamphlet No. 17, *Cylinder and Ton Container Procedure for Chlorine Packaging,* includes procedures for internal examination of chlorine cylinders at the time of each filling.

A good inspection light of sufficient intensity to clearly illuminate the interior walls is mandatory for internal inspection. Flammable gas cylinders must be purged with inert gas or water before being examined with a light. Liquefied gases can cause problems if the cylinder contents are not completely removed prior to purging. Cylinders containing other types of hazardous materials must also be purged to remove residual gas or liquid before being examined with a light. Upon venting, these gases can cool down enough to reduce the gauge pressure such that cylinders will appear to be empty. Warming and/or inverting the cylinder will assist in the removal of residual liquefied gases prior to purging.

General Corrosion

Interior corrosion is best evaluated by a hydrostatic test combined with careful visual inspection. Thickness-measuring and flaw detection devices of the ultrasonic type may be used to evaluate specific conditions. Basic corrosion limits for both low and high pressure steel cylinders were discussed under the External Inspection section of this chapter.

Localized Pitting or Line Corrosion

Localized pitting or line corrosion may not be detected by the hydrostatic test. These types of corrosion cause significant localized stresses. These stresses are detrimental, and the cylinder must be removed from service for further evaluation with respect to corrosion limits as discussed earlier in this chapter.

Internal Defects Other than Corrosion

Any cylinder must be rejected and further evaluated when doubt exists as to its suitability for continued service. Where the bottom of the defect cannot be seen or where its extent cannot be measured by various inspection instruments, the cylinder must be condemned. Examples of such internal defects are cuts, mechanical abrasions, and fabrication irregularities.

HYDROSTATIC TESTING OF CYLINDERS AND TUBES

Introduction

U.S. Department of Transportation regulations in Title 49 of the U.S. *Code of Federal Regulations* at 49 CFR 173.34 state requirements concerning the hydrostatic testing of compressed gas cylinders by water jacket or other suitable methods that are operated in a manner that provides accurate data. [1] In Canada, the regulations of Transport Canada require similar methods. [2] This section briefly describes the methods in use. In general, the methods for hydrostatic testing are used to determine the total expansion, permanent expansion, and percent expansion of cylinders under specified test pressure conditions. These parameters must be determined unless otherwise noted by regulation or specification. The test apparatus must be approved with respect to type and operation by DOT or TC, as applicable.

Detailed regulations concerning the retesting of cylinders and tubes are to be found in 49 CFR 173.34 and equivalent Canadian regulations. CGA C-1, *Methods for Hydrostatic Testing of Compressed Gas Cylinders,* provides a thorough presentation on this topic. [18] In Canada, CSA B338, *Highway Tanks and Portable Tanks for the Transportation of Dangerous Goods,* should be consulted. [22] The DOT regulations also outline the procedure for applying for approval of test equipment. The presentation provided here is only a brief overview of the methods for hydrostatic testing. Those who require more thorough information must refer to the appropriate DOT or TC regulations and to CGA C-1. [1], [2], and [18]

Methods used for hydrostatic testing of compressed gas cylinders and tubes are as follows:

- Water jacket volumetric expansion method
- Direct expansion method
- Pressure recession method
- Proof pressure method

For any of the methods for hydrostatic testing, while the design and details of the apparatus may be adapted to suit individual requirements for a particular installation, the safety recommendations outlined in CGA C-1, *Methods for Hydrostatic Testing of Compressed Gas Cylinders,* should be followed. [18] Only properly trained personnel should be employed to perform the hydrostatic testing.

Inspection of Cylinders

Regardless of the type of hydrostatic testing method used, DOT and TC regulations both specify that the periodic retest must include an external and internal visual examination of the cylinder. It is recommended that these inspections be conducted prior to the hydrostatic retest in accordance with the applicable Compressed Gas Association standards. See references [13] through [16] and [20]. External and internal inspection considerations were discussed more thoroughly in previous sections of this chapter.

Experience in the inspection of cylinders is an important factor in determining the acceptability of a given cylinder for continued service. Those lacking this experience and having cylinders of doubtful acceptability should return them to a manufacturer of the same type of cylinders or to a competent requalification agency for reinspection.

Each high pressure steel cylinder, standing alone, must be given a hammer test by tapping lightly with a metallic object, such as a $\frac{1}{2}$-lb (0.23-kg) machinist hammer, wrench, or equivalent. Any cylinder which has a dull or dead ring must be cleaned, and if the dull or dead ring persists, the cylinder must be condemned. The hammer test is not applicable to aluminum, low pressure, or fiberglass-reinforced plastic (FRP) cylinders.

A careful internal inspection must be made with a suitable light. If the cylinder is seriously corroded, it must be rejected. To avoid a possible flash, cylinders used in a flammable gas service must be thoroughly purged before the interior inspection light is inserted.

Water Jacket Volumetric Expansion Method

Water jacket volumetric expansion is the standard method of testing high pressure cylinders in the compressed gas industries. This method is applicable to all hydrostatic tests when volumetric expansion determinations are required, that is, when expansion, elastic expansion, permanent expansion, and percent permanent expansion measurements are required. It consists of enclosing the cylinder, full of water, in a vessel completely filled with water. The measurement is done with a suitable device, such as a leveling burette, attached to the jacket vessel. This measures the volume of water forced from the jacket upon the application of pressure to the interior of the cylinder, which causes expansion of the cylinder, and the volume remaining displaced upon release of the pressure. These volumes represent the total and permanent expansions of the cylinder, respectively.

This method is also used to determine accurately the elastic expansion, which is directly related to the average wall thickness of a cylinder. In general, an increase in elastic expansion indicates reduction of average wall thickness. A cylinder that is handled properly will retain its original condition unless physically damaged or attacked by corrosion.

A schematic diagram for the testing equipment is shown in Fig. 9-9. Data from a typical water jacket test are shown in Table 9-2.

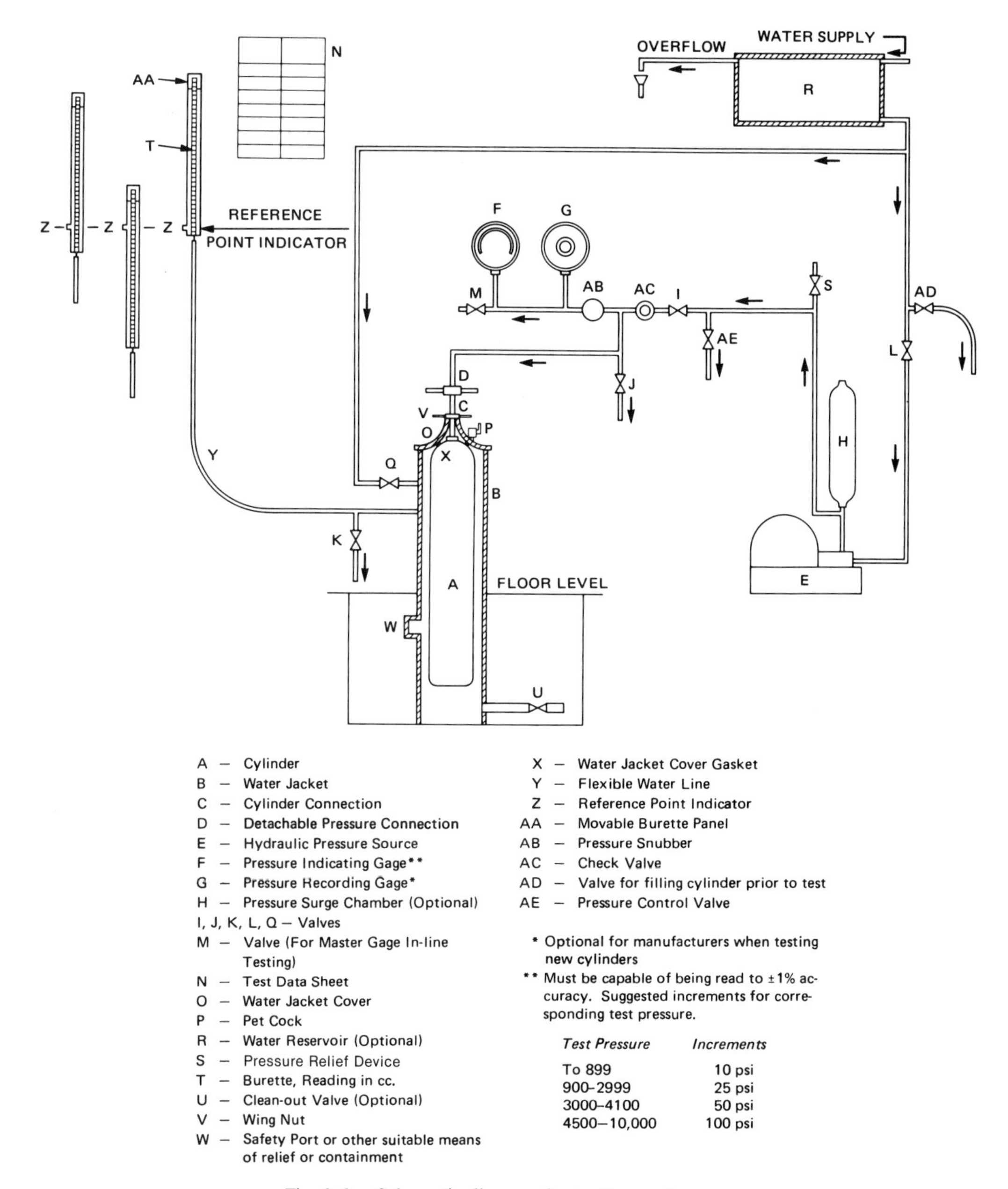

Fig. 9-9. Schematic diagram for testing equipment.

Direct Expansion Method

The direct expansion method is applicable to all hydrostatic tests when volumetric expansion determinations are required. However, it has practical limitations in its use. Although the elastic expansion is also measured in this method, regulations of both the United States and Canada forbid this method to be used to qualify those cylinders that may be charged to 10 percent in excess of marked service pressure as provided for in 49 CFR 173.302, [1], and in Canada, for "+" marked cylinders or TC specification cylinders tested at one and one-half times the service pressure.

The direct expansion method determines

TABLE 9-2. DATA FROM A TYPICAL WATER JACKET TEST.

B.E. APPROVAL NO. 1234
ASSIGNED RETEST SYMBOL ZYX

XYZ OXYGEN CO.
THIS TOWN, THAT STATE 00000

PLANT CHART NO. 10
TEST DATE(S) 2-75

	Serial Number	Identifying Symbol	Size	DOT/ICC Rating	Test Pressure	Volumetric Expansion			Visual Insp.	Disposition*	Tested By	Remarks
						Total	Perm.	Elastic				
1	Calibrated Cylinder				3360						SS	
2	X-1234	XYZ	9 x 51	3A2015	3360	157	1	156	OK	a		
3	X-5432	XYZ	"	3AA2015	"	—	—	—	OK	b		Neck Leaks
4	123456	ABC	"	"	—	—	—	—	C	c		Excessive Corrosion
5	C-6789	CDE	"	3A2015	3360	195	21	174	OK	c		P.E. over 10% of T.E.
6	236567	GHI	"	"	"	188	2	186	OK	a		Not Plus Marked
7	K-1456	ABC	7 x 32	"	"	58	2	56	OK	a		
8	65432R	XYZ	7 x 43	"	"	71	0	71	OK	a		
9	X-5432	XYZ	9 x 51	3AA2015	3460	184	1	183	OK	a		Passed 2nd Test
10	D-2345	ZYX	4 x 17	3AA2015	3360	12.7	.3	12.4	OK	a		
11	Calibrated Cylinder				4000						NN	
12	Z-492	XYZ	9½ x 56	3AA2400	4000	225	1	224	OK	a		
13	Z-495	"	"	"	"	227	2	225	OK	a		
14	Calibrated Cylinder				3775							
15	Y10555	XYZ	9 x 51	3AA2265	"	200	2	198	OK	a		
16	Y10554	XYZ	"	"	"	207	0	207	OK	a		
17	K14999	ABC	7 x 32	3A2015	—	—	—	—	C	c		Failed Hammer Test

*DISPOSITION CODE
a — Return to service
b — Set aside for further tests
c — Scrap
d — Set aside for heat treatment
e — Other (specify) ____________

I hereby certify that all the above tests were made under my supervision and in accordance with DOT Regulations.

(Signed)____________

the total expansion by measuring the amount of water forced into a cylinder to pressurize it to test pressure. The permanent expansion is measured by the amount of water expelled from the cylinder when the pressure is released. This method consists of forcing a measurable volume of water into a cylinder filled with a known weight of water at a known temperature, and measuring the volume of water expelled from the cylinder when the pressure is released.

The permanent volumetric expansion of the cylinder is calculated by subtracting the volume of water expelled from the volume of water forced into the cylinder. The total volumetric expansion of the cylinder is calculated by subtracting the compressibility of the volume of water forced into the cylinder to raise the pressure to the desired test pressure.

A schematic diagram for this type of testing equipment is shown in Fig. 9-10.

Pressure Recession Method

The pressure recession method consists of subjecting the cylinder rapidly to hydrostatic test pressure, then immediately cutting off the pressure supply and observing the recession of pressure in the cylinder due to permanent expansion. Since the expansion of the cylinder is not measured by burette, this method must not be used to qualify cylinders that may be charged to 10 percent in excess of marked service pressure in accordance with 49 CFR 173.302, [1], or where, in Canada, TC specification cylinders are tested to one and one-half times the service pressure or equivalent specification cylinders are marked with a "+" sign.

The pressure recession method may be used for testing cylinders that require a test pressure of 2000 psi (13 789 kPa) or more. The method consists essentially in subjecting the cylinder to the required hydrostatic test pressure, then immediately cutting off further pressure supply and observing for at least two minutes whether or not there occurs a recession of the pressure in the cylinder. If the pressure does not recede, this indicates that at the test pressure the cylinder does not show any permanent expansion.

A schematic diagram for this type of testing equipment is shown in Fig. 9-11.

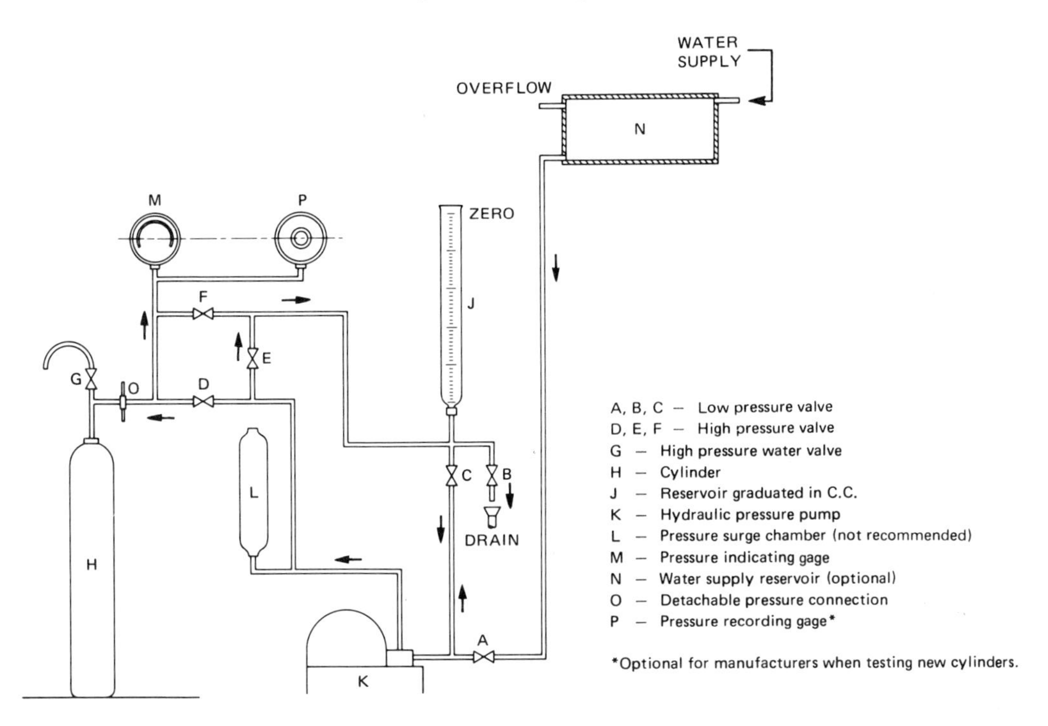

Fig. 9-10. Typical schematic diagram for direct expansion testing equipment.

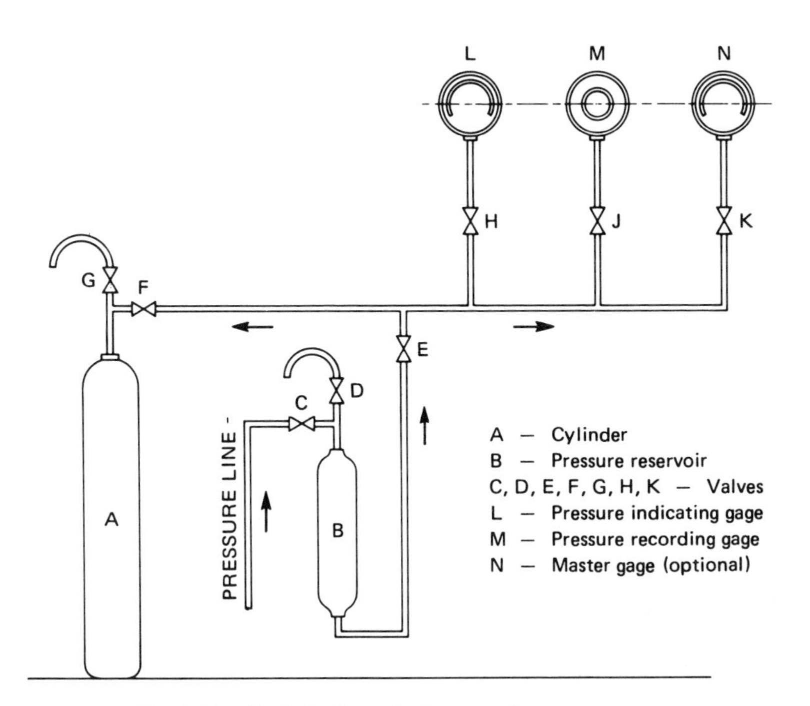

Fig. 9-11. Typical schematic diagram of pressure recession method testing equipment.

Proof Pressure Method

The proof pressure method is permitted where DOT or TC regulations and specifications do not require the determination of total and permanent volumetric expansion of the cylinder. It consists of examining the cylinder under test for leaks and defects.

The test consists essentially of applying an internal pressure equal to that stipulated for the volumetric expansion test and determining whether any weaknesses or leakage exists.

A schematic diagram for this type of testing equipment is shown in Fig. 9–12.

Other Methods

Acoustic emission testing using hydrostatic and/or pneumatic pressure is presently allowed by DOT exemption only and therefore will be addressed only by definition in this chapter as follows: Acoustic emission is defined as the transient elastic waves generated by the rapid release of energy within a material. Acoustic emission technology is simply a method that listens to abnormal sounds within a material, structure, or process, via piezoelectric sensors. The sound is analyzed, processed, displayed, and saved. Criteria are established to evaluate the emission located by this method.

FILLING COMPRESSED GAS CYLINDERS

Prefill Inspection of Compressed Gas Cylinders

Due to the vast diversity of uses and requirements associated with compressed gas cylinders, a prefill inspection of each cylinder is a necessary precaution. This should indicate any corrections or repairs that are required prior to filling. The following paragraphs describe operations that should be included in the prefill process.

DOT and TC regulations specify that cylinders cannot be filled without first determining the ownership of the cylinder. This can be established by means of the cylinder marking and/or written permission of the owner on file at the filling location.

The cylinder markings that require inspec-

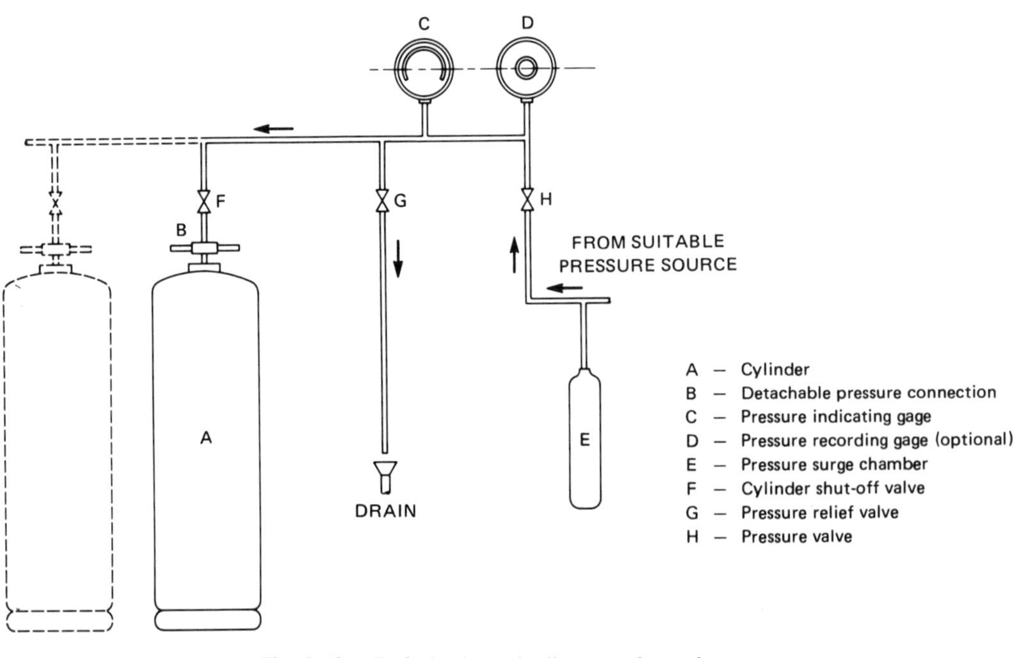

Fig. 9–12. Typical schematic diagram of proof pressure method testing equipment.

tion prior to each filling include the cylinder specification (e.g., ICC, DOT, and CTC or TC), marked service pressure, and the latest and original test dates. The latest retest date must be checked to assure that the cylinder does not require retesting prior to filling. The original test date provides information on the required frequency of testing, which may change after 35 years.

Retest periods may vary depending upon use, material of construction, and the cylinder specification. Many steel cylinders must be retested every five years. DOT and TC regulations allow for Specification 3A and 3AA cylinders to be retested every 10 years provided the cylinder meets the regulatory criteria in 49 CFR 173.34, which are summarized in CGA P-15, *Filling of Industrial and Medical Nonflammable Compressed Gas Cylinders* [1] and [23] Aluminum cylinders must be retested every five years and fiber-reinforced high pressure cylinders every three years.

An external damage inspection must be performed to ensure the cylinder is free of arc burns, evidence of exposure to fire or excessive heat, cuts, digs, gouges, dents, corrosion, and pitting. Also, cylinders must not be bulged, out of round, bowed, and so on. If the cylinder has a flat bottom it must stand upright without leaning. The criteria for steel cylinders is given in CGA C-6, *Standards for Visual Inspection of Steel Compressed Gas Cylinders.* [13] Other CGA publications provide criteria for other types of cylinders. [14], [15], and [16]

A cylinder valve inspection must be performed on each cylinder to ascertain that the valve is suitable for use and the intended gas service. The valve should be free of oil or other hydrocarbon substances that will present a filling hazard, particularly with oxidizing gases. The valve should operate properly and smoothly. The outlet should be in good condition with undamaged threads and seat.

Valve parts such as valve handwheels and valve stems should not be broken or bent and must be intact. Valves with tapered threads at the cylinder connection should have a portion of threads exposed to ensure the valve has not reached the end of its service life. See Chapter 8 for information on proper valving procedures. Each outlet should be the proper connection for the intended gas service, as prescribed in the current edition of CGA V-1, *American National Canadian, and Compressed Gas Association Standard for Compressed Gas Cylinder Valve Outlet and Inlet Connections.* [21]

Each pressure relief device assembly must be examined to ensure it is free of visual indications of abuse, damage, extrusion of fusible metal, or a disk having burst, and that it is intact. Each must be the appropriate type and pressure rating for the gas and cylinder service as prescribed in DOT and TC regulations and CGA S-1.1, *Pressure Relief Device Standards—Part 1—Cylinders for Compressed Gases.* [1], [2], and [24]

Cylinders with neckrings must be examined to determine that the threads are not excessively worn or damaged and that the neckring is firmly attached to each cylinder.

An examination for cleanliness of each valve and the exterior wall of cylinders must be performed to ascertain the absence of oil, grease, or any other hydrocarbon substance. This is particularly important for oxidizing gases to prevent a fire or explosion that could result if a hydrocarbon substance came in contact with an oxidizer.

A test for internal contamination of any residual gas must be performed on each oxygen and breathing gas cylinder prior to filling. The most common type of test is one described as a "sniff test" performed by the fill operator. Alternate detection methods may also be used. Sniff test procedures are described in CGA P-15, *Filling of Industrial and Medical Nonflammable Compressed Gas Cylinders.* [23] Any cylinder which has an odor detected during this inspection must not be filled, but set aside for removal of the contaminant.

Danger: Cylinders which may contain nitrous oxide, toxic, or flammable gases should not be subjected to a sniff test.

The hammer test, or "dead-ring" test, is one of the requirements of DOT and TC regulations for ten-year retest criteria for Specification 3A and 3AA cylinders. The hammer test is a valuable indicator of internal corrosion and may indicate accumulation of liquid and foreign material in the cylinder. This test is generally performed by using a half-pound (0.23-kg) ball-peen hammer or equivalent to tap the cylinder sidewall with a light blow. Cylinders which produce a dull ring must be marked and set aside for further evaluation. A hammer test should not be performed on low pressure steel cylinders or high or low pressure aluminum or composite cylinders, and it is not valid for clustered cylinders.

Filling Nonflammable, Nonliquefied Compressed Gases

Among the gases in the nonflammable, nonliquefied category are noncryogenic nitrogen, oxygen, helium, and argon. Cylinders that have passed the prefill inspection and are to be filled with such gases are connected to the proper manifold. During the connection process, the manifold pigtails, flexible hoses, and station valves are inspected to ensure they are clean and undamaged.

The cylinder content must be identified by labeling. Inappropriate, obsolete, illegible, or damaged labels are removed. When required, new product labels may be applied to each cylinder before, during, or immediately after filling, but each cylinder must be labeled before being removed from the filling manifold area. For recommended product labeling information, refer to CGA C-7, *Guide to the Preparation of Precautionary Labeling and Marking of Compressed Gas Containers.* [7]

Residual cylinder content should be confirmed or disposed of. The method of disposal of residual content will depend on the product and applicable environmental regulations. As necessary, an appropriate disposal system should be set up. The following procedure would generally apply to cylinders containing atmospheric gases. Each cylinder should be vented unless the identification and purity have been confirmed by analysis. Residual gas in all medical gas cylinders must be vented. To vent, each cylinder valve and the manifold vent valve are opened and all residual gas is vented to the atmosphere outside the building, away from the building air intakes. The manifold vent valve is closed when the pressure gauge on the manifold indicates the cylinders are empty. The vacuum pump is started (unless the pump is continuously running), and the manifold vacuum valve is opened slowly.

Warning: Vacuum pumps used for oxygen service must be approved for oxygen service. This includes pumps equipped with fluorocarbon oils or water aspirators. Never evacuate oxygen with a hydrocarbon-filled vacuum pump.

The cylinders are evacuated to a minimum vacuum of approximately 25 inches of mercury (Hg) or 635 mm Hg. The vacuum valve is closed and then the pump shut down (unless the pump is designed for continuous running). The manifold supply valve is opened and filling proceeds. Filling should be based on the pressure/temperature tables found in Appendix 1 of CGA P-15, *Filling of Industrial and Medical Nonflammable Compressed Gas Cylinders.* [23]

Note: DOT and TC Specification 3A and 3AA steel cylinders having a plus rating ("+" after the most recent retest date), which do not contain a pressure relief assembly with a fusible metal-backed frangible disk, may be filled to 110 percent of the marked service pressure, when filled with an atmospheric gas.

During the filling process, there should be a check for the absence of heat of compression. This is performed by placing the bare hand on each cylinder sidewall. A cylinder's sidewall that is colder than other cylinders on the manifold indicates the cylinder is not being filled at the same rate as the other cylinders. The valve must be closed and not reopened on any cylinder that does not warm up from heat of compression. The cylinder should be marked and set aside for further evaluation upon completion of the filling operation.

At approximately 300 psig (2068 kPa), the

fill connections and cylinder valves should be inspected for leaks. Leaks are detected by utilizing a compatible leak detection solution or leak detection instrument. A cylinder that leaks should be marked and isolated for further evaluation. When the correct pressure is attained based on cylinder temperature, the manifold supply valve is closed, as is each cylinder valve. The vent valve is opened to release manifold pressure. Each cylinder is disconnected from the manifold. A leak check of each cylinder valve, including the outlet and pressure relief device, is again performed.

Filling Nonflammable, Liquefied Compressed Gases

Gases in the nonflammable, liquefied compressed gas category include carbon dioxide, nitrous oxide, and others. Liquefied gases must be filled by weight, not by pressure. Note that this section does not apply to Specification TC/DOT-4L cylinders.

Prior to filling, scales must be calibrated according to applicable regulations, or at least once a year. They should be checked for accuracy each day using a "check weight." A record documenting daily accuracy checks should be maintained. The scale(s) must have sufficient range and sensitivity for the cylinder sizes being weighed.

Cylinder content must be identified by labeling. Any inappropriate, obsolete, illegible, or damaged labels must be removed. New product labels, if required, should be applied before filling. For recommended product-labeling information, refer to CGA C-7, *Guide to the Preparation of Precautionary Labeling and Marking of Compressed Gas Containers.* [7]

Cylinders in this service should have the tare weight stamped on each cylinder. After passing the prefill inspection, each cylinder is weighed (without cylinder cap) to determine if the cylinder weight matches the tare weight, within specified limits. An overweight cylinder indicates possible contamination or the presence of residual product. An underweight cylinder may indicate excessive corrosion. A cylinder with either condition requires further examination before filling.

Residual cylinder content should be confirmed or disposed of. The method of disposal of residual content will depend on the product and applicable environmental regulations. As necessary, an appropriate disposal system should be set up. Each cylinder should be vented unless the identification and purity have been confirmed by analysis. Venting should be directed outside the building, away from building air intakes. Residual gas or liquid in all medical cylinders must be vented. Each vented cylinder is evacuated to a minimum vacuum of approximately 25 inches of mercury (Hg) or 635 mm Hg. The vacuum valve and each cylinder valve are closed, and the pump is then shut down (unless the pump is designed for continuous running).

Warning: Vacuum pumps used for an oxidizing gas service must be approved for oxidizing gases. This includes pumps equipped with fluorocarbon oils or water aspirators. Never evacuate an oxidizer with a hydrocarbon-filled vacuum pump.

Cylinders are connected one at a time to the proper filling pigtail or transfer hose. During the connection process, the connection apparatus is inspected to ascertain that it is clean and undamaged. The cylinder is weighed after connection to ensure the weight of the connection apparatus is taken into account in the weighing operation. With the cylinder valve closed, the pigtail vent valve (if provided) is opened and the pigtail shutoff valve is cracked open to purge the line; then the pigtail vent valve is closed. The cylinder valve is opened and filling proceeds. Filling is based on the applicable filling density of the product to be filled. Applicable filling densities are specified in 49 CFR 173.304 or equivalent Canadian regulations. [1] and [2]

When the scale indicates the cylinder has reached the final fill weight, the valve on the cylinder is closed. The pigtail shutoff valve is closed and the pigtail vent valve (if provided) is opened before the cylinder is disconnected. The cylinder is disconnected and the weight of the filled cylinder checked. If the cylinder is overfilled, excess product must be immediately vented.

Caution: Liquefied gas cylinders, if overfilled (''liquid full''), can rupture due to hydrostatic pressure. Never exceed the calculated legal fill weight.

Caution: Liquefied gases remaining in the pigtail or transfer hose, if not properly vented, will expand and may develop excessive pressure, causing rupture and/or a whipping action of the pigtail or transfer hose. Do not allow liquefied gases to remain in the pigtail or transfer hose.

The cylinder valve outlet, pressure relief device, and cylinder-to-valve connection are leak tested. Leaks are detected using a compatible leak detection solution or leak detection instrument. A cylinder that leaks should be marked and set aside for further evaluation.

Quality Control

Product analysis should be performed to make sure the product meets required specifications. A written procedure explaining analytical requirements, minimum specifications, and necessary documentation should be available. Records should be retained as necessary.

Analytical testing and record-keeping requirements for medical gas products are more detailed than above. They are subject to the requirements of the United States Food and Drug Administration or Health and Welfare Canada, and must comply with good manufacturing practices. Medical gas product may be declared unsuitable for shipment in the United States if any of the required records thereof are missing, even if the final product passes the required specifications.

DISPOSITION OF UNSERVICEABLE CYLINDERS WITH KNOWN CONTENTS

This section discusses the recommendations for disposal of compressed gas cylinders which are no longer considered to be serviceable. Some of these unserviceable cylinders failed to qualify for further use under the maintenance requirements of the U.S. Department of Transportation or those of Transport Canada. In other cases, cylinders are occasionally found that appear to have been out of service for a long time, are inadequately marked, and are considered unsafe for further use. In the latter case, the cylinders may be either empty or charged with gas.

Information on how to handle these types of cylinders can be found in CGA C-2, *Recommendations for the Disposition of Unserviceable Compressed Gas Cylinders with Known Contents.* [25] Readers who may encounter unserviceable cylinders are encouraged to study and understand the complete text of CGA C-2. Users who have unserviceable cylinders are cautioned against disposing of cylinders that either are empty or contain product. All possible means should be taken to determine the supplier of the cylinder and return the cylinder to the supplier.

Nonreusable (nonrefillable) containers such as those made to DOT or TC Specifications 39, 40, and 41 are considered by DOT and TC regulations to be unserviceable after one use and therefore must not be refilled or reused for any purpose. Disposal of such cylinders should be in accordance with recommendations obtained from the suppliers who initially filled the cylinders. Such recommendations may sometimes be included in the labeling of the cylinders.

Disposal Should Be by Qualified Personnel Only

The proper and safe disposition of unserviceable compressed gas cylinders is important as very substantial potential hazards may exist that must be recognized and evaluated by those who attempt to dispose of them. Where the contents of the cylinder are unknown and there is no ready means for identifying its properties, the hazard is especially great. The disposal of cylinders with unknown contents is a very complex subject and should only be attempted by specially trained experts. For cylinders in this category, the supplier should be contacted immediately. ***The user should never attempt to dispose of a cylinder with unknown contents.*** The disposal of unserviceable compressed gas cylinders is potentially hazardous because they may contain:

(a) Gas under pressure
(b) Flammable gas
(c) Explosive mixtures
(d) Poisonous or toxic materials
(e) Corrosive, oxidizing, or reactive materials

Individuals or organizations such as scrap dealers, fire services, military organizations, and others who may have reason to dispose of an unserviceable compressed gas cylinder should first attempt to locate the supplier of the cylinder and have the supplier assume responsibility for disposal. If the supplier cannot be determined, such organizations should acquaint themselves with the names and addresses of the nearest manufacturers or distributors of the type of compressed gas or gases in question. The manufacturer or distributor should be requested to remove the cylinders for appropriate disposal. If this is not feasible, the manufacturer or distributor should be requested to provide appropriate instructions and supervision for the safe disposition of unserviceable cylinders with known contents.

The recommendations in CGA C-2 were primarily developed from a safety standpoint and are not designed to assure compliance with applicable federal, state, provincial, or local laws and regulations, including environmental and other standards, which may vary from place to place and over time. Full responsibility for determining applicable regulatory requirements and ensuring compliance therewith rests with the user.

In general, the disposal of all unserviceable compressed gas cylinders requires adherence to the following procedures:

(1) Identify the contents.
(2) Safely purge the contents.
(3) Remove the valve.
(4) Obliterate all markings.
(5) Destroy the cylinder.
(6) Properly discard the cylinder.

It is absolutely essential that the cylinder contents be identified before steps for the disposition of the cylinder are taken.

Suppliers of compressed gases mark their cylinders with product names. These names may be either the proper chemical names or a commonly accepted name such as a trade name. Refer to CGA C-4, *American National Standard Method of Marking Portable Compressed Gas Containers to Identify the Material Contained.* [6] Do not rely upon color coding to determine contents. If the contents cannot be determined, the cylinder should be returned to the manufacturer or distributor. When a cylinder bears adequate product labeling it is reasonable to rely upon the labels for the identity of the cylinder contents.

In order to determine whether there is gas in the cylinder under pressure, the valve should be opened slightly and immediately closed to determine by sound, soap suds, or pressure gauge whether any gas has escaped. ***Due consideration should be taken regarding the hazardous characteristics of the contained gas.***

This should only be done in a safe area, utilizing proper precautions for the gas discharged, and with the cylinder properly supported. This test should be made with caution and only a minimum amount of gas should be permitted to escape.

When opening the valve, the outlet should be pointed away from anyone in the vicinity. The absence of escaping gas does not necessarily mean the cylinder is empty, as the valve may be inoperative or the cylinder may contain a low pressure liquid. Cylinder weight can be used as a possible indicator of residual liquid.

Disposition of Cylinder Contents

The contents of unserviceable cylinders with known contents should be removed by the methods appropriate for the contents, and these must also comply with applicable regulatory requirements. These methods should not be attempted unless someone experienced in a proper method of removal is personally directing the procedure. If this cannot be done, all details on procedures and precautions to be employed must be obtained from a person so qualified.

The means of disposal in general will be related to the properties of the cylinder contents. Oxygen or inert gases can be safely vented to the atmosphere. Flammable gases can be disposed by burning or by other suitable methods. Toxic, reactive, poisonous, or irritating gases are disposed of by methods appropriate to the properties of the contents, that is, incineration or chemical reaction.

If it is suspected that the valve will not properly operate so as to control the release of the contents, precautions must be taken to remove the contents at a controlled rate. Under such circumstances, it is recommended that the cylinder be firmly secured in a restraining device or by other suitable means to minimize the danger of the cylinder being tossed about out of control. A needle valve or regulator may be attached to the discharge connection of the cylinder for adequate control and to prevent excessive discharge rates. If the cylinder contains a chemically reactive material, no device should be so attached, as a rapid buildup of pressure into the small confined volume between the cylinder valve and the attachment could initiate a dangerous explosion.

If a cylinder valve is damaged in a way that prevents the discharge of the cylinder contents in a normal manner, it may be possible to release the pressure in the cylinder through the pressure relief device. However, only qualified personnel, familiar with gas cylinders and their pressure relief devices, should attempt this procedure. It should not be attempted where the gas content may be noxious or where the ejection of the pressure relief device by high cylinder pressure could be hazardous. Further, it should not be attempted without proper personal protection equipment or venting procedures.

If a cylinder valve is damaged to such an extent that it cannot be used to release the gas, do not attempt to remove the gas. Return the cylinder to the supplier, or obtain professional assistance from the supplier or other acknowledged expert.

Disposition of Empty Cylinders

After releasing of the pressure and discharging of the contents from an unserviceable cylinder, the cylinder should be purged if it previously contained a flammable, oxidizing, reactive, or toxic material, before any attempt is made to destroy it with a cutting torch. Purging can be accomplished with the use of inert gases or steam, or by filling the cylinder with water. In certain cases, additional cleaning may be required. After removal of the gas contents of a cylinder, and after purging and cleaning, the empty cylinders should be destroyed with a cutting torch or by other appropriate means that will make it unusable as a pressure vessel. This should be done in such a way that the cylinder cannot be readily repaired. It is further recommended that the specification markings on the cylinder be destroyed.

Disposition of Unserviceable Acetylene Cylinders

Cylinders marked ICC-8, DOT-8, CRC-8, BTC-8, CTC-8, ICC-8AL, DOT-8AL, CRC-8AL, BTC-8AL, or CTC-8AL are authorized for acetylene only. These cylinders are filled with a porous mass which distributes the absorbent that is utilized to retain acetylene. Before disposing of one of these cylinders, it is important that every precaution be taken to handle the cylinder safely. This work should be done by personnel completely familiar with these cylinders and includes:

(1) Cylinder preparation
(2) Cylinder dismantling
(3) Preparation for cylinder destruction
(4) Cylinder destruction

In addition, there are precautions regarding the disposal by cutting and storing of scrapped acetylene cylinders or fillers which must be adhered to. Further details may be found in CGA C-2, *Recommendations for the Disposition of Unserviceable Compressed Gas Cylinders with Known Contents.* [25]

REFERENCES

[1] *Code of Federal Regulations,* Title 49 CFR Parts 100–199, (Transportation), U.S. Department of Transportation, Superintendent of Documents, U.S. Government Printing Office, Washington, DC 20402.

[2] *Transportation of Dangerous Goods Regulations,* Transport Canada, Canadian Government Publishing Centre, Supply and Services Canada, Ottawa, Ontario, Canada K1A 0S9.

[3] CSA B339, *Cylinders, Spheres, and Tubes for the Transportation of Dangerous Goods,* Canadian Standards Association, 178 Rexdale Blvd., Rexdale, Ontario, Canada M9W 1R3.

[4] CSA B340, *Selection and Use of Cylinders, Spheres, Tubes and Other Containers for the Transportation of Dangerous Goods,* Canadian Standards Association, 178 Rexdale Blvd., Rexdale, Ontario, Canada M9W 1R3.

[5] *ASME Boiler and Pressure Vessel Code,* American Society of Mechanical Engineers, 345 East 47th Street, New York, NY 10017.

[6] CGA C-4, *American National Standard Method of Marking Portable Compressed Gas Containers to Identify the Material Contained* (ANSI/CGA C-4), Compressed Gas Association, Inc., 1235 Jefferson Davis Highway, Arlington, VA 22202.

[7] CGA C-7, *Guide to the Preparation of Precautionary Labeling and Marking of Compressed Gas Containers,* Compressed Gas Association, Inc., 1235 Jefferson Davis Highway, Arlington, VA 22202.

[8] ISO 448, *Gas Cylinders for Industrial Use—Marking for Identification of Content,* International Organization for Standardization (Central Secretariat), 1, Rue de Varembe, Case Postale 56, CH-1211 Geneve 20, Switzerland/Suisse.

[9] *Code of Federal Regulations,* Title 21 CFR, Parts 200, 201, 207, 210, and 211 (Food and Drugs), Food and Drug Administration, Superintendent of Documents, U.S. Government Printing Office, Washington, DC 20402.

[10] *Code of Federal Regulations,* Title 29 CFR, Parts 1900–1910 (Labor), U.S. Department of Labor, Superintendent of Documents, U.S. Government Printing Office, Washington, DC 20402.

[11] *Regulations for the Transportation of Dangerous Commodities by Rail,* Canadian Government Publishing Centre, Supply and Services Canada, Ottawa, Ontario, Canada K1A 0S9.

[12] ANSI Z129.1, *American National Standard for Precautionary Labeling of Hazardous Industrial Chemicals,* American National Standards Institute, Inc., 1430 Broadway, New York, NY 10018.

[13] CGA C-6, *Standards for Visual Inspection of Steel Compressed Gas Cylinders,* Compressed Gas Association, Inc., 1235 Jefferson Davis Highway, Arlington, VA 22202.

[14] CGA C-6.1, *Standards for Visual Inspection of High Pressure Aluminum Compressed Gas Cylinders,* Compressed Gas Association, Inc., 1235 Jefferson Davis Highway, Arlington, VA 22202.

[15] CGA C-6.2, *Guidelines for Visual Inspection and Requalification of Composite High Pressure Cylinders,* Compressed Gas Association, Inc., 1235 Jefferson Davis Highway, Arlington, VA 22202.

[16] CGA C-6.3, *Reinspection of Low Pressure Aluminum Compressed Gas Cylinders,* Compressed Gas Association, Inc., 1235 Jefferson Davis Highway, Arlington, VA 22202.

[17] CGA C-5, *Cylinder Service Life—Seamless, Steel, High Pressure Cylinders,* Compressed Gas Association, Inc., 1235 Jefferson Davis Highway, Arlington, VA 22202.

[18] CGA C-1, *Methods for Hydrostatic Testing of Compressed Gas Cylinders,* Compressed Gas Association, Inc., 1235 Jefferson Davis Highway, Arlington, VA 22202.

[19] CGA C-13, *Guidelines for Periodic Visual Inspection and Requalification of Acetylene Cylinders,* Compressed Gas Association, Inc., 1235 Jefferson Davis Highway, Arlington, VA 22202.

[20] CGA C-8, *Standard for Requalification of DOT-3HT Seamless Steel Cylinders,* Compressed Gas Association, Inc., 1235 Jefferson Davis Highway, Arlington, VA 22202.

[21] CGA V-1, *American National, Canadian, and Compressed Gas Association Standard for Compressed Gas Cylinder Valve Outlet and Inlet Connections* (ANSI/CSA/CGA V-1), Compressed Gas Association, Inc., 1235 Jefferson Davis Highway, Arlington, VA 22202.

[22] CSA B338, *Highway Tanks and Portable Tanks for the Transportation of Dangerous Goods,* Canadian Standards Association, 178 Rexdale Blvd., Rexdale, Ontario, Canada M9W 1R3.

[23] CGA P-15, *Filling of Industrial and Medical Nonflammable Compressed Gas Cylinders,* Compressed Gas Association, Inc., 1235 Jefferson Davis Highway, Arlington, VA 22202.

[24] CGA S-1.1, *Pressure Relief Device Standards—Part 1—Cylinders for Compressed Gases,* Compressed Gas Association, Inc., 1235 Jefferson Davis Highway, Arlington, VA 22202.

[25] CGA C-2, *Recommendations for the Disposition of Unserviceable Compressed Gas Cylinders with Known Centents,* Compressed Gas Association, Inc., 1235 Jefferson Davis Highway, Arlington, VA 22202.

CHAPTER 10

Cleaning Components, Equipment, and Systems for Oxygen Service

INTRODUCTION

The cleaning methods described in this chapter are intended for cleaning components, equipment, and systems used in the production, storage, distribution, and use of liquid and gaseous oxygen. Some examples of these (illustrative of the primary intent of this chapter's coverage) are: stationary storage tanks, tank trucks, and tank cars; pressure vessels such as heat exchangers and rectification columns; and associated piping, valves, and instrumentation. The cleaning methods also may be utilized in cleaning other oxygen service equipment such as cylinders, cylinder valves, regulators, welding torches, pipelines, compressors, and pumps.

Oxygen equipment and systems, including all components and parts thereof, must be adequately cleaned to remove harmful contamination prior to the introduction of oxygen. Harmful contamination would include both organic and inorganic materials such as oils, greases, paper, fiber, rags, wood pieces, coal dust, solvents, weld slag, rust, sand, and dirt, which if not removed could cause a violent combustion reaction or even an explosion in an oxygen atmosphere. At the very least, contamination could have deleterious effects on systems and components thereof, their operation, service life, and reliability, and be detrimental to product purity.

OBJECTIVES

This chapter presents methods for cleaning oxygen service equipment. When properly used, these cleaning methods and subsequent inspections will result in the degree of cleanliness required for the safe operation of oxygen service equipment and the necessary product purity required in CGA G-4.3, *Commodity Specification for Oxygen.* [1] Suggested limits of contamination and ways of determining if a component or system is sufficiently clean to be used in oxygen service are given, along with procedures for keeping such equipment clean before being placed in service.

Cleaning a component or system for oxygen service involves the removal of combustible contaminants including the surface residue from manufacturing, hot work, and assembly operations. It also involves the removal of all cleaning agents and the prevention of recontamination before final assembly, installation, and use. These cleaning agents and contaminants include solvents, acids, alkalies, water, moisture, corrosion products, thread lubricants, filings, dirt, scale, slag, weld splatter, organic material (such as oil, grease, crayon, and paint), lint, and other foreign materials.

Injurious contaminants can be removed by cleaning all parts and maintaining this condition during construction, by completely cleaning the system after construction, or by a combination of the two. The methods of cleaning described herein may involve the use of hazardous chemicals and operations. Although some safety measures have been described, this chapter does not purport to address all physical and health hazards associated with any particular cleaning method. Accordingly, users should investigate and implement appropriate safety procedures.

Types of Contaminants

A contaminant is universally defined as a foreign or unwanted substance that can have deleterious effects on a system's operation, life, or reliability. Solid and fluid contaminants are classified into three major categories: organics, inorganics, and particulates of organics and inorganics. The definition of each respective category is given in the following paragraphs.

A list of common contaminants is given in Table 10-1, and their approximate size is shown graphically in Fig. 10-1.

Organics. Chemical compounds of carbon chains or rings which may contain hydrogen with or without oxygen, nitrogen, or other elements. Examples of organics include greases, oils, hair, wood, and nonpolar resin.

TABLE 10-1. EXAMPLES OF COMMON CONTAMINANTS

Organic	Inorganic
Hair	Burrs
Bacteria	Chips
Dandruff	Skin flakes
Pollen	Plating flakes
Wood	Grinding dust
Saliva	Fines
Paper	Welding slag
Insects	Fingerprints
Strings	Ceramic chips
Lint	Sand
Plastic particles	Dust
Lubricants	Glass particles

Inorganics. Chemical compounds that do not contain carbon as the principal element (excepting carbonates, cyanides, and cyanates) that is, matter other than plant or animal. Examples include water-soluble salts (polar), metal, plastic, dust, and dirt.

Particulates. A general term used to describe a finely divided solid of organic or inorganic matter. These solids are usually reported as contaminants by the population of a specific micrometer size. See ASTM F312, *Method for Microscopical Sizing and Counting Particles from Aerospace Fluids on Membrane Filters.* [2]

Nonmetallics

The term *nonmetallics* is used to denote nonmetallic, nondurable materials including elastic and plastic polymers, organics, wood, and cloth products. Such materials may form component parts of the oxygen system to be cleaned, and the compatibility of cleaning solutions with exposure to such components must be considered.

FACTORS FOR SELECTING A CLEANING METHOD

The cleaning of an oxygen system should begin with disassembly to the elemental or piece part level. If cleaning is attempted by flowing a cleaning solution through a component, vulnerable internal elements may be damaged by the strength of the solution required to clean the major elements of the component. Also, contaminants and cleaning solutions may become entrapped in component recesses and may ultimately react with oxygen. After the component has been disassembled, the parts should be grouped according to the method of cleaning to be used. If sealing surfaces are damaged or cracks are observed in the metallic parts, the component must be repaired or replaced. While the methods described here are applicable to all metallic materials, they do not necessarily apply to nonmetallic products.

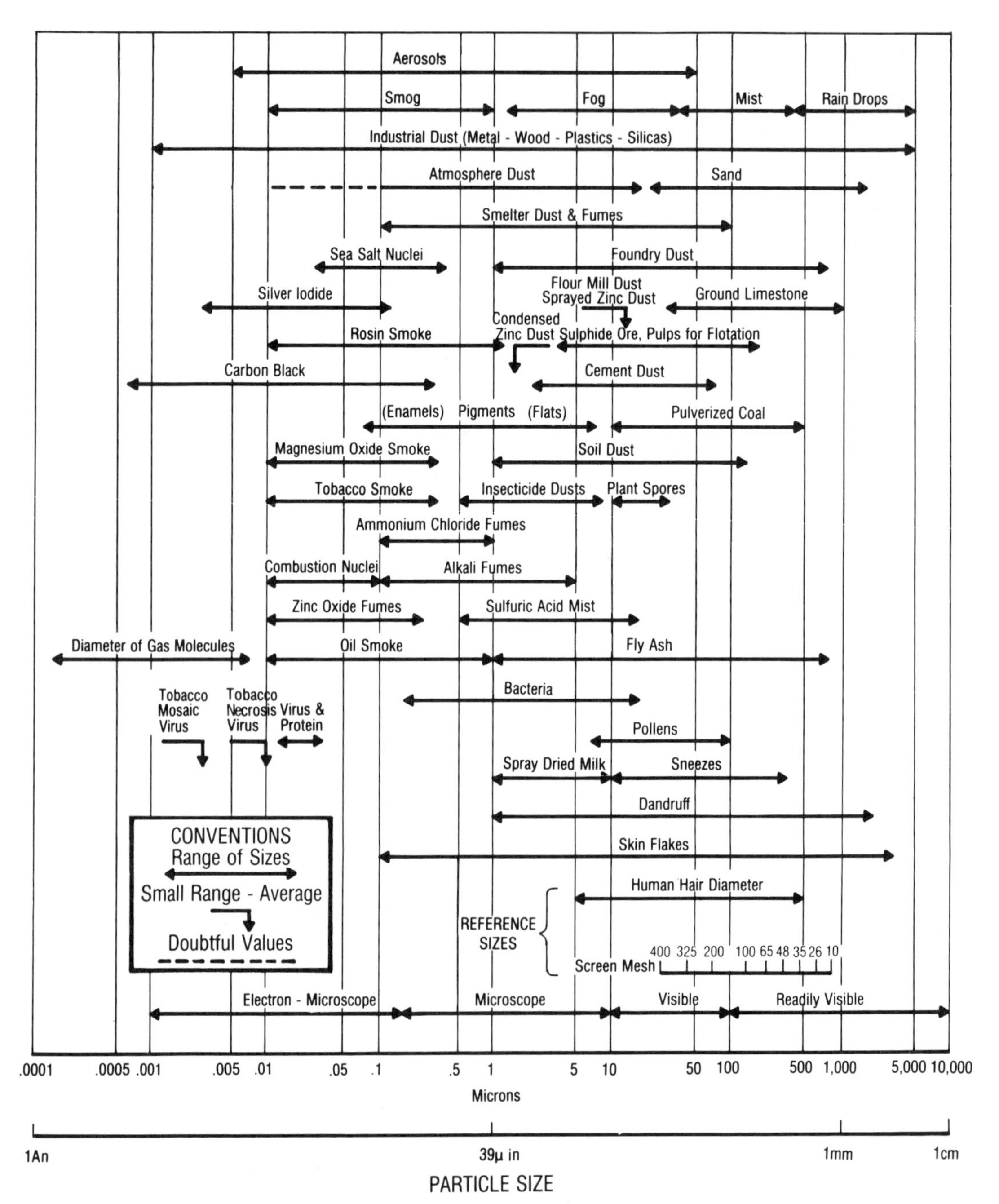

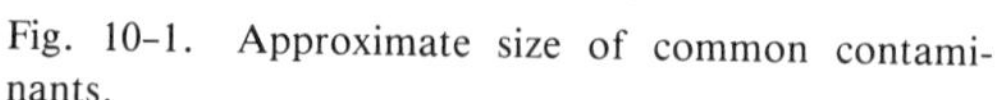
Fig. 10-1. Approximate size of common contaminants.

Special attention should be directed to the cleaning of component nonmetallics. Readers of this chapter are encouraged to consult the supplier of the component nonmetallics for the appropriate cleaning instructions and practices.

Selecting Procedures

The type, possible location, and degree of contamination should initially be estimated in order to decide on the most practical method of cleaning, inspecting, and testing.

In addition, the arrangement of passages must be studied so that cleaning, washing, or draining practices can be adjusted to make sure that dead-end passages and possible traps are adequately cleaned.

Factors in Selecting a Cleaning Method

Factors to be considered in selecting a cleaning method include:

- The type of contaminant (i.e., inorganic, organic, particulate)
- The base material and/or coating of the part to be cleaned
- Initial condition of the part to be cleaned
- The desired final degree of cleanliness of the part
- Environmental impact of the required cleaning method
- Cost effectiveness of the required cleaning method
- Effects of the selected cleaning methods on the part to be cleaned, such as mechanical, chemical, and thermal

Mechanical effects can be produced by mechanical cleaning methods. Wire brushing a surface to remove contaminants may work-harden the part's surface. Tumbling operations may result in a shot-peen effect dependent upon the material, or it may remove plated surfaces.

Chemical effects can manifest themselves as an attack on the part material, directly or indirectly. This in turn may produce a non-desired coating, such as an oxide, which may inhibit further processing or design functioning.

Thermal effects can result because some cleaning methods may require elevated temperatures. Such temperatures may degrade the component, equipment (or part thereof), or system being cleaned.

The following cleaning methods can be used alone or in combination:

(1) Steam cleaning (including hot water and detergents)
(2) Vapor degreasing
(3) Solvent washing (including ultrasonics)
(4) Alkaline (caustic) cleaning
(5) Acid cleaning
(6) Mechanical cleaning (blast cleaning, wire brushing, etc.)
(7) Purging

CLEANING OPERATIONS

In general, cleaning methods can be divided into three major categories: (1) precleaning, (2) intermediate cleaning, and (3) precision cleaning. Depending on the factors previously mentioned, one or more cleaning methods may be necessary in each category. Each of the three categories and the cleaning methods employed are described in subsequent sections.

Precleaning

Precleaning is necessary when the part to be cleaned is grossly contaminated. Examples of gross contamination include: excessive oxide or scale buildup, large quantities of oil and greases, inorganic particulates, scale, dirt, grit, solid objects, hydrocarbons, and fluorocarbons. Precleaning reduces excessive contaminants, thereby increasing the useful life and effectiveness of the cleaning solutions used in subsequent cleaning operations.

The cleaning environment and the handling procedure used for all precleaning operations are not critical, but users are encouraged to be cognizant of and to follow all related safety practices. Precleaning can be divided into two subcategories, mechanical and chemical precleaning, or a combination of the two.

Intermediate Cleaning

Intermediate cleaning generally consists of subjecting the part to both caustic and acid cleaning solutions designed to remove solvent residues and residual contaminants which have been conditioned by precleaning. The cleaning environment and handling procedures used for intermediate cleaning operations are more restrictive than those used for precleaning. The cleaning environment must be controlled in order to minimize introducing contaminants and compromising subsequent final or precision cleaning operations.

Alkali cleaning solutions are caustic and can cause personal injury if safety precautions, such as protective clothing and ventilation, are not rigidly followed. A list of common alkaline salts and detergents is given in Table 10-2.

Precision or Final Cleaning

Precision or final cleaning is the final process to which parts are subjected in order to meet strict levels of cleanliness. Certain nuclear, space, and other critical applications may require that only very-high-purity precision cleaning agents be used. Precision or final cleaning is normally performed by exposing the part to a final cleaning solvent either by vapor degreasing, ultrasonic cleaning, or direct rinsing.

Precision cleaning involves maintenance of the most critical level of cleaning environment and handling control. Precautions should be taken to minimize the recontamination of the part by the final cleaning environment or by the handling procedures. Certain end use applications of the part may require strict cleaning environments, such as classed clean rooms. Fluorocarbon solvents and their blends, or chlorinated solvents, are normally used during the precision cleaning process.

Some equipment components used for vapor degreasing are similar to those used in ultrasonic cleaning. Ultrasonic cleaning equipment consists of an immersion tank, high frequency sound generator, and heater. Vapor degreasing and ultrasonic equipment are commercially available as separate units or combined in one unit which may include a spray wand.

A sample of new wash solvent should be taken for control purposes when required. If solvent monitoring is desired, a representative sample of the used solvent should be taken to determine its contaminant level. Additional solvent should be added to dilute the used solvent to an acceptable contamination level. The basic procedure for precision cleaning parts may include the following steps:

- Suspension of the part in the vapor of the solvent
- Immersion and ultrasonic cleaning in the liquid solvent
- Spray rinsing of the part with filtered solvent

The solvent must be discarded when the cleaning operation does not yield acceptably cleaned surfaces or the solvent fails to meet the acceptance standards.

After the precision cleaning operation, the parts should be dried by purging or blowing with dry, oil-free nitrogen or air to remove entrapped or residual solvents.

TABLE 10-2. COMMON ALKALINE SALTS AND DETERGENTS.

Alkaline Salts	
Sodium hydroxide, NaOH	
Sodium metasilicate, ortho, or trisilicate, Na_2O_3Si	
Soda ash, Na_2CO_3	
Sodium tetraborate, $Na_2B_4O_7$	
Trisodium phosphate (TSP)	
Sodium pyrophosphate, NaP_2O_7	
Sodium polyphosphate	
Alkaline Detergents	
Saponifiers	—Solubilize fats
Wetting agents	—Reduce surface tension
Deflocculents	—Prevent particle agglomeration
Water softeners	—Reduce hardness
Buffering agents	—Maintain pH

CLEANING METHODS

Various cleaning methods are described in the following paragraphs. The two main divisions for grouping types of cleaning methods are mechanical and chemical. For additional information on cleaning methods, design criteria, and materials compatibility for oxygen service, consult ASTM publications [3] through [5] listed in the reference section. Reference [6] is an ASTM Special Technical Publication series of books wherein numerous symposia papers are published covering a wide range of technical studies on these topics.

MECHANICAL CLEANING

Mechanical cleaning may be accomplished by methods such as abrasive blast cleaning, wire brushing, grinding, barrel or mass cleaning, steam cleaning, or hot water cleaning. Details are contained in *Metals Handbook,* Volume 5, "Surface Cleaning, Finishing and Coating." [7]

Abrasive Blast Cleaning

Abrasive blast cleaning may be described as the use of an abrasive propelled through nozzles against the surface of pipe, fittings, or containers to remove mill scale, rust, varnish, paint, or other foreign matter. The medium propelling the abrasive must be oil-free unless the oil is to be removed by subsequent cleaning. The specific abrasive materials used must be suitable for performing the cleaning without depositing contaminants that cannot be removed by subsequent cleaning. Care needs to be taken when blast cleaning so as not to remove an excessive amount of parent metal.

Wire Brushing or Grinding

Accessible surfaces may be wire brushed. Welds may be ground and wire brushed to remove slag, grit, or excess weld material. Carbon steel wire brushes must not be used on aluminum or stainless steel surfaces. Any wire brushes previously used on carbon steel cannot be used on aluminum or stainless steel surfaces.

Barrel or Mass Cleaning

Barrel or mass cleaning can be described as a method that uses a quantity of hard abrasive material placed in a container to clean internal surfaces. The container is rotated or tumbled, imparting relative motion between the components within the container, the abrasive material, and the container.

Swab, Spray, and Dip Cleaning

Swab, spray, and dip cleaning are three methods of delivering solutions to the component surfaces. Each method has its particular advantages. Swabbing is generally used on parts or components to clean small select areas only. Spraying and dipping are used for overall cleaning. These methods are generally employed with alkaline or acid cleaning methods, which are discussed elsewhere in this chapter.

Vacuuming and Blowing

Vacuuming and blowing remove the contaminant from the component by negative or positive air currents. These methods may be used to remove loose dirt, slag, scale, and various particles, but are not suited for the removal of surface oxides, greases, and oils.

Steam Cleaning

The equipment used for steam cleaning may consist of a steam and water supply, a length of hose, and a steam lance with or without a spray nozzle. A detergent is generally incorporated into the steam spray.

Either plant steam or steam from a portable steam generator can be used. If a steam lance is used, the detergent solution may enter the steam gun by venturi action and mix with the steam. Steam removes oil, greases, and soaps by first "thinning" them with the heat. Dispersion and emulsification of the oils then occurs, followed by dilution with the condensed steam. The system should provide control over the steam, water, and detergent flows so that the full effects of the detergent's chemical action, the heat of the steam, and the "abrasive" action of the pressure jet are attained for maximum cleaning efficiency.

If the steam is clean and free of organic material, a secondary cleaning operation with a solvent or alkaline degreaser may not be required in cases where the initial contamination is not heavy or is readily removed with steam.

Hot Water Cleaning

Cleaning with a hot detergent solution may utilize a spray system or a cleaning vat with suitable agitation of either the solution or the parts to be cleaned. Hot detergent solution cleaning can be used where a steam temperature is not necessary to free and fluidize contaminants. Proper consideration must be given to the size, shape, and number of parts to be cleaned so as to ensure adequate contact between the surfaces to be cleaned and the detergent solution. The solution temperature should be in accordance with the recommendation of the manufacturer of the cleaning agent.

Most detergents are water soluble and are best removed by prompt flushing with sufficient quantities of hot or cold clean water, as appropriate, before the cleaning agents have time to precipitate. The equipment is then dried by purging with dry, oil-free air or nitrogen which may be heated to shorten the drying time. Blowing or purging with dry, oil-free air or nitrogen should be used to remove small particles which may be present as a result of any previous cleaning methods employed. Purging should also be used as a means to isolate cleaned surfaces to prevent cross contamination of parts between sequential mechanical cleaning methods or recontamination before packaging.

Purging

It is very important to purge the component to ensure that all residuals from the previous cleaning operations are removed before subsequent cleaning operations or final packaging occur. This can be accomplished by rinsing, drying, and blowing. Rinsing may depend upon the cleaning solutions used, but in general, filtered water may be used. Drying may be done by the application of heat to the component by ovens and infrared lights, or by blowing with clean, oil-free, dry air (heated or unheated).

A more critical purging is performed using clean, dry, oil-free nitrogen gas. Factors such as the duration of the purge, the number of purging operations, and the type of purging operations are dependent upon the component to be cleaned, the cleaning methods employed, and the final application.

CHEMICAL CLEANING

Caustic Cleaning

Caustic cleaning is cleaning with solutions of high alkalinity for the removal of heavy or tenacious surface contamination followed by a rinsing operation. There are many effective materials available for caustic cleaning. They are basically alkalies and are water soluble and nonflammable, and may be harmful in contact with the skin or eyes or if swallowed. The cleaning agents should be chosen so that they do not react chemically with the materials being cleaned.

The water that is used for rinsing should be free of oil and other hydrocarbons and should contain no particles larger than those acceptable on the cleaned surface. Filtration may be required. It may be desirable to analyze the water to determine the type and quantity of impurities. Some impurities may cause undesirable products or reactions with the particular caustic cleaner used.

The cleaning solution can be applied by spraying, immersion flushing, or hand swabbing:

(1) Spraying works well, but requires a method whereby the cleaning solution reaches all areas of the surface. It is also desirable to have provisions for draining the solution faster than it is introduced to avoid accumulation.
(2) Immersion or flushing should be total rather than partial since the solution tends to dry on the surface that is exposed to air.
(3) Hand-swabbed surfaces should be rinsed before the cleaning solution dries.

Generally, cleaning solutions perform better when warm. Depending upon the particular solution, this temperature will be in the range of 100°F to 180°F (37.8°C to 82.2°C). The cleaning solution can be reused until it is too weak or too contaminated as determined by pH or concentration analysis. Both decrease as the solution weakens. Experience will establish when a cleaning solution has become too weak or too contaminated to effectively clean contaminated surfaces.

The degree of cleanliness achieved will ultimately depend on the thoroughness of the rinsing procedures. Some of the contaminants may be held in suspension in the cleaning solution. Therefore, if the cleaning solution is not completely flushed from the surface being cleaned, the contaminant in any remaining solution will redeposit on the surface during the drying operation. The surface must not be allowed to dry between the cleaning phase and the rinsing phase. If this happens, it is very likely that the film or residue will not be adequately removed during the rinsing phase. Frequently, some type of agitation during rinsing is required. This may be by mechanical brushing, fluid impingement, agitation of the parts being cleaned, etc.

The water rinse is often warmed to help remove the cleaning solution and aid in the drying process. A method of determining when the rinsing is complete is to monitor the pH of the outlet rinse water. The pH approaches that of the original rinse water as the rinsing progresses. If drying is not completed with the residual heat in the metal, it can be completed with dry, oil-free air or nitrogen (heated or unheated). If it is desirable that the equipment be maintained in a dry atmosphere before installation or use, the dew point of the contained atmosphere should not be above −30°F (−34.4°C).

Acid Cleaning

The acid cleaning procedure removes oxides and other contaminants by immersion in a suitable acid solution, usually at room temperature. The type of cleaning agent selected will depend, in most cases, on the material to be cleaned. The following general guidelines can be used:

(1) Cleaning agents formulated using phosphoric acid can be used for all metals. These agents will remove oxides, light rust, light oils, and fluxes.

(2) Cleaning agents formulated using hydrochloric acid are recommended for carbon and low-alloy steels only. These agents will remove rust, scale, and oxide coatings and will strip chromium, zinc, and cadmium platings. Certain acid solutions, including hydrochloric or nitric acids, should contain an inhibitor to prevent harmful attacks on base metals. Hydrochloric acid should not be used on stainless steel since it may cause stress corrosion.

(3) Aluminum, copper, and their alloys can be cleaned using solutions based on chromic or nitric acids. These agents are not true cleaning agents but are used for deoxidizing, brightening, and removing black smut which forms during cleaning with an alkaline solution. Some agents are available as liquids and others as powders, and are mixed in concentrations of 5 to 50 percent in water, depending on the cleaning agent and the amount of oxide or scale to be removed.

A storage or immersion tank, acid-resistant recirculation pump, and associated piping and valving compatible with the acid solution are required. Common methods of applying acid cleaning agents used for cleaning metals are:

(1) Large areas may be flushed with an appropriate acid solution.
(2) Small parts may be immersed and scrubbed or agitated in the solution.

Caution: Acid cleaning agents should not be used unless their application and performance are known or are discussed with the cleaning agent manufacturer. The manufacturer's recommendations regarding concentration, temperature, and personnel protective equipment should be followed for safe handling and use of the cleaning agent.

After caustic or acid cleaning, thorough

rinsing of the equipment is performed using cold water. Rinsing must begin as soon as practicable after cleaning to prevent excessive attack on the material being cleaned by the acid cleaning solution. If there is a chance of any cleaning solution becoming trapped in the equipment being cleaned, a dilute alkaline neutralizing solution can be applied, followed by water rinsing. If drying is not completed with the residual heat in the metal, it can be completed with dry, oil-free air or nitrogen (heated or unheated). If it is desirable that the equipment be maintained in a dry atmosphere before installation or use, the dew point of the contained atmosphere should not be above −30°F (−34.4°C).

Solvent Washing

Solvent washing is the removal of organic contaminants from the surface to be cleaned by the use of chlorinated hydrocarbons or other suitable solvents. The solvents frequently used for solvent washing and ultrasonic cleaning are methylene chloride; refrigerant 11 (trichlorofluoromethane); refrigerant 113 (trichlorotrifluoroethane); perchloroethylene; 1,1,1-trichloroethane (methyl chloroform); or trichloroethylene. Suitable corrosion inhibitors and stabilizers should be included in the formulation for the solvents.

Warning: Carbon tetrachloride is not to be used because of its high toxicity, that is, its low threshold limit value (TLV). Trichloroethylene should be used only if absolutely necessary since it is more toxic than 1,1,1-trichloroethane (methyl chloroform).

Caution: Solvent washing agents should not be used unless their application and performance are known or are discussed with the cleaning agent manufacturer. The manufacturer's recommendations regarding concentration, temperature, and personnel protective equipment should be followed for safe handling and use of the cleaning agent.

The boiling points, freezing points, threshold limit values, and Kauri-Butanol numbers of these solvents are listed in Table 10-3. Washing equipment may consist of a recirculating system for the solvent or a closed container for immersing parts. Auxiliary control and test equipment might include the following: space heaters, halogen detectors, thermometers, a utility container, funnel and strainer, an Imhoff cone, dry, oil-free air or nitrogen, and a siphon pump. For ultrasonic cleaning, a high frequency sound generator and container are substituted for the recirculation system.

Caution: Some plastic tubing, including polyvinylchloride (PVC), may have its plasticizer extracted by the solvent and deposited on the surface being cleaned. For this same reason, rubber, neoprene, and nylon tubing should not be used with these solvents when cleaning oxygen equipment. Polyethylene, polypropylene, and polytetrafluoroethylene (PTFE) tubing are satisfactory with the frequently used solvents.

Before a new batch of solvent is used for any cleaning operation, a sample of it should be taken for reference purposes. This sample should be stored in a clean container made of materials that will not contaminate it. The degree of cleanliness of the solvent after a period of use can be determined by comparing it to the reference sample in one of several ways: (1) by comparing its color to that of the reference sample, (2) by an analysis, or (3) by an evaporation procedure.

In the color comparison, it is assumed that the solvent is still sufficiently clean to use if it shows no distinct color change from the reference sample. The color change can be determined visually or by instrument, comparing the simultaneous light transmission through both samples. This should be verified by analytical tests to detect probable contaminants or by calculation of the amount of residue deposited by evaporation of contaminated solvent. ASTM D-2108-85, *Test Method for Color Halogenated Organic Solvents and Admixtures,* may be used if a scale of color changes is to be established for one or more contaminants. [8]

Analytical techniques (e.g., infrared spectroscopy or chromatography), although somewhat more time consuming, can mea-

TABLE 10–3. CHEMICAL AND PHYSICAL PROPERTIES OF CLEANING SOLVENTS

Solvent	Formula	Molecular Weight	Boiling Point		Freezing Point		Density at 68°F (20°C)		Latent Heat of Vaporization at Boiling Point	Evaporation Rate	TLV	Kauri-butanol Number[b] at 77°F
			°F	(°C)	°F	(°C)	lb/ft³	(kg/m³)	(Btu/lb)	(Ether = 100)	(ppm)[a]	(25°C)
1,1,1-Trichloroethane[c]	$C_2H_3Cl_3$	133.42	165.0	(73.89)	−36.0	(−37.78)	82.10[f]	(1315.12)	102.0	37	350	124
Methylene chloride[d]	CH_2Cl_2	84.94	103.6	(39.78)	−142.1	(−96.72)	83.37	(1335.46)	141.7	62	50	136
Perchloroethylene	C_2Cl_4	165.85	250.2	(121.22)	−8.2	(−22.33)	101.50	(1625.87)	90.0	12	50	92
Refrigerant 11 (trichlorofluoromethane)	CCl_3F	137.40	74.8	(23.78)	−168.0	(−111.11)	92.70[g]	(1484.91)	78.31	81	1000	60
Refrigerant 113 (trichlorotrifluoroethane)	$Cl_2FC_2ClF_2$	187.40	117.6	(47.56)	−31.0	(−35.00)	94.29[h]	(1510.38)	63.12	126	1000	31
Trichloroethylene[e]	C_2HCl_3	131.40	188.6	(87.00)	−122.8	(−86.00)	91.42	(1464.41)	103.0	30	50	129

[a]Threshold limit values (time-weighted average) adopted by American Conference of Governmental Industrial Hygienists, 1989–90.
[b]The higher the Kauri-Butanol number, the greater the dissolving power for certain gums.
[c]1,1,1-Trichloroethane has neither a flash nor fire point. It has flammable limits of 7.5% to 15% in air at 77°F (25°C).
[d]Methylene chloride has neither a flash nor fire point. It has flammable limits of 12% to 22% in air at 77°F (25°C).
[e]Trichloroethylene has a listed flash point of 90°F (32.2°C) and flammable limits of 8% to 10.5% in air at 77°F (25°C).
[f]This value is listed for 77°F (25°C).
[g]This value is listed for 70°F (21.1°C).
[h]This value is listed for 117.5°F (47.6°C).

sure quite exactly the extent of solvent contamination with a known contaminant, for example, a particular cutting oil used to machine parts. However, if one or several unknown contaminants are present, results might be more difficult to quantify.

Contamination can be checked by calculating the amount of residue deposited by evaporation of a measured amount of contaminated solvent. This calculation depends on the nonvolatility of any contaminants. However, the vapor pressures of most oils are sufficiently high that significant amounts may evaporate with a large volume of solvent. Therefore, an evaporation determination may give only a lower limit to the amount of dissolved contaminant.

A test for contamination should be run periodically on the solvent used for immersion cleaning of components. If a large vessel or piping system is cleaned by circulating solvent through it, the solvent should be tested at the end of the cleaning period. If the solvent is contaminated, as shown by that test, it must be drained from the equipment and replaced with a batch of clean solvent. After further circulation, this solvent must be similarly tested, and it must either pass or be replaced with clean solvent. A vessel can be considered clean when no distinct color difference exists between the ingoing and drained samples.

After a part is removed from or drained of solvent, techniques such as heating and monitoring the exit purge gas for solvent, for example, by halogen detector, should be used to ensure that all solvent has been removed from the component.

Dirty solvent may be reclaimed by appropriate procedures, reused as is for initial cleaning, or discarded. Disposal must comply with applicable federal, state, municipal, and provincial laws and regulations, including environmental and other standards which might apply.

Caution: Use proper solvent transfer containers (precleaned glass or metal) with no seals that can be dissolved by the solvent.

Removal of solvents is important. After the oil and grease contaminants have been removed or dissolved and the solvent drained, allow the part to dry, purging any cavity, piping, or closed vessel with dry, oil-free air or nitrogen to remove liquid by entrainment. Then circulate the purge gas until the final traces of the solvent have been removed. Purging can be considered complete when the solvent cannot be detected by appropriate methods in the gas venting from the vessel, piping, or component being purged.

If the odor of solvent vapors is detected in the vicinity of the effluent purge gas, the equipment requires additional purging. A halogen leak detector may be used with chlorinated solvents for determining when a vessel, piping, or component is adequately purged. The test method should be agreed upon by the manufacturer and the purchaser.

For equipment being used in oxygen service, it may be desirable to estimate the total quantity of oil or grease removed to justify future extension of operating periods between washing or omission of washing operations.

Solvent cleaning is sometimes enhanced by the use of ultrasonic agitation. Ultrasonic cleaning is the loosening of oil and grease or other contamination from metal surfaces by the immersion of parts in a solvent or detergent solution in the presence of high frequency vibrational energy.

Vapor Degreasing

Vapor degreasing is the removal of soluble organic materials from the surfaces of equipment by the continuous condensation of solvent vapors and their subsequent washing action. Commercial degreasers are available for cleaning metals at room temperatures. Vapor degreasing equipment consists essentially of a vaporizer for generating clean vapors from a contaminated solvent and a vessel for holding the parts to be cleaned in the vapor space.

The solvents frequently used for vapor degreasing are methylene chloride; refrigerant 11 (trichlorofluoromethane); refrigerant 113 (trichlorotrifluoroethane); perchloroethylene; 1,1,1-trichloroethane (methyl chloroform); or trichloroethylene. Suitable corrosion inhibitors and stabilizers should be included in the formulation for the solvents. Some of these solvents are flammable in air under certain conditions and have varying degrees of toxicity. Caution should be exercised in their use. Dry, oil-free air or nitrogen should be available for purging.

The procedure described here is useful for cleaning cold or cryogenic equipment. The temperature of a component must be between the freezing and boiling points of the solvent so that the solvent vapors will condense and wash down by gravity over the equipment surfaces.

This cleaning procedure requires that the solvent be boiled in a vaporizer and the hot solvent vapors then contact a cooler component on whose surface the vapors condense and over which they wash, carrying away soluble contaminants. This action can occur by placing components inside a vapor degreaser chamber into which the solvent vapor rises from a vaporizer chamber. Alternatively, if a vessel is to be cleaned, hot solvent vapor can be piped from a vaporizer into the vessel, on whose inner walls condensation and cleaning will occur. In this case, the equipment should be positioned and connected so that the condensate can be thoroughly drained from the system. Continuous removal of the condensate and its transport back into the vaporizer will carry the dissolved impurities into the vaporizer where they remain, as fresh pure vapors are released to continue the degreasing operation. Cleaning can be considered complete when the returning condensate is as clean as the new solvent.

Note: The vapor degreasing action will stop when the temperature of the vessel reaches the boiling point of the solvent.

The solvent should be removed as described in the solvent washing section of this chapter.

Caution: Vapor degreasing solvents should not be used unless their application and performance are known or are discussed with the cleaning agent manufacturer. The manufacturer's recommendations regarding concentration, temperature, and personnel protective equipment should be followed for safe handling and use of the cleaning agent.

INSPECTION PROCEDURES

Detailed cleaning and quality control procedures should be agreed upon by the manufacturer and the purchaser. A source inspection by the purchaser's representative at the manufacturer's location is desirable. The purchaser should initially and periodically inspect the manufacturer's facilities and audit the cleaning and quality control procedures.

Some industries have found a contamination level equal to or below 500 mg/m^2 (47.5 mg/ft^2) to be the maximum level of hydrocarbon contamination tolerable for components, equipment, and systems in oxygen service. The actual level depends on the specific application (state of fluid, temperature, and pressure).

Likewise, the requirement for limiting the particle and fiber contamination is necessarily dependent upon actual service. Therefore, the user is urged to review component, equipment, and system requirements. If the purchaser's requirement does include a particle and fiber count, some industries have found that a representative square-foot section of surface must show no particle larger than 1000 microns and no more than 20 particles per square foot (215 particles/m^2) between 500 and 1000 microns. Isolated fibers of lint should be no longer than 2000 microns and there should be no accumulation of lint fibers. In some cases, lower particle size and populations may be necessary, depending upon actual service.

Any one or combination of the following tests can be used to assess the degree of cleanliness of a piece of equipment. The degree of cleanliness assessment is limited to the precision and bias of the verification technique. Failure to pass any of the specified tests requires recleaning and reinspection and may require reevaluation of the cleaning procedures. In-process inspections to ensure the adequacy of cleaning procedures may be desirable.

Direct Visual Inspection (White Light)

Direct visual inspection by white light is the most common test used to detect the presence of contaminants such as oils, greases, preservatives, moisture, corrosion products, weld slag, scale, filings, chips, and other foreign matter. The item is observed (20/20 vision without magnification) for the absence of contaminants under strong white light and for the absence of accumulations of lint fibers. This method will detect particulate matter in excess of 50 microns (0.002 inch) and moisture, oils, greases, and so forth, in relatively large amounts. The item being examined must be recleaned if an unacceptable amount of foreign material is detected by this inspection method.

Direct Visual Inspection (Ultraviolet Light)

Ultraviolet light causes many common hydrocarbon or organic oils or greases to fluoresce when they may not otherwise be detectable by other visual means. The surface is observed in darkness or subdued light using an ultraviolet light radiating at wavelengths between 2500 and 3700 angstrom units. Ultraviolet (black light) inspection will indicate if cleaned surfaces are free of any hydrocarbon fluorescence. Accumulations of lint or dust that may be visible under the black light must be removed by blowing with dry, oil-free air or nitrogen, wiping with a clean lint-free cloth, or vacuuming. Not all organic oils fluoresce to the same degree, and for this reason ultraviolet inspection alone cannot be relied upon as a test for degree of cleanliness. Some materials that fluoresce, such as cotton lint, are acceptable unless present in excessive amounts. If fluorescence shows up as a blotch, smear, smudge, or film, the fluorescing area must be recleaned.

Wipe Test

The wipe test is used to detect contaminants on visually inaccessible areas as an aid in complementing the above visual inspections. The surface is rubbed lightly with a clean white paper or lint-free cloth which is then examined under white and ultraviolet light. The area should not be rubbed hard enough to remove any oxide film, as this could be confused with actual surface contamination. The item being examined must be recleaned if an unacceptable amount of foreign material is detected by this inspection method.

Water Break Test

The water break test may be used to detect oily residues not found by other means. The surface is wetted with a spray of clean water. The water should form a thin layer and remain unbroken for at least five seconds. Beading of the water droplets indicates the presence of oil contaminants, and recleaning is required. This method is generally limited to horizontal surfaces.

Solvent Extraction Test

The solvent extraction test may be used to supplement visual techniques or to check inaccessible surfaces by using a solvent to extract contaminants for inspection. The surface is flushed, rinsed, or immersed in a low-residue solvent. Solvent extraction is limited by the ability of the procedure to reach and dissolve the contaminants present and by the loss of contaminant during solvent evaporation. The equipment tested may also contain materials, such as polymers or elastomers, which would be attacked by the solvent and give erroneous results.

The used solvent may be checked to determine the amount of nonvolatile residue by the following procedure. A known quantity of a representative sample of used solvent, which has been filtered, is evaporated almost to dryness, then transferred to a small weighed beaker for final evaporation, with care taken not to overheat the residue. In the same manner, the weight of residue from a similar quantity of clean solvent is determined. The difference in weight of the two residues and the quantity of solvent used should be used to compute the amount of contaminant extracted per square foot (meter) of surface area cleaned.

In a similar manner, a one liter representative sample of the unfiltered used solvent can be placed in an Imhoff cone and evaporated to dryness. The volume of residue can be measured directly and used to compute the amount of contaminant extracted per square foot (meter) of surface area cleaned. Greater sensitivity can be achieved by evaporating successive liters of solvent in the same Imhoff cone.

Another method is to take a representative sample of known quantity of the used solvent and compare it to a similar sample of new solvent by comparing light transmission through the two samples simultaneously. There should be little, if any, difference in color of the solvents and very few particles.

PACKAGING AND LABELING

Once a piece of equipment has been cleaned for oxygen service and the cleaning agent completely removed from the equipment, it should be suitably protected as soon as practicable to prevent recontamination during storage and prior to being placed in service. Following are several ways in which this can be done. The protection provided will depend on a number of factors such as the type of equipment, length of storage, and atmospheric conditions. The type of protection required should be specified by the purchaser of the equipment.

Protection of Openings

Equipment or parts having small openings may be protected by caps or plugs. Small to medium-sized components may be sealed in plastic bags or protected by other appropriate means. Openings on large equipment may be sealed, preferably with caps, plugs, or blind flanges where appropriate. Taped solid board blanks, or other durable covers which cannot introduce contamination into the equipment when removed, can also be used to seal such openings.

Pressurization

Equipment with large internal volumes may be filled with dry, oil-free air or nitrogen after all openings have been sealed and valves closed. Parts in suitable plastic bags may be purged with inert gas or evacuated and sealed.

Where the purchaser's requirements include labeling to show that oxygen service cleaning of parts or equipment has been performed, a statement such as "Cleaned for oxygen service" or other suitable wording should appear on the part or package as applicable. Additional information which may be included is as follows:

(1) A statement, "This equipment is cleaned in accordance with Oxygen Cleaning Specification No. ______"
(2) Date of inspection and the inspector's stamp or marking
(3) Description of the part, including part number if available
(4) A statement, "Do not open until ready for use"

PERSONNEL SAFETY

Cleaning operations for oxygen service equipment must be carried out in a manner which provides for the safety of personnel performing the work and must also conform to local ordinances and federal, state, and provincial regulations.

Operators must be instructed in the safe use of the cleaning agents employed, including any hazards associated with the use of these agents. Written instructions are to be issued whenever special safety considerations are involved. A responsible individual should direct oxygen cleaning operations.

Dangerous Chemicals

Do not use highly toxic chemicals. For example, carbon tetrachloride must not be employed in any cleaning operation. The health hazards associated with the use of any solvent must be considered in its selection. The time-weighted average threshold limit value (TLV) must not be exceeded for a specific solvent; some chlorinated solvents are suspected of being carcinogenic. Breathing of solvent fumes and liquid contact with the skin should be avoided. Material Safety Data Sheets for solvents should be obtained from the solvent manufacturer in accordance with the requirements of 29 CFR 1910.1200(g). [9]

Caution must be exercised in using solvents commonly referred to as nonflammable but which could become flammable in air under certain conditions. The concentrations creating a flammable mixture in air are usually well in excess of the concentrations that cause physiological harm. Therefore, on removing solvents to the extent necessary to protect personnel from respiratory harm, it must not be forgotten that purging with air may create a flammable mixture. Also, failure to purge adequately can leave a flammable mixture which in the presence of heat, flame, or sparks may result in a dangerous energy release.

Following appropriately prescribed procedures for mixing and handling acids and caustics also contributes to the avoidance of injuries. Special consideration should be given to the safe disposal of waste cleaning solution.

Protective Equipment

Face shields or goggles must be provided for face or eye protection from cleaning solutions. Safety glasses with side protection are required for protection from injuries due to flying particles. Protective clothing must be used when required to prevent cleaning solutions from contacting the skin.

Self-contained breathing apparatus (see ANSI Z88.2, *Practices for Respiratory Protection* [10]) must be provided wherever there is a possibility of a deficiency of oxygen due to the use of an inert gas purge, or if there is any possibility of exceeding allowable TLV values. All areas where cleaning compounds and solvents are used should be adequately ventilated. In outdoor operations, locate cleaning operations so that operators can work upwind of solvent vapor accumulations.

Special Situations

Entering Vessels. Work should not be performed inside a vessel or confined area until the vessel or confined area has been properly prepared and work procedures have been established that will ensure the safety of workers. A Hazardous Work Permit (HWP) is an instrument widely used in industry for ensuring safe working conditions, and its use is strongly recommended. The HWP should consider at least the following seven items before anyone enters a vessel or confined space:

(1) *Isolation:* All lines to a vessel should be suitably isolated to prevent the entry of foreign materials, in particular the inert gases (nitrogen, argon, or the rare gases) that cause asphyxiation by oxygen deficiency. Oxygen enrichment is also to be avoided because of the increased fire hazard. Acceptable means of isolating vessels are blanking, double block and bleed valves, or disconnection of all lines from the vessel.

(2) *Periodic monitoring:* The need for periodic monitoring of the atmosphere in any vessel or confined space must be considered before any work is performed.

(3) *Ventilation:* A fresh air supply suitable for breathing is normally supplied to the vessel when personnel are inside.

(4) *Atmospheric analysis:* The atmosphere in a vessel that has been in service or that

has been purged must always be analyzed at appropriate sampling points before entering to determine that the vessel or confined area has been adequately ventilated with fresh air and is safe for entry.

(5) *Rescue procedure:* A reliable procedure for removing personnel from any vessel or confined work space must be available and all workers trained before any work begins. A portable air breathing supply must be available for each worker and used when entering the vessel or confined work space in such a procedure. An appropriate lifeline must be used when necessary and when required by law.

(6) *Work procedure:* When cleaning operations are performed inside oxygen vessels or other such confined spaces, a reliable preplanned procedure for quickly removing or protecting personnel in cases of emergency must be established and understood by all workers before work begins.

(7) *Watcher:* When toxic cleaning agents are used, it is recommended that a watcher be stationed immediately outside a vessel or confined space to ensure the safety of those working within. A portable air breathing supply must be immediately available. If the worker must use a self-contained breathing apparatus or a breathing mask, a "watcher" must be present per OSHA requirements at 29 CFR 1910.134(d). [9]

Other considerations may be required, depending on the type of work being performed. For example, a vessel should not be entered until its temperature is at or near the ambient temperature. All workers involved with any vessel entry should be fully apprised of the total operation prior to tank entry.

Personnel without self-contained breathing apparatus must not enter any vessel unless its atmosphere has a normal air composition. Normal atmospheric air has 21 percent oxyen by volume. However, it is permissible to work in atmospheres having oxygen concentrations in the range of 19 to 23 percent if the other gases present do not exceed their threshold limit values. In the event that the oxygen concentration deviates from 21 percent, a review of the system is required to ensure that excessive oxygen or an asphyxiant gas is not entering the vessel. See also CGA P-14, *Accident Prevention in Oxygen-Rich and Oxygen-Deficient Atmospheres.* [11]

Heating Solvents. Chlorinated solvents can, upon heating, break down into dangerous compounds. A commonly used solvent, trichloroethylene, decomposes at temperatures not far above the boiling point of water. Ventilation must be adequate to prevent breathing excessive amounts of the solvent vapors or their decomposition products. Air respirators must be used in situations where the concentration of solvent vapors or any other foreign material in the atmosphere exceeds their threshold limit value (TLV).

Welding Near Solvents. It is important to ensure that parts to be welded are free of cleaning solvents. Ultraviolet rays and heat from welding can decompose certain chlorinated solvents to produce poisonous phosgene gas or potentially explosive gas mixtures. Accordingly, the atmosphere in the vicinity of such operations must be free from chlorinated solvent vapors.

REFERENCES

[1] CGA G-4.3, *Commodity Specification for Oxygen,* Compressed Gas Association, Inc., 1235 Jefferson Davis Highway, Arlington, VA 22202.

[2] ASTM F312, *Method for Microscopical Sizing and Counting Particles from Aerospace Fluids on Membrane Filters,* ASTM, 1916 Race Street, Philadelphia, PA 19103.

[3] ASTM G-93, *Practice for Cleaning Methods for Material and Equipment Used in Oxygen-Enriched Environments,* ASTM, 1916 Race Street, Philadelphia, PA 19103.

[4] ASTM G-88, *Guide for Designing Systems for Oxygen Service,* ASTM, 1916 Race Street, Philadelphia, PA 19103.

[5] ASTM G-63, *Guide for Evaluating Nonmetallic Materials for Oxygen Service,* ASTM, 1916 Race Street, Philadelphia, PA 19103.

[6] ASTM STP-812, 910, 986, 1040, *Flammability and*

Sensitivity of Materials in Oxygen-Enriched Atmospheres (ASTM Special Technical Publication Series), ASTM, 1916 Race Street, Philadelphia, PA 19103.

[7] *Metals Handbook,* Volume 5, 9th edition (1982), American Society for Metals, Metals Park, OH 44073.

[8] ASTM D-2108-85, *Test Method for Color Halogenated Organic Solvents and Admixtures,* ASTM, 1916 Race Street, Philadelphia, PA 19103.

[9] *Code of Federal Regulations,* Title 29 CFR Part 1910, Occupational Safety and Health Administration, U.S. Department of Labor, U.S. Government Printing Office, Superintendent of Documents, Washington, DC 20402.

[10] ANSI Z88.2, *Practices for Respiratory Protection,* American National Standards Institute, 1430 Broadway, New York, NY 10018.

[11] CGA P-14, *Accident Prevention in Oxygen-Rich and Oxygen-Deficient Atmospheres,* Compressed Gas Association, Inc., 1235 Jefferson Davis Highway, Arlington, VA 22202.

ADDITIONAL REFERENCES

CGA G-4, *Oxygen,* Compressed Gas Association, Inc., 1235 Jefferson Davis Highway, Arlington, VA 22202.

CGA G-4.1, *Cleaning Equipment for Oxygen Service,* Compressed Gas Association, Inc., 1235 Jefferson Davis Highway, Arlington, VA 22202.

PART III

Compressed Gases and Gas Mixtures: Properties, Manufacture, Uses, and Special Requirements for Safe Handling

Part III provides information on properties, uses, and handling for 44 gases (or gas groups) that are of current commercial importance. Gases produced and used only in small laboratory quantities are not included.

Gases are treated individually in separate monographs in all but four cases where closely related gases have been grouped together. The monographs are arranged in alphabetical order for easy reference. An extensive discussion of gas mixtures, an increasingly important part of the gas industry, appears at the end of Part III.

Each gas monograph opens with basic identifying information: the generally accepted chemical name of the gas, its chemical symbol, other names by which the gas is known, its Chemical Abstracts Service (CAS) registry number, its DOT and Transport Canada classification, and its assigned UN number.

The text of each monograph is divided into the following main subsections:

- Identifying Information
- Physical Constants
- Description
- Grades Available
- Uses
- Physiological Effects
- Materials of Construction
- Safe Storage, Handling, and Use

- Disposal
- Handling Leaks and Emergencies
- Methods of Shipment
- Containers
- Methods of Manufacture
- References

Physical constants appear in tabular form in the section. Data are given in customary U.S. units and equivalent SI metric units. Included are:

- Chemical formula
- Molecular weight
- Vapor pressure
- Density of the gas
- Specific gravity (compared to air)
- Specific volume
- Density as a liquid
- Boiling point
- Melting point
- Critical temperature
- Critical pressure
- Critical density
- Triple point
- Latent heat of vaporization
- Latent heat of fusion
- Specific heat
- Ratio of specific heats
- Solubility in water
- Weight of liquid

Physical constants in addition to those listed above are given when they are important to the safe handling and use of a particular gas.

QUALIFICATIONS ON PHYSICAL CONSTANTS AND OTHER MATERIAL

Data given on the physical constants in the gas monographs are based on authoritative scientific and industrial sources, as are all other matters of factual information and recommended practice. In presenting this material, the publisher and the Compressed Gas Association, Inc., assume no legal responsibility whatever for any losses or injuries sustained, or liabilities incurred, by persons or organizations acting in any way on the basis of any part of the material. However, the information and recommendations are believed to be accurate and sound to the best knowledge of the CGA.

The data given on physical constants generally represent the properties of pure commodities rather than those of commercial grades of the gases. The properties of commercial grades should be expected to differ somewhat from the values for pure grades presented here.

For those gases in Part III for which vapor pressure curves are included, the curves were generated from values available in published literature. Curve-smoothing techniques were used as necessary to compensate for graphic anomalies resulting from a shortage of available data points over a given range.

Summaries of shipping regulations in the Part III gas monographs are of course based on the regulations in effect while the *Handbook* was being prepared. For the full and current regulations, the reader is strongly ad-

MOISTURE CONVERSION DATA TABLE.

Dew Point °F	Dew Point °C	Moisture Content ppm (v/v)	Moisture Content mg/L
−110	−78.9	0.58	0.00043
−105	−76.1	0.93	0.00069
−100	−73.3	1.5	0.0011
− 95	−70.5	2.3	0.0017
− 90	−67.8	3.5	0.0026
− 85	−65.0	5.3	0.0040
− 80	−62.2	7.8	0.0058
− 75	−59.4	11.4	0.0085
− 70	−56.7	16.2	0.012
− 65	−53.9	23.0	0.017
− 60	−51.1	32.0	0.024
− 55	−48.3	45.0	0.034
− 50	−45.6	63.0	0.047
− 45	−42.8	87.0	0.065
− 40	−40.0	120	0.089
− 35	−37.2	165	0.12
− 30	−34.4	225	0.17
− 25	−31.6	305	0.23
− 20	−28.9	400	0.30
− 15	−26.1	525	0.39
− 10	−23.3	690	0.51
− 5	−20.5	895	0.67
0	−17.8	1180	0.88

vised to consult current editions of their published forms as identified and referenced.

It is also important for readers to understand that the authorized service pressures given in gas sections for cylinders and other containers are only the minimum service pressures cited in U.S. Department of Transportation or Transport Canada regulations. For example, if cylinders meeting DOT or TC specifications 3AA150 are noted in a gas section as authorized for a given gas, then any other 3AA cylinders with higher service pressures are also authorized (such as 3AA1000, 3AA2000, etc.).

MOISTURE CONVERSION DATA

Water/Dew Point is expressed in ppm (volume/volume) and degrees F at one atmosphere absolute, 14.696 psia (101.325 kPa abs: 760 mm Hg). The Moisture Conversion Table can be used to convert the moisture content (water content) in any gas from ppm (v/v) to mg/L or vice versa.

Acetylene

Chemical Symbol: C_2H_2
Synonyms: Ethine, Ethyne
CAS Registry Number: 74-86-2
DOT Classification: Flammable gas
DOT Label: Flammable or Flammable gas
Transport Canada Classification: 2.1
UN Number: UN 1001

PHYSICAL CONSTANTS

	U.S. Units	SI Units
Chemical formula	C_2H_2	C_2H_2
Molecular weight	26.04	26.04
Vapor pressure[a] at 70°F (21.1°C)	635 psig	4378 kPa
Density of the gas at 32°F (0°C) and 1 atm	0.07314 lb/ft^3	1.1716 kg/m^3
Specific gravity of the gas at 32°F and 1 atm (air = 1)	0.906	0.906
Specific volume of the gas[b] at 70°F (21.1°C) and 1 atm	14.7 ft^3/lb	0.918 m^3/kg
Specific gravity of the liquid at −112°F (−80°C)	0.613	0.613
Density of the liquid at 70°F (21.1°C)	24.0 lb/ft3	384 kg/m3
Boiling point at 10 psig[c]	−103°F	−75.0°C
Melting point at 10 psig[c]	−116°F	−82.2°C
Critical temperature	96.8°F	36.0°C
Critical pressure	907 psia	6250 kPa abs

[a]In this monograph, as throughout the *Handbook,* gauge pressure is denoted by the terms *psig* and *kPa,* whereas absolute pressure is denoted by the terms *psia* and *kPa abs.*

[b]Based on 1.171 g/liter at 32°F (0°C) and 1 atm.

[c]Reported at 10 psig instead of at 1 atm, because at 1 atm, acetylene sublimes directly from the solid to the gaseous state without entering the liquid state. Its sublimation point at 1 atm is −118°F (−83.3°C).

	U.S. Units	SI Units
Critical density	14.4 lb/ft^3	231 kg/m^3
Triple point	−116°F at 17.7 psia	−82.2°C at 122 kPa abs
Latent heat of vaporization at triple point	264 Btu/lb	614 kJ/kg
Latent heat of fusion at −114.7°F (−81.5°C)	41.56 Btu/lb	96.67 kJ/kg
Specific heat of the gas at 60°F (15.5°C) and 1 atm		
C_p	0.383 Btu/(lb)(°F)	1.60 kJ/(kg)(°C)
C_v	0.304 Btu/(lb)(°F)	1.27 kJ/(kg)(°C)
Ratio of specific heats	1.26	1.26
Solubility in water, vol/vol at 60°F (15.6°C)	1.1	1.1
Specific volume of the gas at 60°F (15.6°C) and 1 atm	14.5 ft^3/lb	0.905 m^3/kg
Solubility in water, vol/vol at 32°F (0°C) and 1 atm	1.7	1.7

DESCRIPTION

Acetylene is a compound of carbon and hydrogen in proportions by weight of about 12 parts carbon to 1 part hydrogen (92.3 to 7.7 percent). A colorless, flammable gas, it is slightly lighter than air. Acetylene of 100 percent purity is odorless, but acetylene of ordinary commercial purity generated from calcium carbide has a distinctive, garliclike odor.

Acetylene burns in air with an intensely hot, luminous, and smoky flame. The ignition temperatures of acetylene and of acetylene-air and acetylene-oxygen mixtures vary according to composition, initial pressure, initial temperature, and water vapor content. As a typical example, an air mixture containing 30 percent acetylene by volume at atmospheric pressure can be ignited at about 581°F (305°C). The flammable limits of acetylene-air and acetylene-oxygen mixtures similarly depend on initial pressure, temperature, and water vapor content. In air at atmospheric pressure, the upper flammable limit is about 80 percent acetylene by volume and the lower limit is 2.5 percent acetylene. Some references list the upper flammable limit as 100 percent, which is due to the decomposition of acetylene. If an ignition source is present, 100 percent acetylene under pressure as low as 6 psig (41 kPa) will decompose with violence under certain conditions of container size and shape.

Acetylene can be liquefied and solidified with relative ease. However, in both the liquid and solid states, acetylene is shock sensitive and explodes with extreme violence when ignited. For this reason, DOT and TC regulations prohibit the shipment of liquid or solidified acetylene. A mixture of gaseous acetylene with air or oxygen in certain proportions explodes if ignited. Gaseous acetylene under pressure may also decompose with explosive force under certain conditions, but experience indicates that 15 psig (103 kPa) is generally acceptable as a safe upper pressure limit when proper equipment and procedures are utilized. Generation, distribution through hose or pipe, or utilization of acetylene at pressures in excess of 15 psi gauge pressure (103 kPa) or 30 psi absolute pressure (207 kPa abs) for welding and allied purposes should be prohibited.

Pressure exceeding 15 psig (103 kPa) can be employed provided specialized equipment is used. Where acetylene is to be utilized for chemical synthesis at pressures in excess of

15 psig (103 kPa), or transported through large-diameter pipelines, means to prevent propagation, should a decomposition reaction occur, must be employed. Packing large-diameter pipe with small-diameter pipes as a protection against exposure to fires is recommended.

Acetylene cylinders avoid the decomposition characteristics of the gas by providing a porous-mass filler material having minute cellular spaces so that no pockets of appreciable size remain where "free" acetylene in gaseous form can collect. This porous mass is saturated with acetone or another suitable solvent into which acetylene dissolves. The combination of these two features—porous filler and solvent—allows acetylene to be contained in such cylinders at moderate pressure without danger of explosive decomposition (the maximum authorized cylinder pressure is 250 psig (1724 kPa) at 70°F (21.1°C), with a variation of about 2.5 psig (17 kPa) rise or fall per degree Fahrenheit or 31 kPa rise or fall per degree Celsius of temperature change.

Refer to CGA G-1, *Acetylene,* for a more thorough discussion on the properties of acetylene and the safe use of acetylene cylinders. [1]

GRADES AVAILABLE

Table 1, from CGA G-1.1, *Commodity Specification for Acetylene*, presents component maxima in parts per million, ppm (mole/mole) unless otherwise shown, for the grades (also denoted as quality verification levels) of acetylene. [2] A blank indicates no maximum limiting characteristic. The absence of a listed value in a quality verification level does not mean to imply that the limit-characteristic is or is not present but merely indicates that the test is not required for compliance with the specification.

USES

Approximately 80 percent of the acetylene produced annually in the United States is used for chemical synthesis. It is possible to use acetylene for an almost infinite number of organic chemical syntheses, but this use in North America has been less extensive than in Europe owing to the ready availability of petroleum from which competitive synthesis routes are often possible. Nevertheless, acetylene has come into increasing prominence as the raw material for a whole series of organic compounds, among them acetaldehyde, acetic acid, acetic anhydride, acetone, and vinyl chloride. These compounds may be used in turn to produce a diverse group of products including plastics, synthetic rubber, dyestuffs, solvents, and pharmaceuticals. Acetylene is also utilized to manufacture carbon black. The remaining 20 percent of annual United States acetylene production is used principally for oxyacetylene welding,

TABLE 1. ACETYLENE GRADES AVAILABLE.
(Units in ppm (mole/mole) unless otherwise stated)

Limiting Characteristics	Grades						
	A	B	C	D	E	F	H
Acetylene min. percent assay	95	98	98	98	99.5	99.5	99.6
Phosphine and Arsine (2) (ppm)			500	50	500	50	25
Hydrogen Sulfide (2) (ppm)			500	50	500	50	25

Note 1. Cylinder acetylene contains variable percentage quantities of a solvent (normally acetone), the amount of solvent present in the expelled gas being dependent upon the vapor pressure of the solvent, the conditions of the cylinder, and the conditions of withdrawal. The purities listed in Table 1 are given on a solvent-free basis.

Note 2. Acetylene manufactured from hydrocarbon feedstock is inherently free from phosphine, arsine, and hydrogen sulfide.

cutting, heat treating, and so on. Small amounts are utilized for lighting purposes in buoys, beacons, and similar devices.

PHYSIOLOGICAL EFFECTS

Acetylene can be inhaled in rather high concentrations without chronic effects. When mixed with oxygen in high percentages, it acts as a narcotic and has been used in anesthesia. However, because of the extreme hazard in producing and using such a mixture, it has not been used for this purpose since 1930.

Acetylene acts as an asphyxiant by diluting the oxygen in the air to a level which will not support life. However, prior to reaching a level where suffocation could occur, the lower flammable limit will have been reached, and this, of course, constitutes a most serious hazard.

MATERIALS OF CONSTRUCTION

Only steel or wrought-iron pipe should be used for acetylene piping systems. Joints in piping must be welded or made with threaded or flanged fittings. Heavier wall thickness pipe must be used when threaded piping is used. The materials for fittings can be rolled, forged, or cast steel, or malleable iron. Cast-iron fittings are not permitted. Under certain conditions acetylene forms readily explosive acetylide compounds when in contact with copper, silver, and mercury. For this reason, acetylene and the use of these metals, or their salts, compounds, and high-concentration alloys is to be avoided.

It is generally accepted that brass containing less than 65 percent copper in the alloy, and certain nickel alloys, are suitable for use in acetylene service under normal conditions. Conditions involving contact with highly caustic salts or solutions, or contact with other materials corrosive to copper or copper alloys, can render the above generally acceptable alloys unsatisfactory for this service. The presence of moisture, certain acids, or alkaline materials tends to enhance the formation of copper acetylides. Further information on metallic acetylides can be found by consulting a number of the Additional References given at the end of this monograph. Bulk plant or chemical plant piping should be in accordance with ASME B31.3, *Chemical Plant and Petroleum Refinery Piping*. [3]

Acetylene customers should not attempt to install acetylene piping systems without specific knowledge of the unique properties of acetylene. ASME B31.3 provides general requirements for chemical plant process piping. NFPA 51, *Standard for the Design and Installation of Oxygen-Fuel Gas Systems for Welding, Cutting and Allied Processes*, published by the National Fire Protection Association, provides standards for installation of multiple cylinder systems. [4] For further recommendations on acetylene cylinder discharge manifolds and shop piping, users should consult their supplier and recognized safety authorities such as the Underwriters' Laboratories, Inc., the Associated Factory Mutual Fire Insurance Companies, and the Compressed Gas Association, Inc.

SAFE STORAGE, HANDLING, AND USE

Acetylene is primarily stored in cylinders. In storing acetylene cylinders, the user should comply with all local, municipal, and state or provincial regulations, and with NFPA 51, *Standard for the Design and Installation of Oxygen-Fuel Gas Systems for Welding, Cutting and Allied Processes.* [4]

Inside all buildings, acetylene cylinders should not be stored near oxygen cylinders. Unless they are well separated, there should be a noncombustible partition at least 5 ft (1.5 m) high with a fire resistive rating of one-half hour between acetylene cylinders and oxygen cylinders.

In the United States, acetylene cylinders stored inside a building at user locations must be limited to a total capacity of 2500 ft^3 (70 m^3) of gas, exclusive of cylinders in use or attached for use. Quantities exceeding this total must be stored in a special building or in a separate room as required by NFPA 51. [4]

In Canada, regulations limit the capacity of acetylene cylinders stored inside a building at user locations to a total capacity of 2160 ft^2 (60 m^3) of gas in unsprinklered, combustible structures and limit the quantity to 6130 ft^3 (170 m^3) in sprinklered buildings of combustible or noncombustible construction before a special room or building is required. Conspicuous signs must be posted in the storage area forbidding smoking or the carrying of open lights.

While storage in a horizontal position does not make the acetylene in cylinders less stable or less safe, it does increase the likelihood of solvent loss, which will result in a lower flame quality when the cylinder is used. There can also be a greater hazard with such storage, owing to flame impingement that would result from fuse plug release with ignition, which might cause violent rupture of any impinged nearby cylinders. A release from a fuse plug opening can project flame up to 15 ft (4.6 m). Therefore, it is always preferable to store acetylene cylinders in an upright position. Acetylene cylinders should not be stored where they can be struck or knocked over. Cylinders at user locations should be tightly "nested" or secured with straps or chains.

Handling Acetylene Cylinders

Always call acetylene by its proper name, *acetylene*, to promote recognition of its hazards and the taking of proper precautions. Never refer to acetylene merely as *gas*.

Never attempt to repair or alter cylinders. This should be done only by the cylinder manufacturer. Valve repairs must be performed only by the supplier of the acetylene gas. If a cylinder is leaking, follow the recommendations in the section on Handling Leaks and Emergencies.

Never tamper with pressure relief devices in valves or cylinders. Keep sparks and flames away from acetylene cylinders and under no circumstances allow a torch flame to come in contact with the fusible metal pressure relief devices, which melt at approximately 212°F (100°C). Should the valve outlet of an acetylene cylinder become clogged by ice, thaw with warm but not boiling water.

Never under any circumstances attempt to transfer acetylene from one cylinder to another, to refill acetylene cylinders, or to mix any other gas with acetylene in a cylinder.

In welding shops and industrial plants where both oxyacetylene and electric welding apparatus are used, care must be taken to avoid the handling of this equipment in any manner which may permit the compressed gas cylinders to come in contact with the electric welding apparatus or electrical circuits.

Never use acetylene cylinders as rollers or supports, or for any purpose other than storing acetylene.

Moving Acetylene Cylinders

Cylinders must be protected against dropping when being unloaded from a truck or platform. Special caution is necessary in transporting acetylene cylinders by crane or derrick. Lifting magnets, slings, rope, chain, or any other device in which the cylinders themselves form a part of the carrier must never be used for hoisting acetylene cylinders. Instead, when a crane is used, a platform, cage, or cradle should be provided to protect the cylinders from being damaged by slamming against obstructions and to keep them from falling out. A recommended type of cradle to build for this purpose is shown in Fig. 1.

Horizontal movement of cylinders is easily accomplished by the use of a hand truck; however, when a hand truck is used, some positive method such as chaining should be used to secure a cylinder standing upright in the hand truck. Cylinders must not be transported lying horizontally on hand trucks with the valve overhanging in a position to

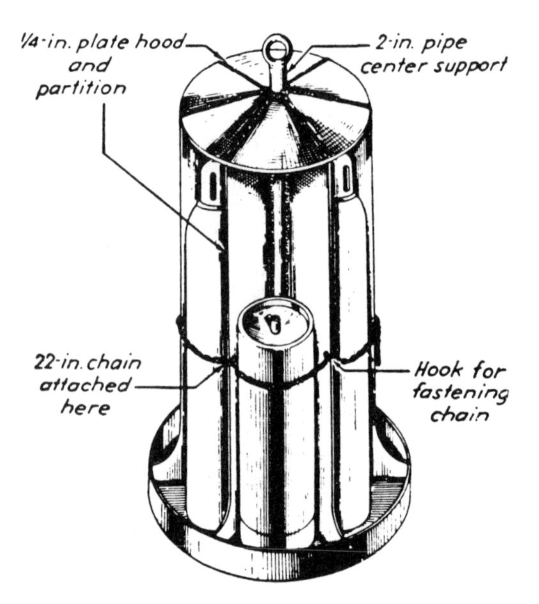

Fig. 1. Recommended type of cradle to hold acetylene cylinders when moved with a crane or derrick.

collide with stationary objects. Cylinders should never be dragged from place to place.

Valves should always be closed before cylinders are moved. Unless cylinders are to be moved while secured in an upright position on a suitable hand truck, pressure regulators should be removed and valve protection caps, if provided for in the cylinder design, should be attached.

Use of Acetylene

When using acetylene for welding or cutting operations with single-cylinder systems, comply with ANSI Z49.1, *Safety in Welding and Cutting.* [5] Also refer to CGA SB-8, *Use of Oxy-Fuel Gas Welding and Cutting Apparatus.* [6]

Never use acetylene through torches or other devices equipped with shutoff valves on the acetylene supply connections without reducing the pressure through a suitable regulator attached to the cylinder valve. Acetylene should never be used in equipment at pressures exceeding 15 psig (103 kPa). Acetylene cylinders should be used in an upright position to avoid loss of solvent and accompanying reduction in flame quality.

In preparing to withdraw acetylene from cylinders, use only wrenches or other tools approved by the manufacturer of the gas for operating cylinder valves. After removing the valve protection cap, partially open (i.e., "crack") the valve for an instant in order to clear the opening of particles of dust or dirt, being careful to stand so that the valve points away from the body. Avoid blowing dangerous amounts of the gas in confined spaces. Do not "crack" an acetylene cylinder valve near welding work, sparks, open flame, or any other possible sources of ignition.

Be sure that all connections are gastight and remain so, and that the connected hose is in good condition and does not have any leaks. Regarding leak detection, see the section on Handling Leaks and Emergencies. Refer also to CGA SB-8, *Use of Oxy-Fuel Gas Welding, and Cutting Apparatus*, for further information. [6]

Always open and close the acetylene cylinder valve slowly to minimize pressure surges. Only use regulators designed for acetylene service. Such gauges are low pressure gauges marked in red above 15 psig to warn against using acetylene at higher pressures. Never use a hammer or mallet in attempting to open or close a valve.

Do not open an acetylene cylinder valve more than one-and-one-half turns. This will provide full flow and minimize the time needed to close the valve in an emergency. Do not stand in front of the regulator and gauge faces when opening the valve.

Do not pile hose, tools, or other objects on top of an acetylene cylinder where they might interfere with quick closing of the valve. On wrench-operated cylinder valves, the wrench used for opening the cylinder valve should always be kept on the valve stem when the cylinder is in use.

Always close the cylinder valve and then bleed pressure from the regulator when the work is finished. Be sure the cylinder valve is closed and all gas is released from the regulator before removing the regulator from a cylinder. When using acetylene in welding and cutting, never allow the regulator and hose to remain pressurized when not in use.

Never apply a torch to the side of a cylinder to raise the pressure. Serious accidents have resulted from violation of this rule.

DISPOSAL

Disposal of any unused cylinder should be handled by the supplier. CGA C-2, *Recommendations for the Disposition of Unserviceable Compressed Gas Cylinders with Known Contents*, provides further information on the proper means to dispose of unserviceable acetylene cylinders. [7]

HANDLING LEAKS AND EMERGENCIES

Because acetylene and air in certain proportions are very flammable and may burn with explosive force, care should be taken to prevent acetylene leakage. Connections must be kept tight and the hose maintained in good condition. Points of suspected leakage should be tested by covering them with soapy water. A leak will be indicated by bubbles of escaping acetylene passing through the soap film. Never test for leaks with an open flame!

If acetylene leaks from around the valve stem when the valve is open, close the valve and tighten the gland nut. This compresses the packing around the stem. If this does not stop the leak, leave the valve closed. Attach a tag to the cylinder stating that the valve is unserviceable and move it to a safe area. *Do not* attempt to tighten the bonnet on O-ring packless valves or on valves having nonrising stems (nonadjustable bonnets). Notify the gas supplier and follow the supplier's instructions for the cylinder's return.

Acetylene cylinders are equipped with fusible metal plugs having a melting point between 208°F and 220°F (212°F nominal; or 97.8°C to 104.4°C, 100°C nominal), which may be located in the top and bottom heads of the cylinders, or in the cylinder valve on 10 (280 L) and 40 ft^3 (1.1 m^3) cylinders.

If acetylene leaks from the valve even when the valve is closed, or if rough handling or other occurrences should cause any fusible plugs to leak, move the cylinder to an open space well away from any possible source of ignition and plainly tag the cylinder as having an unserviceable valve or fuse plug. Open the valve slightly to let the acetylene escape slowly. Place a sign at the cylinder warning "Leaking Flammable Gas/No Smoking" to caution persons against approaching the cylinder with cigarettes or other open lights. When the cylinder is empty, close the valve. Notify the manufacturer immediately of the serial number of the cylinder and the particulars of its defect, as far as known, and await instructions.

For the safe use of acetylene welding and cutting equipment and to prevent fires, the user should comply with NFPA 51B, *Standard for Fire Prevention in Use of Cutting and Welding Processes*. [8] Also, see CGA SB-4, *Handling Acetylene Cylinders in Fire Situations*. [9]

METHODS OF SHIPMENT

Only cylinders are authorized for shipping acetylene. Acetylene producers using the gas for chemical synthesis store acetylene in low-pressure gas holders for which the recommended material is carbon steel.

Under the appropriate regulations, acetylene is authorized for shipment as follows:

By Rail: In cylinders.

By Highway: In cylinders on trucks.

By Water: In cylinders, in the United States via cargo vessels, passenger vessels, passenger or vehicle ferry vessels, and passenger or vehicle railroad car ferry vessels. On barges, in cylinders for barges of U.S. Coast Guard Classes A and C only.

In Canada, cylinders must be stowed according to the requirements of the *International Maritime Dangerous Goods Code*. [10] Cylinders should be shaded from radiant heat, stowed clear of living quarters, and separated from chlorine. For cargo ships or passenger ships which are carrying not more than 25 passengers or 1 per 3 m (10 ft) of length, cylinders should be stowed on deck only. For other passenger ships, cylinders should be stowed on deck in a position not accessible to unauthorized persons.

By Air: In the United States, aboard cargo aircraft only, in cylinders as required up to 300 lb (136 kg) maximum net weight per cylinder. In Canada, acetylene may be shipped aboard cargo aircraft only, in cylinders required up to 15 kg (33 lb) maximum net weight per cylinder.

Acetylene must not be filled to pressures exceeding those specified in 49 CFR 173.303(b) or, in Canada, by CSA B340. [11] and [12]

The maximum filling pressure after "settling back" authorized for acetylene in cylinders that meets the specifications and solvent filling requirements of DOT and Transport Canada is 250 psig at 70°F (1724 kPa at 21.1°C). [11] and [13]

Only cylinders that meet TC/DOT Specifications 8 or 8AL or 8WC, and that also meet requirements for fillings of a porous material and a suitable solvent, can be used for acetylene service. DOT regulations prohibit shipment of cylinders containing acetylene gas unless they were charged by or with the consent of the owner. A periodic requalification of acetylene cylinders is required by DOT and is described in CGA C-13, *Guidelines for Periodic Visual Inspection and Requalification of Acetylene Cylinders*. [14]

Transportation of an acetylene cylinder in a closed passenger vehicle has resulted in accidents due to poor ventilation, heat buildup, and/or improperly secured cylinders. Temperatures in the trunk of a car can reach over 140°F (60°C) on a hot day. An acetylene cylinder can be safely transported after being properly secured in a well-ventilated passenger vehicle such as an open-bed pickup truck. Acetylene cylinders must not be transported in nonventilated compartments of passenger vehicles.

CONTAINERS

Cylinders

Acetylene is most commonly available in cylinders of approximate capacities of 10, 40, 60, 100, 225, 300, 400, and 850 ft^3 (0.2 to 24 m^3). "Lighthouse" type cylinders—those generally used in acetylene-operated automatic aids to marine navigation—are available in larger sizes, the biggest having a capacity of approximately 1400 ft^3 (40 m^3).

Do not attempt to charge acetylene into any cylinders except those constructed for acetylene. Do not charge any other gas but acetylene into an acetylene cylinder. Do not mix any other gas with acetylene in an acetylene cylinder. ***Failure to observe these warnings may result in a serious accident.***

The following marks are required by DOT and TC to be plainly stamped on or near the shoulder or top head of all acetylene cylinders as follows: (1) the U.S. DOT and/or TC specification number: TC/DOT 8 or TC/DOT 8AL or TC 8WC; (2) a serial number and the user's, purchaser's, or maker's identifying symbol (the symbol must be registered with DOT in the United States, Associate Director for Hazardous Materials Regulation (HMR) and/or with Transport Canada in Canada); (3) the date of the test to which it was subjected in manufacture; and (4) the tare weight of the cylinder in pounds and ounces or kilograms.

Note: Where the initials of both regulatory agencies are used, the U.S. Department of Transportation requires the initials DOT to be adjacent to the alpha-numeric specification number.

The markings on cylinders must not be changed except as provided in DOT regulations. Current regulations forbid removal of original markings required by law, but allow for certain additional markings when a detailed application is made to and approval received from the DOT Director for Hazardous Materials Regulation in the United States or the Canadian Transport Commission in Canada. Markings on cylinders must be kept in a readable condition.

Authorized pressure relief devices on acetylene cylinders are Type CG-3 fusible plugs with a nominal yield temperature of 212°F (100°C). [15]

Valve Outlet Connections

The standard valve outlet connection in the United States and Canada for acetylene cylinders over 50 ft^3 (1.41 m^3) is Connection CGA 510. The limited standard connection in the United States and Canada is Connection CGA 300. Small valve series limited standard connections for the United States and Canada are Connections CGA 200 for cylinders of approximately 10 ft^3 (280 L) and Connection 520 for cylinders between 35 and 75 ft^3 (970 L and 2.12 m^3). Additional information may be found in CGA V-1, *American National, Canadian, and Compressed Gas Association Standard for Compressed Gas Cylinder Outlet and Inlet Connections.* [16]

METHODS OF MANUFACTURE

In the United States and Canada, calcium carbide is the principal raw material for acetylene manufacture. Calcium carbide and water may be made to react by several methods to produce acetylene, with calcium hydroxide as a by-produce. Acetylene is also manufactured by the thermal or arc cracking of hydrocarbons and by a process employing the partial combustion of methane with oxygen.

Acetylene manufactured from carbide made in the United States and Canada normally contains less than 0.4 percent impurities other than water vapor. Apart from water, the chief impurity is air, in concentrations of approximately 0.2 and 0.4 percent. The remainder is mostly phosphine, ammonia, hydrogen sulfide, and in some instances, small amounts of carbon dioxide, hydrogen, methane, carbon monoxide, organic sulfur compounds, silicon hydrides, and arsine. Purified cylinder acetylene is substantially free from phosphine, ammonia, hydrogen sulfide, organic sulfur compounds, and arsine. The other impurities are nearly the same as in the original gas.

Acetylene cylinder filling plants should comply with NFPA 51A, *Standard for Acetylene Cylinder Charging Plants.* [17] This standard provides guidance on the design, construction, and operations of these plants. All acetylene plant operations should use CGA C-13, *Guidelines for Periodic Visual Inspection and Requalification of Acetylene Cylinders*, as a reference for prefill cylinder inspection and for periodic requalification of cylinders. [14]

REFERENCES

[1] CGA G-1, *Acetylene*, Compressed Gas Association, Inc., 1235 Jefferson Davis Highway, Arlington, VA 22202.

[2] CGA G-1.1, *Commodity Specification for Acetylene*, Compressed Gas Association, Inc., 1235 Jefferson Davis Highway, Arlington, VA 22202.

[3] ASME B31.3, *Chemical Plant and Petroleum Refinery Piping* (ANSI/ASME B31.3), American Society for Mechanical Engineers, 345 E. 47th Street, New York, NY 10017-2392.

[4] NFPA 51, *Standard for the Design and Installation of Oxygen-Fuel Gas Systems for Welding, Cutting and Allied Processes*, National Fire Protection Association, Batterymarch Park, Quincy, MA 02269.

[5] ANSI Z49.1, *Safety in Welding and Cutting*, American Welding Society, 550 N.W. LeJeune Road, P.O. Box 351040, Miami, FL 33135.

[6] CGA SB-8, *Use of Oxy-Fuel Gas Welding and Cutting Apparatus*, Compressed Gas Association, Inc., 1235 Jefferson Davis Highway, Arlington, VA 22202.

[7] CGA C-2, *Recommendations for the Disposition of Unserviceable Compressed Gas Cylinders with Known Contents*, Compressed Gas Association, Inc., 1235 Jefferson Davis Highway, Arlington, VA 22202.

[8] NFPA 51B, *Standard for Fire Prevention in Use of Cutting and Welding Processes*, National Fire Protection Association, Batterymarch Park, Quincy, MA 02269.

[9] CGA SB-4, *Handling Acetylene Cylinders in Fire Situations*, Compressed Gas Association, Inc.,

1235 Jefferson Davis Highway, Arlington, VA 22202.
[10] *International Maritime Dangerous Goods Code*, International Maritime Organization, 4 Albert Embankment, London, England SE1 7SR.
[11] *Code of Federal Regulations*, Title 49 CFR Parts 100–199, (Transportation), Superintendent of Documents, U.S. Government Printing Office, Washington, DC 20402
[12] CSA B340, *Selection and Use of Cylinders, Spheres, Tubes and Other Containers for the Transportation of Dangerous Goods*, Canadian Standards Association, 178 Rexdale Boulevard, Rexdale (Toronto), Ontario, Canada M9W 1R3.
[13] *Transportation of Dangerous Goods Regulations*, Canadian Government Publishing Centre, Supply and Services Canada, Ottawa, Ontario, Canada K1A 0S9.
[14] CGA C-13, *Guidelines for Periodic Visual Inspection and Requalification of Acetylene Cylinders*, Compressed Gas Association, Inc., 1235 Jefferson Davis Highway, Arlington, VA 22202.
[15] CGA S-1.1, *Pressure Relief Device Standards—Part 1—Cylinders for Compressed Gases*, Compressed Gas Association, Inc., 1235 Jefferson Davis Highway, Arlington, VA 22202.
[16] CGA V-1, *American National, Canadian, and Compressed Gas Association Standard for Compressed Gas Cylinder Valve Outlet and Inlet Connections*, Compressed Gas Association, Inc., 1235 Jefferson Davis Highway, Arlington, VA 22202.
[17] NFPA 51A, *Standard for Acetylene Cylinder Charging Plants*, National Fire Protection Association, Batterymarch Park, Quincy, MA 02269.

ADDITIONAL REFERENCES

Acetylene—Its Properties, Manufacture and Uses, S. A. Miller, London, 1965.

CGA G-1.2, *Recommendations for Chemical Acetylene Metering*, Compressed Gas Association, 1235 Jefferson Davis Highway, Arlington, VA 22202.

CGA G-1.3, *Acetylene Transmission for Chemical Synthesis*, Compressed Gas Association, Inc., 1235 Jefferson Davis Highway, Arlington, VA 22202.

CGA G-1.6, *Recommended Practices for Mobile Acetylene Trailer Systems*, Compressed Gas Association, 1235 Jefferson Davis Highway, Arlington, VA 22202.

Metallic Acetylides

Bramfeld, V. F., Clark, M. T., and Seyfang, A. P., "Copper Acetylides," *J. Soc. Chem. Ind. (London)*, **66**, 346–53 (October 1947).

"Conditions of Formation and Properties of Copper Acetylide," unpublished research paper by L'Air Liquide, Paris, France.

Feitnecht, von H., and Hugi-Carmes', L., "Ueber Bildung und Eigenschafter der Kupferacetylide," *Schweizer Archiv Angew, Wiss. Tech.*, **10**, 23 (1957).

"The Formation and Properties of Acetylides," paper presented by G. Benson, Shawinigan Chemicals, Ltd., at the Compressed Gas Association Canadian Section, September 17, 1950.

Nieuwland and Vogt, *The Chemistry of Acetylene*, ACS Monograph No. 99, Van Nostrand Reinhold, New York.

Air

Synonyms: Compressed air, atmospheric air, the atmosphere (of the earth)
CAS Registry Number: None (for nitrogen, 7727-37-9;
for oxygen, 7782-44-7)
DOT Classification: Nonflammable gas
DOT Label: Nonflammable gas
Transport Canada Classification: 2.2
UN Number: UN 1002 (compressed gas); UN 1003 (refrigerated liquid)

PHYSICAL CONSTANTS

	U.S. Units	SI Units
Chemical name	Air	Air
Molecular weight	28.975	28.975
Density of the gas at 70°F (21.1°C) and 1 atm	0.07493 lb/ft^3	1.2000 kg/m^3
Specific gravity of the gas at 70°F (21.1°C) and 1 atm (air = 1)	1.00	1.00
Specific volume of the gas at 70°F (21.1°C) and 1 atm	13.346 ft^3/lb	0.8333 m^3/kg
Boiling point at 1 atm	−317.8°F	−194.3°C
Freezing point at 1 atm	−357.2°F	−216.2°C
Critical temperature	−221.1°F	−140.6°C
Critical pressure	547 psia	3771 kPa abs
Critical density	21.9 lb/ft^3	351 kg/m^3
Latent heat of vaporization at normal boiling point	88.2 Btu/lb	205 kJ/kg
Specific heat of gas at 70°F (21.1°C) and 1 atm		
C_p	0.241 Btu/(lb)(°F)	1.01 kJ/(kg)(°C)
C_v	0.172 Btu/(lb)(°F)	0.720 kJ/(kg)(°C)
Ratio of specific heats (C_p/C_v)	1.40	1.40
Solubility in water, vol/vol at 32°F (0°C)	0.0292	0.0292

	U.S. Units	SI Units
Weight of liquid at normal boiling point	7.29 lb/gal	874 kg/m^3
Density of liquid at boiling point and 1 atm	54.56 lb/ft^3	874.0 kg/m^3
Gas/liquid ratio (liquid at boiling point, gas at 70°F and 1 atm), vol/vol	728.1	728.1
Thermal conductivity		
at −148°F (−100°C)	0.0095 Btu/(hr)(ft)(°F/ft)	0.0164 W/(m)(°C)
at 32°F (0°C)	0.0140 Btu/(hr)(ft)(°F/ft)	0.0242 W/(m)(°C)
at 212°F (100°C)	0.0183 Btu/(hr)(ft)(°F/ft)	0.0317 W/(m)(°C)

DESCRIPTION

Air is the natural atmosphere of the earth—a nonflammable, colorless, odorless gas that consists of a mixture of gaseous elements (with water vapor, a small amount of carbon dioxide, and traces of many other constituents). Synthesized air is produced by combining pure oxygen and nitrogen and contains between 19.5 and 23.5 percent oxygen, with the balance nitrogen and with a major portion of the other components eliminated. Dry air is noncorrosive. Liquefied air is transparent with a bluish cast and has a milky color when it contains carbon dioxide.

Because air is a mixture, not a compound, it can be separated into its components. The most common method is the liquefaction of air by reducing its temperature to approximately −320°F (−195.6°C), then fractionally distilling to remove each of the constituents as fractions.

Air can be compressed at the point of use for most practical applications. To meet needs for air of special purity or specified composition (as in certain medical, scientific, industrial, fire protection, undersea, and aerospace uses), it is purified or compounded synthetically and shipped in cylinders as a nonliquefied gas at high pressures.

A typical analysis of dry air at sea level has the following composition:

Component	% by Mole	% by Weight
Nitrogen	78.084	75.5
Oxygen	20.946	23.2
Argon	0.934	1.33
Carbon dioxide	0.0335[a]	0.045
Neon	0.001818	—
Helium	0.000524	—
Methane	0.0002	—
Krypton	0.000114	—
Nitrous oxide	0.00005	—
Xenon	0.0000087	—

[a]Variable.

Atmospheric air also contains varying amounts of water vapor. For most practical purposes, the air composition is taken to be 78 percent nitrogen and 21 percent oxygen by volume, and to be 75.5 percent nitrogen and 23.2 percent oxygen by weight. The other "atmospheric trace gases" together comprise less than 1 percent. Trace impurity levels may vary with geographic locations or with proximity to industrial areas or highways carrying dense traffic. This composition remains relatively constant at altitudes up to 70 000 ft (21 336 m).

GRADES AVAILABLE

Table 1, from CGA G-7.1, *Commodity Specification for Air,* presents the components maxima in parts per million (mole/mole), unless shown otherwise, for types and grades of air, which are also sometimes denoted as quality verification levels. [1] A blank indicates no maximum limiting characteristic. The absence of a value in a listed grade does not mean to imply that the limiting characteristic is or is not present, but merely indicates that the test is not required for compliance with the specification. Please note that the grade of air may be provided by compressing atmospheric air or synthetically by mixing pure oxygen and nitrogen. A table which provides a means of converting moisture data into the particular units of interest appears as part of the Introduction to Part III.

USES

Air meeting particular purity specifications has many important applications. Some of these applications are in medical, undersea, aerospace, and atomic energy fields. It is also employed in self-contained breathing apparatus used by industrial, emergency response, and fire-fighting personnel, and as a power source for some kinds of pneumatic equipment.

PHYSIOLOGICAL EFFECTS

Air is nontoxic and nonflammable. Of the constituents which make up air, only oxygen and nitrogen are necessary for life. The other "trace gases," while useful for many industrial and scientific purposes, have, to the best of present-day physiological and medical knowledge, no physiologic role. Only oxygen (O_2) and nitrogen (N_2) are essential in respirable air.

The nitrogen in the air we breathe has no metabolic function, but serves as an inert diluent and has a mechanical function in maintaining inflation of gas-filled body cavities such as the pulmonary alveoli, middle ear, and the sinus cavities. Without nitrogen, as the oxygen is absorbed by the blood, these cavities may tend to contract and to collapse, with painful and possibly serious consequences.

The oxygen contained in the air we breathe is necessary to support the metabolic processes by which our bodies convert our "fuels," the foods we eat (carbohydrates, fats, and proteins), into heat and energy. The average man in the course of a 24-hour day will consume approximately 26 ft^3 (0.74 m^3) of oxygen. The oxygen consumed will weigh about 2.5 lb (1.13 kg), which is approximately equal to the weight of the food consumed during the same period. To obtain this oxygen, the person will breathe approximately 500 ft^3 (14.16 m^3) of air.

Air, like any gas, is capable of being compressed or of being rarefied. Although the fraction or percentage of oxygen in air remains constant, an increase in air pressure results in an increase in the partial pressure of oxygen; conversely, a decrease in pressure results in a reduction in the partial pressure of oxygen. Whenever the partial pressure of oxygen in the atmosphere falls significantly, the average individual is likely to begin to suffer symptoms of "hypoxia" (oxygen deficiency) without warning. Conversely, oxygen at an elevated partial pressure also may be toxic if breathed for extended periods of time.

The breathing of air enriched by the addition of oxygen, or of oxygen alone, is common practice (a) in aviation or in mountain climbing at altitudes above 10 000 ft (3048 m), (b) for inhalation therapy or for resuscitation, and (c) in some types of protective breathing equipment. In air which contains more than the normal 21 percent oxygen, combustible materials are easier to ignite and burn faster. The higher the concentration of oxygen, the greater the fire risk. In a compartment (such as a tunnel, caisson, or

TABLE 1. AIR GRADES AVAILABLE.[1]
(Units in ppm (mole/mole) unless shown otherwise)

Limiting Characteristics	Maxima for Gaseous Air								
	A	K	L	D	E	G	J	M	N[7]
Percent 0_2 Balance predominantly N_2 (Note 2)	atm/ 19.5–23.5	atm/ 19.5–23.5	atm/ 19.5–23.5	atm/ 19.5–23.5	atm/ 20–22	atm/ 19.5–23.5	atm/ 19.5–23.5	atm/ 19.5–23.5	atm/ 19.5–23.5
Water, ppm (v/v) (Note 3)		200	50				1	3	
Dew Point, °F (Note 3)		−33	−54				−104	−92	
Oil (condensed) (mg/m^3 at NTP)				5[4]	5[4]				None[a]
Carbon Monoxide				10[5,6]	10	5	1	1	10
Odor (Note 8)									None
Carbon Dioxide				1000[6]	500	500	0.5	1	500
Total Hydrocarbon Content (as methane)		25			25	15	0.5	1	
Nitrogen Dioxide						2.5	0.1	0.5	2.5
Nitric Oxide									
Sulfur Dioxide						2.5	0.1		5
Halogenated Solvents						10	0.1		
Acetylene							0.05		
Nitrous Oxide							0.1		
USP									Yes

[a]Includes water.

Note 1. The second edition of CGA G-7.1, published in 1973, listed nine quality verification levels of gaseous air lettered A to J and two quality verification levels of liquid air lettered A and B. Some of those letter designations have been dropped since they no longer represent major volume usage by industry. Four new letter designations, K, L, M, and N, have been added to reflect current specifications. To get a listing of quality verification levels dropped, see CGA G-7.1-1973 or contact the Compressed Gas Association.

Note 2. The term *atm* (atmospheric) denotes the oxygen content normally present in atmospheric air; the numerical values denote the oxygen limits for synthesized air.

Note 3. The water content of compressed air required for any particular quality verification level may vary with the intended use from saturated to very dry. For breathing air used in conjunction with self-contained breathing apparatus in extreme cold where moisture can condense and freeze, causing the breathing apparatus to malfunction, a dew point not to exceed −50°F (63 ppm v/v) is recommended. If a specific water limit is required, it should be specified as a limiting concentration in ppm (v/v) or dew point. Dew point is expressed in °F at one atmosphere pressure absolute, 101 kPa abs (760 mm Hg).

Note 4. Test not required for synthesized air whose oxygen and nitrogen components are produced by air liquefaction.

Note 5. Test not required for synthesized air when nitrogen component was previously analyzed and meets the *National Formulary* (NF) specification. [2]

Note 6. Test not required for synthesized air when oxygen component was produced by air liquefaction and meets the *United States Pharmacopeia* (USP) specification. [2]

Note 7. To apply to cylinders only; does not apply to on-stream compressor systems.

Note 8. Specific measurement of odor in gaseous air is impractical. Air normally may have a slight odor. The presence of a pronounced odor should render the air unsatisfactory. Odor is checked by sniffing a moderate flow of air from the container being tested by cupping the hand and bringing some of the gas being vented toward the nose.

chamber) filled with air under pressure, most combustible materials will ignite more readily and burn much more rapidly than they would in air at normal atmospheric pressure, because of the increase in partial pressure of oxygen, even though the air contains only the normal 21 percent of oxygen.

The oxygen content of compressed air for human respiration must be held within the limits given by the appropriate specifications in CGA G-7.1, *Commodity Specification for Air*, so as to provide adequate oxygen content for physiological needs and yet not include an excessive concentration of oxygen which might create a fire or health hazard. [1] See CGA G-7, *Compressed Air for Human Respiration*, for additional information. [3]

MATERIALS OF CONSTRUCTION

Since dry air is noncorrosive, it may be contained in equipment constructed with any common, commercially available metals.

SAFE STORAGE, HANDLING, AND USE

Cylinders and other containers charged with air at high pressure must be handled with all the precautions necessary for safety with any nonflammable compressed gas, recognizing that compressed air is a strong oxidizing agent. See Chapter 5 for general guidelines. Note that with liquid air, the oxygen concentration will change with time from that at the time the container was filled.

DISPOSAL

When disposing of compressed air from cylinders, make sure the cylinder is secure and that appropriate precautions are made with respect to noise levels.

HANDLING LEAKS AND EMERGENCIES

Avoid contact of the skin with liquid air or its cold boil-off gas. Flush liquid air spills with water to accelerate evaporation. Because of the preferential evaporation of nitrogen, initially high concentrations of nitrogen followed subsequently by the presence of a rich oxygen liquid and evolution of an oxygen-rich atmosphere may occur.

METHODS OF SHIPMENT

Under the appropriate regulations, air is authorized for shipment as follows: [4], [5]

By Rail: In cylinders as a compressed gas, and as a liquid in special containers for cryogenic gas.

By Highway: In cylinders as a compressed gas, and as a liquid in special containers for liquefied cryogenic gas.

By Water: In cylinders on passenger vessels, cargo vessels, and ferry and railroad car ferry vessels (passenger or vehicle).

By Air: In cylinders as a compressed gas, aboard passenger aircraft up to 150 lb (68 kg), or aboard cargo aircraft up to 300 lb (136 kg), maximum net weight per cylinder. In insulated containers meeting specified requirements, liquid air, either low pressure or pressurized, in cargo aircraft only up to 300 lb (136 kg) maximum net weight per container. Nonpressurized liquid air is not accepted for shipment aboard cargo or passenger aircraft.

CONTAINERS

In the United States, Department of Transportation regulations describe how containers shall be in compliance with current DOT regulations. See Title 49 of the U.S. *Code of Federal Regulations* Parts 100–179 and CGA C-4, *American National Standard Method of Marking Portable Compressed Gas Containers to Identify the Material Contained*. [4] and [6]

In Canada, the *Transportation of Dangerous Goods Regulations* of Transport Canada will apply. [5] Identification of the contents of portable medical gas containers must be as laid out in the *Canadian General Standards Board Standard for Identification of Medical Gas Containers, Pipelines, and Valves*. [7]

Filling Limits. The maximum filling limits at 70°F (21.1°C) for compressed air are the authorized service pressures marked on the cylinders. Authorized cylinders meeting special requirements may be filled to a limit of up to 110 percent of their marked service pressures. See 49 CFR 173.302 (c). [4]

Compressed air may be shipped in qualified cylinders authorized by the DOT or TC for nonliquefied compressed gas. (These include cylinders meeting TC/DOT specifications 3A, 3AA, 3AX, 3AAX, 3AL, 3B, 3E, 4B, 4BA, and 4BW in addition, continued use of cylinders meeting DOT specifications 3, 3C, 3D, 4, 4A, 4C, 25, 26, 33, and 38 is authorized, but new construction is not authorized.)

All cylinders authorized for compressed air service must be requalified by hydrostatic retest every 5 or 10 years under regulations at the time of publication, with the following exceptions: TC/DOT-4 cylinders, every 10 years; no periodic retest is required for cylinders of types 3C, 3E, 4C, and 4L.

Container Preparation

Container preparation shall be as necessary to assure that the container contents meet the requirements of the specified grade of air.

Containers for Air Intended for Human Respiration

Breathing air containers must be processed by a method which encompasses inspection, evacuation, purging, or cleaning procedures to ensure that the container and component parts are not reactive, additive, or absorptive to an extent that significantly affects the identity, quality, and purity of the specified quality verification level of air being supplied.

Note: See CGA G-7, *Compressed Air for Human Respiration* for information pertaining to air intended for breathing purposes. [3] In Canada, refer to Canadian Standards Association (CSA) Standard Z180.1-M1978, *Compressed Breathing Air.* [8]

Valve Outlet Connections

Cylinder valve inlet and outlet connections on compressed air containers offered in transportation must conform to CGA V-1, *American National, Canadian, and Compressed Gas Association Standard for Compressed Gas Cylinder Valve Outlet and Inlet Connections*, and CGA V-7, *Standard Method of Determining Cylinder Valve Outlet Connections for Industrial Gas Mixtures.* [9] and [10]

Connection CGA 346 is standard for all air, both industrial and for breathing purposes. Connection CGA 590 has been retained as a "limited standard" for industrial air.

Connections CGA 346 and CGA 590 are assigned for compressed air up to 3000 psig (20 684 kPa). These connections must not be used for pressures that exceed 3000 psig (20 684 kPa) at 120°F (48.9°C). Cylinders filled to 2640 psig (18 200 kPa) when at 70°F (21.1°C) shall be deemed to have met this requirement.

Connection CGA 347 is the standard for intermediate high pressure air (3001 to 5500 psig, 20 691 to 37 921 kPa) as measured at 120°F (48.9°C).

Connection CGA 702 is the standard for high pressure air (5501 to 7500 psig, 37 928 to 51 711 kPa) as measured at 120°F (48.9°C). Connection CGA 677 (which was the temporary high pressure standard for air) became obsolete for that purpose with the publication of the sixth edition of CGA V-1.

Connection CGA 950, the pin-indexed yoke connection, is the assigned standard for compressed breathing air in small-size medical cylinders at pressures up to 3000 psig (20 684 kPa). Connections CGA 850 and 855 (formerly CGA 1310) are assigned to SCUBA (self-contained underwater breathing apparatus) at pressures up to 3000 psig (20 684 kPa).

The pressure relief devices on cylinder valves on air cylinders must conform to those authorized in CGA S-1.1, *Pressure Relief Device Standards—Part 1—Cylinders for Compressed Gases.* [11]

Similarly, pressure relief devices on air cargo and portable tanks must conform with the requirements of CGA S-1.2, *Pressure Relief Device Standards—Part 2—Cargo and Portable Tanks for Compressed Gases*, and on air storage containers, pressure relief devices must be in accordance with CGA S-1.3, *Pressure Relief Device Standards—Part 3—Compressed Gas Storage Containers.* [12] and [13]

METHODS OF MANUFACTURE

Air may be compressed from the atmosphere and purified by chemical and mechanical means. It may also be synthetically produced from the already purified major components nitrogen and oxygen.

REFERENCES

[1] CGA G-7.1, *Commodity Specification for Air*, Compressed Gas Association, Inc., 1235 Jefferson Davis Highway, Arlington, VA 22202.

[2] *United States Pharmacopoeia/National Formulary* (current edition), United States Pharmacopeial Convention, Inc., 12601 Twinbrook Parkway, Rockville, MD 20852.

[3] CGA G-7, *Compressed Air for Human Respiration*, Compressed Gas Association, Inc., 1235 Jefferson Davis Highway, Arlington, VA 22202.

[4] *Code of Federal Regulations*, Title 49 CFR Parts 100-199 (Transportation), Superintendent of Documents, U.S. Government Printing Office, Washington, DC 20402.

[5] *Transportation of Dangerous Goods Regulations*, Canadian Government Publishing Centre, Supply and Services Canada, Ottawa, Ontario, Canada K1A 0S9.

[6] CGA C-4, *American National Standard Method of Marking Portable Compressed Gas Containers to Identify the Material Contained*, Compressed Gas Association, Inc., 1235 Jefferson Davis Highway, Arlington, VA 22202.

[7] *Canadian General Standards Board Standard for Identification of Medical Gas Containers, Pipelines, and Valves*, Canadian Government Publishing Centre, Supply and Services Canada, Ottawa, Ontario, Canada K1A OS9.

[8] CSA Z180.1-M1978, *Compressed Breathing Air*, Canadian Standards Association, 178 Rexdale Boulevard, Rexdale (Toronto), Ontario, Canada M9W 1R3.

[9] CGA V-1, *American National, Canadian and Compressed Gas Association Standard for Compressed Gas Cylinder Valve Outlet and Inlet Connections* (ANSI/CSA, CGA V-1), Compressed Gas Association, Inc., 1235 Jefferson Davis Highway, Arlington, VA 22202.

[10] CGA V-7, *Standard Method of Determining Cylinder Valve Outlet Connections for Industrial Gas Mixtures*, Compressed Gas Association, Inc., 1235 Jefferson Davis Highway, Arlington, VA 22202.

[11] CGA S-1.1, *Pressure Relief Device Standards—Part 1—Cylinders for Compressed Gases*, Compressed Gas Association, Inc., 1235 Jefferson Davis Highway, Arlington, VA 22202.

[12] CGA S-1.2, *Pressure Relief Device Standards—Part 2—Cargo and Portable Tanks for Compressed Gases*, Compressed Gas Association, Inc., 1235 Jefferson Davis Highway, Arlington, VA 22202.

[13] CGA S-1.3, *Pressure Relief Device Standards—Part 3—Compressed Gas Storage Containers*, Compressed Gas Association, Inc., 1235 Jefferson Davis Highway, Arlington, VA 22202.

ADDITIONAL REFERENCES

Federal Specification BB-A-1034A, with amendment (1) of 12-15-70, *Air, Compressed, for Breathing Purposes*, General Services Administration, 7th and D Streets, SW, Room 6039, Washington, DC 20407.

Military Specification MIL-A-27420, *Air, Liquid Breathing*, 4-30-70, Naval Publications and Forms Center, 5801 Tabor Ave., Attn.: Customer Service, Code 1052, Philadelphia, PA 19120.

Ammonia (Anhydrous)

Chemical Symbol: NH_3
Synonyms: Anhydrous Ammonia
CAS Registry Number: 7664-41-7
DOT Classification: Nonflammable gas
DOT Label: Nonflammable gas
Transport Canada Classification: 2.4
UN Number: UN 1005
IMO Number: 2.3
ASHRAE Refrigerant Number: R-717

PHYSICAL CONSTANTS

	U.S. Units	SI Units
Chemical formula	NH_3	NH_3
Molecular weight	17.031	17.031
Vapor pressure		
at 70°F (21.1°C)	114.1 psig	786.7 kPa
at 105°F (40.6°C)	214.2 psig	1476.8 kPa
at 115°F (46.1°C)	251.5 psig	1734.0 kPa
at 130°F (54.4°C)	315.6 psig	2176.0 kPa
Density of the gas		
at 32°F (0°C) and 1 atm	0.0481 lb/ft^3	0.771 kg/m^3
Specific gravity of the gas		
at 32°F and 1 atm (air = 1)	0.5970	0.5970
Specific volume of the gas		
at 32°F (0°C) and 1 atm	20.78 ft^3/lb	1.297 m^3/kg
Density of the liquid		
at 70°F (21.1°C)	38.00 lb/ft^3	608.7 kg/m^3
at 105°F (40.6°C)	36.12 lb/ft^3	578.6 kg/m^3
at 115°F (46.1°C)	35.55 lb/ft^3	569.4 kg/m^3
at 130°F (54.4°C)	34.66 lb/ft^3	555.2 kg/m^3
Boiling point at 1 atm	−28°F	−33.3°C
Freezing point at 1 atm	−107.9°F	−77.7°C
Critical temperature	271.4°F	133.0°C
Critical pressure	1657 psia	11.425 kPa
Critical density	14.7 lb/ft^3	235 kg/m^3

	U.S. Units	SI Units
Triple point	−107.86°F at 0.88 psia	−77.70°C at 6.1 kPa abs
Latent heat of vaporization at boiling point and 1 atm	589.3 Btu/lb	13.71 × 10^5 J/kg
Latent heat of fusion at −107.9°F (−77.72°C)	142.8 Btu/lb	332.2 kJ/kg
Specific heat of gas at 59°F (15°C) and 1 atm		
C_p	0.5232 Btu/(lb)(°F)	2.191 kJ/(kg)(°C)
C_v	0.3995 Btu/(lb)(°F)	1.673 kJ/(kg)(°C)
Ratio of specific heats (C_p/C_v)	1.3096	1.3096
Solubility in water vol(liq.)/vol(liq.) at 68°F (15.5°C)	0.848	0.848
Weight of the liquid per gallon at 60°F (15.5°C)	5.147 lb/gal	616.7 kg/m^3
Vapor density at −28°F (−33°C) and 1 atm	0.0555 lb/ft^3	0.8898 kg/m^3
Specific gravity of the liquid at −28°F (−33°C) compared to water at 39.2°F (4°C)	0.6819	0.6819
Liquid density at −28°F (−33°C) and 1 atm	42.57 lb/ft^3	681.9 kg/m^3
Flammable limits (percent in air, by volume)	16–25%	16–25%
Ignition temperature	1562°F	850.0°C
Heat of solution extrapolated to 0% concentration by weight	347.4 Btu/lb	8.081 × 10^5 J/kg
Heat of solution at 28% concentration by weight	214.9 Btu/lb	4.999 × 10^5 J/kg

Note: One atmosphere = 14.696 psia, 101.325 kPa (abs), 760 mm of mercury, or 1.01325 bars.

DESCRIPTION

Ammonia is the compound formed by the chemical combination of the two gaseous elements, nitrogen and hydrogen, in the molar proportion of one part nitrogen to three parts hydrogen. This relationship is shown in the chemical symbol for ammonia, NH_3. On a weight basis the ratio is 14 parts nitrogen to 3 parts hydrogen or approximately 82 percent nitrogen to 18 percent hydrogen.

The term *ammonia* as used throughout this monograph is the name of the chemical compound, NH_3, which is commonly called anhydrous ammonia. *Anhydrous* means "without water" and when used with *ammo-*

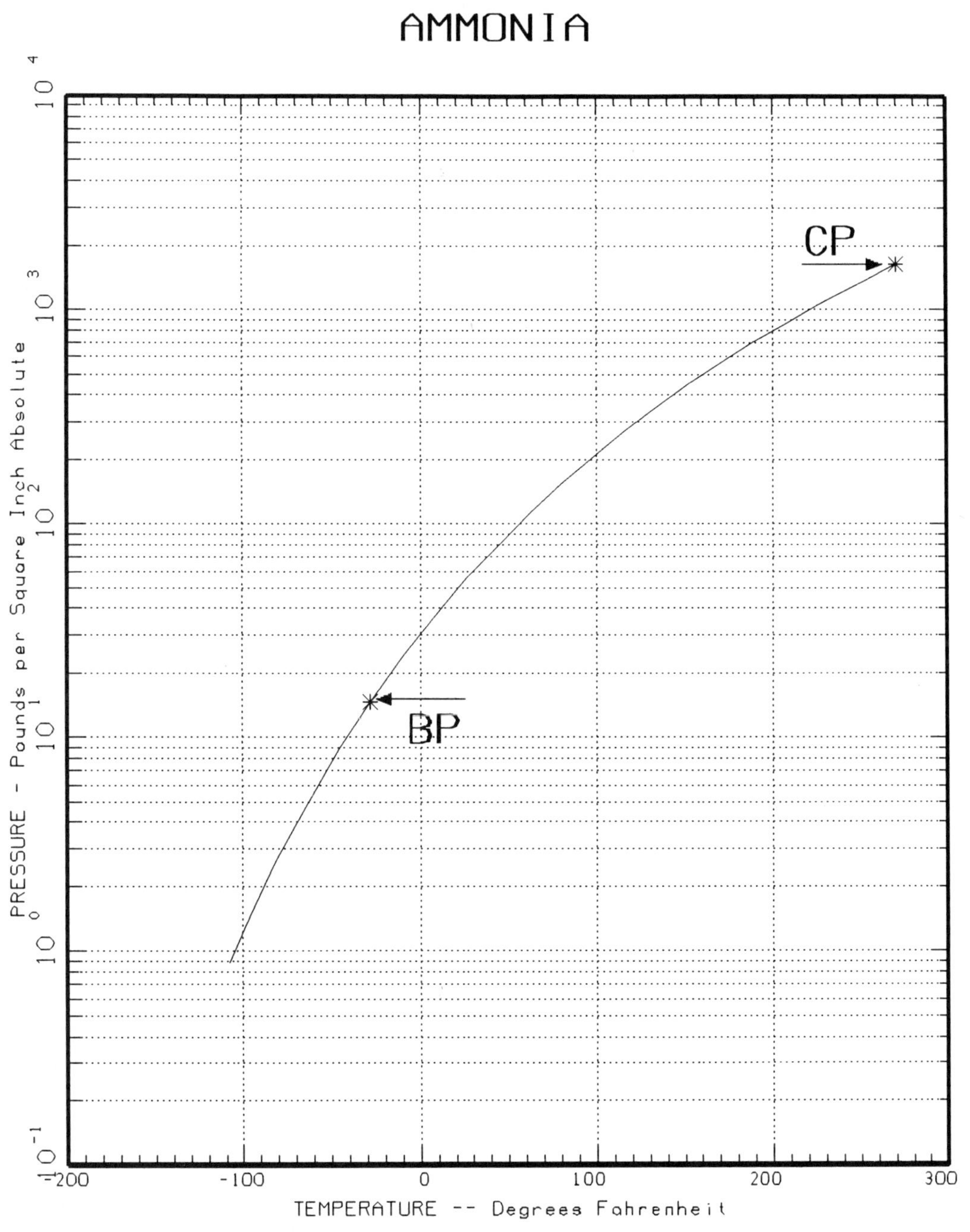

Fig. 1. Vapor Pressure Curve for Ammonia.

nia indicates that the water content is less than 0.2 percent. This differentiates it from the various widely used aqueous solutions of ammonia.

At room temperature and atmospheric pressure, ammonia is a pungent, colorless gas. It may be compressed and cooled to a colorless liquid. Between the melting and critical points, liquid ammonia exerts a vapor pressure which increases with rising temperature. When anhydrous ammonia in a closed container is in equilibrium with anhydrous ammonia vapor, the pressure within the container bears a definite relationship to the temperature as shown by the curve in Fig. 1.

Liquid ammonia is lighter than water, having a density of 42.57 lb/ft^3 (681.9 kg/m^3) at −28°F (−33.3°C); as a gas, ammonia is lighter than air, its relative density being 0.597 compared to air at a pressure of 1 atm and a temperature of 32°F (0.0°C). Under the latter conditions, 1 lb (0.4536 kg) of ammonia vapor occupies a volume of 20.78 ft^3 (0.5884 m^3).

The relationships of temperature to vapor

TABLE 1. PROPERTIES OF LIQUID AMMONIA AT VARIOUS TEMPERATURES.[a]

Temperature		Vapor Pressure (1)		Liquid Density (2)		Pounds per U.S. Gallon (3)	Specific Gravity of Liquid (Compared to Water at 4°C) (4)	Latent Heat (5)	
(°F)	(°C)	psig	(kPa)	lb/ft^3	kg/m^3			Btu/lb	(J/g)
−28	(−33.3)	0.0	(0.0)	42.5	(681.9)	5.69	0.682	589.3	(1370.7)
−20	(−28.9)	3.6	(24.8)	42.22	(676.3)	5.64	0.676	583.6	(1357.5)
−10	(−23.3)	9.0	(62.1)	41.78	(669.3)	5.59	0.669	576.4	(1340.7)
0	(−17.8)	15.7	(108.2)	41.34	(662.2)	5.53	0.662	568.9	(1323.3)
10	(−12.2)	23.8	(164.1)	40.89	(655.0)	5.47	0.655	561.1	(1305.1)
20	(−6.7)	33.5	(231.0)	40.43	(647.6)	5.40	0.647	553.1	(1286.5)
30	(−1.1)	45.0	(310.3)	39.96	(640.1)	5.34	0.640	544.8	(1267.2)
40	(4.4)	58.6	(404.0)	39.49	(632.6)	5.28	0.633	536.2	(1247.2)
50	(10.0)	74.5	(513.7)	39.00	(624.7)	5.21	0.625	527.3	(1226.5
60	(15.6)	92.9	(640.5)	38.50	(616.7)	5.15	0.617	518.1	(1205.1)
65	(18.3)	103.1	(710.8)	38.25	(612.7)	5.11	0.613	513.4	(1194.2)
70	(21.1)	114.1	(786.7)	38.00	(608.7)	5.08	0.609	508.6	(1183.0)
75	(23.9)	125.8	(867.4)	37.74	(604.5)	5.05	0.605	503.7	(1171.6)
80	(26.7)	138.3	(953.5)	37.48	(600.4)	5.01	0.600	498.7	(1160.0)
85	(29.4)	151.7	(1045.9)	37.21	(596.0)	4.97	0.596	493.6	(1148.1)
90	(32.2)	165.9	(1143.8)	36.95	(591.9)	4.94	0.592	488.5	(1136.3)
95	(35.0)	181.1	(1248.6)	36.67	(587.4)	4.90	0.587	483.2	(1123.9)
100	(37.8)	197.2	(1359.6)	36.40	(583.1)	4.87	0.583	477.8	(1111.4)
105	(40.6)	214.2	(1476.9)	36.12	(578.6)	4.83	0.579	472.3	(1098.6)
110	(43.3)	232.3	(1601.7)	35.84	(574.1)	4.79	0.574	466.7	(1084.8)
115	(46.1)	251.5	(1476.8)	35.55	(569.5)	4.75	0.569	460.9	(1072.1)
120	(48.9)	271.7	(1873.3)	35.26	(564.8)	4.71	0.565	455.0	(1058.3)
125	(51.7)	293.1	(2020.9)	34.96	(560.0)	4.67	0.560	488.9	(1044.1)
130	(54.4)	315.6	(2176.0)	34.66	(555.2)	4.63	0.555	(443)[b]	(1030.4)
135	(57.2)	339.4	(2340.1)	34.35	(550.2)	4.59	0.550	(436)[b]	(1014.1)
140	(60.0)	364.4	(2512.4)	34.04	(545.3)	4.55	0.545	(430)[b]	(1000.2)

[a]Data for customary U.S. units in columns 1, 2, and 5 taken from U.S. Bureau of Standards Circular No. 142, "Table of Thermodynamic Properties of Ammonia," 1st ed., April 16, 1923. Values for columns 3 and 4 calculated from column 2. Metric conversions in parentheses have been rounded.

[b]The figure in parentheses was calculated from empirical equations given in U.S. Bureau of Standards Scientific Papers Nos. 313 and 315 and represents a value obtained by extrapolation beyond the range covered in the experimental work.

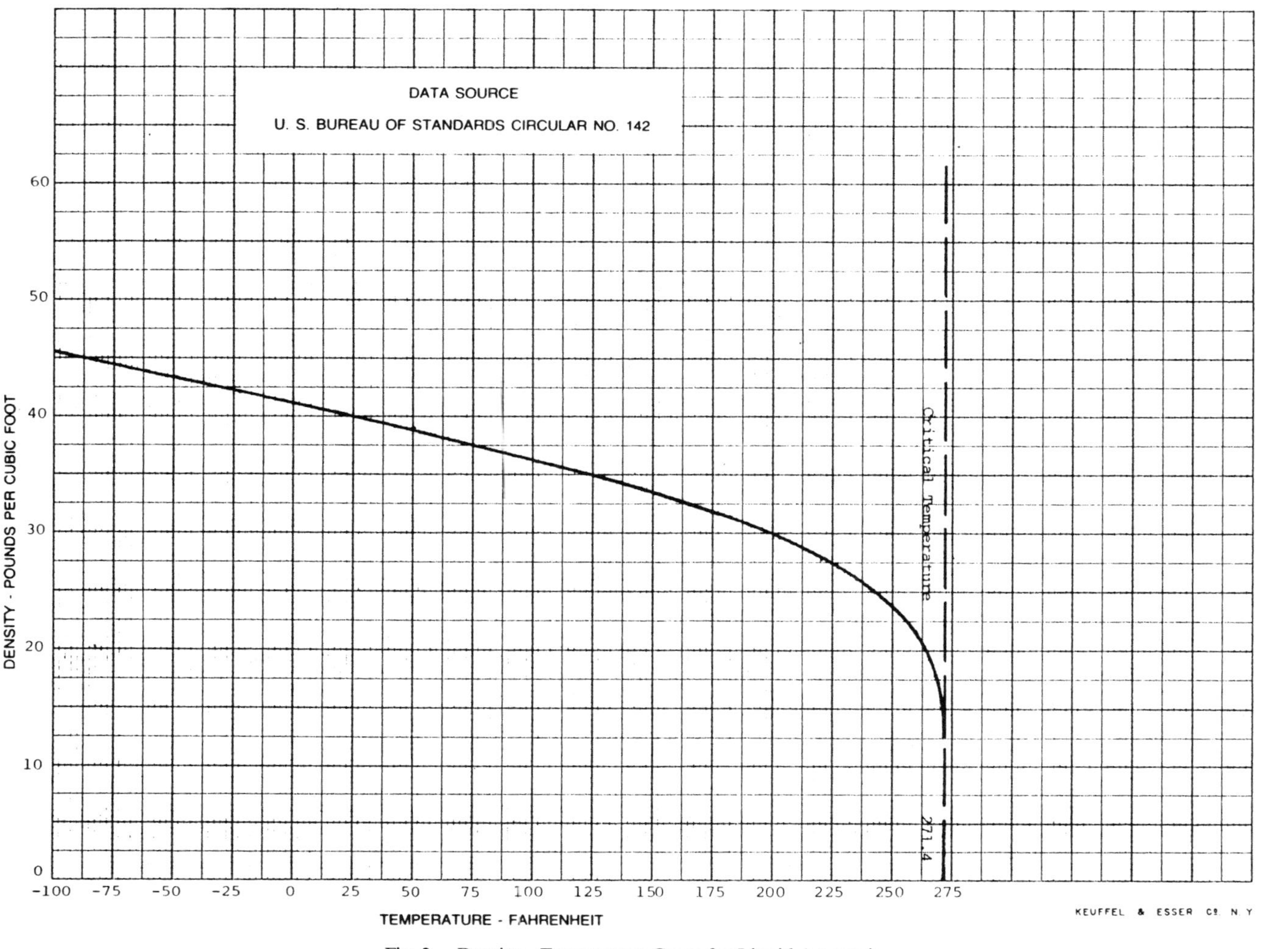

Fig. 2. Density—Temperature Curve for Liquid Ammonia.

TABLE 2. THERMODYNAMIC PROPERTIES OF SATURATED AMMONIA LIQUID AND VAPOR.

Temperature		Pressure		Entropy J/(mol-K)		Enthalpy kJ/mol		Latent Heat of Vapor-ization	Specific Volume cm^3/mol		Density g/cm^3	
K	°F	kPa	atm	Liquid	Vapor	Liquid	Vapor	kJ/mol	Liquid	Vapor	Liquid	Vapor
195.42	−107.9	6.076 5	0.059 97	71.59	199.95	10.435	35.522	25.087				
200	−99.7	8.666 3	0.085 53	73.30	198.41	10.770	35.794	25.024	23.37	189 730	0.729	0.000 09
210	−81.7	17.793	0.175 6	76.90	194.26	11.510	36.158	24.648	23.74	96 858	0.717	0.000 18
220	−63.7	33.91	0.334.7	80.33	190.54	12.255	36.497	24.242	23.13	53 070	0.706	0.000 32
230	−45.7	60.61	0.598 2	83.68	187.11	13.008	36.794	23.786	24.55	30 895	0.694	0.000 55
240	−27.7	102.58	1.012 4	86.86	183.89	13.770	37.054	23.284	24.99	18 905	0.682	0.000 90
250	−9.7	165.4	1.632 0	90.00	181.13	14.535	37.317	22.782	25.46	12 088	0.669	0.001 41
260	8.3	255.9	2.526	93.01	178.57	15.309	37.560	22.251	25.96	8 029	0.656	0.002 12
270	26.3	381.9	3.769	95.94	176.23	16.096	37.773	21.677	26.50	5 505	0.643	0.003 09
280	44.3	551.8	5.446	98.78	174.05	16.887	37.961	21.074	27.07	3 882	0.629	0.004 39
290	62.3	775.3	7.652	101.59	172.00	17.694	38.120	20.426	27.70	2 803	0.615	0.006 08
300	80.3	1 062.4	10.485	104.31	170.08	18.506	38.238	19.732	28.39	2 006	0.600	0.008 24
310	98.3	1 424.9	14.063	106.98	168.24	19.330	38.317	18.987	29.14	1 545	0.584	0.011 02
320	116.3	1 873	18.49	109.62	166.40	20.171	38.388	18.217	29.98	1 174	0.568	0.014 51
330	134.3	2 420	23.90	112.21	164.77	21.033	38.380	17.347	30.92	900.1	0.551	0.018 92
340	152.3	3 082	30.42	114.77	163.01	21.924	38.305	16.381	31.99	696.1	0.532	0.024 47
350	170.3	3 870	38.19	117.32	161.00	22.840	38.137	15.297	33.24	540.4	0.512	0.031 52
360	188.3	4 800	47.40	119.87	158.99	23.799	37.878	14.079	34.73	419.3	0.490	0.040 62
370	206.3	5 891	58.14	122.47	156.73	24.807	37.472	12.665	36.58	323.9	0.466	0.052 58
380	224.3	7 150	70.60	125.27	153.97	25.832	36.807	10.975	39.03	247.0	0.436	0.068 95
390	242.3	8 607	84.94	128.28	150.58	27.087	35.736	8.649	42.61	182.6	0.400	0.093 27
400	260.3	10 280	101.5	130.00	145.10	28.514	33.731	5.217	49.45	124.5	0.344	0.136 80
405.6	270.4	11 300	111.5	138.24		31.066		0.0	72.47		0.235	

Source: P. Davies, in F. Din, Ed., *Thermodynamic Functions of Gases,* Volume 1, pg. 89, 1962, Butterworth, Inc., Washington, DC.

pressure, density, specific gravity, and latent heat for liquid ammonia are shown in Table 1. Vapor pressure–temperature and density-temperature curves are shown in Figs. 1 and 2, respectively. Values of thermodynamic properties of liquid ammonia and gaseous ammonia are given in Table 2.

As a chemical compound, ammonia is highly associated and stable. Dissociation begins to occur at 840°F to 930°F (449°C to 499°C) and atmospheric pressure, with the products being nitrogen and hydrogen. Experiments conducted by the Underwriters' Laboratories indicate that an ammonia-air mixture in a standard quartz bomb will not ignite at temperatures below 1560°F (850°C). When an iron bomb was used, the ignition temperature was 1200°F (651°C) due to a catalytic effect. Ammonia gas burns at atmospheric pressure, but only within the limited range of 16 to 25 percent by volume of ammonia in air.

Ammonia is a highly reactive chemical, forming ammonium salts in reactions with inorganic and organic acids; amides in reactions with esters, acid anhydrides, acyl halides, carbon dioxide, or sulfonyl chlorides; and amines in reactions with halogen compounds or oxygen-containing compounds such as polyhydric phenols, alcohols, aldehydes, and aliphatic ring oxides.

GRADES AVAILABLE

No commodity grade specifications for ammonia have been published as standard for the industry. However, generally accepted grade designations are shown below.

Grade	Ammonia, Percent Weight Minimum
Commercial	99.5
Agricultural	99.7[a]
Refrigeration	99.95
Technical	99.98
Metallurgical	99.995
Electronic	99.998

[a]When 82.0% weight minimum nitrogen is guaranteed.

In these grade designations, the assay values are established by difference following the determination of specified impurities, usually water and oil. Federal government specifications for technical grade ammonia may be found in Federal Specification 0-A-445B. [1]

USES

About 80 percent of all ammonia produced in the United States is used in agriculture as a source of nitrogen, which is essential for plant growth. Nitrogen makes up about 16 percent of plant protein. When a fruit, vegetable, or grain crop is grown and harvested, nitrogen is removed from the soil. If the fertility of the land is to be maintained, nitrogen and other elements essential to plant growth, such as potassium and phosphorus, must be restored to the soil by fertilization. Depending upon the particular crop, up to 200 lb of nitrogen may be economically applied per acre.

About 4 million tons of ammonia containing 82 percent nitrogen are applied directly to the soil each year in the United States. It can be injected at a depth of several inches below the surface of the soil by specially designed equipment, or it can be dissolved in irrigation water.

Ammonia is utilized extensively in the fertilizer industry to produce solid material such as ammonium salts, nitrate salts, and urea. Ammonium sulfate, ammonium nitrate, and ammonium phosphate are made directly by neutralizing the corresponding acids—sulfuric acid, nitric acid, and phosphoric acid—with ammonia. Urea is an organic compound formed by combining ammonia and carbon dioxide. Ammonium sulfate, ammonium nitrate, ammonium phosphate, and urea are used for direct application to the soil in dry form and in combination with other phosphate and potassium salts.

Ammonia is also used in the production of nitrogen fertilizer solutions which consist of ammonia, ammonium nitrate, urea, and water in various combinations. Some are pressure solutions and others are not. Nonpressure and low pressure solutions are widely used for direct application to the soil. Pressure solutions containing free ammonia are used in the manufacture of high-analysis mixed fertilizers.

In addition to their use as fertilizers, ammonia and urea are used as a source of protein in ruminant livestock feeds. Urea is used in mixed feed supplements to supply the nitrogen needed for the biosynthesis of proteins by the microorganisms in ruminating animals such as cattle, sheep, and goats.

Ammonia is oxidized in the production of nitric acid, the principal ammonia derivative used in making explosives. Both industrial and military explosives are divided into two main types: high explosives such as dynamite, nitroglycerine, and TNT, which detonate rapidly to give a shattering blast for demolition purposes; and low explosives such as nitrocellulose, which detonate slowly to give a heaving/pushing effect for propellant or blasting applications. Dynamite, a general term for high explosives used in mining and construction, contains nitroglycerine or other organic nitrogen compounds absorbed in a combustible material. In ammonia dynamites, ammonium nitrate, made by reacting ammonia and nitric acid, replaces all of the nitroglycerine. Blasting-gelatin dynamites consist of a colloidal mixture of nitroglycerine and nitrocellulose. The latter is made by treating cellulose with a mixture of nitric and sulfuric acids.

Ammonium nitrate is the principal base material in slurry explosives and lower-cost blasting agents. It can be converted to an effective blasting agent by proper mixing with a carboniferous material such as fuel oil. Ammonium nitrate/fuel oil (ANFO) mixtures are used extensively in open-pit mining and outdoor construction work because of ease of handling, availability, low cost, and safety.

Ammonia is required for the synthesis of ammonium salts and certain alkalies, dyes, pharmaceuticals, synthetic textile fibers, and plastics.

Used in both absorption and compression type systems, ammonia is the oldest, most efficient and economical mechanical refrigerant known.

Ammonia or dissociated ammonia is used in such metal-treating operations as nitriding, carbo-nitriding, bright annealing, furnace brazing, sintering, and other applications where protective atmospheres are required.

Ammonia is used in pH control, in mineral beneficiation, in the neutralization of acidic components during petroleum refining, and in the treatment of acidic wastes.

Dissociated ammonia provides a convenient source of hydrogen for hydrogenation and other applications.

Ammonia is used in extracting certain metals such as copper, nickel, and molybdenum from their ores.

Ammonia vapor is utilized as the developing agent for diazonium salts (white printing) and is also employed in the production of diazotype microfilm duplicates.

Ammonia is used in scrubbers to neutralize sulfur oxides in their removal from stack gases in electric power generation and other furnace operations such as in smelting. It is also used to improve the efficiency of electrostatic precipitators in the removal of particulate matter.

Ammonia is highly soluble in water, forming aqueous ammonia (ammonium hydroxide or aqua ammonia), which has many applications. In a very dilute solution (2 to 5 percent ammonia) it is available as "household ammonia." The solubility of ammonia in water at various temperatures is shown in Fig. 3.

Ammonia is used in water treatment in conjunction with chlorine. Ammonia is used to remove trihalomethanes from water. Water solutions of ammonia are used to regenerate weak anion exchange resins.

PHYSIOLOGICAL EFFECTS

Persons having chronic respiratory disease or persons who have shown evidence of undue sensitivity to ammonia should not be employed where they will be exposed to ammonia.

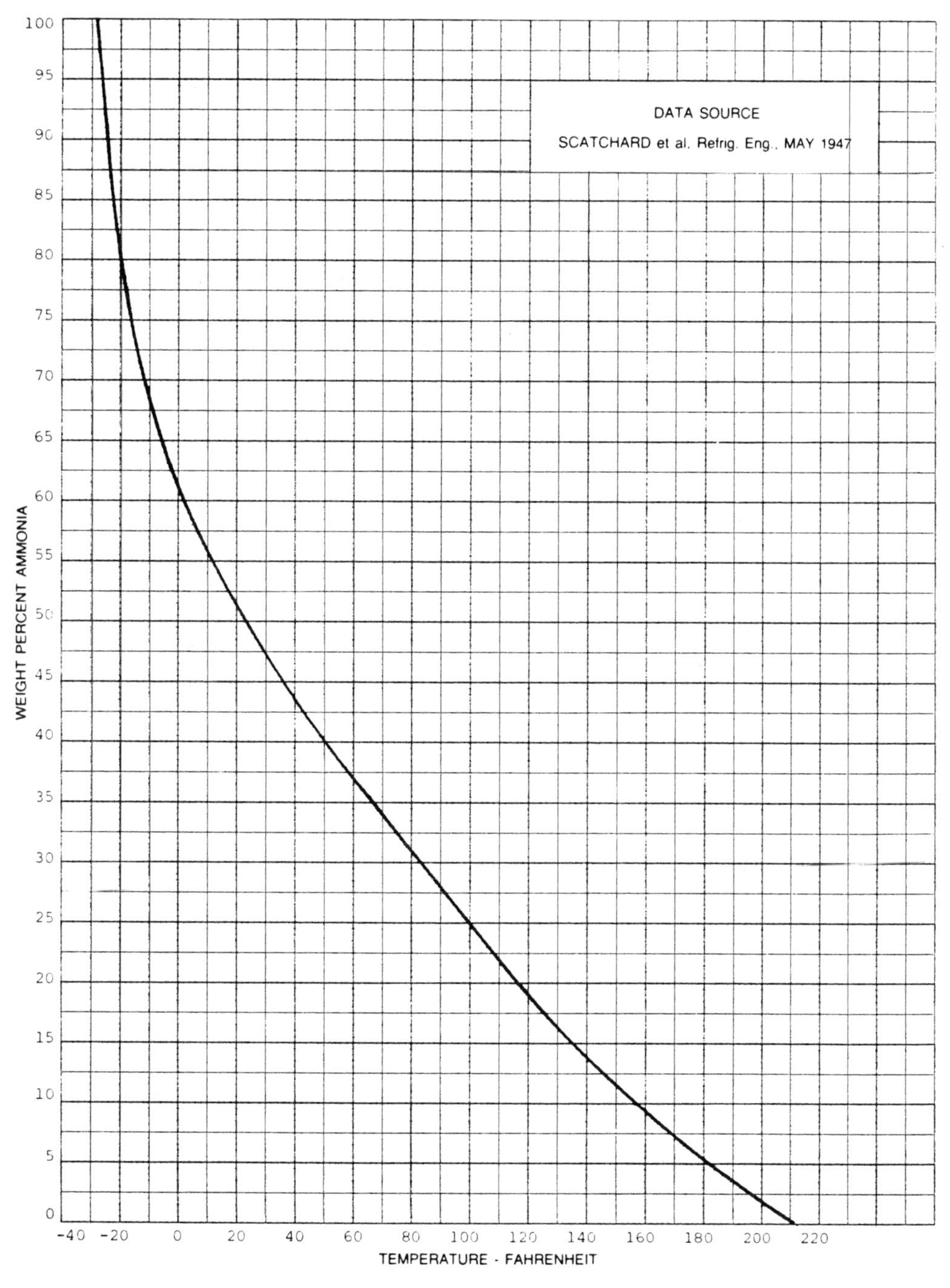

Fig. 3. Solubility Curve for Ammonia.

Ammonia is not a cumulative metabolic poison; ammonium ions are actually important constituents of living systems. However, ammonia in the ambient atmosphere has an intense irritating effect on the mucous membranes of the eyes, nose, throat, and lungs. High levels of ammonia can produce corrosive effects on tissues and can cause laryngeal and bronchial spasm and edema so as to obstruct breathing. The pungent odor of ammonia affords a protective warning, and conscious persons avoid breathing significantly contaminated air. Unconscious persons, however, are not similarly protected.

Table 3 indicates human physiological response to various concentrations of ammonia in air. Individuals differ in their sensitivity to ammonia, some persons being highly reactive to relatively low concentrations and others showing a significant tolerance to the irritative effects.

In accordance with U.S. Department of Labor regulations, an employee's exposure to ammonia must be limited to a concentration not to exceed 35 ppm (27 mg/m^3) of ammonia in air by volume based upon a 15-minute time-weighted average. [2] Concentrations in the range of 20 to 50 ppm are readily perceptible, and it is therefore unlikely that any individual would become overexposed unknowingly.

Since liquid ammonia vaporizes readily and has a great affinity for water, it may cause severe injury to the skin by freezing the tissue and subjecting it to caustic action and dehydration. A chemical burn, which may be severe, will result.

MATERIALS OF CONSTRUCTION

Most common metals are not affected by dry ammonia. However, when combined with water vapor, ammonia will attack copper, zinc, or alloys containing copper as a major alloying element. Therefore, these materials should not be used in contact with ammonia. Certain high-tensile-strength steel have developed stress-corrosion cracking in ammonia service, but such cracking can be prevented by the use of 0.2 percent water by weight in the ammonia as an inhibitor. Ammonia storage tanks and their valves and fittings are usually made of steel.

DOT regulations covering the construction of ammonia containers prohibit the use of copper, silver, zinc, or alloys of these materials. For details regarding materials of construction recommended for piping, tubing, fittings, and hose, refer to the latest edition of ANSI K61.1, *American National Standard Safety Requirements for the Storage and Handling of Anhydrous Ammonia.* [3]

Piping and Equipment

Moist ammonia corrodes copper, zinc, and their alloys. It is customary, therefore, that

TABLE 3. PHYSIOLOGICAL RESPONSE TO AMMONIA.

Response	Concentration
Least perceptible odor	5 ppm
Readily detectable odor	20–50 ppm
No discomfort or impairment of health for prolonged exposure	50–100 ppm
General discomfort and eye tearing; no lasting effect on short exposure	150–200 ppm
Severe irritation of eyes, ears, nose, and throat; no lasting effect on short exposure	400–700 ppm
Coughing, bronchial spasms	1700 ppm
Dangerous, less than one-half hour exposure may be fatal	2000–3000 ppm
Serious edema, strangulation, asphyxia, rapidly fatal	5000–10 000 ppm
Immediately fatal	10 000 ppm

Note: Concentrations are for ammonia in air by volume. Exposure levels which are tolerated by normal persons may produce coughing and bronchospasm in others.

iron or steel can be used in piping and fittings. Certain aluminum alloys are used also in appurtenances and can be used in other parts of systems for ammonia. Copper, brass, or galvanized fittings and pipe should not be used. Unions, valves, gauges, pressure regulators, and relief valves or other fittings having copper, brass, or bronze parts are not suitable for ammonia service.

Metallic and nonmetallic gasket materials, such as carbon steel or stainless steel spiral-wound and aluminum are suitable for ammonia service. Some fluorocarbon plastic materials and neoprene have also been found suitable. Check with the gasket manufacturer to ensure compatibility with ammonia.

All ammonia piping should be extra-heavy (Schedule 80) steel when threaded joints are used. Standard-weight (Schedule 40) steel may be used when joints are either welded or joined by welding-type flanges. Flow should not be restricted by an excessive number of elbows and bends. All refrigeration system piping should conform to the latest edition of ANSI B31.5, *American National Standard for Refrigeration Piping*. [4] Also see the latest edition of ANSI K61.1 requirements. [3]

Welded instead of threaded joints should be used whenever possible, particularly on larger pipe sizes. When threaded joints are used, a suitable pipe thread luting compound or perfluorinated plastic tape should be used. Piping should be well supported, and provisions should be made to protect all piping against the effects of expansion, contraction, jarring, vibration, settling, and external corrosion.

All piping and tubing should be tested after assembly at a pressure not less than the normal operating pressure of the system. Any leaks found should be eliminated.

Threaded nipples, when used, should be cut from Schedule 80 steel pipe.

All fittings must be at least extra heavy. Nonmalleable metals, such as cast iron, must not be used. Forged or cast steel valves and fittings should be used in any service subject to significant strain or vibration.

Hose should be manufactured specifically for ammonia service and should have a minimum burst pressure not less than 1750 psig (12 066 kPa). Hose couplings should be made of steel. Each hose together with its internal couplings should be hydrostatically tested and should be leak-free at the maximum rated working pressure of the hose assembly including couplings. Hose assemblies should be inspected visually at regular intervals thereafter, depending upon the frequency and severity of service, and replaced as necessary. Refer to CGA P-7, *Standard for Requalification of Cargo Tank Hose Used in the Transfer of Compressed Gases.* [5]

SAFE STORAGE, HANDLING, AND USE

Personnel working with anhydrous ammonia should be thoroughly familiar with safety precautions for handling a gas corrosive to human tissue as well as measures for handling emergencies.

Use of welding or flamecutting equipment on or in an ammonia container is not recommended unless all ammonia has been purged and all of the residue has been removed.

See Chapter 5 for general guidelines concerning proper storage, handling, and use of compressed gases. For more specific and comprehensive information regarding the safe storage, handling, and use of anhydrous ammonia, refer to CGA G-2, *Anhydrous Ammonia*, and ANSI K61.1, *American National Standard Safety Requirements for the Storage and Handling of Anhydrous Ammonia.* [6] and [3] These are complementary publications. CGA G-2 focuses primarily on the properties of anhydrous ammonia; the types of containers, piping, and other equipment used with anhydrous ammonia; and the regulations and other requirements pertinent to transportation and use. CGA G-2 also contains some guidelines regarding emergency actions. As the title implies, the primary focus of ANSI K61.1 is storage and handling procedures pertinent to the safe use of anhydrous ammonia.

Withdrawing Ammonia as a Liquid from a Cylinder

The cylinder should be positioned horizontally with the valve outlet pointing up. In this position the dip tube will point down and be in the liquid phase.

Before an empty cylinder is disconnected, the cylinder valve should be closed and the bleed line valve opened to exhaust the residual ammonia from the connecting hose. If frost forms on the hose, the presence of liquid ammonia is indicated and the connector should not be removed until the frost has melted.

The practice of manifolding to withdraw liquid ammonia from two or more cylinders simultaneously is hazardous because, under certain temperature conditions, it is possible for liquid to flow from one cylinder into another cylinder until it is completely filled. If the valve of this completely filled cylinder were then closed, any rise in temperature would result in the development of hydrostatic pressure due to liquid ammonia expanding and could result in the rupture of the cylinder. The practice of manifolding cylinders to withdraw liquid ammonia should be avoided, but if necessary, check valves should be installed in the piping from each cylinder to prevent reverse flow.

Withdrawing Ammonia as Vapor from a Cylinder

When a cylinder is vertical, the dip tube is normally in the vapor phase. When placed horizontally for vapor withdrawal, a cylinder should have its valve outlet pointing downward, which points the dip tube upward into the vapor phase.

A full cylinder in a vertical position may discharge liquid ammonia if it has been stored in a location warm enough to cause the liquid level to rise above the lower end of the dip tube.

The flow rate of ammonia vapor from a cylinder is dependent on the rate of vaporization of the liquid ammonia. Since heat transfer from the surrounding atmosphere through the cylinder wall is greatest through the portion of the wall that is wet with liquid internally, the rate of vaporization will diminish as the liquid level in the cylinder decreases. A cylinder of 150 lb (68 kg) size at room temperature standing vertically may be expected to deliver up to 30 ft^3 (850 L) of ammonia vapor per hour continuously until empty.

Frost forming on a cylinder indicates that the refrigeration effect of vaporization has cooled the liquid to 32°F (0°C) or less, and as a result, the flow rate and pressure may become inadequate. Manifolding with additional cylinders will increase the heat transfer surface. Shutting off the valve and allowing a frosted cylinder to warm up will restore pressure and flow rate. The procedure can be repeated until the cylinder is empty.

It should be emphasized that a cylinder must never be warmed directly by a water bath, steam, or flame. These are very dangerous practices which might create pressure sufficient to rupture the cylinder.

Determining When Cylinders Are Empty

The best way to determine if an ammonia cylinder is empty is to weigh it, without cap, and compare the weight with the tare weight stamped on the cylinder.

A pressure gauge should be installed in a system utilizing vapor supplied from cylinders. A gauge pressure reading of less than 25 psi (172 kPa) without frost formation on the cylinder walls indicates that the cylinder is essentially empty. The ammonia vapor content of a 150-lb (68-kg) cylinder at 25 psig (172 kPa) is approximately 0.54 lb (0.24 kg).

It is desirable to install a pressure-actuated switch to operate an alarm when ammonia pressure has dropped to 25 psig (172 kPa) or to the pressure of the system into which ammonia is being introduced. On cylinder installations supplying ammonia vapor, this type of alarm will give notice to switch cylinders to maintain flow.

DISPOSAL

Ammonia may be disposed of by discharge into water of sufficient volume to absorb it. A ratio of 10 parts water to one part ammonia may be considered sufficient. Disposal of the resultant ammonium hydroxide, including and subsequent neutralization products, must be done in an environmentally safe manner that, for example, will not be harmful to aquatic life, and must be performed in compliance with applicable federal, state, provincial, and local regulations.

HANDLING LEAKS AND EMERGENCIES

A leak in an ammonia system can be detected by odor. The location of the leak may be determined with moist red litmus paper or moist filter paper impregnated with phenolphthalein or detection instruments. These chemical test papers change color in ammonia vapor. Other means of detection involve the use of sulfur dioxide, which forms a white fog in contact with ammonia vapor.

Only personnel trained for and designated to handle emergencies should attempt to stop a leak. Respiratory equipment of a type suitable for ammonia must be worn. All persons not so equipped must leave the affected area until the leak has been stopped.

If ammonia vapor is released, the irritating effect of the vapor typically will force personnel to leave the area before they have been exposed to dangerous concentrations. To facilitate their rapid evacuation there should be sufficient well-marked and easily accessible exits. If, despite all precautions, a person should become trapped in an ammonia atmosphere, he or she should breathe as little as possible and open his or her eyes only when necessary. Depending upon the concentration of ammonia, partial protection may be gained by holding a wet cloth over the nose and mouth. Since ammonia vapor in air will rise, a trapped person should remain close to the floor to take advantage of the lower vapor concentrations at that level.

With good ventilation or rapidly moving air currents, ammonia vapor, being lighter than air, can be expected to dissipate readily to the upper atmosphere without further action being necessary. Lacking these conditions, the concentration of ammonia vapor in air can be reduced effectively by the use of adequate volumes of water applied through spray or fog nozzles. Do not put water on a liquid ammonia spill unless sufficient water is available. Sufficient water may be considered as 100 parts of water to one part of ammonia.

Under some circumstances ammonia in a container is colder than the available water supply. At such times water must not be sprayed on the container walls since it would heat the ammonia and aggravate any gas leak. If it is found necessary to dispose of ammonia, as from a leaking container, liquid ammonia may be discharged into a vessel containing water sufficient to absorb it. Sufficient water may be considered as ten parts of water to one part of ammonia. The ammonia must be injected into the water as near the bottom of the vessel as practical.

A leak at a valve stem on an ammonia cylinder in service can usually be stopped by tightening the packing nut which has a left-hand thread. If this fails, the valve should be closed. A cylinder which continues to leak should be removed from the building to a safe area and the supplier notified. For further information, see *Emergency Services Guide for Selected Hazardous Materials*, published by the U.S. Department of Transportation [7], or the *Emergency Response Guide* from Transport Canada. [8]

Fire Exposure

An ammonia container exposed to a fire should be removed if it can be done safely. If for any reason the container cannot be removed, it should be kept cool with water spray from a safe distance until well after the fire is out. Fire-fighting personnel should be equipped with protective clothing and respiratory equipment. Information on such safety equipment, and on general procedures for securing the area, is given in CGA G-2, *Anhydrous Ammonia*. [6]

When ammonia is discharged into the environment accidentally or intentionally in an amount equal to or exceeding the reportable quantity, 100 lb (45.415 kg) during a 24-hour period (as shown in 49 CFR 172.101), [9], such discharge must be reported to the National Response Center, (800) 424-8802, in Washington, DC; alternate number is (202) 267-2675. Refer to Section 103(a) and 103(b) of the "Comprehensive Environmental Response, Compensation, and Liability Act of 1980" (Superfund Law). For further information see *Emergency Response Guidebook*, DOT P5800, published by the Department of Transportation. [10]

Exposure and Emergency Actions

Call a physician immediately for any person who has been burned or overcome by ammonia. The patient should be removed to an area free from fumes, and preferably a warm room. The patient should be placed in a reclining position with head and shoulders elevated and kept warm by the use of blankets or other cover, if necessary.

Prior to medical aid by the physician, first aid measures should be taken. Those presented herein are based upon what is believed to be common practice in industry. Their adoption in any specific case should, of course, be subject to prior endorsement by a competent medical advisor.

First Aid

Any conscious person who has inhaled ammonia causing irritation should be assisted to an uncontaminated area and inhale fresh air. A person overcome by ammonia should immediately be carried to an uncontaminated area. If breathing has ceased, artificial respiration must be started immediately, preferably by trained personnel. If breathing is weak or has been restored by artificial respiration, oxygen may be administered.

If contacted by ammonia, the eyes must be flooded immediately with copious quantities of clean water. Speed is essential. In isolated areas, water in a squeeze bottle which can be carried in the pocket is helpful for emergency irrigation purposes. Eye fountains should be used, but if they are not available, water may be poured over the eyes. In any case, the eyelids must be held open and irrigation must continue for at least 15 minutes. The patient must receive prompt attention from a physician, preferably an ophthalmologist. Persons subject to ammonia exposure should not wear contact lenses.

If liquid ammonia contacts the skin, the area affected should be immediately flooded with water. If no safety shower is available, immerse in any available water of acceptable temperature. Water will have the effect of thawing out clothing which may be frozen to the skin. Such clothing should be removed and flooding with water continued for at least 15 minutes. Do not apply salves or ointments or cover burns with dressing; however, protect the injured area with a clean cloth prior to medical care. Do not attempt to neutralize the ammonia. If ammonia has entered the nose or throat and the patient can swallow, have him or her drink large quantities of water. ***Never give anything by mouth to an unconscious person.***

METHODS OF SHIPMENT

Under the appropriate regulations, anhydrous ammonia is authorized for shipment as follows: [9] and [11]

By Rail: In insulated and uninsulated tank cars (shipment by rail in cylinders and in portable tanks is also authorized, as is the currently seldom used method of shipment by rail in multi-unit tank car tanks.

By Highway: In cargo tank motor vehicles, including semitrailers and full trailers (shipment by highway in cylinders and portable tanks is also authorized, as is the currently seldom used method of shipment by highway in multi-unit tank car tanks).

By Water: In barges on inland waterways and tankers or vessels on oceangoing routes, stowed in a well-ventilated space.

By Air: On cargo aircraft only up to 300 lb (136 kg) maximum net weight per container; forbidden on passenger aircraft.

By Pipeline: In interstate pipelines authorized for liquid service, not gaseous service.

CONTAINERS

Under the appropriate regulations, anhydrous ammonia is transported as a liquefied compressed gas in cylinders, insulated and uninsulated tank cars (and multi-unit tank car tanks), barges, and tankers. It is stored in bulk in large-capacity containers installed above or below ground. Normal aboveground storage is in uninsulated, pressure storage tanks. Very large aboveground containers are often low pressure, refrigerated, and consequently insulated tanks.

Filling Density

Maximum allowable filling densities for shipping containers in common usage are shown in Table 4.

Cylinders

Cylinders meeting the requirements of the following TC/DOT specifications are authorized for ammonia service: 3A480, 3A480X,

TABLE 4. MAXIMUM ALLOWABLE FILLING DENSITIES FOR AMMONIA SHIPPING CONTAINERS.

Type Container	DOT Specification Number		Maximum Allowable Filling Density Percent by Weight
Cylinders	(See Table 5)		54
Portable Tanks	51	(Notes 1 and 7)	56
Nurse Tanks	(Nonspecification)	(Notes 2 and 7)	56
Cargo Tanks	MC-330	(Notes 1 and 7)	56
	MC-331 or TC-331	(Notes 1 and 7)	56
Multi-unit Tank Car Tanks	106A500-X	(Note 3)	50
Single-unit Tank Cars	105A300-W	(Note 4)	57
	112S340-W	(Notes 5 and 6)	57
	112S400-F	(Notes 5 and 6)	57
	114S340-W	(Notes 5 and 6)	57

Note 1. Uninsulated cargo and portable tanks may be filled to 87.5% by volume provided the temperature of the ammonia is not less than 30°F (−1.1°C).

Note 2. Must be operated by a private carrier exclusively for agricultural purposes, have a minimum design of 250 psig, meet ASME Code requirements, and meet other requirements stipulated by 49 CFR 173.315(m).

Note 3. Authorized for rail, highway, and cargo vessel transportation, but currently rarely used.

Note 4. Each Specification 105 tank car built after August 31, 1981 shall conform to class DOT 105S. After December 31, 1986, each Specification 105 tank car built before September 1, 1981, and with a water capacity (shell full volume including manways) exceeding 18 500 U.S. gal (70.03 m^3), shall conform to class DOT-105S.

Note 5. Filling density of 58.8% permitted November through March inclusive. Storage in transit prohibited.

Note 6. When class DOT-112S or 114S tank cars are prescribed, class DOT-112T, 114T, 112J, and 114J tank cars may also be used.

Note 7. These containers are normally filled by liquid volume measurement techniques. A 56% filling density is equivalent to 82 liquid volume percent for ammonia.

3AA480, 3AL480, 3E1800, 4, 4A480, and 4AA480. [9] and [11] The use of existing DOT-3A480X and DOT-4 cylinders is authorized, but new construction is not permitted, having been deleted from 49 CFR 178, "Shipping Container Specifications," on September 11, 1980. [9]

Ammonia cylinders are available in seamless (one-piece) or welded (two-piece) construction and in a number of convenient sizes, the two most widely used having capacities of 100 and 150 lb (45.4 and 68.0 kg) of ammonia. Dimensions and weights of some typical cylinders are listed in Table 5.

The valve of each ammonia cylinder having a capacity of 15 lb (6.8 kg) or more has an internal dip tube connected to it. This makes it possible to withdraw either liquid or vapor by positioning the cylinder so that the end of the dip tube is in the desired phase. Some ammonia cylinders are furnished by certain suppliers on request with full-length dip tubes to provide liquid withdrawal with the cylinder in the vertical (valve up) position.

Cylinders made in compliance with TC/DOT specifications 4, 3A, 3AA, 3A480X, 4A, or 4AA480 used exclusively for ammonia of at least 99.95 percent purity may, in lieu of hydrostatic retest, be given a complete external visual inspection at the time such periodic retest becomes due. See 49 CFR 173.34(e)(10). [9]

Pressure relief devices authorized for use with ammonia cylinders are the Type CG-2 fusible plug device with a nominal yield temperature of 165°F (73.9°C). No pressure relief device is required for cylinders under 165 lb (74.8 kg). [12]

Requalification

In addition to hydrostatic testing of cylinders at the time of manufacture, the U.S. Department of Transportation requires the owner or his authorized agent to fulfill periodic requalification requirements for his cylinders. See 49 CFR 173.34(e). [9] Hydrostatic retest or visual inspection under certain conditions are the accepted methods for requalifying cylinders. Requalifying periods and test pressures for cylinders in ammonia service are shown in Table 6.

Cylinders exposed to fire must be properly heat treated and tested before being returned to service. See 49 CFR 173.34(f). [9]

If, as a result of hydrostatic tests, a cylinder leaks or shows a permanent expansion which exceeds 10 percent of the total expansion, it must be condemned. Except for TC/DOT-3AL aluminum cylinders approved for ammonia service, cylinders condemned because of excessive permanent expansion may be requalified if heat-treated and then hydrostatically retested. See 49 CFR 173.34(e)(4). [9]

Valve Outlet Connections

The standard valve outlet connections in the United States and Canada for ammonia cylinders are Connections CGA 240 and 705. Standard yoke connections in the United States and Canada are Connections CGA 800 and 845. Connection CGA 660 is a limited standard connection for ammonia and may be used for laboratory applications of electronic grade ammonia. Lecture bottles are supplied with Connection CGA 110 or 180. See ANSI/CGA V-1, *American National, Canadian, and Compressed Gas Association Standard for Compressed Gas Cylinder Valve Outlet and Inlet Connections.* [13]

Portable Tanks

Portable tanks used for transportation of ammonia must comply with TC/DOT Specification 51. [9] and [11] They are designed to be temporarily attached to a motor vehicle, other vehicle, railroad car other than tank car, or vessel, and are equipped with skids, mountings, or accessories to facilitate handling of the tank by mechanical means. If a portable tank is used as a cargo tank, it must comply with all the requirements prescribed for cargo tanks.

Portable tanks complying with TC/DOT Specification 51 must have a capacity in ex-

TABLE 5. APPROXIMATE DIMENSIONS AND WEIGHTS OF TYPICAL AMMONIA CYLINDERS.

DOT Cylinder	Ammonia Capacity		Average Tare Weight pounds (kg)		Overall Length		Outside Diameter		Wall Thickness		Minimum Volume	
Spec. No.	Pounds	(kg)	Less Cap	With Cap	Inches	(mm)	Inches	(mm)	Inches	(mm)	Cu. In.	(L)
3AA1800	2	(0.91)	5 (2.28)	NA	16	(406)	3.50	(89)	0.070	(1.78)	108	(1.77)
3A480	100	(45.4)	134 (60.8)	137 (62.1)	59	(1499)	12.50	(317)	0.176	(4.47)	5158	(84.59)
3A480	150	(68.0)	195 (88.4)	198 (89.8)	60	(1524)	15.00	(381)	0.212	(5.38)	7710	(126.44)
3A480X	100	(45.4)	88 (39.9)	91 (41.3)	57	(1448)	12.25	(311)	0.120	(3.05)	5158	(84.59)
3A480X	150	(68.0)	135 (61.2)	138 (62.6)	58	(1473)	14.75	(375)	0.125	(3.17)	7710	(126.44)
4AA480	25	(11.3)	43 (19.5)	NA	30	(508)	12.25	(311)	0.153	(3.89)	1344	(22.04)
4AA480	100	(45.4)	114 (51.7)	117 (53.1)	56	(1422)	12.25	(311)	0.153	(3.89)	5158	(84.59)
4AA480	150	(68.0)	158 (71.7)	161 (73.0)	58	(1473)	14.75	(375)	0.185	(4.70)	7710	(126.44)

TABLE 6. REQUALIFYING PERIODS AND TEST PERIODS FOR TYPICAL CYLINDERS IN AMMONIA SERVICE.

Specification	Visual Inspection Period (years)	Hydrostatic Test Period (years)	Test Pressure (psig)	Test Pressure (kPa)
TC/DOT-4[a]	5	10	700	4 830
TC/DOT-3A480	5	10	800	5 516
TC/DOT-3A480X	5	10	800	5 516
TC/DOT-3AA480	5	5	800	5 516
TC/DOT-3AA1800	5	5	3000	20 685
TC/DOT-4A480	5	5	800	5 516
TC/DOT-4AA480	5	10	960	6 410

[a]Also see 49 CFR 173.34(e)(2). [9]

cess of 1000 lb (454 kg) of water. If shipped on vessels under the jurisdiction of the United States Coast Guard, the gross weight must not exceed 55 000 lb (24 948 kg). [14]

TC/DOT Specification 51 prescribes fabrication from steel in accordance with the *ASME Boiler and Pressure Vessel Code*, Section VIII, and a minimum design pressure of 265 psig (1827 kPa) is required. [15] Steel of a thickness less than 3/16 inch (4.763 mm) must not be used for the shell, heads, or protective housings.

All valves, fittings, accessories, pressure relief and safety devices, gauging devices, and the like must be adequately protected against mechanical damage by a protective housing or must be recessed and under a cover plate.

Systems Mounted on Farm Wagons

Special farm vehicles have been developed for the handling of ammonia as a fertilizer in agriculture. Containers mounted on farm equipment and used as mobile storage for supplying ammonia as it is applied to the soil in the field are called applicator tanks. Containers mounted on farm wagons and used to transport ammonia over the highways and to replenish applicator tanks are called nurse tanks. Sometimes nurse tanks are used as applicator tanks. State, provincial, and local safety and design regulations should be consulted regarding the operation of these vehicles. Nurse tanks are regulated by the DOT under the provisions of 49 CFR 173.315(m). [9]

Further information on appurtenances and markings for these tanks, and on other requirements for nurse tanks and applicator tanks, is given in CGA G-2, *Anhydrous Ammonia.* [6] See also ANSI K61.1, *American National Standard Safety Requirements for the Storage and Handling of Anhydrous Ammonia.* [3]

Multi-Unit Tank Car Tanks

The DOT authorizes the transportation of ammonia in DOT Specification 106A500X containers by rail, highway, or water modes. These containers are uninsulated, have a test pressure of 500 psig (3447 kPa), and have an ammonia capacity of approximately 800 lb (363 kg) each.

In the highway mode, these containers may be transported on trucks or semitrailers provided they are securely chocked or clamped in place to prevent shifting and adequate facilities are available for safe handling of the containers. With limited exception, DOT

106A500X multi-unit tank car tanks are not used today for the transportation of ammonia.

Cargo Tanks

A cargo tank is a container permanently attached to or forming a part of any motor vehicle such as a tank truck, trailer, or semitrailer. An ammonia cargo tank generally complies with DOT Specification MC-330 (before September 1, 1967) or MC-331 (later construction) with a minimum design pressure of 265 psig (1827 kPa). A straight truck mounted ammonia cargo tank is referred to as a tank truck or a bob-tail. The larger trailer or semitrailer mounted ammonia cargo tank is referred to as a transport or tank wagon. See 49 CFR 173.33, 173.315(a), and 178.337. [9] Early in 1982, the DOT authorized the intrastate transportation of ammonia in nonspecification cargo tanks and specification cargo tanks having a minimum design pressure of not less than 250 psig (1724 kPa) and meeting certain limiting conditions. See 49 CFR 173.315(a)(1) Note 17. [9]

The DOT does not limit the size of cargo tanks, except if shipped on vessels under the jurisdiction of the United States Coast Guard, where the gross weight must not exceed 55 000 lb (24 948 kg). [14] State regulations limit the gross weight of the vehicle.

DOT Specification MC-331 prescribes fabrication from steel under Section VIII of the *ASME Boiler and Pressure Vessel Code*, and a minimum design pressure of 265 psig (1827 kPa) is required. [15] Steel of a thickness less than 3/16 inch (4.763 mm) must not be used for the shell, heads, or protective housings. Each cargo tank used for ammonia must be post-weld heat treated. The post-weld heat treatment must be performed as prescribed by the ASME Code. See 49 CFR 178.337-1. [9]

All containers, except those filled by weight, must be equipped with gauging devices to indicate the maximum permitted liquid level. Permitted types for ammonia are the rotary tube, adjustable slip tube, and fixed-length dip tube. If other gauging devices are installed, they must not be used as primary controls for filling cargo tanks. See 49 CFR 173.315(h). [9]

Tank Cars

Ammonia is authorized for shipment in special tank cars which may be either insulated or uninsulated. After December 31, 1986, however, Specification 105 tank cars built before September 1, 1981, and having a water capacity exceeding 18 500 gallons (70 m^3) must conform to class DOT-105S. Insulated tank cars built after August 31, 1981, must be constructed to DOT Specification 105S300-W, although at the time of this publication, DOT Specification 105A300-W tank cars were still authorized to transport ammonia. Uninsulated tank cars must be constructed to DOT Specification 112S340-W, 112S400-F, or 114S340-W. LP-gas tank cars of class 105J, 112J, 112T, 114J, and 114T are authorized for dual service transporting of either LP-gas or ammonia. Table 5 lists the tank cars authorized to transport ammonia at the time of this publication, but cars of the same class having higher marked test pressures may also be used. Shippers should refer to 49 CFR 173.314 for tank car specifications currently authorized. [9]

Many older tank cars have a nominal water capacity of 11 000 gallons (42 m^3) and hold approximately 26 short tons (24 000 kg) of ammonia. The majority of these small tank cars are of the insulated class 105A variety. Most of the tank cars built since the early 1960s are large uninsulated tank cars of up to 33 500 gallons (127 m^3) water capacity and hold about 80 short tons (73 000 kg) of ammonia. These are class 112 or 114 tank cars and are referred to as jumbo cars. Some still refer to the smaller tank cars as standard cars. Most new class DOT 105 tank cars are of the jumbo size.

Before loading or unloading, an external visual inspection of each ammonia tank car should be performed. If the tank car tank, or jacket if the tank is insulated, shows evidence of abrasion, dents, gouges, severe corrosion, leakage, or other defects, including damage to any part of the tank car structure, such condition should be reported promptly to the owner or shipper to determine whether or not the integrity of the tank car is satisfactory to conduct normal loading or unloading procedures.

A tank car used in seasonal service must be suitable for the commodities transported and should be purged completely of the previous commodity when service is changed. See 49 CFR 173.314. [9]

An ammonia tank car must be consigned for delivery and unloaded on a private track. See 49 CFR 171.8 and 174.204(a). [9] If a private track is unavailable, an ammonia tank car equipped with check valves may be consigned for delivery and unloaded on a carrier track provided it is unloaded into permanent storage of sufficient capacity to receive the entire contents of the car. See 49 CFR 174.204(a)(2). [9]

Although federal regulations do not require an ammonia tank car on a private track to be unloaded into permanent storage of sufficient capacity to receive the entire contents of the car, some state regulations do impose this requirement, and state authorities should be consulted to determine if such provisions are applicable at a given location.

Barges and Tankers

Barges and tankers for the transportation of ammonia are specially designed and fabricated for this service. Barges are widely used on inland waterways, while tankers are limited to oceangoing routes. Both require suitable terminals for loading and unloading.

Most ammonia transported by water is refrigerated to −28°F (−33.3°C). Older barges were not refrigerated and required pressure storage at the receiving terminal. After unloading into high pressure storage, the ammonia was refrigerated and transferred to low pressure storage.

Most barges for transporting ammonia have two refrigerated tanks with a total capacity of 2500 tons (2.3 gigagrams). There is refrigeration equipment on board to maintain the ammonia at −28°F (−33°C). Oceangoing tankers have capacities of up to 56 000 tons (50.8 Gg), but most are 9000 to 40 000 tons (8.2 Gg to 36.3 Gg).

The design and fabrication of a barge or a tanker for ammonia service must be approved by the authorities in the country where the vessel is fabricated. Any vessel operating in U.S. waters must be approved by the U.S. Coast Guard. Any modification after fabrication must also be approved.

Pipelines

Ammonia is transported through 6- to 8-inch steel pipelines as a liquefied compressed gas. These pipelines are operated by common carrier companies and are regulated by the U.S. Department of Transportation through the Office of Pipeline Safety Regulation of the Research and Special Programs Administration. See 49 CFR Part 195, "Transportation of Liquids by Pipelines." [9]

Ammonia transported by pipeline is at ambient temperature. Relatively small terminals using pressure storage can take ammonia directly from the pipelines. However, most of the terminals operated in conjunction with pipelines utilize large-scale, refrigerated storage.

Pipelines can connect multiple producing plants with multiple storage locations. The storage location may be hundreds of miles away from the producing plant.

Receipt of ammonia by pipeline requires careful, joint study and planning between producer, pipelines operator, and terminal operator.

Steels for pipelines transporting ammonia are specified by ANSI/ASME B31.4, *American National Standard for Liquid Petroleum Transportation Piping Systems*, which also requires that the ammonia be inhibited with at least 0.2 percent water by weight. [16]

Stationary Containers

This section covers pressure containers mounted on foundation piers and used primarily as ammonia storage tanks. Such tanks may be insulated, although most are not. Size limitations may be imposed by regulations or local conditions. The most recently constructed sizes range from 500 gallons to 45 000 gallons (2 m^3 to 170 m^3).

Ammonia storage containers must be designed for at least 250 psig (1724 kPa), constructed in accordance with the *ASME Boiler and Pressure Vessel Code*, Section VIII, and should be stress relieved after fabrication. [15] Since it is possible that the design of the container, its capacity, and its location may be influenced by state, provincial, or municipal regulations and insurance restrictions, a thorough investigation of pertinent requirements should be made prior to fabrication. All containers must be inspected by a National Board Inspector who shall certify the tank as being in compliance with applicable requirements of the ASME Code and registered with the National Board of Boiler and Pressure Vessel Inspectors.

METHODS OF MANUFACTURE

Ammonia was first produced in the United States on a commercial basis around 1890 as a by-product of the destructive distillation of coal in the manufacture of coke and coal gas. The first commercially successful synthetic ammonia plant in the United States was put into operation in 1921. It utilized the Haber-Bosch process, in which a preheated mixture of nitrogen and hydrogen was subjected to pressure in the presence of a contact catalyst.

Most ammonia produced commercially today is manufactured by processes which are modifications of the Haber-Bosch process. Several sources of hydrogen have been used, including natural gas, refinery gas, or coke-oven gas. Nitrogen may be supplied by introducing compressed air into the process stream or by introducing nitrogen from an air separation unit. In modern plants, natural gas, air, and steam are reacted at high temperatures in the presence of catalysts to yield a mixture of hydrogen, nitrogen, and oxides of carbon. Following conversion or removal of the oxides, the remaining hydrogen-nitrogen mixture is compressed and passed over catalysts, where ammonia is synthesized at elevated temperatures.

During 1988, production of ammonia in the United States was estimated at 17 million short tons which is 13 percent of the estimated 131 million short tons of worldwide production.

REFERENCES

[1] Federal Specification O-A-445B, *Ammonia Technical*, February 25, 1975, General Services Administration Regional Office in Washington, DC, and other cities.

[2] *Code of Federal Regulations*, Title 29 (Labor) Part 1910, Section 1000, U.S. Government Printing Office, Washington, DC 20402.

[3] ANSI K61.1 (CGA G-2.1), *American National Standard Safety Requirements for the Storage and Handling of Anhydrous Ammonia*, American National Standards Institute, Inc., 1430 Broadway, New York, NY 10018, or Compressed Gas Association, Inc., 1235 Jefferson Davis Highway, Arlington, VA 22202.

[4] ANSI B31.5, *American National Standard for Refrigeration Piping*, American National Standards Institute, Inc., 1430 Broadway, New York, NY 10018.

[5] CGA P-7, *Standard for Requalification of Cargo Tank Hose Used in the Transfer of Compressed Gases*, Compressed Gas Association, Inc., 1235 Jefferson Davis Highway, Arlington, VA 22202.

[6] CGA G-2, *Anhydrous Ammonia*, Compressed Gas Association, Inc., 1235 Jefferson Davis Highway, Arlington, VA 22202.

[7] *Emergency Services Guide For Selected Hazardous Materials*, U.S. Department of Transportation, Superintendent of Documents, U.S. Government Printing Office, Washington, DC 20402.

[8] *Emergency Response Guide* (Transport Canada), Canadian Government Publishing Centre, Supply and Services Canada, Ottawa, Ontario, Canada K1A 0S9.

[9] *Code of Federal Regulations*, Title 49 CFR Parts 100–199 (Transportation), Superintendent of Documents, U.S. Government Printing Office, Washington, DC 20402.

[10] DOT P5800, *Emergency Response Guidebook* (U.S. Department of Transportation), Superintendent of Documents, U.S. Government Printing Office, Washington, DC 20402.

[11] *Transportation of Dangerous Goods Regulations*, Canadian Government Publishing Centre, Supply

and Services Canada, Ottawa, Ontario, Canada K1A 0S9.

[12] CGA S-1.1, *Pressure Relief Device Standards—Part 1—Cylinders for Compressed Gases*, Compressed Gas Association, Inc., 1235 Jefferson Davis Highway, Arlington, VA 22202.

[13] CGA V-1, *Standard for Compressed Gas Cylinder Valve Outlet and Inlet Connections* (ANSI/CSA/CGA V-1), Compressed Gas Association, Inc., 1235 Jefferson Davis Highway, Arlington, VA 22202.

[14] *Code of Federal Regulations*, Title 46 CFR Parts 0–199 (Shipping), Superintendent of Documents, U.S. Government Printing Office, Washington, DC 20402.

[15] *ASME Boiler and Pressure Vessel Code* (Section VIII), American Society of Mechanical Engineers, 345 East 47th Street, New York, NY 10017.

[16] ANSI/ASME B31.4, *American National Standard for Liquid Petroleum Transportation Piping Systems*, American National Standards Institute, Inc., 1430 Broadway, New York, NY 10018.

Argon

Chemical Symbol: Ar
Synonym: LAR (liquid only)
CAS Registry Number: 7440-37-1
DOT Classification: Nonflammable gas
DOT Label: Nonflammable gas
Transport Canada Classification: 2.2
UN Number: UN 1006 (compressed gas); UN 1951 (refrigerated liquid)

PHYSICAL CONSTANTS

	U.S. Units	SI Units
Chemical formula	Ar	Ar
Molecular weight	39.95	39.95
Density of the gas at 70°F (21.1°C) and 1 atm	0.103 lb/ft^3	1.650 kg/m^3
Specific gravity of the gas at 70°F (21.1°C) and 1 atm	1.38	1.38
Specific volume of the gas at 70°F (21.1°C) and 1 atm	9.71 ft^3/lb	0.606 m^3/kg
Density of the liquid at boiling point and 1 atm	87.02 lb/ft^3	1394 kg/m^3
Boiling point at 1 atm	−302.6°F	−185.9°C
Melting point at 1 atm	−308.6°F	−189.2°C
Critical Temperature	−188.1°F	−122.3°C
Critical Pressure	711.5 psia	4905 kPa abs
Critical Density	33.44 lb/ft^3	535.6 kg/m^3
Triple Point	−308.8°F at 9.99 psia	−199.3°C at 68.9 kPa abs
Latent heat of vaporization at boiling point and 1 atm	69.8 Btu/lb	162.3 kJ/kg
Latent heat of fusion at triple point	12.8 Btu/lb	29.6 kJ/kg
Specific heat of the gas at 70°F (21.1°C) and 1 atm		
C_p	0.125 Btu/(lb)(°F)	0.523 kJ/(kg)(°C)
C_v	0.075 Btu/(lb)(°F)	0.314 kJ/(kg)(°C)

	U.S. Units	SI Units
Ratio of specific heats	1.67	1.67
Solubility in water at 32°F (0°C) vol/vol	0.056	0.056
Weight of the liquid at boiling point	11.63 lb/gal	1394.0 kg/m^3

DESCRIPTION

Argon belongs to the family of inert, rare gases of the atmosphere. It is plentiful compared to the other rare atmospheric gases; 1 million ft^3 (28 317 m^3) of dry air contains 9300 ft^3 (263 m^3) of argon. Argon is colorless, odorless, tasteless, and nontoxic. It is extremely inert and forms no known chemical compounds. It is slightly soluble in water.

GRADES AVAILABLE

Table 1, from CGA G-11.1, *Commodity Specification for Argon*, presents the component maxima in parts per million, ppm (mole/mole), unless otherwise shown, for the grades of argon, which are also sometimes denoted as quality verification levels. [1] Gaseous argon is referred to as Type I and liquid argon as Type II. A blank indi-

TABLE 1. ARGON GRADES AVAILABLE.
(Units in ppm (mole/mole) unless shown otherwise)

<table>
<tr><th rowspan="2">Limiting Characteristics</th><th colspan="6">Maxima for Gaseous and Liquid Nitrogen</th></tr>
<tr><th>A</th><th>B</th><th>C</th><th>D</th><th>E</th><th>F</th></tr>
<tr><td>Argon min. % (mole/mole)</td><td>99.985</td><td>99.996</td><td>99.997</td><td>99.998</td><td>99.999</td><td>99.9985</td></tr>
<tr><td>Water (v/v)</td><td>23.0</td><td>14.3</td><td>10.5</td><td>3.5</td><td>1.5</td><td>1</td></tr>
<tr><td>Dew Point, °F</td><td>−65</td><td>−72</td><td>−76</td><td>−90</td><td>−100</td><td>−104</td></tr>
<tr><td>Oxygen</td><td>50</td><td>7</td><td>5</td><td>2</td><td>1</td><td>2</td></tr>
<tr><td>Nitrogen</td><td>50</td><td>15</td><td>20</td><td>10</td><td>5</td><td>10</td></tr>
<tr><td>Hydrogen</td><td>50</td><td>1</td><td>1</td><td>1</td><td>1</td><td>1</td></tr>
<tr><td>Total Hydrocarbon Content (as methane)</td><td></td><td rowspan="2">5
Note 1</td><td rowspan="2">3
Note 1</td><td>0.5</td><td>0.5</td><td>0.5</td></tr>
<tr><td>Carbon Dioxide</td><td></td><td>0.5</td><td>0.5</td><td rowspan="2">0.5
Note 2</td></tr>
<tr><td>Carbon Monoxide</td><td></td><td></td><td></td><td></td><td></td></tr>
<tr><td>Permanent Particulates</td><td colspan="5">Type II may require filtering.</td><td>Note 3</td></tr>
</table>

Note 1. Value noted is combined total hydrocarbons and carbon dioxide.
Note 2. Value noted is combined carbon dioxide and carbon monoxide.
Note 3. To be determined between supplier and customer.

cates no maximum limiting characteristic. The absence of a value in a listed grade does not mean to imply that the limiting characteristic is or is not present, but merely indicates that the test is not required for compliance with the specification.

USES

Argon is extensively used in filling incandescent and fluorescent lamps and electronic tubes; as an inert gas shield for arc welding and cutting; as a blanket in the production of titanium, zirconium, and other reactive metals; to flush molten metals to eliminate porosity in castings; and to provide a protective shield for growing silicon and germanium crystals.

PHYSIOLOGICAL EFFECTS

Argon is nontoxic and largely inert. It can act as a simple asphyxiant by diluting the concentration of oxygen in air below levels necessary to support life. Inhalation of it in excessive concentrations can result in dizziness, nausea, vomiting, loss of consciousness, and death. Death may result from errors in judgment, confusion, or loss of consciousness, which prevents self-rescue. At low oxygen concentrations, unconsciousness and death may occur in seconds without warning.

Gaseous argon must be handled with all the precautions necessary for safety with any nonflammable, nontoxic compressed gas. All precautions necessary for the safe handling of any gas liquefied at very low temperatures must be observed with liquid argon. Extensive tissue damage or burns can result from exposure to liquid argon or cold argon vapors.

MATERIALS OF CONSTRUCTION

Gaseous argon is noncorrosive and inert, and may consequently be contained in systems constructed of any common metals and designed to safely withstand the pressures involved. At the temperature of liquid argon, ordinary carbon steels and most alloy steels lose their ductility and are considered unsafe for liquid argon service. Satisfactory materials for use with liquid argon include Type 18-8 stainless steel and other austenitic nickel-chromium alloys, copper, Monel, brass, and aluminum.

SAFE STORAGE, HANDLING, AND USE

Gaseous argon is commonly stored in high pressure cylinders, tubes, or tube trailers. Liquid argon is commonly stored at the consumer site in cryogenic liquid cylinders and specially designed vacuum-insulated storage tanks.

All of the precautions necessary for the handling of any nonflammable gas or cryogenic liquid must be taken. For additional details, see Chapter 5 and the compressed gas manufacturer's Material Safety Data Sheet.

Liquid and gaseous systems should be designed and installed only under the direction of personnel thoroughly familiar with liquid and gaseous argon equipment and in compliance with state, provincial, and local requirements.

DISPOSAL

Vent argon gas slowly to a well-ventilated outdoor location remote from personnel work areas and building air intakes. Return cylinders to the supplier with residual pressure, the cylinder valve tightly closed and the valve caps in place. Allow liquid argon to evaporate in well-ventilated outdoor locations which are remote from work areas and building air intakes.

HANDLING LEAKS AND EMERGENCIES

Ventilate adjacent enclosed areas to prevent the formation of oxygen-deficient atmospheres caused by the release of gaseous argon or by the evaporation of liquid argon. Personnel, including rescue workers, should not enter areas where the oxygen concentration is below 19 percent, unless provided with a self-contained breathing apparatus or air-line respirator.

Avoid contact of the skin with liquid argon or its cold boil-off gas. Flush liquid argon spills with water to accelerate evaporation.

METHODS OF SHIPMENT

Under the appropriate regulations, argon is authorized for shipment as follows (argon gas, except where liquid argon is indicated): [2] and [3]

By Rail: In cylinders and in tube tank cars. Liquid argon, in vacuum-insulated cylinders and tank cars.

By Highway: In cylinders on trucks, and in tube trailers. Liquid argon, in vacuum-insulated cylinders and cargo tanks.

By Water: In cylinders on cargo and passenger vehicles, and on ferry and railroad car ferry vessels (passenger or vehicle). In authorized cargo tanks on cargo vessels only. Liquid argon, in pressurized cylinders on cargo and passenger vessels and ferry and railroad car ferry vessels (passenger or vehicle).

By Air: In gaseous cylinders aboard passenger aircraft up to 150 lb (68 kg), and aboard cargo aircraft up to 300 lb (136 kg), maximum net weight per cylinder. Liquid argon, aboard passenger aircraft up to 10 lb (45 kg) maximum net weight per container; liquid argon aboard cargo aircraft up to 1100 lb (499 kg) maximum net weight per container.

CONTAINERS

Argon gas is authorized for shipment in cylinders, tube tank cars, and tube trailers. Liquid argon is shipped as a cryogenic fluid in vacuum-insulated cylinders, insulated portable tanks, insulated tank trucks, and tank cars.

Filling Limits

For gaseous argon, the maximum filling limits authorized are as follows: Cylinders and tube trailers may be filled to the authorized service pressure marked on the cylinder or tube assemblies at 70°F (21.1°C). In the case of cylinders of Specifications 3A, 3AA, 3AX, 3AAX, and 3T that meet special requirements, they may be filled up to 10 percent in excess of their marked service pressures. See 49 CFR 173.302 (c). [2] Tube tank cars (uninsulated cars of the TC/DOT-107A type) are authorized to be filled to not more than seven-tenths of the marked test pressure at 130°F (54.4°C).

For liquid argon, the maximum filling limits authorized are: Specification TC/DOT-4L cylinders are authorized for the transportation of liquid argon when carried in the vertical position. The filling density must be in accordance with Table 2.

Cylinders

Cylinders which comply with TC/DOT specifications 3A and 3AA are the types usually used to ship gaseous argon, but it is au-

TABLE 2. FILLING DENSITY OF ARGON IN TC/DOT-4L LIQUID CYLINDERS.

Pressure Control Valve Setting (maximum start-to-discharge pressure)		Maximum Permitted Filling Density (Percent by weight)
psig	(kPa)	
45	(310)	133
75	(517)	130
105	(724)	127
170	(1172)	122
230	(1585)	119
295	(2034)	115
360	(2482)	113
450	(3103)	111
540	(3723)	107
625	(4309)	104

thorized for shipment in any cylinders approved for nonliquefied compressed gas. These include cylinders meeting Specifications 3A, 3AA, 3AX, 3AAX, 3B, 3E, 3T, 3AL, 39, 4B, 4BA, and 4BW; in addition, continued use of cylinders complying with Specifications 3C, 3D, 4, 4A, 4C, 3, 25, 26, 33, and 38 is authorized, but new construction is not authorized.

Liquid argon is authorized for shipment in cylinders which meet DOT specification 4L. For liquid argon at pressures under 25.3 psig (174 kPa), the container specification is not regulated by the DOT. However, in Canada, compressed gases and refrigerated liquids are regulated regardless of pressure.

All cylinders authorized for gaseous argon service must be requalified by hydrostatic retest every five or ten years under present regulations, with the following exceptions: DOT-4 cylinder, every ten years; and no periodic retest is required for cylinders of types 3C, 3E, and 4C. Also, for cylinders of the 4L type authorized for liquid argon service, no periodic retest is required for requalification.

Pressure relief devices authorized for use on argon cylinders are the Type CG-1 rupture disk, CG-4 and CG-5 combination rupture disk/fusible plug devices, and the CG-7 pressure relief valve. Refer to CGA S-1.1, *Pressure Relief Device Standards—Part I—Cylinders for Compressed Gases*, for further information. [4]

Valve Outlet Connections

Standard connections in the United States and Canada for argon cylinders are as follows: for service pressure up to 3000 psig (20 684 kPa), Connection CGA 580; for 3001 to 5500 psig (20 691 to 37 921 kPa), Connection CGA 680; for 5501 to 7500 psig (37 928 to 51 711 kPa), Connection CGA 677; and for cryogenic liquid withdrawal, Connection CGA 295. The standard connections for cylinders are shown in CGA V-1, *American National, Canadian, and Compressed Gas Association Standard for Compressed Gas Cylinder Valve Outlet and Inlet Connections*. [5]

Tank Cars

Gaseous argon is authorized for rail shipment in tank cars that comply with TC/DOT specifications 107A. TC/DOT regulations require that the pressure to which the containers are charged must not exceed seven-tenths of the marked test pressure at 130°F (54.4°C). Liquid argon is also shipped in vacuum-insulated tank cars (AAR204W) at pressures less than 25.3 psig (174 kPa).

Tube Trailers

Gaseous argon is shipped in tube trailers with capacities ranging to more than 40 000 ft^3 (1133 m^3). These trailers are built to comply with DOT cylinder specifications 3A, 3AA, 3AX, 3AAX, or 3T. The trailers commonly serve as the storage supply for the user, with the supplier replacing trailers as they are emptied.

Tank Trailers (Cargo Tanks)

Liquid argon is shipped in bulk at pressures below 25.3 psig (174 kPa) in special insulated tank trailers, with capacities in excess of 400 000 ft^3 (11 327 m^3).

Small Portable Containers

Liquid argon is shipped and stored in small portable containers (dewar flasks) which hold quantities ranging from 1 to 25 gallons (4 to 95 L) or more. These containers are encased in shells and are heavily insulated; they maintain the liquid at atmospheric pressure.

METHODS OF MANUFACTURE

Argon is manufactured in air separation plants by means of fractional distillation after the liquefaction of air.

REFERENCES

[1] CGA G-11.1, *Commodity Specification for Argon*, Compressed Gas Association, Inc., 1235 Jefferson Davis Highway, Arlington, VA 22202.

[2] *Code of Federal Regulations*, Title 49 CFR Parts 100–199 (Transportation), Superintendent of Documents, U.S. Government Printing Office, Washington, DC 20402.

[3] *Transportation of Dangerous Goods Regulations*, Canadian Government Publishing Centre, Supply and Services Canada, Ottawa, Ontario, Canada K1A 0S9.

[4] CGA S-1.1, *Pressure Relief Device Standards—Part 1—Cylinders for Compressed Gases*, Compressed Gas Association, Inc., 1235 Jefferson Davis Highway, Arlington, VA 22202.

[5] CGA V-1, *American National, Canadian, and Compressed Gas Association Standard for Compressed Gas Cylinder Valve Outlet and Inlet Connections* (ANSI/CSA/CGA V-1), Compressed Gas Association, Inc., 1235 Jefferson Davis Highway, Arlington, VA 22202.

ADDITIONAL REFERENCES

CGA P-9, *The Inert Gases, Argon, Nitrogen and Helium*, Compressed Gas Association, Inc., 1235 Jefferson Davis Highway, Arlington, VA 22202.

CGA P-12, *Safe Handling of Cryogenic Liquids*, Compressed Gas Association, Inc., 1235 Jefferson Davis Highway, Arlington, VA 22202.

CGA P-14, *Accident Prevention in Oxygen-Rich and Oxygen-Deficient Atmospheres*, Compressed Gas Association, Inc., 1235 Jefferson Davis Highway, Arlington, VA 22202.

Saturated Liquid Densities of Oxygen, Nitrogen, Argon and Para-Hydrogen (National Bureau of Standards Technical Note 361), National Institute of Standards and Technology, Gaithersburg, MD 20899.

McCarty, R. D., *Interactive Fortran IV Computer Programs for the Thermodynamic and Transport Properties of Selected Cryogens (Fluid Pack)* (National Bureau of Standards Technical Note 1025, October 1980), National Institute of Standards and Technology, Gaithersburg, MD 20899.

Younglove, B. A., *Interactive Fortran Program to Calculate Thermophysical Properties of Six Fluids* (National Bureau of Standards Technical Note 1048, July 1982), National Institute of Standards and Technology, Gaithersburg, MD 20899.

Arsine

Chemical Symbol: AsH_3
Synonyms: Arsenic hydride, arsenic trihydride, arseniuretted hydrogen, arsenious hydride, hydrogen arsenide
CAS Registry Number: 7784-42-1
DOT Classification: Poison A
DOT Label: Poison gas, Flammable gas
Transport Canada Classification: 2.3 (2.1)
UN Number: UN 2188

PHYSICAL CONSTANTS

	U.S. Units	SI Units
Chemical formula	AsH_3	AsH_3
Molecular weight	77.95	77.95
Vapor pressure		
at 70°F (21.1°C)	219.7 psia	1514.7 kPa abs
at 105°F (40.6°C)	319 psia	2199.4 kPa abs
at 115°F (46.1°C)	363 psia	2502.8 kPa abs
at 130°F (54.4°C)	435 psia	2999.2 kPa abs
Density of the gas		
at 68°F (20°C) and 1 atm	0.2025 lb/ft^3	3.24 kg/m^3
Specific gravity of the gas		
at 70°F (21.1°C) and 1 atm		
(air = 1)	2.69	2.69
Specific volume of the gas		
at 70°F (21.1°C) and 1 atm	5.0 ft^3/lb	0.312 m^3/kg
Density of liquid		
at 68°F (20°C)	83.59 lb/ft^3	1.339 kg/m^3
at 70°F (21.1°C)	83.55 lb/ft^3	1.335 kg/m^3
at 105°F (40.6°C)	78.60 lb/ft^3	1.259 kg/m^3
at 115°F (46.1°C)	77.22 lb/ft^3	1.237 kg/m^3
at 130°F (54.4°C)	75.16 lb/ft^3	1.204 kg/m^3
Boiling point at 1 atm	−80.5°F	−62.5°C
Freezing point at 1 atm	−178.4°F	−116.9°C
Critical temperature	211.8°F	99.9°C
Critical pressure	957 psia	6598 kPa abs

	U.S. Units	SI Units
Critical density	—	—
Latent heat of vaporization at −80.46°F (−62.48°C)	92 Btu/lb	214 kJ/kg
Heat capacity at 77°F (25°C)	0.2125 Btu/lb	0.494 kJ/kg
Solubility in water	0.23 in.^{3}/1 in.^{3}	0.23 cm^3/1 cm^3
Flammable limits in air (percent by volume)	4.5–64%	4.5–64%

DESCRIPTION

Arsine is a colorless, extremely toxic, flammable gas at room temperature and atmospheric pressure and is heavier than air. It has a mild garliclike odor and acts as a blood and nerve poison. It can be fatal if inhaled in sufficient quantity and can form flammable mixtures with air.

Arsine is shipped as a liquefied compressed gas in steel cylinders under its own vapor pressure of 219.7 psia (1515 kPa abs). Arsine is slightly soluble in both water and organic solvents. It reacts readily with agents such as potassium permanganate, bromine, and sodium hypochlorite to form arsenic compounds. Arsine is stable at room temperature, but begins to decompose into its elements around 446–464°F (230–240°C).

GRADES AVAILABLE

Arsine is usually sold in ultrahigh-purity grades (99.9995+ percent) primarily for use in the electronic industry. A typical specification for an electronic grade arsine is as follows:

Oxygen	1 ppm
Nitrogen	1 ppm
Carbon monoxide	1 ppm
Carbon dioxide	1 ppm
Moisture	1 ppm
C_1-C_5 Hydrocarbons	1 ppm
Arsine	99.9995%

Manufacturers supply various grades of product depending on the application and purity requirements such as:

- Electronic Grade
- Ultra Large Scale Integration (ULSI) Grade
- Metal Organic Chemical Vapor Deposition (MOCVD) Grade
- Semiconductor Grade

USES

While some arsine is commercially produced for use in organic synthesis, it is mainly used in the electronics industry for the production of semiconductors. Examples of the applications include:

- N-type doping of epitaxial silicon
- N-type diffusions in silicon
- Ion-Implantation
- Growth of gallium arsenide (GaAs) and gallium arsenide phosphide (GaAsP) and other compounds formed with Group III/V elements from the periodic table

PHYSIOLOGICAL EFFECTS

Arsine is an extremely toxic gas which attacks the nervous and circulatory systems and can be fatal if inhaled in sufficient quantity. It is a powerful hemolytic agent, and victims may have delayed symptoms for up to 24 hours. Both chronic and acute exposures to arsine are dangerous. Effects of a single (acute) inhalation exposure include rapid intravascular hemolysis, hemoglobinuria with accompanying dark urine, malaise, dizziness, headache, nausea, vomiting, abdominal pain, diarrhea, and collapse. Pulmonary edema may occur following overexposure. There may be a delay of several hours before the onset of signs or symptoms.

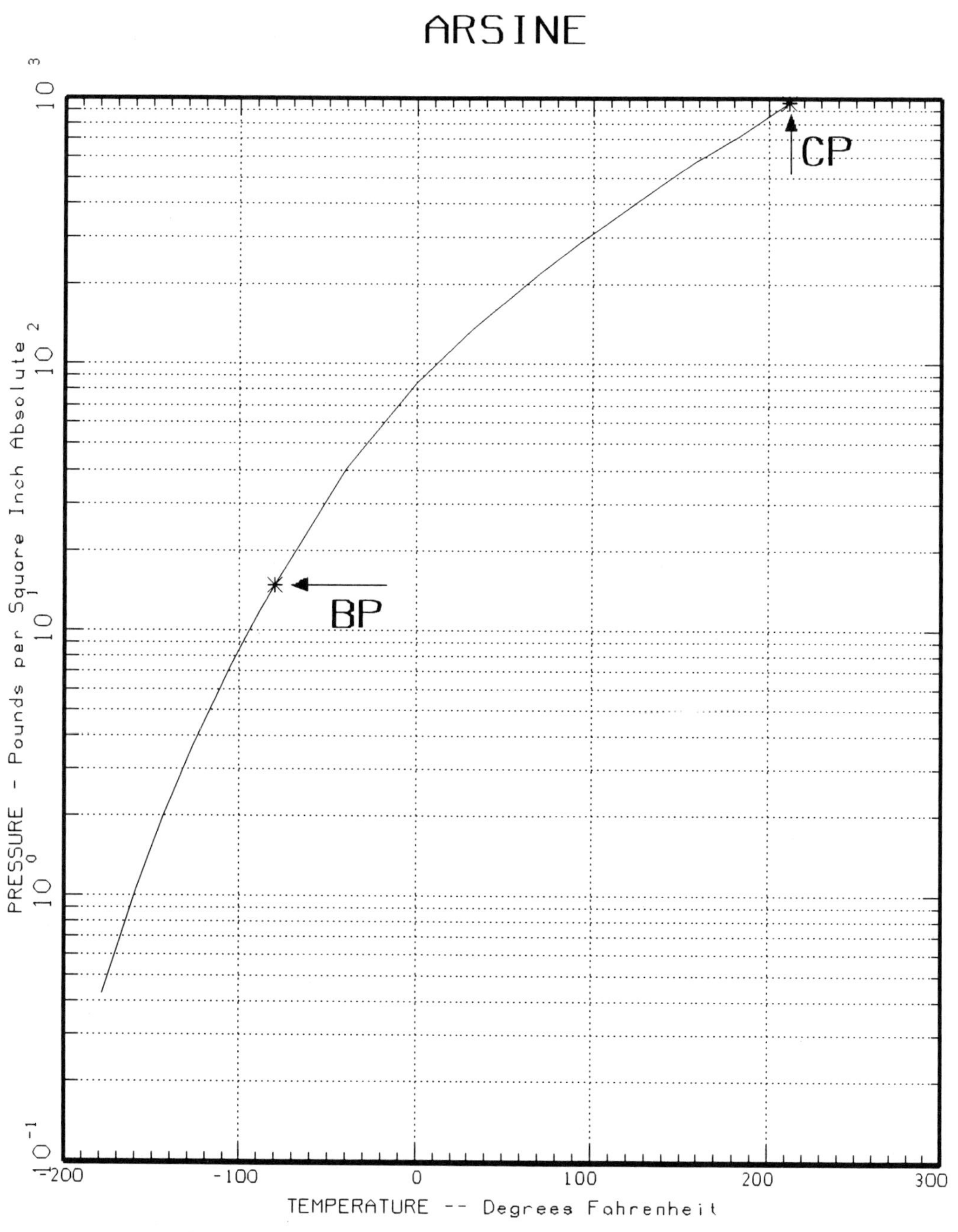

Fig. 1. Vapor Pressure Curve for Arsine.

Effects of repeated (chronic) overexposure may result in peripheral neuropathy, hyperpigmentation, keratosis, cardiovascular disease, and progressive anemia. Other effects of overexposure include pulmonary edema, jaundice, and severe hemolytic anemia. Severe kidney damage may occur with oliguria or anuria leading to uremia and death. Severe liver and cardiac damage may also occur. There is evidence (published in reports of the National Toxicology Program and the International Agency for Research on Cancer) that inorganic arsenic compounds are skin and lung carcinogens in humans.

EFFECTS ON HUMANS FROM CONCENTRATION AND DURATION OF ARSINE EXPOSURE

Concentration	Exposure	Effects on Humans
500 ppm	few minutes	lethal
250 ppm	30 minutes	dangerous to life
6–16 ppm	30–60 minutes	dangerous to life

The American Conference of Governmental Industrial Hygienists has adopted a time-weighted average threshold limit value (TLV-TWA) for arsine of 0.05 ppm. [1] The U.S. Occupational Safety and Health Administration has established an eight-hour time-weighted average exposure limit of 0.05 ppm (0.2 mg/m^3) for arsine. [2]

MATERIALS OF CONSTRUCTION

Arsine is noncorrosive and may, therefore, be used with most of the commercially available metals. However, since arsine is mainly used for the electronics industry, stainless steel is recommended for the gas delivery systems. Stainless steel regulators should be utilized for all high-purity applications with arsine and arsine mixtures.

SAFE STORAGE, HANDLING, AND USE

Since arsine is an extremely toxic and flammable gas, appropriate precautions must be taken in its storage and handling. Store and use arsine and arsine mixtures only in ventilated gas cabinets, exhaust hoods, or highly ventilated rooms that supply a large volume of forced air ventilation. Explosion-proof forced draft gas cabinets or fume hoods are recommended. Use piping and equipment adequately designed to withstand the pressures to be encountered.

Since arsine may form explosive mixtures with air, keep it away from heat and all ignition sources such as flames and sparks. All lines, connections, equipment, and so on, must be thoroughly checked for leaks and grounded prior to use. Only use spark-proof tools and explosion-proof equipment. The compatibility with plastics and elastomers should be confirmed.

For basic safety information on the handling of compressed gas cylinders, refer to CGA P-1, *Safe Handling of Compressed Gases in Containers*. [3] Also see Chapter 5. The storage, handling, and use of arsine in certain areas are regulated by fire codes such as Article 80 of the *Uniform Fire Code*. [4]

DISPOSAL

Low concentrations of arsine can be removed from waste gas streams through the use of liquid scrubbers containing oxidizing solutions such as potassium permanganate, or through the use of dry scrubbers containing oxidizing or absorbing agents. Incineration is also used but requires secondary removal of resulting arsenic trioxide particulates. Disposal of pure arsine should not be undertaken except by experienced personnel.

Prevent waste from contaminating the surrounding area. Discard any product, residue, disposable container, or liner in an environmentally acceptable manner, in full compliance with federal, state, provincial, and local regulations.

HANDLING LEAKS AND EMERGENCIES

Systems designed to handle arsine must be pressure and leak tested with nitrogen or another inert gas prior to being placed in service. Any leak or spill situation with arsine has the potential of serious toxic exposure, as well as fire and explosion.

All areas where arsine is used or stored should be monitored with sensitive gas detection instruments. Detection of concentrations below the TLV level (0.05 ppm) should initiate corrective action, while higher concentrations should initiate an alarm calling for evacuation of all potentially exposed personnel.

If a leak occurs sufficient to cause dangerous levels of arsine, immediately evacuate all personnel from the danger area. Use a self-contained breathing apparatus and full-coverage (Class A) protective clothing to enter the area of the leak. Corrective action to stop the leak should be initiated. Where concentrations in the flammable range may exist, remove all sources of ignition, if without risk. Ventilate the area of a leak or move the leaking container to a well-ventilated area sufficiently far away to avoid exposure to personnel or the local population.

First Aid

If arsine is inhaled or suspected of being inhaled, rescuers, after donning proper protective clothing and equipment, should immediately remove the victim to fresh air. Arsine is known to be skin absorbable. In all cases of exposure, immediate treatment by a physician knowledgeable of the toxic properties of this gas is required, even if no symptoms are present. Specific medical information is available in *Effects of Exposure to Toxic Gases—First Aid and Medical Treatment.* [5]

METHODS OF SHIPMENT

Under the appropriate regulations, arsine is authorized for shipment as follows: [6] and [7]

By Rail: In cylinders (via freight only)

By Highway: In cylinders on trucks

By Water: In cylinders on cargo vessels on deck only and away from living quarters; forbidden on passenger vessels

By Air: Forbidden

See 49 CFR 172.101, "Hazardous Materials Table," for specific requirements. [6]

CONTAINERS

Arsine is authorized for shipment in cylinders manufactured in compliance with TC/DOT Specification 3A, 3AA, 3E, or 3AL and must not exceed 125 lb water capacity (nominal). Specification 3AL cylinders containing arsine may only be transported by highway and rail. Cylinder valves must be protected and sealed from leaks in accordance with 49 CFR 173.327. [6]

The Specification 3A, 3AA, and 3AL type cylinders authorized for arsine service must be requalified by hydrostatic retest every five years under present regulations. For Specification 3E 1800 cylinders, no hydrostatic test is required.

Pressure relief devices are prohibited for use on arsine cylinders.

Valve Outlet Connections

The standard valve outlet connection in the United States and Canada for arsine cylinders is Connection CGA 350.

METHODS OF MANUFACTURE

There are several processes which may be used to commercially produce arsine. The selection of any particular process depends on the reagents required, yield, and disposal of waste products. The processes include:

- Reaction of lithium aluminum hydride with arsenic trichloride
- Reaction of metallic arsenides with acid
- Reaction of arsenious oxide with potassium borohydride in sodium hydroxide solution
- Reaction of sodium arsenide with ammonium bromide in liquid ammonia

REFERENCES

[1] *Threshold Limit Values and Biological Exposure Indices* (1989–90 ed.), American Conference of Governmental Industrial Hygienists, 6500 Glenway Avenue, Bldg. D-7, Cincinnati, OH 45211- 4438.

[2] CGA P-1, *Safe Handling of Compressed Gases in Containers*, Compressed Gas Association, 1235 Jefferson Davis Highway, Arlington, VA 22202.

[3] *Uniform Fire Code*, Western Fire Chiefs Association and International Conference of Building Officials, 5360 South Workman Mill Road, Whittier, CA 90601.

[4] *Effects of Exposure to Toxic Gases—First Aid and Medical Treatment*, Matheson Gas Products, Inc., Secaucus, NJ 07094 (3rd ed., 1988).

[5] *Code of Federal Regulations*, Title 49 CFR Parts 100–199, (Transportation), Superintendent of Documents, U.S. Government Printing Office, Washington, DC 20402.

[6] *Transportation of Dangerous Goods Regulations*, Canadian Government Publishing Centre, Supply and Services Canada, Ottawa, Ontario, Canada K1A 0S9.

ADDITIONAL REFERENCES

Arsine Product Information Sheet (F-3892B), Union Carbide Corporation, Linde Division, 39 Old Ridgebury Road, Danbury, CT 06817.

DHEW (NIOSH) Intelligence Bulletin No. 32 (1979), *Arsine (Arsenic Hydride) Poisoning in the Workplace*, National Institute for Occupational Safety and Health, NIOSH Mail Stop C-13, 4676 Columbia Parkway, Cincinnati, OH 45226.

Encyclopedie Des Gaz, L'Air Liquide. Elsevier/North-Holland Inc., 52 Vanderbilt Avenue, New York, NY 10017, 1976.

Material Safety Data Sheet for Arsine (L4565-A), Union Carbide Corporation, Linde Division, 39 Old Ridgebury Road, Danbury, CT 06817.

Matheson Gas Data Book, 6th ed., Matheson Gas Products, Inc., Secaucus, NJ 07094, 1980.

Matheson Unabridged Gas Data Book, 5th ed., Matheson Gas Products, Inc., Secaucus, NJ 07094, 1977.

Boron Trichloride

Chemical Symbol: BCl_3
Synonym: Boron chloride
CAS Registry Number: 10294-34-5
DOT Classification: Corrosive material
DOT Label: Corrosive
Transport Canada Classification: 2.3 (8)
UN Number: UN 1741

PHYSICAL CONSTANTS

	U.S. Units	SI Units
Chemical formula	BCl_3	BCl_3
Molecular weight	117.17	117.17
Density of the gas at 70°F (21.1°C) and 1 atm	0.303 lb/ft^3	4.85 kg/m^3
Specific volume of the gas	3.3 ft^3/lb	0.206 m^3/kg
Specific gravity at 70°F (21.1°C) and 1 atm (air = 1)	4.045	4.045
Vapor pressure at 70°F (21.1°C)	19.1 psia	131.7 kPa abs
Liquid density at 70°F (21.1°C)	82.7 lb/ft^3	1324 kg/m^3
Critical temperature	353.8°F	178.8°C
Critical pressure	561.3 psia	3870 kPa abs
Critical density	16.8 lb/ft^3	269 kg/m^3
Boiling point	54.5°F	12.5°C
Melting point	−161.1°F	−107.3°C
Latent heat of vaporization at 70°F (21.1°C)	85.9 Btu/lb	199.8 kJ/kg
Specific heat of the liquid at 70°F (21.1°C) C_p	0.229 Btu/(lb)(°F)	0.962 kJ/(kg)(°C)
Specific heat of the gas at 70°F (21.1°C) C_p	0.127 Btu/(lb)(°F)	0.533 kJ/(kg)(°C)

DESCRIPTION

Boron trichloride is a colorless, acid gas that fumes in the presence of moist air. It is packaged in steel cylinders as a liquid under its own vapor pressure of 19.1 psia (131.7 kPa abs) at 70°F (21.1°C). It reacts with water or moist air to produce hydrochloric and boric acid.

GRADES AVAILABLE

Boron trichloride is available for commercial and industrial purposes in C.P. grade with a minimum purity of 99.9 percent by weight. A typical commercial (C.P.) grade analysis by weight is:

Boron trichloride	99.95%
Free chlorine as Cl_2	0.01%
Silicon as Si	0.0015%
Phosgene	0.06%

It is also available to the electronic industry in Electronic and VLSI Echant grade. The impurity specification for the latter is:

Aluminum	0.5 ppm
Calcium	0.5 ppm
Chlorine	1.0 ppm
Copper	0.5 ppm
Iron	0.5 ppm
Nickel	0.5 ppm
Phosgene	1.0 ppm
Potassium	0.5 ppm
Silicon	1.0 ppm
Sodium	0.5 ppm

USES

Boron trichloride is used in the refining of aluminum, copper, magnesium, and zinc to remove oxides, nitrides, and carbides from the molten metal. Carbon monoxide, hydrogen, and nitrogen can be removed from an aluminum melt by treating with boron trichloride. It also improves the tensile strength of aluminum and will allow remelting without a major change in the grain structure.

The electronic industry benefits from boron trichloride in many applications. It is used in the production of optical fibers, as a p-type dopant for thermal diffusion in silicon, and for ion implantation.

PHYSIOLOGICAL EFFECTS

Boron trichloride is irritating and corrosive to all living tissue. Exposure of skin tissue to higher concentrations of boron trichloride or the liquid can cause hydrochloric acid burns and skin lesions resulting in tissue destruction and scarring. Chemical pneumonitis (deep lung inflammation) and pulmonary edema (abnormal fluid buildup in the lungs) result from excessive exposure to the lower respiratory tract and deep lung. Burns to the eyes result in lesions and possible loss of vision. Symptoms of exposure include tearing of eyes, coughing, labored breathing, and excessive salivary and sputum formation.

The American Conference of Governmental Industrial Hygienists (ACGIH) has not established a threshold limit value (TLV) for boron trichloride. It is recommended that compliance with the 5 ppm ceiling limit for hydrogen chloride be used. [1]

MATERIALS OF CONSTRUCTION

Piping, valves, and other equipment used in direct contact with anhydrous boron trichloride should be of stainless steel. Carbon steel may be used in some areas if the temperature remains below 265°F (129.4°C). Monel and other nickel alloys also offer good resistance. In the presence of moisture, the formation of hydrochloric acid will cause most metals to corrode. Platinum, silver, or tantalum offer the best resistance.

SAFE STORAGE, HANDLING, AND USE

Boron trichloride should be stored in a cool, dry, well-ventilated area. Cylinders should be firmly secured and protected from damage. Full and empty cylinders should be stored separately. Inventory should be rotated to prevent prolonged storage of a full cylinder. The temperature in the storage area should not exceed 125°F (51.7°C). Personnel

Vapor Pressure

BORON TRICHLORIDE

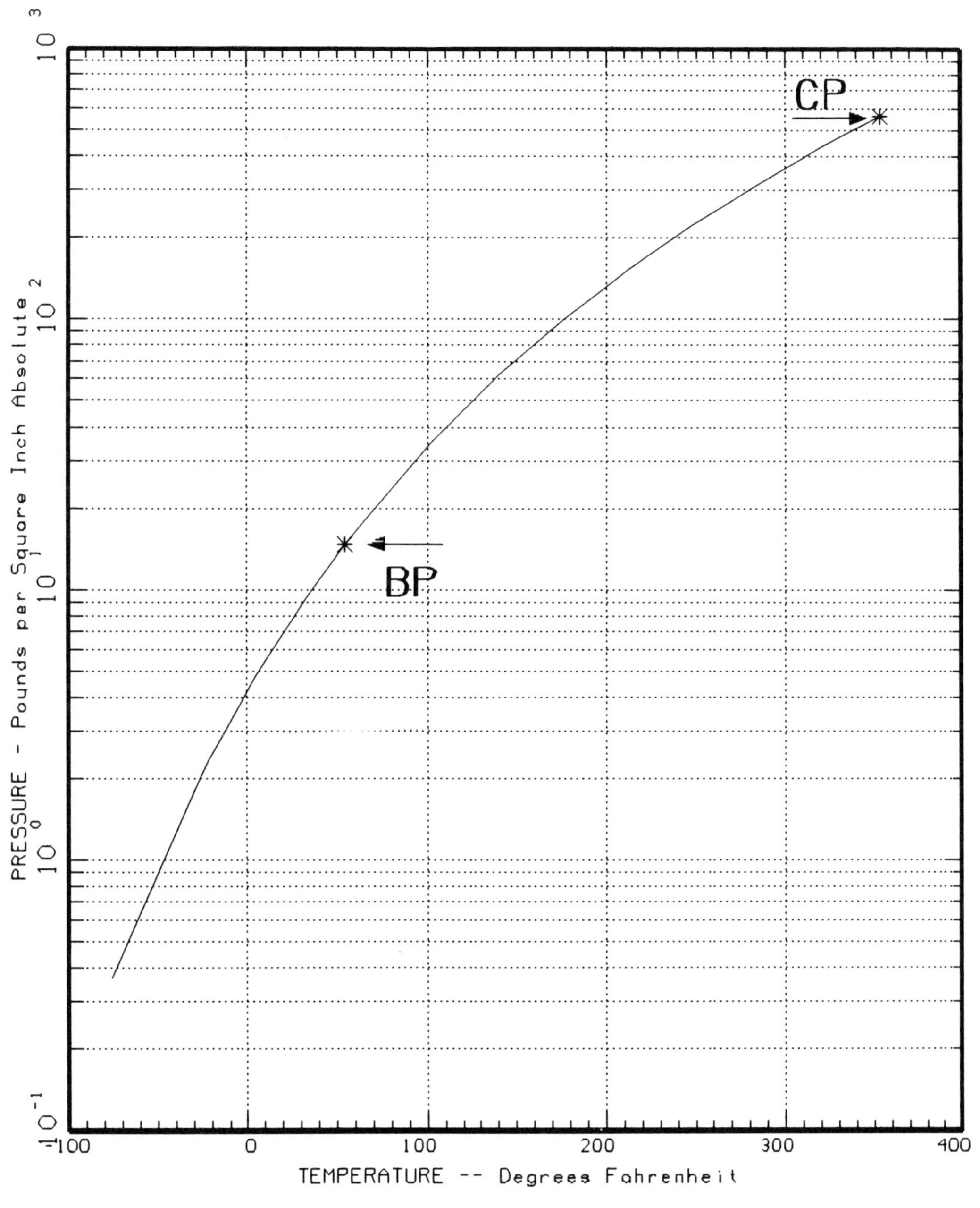

Fig. 1. Vapor Pressure Curve for Boron Trichloride.

handling boron trichloride should be equipped with rubber gloves, chemical-resistant overgarments, chemical goggles or safety glasses, and face shield. They should also have access to full-face respirators and/or self-contained breathing apparatus, eye wash, and safety shower. The work area should be adequately ventilated to prevent the accumulation of gas in case of a release.

The valve protection cap should remain in place unless the container is secured with the valve outlet piped to the point of use. Do not heat the cylinder by any means to increase the discharge rate. Keep the system scrupulously dry. In the presence of trace amounts of water, boron trichloride forms hydrochloric acid. Hydrochloric acid is very corrosive to most common materials of construction. Systems should be purged with a dry inert gas before and after use.

When the system is not in use it should be closed and filled with approximately 5–10 psig of inert gas head pressure. This will keep moisture from entering the system.

DISPOSAL

Disposal of boron trichloride may be accomplished by slowly discharging the gas into a counter-current flow scrubber or other suitable vessel containing approximately 15 percent sodium hydroxide or other alkali and water. It is necessary to use a vacuum break or reverse flow trap to prevent the caustic solution from drawing back into the system or cylinder.

HANDLING LEAKS AND EMERGENCIES

Boron trichloride leaks are easily detected by the presence of fuming vapors. Systems should be vented and purged with inert gas before making repairs. If a leak is in the cylinder or cylinder valve, the supplier should be contacted for assistance. Leaks from the upper part of the valve can often be stopped by simply tightening the valve packing nut (by turning clockwise as viewed from above).

First Aid

Inhalation. Remove personnel from exposure and administer oxygen if chest symptoms occur. Airway obstruction, as indicated by the presence of laryngeal stridor, may require the placement of an airway by trained emergency medical technicians. Chest symptoms may be treated by responding emergency medical technicians, by the administration of anticholinergic inhalants (such as Atrovent) and bronchodilators (such as albuterol).

Skin Contact. Washing has been shown to be effective for boron trichloride skin burns, when continued for 10–30 minutes, possibly because of the associated decrease in skin temperature. After the initial washing, affected areas should be treated with ice water poultices for 30 minutes.

Eye Contact. Any contact of boron trichloride with the eyes should be regarded as serious. Severe irritation of the eyes and eyelids may occur if the eyes are not washed out promptly, and prolonged contact may result in impairment or loss of vision.

If the eyes are affected, they should be irrigated without delay with copious quantities of running water. The eyelids should be spread apart with the fingers to ensure contact with all accessible tissues. Irrigation should continue for at least 15 minutes. Two or three drops of pontocaine (0.5 percent aqueous solution) may be instilled into the eye to facilitate irrigation.

See *Effects of Exposure to Toxic Gases—First Aid and Medical Treatment* for further information. [2]

METHODS OF SHIPMENT

Under appropriate regulations, boron trichloride is authorized for shipment as follows: [3] and [4]

By Rail: In cylinders and tank cars.

By Highway: In cylinders on trucks.

By Water: In cylinders via cargo vessels, ferry, and railroad car ferry. Forbidden on passenger vessels. May be stored above or under deck. However, under deck is preferred.

By Air: Cargo craft only; maximum one quart per package.

CONTAINERS

Boron trichloride is shipped as a compressed liquefied gas under its own vapor pressure (20.6 psia at 70°F; 142 kPa abs at 21.1°C) in cylinders and tank cars.

Cylinders

Boron trichloride is approved for filling in specification steel or nickel cylinders as prescribed for any compressed gas, except acetylene.

The Type CG-2 fusible plug is the authorized pressure relief device for cylinders of boron trichloride. It has a nominal yield temperature of 165°F (73.9°C). For additional information, see CGA S-1.1, *Pressure Relief Device Standards—Part I—Cylinders for Compressed Gases*. [5]

Valve Outlet Connections

The standard valve outlet connection in the United States and Canada for boron trichloride cylinders is Connection CGA 660. [6]

Tank Cars

Boron trichloride is approved for filling in Specification 105A300W tank cars or Specification 106A500X multi-unit tank cars.

METHODS OF MANUFACTURE

Boron trichloride is prepared by chlorinating a mixture of finely divided carbon and boric oxide at 1600–1800°F (871–982°C). It can also be prepared by heating boric oxide with sodium, potassium, or lithium chloride at 1472–1832°F (800–1000°C) or by heating sodium borofluoride with magnesium chloride at 932–1832°F (500–1000°C). [7]

REFERENCES

[1] *Threshold Limit Values and Biological Exposure Indices*, 1989-90 ed., American Conference of Governmental Industrial Hygienists, 6500 Glenway Avenue, Bldg. D-7, Cincinnati, OH 45211-4438.

[2] *Effects of Exposure to Toxic Gases—First Aid and Medical Treatment*, 3rd ed., Matheson Gas Products, Inc., Secaucus, NJ 07094. (1988)

[3] *Code of Federal Regulations*, Title 49 CFR Parts 100–199, (Transportation), Superintendent of Documents, U.S. Government Printing Office, Washington, DC 20402.

[4] *Transportation of Dangerous Goods Regulations*, Canadian Government Publishing Centre, Supply and Services Canada, Ottawa, Ontario, Canada K1A OS9.

[5] CGA S-1.1, *Pressure Relief Device Standards—Part 1—Cylinders for Compressed Gases*, Compressed Gas Association, Inc., 1235 Jefferson Davis Highway, Arlington, VA 22202.

[6] CGA V-1, *American National, Canadian, and Compressed Gas Association Standard for Compressed Gas Cylinder Valve Outlet and Inlet Connections* (ANSI/CSA/CGA V-1), Compressed Gas Association, Inc., 1235 Jefferson Davis Highway, Arlington, VA 22202.

[7] *Matheson Gas Data Book*, 6th ed., Matheson Gas Products, Inc., Secaucus, NJ 07094. (1980)

ADDITIONAL REFERENCES

Encyclopedia Des Gaz, L'Air Liquide. Elsevier/North-Holland Inc., 52 Vanderbilt Avenue, New York, NY 10017.

Material Safety Data Sheet for Boron Trichloride (1987), Air Products and Chemicals, Inc., Allentown, PA 18195.

Boron Trifluoride

Chemical Symbol: BF_3
Synonym: Boron fluoride
CAS Registry Number: 7637-07-2
DOT Classification: Nonflammable gas
DOT Label: Nonflammable gas, Poison
Transport Canada Classification: 2.3 (8)
UN Number: UN 1008

PHYSICAL CONSTANTS

	U.S. Units	SI Units
Chemical formula	BF_3	BF_3
Molecular weight	67.81	67.81
Density of the gas at 32°F (0°C) and 1 atm	0.192 lb/ft^3	3.08 kg/m^3
Specific gravity of the gas at 70°F (21.1°C) and 1 atm (air = 1)	2.32	2.32
Specific gravity of liquid at boiling point (water = 1)	1.57	1.57
Specific volume of the gas at 70°F (21.1°C) and 1 atm	5.6 ft^3/lb	35 m^3/kg
Density of the liquid at −196.8°F (127.1°C)	99.2 lb/ft^3	1589.0 kg/m^3
Boiling point at 1 atm	−148.5°F	−100.3°C
Melting point at 1 atm	−198.5°F	−128°C
Critical temperature	9.95°F	−12.3°C
Critical pressure	723 psia	4985 kPa abs
Critical density	36.9 lb/ft^3	591.1 kg/m^3
Triple point at 1 atm	−196.8	−127.1°C
Latent heat of vaporization at 149.6°F (−100.9°C)	122.6 Btu/lb	285.2 kJ/kg
Latent heat of fusion at −199.7°F (−128.7°C)	26.4 Btu/lb	61.4 kJ/kg
Specific heat of gas at 78°F (26.6°C) and 1 atm C_p	0.178 Btu/(lb)(°F)	0.745 kJ/(kg)(°C)

	U.S. Units	SI Units
Solubility in water, weight percent		
at 32°F (0°C) and 1 atm	322%	322%
Weight of the liquid		
at melting point and 1 atm	13.6 lb/gal	1629.6 kg/m^3

DESCRIPTION

Boron trifluoride is a colorless gas which has a persistent, irritating, acidic odor and which hydrolyzes in moist air to form dense white fumes. It is shipped as a nonliquefied compressed gas at varying pressures up to 2000 psig (13 790 kPa). Boron trifluoride reacts readily with water with the evolution of heat to form the hydrates $BF_3 \cdot H_2O$ and $BF_3 \cdot 2H_2O$, which are relatively strong acids. Inhalation of the gas irritates the respiratory system, and high concentrations in contact with the skin can cause burns similar to hydrogen fluoride but less severe. Boron trifluoride readily dissolves in water, evolving large quantities of heat, and in organic compounds containing oxygen or nitrogen. The reaction is so rapid that if boron trifluoride from a cylinder is fed beneath the surface of these liquids, there is danger of backflow into the cylinder unless a vacuum break or trap is provided in the feed line. Boron trifluoride catalyzes a variety of reactions and forms a great many addition compounds.

GRADES AVAILABLE

Boron trifluoride is available for commercial and industrial use in technical grades having much the same component proportions from one producer to another. The specification for a typical technical grade is as follows:

Component	Specification (percent by weight)
Boron trifluoride	99.0 minimum
Sulfur dioxide	0.04 maximum
Sulfate	0.05 maximum
Silicon tetrafluoride	0.1 maximum
Noncondensables (air)	0.6 maximum

USES

Boron trifluoride is used as a catalyst for polymerizations, alkylations, and condensation reactions; as a gas flux for internal soldering or brazing; as an extinguisher for magnesium fires; and as a source of B^{10} isotope.

PHYSIOLOGICAL EFFECTS

Boron trifluoride irritates the nose, mucous membranes, and other parts of the respiratory system. Concentrations as low as 1 ppm in air can be detected by the sense of smell.

A TLV ceiling of 1 ppm (3 mg/m^3) for exposures to boron trifluoride has been adopted by the American Conference of Governmental Industrial Hygienists. [1] The U.S. Occupational Safety and Health Administration similarly has adopted a ceiling limit of 1ppm (3mg/m^3) for exposures to boron trifluoride. [2]

The irritating sensation and white fumes produced by boron trifluoride in air give easily noticed warning of the escape of even small amounts of the gas; moreover, personnel do not get used to its odor, and tend to seek fresh air if traces of it are inhaled. High concentrations are not only injurious if inhaled but in contact with the skin can cause dehydrating burns similar to those inflicted by acids. Contact of the vapor or liquid with the eyes should also be avoided. In case of burns or other serious exposures, call a physician immediately, and administer first aid.

Protective clothing and equipment required for personnel working with boron trifluoride includes at least rubber gloves, safety glasses, and face shields worn as minimum protection. It is advisable as well that a long-sleeve shirt buttoned at the wrists be worn.

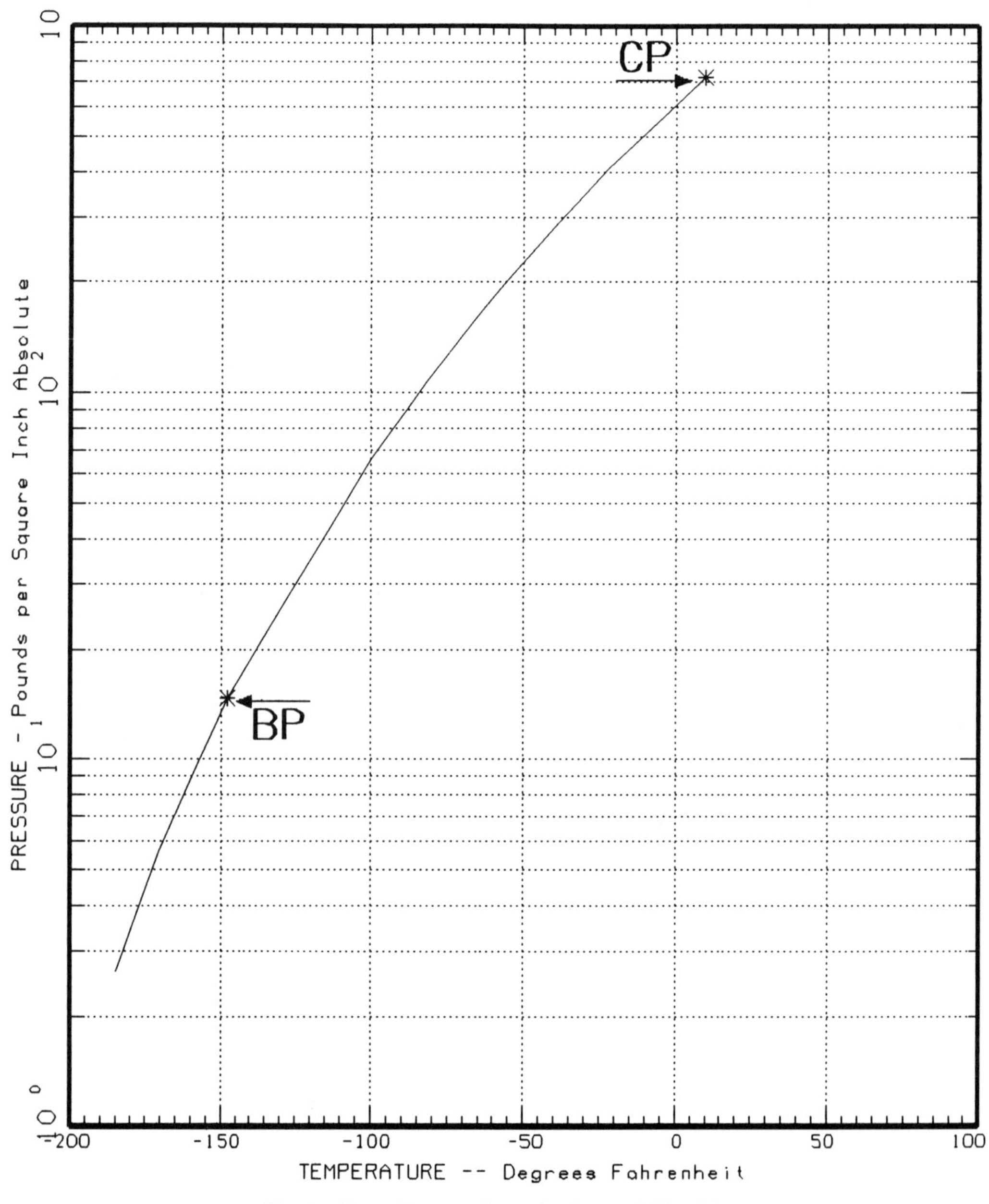

Fig. 1. Vapor Pressure Curve for Boron Trifluoride.

MATERIALS OF CONSTRUCTION

Dry boron trifluoride does not react with the common metals of construction, but if moisture is present, the acidic hydrates formed, $BF_3 \cdot H_2O$ and $BF_3 \cdot 2H_2O$, can corrode many common metals rapidly. Consequently, lines, pressure regulators, and valves in boron trifluoride service must be well protected from the entrance of moist air between periods of use. Cast iron must not be used because active fluorides attack its structure. If steel piping is used for boron trifluoride, forged-steel fittings must be used instead of cast-iron fittings. Stainless steel, Monel, nickel, and Inconel are good materials of construction.

Among materials suitable for gaskets are Teflon and other appropriate fluorocarbon or chlorofluorocarbon plastics. Most plastics become embrittled in boron trifluoride service, but tubing of neoprene or butyl rubber can be used temporarily where positive pressure is not involved. The use of polyvinyl chloride should be avoided.

SAFE STORAGE, HANDLING, AND USE

Boron trifluoride should be stored in a dry, clean area and protected against corrosive materials and potential physical damage. Protect against high temperatures; the maximum temperature to which boron trifluoride containers are exposed should be no higher than 125°F (51.7°C). Users of smaller quantities usually transfer the gas to process through seamless tubing, with a pressure-reducing regulator or needle valve for control. Materials may be steel or stainless steel.

The high pressure system should be designed for a working pressure equal to or greater than the assigned working pressure of the supply cylinder. The system should be cleaned and degreased with a solvent (1,1,1, trichloroethane) and purged with a dry inert gas (nitrogen) prior to introducing the boron trifluoride.

Larger systems should be designed using recommended materials of construction.

Any boron trifluoride transfer system must include vacuum breaks or effective check valves to prevent backflow of process materials into cylinders or tubes supplying the gas. Care should be taken to ensure that blockage by corrosion products does not give a false indication of an empty cylinder or system.

DISPOSAL

Disposal of boron trifluoride may be accomplished by slowly discharging the gas into a countercurrent flow caustic scrubber or other suitable vessel containing approximately 15–35 percent potassium hydroxide or other caustic solution. The system must contain a vacuum break, reverse flow trap, or check valve to prevent caustic solution from drawing back into the cylinder.

HANDLING LEAKS AND EMERGENCIES

Leaks can very readily be detected by the dense white fumes formed by boron trifluoride in moist air, or by its sharp and irritating odor. Emergency personnel equipped with complete breathing, eye, and skin protective equipment are the only ones who should respond to correct leaks; others should leave the area at once until the leak(s) have been stopped and the area has been thoroughly purged of vapors and secured. Emergency procedures for dealing with leaks should be fully established and practiced in advance.

Full-protection (Class A) rubber or plastic garments and breathing apparatus with positive-pressure self-contained air supplies must be available for emergencies that may make it necessary for personnel to enter an area containing a high concentration of boron trifluoride.

First Aid

Inhalation. Remove personnel from exposure and administer oxygen if chest symptoms occur. Airway obstruction, as indicated by the presence of laryngeal stridor, may require the placement of an airway by trained emergency medical technicians. Chest symptoms may be treated by responding emergency medical technicians, by the administration of anticholinergic inhalants (such as Atrovent) and bronchodilators (such as albuterol).

Skin Contact. Washing has been shown to be effective for boron trifluoride skin burns, when continued for 10–30 minutes, possibly because of the associated decrease in skin temperature. After the initial washing, affected areas should be treated with ice water poultices for 30 minutes.

Eye Contact. Any contact of boron trifluoride with the eyes should be regarded as serious. Severe irritation of the eyes and eyelids may occur if the eyes are not washed out promptly, and prolonged contact may result in impairment or loss of vision.

If the eyes are affected, they should be irrigated without delay with copious quantities of running water, as with an eyewash fountain. The eyelids should be spread apart with the fingers to ensure contact with all accessible tissues. Irrigation should continue for at least 15 minutes. Two or three drops of pontocaine (0.5 percent aqueous solution) may be instilled into the eye to facilitate irrigation.

See *Effects of Exposure to Toxic Gases—First Aid and Medical Treatment* for further information. [2]

METHODS OF SHIPMENT

Under the appropriate regulations, boron trifluoride is authorized for shipment as follows: [3] and [4]

By Rail: In cylinders.

By Highway: In cylinders on trucks, and in tube trailers with tubes meeting cylinder specifications.

By Water: In cylinders stowed on deck aboard cargo vessels only.

By Air: Forbidden.

CONTAINERS

Boron trifluoride is authorized for shipment in cylinders, and is also shipped in motor vehicle tube trailers with tubes built to comply with cylinder specifications.

Filling Limits

The maximum filling limit at 70°F (21.1°C) permitted for boron trifluoride is the service pressure of the container.

Cylinders

Boron trifluoride is frequently shipped and used in cylinders meeting TC/DOT Specifications 3A and 3AA and having service pressures ranging from 1800 to 2400 psig (12 410 to (16 547) kPa). Common commercial cylinder sizes have capacities from 6 to 62 lb (34 to 345 ft^3)(2.7 to 28 kg, or up to about 9769 L) of gas.

Under DOT regulations, boron trifluoride is also authorized for shipment in any other cylinders specified as appropriate for nonliquefied compressed gas (which include cylinders meeting DOT Specifications 3B, 3E, 4B, 4BA, and 4BW; cylinders meeting Specifications 3, 3C, 3D, 4, 4A, 4C, 7, 25, 26, 33, and 38 may also be continued in boron trifluoride service, but new construction is not authorized). [3] Cylinders of the 3A and 3AA type used in boron trifluoride service must be requalified by hydrostatic retest every five years under present regulations. All other cylinders authorized for boron trifluoride must similarly be requalified by hydrostatic retest every five years, with the following exceptions: DOT-4 cylinders, every ten years; no periodic retest is required for cylinders of types 3C, 3E, 4C, and 7.

Authorized pressure relief devices for use on boron trifluoride cylinders are Type CG-4 and CG-5 combination rupture disk/fusible plug devices. See CGA S-1.1, *Pressure Relief Device Standards—Part I—Cylinders for Compressed Gasses,* for further information and requirements. [5]

Valve Outlet Connections

The standard valve outlet connection in the United States and Canada for boron trifluoride cylinders is Connection CGA 330. [6]

Tube Trailers

Bulk shipment of boron trifluoride is made in high-pressure tube trailers, for which the tubes have been built to comply with DOT cylinder specifications 3A and 3AA and to have service pressures of around 2000 psig (13 790 kPa). Common capacities of these trailers vary from 9000 to 15 000 lb (4081 to 6802 kg) of gas.

METHODS OF MANUFACTURE

Basically all methods of preparation of boron trifluoride entail the reaction of a boron compound with a fluorine-containing compound in the presence of an acid. In one major method of commercial preparation, boron trifluoride is produced from boric oxide and hydrofluoric acid (in the reaction $B_2O_3 + 6HF \rightarrow 2BF_3 + 3H_2O$), the product being purified and compressed before packaging in cylinders. Other methods of manufacture employ reactions between boron trichloride or borax and hydrofluoric acid.

REFERENCES

[1] *Threshold Limit Values and Biological Exposure Indices,* 1989-90 ed., American Conference of Governmental Industrial Hygienists, 6500 Glenway Ave, Bldg. D-7, Cincinnati, OH 45211-4438.

[2] *Effects of Exposure to Toxic Gases—First Aid and Medical Treatment,* 3rd ed., Matheson Gas Products, Inc., Secaucus, NJ 07094. (1988)

[3] *Code of Federal Regulations,* Title 49 CFR Parts 100–199, (Transportation), Superintendent of Documents, U.S. Government Printing Office, Washington DC 20402.

[4] *Transportation of Dangerous Goods Regulations,* Canadian Government Publishing Centre, Supply and Services Canada, Ottawa, Ontario, Canada K1A 0S9.

[5] CGA S-1.1, *Pressure Relief Device Standards—Part 1—Cylinders for Compressed Gases,* Compressed Gas Association, Inc., 1235 Jefferson Davis Highway, Arlington, VA 22202.

[6] CGA V-1, *American National, Canadian, and Compressed Gas Association Standard for Compressed Gas Cylinder Valve Outlet and Inlet Connections* (ANSI/CSA/CGA V-1), Compressed Gas Association, Inc., 1235 Jefferson Davis Highway, Arlington, VA 22202.

ADDITIONAL REFERENCES

Booth, H. S., and Martin, D. R., *Boron Trifluoride and Its Derivatives,* Wiley, New York, 1949.

Booth, H. S., and Martin, D. R., *J. Am. Chem. Soc.,* **64**, 2198–2205 (1942).

Euken, A., and Schroder, E., *Z. Physik. Chem.,* **B41**, 307–319 (1938).

Fischer, W., and Weidemann, W., *Z. Anorg. Allgem. Chem.,* **213**, 106–114 (1933).

LeBoucher, L., Fischer, W., and Blitz, W., *Z. Anorg. Allgem Chem.,* **207**, 61–72 (1932).

Topchiev, A. V., Zavgorodnii, S. V., and Paushkin, Y. M. (Greaves, J. T., trans.), *Boron Fluoride and Its Compounds as Catalysts in Organic Chemistry,* New York, Pergamon Press, 1959.

1,3-Butadiene (Butadiene)

Chemical Symbol: $H_2C{:}CHCH{:}CH_2$ (or $CH_2{:}CHCH{:}CH_2$)
CAS Registry Number: 106-99-0
Synonyms: Vinylethylene, biethylene, erythrene, bivynl, divynl B
DOT Classification: Flammable gas
DOT Label: Flammable gas
Transport Canada Classification: 2.1
UN Number: UN 1010

PHYSICAL CONSTANTS

	U.S. Units	SI Units
Chemical formula	$H_2C{:}CHCH{:}CH_2$	$H_2C{:}CHCH{:}CH_2$
Molecular weight	54.092	54.092
Vapor pressure		
at 70°F (21.1°C)	21.35 psig	147.2 kPa
at 100°F (37.8°C)	35.8 psig	247 kPa
at 115°F (46.1°C)	54.8 psig	378 kPa
at 130°F (54.4°C)	77 psig	530 kPa
Density of the gas		
at 70°F (21.1°C) and 1 atm	0.37 lb/ft^3	5.9 kg/m^3
Density of the liquid		
at 60°F (15.6°C)	39.05 lb/ft^3	625.5 kg/m^3
at 70°F (21.1°C)	38.69 lb/ft^3	619.8 kg/m^3
at 105°F (40.6°C)	37.0 lb/ft^3	593 kg/m^3
at 115°F (46.1°C)	36.57 lb/ft^3	585.8 kg/m^3
at 130°F (54.4°C)	35.69 lb/ft^3	571.7 kg/m^3
Specific gravity of the gas at 60°F (15.6°C) and 1 atm (air = 1)	1.9153	1.9153
Specific gravity of the liquid at 60°F (15.6°C), at saturation pressure (absolute value from weights in vacuum for the air-saturated liquid)	0.6272	0.6272

	U.S. Units	SI Units
Specific volume of the gas at 70°F (21.1°C) and 1 atm	6.9 ft^3/lb	0.43 m^3/kg
Specific heat of liquid at 1 atm	0.5079 Btu/(lb)(°F)	2.126 kJ/(kg)(°C)
Specific heat of ideal gas at 60°F (15.6°C) and 1 atm		
C_p	0.3412 Btu/(lb)(°F)	1.429 kJ/(kg)(°C)
C_v	0.3045 Btu/(lb)(°F)	1.275 kJ/(kg)(°C)
Ratio of specific heats (C_p/C_v)	1.121	1.121
Boiling point at 1 atm	24.046°F	−4.411°C
Melting point at 1 atm	−164.05°F	−108.92°C
Critical temperature	306°F	152°C
Critical pressure	628 psia	4330 kPa abs
Critical density	15.3 lb/ft^3	245 kg/m^3
Triple point	−164.05°F at 0.010 psia	−108.92°C at 0.069 kPa abs
Solubility in water, at 74°F (23.3°C) and 1 atm weight percent	0.0501%	0.0501%
Weight of liquid at 70°F (21.1°C)	5.172 lb/gal	619.7 kg/m^3
Latent heat of vaporization at boiling point	179.6 Btu/lb	417.7 kJ/kg
Latent heat of fusion at melting point	63.5 Btu/lb	148 kJ/kg
Gross heat of combustion		
Real gas at 60°F (15.6°C) and 1 atm	2954.8 Btu/ft^3	110 090 kJ/m^3
Liquid at 60°F (15.6°C) and saturation pressure	20 095 Btu/lb	46 741 kJ/kg
Liquid at 60°F (15.6°C) and saturation pressure	104 898 Btu/gal	29.237 × 10^6 kJ/m^3
Flammability limits (percent, in air by volume)	2–11.5%	2–11.5%
Net heat of combustion		
Real gas at 60°F (15.6°C) and 1 atm	2800 Btu/ft^3	104 300 kJ/m^3
Liquid at 60°F (15.6°C) and saturation pressure	19 035 Btu/lb	44 275 kJ/kg
Liquid at 60°F (15.6°C) and saturation pressure	99 364 Btu/gal	27 695 × 10^6 kJ/m^3

DESCRIPTION

1,3-Butadiene (butadiene) is a flammable, colorless gas with a mild aromatic odor. (Only 1,3-butadiene is treated in this section; the information given here should not be assumed to apply to 1,2-butadiene or to other butadienes.)

1,3-Butadiene is highly reactive and readily polymerizes, and it is authorized for shipment only if inhibited. Among inhibitors often used are tertiary butylcathechol, di-n-butylamine, and phenyl-betanaphthylamine. Inhibited 1,3-butadiene is shipped as a liquefied compressed gas under its own low vapor pressure of about 21 psig at 70°F (145 kPa at 21.1°C).

GRADES AVAILABLE

1,3-Butadiene is available for commercial and industrial use in various grades having much the same component proportions from one producer to another.

All grades contain approximately 115 ppm of a polymerization inhibitor, such as tertiary-butylcathechol. Distillation or washing with dilute caustic solution is employed for removing the inhibitor when necessary.

USES

One major use of 1,3-butadiene has been in the making of synthetic rubber. Among the types of synthetic rubber made with 1,3-butadiene are styrene-butadiene and nitrile-butadiene rubbers. *Cis*-polybutadiene is also an extender and substitute for rubber, and *trans*-polybutadiene is a type of rubber with unusual properties.

1,3-Butadiene is also used extensively for various polymerizations in manufacturing plastics. It combines with activated olefins in the Diels-Alder reaction to give hydroaromatic hydrocarbons. 1,3-Butadiene undergoes 1,4 cyclization with reactants containing sulfur, oxygen, and nitrogen. [1]

Copolymers with high proportions of styrene have found applications as stiffening resins for rubber, in water-base and other paints, and in high-impact plastics. Butadiene also serves as a starting material for Nylon 66 (adiponitrile) and an ingredient in rocket fuel (butadiene-acrylonitrile polymer).

PHYSIOLOGICAL EFFECTS

If inhaled in high concentrations, 1,3-butadiene has an anesthetic or mild narcotic action which appears to vary with individuals. Inhalation of a 1 percent concentration in air has been reported to have had no effect on the respiration or blood pressure of individuals, but such exposure may cause the pulse rate to quicken and give a sensation of prickling and dryness in the nose and mouth. Inhalation in higher concentrations has brought on blurring of vision and nausea in some persons. Inhalation in excessive amounts leads to progressive anesthesia, and exposure to a 25 percent concentration for 23 minutes proved fatal in one instance. No cumulative action on the blood, lungs, liver, or kidneys has been evidenced.

The potential carcinogenic effect of 1,3-butadiene is presently under study, especially with respect to its metabolism to an epoxide, 1,2:3,4-diepoxy butane. [2] The American Conference of Governmental Industrial Hygienists (ACGIH) has classified 1,3-butadiene as a ''suspected human carcinogen'' [3] The National Toxicology Program has classified 1,3-butadiene as showing clear evidence of carcinogenicity.

The maximum allowable exposure in an 8-hour period must not exceed the 8-hour time-weighted average of 1000 ppm as specified by OSHA (29 CFR 1910.1000). [4] At the time of publication, OSHA had a rule-making pending to lower the permissible exposure level. ACGIH has adopted a time-weighted threshold limit value (TLV-TWA) of 10 ppm (22 mg/m^3) for 1,3-butadiene. [3] The National Institute for Occupational Safety and Health (NIOSH) recommends that 1,3-butadiene be regarded as a potential occupational carcinogen, teratogen, and

Vapor Pressure

1,3-BUTADIENE

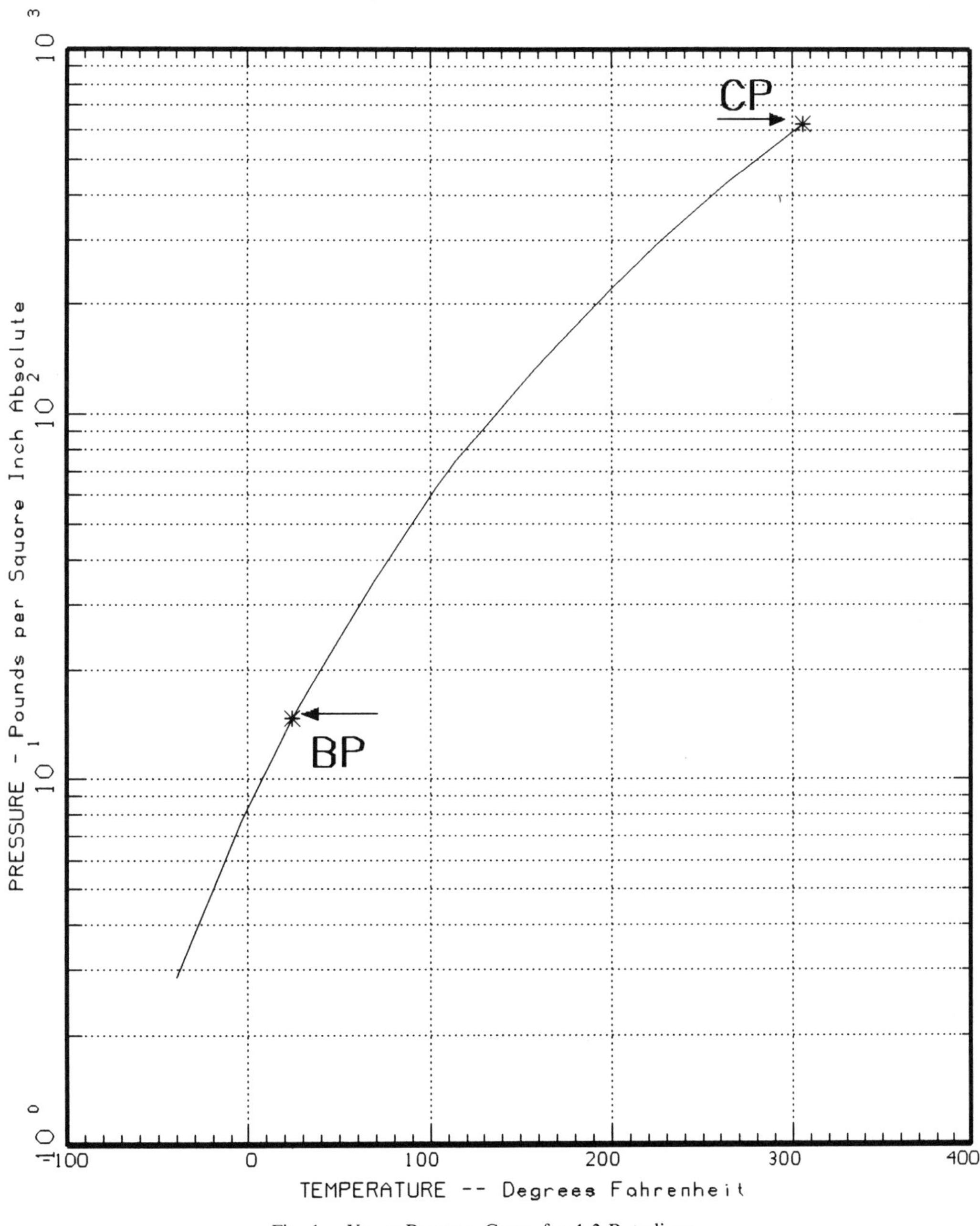

Fig. 1. Vapor Pressure Curve for 1,3-Butadiene.

possible reproduction hazard. NIOSH has recommended that exposure be at the lowest feasible level.

Contact with excessive concentrations of butadiene vapors also irritates the eyes, lungs, and nasal passages. Prolonged contact between liquid butadiene and the skin causes freezing of the tissue. Butadiene liquid evaporates rapidly, and delayed skin burns may result if liquid butadiene is allowed to remain trapped in clothing or in shoes.

MATERIALS OF CONSTRUCTION

1,3-Butadiene is noncorrosive and may be used with any common metals. Steel is recommended for tanks and piping in butadiene service by some authorities. If used with plastics, compatibility must be confirmed. Welded rather than threaded connections are similarly recommended because 1,3-butadiene tends to leak through even extremely small openings. If threaded connections are used, Schedule 80 pipe must be used. Before being exposed to 1,3-butadiene that is not inhibited, iron surfaces should be treated with an appropriate reducing agent such as sodium nitrite because polymerization is accelerated by oxygen (even if present as in ferrous oxide) as well as by heat.

SAFE STORAGE, HANDLING, AND USE

All the precautions necessary for the safe handling of a flammable compressed gas must also be observed with 1,3-butadiene, and special precautions must be taken against its possible polymerization.

1,3-Butadiene must be kept inhibited in storage to prevent polymerization and the formation of spontaneously flammable peroxides. The inhibitor content of butadiene stored for any appreciable period should be regularly measured and maintained at safe levels. Ignition within a storage tank can be prevented by diluting the vapor phase with a sufficient proportion of inert gas. Because of its high volatility, 1,3-butadiene is usually stored under pressure, or in insulated tanks at reduced temperatures, preferably below 35°F (1.6°C).

Cylinders of 1,3-butadiene should be stored upright in a cool, dry, well-ventilated location away from sources of ignition. Reserve stock of this material should be segregated from cylinders containing oxygen, other oxidizing chemicals and gases, and combustible materials. Because of the tendency of 1,3-butadiene to dimerize, the material should be kept as cool as possible. Keep away from heat, sparks and open flame. Store and use with adequate ventilation at all times. Use only in a closed system. Close valve when empty.

Use piping and equipment adequately designed to withstand pressures to be encountered. Ground all equipment. Only use spark-proof tools and explosion-proof equipment.

All precautions necessary for storage tanks containing flammable compressed gas must be taken with 1,3-butadiene storage installations. Tanks should be located outdoors, isolated from boiler houses and other possible sources of ignition, and provided with adequate diking or drainage to confine or discard the contents should the tank rupture. Since this gas is heavier than air and tends to seek lower levels, storage pits and depressions in basements should be avoided. Processes employing 1,3-butadiene should be designed so that personnel are not exposed to the vapor or liquid. Installations must comply with all local regulations and should be designed with the help of authorities thoroughly familiar with butadiene.

DISPOSAL

Discard any product, residue, disposable container, or liner in an environmentally acceptable manner in full compliance with federal, state, provincial, and local regulations.

HANDLING LEAKS AND EMERGENCIES

Due to its high flammability and extreme reactivity, 1,3-butadiene should be stored in a cool, dry, well-ventilated area away from oxidizers, sources of heat, open flames, or sparks. Should a leak occur, evacuate personnel from the area, shut off all ignition sources, ventilate the area, and reduce vapors with fog or fine water spray. Shut off the leak if this can be done without risk to personnel. As flammable vapors may spread, check atmosphere with appropriate devices while using self-contained breathing apparatus. For controlling large flows, personnel may have to wear approach type protective suits and positive pressure self-contained breathing apparatus.

Butadiene forms explosive mixtures with air and oxidizing agents. It may be self-igniting. Should a fire occur, evacuate all personnel from the danger area. Do not extinguish flames due to the possibility of explosive re-ignition. Immediately cool containers with water spray from a maximum distance, taking care not to extinguish flames, and allow fire to burn out. If flames are accidentally extinguished, keep personnel out of the area as explosive re-ignition may occur. If ignition sources can be removed without risk, do so while using self-contained breathing apparatus.

The inhibitor in 1,3-butadiene will protect only the liquid in which it is dissolved; exposure to air will cause peroxides to form, which in turn will affect the rate of polymer formation. If polymerization occurs in a cylinder or tank car, the cylinder or tank car may rupture violently. [1] and [5]

First Aid

Inhalation: 1,3-butadiene has a low order of inhalation toxicity. However, its vapors are mildly irritating to mucus membranes. Any person who becomes aware of symptoms of overexposure should go promptly to an uncontaminated area and inhale fresh air or oxygen. In the event of a massive overexposure, the victim may become unconscious. If this occurs, the victim should be removed to an uncontaminated area where artificial respiration can be administered, followed by administration of oxygen. A physician should be called.

Eye Contact: If liquid 1,3-butadiene comes in contact with the eyes, they should be flushed with tap water for at least 15 minutes. If irritation persists, the victim should be promptly referred to a physician.

Skin Contact: Contact with 1,3-butadiene in liquid form can cause frostbite. If this occurs, cover the frostbitten part with a warm woolen material, immerse the frostbitten part in warm water, and warm slowly until normal circulation is restored. Call a physician.

METHODS OF SHIPMENT

Under the appropriate regulations, inhibited 1,3-butadiene is authorized for shipment as follows: [6] and [7]

By Rail: In cylinders, single-unit tank cars, and TMU (ton multi-unit) tank cars.

By Highway: In cylinders on trucks, and in cargo tanks and portable tanks.

By Water: In cylinders and portable tanks (tanks meeting Specification DOT-51 and not over 20 000 lb (9072 kg) gross weight aboard cargo vessels only). In authorized tank cars aboard trainships only, and in authorized motor vehicle tank trucks aboard trailerships and trainships only. In cylinders on barges of U.S. Coast Guard classes A, CA, and CB only. In cargo tanks aboard tankships and tank barges (to maximum filling densities by specific gravity as stated in Coast Guard regulations). [8]

By Air: In cylinders on cargo airlift only up to 300 lb (136 kg) maximum net weight per cylinder. Forbidden on passenger aircraft.

CONTAINERS

Inhibited 1,3-butadiene is authorized for shipment in cylinders, single-unit tank cars. TMU tank cars, and motor vehicle cargo tanks and portable tanks.

Filling Limits

The maximum filling limits authorized for inhibited butadiene are as follows:

In cylinders—not in excess of the cylinder service pressure at 70°F (21.1°C) or in excess of 5/4 of the service pressure at 130°F (54.4°C). In addition, the liquid portion of the content at 130°F (54.4°C) must not completely fill the container.

In other authorized containers—filling limits as with liquefied petroleum gas; these maximum filling densities are prescribed according to the specific gravity of the liquid material at 60°F (15.6°C) in detailed tables that are part of the DOT regulations. Producers and suppliers who charge inhibited 1,3-butadiene containers other than cylinders should consult these tables in the current regulations. [6] and [7]

The lower and upper limits and the maximum filling densities authorized in Title 49 of the U.S. *Code of Federal Regulations* for such containers are as follows (percent water capacity by weight):

In single-unit tank cars and TMU tank cars—from 45.500 percent (insulated tanks, April through October) for 0.500 specific gravity, to 61.57 percent (uninsulated tanks, November through March, with no storage in transit) for 0.635 specific gravity. In addition, filling must not exceed various specified limits of pressure and liquid content at temperatures of 105°F (40.6°C), 115°F (46.1°C), 130°F (54.4°C), as given in DOT regulations. See 49 CFR 173.314(f). [6]

In cargo tanks and portable tanks—from thirty-eight percent (tanks of 1200 gallons or 4.542 m^3 capacity or less) for 0.473 to 0.480 specific gravity, to 60 percent (tanks of over 1200 gallons or 4.542 m^3 capacity) for 0.627 specific gravity and over (except when using fixed-length dip tubes or other fixed maximum liquid level indicators). Moreover, the tank must be not liquid full at 105°F (40.6°C) if insulated or at 115°F (46.1°C) if uninsulated, and the gauge vapor pressure at 115°F (46.1°C) must not exceed the tank's design pressure. See 49 CFR 173.315(b). [6]

Cylinders

Inhibited 1,3-butadiene is authorized by the DOT for shipment in any cylinders specified for liquefied compressed gas; such cylinders include those that meet the following TC/DOT specifications; 3A, 3AA, 3B, 3BN, 3E, 4B, 4BA, 4B-ET, 4BW, and 39. Cylinders complying with DOT specifications 3, 3D, 4, 4A, 9, 25, 26, 38, 40, and 41 may also be continued in butadiene service, but new construction is not authorized.

Cylinders of all types authorized for inhibited 1,3-butadiene service must be requalified by hydrostatic retest every five years under present regulations with the following exceptions:

(1) No periodic retest is required for 3E cylinders.

(2) Ten years is the required retest interval for Specification 4 cylinders.

(3) External visual inspection may be used in lieu of hydrostatic retest for cylinders that are used exclusively for inhibited 1,3-butadiene which is commercially free from corroding components and that are of the following types (including cylinders of these types with higher service pressure): DOT 3A, 3AA, 3A450X, 3B, 4B, 4BA, 4BW. Continued use of existing cylinders of the ICC-26-240 and 26-300 types is authorized, but no new construction is authorized.

Authorized pressure relief devices for cylinders of inhibited 1,3-butadiene are Type CG-7 pressure relief devices. See CGA S-1.1, *Pressure Relief Device Standards—Part 1—Cylinders for Compressed Gases,* for additional requirements and information. [9]

Valve Outlet Connections

The standard valve outlet connection in the United States and Canada for inhibited 1,3-butadiene cylinders is Connection CGA 510. [10]

Cargo Tanks and Portable Tanks

Inhibited 1,3-butadiene may be shipped by motor vehicle under DOT regulations in cargo tanks meeting DOT specifications MC-330 or MC-331, and in portable tanks complying with DOT-51 specifications. The minimum design pressure required for these tanks is 100 psig (689 kPa).

Single-Unit Tank Cars and TMU Tank Cars

Single-unit tank cars and TMU tank cars are authorized by the DOT for the shipment of inhibited 1,3-butadiene as follows provided that interior pipes of loading and unloading lines are equipped with excess flow valves:

At pressures not exceeding 75 psig (517 kPa) at 105°F (40.6°C) in single-unit tank cars which comply with DOT specifications 105J100W or 111J100W4 (built before March 1, 1984); also, in TMU tanks cars which meet DOT specifications 106A500-X.

At pressures not exceeding 255 psig (1758 kPa) at 115°F (46.1°C) in single-unit tank cars which comply with DOT specifications 112T340W, 112J340W, 114T340W, 114J340W.

At pressures not exceeding 300 psig at 115°F (46.1°C) in single-unit tank cars which comply with DOT specifications 112T400W, 112J400W, 114T400W, 114J400W.

Refer to 49 CFR 173.314 for further information and limitations. [6]

METHODS OF MANUFACTURE

1,3-Butadiene is made commercially by dehydrogenating butanes or butenes in the presence of a catalyst, by reacting ethanol and acetaldehyde, and by the cracking of naphtha and light oil. It is also derived as a by-product in ethylene production.

REFERENCES

[1] *Matheson Gas Data Book,* 6th ed., Matheson Gas Products, Inc., Secaucus, NJ 07094. (1980)

[2] *Effects of Exposure to Toxic Gases—First Aid and Medical Treatment,* 3rd ed., Matheson Gas Products, Inc., Secaucus, NJ 07094. (1988)

[3] *Threshold Limit Values and Biological Exposure Indices,* (1988–89 ed.), American Conference of Governmental Industrial Hygienists, 6500 Glenway Avenue, Bldg. D-7, Cincinnati, OH 45211-4438.

[4] *Code of Federal Regulations,* Title 29 CFR Parts 1900–1910 (Labor), Superintendent of Documents, U.S. Government Printing Office, Washington, DC 20402.

[5] *Emergency Handling of Hazardous Materials in Surface Transportation,* Association of American Railroads, Bureau of Explosives (BOE), 50 F Street, N.W., Washington, DC 20001.

[6] *Code of Federal Regulations,* Title 49 CFR Parts 100–199 (Transportation), Superintendent of Documents, U.S. Government Printing Office, Washington, DC 20402.

[7] *Transportation of Dangerous Goods Regulations,* Canadian Government Publishing Centre, Supply and Services Canada, Ottawa, Ontario, Canada K1A 0S9.

[8] *Code of Federal Regulations,* Title 46 CFR Parts 0-199 (Shipping), Superintendent of Documents, U.S. Government Printing Office, Washington, DC 20402.

[9] CGA S-1.1, *Pressure Relief Device Standards—Part 1—Cylinders for Compressed Gases,* Compressed Gas Association, Inc., 1235 Jefferson Davis Highway, Arlington, VA 22202.

[10] CGA V-1, *American National, Canadian, and Compressed Gas Association Standard for Compressed Gas Cylinder Valve Outlet and Inlet Connections* (ANSI/CSA/CGA V-1), Compressed Gas Association, Inc., 1235 Jefferson Davis Highway, Arlington, VA 22202.

Carbon Dioxide

Chemical Symbol: CO_2
Synonym: Carbon anhydride
CAS Registry Number: 124-38-9
DOT Classification: Nonflammable gas
DOT Label: Nonflammable gas
Transport Canada Classification: 2.2
UN Number: UN 1013 (compressed gas), UN 2187 (refrigerated liquid), UN 1845 (solid)

PHYSICAL CONSTANTS

	U.S. Units	SI Units
Chemical formula	CO_2	CO_2
Molecular weight	44.01	44.01
Vapor pressure		
at 70°F (21.1°C)	838 psig	5778 kPa
at 32°F (0°C)	491 psig	3385 kPa
at 2°F (−16.7°C)	302 psig	2082 kPa
at −20°F (−28.9°C)	200 psig	1379 kPa
at −69.9°F (−56.5°C)	60.4 psig	416 kPa
at −109.3°F (−78.5°C)	0 psig	0 kPa
Density of the gas		
at 70°F (21.1°C) and 1 atm	0.1144 lb/ft^3	1.833 kg/m^3
at 32°F (0°C) and 1 atm	0.1234 lb/ft^3	1.977 kg/m^3
Specific gravity of the gas		
at 70°F (21.1°C) and 1 atm (air = 1)	1.522	1.522
at 32°F (0°C) and 1 atm (air = 1)	1.524	1.524
Specific volume of the gas		
at 70°F (21.1°C) and 1 atm	8.741 ft^3/lb	0.5457 m^3/kg
at 32°F (0°C) and 1 atm	8.104 ft^3/lb	0.5059 m^3/kg
Density of the liquid, saturated		
at 70°F (21.1°C)	47.6 lb/ft^3	762 kg/m^3

	U.S. Units	SI Units
at 32°F (0°C)	58.0 lb/ft³	929 kg/m³
at 2°F (−16.7°C)	63.3 lb/ft³	1014 kg/m³
at −20°F (28.9°C)	66.8 lb/ft³	1070 kg/m³
at −69.9°F (−56.6°C)	73.5 lb/ft³	1177 kg/m³
Sublimation temperature at 1 atm	−109.3°F	−78.5°C
Critical temperature	87.9°F	31.1°C
Critical pressure	1070.6 psia	7382 kPa abs
Critical density	29.2 lb/ft³	468 kg/m³
Triple point	−69.9°F at 60.4 psig	−56.6°C at 416 kPa
Latent heat of vaporization		
at 32°F (0°C)	100.8 Btu/lb	234.5 kJ/kg
at 2°F (−16.7°C)	119.0 Btu/lb	276.8 kJ/kg
at −20°F (−28.9°C)	129.7 Btu/lb	301.7 kJ/kg
Latent heat of fusion at −69.9°F (−56.6°C)	85.6 Btu/lb	199 kJ/kg
Specific heat of the gas at 77°F (25°C) and 1 atm		
C_p	0.203 Btu/(lb)(°F)	0.850 kJ/(kg)(°C)
C_v	0.157 Btu/(lb)(°F)	0.657 kJ/(kg)(°C)
Ratio of specific heats (C_p/C_v) at 59°F (15°C)	1.304	1.304
Solubility in water, vol/vol at 68°F (20°C)	0.90	0.90
Weight of the liquid at 2°F (−16.7°C)	8.46 lb/gal	1014 kg/m³
Cylinder pressure at 68% filling density (42.5 lb/ft; 681 kg/m³		
at 70°F (21.1°C)	838 psig	5778 kPa
at 100°F (37.8°C)	1450 psig	9997 kPa
at 130°F (54.4°C)	2250 psig	15 513 kPa
Latent heat of sublimation at −109.3°F (−78.5°C)	245.5 Btu/lb	571.3 kJ/kg
Viscosity of saturated liquid at 2°F (−16.7°C)	0.287 lb/(ft)(hr)	0.000119 Pa·s

DESCRIPTION

Carbon dioxide is a compound of carbon and oxygen in proportions by weight of about 27.3 percent carbon to 72.7 percent oxygen. A gas at normal atmospheric temperatures and pressures, carbon dioxide is colorless, odorless, and about 1.5 times as heavy as air. A slightly acid gas, it is felt by some persons to have a slight pungent odor and biting taste.

Carbon dioxide gas is relatively nonreactive and nontoxic. It will not burn, and it will not support combustion or life. When dissolved in water, carbonic acid (H_2CO_3) is formed. The pH of carbonic acid varies from 3.7 at atmospheric pressure to 3.2 at 23.4 atm.

Carbon dioxide may exist simultaneously as a solid, liquid, and gas at a temperature of −69.9°F (−56.6°C) and a pressure of

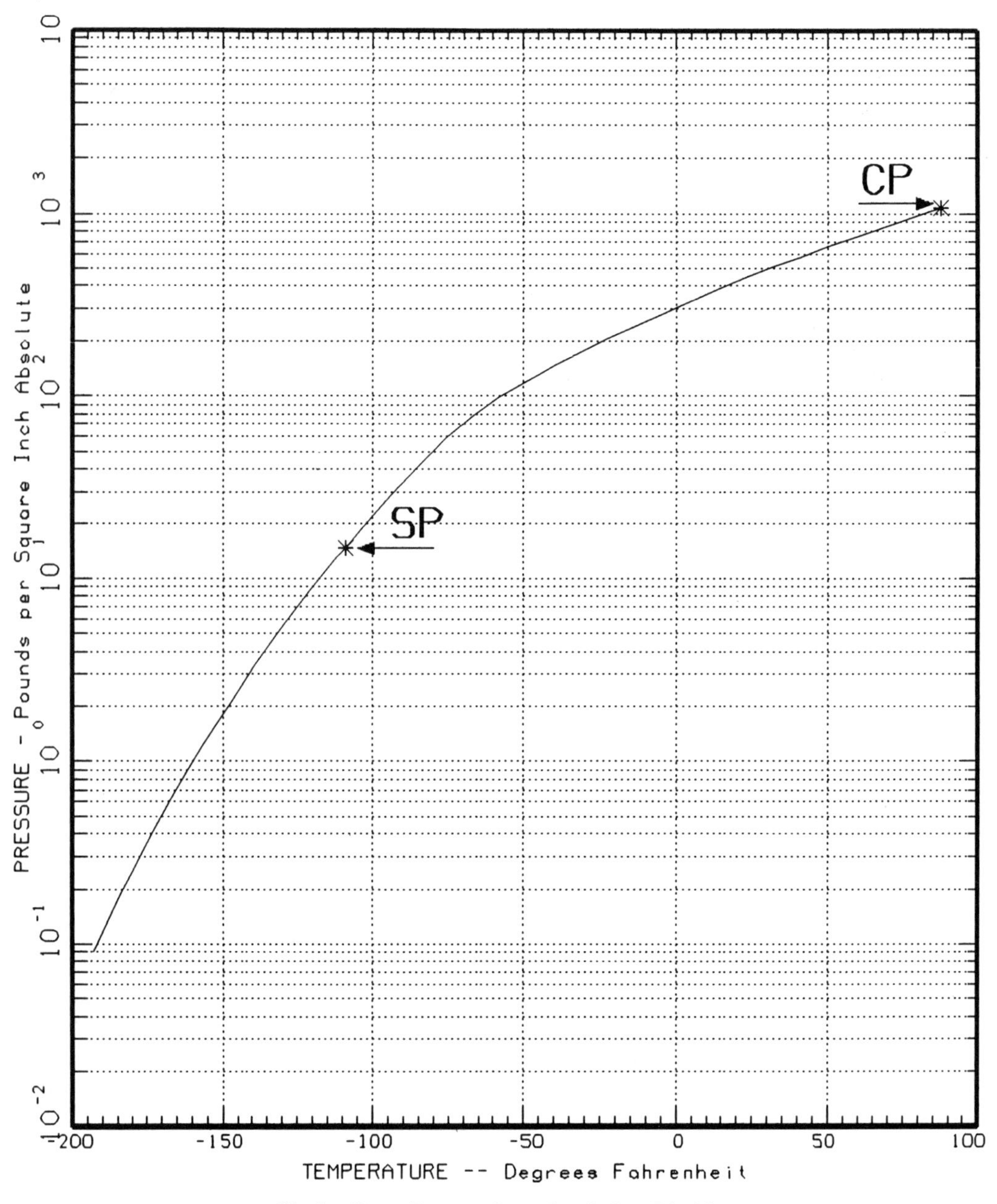

Fig. 1. Vapor Pressure Curve for Carbon Dioxide.

60.4 psig (416 kPa), its triple point. Figure 1 is the vapor pressure curve for carbon dioxide. Figure 2 shows the triple point and full equilibrium curve for carbon dioxide.

At temperatures and pressures below the triple point, carbon dioxide may be either a solid ("dry ice") or a gas, depending upon temperature conditions. Solid carbon dioxide at a temperature of −109°F (−78.5°C) and atmospheric pressure transforms directly to a gas (sublimes without passing through the liquid phase). Lower temperatures will result if solid carbon dioxide sublimes at pressures less than atmospheric. Thermodynamic properties of saturated carbon dioxide in the solid, liquid, and vapor phase are given in Tables 1A and 1B.

At temperatures and pressures above the triple point and below 87.9°F (31.1°C), carbon dioxide liquid and gas may exist in equilibrium in a closed container. Within this temperature range, the vapor pressure in a closed container holding carbon dioxide liquid and gas in equilibrium bears a definite relationship to the temperature. Above the critical temperature, which is 87.9°F (31.1°C), carbon dioxide cannot exist as a liquid regardless of the pressure.

GRADES AVAILABLE

Table 2, from CGA G-6.2, *Commodity Specification for Carbon Dioxide,* presents the component maxima in parts per million, ppm (mole/mole) unless otherwise shown, for the grades (also denoted as quality verification levels) of carbon dioxide. [1] A blank indicates no maximum limiting characteristic. The absence of a listed value in a quality verification level does not mean to imply that the limiting characteristic is or is not present, but merely indicates that the test is not required for compliance with the specification.

QVLs E, F, and H generally refer to carbon dioxide in cylinders.

Note: A gas/liquid equilibrium must be established in the cylinder prior to verification of the limiting characteristics.

QVLs E and H also generally refer to carbon dioxide in bulk liquid storage and transport containers. QVL J refers to solidified carbon dioxide (dry ice).

USES

Solid carbon dioxide is used quite extensively to refrigerate dairy products, meat products, frozen foods, and other perishable foods while in transit. It is also used as a cooling agent in many industrial processes, such as grinding heat-sensitive materials, rubber tumbling, cold-treating metals, shrink fitting of machinery parts, vacuum cold traps, and so on.

Gaseous carbon dioxide is used to carbonate soft drinks, for pH control in water treatment, in chemical processing, as a food preservative, as an inert "blanket" in chemical and food processing and metal welding, as a growth stimulant for plant life, for hardening molds and cores in foundries, and in pneumatic devices.

Liquid carbon dioxide is used as an expendable refrigerant for freezing and chilling food products; for low temperature testing of aviation, missile, and electronic components; for stimulation of oil and gas wells; for rubber tumbling; and for controlling chemical reactions. Liquid carbon dioxide is also used as a fire-extinguishing agent in portable and built-in fire-extinguishing systems.

PHYSIOLOGICAL EFFECTS

Carbon dioxide is normally present in the atmosphere at about 0.035 percent by volume. It is also a normal end product of human and animal metabolism. The exhaled breath contains up to 5.6 percent carbon dioxide. The greatest physiological effect of carbon dioxide is to stimulate the respiratory center, thereby controlling the volume and rate of respiration. It is able to cause dilation and constriction of blood vessels and is a vital constituent of the acid-base mechanism that controls the pH of the blood.

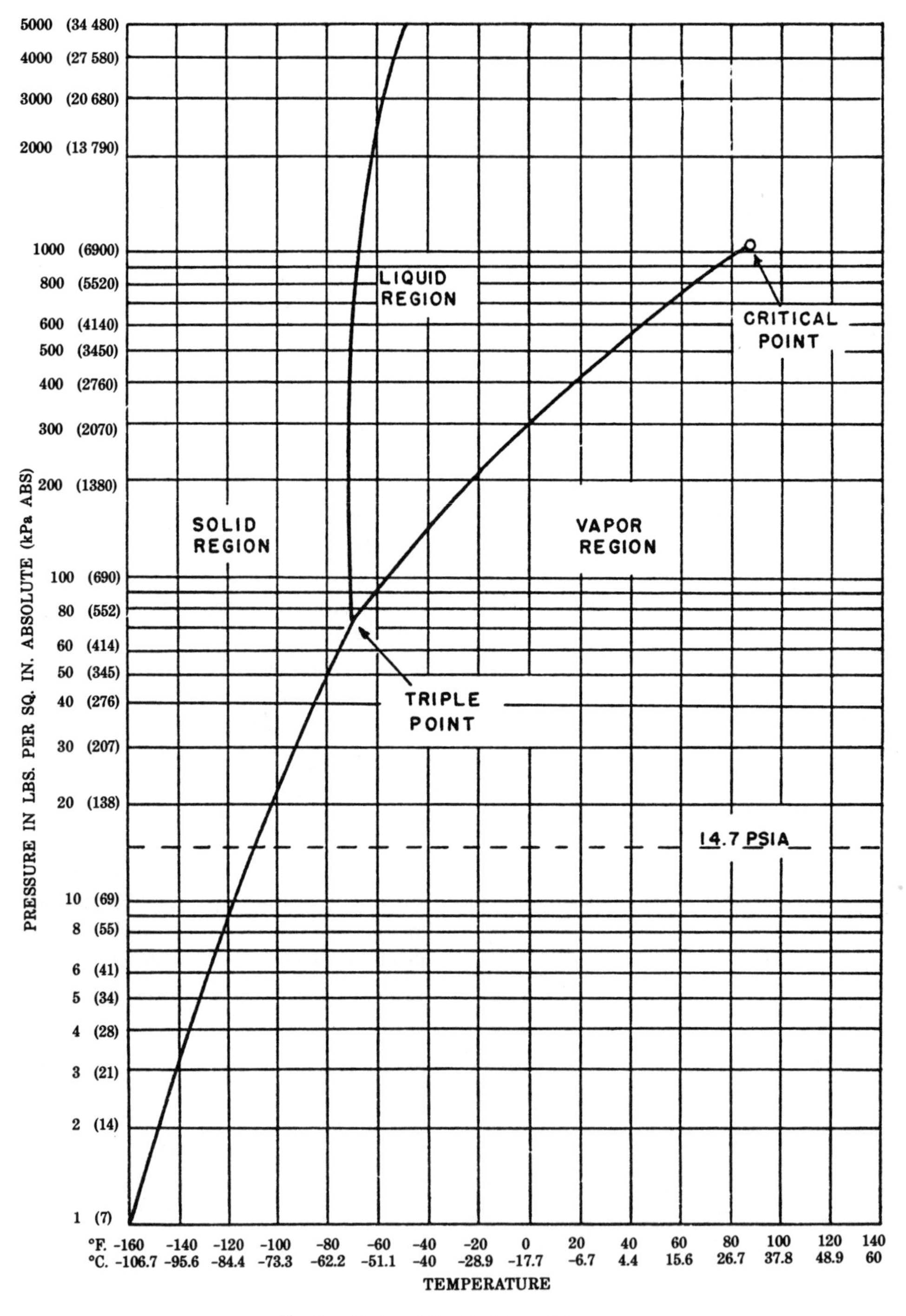

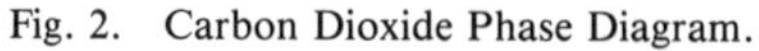
Fig. 2. Carbon Dioxide Phase Diagram.

Table 1A. Thermodynamic Properties of Saturated Carbon Dioxide Solid, Liquid, and Vapor Phases (Customary SI units).

Phase	Temp C	Pressure kPa absolute	Pressure kPa gauge	Density kg/m³ Solid or Liquid	Volume m³/kg x 10⁻³ Vapor	Enthalpy* kJ/kg Solid or Liquid	Enthalpy* kJ/kg Vapor	Entropy* kJ/(kg)(K) Solid or Liquid	Entropy* kJ/(kg)(K) Vapor
Solid and Vapor	-102	11.36	- 89.97	1597	2837	123.5	710.1	1.415	4.841
	-100	13.97	- 87.36	1595	2327	125.8	711.3	1.428	4.809
	- 98	17.15	- 84.18	1593	1916	128.2	712.4	1.442	4.777
	- 96	20.95	- 80.38	1591	1583	130.5	713.6	1.455	4.746
	- 94	25.49	- 75.84	1588	1314	132.9	714.8	1.469	4.716
	- 92	30.89	- 70.44	1585	1095	135.3	715.9	1.482	4.687
	- 90	37.27	- 64.06	1582	917.3	137.7	717.1	1.495	4.658
	- 88	44.76	- 56.57	1579	771.9	140.2	718.2	1.508	4.630
	- 86	53.53	- 47.80	1576	651.3	142.6	719.3	1.521	4.603
	- 84	63.77	- 37.56	1573	550.7	145.1	720.4	1.534	4.376
	- 82	75.72	- 25.61	1569	467.1	147.6	721.4	1.548	4.550
	- 80	89.62	- 11.71	1565	397.7	150.1	722.4	1.561	4.523
	- 78.5	101.3	0	1562	354.7	152.1	723.1	1.569	4.504
	- 78	105.7	4.4	1561	339.8	152.7	723.4	1.574	4.498
	- 76	124.2	22.9	1558	291.1	155.3	724.4	1.586	4.473
	- 74	145.6	44.3	1554	249.9	157.9	725.4	1.599	4.449
	- 72	170.0	68.7	1549	215.1	160.5	726.3	1.612	4.425
	- 70	198.1	96.8	1545	185.7	163.1	727.1	1.625	4.402
	- 68	230.2	128.9	1541	160.8	165.8	727.7	1.638	4.378
	- 66	267.0	165.7	1536	139.5	168.4	728.1	1.651	4.353
	- 64	308.9	207.6	1532	121.1	171.1	728.4	1.664	4.328
	- 62	356.7	255.4	1527	105.1	173.9	728.6	1.677	4.304
	- 60	409.8	308.5	1522	91.23	176.7	728.7	1.690	4.281
	- 58	467.1	365.8	1517	81.00	179.5	728.8	1.703	4.262
	- 56.6	518.0	416.7	1513	72.22	181.4	729.0	1.722	4.250
Triple Point									
Liquid and Vapor	- 56.6	518.0	416.7	1178	72.22	380.5	729.0	2.641	4.250
	- 56	531.7	430.4	1176	71.10	381.5	729.1	2.643	4.244
	- 54	578.9	477.6	1168	64.72	385.2	729.5	2.659	4.229
	- 52	629.5	528.2	1161	59.78	388.9	729.8	2.675	4.215
	- 50	683.6	582.3	1154	55.41	392.5	730.2	2.691	4.203
	- 48	741.0	639.7	1146	51.36	396.2	730.6	2.707	4.192
	- 46	801.9	700.6	1139	47.63	399.9	731.1	2.723	4.181
	- 44	866.3	865.0	1131	44.20	403.7	731.5	2.739	4.170
	- 42	934.3	833.0	1123	41.05	407.5	732.0	2.756	4.160
	- 40	1006	904.7	1115	38.16	411.3	732.4	2.772	4.149

Phase	Temp C	Pressure kPa absolute	Pressure kPa gauge	Density kg/m³ Solid or Liquid	Volume m³/kg x 10⁻³ Vapor	Enthalpy* kJ/kg Solid or Liquid	Enthalpy* kJ/kg Vapor	Entropy* kJ/(kg)(K) Solid or Liquid	Entropy* kJ/(kg)(K) Vapor
Liquid and Vapor	-38	1082	981	1107	35.52	415.1	732.9	2.788	4.139
	-36	1162	1061	1099	33.11	419.0	733.3	2.803	4.129
	-34	1246	1145	1091	30.90	422.9	733.7	2.819	4.119
	-32	1335	1234	1083	28.87	426.8	734.1	2.835	4.109
	-30	1429	1328	1074	27.00	430.8	734.4	2.851	4.100
	-28	1527	1426	1066	25.27	434.8	734.7	2.867	4.091
	-26	1630	1529	1057	23.66	438.8	734.9	2.883	4.081
	-24	1739	1638	1048	22.16	442.8	735.0	2.899	4.072
	-22	1852	1751	1039	20.76	446.9	735.0	2.915	4.062
	-20	1971	1870	1030	19.45	451.0	735.0	2.931	4.053
	-18	2095	1994	1021	18.24	455.1	734.9	2.947	4.044
	-16	2226	2125	1011	17.13	459.3	734.7	2.963	4.034
	-14	2362	2261	1002	16.09	463.6	734.4	2.979	4.024
	-12	2503	2402	991.9	15.11	467.8	734.1	2.994	4.014
	-10	2649	2548	982.0	14.19	472.2	733.6	3.010	4.004
	- 8	2804	2703	971.8	13.34	476.6	733.0	3.027	3.993
	- 6	2964	2863	961.5	12.54	481.1	732.2	3.043	3.983
	- 4	3131	3030	951.5	11.79	485.6	731.4	3.059	3.973
	- 2	3305	3204	940.7	11.07	490.3	730.5	3.076	3.962
	0	3485	3384	929.4	10.38	495.0	729.4	3.092	3.950
	2	3673	3572	917.4	9.703	499.8	727.7	3.109	3.937
	4	3869	3768	905.0	9.046	504.7	725.4	3.126	3.923
	6	4071	3970	892.1	8.435	509.8	723.1	3.143	3.908
	8	4282	4181	878.0	7.878	515.0	720.8	3.161	3.894
	10	4501	4400	863.6	7.375	520.4	718.5	3.179	3.879
	12	4730	4629	848.2	6.900	525.9	716.1	3.198	3.864
	14	4966	4865	831.9	6.446	531.6	713.5	3.217	3.849
	16	5210	5109	814.3	6.006	537.6	710.5	3.236	3.833
	18	5464	5363	795.5	5.577	543.8	707.2	3.256	3.817
	20	5727	5626	775.2	5.157	550.4	703.5	3.278	3.800
	22	6001	5900	753.6	4.745	557.8	699.4	3.303	3.783
	24	6285	6184	728.9	4.337	566.0	694.6	3.331	3.765
	26	6581	6480	696.4	3.914	575.4	688.2	3.364	3.745
	28	6890	6789	655.7	3.460	586.3	679.1	3.403	3.710
	30	7211	7110	593.1	2.910	602.5	664.4	3.454	3.658
	31.1	7382	7281	467.9	2.137	634.3	634.3	3.552	3.552

*Based on 0 for the perfect crystal at 0 degrees kelvin (-273.15C)

TABLE 1B. THERMODYNAMIC PROPERTIES OF SATURATED CARBON DIOXIDE SOLID, LIQUID, AND VAPOR PHASES (SI UNITS).

Temp F	Pressure		Volume cu ft/lb	Density lb/cu ft	Enthalpy** Btu/lb		Entropy** Btu/(lb) (°R)	
	psia	psig	Vapor	Solid or Liquid	Solid or Liquid	Vapor	Solid or Liquid	Vapor
Solid and Vapor								
-150	1.793	26.26*	41.81	99.7	53.55	305.50	0.3394	1.1530
-140	3.171	23.46*	24.41	99.2	56.36	306.90	0.3483	1.1318
-130	5.405	18.92*	14.69	98.8	59.22	308.30	0.3571	1.1126
-120	8.923	11.75*	9.131	98.3	62.15	309.62	0.3658	1.0945
-110	14.34	7.725*	5.829	97.5	65.16	310.81	0.3743	1.0770
-109.3	14.70	0	5.683	97.5	65.38	310.90	0.3748	1.0758
-105	17.94	3.244	4.708	97.2	66.69	311.45	0.3786	1.0687
-100	22.28	7.584	3.814	96.9	68.24	312.01	0.3829	1.0606
- 95	27.67	12.97	3.103	96.6	69.80	312.50	0.3872	1.0527
- 90	34.14	19.44	2.531	96.2	71.41	312.89	0.3915	1.0449
- 85	41.63	26.93	2.074	95.7	73.01	313.11	0.3959	1.0372
- 80	50.58	35.88	1.714	95.3	74.63	313.22	0.4002	1.0292
- 75	61.72	47.02	1.418	94.9	76.28	313.29	0.4045	1.0215
- 70	74.76	60.06	1.182	94.4	77.96	313.42	0.4089	1.0151
- 69.9	75.13	60.43	1.157	94.4	78.01	313.42	0.4112	1.0150
Triple Point								
Liquid and Vapor								
- 69.9	75.13	60.43	1.157	73.53	163.6	313.42	0.6308	1.0150
- 68	78.48	63.78	1.117	73.30	164.4	313.49	0.6318	1.0128
- 66	82.34	67.64	1.060	73.05	165.3	313.58	0.6340	1.0107
- 64	86.35	71.65	1.011	72.79	166.1	313.67	0.6362	1.0088
- 62	90.50	75.80	0.9659	72.54	167.0	313.76	0.6384	1.0070
- 60	94.76	80.06	0.9254	72.28	167.9	313.85	0.6406	1.0053
- 58	99.15	84.45	0.8876	72.01	168.8	313.94	0.6427	1.0038
- 56	103.7	89.00	0.8501	71.76	169.6	314.03	0.6448	1.0024
- 54	108.4	93.70	0.8145	71.49	170.5	314.12	0.6469	1.0010
- 52	113.2	98.50	0.7809	71.23	171.4	314.23	0.6490	0.9996
- 50	118.2	103.5	0.7489	70.97	172.3	314.33	0.6511	0.9982
- 48	123.4	108.7	0.7184	70.71	173.2	314.44	0.6533	0.9967
- 46	128.8	114.1	0.6896	70.44	174.1	314.57	0.6555	0.9952
- 44	134.3	119.6	0.6622	70.16	175.0	314.68	0.6577	0.9938
- 42	140.0	125.3	0.6361	69.88	175.9	314.78	0.6599	0.9924
- 40	145.9	131.2	0.6113	69.60	176.8	314.89	0.6621	0.9910
- 38	152.0	137.3	0.5876	69.33	177.7	314.98	0.6642	0.9897
- 36	158.1	143.4	0.5648	69.04	178.7	315.09	0.6663	0.9883
- 34	164.6	149.9	0.5432	68.76	179.6	315.20	0.6684	0.9869
- 32	171.2	156.5	0.5227	68.48	180.5	315.31	0.6704	0.9855
- 30	178.0	163.3	0.5031	68.20	181.4	315.40	0.6725	0.9842
- 28	184.9	170.2	0.4844	67.92	182.3	315.49	0.6746	0.9829
- 26	192.1	177.4	0.4665	67.64	183.2	315.58	0.6767	0.9817
- 24	199.6	184.9	0.4492	67.35	184.3	315.67	0.6788	0.9805
- 22	207.2	192.5	0.4325	67.06	185.2	315.74	0.6810	0.9793
- 20	215.0	200.3	0.4166	66.77	186.1	315.81	0.6831	0.9781
- 18	223.1	208.4	0.4014	66.47	187.0	315.86	0.6852	0.9769
- 16	231.3	216.6	0.3868	66.17	188.1	315.92	0.6873	0.9756
- 14	239.8	225.1	0.3731	65.87	189.0	315.95	0.6894	0.9743
- 12	248.7	234.0	0.3599	65.56	189.9	315.99	0.6916	0.9730
- 10	257.6	242.9	0.3473	65.25	191.0	316.01	0.6937	0.9717
- 8	266.9	252.2	0.3351	64.93	191.9	316.01	0.6958	0.9704

*Inches of mercury below atmospheric pressure.

Liquid and Vapor

Temp F	Pressure		Volume cu ft/lb	Density lb/cu ft	Enthalpy** Btu/lb		Entropy** Btu/(lb) (°R)	
	psia	psig	Vapor	Solid or Liquid	Solid or Liquid	Vapor	Solid or Liquid	Vapor
-6	276.3	261.6	0.3233	64.62	193.0	316.01	0.6979	0.9691
-4	285.8	271.1	0.3119	64.29	193.9	315.99	0.7000	0.9678
-2	295.7	281.0	0.3010	63.96	194.9	315.95	0.7022	0.9666
+ 0	305.8	291.1	0.2906	63.63	195.8	315.92	0.7043	0.9654
2	316.3	301.6	0.2805	63.30	196.9	315.88	0.7064	0.9642
4	327.0	312.3	0.2707	62.97	198.0	315.83	0.7085	0.9629
6	337.9	323.2	0.2613	62.64	198.9	315.76	0.7106	0.9616
8	349.0	334.3	0.2523	62.30	200.0	315.68	0.7127	0.9603
10	360.5	345.8	0.2436	61.99	200.9	315.59	0.7148	0.9589
12	372.2	357.5	0.2353	61.69	202.0	315.50	0.7169	0.9575
14	384.3	369.6	0.2273	61.32	203.0	315.40	0.7190	0.9561
16	396.5	381.8	0.2196	61.02	204.1	315.27	0.7211	0.9547
18	409.0	394.3	0.2122	60.67	205.2	315.13	0.7232	0.9533
20	421.9	407.2	0.2050	60.32	206.3	314.96	0.7253	0.9520
22	435.1	420.4	0.1980	59.91	207.4	314.80	0.7275	0.9507
24	448.7	434.0	0.1911	59.57	208.4	314.62	0.7297	0.9493
26	462.5	417.8	0.1845	59.17	209.5	314.42	0.7319	0.9479
28	476.6	461.9	0.1783	58.78	210.6	314.19	0.7341	0.9465
30	490.8	476.1	0.1722	58.40	211.7	313.90	0.7363	0.9450
32	505.5	490.8	0.1663	58.02	212.8	313.58	0.7385	0.9434
34	520.5	505.8	0.1602	57.59	214.0	313.20	0.7407	0.9417
36	536.0	521.3	0.1542	57.12	215.1	312.77	0.7429	0.9399
38	551.7	537.0	0.1482	56.70	216.4	312.28	0.7452	0.9380
40	567.7	553.0	0.1425	56.29	217.4	311.76	0.7475	0.9360
42	584.0	569.3	0.1372	55.89	218.7	311.20	0.7598	0.9340
44	600.8	586.1	0.1321	55.44	220.0	310.63	0.7521	0.9321
46	617.8	603.1	0.1273	54.95	221.2	310.05	0.7544	0.9302
48	635.2	620.5	0.1226	54.43	222.5	309.47	0.7568	0.9283
50	652.9	638.2	0.1181	53.91	223.7	308.90	0.7593	0.9264
52	671.2	656.5	0.1138	53.45	225.0	308.32	0.7618	0.9246
54	689.7	675.0	0.1095	52.95	226.4	307.75	0.7643	0.9227
56	708.6	693.9	0.1054	52.37	227.7	307.13	0.7668	0.9207
58	727.9	713.2	0.1014	51.81	229.1	306.49	0.7694	0.9187
60	747.6	732.9	0.09752	51.17	230.6	305.78	0.7720	0.9166
62	767.7	753.0	0.09372	50.47	232.0	305.03	0.7746	0.9145
64	788.3	773.6	0.08999	49.78	233.5	304.22	0.7773	0.9123
66	809.3	794.6	0.08631	49.08	235.1	303.35	0.7801	0.9100
68	830.8	816.1	0.08261	48.39	236.7	302.45	0.7830	0.9077
70	852.7	838.0	0.07894	47.62	238.3	301.52	0.7861	0.9053
72	875.0	860.3	0.07535	46.80	240.3	300.51	0.7894	0.9030
74	897.8	883.1	0.07173	45.90	242.1	299.39	0.7930	0.9006
76	921.1	906.4	0.06811	44.94	244.3	298.10	0.7970	0.8982
78	945.1	930.4	0.06411	43.90	246.4	296.57	0.8013	0.8957
80	969.5	954.8	0.06013	42.67	248.9	294.75	0.8060	0.8924
82	994.5	979.8	0.05603	41.23	251.5	292.46	0.8112	0.8881
84	1020	1005	0.05171	39.59	254.7	289.67	0.8170	0.8821
86	1046	1031	0.04711	37.03	259.0	285.64	0.8249	0.8737
87.9	1071	1056	0.03423	29.21	272.7	272.70	0.8483	0.8483

**Based on 0 for the perfect crystal at absolute zero of temperature, -459.67F (-273.15C)

TABLE 2. CARBON DIOXIDE GRADES AVAILABLE (Units in ppm (mole/mole) unless otherwise stated).

Limiting Characteristics	Quality Verification Levels				
	E	F	G	H	I
Carbon Dioxide, min. % (mole/mole)	99.0	99.5	99.5	99.8	
Water, ppm v/v (Vapor)	200	120	32	32	
Dew Point, °F	−32	−40	−60	−60	
Total Hydrocarbons (as Methane)			50	50	
Oxygen				30	
Carbon Monoxide	10 (vapor)			10	
Hydrogen Sulfide	1 (vapor)			0.5 (Note 3)	
Nitric Oxide	2.5 (vapor)			2.5	
Nitrogen Dioxide	2.5 (liquid)			2.5	
Ammonia	25 (liquid)				
Sulfur Dioxide	5 (liquid)			5	
Carbonyl Sulfide				0.5 (Note 3)	
Nonvolatile Residues, ppm (wt/wt)		10	10	10	500
Odor	Free of Foreign Odor, Note 1				
Color					White Opaque
U.S.P., Note 2	Yes				

Note 1. In gas phase or water solution.

Note 2. For *United States Pharmacopeia* (USP) quality verification, the verification tests must be performed in the following sequence: carbon monoxide, hydrogen sulfide, nitric oxide, nitrogen dioxide, ammonia, sulfur dioxide, water, percent purity (after the filled cylinder has warmed to ambient temperature). A gas/liquid equilibrium must be established in the cylinder. For a "vapor" sample, test carbon dioxide from the vapor phase of the cylinder; for a "liquid" sample, arrange the cylinder so that when its valve is opened, a portion of the carbon dioxide liquid phase is released through a regulator or a piece of pressure tubing of sufficient length to allow all liquid to vaporize.

Note 3. Total sulfides in QVL H shall not exceed 0.5 ppm as hydrogen sulfide.

Note 4. Unless otherwise specified, the sample shall be obtained from the vapor phase of the container, or as agreed upon between the supplier and the customer.

Note 5. The 1973 edition of G-6.2 listed four quality verification levels lettered A to D for gaseous and liquid carbon dioxide, and one quality verification level of solid carbon dioxide. These letter designations have since been dropped because they no longer represent major volume of usage by industry. To get the specifications of quality verification levels dropped, see CGA G-6.2-1973 or contact the Compressed Gas Association.

Carbon dioxide acts as a stimulant and a depressant on the central nervous system. Increases in heart rate and blood pressure have been noted at a concentration of 7.6 percent, and dyspnea (labored breathing), headache, dizziness, and sweating occur if exposure at that level is prolonged. At concentrations of 10 percent and above, unconsciousness can result in one minute or less. Impairment in performance has been noted during prolonged exposure to concentrations of 3 percent carbon dioxide even when the oxygen concentration was 21 percent.

Table 3 shows the tolerance time of a healthy male under exercising conditions.

Inhalation of gaseous carbon dioxide can adversely affect body function. Skin, eye, or mouth contact with dry ice or compressed carbon dioxide can cause tissue damage or burns.

Effects of Overexposure

Gaseous carbon dioxide is an asphyxiant. Concentrations of 10 percent or more can produce unconsciousness or death. Lower concentrations may cause headache, sweating, rapid breathing, increased heartbeat, shortness of breath, dizziness, mental depression, visual disturbances, and shaking.

TABLE 3. PHYSIOLOGICAL TOLERANCE TIME FOR VARIOUS CARBON DIOXIDE CONCENTRATIONS.

CO_2 in Air, Percent by Volume	Maximum Exposure Limit, Minutes
0.5	indefinite
1.0	indefinite
1.5	480
2.0	60
3.0	20
4.0	10
5.0	7
6.0	5
7.0	less than 3

Note: For healthy male under exercising conditions.

The seriousness of the latter manifestations are dependent upon the concentration of carbon dioxide and the length of time the individual is exposed.

MATERIALS OF CONSTRUCTION

The common commercially available metals can be used for carbon dioxide installations (those not handling carbon dioxide in aqueous solutions). Any carbon dioxide system at the user's site must be designed to safely contain the pressures involved and must conform with all state and local regulations. See also CGA G-6.1, *Standard for Low Pressure Carbon Dioxide Systems at Consumer Sites.* [2] For low-pressure carbon dioxide systems (up to 400 psig or 2758 kPa), containers and related equipment should have design pressures rated at least 10 percent above the normal maximum operating pressure.

For such systems, Schedule 80 threaded steel pipe with forged steel fittings rated at 2000 psi (13 790 kPa) or seamless Schedule 40 steel pipe with welded joints is recommended; alternate recommendations include stainless steel, copper, or brass pipe and stainless steel or copper tubing. Special materials and construction are required for containers operating at temperatures below −20°F (−28.9°C).

Since carbon dioxide forms carbonic acid when dissolved in water, systems handling carbon dioxide in aqueous solutions must be fabricated from such acid-resistant materials as certain stainless steels, Hastelloy metals, or Monel metal.

SAFE HANDLING, STORAGE, AND USE

Carbon dioxide is contained, shipped, and stored in either liquefied or solid form. Additional information regarding the safe storage, handling, and use of carbon dioxide can be found in CGA G-6, *Carbon Dioxide.* [3] Applications using gaseous carbon dioxide are supplied by gas converted from liquid or solid carbon dioxide. Liquid made from dry ice in converters must not be used to fill cylinders.

Being more dense than air, carbon dioxide gas may accumulate in low or confined areas under certain conditions of use or storage. Precautions with regard to ventilation are required. Appropriate warning signs should be affixed outside of those areas where high concentrations of carbon dioxide gas may accumulate. Suggested wording for such a sign is:

CAUTION
CARBON DIOXIDE GAS
Ventilate before entering.
A high CO_2 gas concentration may occur in this area and may cause suffocation.

For information on current recommended labeling practices concerning carbon dioxide containers, refer to CGA C-7, *Guide to the Preparation of Precautionary Labeling and Marking of Compressed Gas Containers.* [4]

When entering low or confined areas where a high concentration of carbon dioxide gas is present, do not use air-breathing or filter-type gas masks. Gas masks of the self-contained type, or the type which feeds clean outside air to the breathing mask, are required.

Liquefied Carbon Dioxide

Users of liquefied carbon dioxide must comply with all state, provincial, municipal and other local regulations.

Except for bulk low pressure containers, storage containers of liquefied carbon dioxide are noninsulated and nonrefrigerated. The contained carbon dioxide is, therefore, at ambient temperature and relatively high pressure.

Caution must be taken that cylinders and high pressure tubes charged with liquid carbon dioxide not be allowed to reach a temperature exceeding 125°F (51.7°C). Storage should never be near furnaces, radiators, or any other source of heat. Storage areas should be designed to prevent heavy objects from accidentally shearing off piping, valves, or pressure relief devices.

Pressure relief devices on converters located in confined areas where the carbon dioxide discharged cannot be dissipated must be vented outdoors away from personnel and building air intakes. Such piping must not be capped at the outlet end or equipped with valves or other means of stopping or restricting the flow of the gas.

Liquid Carbon Dioxide in Bulk Containers at Low Pressures

Bulk containers for storing liquid carbon dioxide at low pressures are well insulated and equipped with a means, usually mechanical refrigeration, to control and limit internal temperatures and pressures. Storage temperatures are maintained well below ambient temperature, usually in the range of −20°F to −10°F (−28.9°C to −23.3°C) with corresponding carbon dioxide pressures of 200 to 345 psig (1379 to 2379 kPa).

Bulk containers used for storing liquid carbon dioxide at low pressures should be protected from tampering by unauthorized individuals. If a small enclosed location is used, the outlet from the pressure relief valves must be piped to the outside away from personnel, building air intakes, or other points such that discharge of the carbon dioxide vapor will not result in a high concentration of carbon dioxide. Such piping must not be equipped with valves or other means of stopping the flow of gas.

The storage container should preferably be located in an area that is not subject to unduly high temperatures. If the ambient temperature is above 110°F (43.3°C) for long periods of time, it may be necessary to provide additional refrigeration capacity. If the ambient temperature falls below 0°F (−17.8°C) for a prolonged period no harm will result, but the carbon dioxide pressure may fall below the desired range.

Dusty, oily locations should be avoided because of the tendency of dust and oil to collect on the refrigerator condenser and thus reduce its efficiency. A dry, well-ventilated location is preferable.

The storage container should not be located in an area where it might be struck by heavy moving or falling objects. A break or tear in the outer shell or covering of the insulation will destroy the vapor seal and allow water vapor to enter the insulation with eventual losses of insulation efficiency. Whenever liquid carbon dioxide is discharged directly to the atmosphere, as in those cases where sudden cooling is desired, extreme caution should be exercised to guard against and counteract the heavy recoil inherent with the discharge of a dense liquid (weighing more than water) under high pressure. Lines should be anchored firmly against this recoil by means of positive mechanical devices installed prior to use.

All lines from a bulk carbon dioxide container should be anchored in such a fashion that shrinkage of the piping or tubing due to the passage of the cold liquid carbon dioxide through them will not tear them loose. Such piping must be protected with adequate pressure relief devices to prevent unduc pressure buildup from entrapped liquid carbon dioxide. Valves used in such lines should have a design pressure not less that 350 psig (2413 kPa.)

The rapid discharge of liquid carbon dioxide through a line which is not grounded will result in a buildup of static electricity potential, which may be dangerous to operating personnel. Such lines should, therefore, be grounded before use.

Flexible hoses used with liquid carbon dioxide should have a minimum working pressure of 500 psig (3447 kPa) for low temperature operation. They should be reinforced with wire braid and attached to couplings of sufficient strength to resist separation from the hose at pressures below the minimum bursting pressure of the hose. Hose lines must be equipped with pressure relief valves if liquid carbon dioxide can become trapped between two valves.

After use of such flexible hose, it becomes quite rigid and may contain loose dry ice snow. Do not fold, bend, or distort the hose, or point it in any direction where pressure buildup within the hose will eject the dry ice snow so as to endanger personnel. The hoses should be examined periodically for wear and damage to both the hose body and couplings. Any hose showing evidence of weakening should be removed from service immediately.

Whenever liquid carbon dioxide is discharged into confined spaces to reduce temperature rapidly, large volumes of carbon dioxide gas will be evolved, amounting to some 8.5 ft^3 of gas per pound of liquid (0.53 m^3/kg). These spaces must therefore have some provision incorporated to vent this vapor to the atmosphere to prevent pressure rise within the space.

Carbon Dioxide Cylinder Manifolding Systems

Piping from cylinders to the point of use in carbon dioxide cylinder manifolding systems must be of correct high pressure design (of at least Schedule 80 high pressure steel pipe or type "K" copper tubing, with provisions made in the piping for adequate pressure relief devices). Piping should be adequately braced.

Transfer of liquid carbon dioxide from one high pressure carbon dioxide cylinder to another may be accomplished by several means, including direct transfer by pressure differential or, more usually, by means of a pump. For refilling fire extinguishers or any other carbon dioxide cylinder, consult the supplier and follow the supplier's recommendations.

When a depleted carbon dioxide cylinder is removed from a manifold supply line, close the valve first and leave it closed to prevent air from entering the "empty" cylinder.

Outlets from pressure relief valves should

be piped to the outside away from personnel and building air intakes to prevent accumulation of heavy carbon dioxide vapors. Such piping must not be capped on the outlet end or equipped with valves or other means of stopping the flow of gas.

No attempt should be made to use carbon dioxide vapor without a pressure-reducing regulator of suitable design and in good condition.

Solid Carbon Dioxide

Solid carbon dioxide (dry ice) has a temperature of −109.3°F (−78.5°C) and must be protected during storage with thermal insulation in order to minimize loss through sublimation. Dry ice should be stored in well-insulated storage containers, preferably in a cool, nonconfined, or ventilated area.

Do not handle dry ice with bare hands. Use heavy gloves or dry ice tongs. Handle blocks of dry ice carefully, as injuries can occur if one is accidentally dropped on the feet.

A suggested wording for a caution label for dry ice follows:

SOLID CARBON DIOXIDE (DRY ICE)
WARNING—EXTREMELY COLD
(−109°F)

Avoid contact with skin and eyes; use gloves. Do not taste. Keep out of children's reach. Liberates gas which may cause suffocation. Do not put in stoppered or closed containers.

Solid Carbon Dioxide and Converters

Solid carbon dioxide may be transformed into a liquid in dry ice converters. These converters should be located in areas where they will never be subjected to temperatures of more than 125°F (51.7°C). Converter locations must be protected so that unauthorized persons cannot tamper with fittings and valves. Adequate protection should be provided to prevent heavy objects from shearing off piping, valves, or pressure relief devices.

The same precautions for handling high pressure gas or liquid carbon dioxide apply equally to solid carbon dioxide converters. The only differences between liquid and solid storage are the method of filling and the amount of liquefied carbon dioxide contained. Dry ice converters are charged with dry ice, and the amount charged must not exceed the rated capacity of the converters. Do not overfill. When charging a dry ice converter, do not leave it open unnecessarily. This will minimize condensation within the converter and thus limit the possibility of corrosion.

Pressure relief devices on converters located in confined areas where the carbon dioxide discharged cannot be dissipated must be vented outdoors remote from personnel and building air intakes. Such piping must not be capped on the outlet end or equipped with valves or other means of stopping or restricting the flow of the gas.

DISPOSAL

Disposal of Refrigerated Liquid CO_2

Vent the carbon dioxide slowly to a well-ventilated outdoor location remote from inhabited areas and building air intakes. Venting liquid rather than vapor minimizes the chance of cooling vessels and piping to hazardously low temperatures as the internal pressure is reduced. Be alert to any resultant fog that may drift across highways, sidewalks, and areas where work may need to be done. If possible, venting should be done on a clear day with at least moderate winds.

Disposal of Cylinder CO_2

Vent slowly from the liquid phase of the cylinder to a well-ventilated outdoor location away from personnel or work areas.

Venting vapor could self-refrigerate the cylinders to a hazardously low temperature. Return the emptied cylinders to the supplier with some residual pressure, with the cylinder valve tightly closed and with the valve protective cap(s) in place.

Disposal of Dry Ice

Solid carbon dioxide (dry ice) block or snow should be placed in a remote area and allowed to sublime until no solid is left. The disposal area should be protected from unauthorized personnel and located such that hazardous concentrations of carbon dioxide vapors will not build up near any work space or otherwise populated area. Alternatively, if it is not practical to remove the dry ice to a remote location, guard any areas in which dry ice has accumulated from unauthorized and unprotected persons until it has all sublimed.

HANDLING LEAKS AND EMERGENCIES

Ventilate adjacent enclosed areas to prevent the formation of toxic concentrations of carbon dioxide. Personnel, including rescue workers, should not enter areas in which the carbon dioxide content exceeds about 3 percent by measurement unless wearing self-contained breathing apparatus or air-line respirators.

Avoid contact of the skin or eyes with solid carbon dioxide (dry ice) or objects cooled by solid carbon dioxide.

First Aid

If a person has inhaled large amounts of carbon dioxide and is exhibiting adverse effects, move the exposed individual to fresh air at once. If breathing has stopped, perform artificial respiration. Keep the person warm and at rest. Get medical attention at once. Fresh air and assisted breathing are appropriate for all cases of overexposure to gaseous carbon dioxide. Recovery is usually complete and uneventful.

If solid carbon dioxide (dry ice) or compressed CO_2 gas comes in contact with the body, stop the exposure at once. If frostbite has occurred, obtain medical attention. Immerse in warm (about 107°F) water. Do not rub.

If the eyes are involved, obtain prompt medical attention. The only appropriate first aid measure is a soft sterile pad held in place over the affected eye.

METHODS OF SHIPMENT

Under the appropriate regulations, liquid or gaseous carbon dioxide is authorized for shipment as follows: [5] and [6]

By Rail: In cylinders and in insulated portable tanks and tank cars.

By Highway: In cylinders, in insulated cargo tanks, and in insulated portable tanks.

By Water: In cylinders abroad passenger vessels, cargo vessels, and all types of ferry vessels. In authorized tank cars, tank trucks, and portable tanks (maximum 20 000 lb or 9072 kg gross weight) aboard cargo vessels and railroad car ferry vessels (passenger or vehicle); and in tank trucks (cargo tanks) aboard passenger or vehicle ferry vessels.

By Air: In cylinders aboard passenger aircraft up to 150 lb (68 kg) maximum net weight per cylinder, and in cylinders aboard cargo aircraft up to 300 lb (136 kg) maximum net weight per cylinder.

Shipment of solid carbon dioxide by rail or highway is not subject to DOT regulations (under which it is not designated as a dangerous article), while shipment by water and air must meet only certain labeling and packaging requirements. See Title 49 of the U.S. *Code of Federal Regulations.* [5] In Canada, however, transportation of carbon dioxide is regulated by the *Transportation of Dangerous Goods Regulations.* [6]

CONTAINERS

Liquefied carbon dioxide is shipped in cylinders, insulated portable tanks, insulated cargo tanks, and insulated tank cars. In high pressure supply systems, it may be stored and used from single or manifolded cylinders. In low pressure systems, it is stored in insulated pressure vessels with controlled refrigeration systems and auxiliary heating when required.

Normally, solid blocks of carbon dioxide weighing 50–60 lb (22.7–27.2 kg) are wrapped in heavy kraft paper bags. They are stored and shipped in insulated containers and storage boxes of varying size. Extruded dry ice pellets are normally stored and shipped in bulk or bags in insulated containers.

Filling Limits

The maximum allowable filling densities authorized for carbon dioxide are:

In cylinders—the weight of carbon dioxide in the cylinder must not exceed 68 percent of the weight of water the cylinder will hold at 60°F (15.6°C).

In tank cars—the liquid portion of the gas does not completely fill the tank at 0°F (−17.8°C) and in no case exceeds the maximum fill density of 95 percent by volume.

In cargo tanks and portable tanks—the liquid portion of the gas does not exceed 95 percent by volume.

Cylinders

There are two kinds of liquefied carbon dioxide cylinders in commercial use: the standard type and the siphon type. The standard cylinder, in an upright position, discharges gas; inverted, it discharges liquid. The valve on the siphon cylinder is equipped with an eductor tube, also known as a dip tube, extending to the bottom of the cylinder. It discharges liquid when the cylinder is in the upright position. With the exception of fire extinguisher cylinders, all siphon-type cylinders are clearly identified by the word *siphon, dip tube,* or other descriptive phrase. *Under no circumstances shall a pressure regulator be attached to a siphon cylinder.*

Cylinders containing liquid carbon dioxide under balanced thermal conditions will have a pressure of 733 psig (5054 kPa) at a temperature of 60°F (15.6°C).

Cylinders that meet the following TC/DOT specifications are authorized for liquefied carbon dioxide service: 3A1800, 3HT2000, 3AA1800, 3E1800, 3T1800, 3AX-1800 AL1800, 3AL1800, and 39; other cylinders may be used by special permit. DOT-3 cylinders may also be continued in carbon dioxide service, but new construction is not authorized.

Under present regulations, cylinders authorized for liquefied carbon dioxide service including fire extinguishers must be requalified by hydrostatic retest every five years, with three exceptions: 3HT cylinders must be requalified by retest every three years; no periodic retest is required for 3E cylinders; and specification 39 cylinders are nonreuseable and nonrefillable.

Pressure relief devices authorized for use on carbon dioxide cylinders include the Type CG-1 rupture disk and the Type CG-7 pressure relief valve. For specific requirements, see CGA S-1.1, *Pressure Relief Device Standards—Part 1—Cylinders for Compressed Gases.* [7]

Valve Outlet Connections

The standard valve outlet connection in the United States and Canada for carbon dioxide cylinders is Connection CGA 320. The standard yoke connection in the United States and Canada (for medical use of the gas) is Connection CGA 940. [8]

As previously stated, the valve must be equipped with a Type CG-1 pressure relief device equipped with a rupture disk. A "backed" pressure relief device, that is, a combination rupture disk/fusible plug device, *must not* be used on a carbon dioxide cylinder valve.

Cargo Tanks

Liquefied carbon dioxide is authorized for shipment in cargo tanks complying with DOT specifications MC-330 and MC-331. The minimum design pressure for these tanks must be 200 psig (1379 kPa) (or, if built to requirements in "Low Temperature Operation of the *ASME Boiler and Pressure Vessel Code,* Section VIII, "Unfired Pressure Vessels," the design pressure may be reduced to 100 psig (689 kPa) or the controlled pressure, whichever is greater). [9] In Canada, tank trailers (highway tanks) for liquefied carbon dioxide (LCO_2) are specified by CSA B620, *Highway Tanks and Portable Tanks for the Transportation of Dangerous Goods,* and CSA B622, *CSA Standard for the Selection and Use of Tank Trucks, Tank Trailers, and Portable Tanks for the Transportation of Dangerous Goods for Class 2 By Road.* [10] and [11]

Portable Tanks

Liquefied carbon dioxide is also authorized for shipment in portable tanks conforming to specification DOT-51. The minimum design pressure is the same as for cargo tanks.

Tank Cars

Tank cars may be used if they meet TC/DOT specification 105A500W and are fitted as required with insulation and pressure-regulating valves. See 49 CFR 173.314. [5]

METHODS OF MANUFACTURE

Unrefined carbon dioxide gas is obtained from the combustion of coal, coke, natural gas, oil, or other carbonaceous fuels; from by-product gases from steam-hydrocarbon reformers, lime kilns, and so on; from fermentation processes; and from gases found in certain wells and natural springs. The gas obtained from these sources is liquefied and purified by several different processes to a purity of about 99.9 percent or better.

In general, the process involved in producing solid carbon dioxide is as follows: First, cold liquid carbon dioxide is piped into either a special hydraulic press or an extruder. As the liquid boils and evaporates, the vapors are vented or pumped off. The remaining liquid cools until it finally freezes into solid carbon dioxide particles (dry ice snow). After the vapor pressure has been reduced to atmospheric pressure by pumping or bleeding the vapors away, the solid carbon dioxide particles are pressed into a block or extruded through dies.

REFERENCES

[1] CGA G-6.2, *Commodity Specification for Carbon Dioxide,* Compressed Gas Association, Inc., 1235 Jefferson Davis Highway, Arlington, VA 22202.

[2] CGA G-6.1, *Standard for Low Pressure Carbon Dioxide Systems at Consumer Sites,* Compressed Gas Association, Inc., 1235 Jefferson Davis Highway, Arlington, VA 22202.

[3] CGA G-6, *Carbon Dioxide,* Compressed Gas Association, Inc., 1235 Jefferson Davis Highway, Arlington, VA 22202.

[4] CGA C-7, *Guide to the Preparation of Precautionary Labeling and Marking of Compressed Gas Containers,* Compressed Gas Association, Inc., 1235 Jefferson Davis Highway, Arlington, VA 22202.

[5] *Code of Regulations,* Title 49 CFR Parts 100–199 (Transportation), Superintendent of Documents, U.S. Government Printing Office, Washington, DC 20402.

[6] *Transportation of Dangerous Goods Regulations,* Canadian Government Publishing Centre, Supply and Services Canada, Ottawa, Ontario, Canada K1A 0S9.

[7] CGA S-1.1, *Pressure Relief Device Standards—Part 1—Cylinders for Compressed Gases,* Compressed Gas Association, Inc., 1235 Jefferson Davis Highway, Arlington, VA 22202.

[8] CGA V-1, *American National, Canadian, and Compressed Gas Association Standard for Compressed Gas Cylinder Valve Outlet and Inlet Connections* (ANSI/CSA/CGA V-1), Compressed Gas Association, Inc., 1235 Jefferson Davis Highway, Arlington, VA 22202.

[9] *ASME Boiler and Pressure Vessel Code,* Section VIII, American Society of Mechanical Engineers, 345 E. 47th Street, New York, NY 10017.

[10] CSA B620, *Highway Tanks and Portable Tanks*

for the Transportation of Dangerous Goods, Canadian Standards Association, 178 Rexdale Blvd., Rexdale (Toronto), Ontario, Canada M9W 1R3.

[11] CSA B622, *CSA Standard for the Selection and Use of Tank Trucks, Tank Trailers, and Portable Tanks for the Transportation of Dangerous Goods for Class 2 By Road,* Canadian Standards Association, 178 Rexdale Blvd., Rexdale (Toronto), Ontario, Canada M9W 1R3.

ADDITIONAL REFERENCES

American Institute of Physics Handbook, 3rd ed., McGraw-Hill, New York, 1972.

ASHRAE Handbook of Fundamentals, American Society of Heating, Refrigerating and Air-Conditioning Engineers, 1791 Tullie Circle, N. E., Atlanta, GA 30329, 1972.

Chemical Engineer's Handbook; 4th ed. McGraw-Hill.

Handbook of Chemistry and Physics, 48th ed., Chemical Rubber Publishing Co., 1967.

Newitt, Pai, Kuloor, and Huggill (Din, F., ed.) *Thermodynamic Functions of Gases,* Vol. 1, Butterworth's, London, 1956.

Vukalovick, M. P., and Altunin, V. V., (Gaunt, D. S., trans. 1968), *Thermophysical Properties of Carbon Dioxide,* Collett's Ltd., London, 1965.

Yaws, C. L., Li, K. Y., and Kuo, C. J. *"Carbon Dioxides: CO and CO2," Chemical Engineering,* September 30, 1974, p. 115.

Carbon Monoxide

Chemical Symbol: CO
CAS Registry Number: 630-08-0
DOT Classification: Flammable gas
DOT Label: Flammable gas
Transport Canada Classification: 2.1 (6.1)
UN Number: UN 1016 (compressed gas); NA 9202 cryogenic liquid

PHYSICAL CONSTANTS

	U.S. Units	SI Units
Chemical formula	CO	CO
Molecular weight	28.01	28.01
Density of the gas at 70°F (21.1°C) and 1 atm	0.0725 lb/ft^3	1.161 kg/m^3
Specific gravity of the gas at 70°F (21.1°C) and 1 atm (air = 1)	0.9676	0.9676
Specific volume of the gas at 70°F (21.1°C) and 1 atm	13.8 ft^3/lb	0.862 m^3/kg
Density of liquid at −312.7°F (−191.5°C)	49.37 lb/ft^3	790.8 kg/m^3
Boiling point at 1 atm	−312.7°F	−191.5°C
Melting point at 1 atm	−340.6°F	−207.0°C
Critical temperature	−220.4°F	−140.2°C
Critical pressure	507.5 psig	3499 kPa abs
Critical density	18.79 lb/ft^3	301 kg/m^3
Triple point	−337.1°F at 2.2 psia	−205.1°C at 15.2 kPa abs
Latent heat of vaporization at −312.7°F (−191.5°C)	92.79 Btu/lb	215.8 kJ/kg
Latent heat of fusion at −340.6°F (−207.0°C)	12.85 Btu/lb	29.89 kJ/kg
Specific heat of gas at 60°F (15.6°C) and 1 atm		
C_p	0.2478 Btu/(lb)(°F)	1.037 kJ/(kg)(°C)
C_v	0.1766 Btu/(lb)(°F)	0.7394 kJ/(kg)(°C)

	U.S. Units	SI Units
Ratio of specific heats (C_p/C_v)	1.403	1.403
Solubility in water, vol/vol at 32°F (0°C)	0.035	0.035
Weight of liquid at −312.7°F (−191.5°C)	6.78 lb/gal	812 kg/m^3
Net heat of combustion at 77°F (25°C)	4343.6 Btu/lb	10 103 kJ/kg
Flammable limits in air (percent by volume)	12.5–74%	12.5–74%
Density of saturated vapor (Nbp)	0.0276 lb/ft^3	4.42 kg/m^3

DESCRIPTION

Carbon monoxide is a colorless, odorless, flammable toxic gas. Liquid carbon monoxide is a cryogenic liquid which exists at a temperature of −313°F (−192°C) and atmospheric pressure. It becomes a flammable vapor upon addition of heat and has a threshold limit value (TLV) of 50 ppm. If inhaled, concentrations of 0.4 percent in air prove fatal in less than an hour, while inhalation of high concentrations can cause sudden collapse with little or no warning. Pure carbon monoxide has a negligible corrosive effect on metals at atmospheric pressures. Impure carbon monoxide, containing water vapor, sulfur compounds, and/or other impurities, causes stress corrosion to ferrous metals at elevated pressures.

Chemically, carbon monoxide is stable with respect to decomposition. At temperatures of 570–2700°F (299–1482°C), it reduces many metal oxides to lower metal oxides, metals, or metal carbides. Hydrogenation of carbon monoxide yields products varying according to catalysts and conditions, which include methane, benzene, olefins, paraffin waxes, hydrocarbon high polymers, methanol, higher alcohols, ethylene glycol, glycerol, and other oxygenated products. Carbon monoxide also combines with the alkali and alkaline earth metals; reacts with chlorine, bromine, sulfur, Grignard reagents, and sodium alkyls; adds to alcohols; and enters into many other reactions.

GRADES AVAILABLE

Carbon monoxide is available for commercial and industrial use in various grades of purity ranging from 98 to 99.99 minimum percent.

USES

Carbon monoxide is used in the chemical industry, to produce such commodities as methanol and phosgene, and in organic synthesis. In metallurgy, it is used to recover high-purity nickel from crude ore in the Mond process and for special steels and reducing oxides. It is also used to obtain powdered metals of high purity, such as zinc white pigments, and to form certain metal catalysts applied in synthesizing hydrocarbons or organic oxygenating compounds. It is used as well in the manufacture of acids, esters, and hydroxy acids, such as acetic and propionic acids, and their methyl esters, and glycolic acid.

PHYSIOLOGICAL EFFECTS

Carbon monoxide is a chemical asphyxiant and acts toxically by combining with the hemoglobin of the red blood cells to form the stable compound carbon monoxide–hemoglobin. It thus prevents the hemoglobin from taking up oxygen, thereby depriving the body of the oxygen needed for metabolic respiration. The affinity of carbon monoxide for hemoglobin is about 300 times the affinity of oxygen for hemoglobin. The inhalation of concentrations as low as 0.04 percent will result in headache and discomfort within 2 to 3 hours. Inhalation of a 0.4 percent concentration in air proves fatal in less than 1 hour. Lacking odor and color, carbon monoxide gives no warning of its presence, and inhalation of heavy concentrations can cause sudden, unexpected collapse.

The eight-hour time-weighted average threshold limit value (TLV) adopted by the American Conference of Governmental Industrial Hygienists is 50 ppm (55 mg/m^3) for exposure to carbon monoxide. [1] The U.S. Occupational Safety and Health Administration has adopted an eight-hour time-weighted average exposure limit of 35 ppm (40 mg/m^3) and a ceiling limit of 200 ppm (229 mg/m^3) for carbon monoxide. [2]

MATERIALS OF CONSTRUCTION

Steel and other common metals are satisfactory for use with dry, sulfur-free carbon monoxide at pressures up to 2000 psig (13 790 kPa). Iron, nickel, and other metals can react with carbon monoxide at elevated pressures to form carbonyls in small quantities. The presence of moisture and sulfur-containing impurities in carbon monoxide appreciably increases its corrosive action on steel at any pressure. High pressure plant equipment is often lined with copper for increased resistance to carbon monoxide attack. Very highly alloyed chrome steels are sufficiently resistant to corrosion by carbon monoxide containing small amounts of sulfur-bearing impurities. Users are strongly urged to make stress corrosion tests of samples of proposed construction materials in order to select one which will withstand the high pressure use of carbon monoxide under actual conditions.

Liquid storage vessels must be well designed and properly insulated to maintain liquid temperature for a reasonable period of time without adverse pressure rise or venting. The inner vessel must be designed, fabricated, inspected, and stamped in accordance with Section VIII, "Unfired Pressure Vessels," of the *ASME Boiler and Pressure Vessel Code*. [3] The Type 300 series stainless steels, 9 percent nickel steel, and aluminum alloys are suitable for inner vessel material construction. Systems for storing and handling carbon monoxide should be designed to meet the appropriate ASME standards and must conform to all applicable state and local regulations.

SAFE STORAGE, HANDLING, AND USE

All precautions necessary for the safe handling of any flammable gas must be observed with carbon monoxide. Among these, special care should be taken to avoid storing carbon monoxide cylinders with cylinders containing oxygen or other highly oxidizing or flammable materials. It is recommended that carbon monoxide cylinders in use be grounded and protected by check valves to prevent the backflow of foreign materials into the cylinders. Areas in which cylinders are being used must be free of all ignition sources and hot surfaces. Carbon monoxide detectors with alarms should be installed in storage areas for personnel protection.

Of all the safety procedures required for the safe handling of carbon monoxide, the "buddy system" must not be overlooked. Always have at least two persons knowledgeable about carbon monoxide in the operating area.

Special precautions must be observed when handling liquid carbon monoxide. Trailer loading and unloading is accomplished through pressure transfer and pump transfer of product. In pressure filling, the pressure of the liquid storage vessel is raised to 10–25 psi above the container being filled, depending upon specific system requirements. In pump filling, it is first necessary to cool the pump sufficiently so that it will prime.

Special rules must be adhered to for liquid trailers when transferring product. Only authorized personnel are allowed in the loading or unloading area. Hose connections must be compatible with the product being transferred. Additional precautions are provided in CGA P-13, *Safe Handling of Liquid Carbon Monoxide*. [4]

DISPOSAL

Carbon monoxide is normally disposed of by incineration using a suitable flare system. Follow local, state, provincial, and federal regulations.

HANDLING LEAKS AND EMERGENCIES

As in the case of any flammable gas, never use a flame in trying to detect carbon monoxide leaks. Portable detection equipment or the formation of bubbles by a soapy solution applied to a suspected area will indicate leaks. Carbon monoxide alarm detectors must be installed in all indoor areas in which the gas in regularly used in more than small laboratory amounts. Colorimetric "sniffer" tubes are also utilized for detection of carbon monoxide. Samples of atmospheric air containing carbon monoxide are aspirated through such tubes.

First Aid Suggestions

Prompt medical attention is of utmost importance. Personnel accidentally overcome by carbon monoxide should be given first aid prior to a physician's arrival. The first aid measures presented here are based upon what is believed to be common practice in industry, but they should be reviewed and amplified into a complete first aid program by a competent medical advisor before adoption in any specific case.

It is extremely important to hasten the elimination of carbon monoxide from the bloodstream, should poisoning occur. Such elimination is best effected by inhalation of breathing oxygen. The inhalation of pure oxygen will eliminate carbon monoxide much more rapidly than inhalation of fresh air.

Any person showing symptoms of carbon monoxide poisoning must be moved immediately to fresh, but not cold, air. Keep the victim warm and on his or her stomach with the face turned to one side. If breathing, the victim should be given breathing oxygen to inhale. If the victim is not breathing, start manual artificial respiration at once with simultaneous administration of oxygen.

If available, a mixture of oxygen containing 7 to 10 percent carbon dioxide by volume will hasten the elimination of carbon monoxide from the bloodstream by acting as a powerful respiratory and cardiac stimulant. The victim will generally breathe more deeply with the administration of this mixture than with pure oxygen. Inhalation of the mixture should continue for at least 15 to 30 minutes.

Drugs are of little value in the treatment of carbon monoxide poisoning. Coffee may be given if the patient is able to hold a cup. Alcohol should not be given.

METHODS OF SHIPMENT

Under the appropriate regulations, carbon monoxide is authorized for shipment as follows: [5] and [6]

By Rail: In cylinders (in one outside container to a maximum of 150 lb or 68 kg).

By Highway: In cylinders on trucks, tube trailers, and cryogenic trailers.

By Water: In cylinders via cargo vessels only in accordance with the IMDG Code. [7] In cylinders on barges of U.S. Coast Guard classes A and C only.

By Air: Aboard cargo aircraft only in appropriate cylinders up to 150 lb (68 kg) maximum net weight per cylinder.

CONTAINERS

Carbon monoxide is presently being shipped in cylinders, tube trailers, and bulk cryogenic containers.

Filling Limits

The maximum pressure authorized for carbon monoxide cylinders is 1000 psig at 70°F

(6895 kPa at 21.1°C) except, if the gas is dry and sulfur-free, the cylinders can be charged to 5/6 the service pressure but never more than 2000 psig at 70°F (13 789 kPa at 21.1°C). See 49 CFR 173.302(f) and CSA B340, *Selection and Use of Cylinders, Spheres, Tubes, and Other Containers for the Transportation of Dangerous Goods.* [5] and [8]

Cylinders

Only cylinders that meet the following TC/DOT specifications and that have service pressures at 70°F (21.1°C) of 1800 psig (12 410 kPa) or higher are authorized for carbon monoxide service: 3A, 3AX, 3AA, 3AAX, 3AL, 3E, 3T, and 3. Flammable gases shipped in 3AL cylinders must be transported by highway, rail, and cargo-only aircraft.

Under present regulations, the cylinders authorized for carbon monoxide service must be requalified by hydrostatic test every five years with the exception of type 3E (for which periodic hydrostatic retest is not required).

Pressure relief devices authorized for use on carbon monoxide cylinders include the CG-4 and CG-5 type devices. These are combination rupture disk/fusible plug devices. For more information on their use, see CGA S-1.1, *Pressure Relief Device Standards—Part 1—Cylinders for Compressed Gases.* [9]

Valve Outlet Connections

The standard valve outlet connection for carbon monoxide in the United States and Canada is Connection CGA 350. [10]

Cryogenic Vessels

Materials selected for handling liquid carbon monoxide must be suitable for use at cryogenic temperatures. Refer to the applicable sections of the *ASME Boiler and Pressure Vessel Code* and ANSI/ASME B31.3, *Chemical Plant and Petroleum Refinery Piping.* [3] and [11] Also see CGA P-12, *Safe Handling of Cryogenic Liquids.* [12]

METHODS OF MANUFACTURE

Pure carbon monoxide is extracted from synthesis gas, blast-furnace gas, or coke-oven gas by the following techniques: (1) absorption of carbon monoxide by an ammoniacal cuprous salt solution at elevated pressure, followed by desorption by pressure release; (2) low temperature condensation and fractionation; (3) liquid methane scrubbing and separation; and (4) permeable membrane.

REFERENCES

[1] *Threshold Limit Values and Biological Exposure Indices,* 1989-90 ed., American Conference of Governmental Industrial Hygienists, 6500 Glenway Avenue, Bldg. D-7, Cincinnati, OH 45211-4438.

[2] *Code of Federal Regulations*, Title 29 CFR Parts 1900–1910 (Labor), Superintendent of Documents, U.S. Government Printing Office, Washington, DC 20402.

[3] *ASME Boiler and Pressure Vessel Code* (Section VIII), American Society of Mechanical Engineers, 345 E. 47th St., New York, NY 10017.

[4] CGA P-13, *Safe Handling of Liquid Carbon Monoxide*, Compressed Gas Association, Inc., 1235 Jefferson Davis Highway, Arlington, VA 22202.

[5] *Code of Federal Regulations*, Title 49 CFR Parts 100–199, (Transportation), Superintendent of Documents, U.S. Government Printing Office, Washington, DC 20402.

[6] *Transportation of Dangerous Goods Regulations.* Canadian Government Publishing Centre, Supply and Services Canada, Ottawa, Ontario, Canada KIA 0S9.

[7] *International Maritime Dangerous Goods Code*, International Maritime Organization, 4 Albert Embankment, London, United Kingdom SE1 7SR.

[8] CSA B340, *Selection and Use of Cylinders, Spheres, Tubes, and Other Containers for the Transportation of Dangerous Goods*, Canadian Standards Association, 178 Rexdale Blvd., Rexdale (Toronto), Ontario, Canada M9W 1R3.

[9] CGA S-1.1, *Pressure Relief Device Standards—Part 1—Cylinders for Compressed Gases*, Compressed Gas Association, Inc., 1235 Jefferson Davis Highway, Arlington, VA 22202.

[10] CGA V-1, *American National, Canadian, and Compressed Gas Association Standard for Compressed Gas Cylinder Valve Outlet and Inlet Con-*

nections (ANSI/CSA/CGA V-1), Compressed Gas Association, Inc., 1235 Jefferson Davis Highway, Arlington, VA 22202.

[11] ANSI/ASME B31.3, *Chemical Plant and Petroleum Refinery Piping*, American National Standards Institute, 1430 Broadway, New York, NY 10018.

[12] CGA P-12, *Safe Handling of Cryogenic Liquids*, Compressed Gas Association, Inc., 1235 Jefferson Davis Highway, Arlington, VA 22202.

Carbon Tetrafluoride

Chemical Symbol: CF_4
Synonyms: Tetrafluoromethane, R-14
CAS Registry Number: 75-73-0
DOT Classification: Nonflammable gas
DOT Label: Nonflammable gas
Transport Canada Classification: 2.2
UN Number: UN 1956
UN Number (Canada): UN 1982

PHYSICAL CONSTANTS

	U.S. Units	SI Units
Chemical formula	CF_4	CF_4
Molecular weight	88.00	88.00
Density of gas		
at 70°F (21.1°C) and 1 atm	0.228 lb/ft^3	3.65 kg/m^3
Specific gravity of the gas		
at 70°F (21.1°C), 1 atm (air = 1)	3.050	3.050
Specific volume of the gas		
at 70°F (21.1°C), 1 atm	4.38 ft^3/lb	0.274 m^3/kg
Boiling point	−198.5°F	−128°C
Melting point	−298.5°F	−183.6°C
Critical temperature	−49.9°F	−45.5°C
Critical pressure	543.2 psia	3745 kPa
Critical density	39.06 lb/ft^3	625.7 kg/m^3
Specific heat of gas		
at 77°F (25°C)		
C_p	0.166 Btu/(lb)(°F)	0.696 kJ/(kg)(°C)
C_v	0.144 Btu/(lb)(°F)	0.602 kJ/(kJ)(°C)
Latent heat of vaporization		
at −198.5°F (−128°C)	58.5 Btu/lb	135.7 kJ/kg

DESCRIPTION

Carbon tetrafluoride is a colorless, odorless, nonflammable gas that is three times as heavy as air at standard temperature and pressure. It is shipped as a nonflammable compressed gas at cylinder pressures up to 2000 psig (13 789 kPa) at 70°F (21.1°C).

GRADES AVAILABLE

Carbon tetrafluoride is available in various grades for industrial and commercial use. It has a minimum purity of 99.7 percent by volume.

There is also a semiconductor grade available with the following impurity specification:

Carbon dioxide	<5.0 ppm
Carbon monoxide	<10 ppm
Nitrogen	<400 ppm
Oxygen and argon	<50 ppm
Sulfur hexafluoride	<5.0 ppm
Nitrous oxide	<5.0 ppm
Total halocarbons	<15 ppm
Total acidity—HF	<0.1 ppm

USES

Carbon tetrafluoride is used as a low temperature refrigerant gas. It is also widely employed by the electronics industry as a dry etchant in microchip manufacture. It is blended with oxygen and used to desmear and etch-back "through holes" on printed circuit boards.

PHYSIOLOGICAL EFFECTS

Carbon tetrafluoride is considered to be nontoxic. The major hazard of overexposure to carbon tetrafluoride is the exclusion of an adequate supply of oxygen; therefore, it should be treated as a simple asphyxiant.

MATERIALS OF CONSTRUCTION

Carbon tetrafluoride is noncorrosive and may be used with any common structural material. Silver- and copper-bearing alloys can act as catalysts for the decomposition of carbon tetrafluoride at high temperatures. Alloys containing more than 2 percent magnesium should not be used if moisture is present.

SAFE STORAGE, HANDLING, AND USE

Use forced ventilation and/or local exhaust to prevent an accumulation of gas that could reduce the oxygen level in the air. Ensure good floor ventilation. Valve protection caps should remain in place unless the cylinder is secured and readied for use. Use a pressure-reducing regulator when connecting a cylinder to a lower-pressure piping system. Do not heat the cylinder. Use a check valve or trap in the discharge line to prevent backflow into the cylinder. Do not store cylinders in areas where the temperature may exceed 125°F (51.7°C).

DISPOSAL

Return all unused quantities of carbon tetrafluoride to the supplier. In the event that small quantities must be disposed of on site, all federal, state, provincial, and local regulations regarding waste disposal should be followed. Controlled venting of this gas to the atmosphere is a potential disposal option. Prior to venting, the approval of state, provincial, and local agencies governing air emissions should be obtained. Provisions should also be made to avoid possible asphyxiating conditions upon venting.

HANDLING LEAKS AND EMERGENCIES

Carbon tetrafluoride may be detected by bubble formation after applying a soap solution to the suspected leak area. Minute leaks may be detected by using a halogen leak detector. If the leak should occur in a cylinder or cylinder valve, move the cylinder to a well-ventilated area and contact the supplier.

METHODS OF SHIPMENT

Under the appropriate regulations, carbon tetrafluoride is authorized for shipment as follows: [1] and [2]

By Rail: In cylinders.

By Highways: In cylinders on trucks and tube trailers.

By Water: In cylinders above or below decks.

By Air: Aboard passenger aircraft: 150 lb (68 kg) maximum net quantity in one package; aboard cargo aircraft: 300 lb (136 kg) maximum net quantity in one package.

CONTAINERS

Carbon tetrafluoride is authorized for shipment in DOT specified cylinders as identified in 49 CFR 173.302 and equivalent Canadian regulations. [1] and [2] Authorized pressure relief devices include the Type CG-1 rupture disk, the Type CG-4 or CG-5 combination rupture disk/fusible plug device, or the Type CG-7 pressure relief valve. [3] The rupture disk must have a bursting pressure not exceeding the minimum prescribed test pressure of the cylinder.

Valve Outlet Connections

The standard valve outlet connection in the United States and Canada for carbon tetrafluoride cylinders is Connection CGA 580. The limited standard valve outlet connection is Connection CGA 320. [4]

METHODS OF MANUFACTURE

Carbon tetrafluoride is prepared by the direct fluorination of carbon or an electrochemical process using acetic acid and hydrogen fluoride.

REFERENCES

[1] *Code of Federal Regulations*, Title 49 CFR Parts 100–199 (Transportation), Superintendent of Documents, U.S. Government Printing Office, Washington, DC 20402.

[2] *Transportation of Dangerous Goods Regulations*, Canadian Government Publishing Centre, Supply and Services Canada, Ottawa, Ontario, Canada KIA 0S9.

[3] CGA S-1.1, *Pressure Relief Device Standards—Part 1—Cylinders for Compresssed Gases*, Compressed Gas Association, Inc., 1235 Jefferson Davis Highway, Arlington, VA 22202.

[4] CGA V-1, *American National, Canadian, and Compressed Gas Association Standard for Compressed Gas Cylinder Valve Outlet and Inlet Connections* (ANSI/CSA/CGA V-1), Compressed Gas Association, Inc., 1235 Jefferson Davis Highway, Arlington, VA 22202.

Chlorine*

Chemical Symbol: Cl_2
CAS Registry Number: 7782-50-5
DOT Classification: Nonflammable gas
DOT Label: Nonflammable gas and Poison, or "Chlorine," as an alternative
Transport Canada Classification: 2.4
UN Number: UN 1017
RTEC Number: F02100000 (Assigned in the U.S. by the National Institute for Occupational Safety and Health)

PHYSICAL CONSTANTS [1]

	U.S. Units	SI Units
Chemical formula	Cl_2	Cl_2
Atomic weight	35.453	35.453
Molecular weight	70.906	70.906
Vapor pressure		
at 0°F (−6.7°C)	28.533 psia	196.73 kPa abs
at 20°F (4.4°C)	42.730 psia	294.61 kPa abs
at 40°F (426.7°C)	61.767 psia	425.87 kPa abs
at 80°F (26.7°C)	118.16 psia	814.71 kPa abs
at 120°F (48.9°C)	205.70 psia	1418.3 kPa abs
Density of the gas		
at 32°F (0°C) and 1 atm	0.20057 lb/ft^3	3.2127 kg/m^3
Specific gravity of the gas		
at 32°F and 1 atm (air = 1)	2.485	2.485
Specific volume of the gas		
at 70°F (21.1°C) and 1 atm	5.3882 ft^3/lb	0.33638 m^3/kg
Density of saturated liquid		
at 32°F (0°C)	91.563 lb/ft^3	1466.7 kg/m^3
at 60°F (15.6°C)	88.765 lb/ft^3	1421.9 kg/m^3
at 80°F (26.7°C)	86.674 lb/ft^3	1388.4 kg/m^3
at 120°F (48.9°C)	82.208 lb/ft^3	1316.8 kg/m^3

*This monograph compliments of the Chlorine Institute.

	U.S. Units	SI Units
at 160°F (71.1°C)	77.231 lb/ft³	1237.1 kg/m³
at 200°F (93.3°C)	71.461 lb/ft³	1144.7 kg/m³
Boiling point at 1 atm	−29.15°F	−33.97°C
Melting point at 1 atm	−149.76°F	−100.98°C
Critical temperature	290.75°F	143.75°C
Critical pressure	1157.0 psia	7977 kPa abs
Critical density	35.8 lb/ft³	573 kg/m³
Triple point	0.20226 psig at −149.76°F	1.3945 kPa abs at −100.98°C
Latent heat of vaporization at boiling point	123.85 Btu/lb	288.08 kJ/kg
Latent heat of fusion at melting point	38.836 Btu/lb	90.331 kJ/kg
Specific heat of dry gas (C_p)		
at 0°C, 1 atm	0.11565 Btu/(lb)(°F)	0.48422 kJ/(kg)(°C)
at 30°C, 1 atm	0.11664 Btu/(lb)(°F)	0.48834 kJ/(kg)(°C)
at 30°C, 101.5 psig (700 kPa)	0.13108 Btu/(lb)(°F)	0.5488 kJ/(kg)(°C)
Ratio of specific heats (C_p/C_v)		
at 0°C, 1 atm	1.3473	1.3473
at 30°C, 1 atm	1.3352	1.3352
at 30°C, 101.5 psig (700 kPa)	1.4240	1.4240
Viscosity at 68°F (20°C) and 1 atm		
Liquid	0.346 cP	0.346 mPa·s
Gas	0.0134 cP	0.0134 mPa·s
Weight of the liquid at 70°F (21.1°C)	11.728 lb/gal	1405.3 kg/m³
Specific gravity of saturated liquid at 32°F and 53.507 psia (0°C and 368.92 kPa abs or 3.641 atm	1.4667	1.4667

DESCRIPTION

Chlorine is a greenish-yellow, nonflammable gas with a distinctive, pungent odor. It is almost 2.5 times as heavy as air. The gas acts as a severe irritant if inhaled. Chlorine liquid has the color of clear amber and is about 1.5 times as heavy as water. It is shipped as a compressed liquefied gas having a vapor pressure of 86.767 psig at 70°F (598.24 kPa at 21.1°C). Chlorine is nonflammable in both gaseous and liquid states. However, like oxygen, it is capable of supporting the combustion of certain substances. Many organic chemicals react readily with chlorine, in some cases with explosive violence. Chlorine usually forms univalent compounds, but it can combine with a valence of 3, 4, 5, or 7.

Chlorine is only slightly soluble in water. When it reacts with pure water, weak solutions of hydrochloric and hypochlorous acids are formed. Chlorine hydrate ($Cl_2 \cdot 8H_2O$) may crystallize below 49.3°F (9.6°C).

Chlorine unites under specific conditions with most of the elements, and these reac-

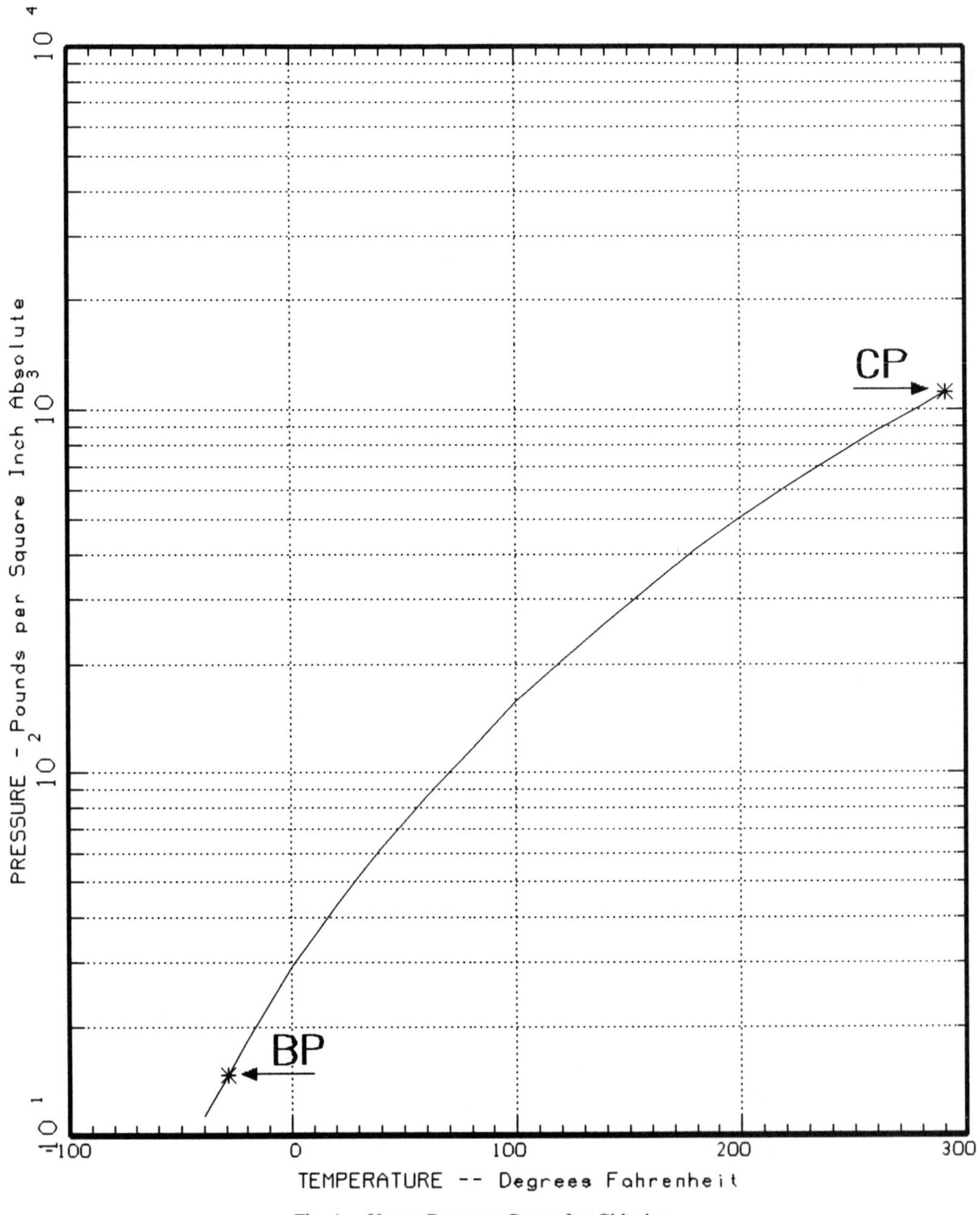

Fig. 1. Vapor Pressure Curve for Chlorine.

tions may be extremely rapid. At its boiling point it reacts with sulfur. It does not react directly with oxygen or nitrogen. The oxides and nitrogen compounds are well known, but can be prepared only by indirect methods. Mixtures of chlorine and hydrogen can react with explosive violence, forming hydrogen chloride. Ignition limits depend on pressure, temperature, and concentration. At 70°–80°F (21°–27°C), the ignition limits range from 3 percent to 93 percent by volume of hydrogen.

The preparation of soda and lime bleaches (sodium and calcium hypochlorite) are typical reactions of chlorine with alkalies and alkaline earth metal hydroxides. The hypochlorites formed are powerful oxidizing agents. Because of its great affinity for hydrogen, chlorine removes hydrogen from some of its compounds such as the reaction with hydrogen sulfide to form hydrochloric acid and sulfur.

Chlorine reacts with organic compounds much the same as with inorganics to form chlorinated derivatives and hydrogen chloride. Some of these reactions can be explosive, including those with hydrocarbons, alcohols, and ethers, and proper methods must be applied in operations in which they are involved.

GRADES AVAILABLE

Chlorine for commercial and industrial use has much the same quality from all producers. High-purity grades (99.9 percent) are available from specialty gas suppliers.

USES

The largest quantities of chlorine are used in manufacturing chemicals. These include:

- Such solvents as carbon tetrachloride, trichloroethylene, 1,1,1-trichloroethane, perchloroethylene, and methylene chloride
- Pesticides and herbicides
- Plastics and fibers such as vinyl chloride and vinylidene chloride
- Refrigerants and propellants such as the halocarbons and methyl chloride.

Chlorine is also an ingredient in the widely used bleach, deodorizer, and disinfectant, sodium hypochlorite. In addition, chlorine is widely used in bleaching pulp, paper, and textiles; for drinking and swimming water purification; in the sanitation of industrial and sewage wastes; and in the degassing of aluminum melts.

PHYSIOLOGICAL EFFECTS

Chlorine gas is primarily a respiratory irritant. It is so intensely irritating that low concentrations in the air (well below 1 ppm) are readily detectable by the normal person. At concentrations above 5 ppm, the severely irritating effects of the gas makes it unlikely that any person will remain in a chlorine-contaminated atmosphere unless he or she is unconscious or trapped.

Liquid chlorine will cause skin and eye burns on contact. In the United States, the eight-hour time weighted exposure limit established by the Occupational Safety and Health Administration is 0.5 ppm (1.5 mg/m^3). OSHA has also adopted a 15-minute short term exposure limit (STEL) of 1 ppm (3 mg/m^3). [2]

The American Conference of Governmental Industrial Hygienists has adopted an 8-hour time-weighted average threshold limit value (TLV-TWA) to 0.5 ppm and a short-term exposure limit (STEL) to 1.0 ppm. [3]

Acute Toxicity

When a sufficient concentration of chlorine gas is present, it will irritate the mucous membranes, the respiratory system, and the skin. Large amounts cause irritation of eyes, coughing, and labored breathing. Exposure to chlorine of excessive duration or concentration results in general excitement of the person affected, accompanied by restlessness, throat irritation, sneezing, and copious salivation. The symptoms of exposure to high concentrations are retching and vomiting, followed by difficult breathing. In extreme cases, the difficulty of breathing may increase to the point where death can occur from suffocation. Liquid chlorine in contact with the eyes or skin will cause local irritation and/or burns. (See Table 1.)

Systemic and Chronic Effects

Chlorine produces no known systemic effects. All symptoms and signs result directly or indirectly from the local irritant action. Low concentrations of chlorine gas in the air may have a minor irritating effect or may produce slight symptoms after several hours' exposure, but careful examination of persons repeatedly exposed to such conditions reportedly have shown no permanent physiological effect.

MATERIALS OF CONSTRUCTION

At ordinary temperatures, dry chlorine, either liquid or gas, does not corrode steel. In the presence of moisture, however, highly corrosive conditions exist due to the formation of hydrochloric and hypochlorous acids. Thus precautions should be taken to keep chlorine and equipment free of moisture. Piping, valves, and containers should be closed or capped when not in use to keep atmospheric moisture out of the system.

Dry Chlorine Systems

In general, steel piping is recommended for handling dry chlorine, defined as containing no more than 150 ppm of water by weight. Stainless steels of the Type 300 series have useful properties for service at low temperatures but can fail due to chloride stress corrosion cracking, particularly in the presence of moisture at ambient or elevated temperatures.

The reaction rate of dry chlorine with most metals increases rapidly above a temperature which is characteristic of the metal. Below 250°F (121°C), iron, copper, steel, lead, nickel, platinum, silver, and tantalum are resistant to dry chlorine gas or liquid. At ordinary temperatures, chlorine reacts with aluminum, arsenic, gold, mercury, selenium, tellurium, and tin. Titanium reacts violently with dry chlorine. Carbon steel ignites in chlorine near 483°F (251°C).

TABLE 1. HAZARD OF CHLORINE AT DIFFERENT ATMOSPHERIC CONCENTRATIONS. [4]

Chlorine Concentration in the Air		
ppm	mg/m³	Degree of Hazard
>1.0	>3	Threshold of odor perception for the average person.
3–5	9–15	Slight irritation of the nose and upper respiratory tract.
5–8	15–24	Irritation of the respiratory tract and eyes.
15–20	45–60	Immediate severe irritation of the respiratory tract, intense coughing, and choking.
30	60	Shortness of breath, chest pain; possibly nausea and vomiting.
40–60	120–180	Development of chemical bronchitis and fluid in the lungs, which may occur after several hours; chemical pneumonia may occur several days later.
Prolonged exposure above 50 ppm		Unconsciousness and death.

Recommendations for materials, pipe fittings, and some miscellaneous equipment suitable for use with dry chlorine are obtainable from the Chlorine Institute. [5]

Moist Chlorine Systems

As already noted, moist chlorine is corrosive to most common metals. Platinum, silver, tantalum, and titanium are resistant. Hastelloy C and Monel are widely used. At low pressures, moist chlorine can be handled in chemical stoneware, glass, or porcelain equipment. Hard rubber, plasticized polyvinyl chloride, glass fiber reinforced polyester, polyvinylidene chloride or fluoride, and fully halogenated fluorocarbon resins have been successfully used. For higher pressure, combinations using resistant lining materials with the common metals for strength generally are used. In general, operations involving moist chlorine require individual study, and materials of construction must be selected with great care.

SAFE STORAGE, HANDLING, AND USE

All precautions necessary for the safe handling of any nonflammable toxic gas must be observed with chlorine. Chlorine equipment and handling systems should be designed by engineers familiar with chlorine. Periodic inspection by knowledgeable persons should be made to ensure that the equipment is used appropriately and that the system is kept in suitable operating condition.

Every installation should have an ongoing safety program. Periodic training sessions and safety inspections should be conducted. Special attention should be paid to the appropriateness of emergency plans and equipment.

Employee Protection and Training

Persons afflicted with asthma, bronchitis, and other chronic lung conditions or irritations of the upper respiratory tract should not be employed in areas where chlorine is handled. All employees working with or around chlorine should be given preemployment and periodic physical examinations, including lung X-rays.

All employees handling or working around chlorine should be trained to handle it properly and safely with special emphasis placed on actions to be taken and equipment to be used in case of emergencies such as leaks. Each employee should be trained in the properties and physiological effects of chlorine [6], the location and proper use of the several types of respiratory equipment, and the conditions under which each type must be used. Each employee should also be trained in first aid procedures, particularly in administering artificial respiration. Quiz sessions on actions to be taken in emergencies, the proper use of respiratory equipment, and first aid measures should be held at regular intervals. [7]

Personal Protective Equipment

Suitable respiratory protective equipment should be available for handling emergencies and should be located outside the probable location of any likely chlorine contamination. Respiratory protection equipment should be routinely inspected and maintained in good condition and should be cleaned after each use and at regular intervals. Equipment used by more than one person should be sanitized after each use. Respiratory equipment must be certified by the National Institute of Occupational Safety and Health or other relevant authority.

Most chlorine releases are at low concentrations where the oxygen content in the air is greater than 19.5 percent and chemical cartridge respirators (up to 10 ppm) or canister masks (25 ppm, maximum) would offer adequate protection. However, without chlorine-monitoring equipment for sampling air in the vicinity of the leak, the use of positive pressure self-contained breathing apparatus (SCBA), with full face piece, is required.

Fit testing and regular maintenance programs for respirator equipment are necessary. Personal respirators for those regularly handling chlorine containers and chlorination equipment are advisable. SCBA should be stored in adjacent areas not likely to be contaminated by chlorine releases.

Protective clothing is not required for performing routine plant operations. Resistant plastic or rubber gloves should be worn by personnel who may come in contact with ferric chloride during maintenance operations. Suitable footgear, respiratory protection equipment, and other protective equipment such as a safety harness and lifeline, should be worn by persons entering tanks. At least one other person should oversee operations from outside the tank at all times. While not specific for chlorine, safety glasses or goggles, hard hats, and safety shoes should be worn or available as dictated by the special hazards of the area or by plant practices.

Oxygen administration equipment should be available either on site or at a nearby facility. Oxygen should be given only under guidance from qualified personnel suitably trained to use oxygen equipment according to medical recommendations.

Shower baths and bubble-type fountains or the equivalent should be installed where contact of the skin, eyes, or clothing with liquid or high concentrations of gaseous chlorine is a possibility.

DISPOSAL

Chlorine facilities should be designed and operated so that chlorine is not released to the environment. If accidental release should occur, the environmental effects, as well as all relevant reporting requirements, must be considered. In the case of chlorine emergencies, or if a chlorine-consuming process involves the discharge of wastes containing chlorine, all government regulations regarding health and safety or the pollution of natural resources must be followed.

Chlorine is only slightly soluble in water and there normally would be little absorption in water from a cloud of chlorine gas. Many forms of aquatic life are adversely affected by chlorine in concentrations well below 0.1 ppm, but harmful concentrations are unlikely unless chlorine or wastes containing chlorine are directly discharged in water.

Atmospheric releases of chlorine, where possible to contain, should be absorbed in an alkali solution. Chlorine affects most vegetation, sometimes retarding growth rate or yield.

HANDLING LEAKS AND EMERGENCIES

Immediate steps should be taken to find and stop chlorine leaks as soon as there is any indication of chlorine in the air. Chlorine leaks never get better; they always get worse, unless promptly corrected. Authorized, trained personnel equipped with suitable personal respiratory protection should investigate whenever a chlorine leak occurs. All other persons should be kept away from the affected area until the cause of the leak has been found and remedied.

If the leak is extensive, all persons in the path of fumes should be evacuated by the proper authorities. If outdoors, keep all persons upwind from the leak. Also, if possible, keep all persons in locations higher than the leak. At sites involving the handling of chlorine outdoors, it is advisable to have a wind sock or weather vane installed in a prominent location to readily determine wind direction. Unless there is a fire-caused updraft, gaseous chlorine tends to lie close to the ground or floor because it is approximately $2\frac{1}{2}$ times as heavy as air.

Finding Leaks

If ammonia vapor is directed at a leak, a white cloud will form indicating the source of the leak. Use a plastic squeeze bottle designed to emit vapors containing aqua ammonia. Avoid contact of aqua ammonia with brass or copper. Commercial 26° Baumé aqua ammonia should be used (household ammonia is not strong enough). Containers, piping, and equipment should be checked for leaks daily.

Emergency Assistance

If a chlorine leak cannot be handled promptly by the user's personnel, the nearest office or plant of the supplier should be called for assistance. If the supplier cannot be immediately reached, CHEMTREC (CANUTEC in Canada) should be called. Refer to the section on Emergency Response in Chapter 5. The Chlorine Institute's CHLOREP system is in operation throughout the United States and Canada to assist anyone who needs help with an emergency involving chlorine. This assistance can be summoned day and night by calling CHEMTREC (toll-free) in the United States and CANUTEC in Canada. The telephone numbers of the supplier and CHEMTREC or CANUTEC should be posted in suitable places and the numbers kept current. When phoning for assistance, give the following:

(1) Your company name, address, telephone number, and the person or persons to contact for further information
(2) Description of the emergency
(3) Travel directions to the site
(4) Type and size of container involved
(5) Corrective measures that are being applied
(6) Other pertinent information such as weather conditions, injuries, etc.
(7) Chlorine supplier's name, address, and telephone number

In Case of Fire

If fire is present or threatened, chlorine containers should be moved away from the fire. If a nonleaking container cannot be moved, it should be kept cool by spraying water on it. If the container is leaking, water should not be used. Chlorine and water react and acids are formed that will corrode the container, and the leak quickly will get worse. However, where several containers are involved and some are leaking, it may be prudent to use water to prevent rupture of the nonleaking containers.

All unauthorized persons should be kept at a safe distance. Never immerse or throw a leaking container into a body of water because the leak will be aggravated and the container may float when still partially full of liquid chlorine, allowing gas evolution at the surface.

Equipment and Piping Leaks

If a leak occurs in equipment in which chlorine is being used, the supply of chlorine should be shut off and the chlorine which is under pressure at the leak should be properly disposed of, such as in an alkaline solution.

Valve Leaks

Leaks around shipping container valve stems usually can be stopped by tightening the packing nut or packing gland. If this does not stop the leak, the container valve should be closed, and the chlorine which is under pressure in the outlet piping should be properly disposed of. If a container valve does not shut off tight, the outlet cap or plug should be applied. Ton containers have two valves; in case of a valve leak, the container should be rolled so the valves are in a vertical plane, with the leaky valve on top.

Container Leaks

If confronted with container leaks other than the valves, take one or more of the following steps:

(1) If possible, turn the container so that gas instead of liquid escapes. The quantity of chlorine that escapes from a gas leak is about 1/15 the amount that escapes from a liquid leak through the same size hole.

(2) Apply appropriate emergency kit device, if available.

(3) Call for emergency assistance.

(4) If practical, reduce pressure in the container by removing the chlorine as a gas (not as a liquid) to process or to a disposal system.

(5) Move the container to an isolated spot remote from personnel.

Leaks in Transit

If a chlorine leak develops in transit through a populated area, it is generally advisable to keep the vehicle or tank car moving until open country is reached in order to minimize the hazards from the escaping gas. Appropriate emergency measures should then be taken as quickly as possible.

If a motor vehicle transporting chlorine containers is damaged, any leaking containers should, if possible, be positioned so that only gas escapes, and if necessary, moved to an isolated area before attempts are made to stop the leaks. If there is any possibility of fire, the containers should be removed from the vehicle.

If a tank car or tank truck is damaged and chlorine is leaking, an appropriate downwind area should be evacuated and emergency clearing operations should not be started until safe working conditions are established.

Preparation for Handling Emergencies

List at least several physicians who could be summoned in the event of an emergency. Request all physicians listed to familiarize themselves with the treatment of persons exposed to chlorine. Provide the telephone numbers of these physicians along with those of the supplier and CHEMTREC or CANUTEC to the plant telephone operator. Post a similar list in the area where chlorine is handled. Also, list and post the telephone numbers of the nearest hospital and fire and police departments. Arrange for a telephone extension in areas where chlorine is handled and keep these extensions open at night, on weekends, and on holidays.

Alkali Absorption

At regular points in areas of chlorine storage and use, provisions should be made for emergency disposal of chlorine from leaking containers. Chlorine may be absorbed in solutions of caustic soda or soda ash. The proportions of alkali and water recommended for this purpose, in the amounts needed to absorb indicated quantities of chlorine, are given in Table 2.

A tank suitable for holding the required alkaline solution should be provided in a convenient location. The alkali should be stored in a form such that a solution can readily be prepared as needed. Never immerse any chlorine container in the solution or a body of water. After the solution is prepared, the chlorine should be passed from the container into the solution through plastic pipe or rubber hose properly weighted to hold it under the surface. When chlorine is absorbed in alkaline solutions, the heat of reaction is substantial. Caustic solutions can cause burns to personnel.

Emergency Kits

Most chlorine suppliers have emergency kits and skilled technicians to use them.

TABLE 2. RECOMMENDED ALKALINE SOLUTIONS FOR ABSORBING CHLORINE.

Chlorine Container Capacity (lb net)	20 Weight Percent Caustic Soda Solution		10 Weight Percent Soda Ash Solution	
	100% NaOH (lb)	Water (gal)	100% Na_2CO_3 (lb)	Water (gal)
100	135	65	359	390
150	203	98	538	585
2000	2708	1300	7176	7800

These kits are designed to control most leaks in chlorine shipping containers. Kit A is for chlorine cylinders; Kit B for ton containers, and Kit C for tank cars and tank trucks. Many customers have found it advisable to purchase kits and train employees in their use. Additional information is available from the Chlorine Institute. [7]

First Aid

Prompt treatment of anyone overcome or seriously exposed to chlorine is of the utmost importance. The patient should be removed from the contaminated area and medical assistance obtained as soon as possible. [8]

Contact with Skin or Mucous Membranes. If the patient has also inhaled chlorine, first aid for inhalation should be given first. If the chlorine has contaminated the skin or clothing, the emergency shower or any other means of washing with copious amounts of water should be used immediately. Contaminated clothing should be removed under the shower and the chlorine should be washed off with very large quantities of water. Skin areas should be washed with large quantities of running water for 15 minutes or longer. Do not attempt neutralization or apply any salves or ointments to damaged skin. Refer to a physician if irritation persists after irrigation or if skin is broken or blistered.

Contact with the Eyes. If even minute quantities of liquid chlorine enter the eyes, or if the eyes have been exposed to strong concentrations of chlorine gas, they should be flushed immediately with copious quantities of running water for at least 15 minutes. Never attempt to neutralize with chemicals. The eyelids should be held apart during this period to ensure contact of water with all accessible tissues of the eyes and lids.

Call a physician, preferably an eye specialist, at once. If a physician is not immediately available, the eye irrigations should be continued for a second period of 15 minutes. No oils or oily ointments, or any medications, should be instilled unless ordered by the physician.

Inhalation. If breathing has not ceased, the patient should be placed in a comfortable position. Firmness and assurance will help alleviate patient anxiety. Slow, deep breathing should be encouraged. Trained personnel should administer oxygen as soon as possible. The victim should be kept warm and remain at rest until medical help arrives. Call a physician immediately.

Caution: Never give anything by mouth to an unconscious or convulsing patient.

If breathing has apparently ceased, the victim must be removed to fresh air. Artificial respiration by acceptable means, such as mouth-to-mouth, should be started immediately. Avoid breathing the exhaled contaminated breath of the victim. If the heart has stopped, CPR should be started by a trained person. Oxygen should be administered by first aid attendants trained in the use of the specific oxygen equipment at hand.

METHODS OF SHIPMENT

Chlorine normally is shipped as a liquefied compressed gas. It is the responsibility of each person shipping or transporting chlorine to know and comply with all applicable regulations. Under the appropriate regulations, chlorine is authorized for shipment as follows: [9] and [10]

By Rail: In cylinders, single-unit tank cars, and TMU tank cars.

By Highway: In cylinders and ton containers (by truck or on special trailers) and cargo tank trucks.

By Water: In cylinders, single-unit tank cars, and ton containers aboard cargo vessels only. In bulk, in steel tank barges. (In an emergency involving life or health, and on application made to the Commandant of the Coast Guard, limited shipments of chlorine may be made under conditions authorized by the Commandant aboard passenger vessels and ferry vessels.) In cylinders on barges of U.S. Coast Guard classes A, BA, CA, and CB.

By Air: In cylinders aboard cargo aircraft only up to 150 lb (68 kg) maximum net weight per cylinder.

CONTAINERS

Chlorine is stored and shipped as a liquefied gas under pressure in steel cylinders, ton containers, tank motor vehicles, tank cars, and tank barges. All containers are equipped with one or more pressure relief devices, except as noted below. Chlorine containers other than barges must comply with the authorized U.S. Department of Transportation or Transport Canada specifications. Containers in transportation must be suitably labeled and placarded as required by regulations.

Filling Limits

The maximum filling density authorized for chlorine containers is 125 percent water capacity by weight at 60°F (15.6°C).

Cylinders

Chlorine is authorized for shipment in cylinders with not over 150 lb (68 kg) capacity which comply with TC/DOT specifications 3A480 or 3AA480. The authorized pressure relief device is a Type CG-2 fusible plug. [11] and [12] Cylinders conforming with some older specifications may still be used. "Lecture bottles" conforming with Specification 3E1800 or sample cylinders conforming with Specification 3BN480 also may be used, and these need not have a pressure relief device.

Storing Cylinders. In addition to the precautions required for the safe handling and storage of any compressed gas cylinders, chlorine cylinders as well as ton containers should be segregated from containers holding other compressed gas (e.g., anhydrous ammonia) and hydrocarbon chemicals or any flammable materials. The storage area must be well ventilated, and storage below ground should be avoided.

Valve Outlet Connections

The standard valve outlet connection for chlorine cylinders in the United States and Canada is Connection CGA 820. The limited standard connection for chlorine cylinders for use in the specialty gas industry only is Connection CGA 660. [13]

Ton Containers

Chlorine is authorized for shipment in ton containers with 2000 lb (907 kg) capacity which comply with DOT specification 106A500X. The pressure relief device is a fusible plug. Ton containers conforming with older specifications may still be used. Valve outlet connections are the same as for cylinders.

Ton containers are authorized for motor vehicle transport as well as rail transport as a part of a TMU tank car. TMU cars (specially built railcars) are designed to carry 15 ton containers, but these cars are nearly obsolete.

Tank Motor Vehicles

Chlorine is authorized for shipment in tank trailers (cargo tanks) complying with DOT specifications MC-331 or MC-330. These are generally 15–22 tons (13 600–20 000 kg) capacity. DOT regulations contain special requirements for chlorine cargo tanks.

Tank Cars

Chlorine is authorized for shipment in tank cars meeting DOT specifications 105 A500W. Cars built to some older specifications may still be used. DOT regulations contain special requirements for chlorine tanks.

Tank Barges

Plans and specifications for the construction of barges and for the tanks must be approved by the U.S. Coast Guard or the Canadian Coast Guard, or both if the barge is to be operated on U.S. and Canadian waterways. The most common barge capacities are 600 tons (540 Mg), configured with four tanks of 150 tons capacity each, and 1100 tons (997 Mg), configured with four tanks of 275 tons capacity each.

Bulk Storage Facilities

Consumers of chlorine in bulk quantities who receive shipment in single-unit tank cars often withdraw the chlorine direct to process, without transferring it to bulk storage facilities of their own.

Bulk chlorine consumers supplied by tank barges and cargo tanks, however, usually unload into their own storage facilities. Whether supplied by tank trailer, tank car, or tank barge, storage facilities must be designed by experienced engineers and must conform to all applicable regulations.

METHODS OF MANUFACTURE

Chlorine is produced predominantly by an electrolysis process in diaphragm, mercury, or membrane cells. In each process, a salt solution, most often sodium chloride, is electrolyzed by the action of direct electric current which converts chloride ions to elemental chlorine. A small amount is produced by other electrolytic processes or nonelectrolytically.

REFERENCES

[1] "Properties of Chlorine in SI Units," The Chlorine Institute, Inc., 2001 L Street, N.W., Washington, DC 20036.

[2] *Code of Federal Regulations*, Title 29 CFR Parts 1900–1910 (Labor), Superintendent of Documents, U.S. Government Printing Office, Washington, DC 20402.

[3] *Threshold Limit Values and Biological Exposure Indices*, 1989–90 ed., American Conference of Governmental Industrial Hygienists, 6500 Glenway Avenue, Bldg. D-7, Cincinnati, OH 45211-4438

[4] *Chlorine and Your Health*, The Chlorine Institute, Inc., 2001 L Street, N.W., Washington, DC 20036

[5] *Piping Systems for Dry Chlorine*, The Chlorine Institute, Inc., 2001 L Street, N.W., Washington, DC 20036.

[6] Material Safety Data Sheet for Chlorine, The Chlorine Institute, Inc., 2001 L Street, N.W., Washington, DC 20036.

[7] *Chlorine Manual*, The Chlorine Institute, Inc., 2001 L Street, N.W., Washington, DC 20036.

[8] *First Aid and Medical Management of Chlorine Exposure*, The Chlorine Institute, Inc., 2001 L Street, N.W., Washington, DC 20036.

[9] *Code of Federal Regulations*, Title 49 CFR Parts 100–199 (Transportation), Superintendent of Documents, U.S. Government Printing Office, Washington, DC 20402.

[10] *Transportation of Dangerous Goods Regulations*, Canadian Government Publishing Centre, Supply and Services Canada, Ottawa, Ontario, Canada K1A 0S9.

[11] CGA S-1.1. *Pressure Relief Device Standards—Part 1—Cylinders for Compressed Gases*, Compressed Gas Association, Inc., 1235 Jefferson Davis Highway, Arlington, VA 22202.

[12] Chlorine Institute Pamphlet No. 17, *Cylinder and Ton Container Procedure for Chlorine Packaging*, The Chlorine Institute, Inc., 2001 L Street, N.W., Washington, DC 20036.

[13] CGA V-1, *American National, Canadian, and Compressed Gas Association Standard for Compressed Gas Cylinder Valve Outlet and Inlet Connections*, (ANSI/CSA/CGA V-1), Compressed Gas Association, Inc., 1235 Jefferson Davis Highway, Arlington, VA 22202.

ADDITIONAL REFERENCES

For a *Catalog* listing numerous technical and safety publications and audiovisuals on chlorine, contact the Chlorine Institute, 2001 L Street, N.W., Washington, DC 20036.

Diborane

Chemical Symbol: B_2H_6
Synonyms: Boroethane, boron hydride
CAS Registry Number: 19287-45-7
DOT Classification: Flammable gas
DOT Label: Flammable gas and Poison
Transport Canada Classification: 2.1 (6.1)
UN Number: UN 1911

PHYSICAL CONSTANTS

	U.S. Units	SI Units
Chemical formula	B_2H_6	B_2H_6
Molecular weight	27.67	27.67
Vapor pressure		
at −220°F (−140°C)	0.522 psia	3.60 kPa abs
at −160°F (−106°C)	5.80 psia	39.99 kPa abs
at −100°F (−73.3°C)	38.22 psia	263.52 kPa abs
at 60°F (15.5°C)	536.55 psia	3699.38 kPa abs
Density of the gas		
at 32°F (0°C) and 1 atm	0.0779 lb/ft^3	1.25 kg/m^3
at 70°F (21.1°C) and 1 atm	0.0712 lb/ft^3	1.14 kg/m^3
Specific gravity of the gas		
at 32°F (0°C) and 1 atm (air = 1)	0.952	0.952
Specific volume of the gas		
at 32°F (0°C) and 1 atm	12.84 ft^3/lb	0.801 m^3/kg
at 70°F (21.1°C) and 1 atm	14.05 ft^3/lb	0.877 m^3/kg
Density of the liquid		
at −220°F (−140°C)	30.9 lb/ft^3	495 kg/m^3
at −184°F (−120°C)	29.6 lb/ft^3	474 kg/m^3
at −112°F (−80°C)	25.9 lb/ft^3	415 kg/m^3
at −40°F (−40°C)	22.2 lb/ft^3	356 kg/m^3
at 32°F (0°C)	16.9 lb/ft^3	271 kg/m^3
Flammable limits (percent in air by volume)	0.9–98%	0.9–98%

	U.S. Units	SI Units
Boiling point at 1 atm	−134.5°F	−92.5°C
Melting point at 1 atm	−265.9°F	−165.5°C
Critical temperature	62°F	16.7°C
Critical pressure	581 psia	4005.8 kPa abs
Critical density	10.3 lb/ft³	165 kg/m³
Latent heat of vaporization at −134.5°F (−92°C)	222.0 Btu/lb	515.9 J/g
Heat capacity of gas at 77°F (25°C) and 14.7 psia C_p	0.4935 Btu/lb°F	2.065 J/g°C
Solubility in water	Decomposes to form boric acid and hydrogen	

DESCRIPTION

Diborane is a colorless gas at room temperature and atmospheric pressure. It has an unpleasant, distinctive, sickly sweet odor. Diborane is a highly flammable gas which forms a flammable mixture with air over a range of 0.9 to 98 percent diborane (at 1 atm). Diborane burns in air (or oxygen) with a blue to green flame. Diborane is considered pyrophoric at room temperature. The gas is easily ignited by a spark or the heat of reaction with moisture in air.

Pure diborane is insensitive to mechanical shock; however, shock and thermally sensitive mixtures may be formed in the presence of impurities such as oxygen, water, halogenated hydrocarbons, and so on. [1] Thermal decomposition of diborane can result in excessive pressure buildup. Vessels for containment of diborane should be designed to contain such resultant decomposition pressure.

GRADES AVAILABLE

Diborane is sold in ambient or refrigerated cylinders with a purity of 99 percent or greater. It is commonly used in the electronics industry, mainly in the form of dilute mixtures. A typical specification for electronic grade diborane is as follows:

Methane	150 ppm
Methyl ether	30 ppm
Higher hydrides	<0.01%
Carbon dioxide	<0.05%
Nitrogen	<0.05%
Hydrogen	>0.1 %
Diborane	balance

USES

Diborane is commonly used in the electronics industry for semiconductor doping by mixing small concentrations with silane in the gas phase prior to decomposition. When reacted with silane and oxygen, diborane also produces the cladding layers of wave guides for fiber optics by chemical vapor deposition.

Other uses of diborane include the preparation of boron nitride by the reaction of diborane with ammonia, as a catalyst for polymerization, and for the conversion of olefins to trialkyl boranes. It is also used in the conversion of amines to amine boranes and as a selective reducing agent with carbonyl compounds such as aldehydes and ketones to form alcohols.

PHYSIOLOGICAL EFFECTS

The U.S. Occupational Safety and Health Administration has established an eight-hour time-weighted average exposure limit of 0.1 ppm (0.1 mg/m³) for diborane which is also the threshold limit value (TLV) adopted by the American Conference of Governmental Industrial Hygienists. [2] and [3]

Inhalation

Diborane acts primarily as an irritant to the respiratory system. Inhalation causes a heat-producing reaction with the moisture in the lungs. Boric acid is produced by hydrolysis. [1] An acute inhalation exposure may lead to respiratory distress—chest tightness, dyspnea, nonproductive cough and wheezing. Subacute exposure to low concentrations of diborane is more likely to produce central nervous system symptoms such as light-headedness, headache, fatigue, and drowsiness. [4]

Skin and Eye Hazard

Skin and eye contact with diborane causes burns. Although there is no evidence of penetration of the skin by pure diborane, higher hydrides resulting from decomposition will penetrate the skin, causing systemic poisoning. [4]

MATERIALS OF CONSTRUCTION

Common metals are suitable as materials of construction. These include the following metals and metal alloys: chrome-molybdenum steel, Type 300 stainless steel, brass, lead, Monel, K-Monel, and nickel. Piping and appurtenances for undiluted diborane must be designed by experienced engineers and safety and fire protection specialists. Saran, polyethylene, Kel-F, Teflon, graphite, and high-vacuum silicone grease are satisfactory for use with diborane.

In addition to the ability of a material to withstand chemical attack, the evaluation of materials compatibility with diborane should also emphasize the effect of the material on diborane stability (as expressed by the decomposition rate). The use of the following materials is ***not recommended:***

- Metal oxides
- Natural rubbers
- Neoprene
- Leak-lock
- Permatex
- Ordinary oil and grease
- Nordel 1145 RPT elastomer, unfilled and SiO_2-filled
- W-970 silicon elastomer, unfilled and SiO_2-filled
- C1S-4 polybutadiene elastomer, unfilled and SiO_2-filled

SAFE STORAGE, HANDLING, AND USE

Diborane should be stored at all times under conditions that will minimize decomposition and avoid contact with air. Normally diborane is stored in TC/DOT cylinders. The cylinder valve should be tightly closed and the valve caps in place. At ambient temperatures, pure diborane stored in a cylinder will undergo 10 to 20 percent decomposition per month. If refrigerated, diborane can be stored for several months with a much reduced rate of decomposition. Because of the potential of cylinder failure under extreme cold conditions, diborane should not be stored below −112°F (−80°C).

Do not immerse cylinders in liquid nitrogen.

Store diborane cylinders with internal pressure above atmospheric pressure to prevent the possibility of air leakage into the cylinder causing the formation of explosive mixtures. Cylinder pressure should never be allowed to fall below 30 psig (207 kPa) at room temperature. Hence, the pressure during storage at −112°F (−80°F) would not be subatmospheric. If a cylinder pressure of less than 30 psig is reached prior to storage, dry inert nitrogen should be added to the diborane cylinder to maintain at least 30 psig.

Open flames, sparks, heat, and all sources of ignition should be kept away from diborane storage areas.

All systems to handle diborane should be designed with these facts in mind:

- The standard valve outlet connection assigned to diborane cylinders is Connection CGA 350.
- For systems downstream of the cylinder valve, only welded and VCR-type fittings should be used in order to prevent

leakage both in and out, under vacuum and pressure.

- All lines and vessels should be cleaned, dried, and purged with dry nitrogen or helium prior to diborane contact.
- Nitrogen-purge taps should be provided at strategic points.
- Subatmospheric operations should be minimized.
- Scrubbers or incinerators should be used for the disposal of vented diborane.
- A metal diaphragm regulator should be used for pressure regulation.

For additional information on the safe storage and handling of diborane, see the Callery Chemical Company publication entitled *Diborane Handling Bulletin.* [1]

DISPOSAL

The preferred method of disposal for diborane is incineration. The gas should be slowly fed to a burn basin which contains a burning solvent such as naphtha or kerosene in accordance with local state, provincial, and federal regulations. The source diborane must be at least 50 feet away from the incinerator. Incinerators with a hydrogen fuel are also available, and wet scrubbers may also be used.

HANDLING LEAKS AND EMERGENCIES

As previously noted, diborane is considered pyrophoric such that the gas is easily ignited by a spark or the heat of reaction with moisture in air. If significant diborane is mixed with air in a restricted space, a delayed ignition may result in an explosive condition. Diborane fires should be allowed to burn out while keeping adjacent areas, especially those containing combustibles, cool with water.

Do not use halogenated fire-extinguishing agents.

Cleaning of an area contaminated by a diborane fire should start with the hosing down of surfaces with water, followed by a 5 percent ammonia and 5 percent trisodium phosphate detergent wash to remove residual products.

If a leak should occur, evacuate all personnel. Responders using self-contained breathing apparatus and protective clothing should attempt to shut off the source of the leak.

First Aid

Call a physician in all cases.

If diborane is inhaled, remove the victim to fresh air. If the victim is not breathing, give artificial respiration, preferably mouth-to-mouth. If breathing is difficult, administer oxygen.

If eyes have been exposed, flush immediately with water for at least 15 minutes while holding eyelids open. If diborane is on skin, flush the exposed area with water for at least 15 minutes, then shower and wash thoroughly afterwards.

METHODS OF SHIPMENT

Under the appropriate regulations, diborane is authorized for shipment as follows:

By Rail: In cylinders, as compressed gas or refrigerated liquid.

By Highway: In cylinders, as compressed gas or refrigerated liquid.

By Water: In cylinders as compressed gas only on cargo vessels stowed above deck.

By Air: Forbidden.

CONTAINERS

Diborane is shipped as a compressed gas in cylinders in 100-g, 1-lb, and 5-lb lots. Refrigerated diborane is shipped as a liquid in cylinders in 1-, 5-, or 40-lb lots. The outside packaging is filled with enough dry ice to provide refrigeration for at least ten days.

The cylinders are typically Specification TC/DOT 3AA. They are usually placed inside a wooden box, drum, or insulated container prior to shipment.

Authorized pressure relief devices for use on diborane cylinders are the CG-4 and CG-5 type, which are combination rupture disk/fusible plug devices. The fusible alloy in the CG-4 type device has a nominal yield temperature of 165°F (74°C); that in the CG-5 type device has a nominal yield temperature of 212°F (100°C). [5]

Valve Outlet Connections

The standard valve outlet connection in the United States and Canada for diborane cylinders is Connection CGA 350. [6]

METHODS OF MANUFACTURE

There are several methods known to prepare diborane. [7] Some chemical equations for the preparation of diborane are:

$$3NaBH_4 + 4BF_3 \xrightarrow{\text{diglyme}} 2B_2H_6 + 3NaBF_4$$

$$[BH_4]^- + H^+ \ (H_3PO_4 \text{ or } H_2SO_4) \longrightarrow \tfrac{1}{2}B_2H_6 + H_2$$

$$2NaBH_4 + I_2 \xrightarrow{\text{diglyme}} 2NaI + B_2H_6 + H_2$$

In addition, stoichiometric amounts of boron trichloride (BCl_3) and sodium hydride (NaH) reacted at room temperature in diglyme will give good yields of diborane. [8]

Boron trichloride reacted with hydrogen in an electric discharge over aluminum at 177–260°F (350–500°C) will also produce diborane. [9] and [10]

REFERENCES

[1] *Diborane Handling Bulletin,* Callery Chemical Company, P.O. Box 429, Pittsburgh, PA 15230. (1982)

[2] *Code of Federal Regulations,* Title 29 CFR Parts 1900–1910 (Labor), Superintendent of Documents, U.S. Government Printing Office, Washington, DC 20402.

[3] *Threshold Limit Values and Biological Exposure Indices,* 1989–90 ed., American Conference of Governmental Industrial Hygienists, 6500 Glenway Avenue, Bldg. D-7, Cincinnati, OH 45211-4438.

[4] *Diborane Health and Safety Bulletin,* Callery Chemical Company, P.O. Box 429, Pittsburgh, PA 15230. (1982)

[5] CGA S-1.1, *Pressure Relief Device Standards—Part 1—Cylinders for Compressed Gases,* Compressed Gas Association, Inc., 1235 Jefferson Davis Highway, Arlington, VA 22202.

[6] CGA V-1, *American National, Canadian, and Compressed Gas Association Standard for Compressed Gas Cylinder Valve Outlet and Inlet Connections* (ANSI/CSA/CGA V-1), Compressed Gas Association, Inc., 1235 Jefferson Davis Highway, Arlington, VA 22202.

[7] *Boranes,* Kirk-Othmer Encyclopedia of Chemical Technology, 3rd ed., Vol. 4, 1980, p. 152.

[8] Lappert, M. F., *Chemical Review,* **56,** p. 959 (1956).

[9] Schroeder, S., and Thiele, K. H., *Anorg. Allg. Chem.,* **428,** (1) (1977).

[10] Borg, A., and Green, A. A., *J. Am. Chem. Soc.* **65**, 1838 (1943).

ADDITIONAL REFERENCES

Code of Federal Regulations, Title 49 CFR Parts 100–199 (Transportation), Superintendent of Documents, U.S. Government Printing Office, Washington, DC 20402.

Transportation of Dangerous Goods Regulations, Canadian Government Publishing Centre, Supply and Services Canada, Ottawa, Ontario, Canada K1A 0S9.

Diborane Technical Bulletin, Callery Chemical Company, P.O. Box 429, Pittsburgh, PA 15230.

Dichlorosilane

Chemical Symbol: SiH_2Cl_2
Synonyms: Dichlorosilicane
CAS Registry Number: 4109-96-0
DOT Shipping Name: Flammable liquid, Corrosive, NOS (Dichlorosilane)
DOT Classification: Flammable liquid
DOT UN Number: UN 2924 (U.S. domestic shipments)
DOT Label: Flammable liquid, Corrosive, Poison
Transport Canada Classification: 2.3 (2.1)
TC Shipping Name: Dichlorosilane
TC, IMO, UN Number: 2189

PHYSICAL CONSTANTS

	U.S. Units	SI Units
Chemical formula	SiH_2Cl_2	SiH_2Cl_2
Molecular weight	101.0	101.0
Vapor pressure		
at 46.7°F (8.2°C)	14.70 psia	101.33 kPa abs
at 68.0°F (20.0°C)	24.25 psia	167.18 kPa abs
at 104.0°F (40.0°C)	44.09 psia	303.98 kPa abs
at 140.0°F (60.0°C)	76.42 psia	526.89 kPa abs
Density of the gas		
at 70°F (21.1°C) and 1 atm	0.2640 lb/ft^3	4.228 kg/m^3
Specific gravity of the gas		
at 70°F (21.1°C) and 1 atm		
(air = 1)	3.48	3.48
Specific gravity of the gas		
at 70°F (21.1°C) and 1 atm	3.83 ft^3/lb	0.239 m^3/kg
Density of the liquid		
at 70°F (21.1°C)	77.10 lb/ft^3	1235 kg/m^3
at 105°F (40.6°C)	74.16 lb/ft^3	1188 kg/m^3
at 115°F (46.1°C)	73.35 lb/ft^3	1175 kg/m^3
at 122°F (50.0°C)	72.79 lb/ft^3	1166 kg/m^3
Boiling point at 1 atm	46.8°F	8.2°C
Melting point at 1 atm	−187.6°F	−122.0°C
Critical temperature	348.8°F	176.0°C
Critical pressure	678.2 psia	4676 kPa abs

	U.S. Units	SI Units
Critical density	28.9 lb/ft^3	463 kg/m^3
Latent heat of vaporization at 46.76°F (8.2°C)	116.0 Btu/lb	269.64 kJ/kg
Heat capacity of gas at 77°F (25°C) and 14.7 psia C_p	0.1467 Btu/(lb)(°F)	0.6143 kJ/(kg)(°C)
Solubility in water	Reacts	Reacts
Flammable limits in air (percent by volume)	4.1–98.8%	4.1–98.8%
Autoignition Temperature*	136.4°F	58°C

DESCRIPTION

Dichlorosilane is a highly flammable, corrosive, and toxic gas at room temperature and atmospheric pressure. It causes severe burns on contact with eyes, skin, and mucous membranes. With water or moisture, it hydrolyzes rapidly to yield silica and silicon oxyhydride along with hydrochloric acid. It is shipped as a liquefied gas in low pressure cylinders at its vapor pressure of 9.1 psig (62.7 kPa) at 70°F (21.1°C). It can form flammable mixtures with air and oxidizing agents.

GRADES AVAILABLE

Dichlorosilane is primarily sold in ultrahigh-purity grades for use in the electronics industry. A typical specification usually quantifies the acceptable levels of hydrocarbons and metals.

USES

Dichlorosilane is primarily used in the electronics industry for such applications as growth of epitaxial or polycrystalline silicon and chemical vapor deposition of silicon dioxide and silicon nitride.

It is an outstanding material for epitaxial deposition. Its silicon content by weight is greater than either trichlorosilane or silicon tetrachloride. Dichlorosilane deposits silicon more efficiently and at lower temperatures than the other chlorosilanes in epitaxial reactors. Dichlorosilane significantly lowers the processing time from that required with silane for deposition of thick layers at reduced temperatures.

The deposition rate of dichlorosilane does not vary appreciably with minor temperature changes inside the reactor, therefore substantially lowering the rejection rate by reducing variations in layer thickness and degree of taper. Since the deposition rate of dichlorosilane is not as temperature sensitive as that of other chlorosilanes, the rate is controlled by adjusting the dichlorosilane concentration in the hydrogen feedstream.

PHYSIOLOGICAL EFFECTS

Dichlorosilane hydrolyzes and oxidizes readily to release hydrogen chloride; therefore, the symptoms, effects, and treatment will be similar to those for hydrogen chloride. Dichlorosilane will cause severe burns on contact with eyes, skin, and mucous membranes.

If dichlorosilane is inhaled, immediately remove the victim to fresh air. If breathing is difficult, give oxygen. Prompt treatment by a physician is required, even if no symptoms of exposure are evident, since the symptoms may be delayed.

Inhalation of low concentrations of vapors will cause irritation of the respiratory tract, producing cough, excess sputum, and chest discomfort. Inhalation of vapors can cause

*Note: The measured autoignition temperature of dichlorosilane is above 212°F (100°C), but accumulated monochlorosilane above bulk liquid may reduce the autoignition temperature to about 136.4°F (58°C).

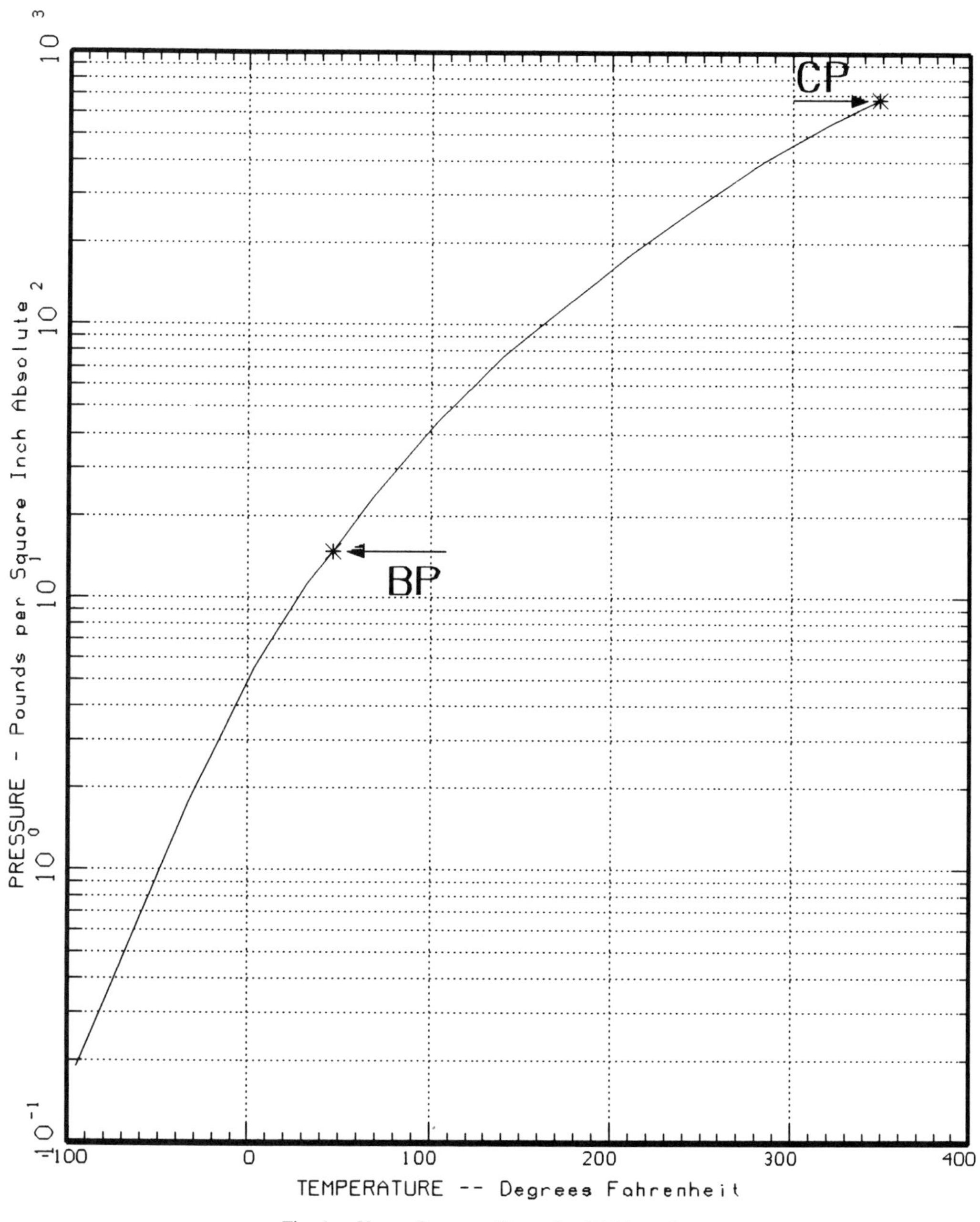

Fig. 1. Vapor Pressure Curve for Dichlorosilane.

severe irritation or burns of moist skin, mucous membranes, and the upper respiratory tract as well as delayed pulmonary edema. Chronic exposure to the vapors may cause discoloration or erosion of the teeth, bleeding of nose and gums, and ulceration of the nasal mucosa.

Vapor contact with the eye will cause severe irritation experienced as pain in the eye, excess lachrymation, closure of the eyelids, and marked excess redness and swelling of the conjunctiva. If high concentrations of hydrogen chloride vapor are formed, then corneal injury can occur. Splash contamination may cause severe conjunctivitis seen as marked excess redness and swelling of the conjunctiva, discharge, iritis, and severe corneal injury. The corneal injury, if untreated, could result in permanent blindness.

Prolonged or widespread skin contact with the liquid may result in the absorption of potentially harmful amounts of material.

MATERIALS OF CONSTRUCTION

Dichlorosilane, in the complete absence of water, can be safely stored in mild steel equipment. In the presence of even small traces of water, dichlorosilane becomes extremely corrosive since the Si-Cl bonds react rapidly with water, generating hydrogen chloride.

Because of reactivity with water, dichlorosilane should always be handled in dry equipment with a dry inert gas, such as nitrogen. For transfer service, dry inert gas is preferred to pumping. Some examples of other common compatible materials used include: Viton, Teflon, Kel-F, nickel, Monel, and some types of stainless steel.

SAFE STORAGE, HANDLING, AND USE

Since dichlorosilane is a highly flammable, corrosive, and toxic liquefied gas, appropriate precautions must be taken in its storage and handling. During the handling of chlorosilanes, the use of such protective equipment as goggles, neoprene or natural rubber gloves, and protective clothing is essential. A self-contained breathing apparatus, as well as both safety showers and eyewash fountains, should be available for emergency use.

Cylinders should be assigned a definite area for storage. The area should be dry, cool, well ventilated, fire resistant, and away from ignition sources. Keep cylinders protected from excessive temperature rise by storing them away from radiators or other heat sources. Storage conditions should comply with local and state regulations.

Cylinders may be stored in the open, but protected against extremes of weather and from the dampness of the ground to prevent rusting. During the summer, cylinders stored in the open should be shaded against the continuous direct rays of the sun in those localities where extreme temperatures prevail.

No part of a cylinder should be subjected to a temperature higher than 125°F (51.7°C). Temperatures in excess of 125°F (51.7°C) may cause a cylinder to become liquid full, resulting in excessive hydrostatic pressure buildup, thereby causing the potential for the cylinder to rupture. Never permit a flame to come in contact with any part of a compressed gas cylinder.

Dichlorosilane has a low autoignition temperature. Exposure to heat from a fire or from a water-dichlorosilane reaction can cause the product to auto-ignite. The acidic decomposition products formed by burning dichlorosilane from leaks may rapidly attack the metal at the leak area, especially if the metal is hot.

Store and use dichlorosilane only in adequately ventilated areas. It should be used only in a closed system constructed with compatible materials and designed to withstand the pressures involved. Keep away from heat and all ignition sources such as flames and sparks, since dichlorosilane will form flammable mixtures with air and other oxidizing agents. All lines, connections, equipment, and so forth, must be thoroughly checked for leaks and grounded prior to use. Use only spark-proof tools and explosion-proof equipment.

Because of the reactivity of dichlorosilane, reverse flow into the cylinder may cause it to rupture. Always use a vacuum break or other protective apparatus in any line or piping from the cylinder to prevent reverse flow.

For general information on safe handling of compressed gas cylinders, see CGA P-1, *Safe Handling of Compressed Gases in Containers.* [1] Also see Chapter 5.

The storage, handling, and use of dichlorosilane are regulated in some areas under fire codes such as Article 80 of the *Uniform Fire Code.* [2]

DISPOSAL

Dichlorosilane should not be discharged directly into surface waters or sewer systems since an acidic waste product is formed. The disposal can be accomplished by controlled introduction of the product into water. The exothermic reactions of dichlorosilane with water (hydrolysis) result in the formation of hydrochloric acid and an insoluble silicon-containing solid or fluid. In order to prevent air pollution, the quantity of water must be sufficient to dissolve all of the hydrogen chloride that will be formed. The ratio of water to dichlorosilane should be at least 10 to 1. The corrosive and exothermic nature of the reaction should be considered in selecting materials of construction for the equipment used in this procedure.

The hydrochloric acid formed should then be neutralized with an alkali agent such as aqueous ammonia, sodium hydroxide, lime slurry, etc., and should be added as an aqueous solution with agitation to the acidic medium. Consideration must be given to the additional heat that will be produced by the neutralization. Silicon-containing solids should be washed to remove residual acid. Discard any product, residue, disposable container, or liner in an environmentally acceptable manner, in full compliance with federal, state, provincial, and local regulations.

HANDLING LEAKS AND EMERGENCIES

Leak Detection

Large leaks of dichlorosilane will be evident by the formation of dense white fumes of hydrogen chloride upon contact with moist air. Small leaks may be detected by holding an open bottle of concentrated ammonium hydroxide solution near the site of the suspected leak (dense white fumes will form). Wet blue litmus paper will turn pink when exposed to a dichlorosilane leak.

Emergencies

In the event dichlorosilane is released or spilled, immediately evacuate all personnel from the danger area. It can form flammable mixtures with air; therefore, remove all sources of ignition, if without risk. Due to its low autoignition temperature, exposure to heat from a fire or from a water-dichlorosilane reaction can cause the product to autoignite. Vapors formed may travel or be moved by air currents and ignited by pilot lights, other flames, smoking, sparks, heaters, electrical equipment, static discharges, or other ignition sources at locations distant from the product-handling point. The flammable atmospheres may linger. Before entering the area, especially a confined area, check the atmosphere with an appropriate detection device. Use self-contained breathing apparatus and protective clothing where needed.

If properly protected with breathing apparatus and impervious protective clothing, personnel can treat releases or spills by flooding with water or a 5 percent sodium bicarbonate solution. Collect treated spills and contain for disposal.

A dichlorosilane fire will usually result if a spill contacts water. If flames are extinguished, explosive reignition may occur. Appropriate measures should be taken, for example, total evacuation of the area. Approach the area with extreme caution. Reverse flow into the cylinder may cause it to rupture. Corrosive vapors may be reduced with a coarse water spray. Shut off the leak if without risk. Ventilate the area, and prevent runoff from contaminating the surrounding environment.

In the event of a small fire, carbon dioxide can be used to try to extinguish it. Reignition will usually occur. If reignition occurs, or the fire cannot be extinguished, then let it burn out unless the flow of dichlorosilane can be stopped safely.

For large fires, use a coarse water spray or all-purpose type foams applied in accordance with manufacturer's recommendations. If the fire persists, cool containers in the affected area with coarse water spray from a maximum distance while allowing the fire to burn. Stop the flow of burning dichlorosilane if without risk, while continuing the cooling water spray.

First Aid

Inhalation. Remove the victim from exposure. Administer oxygen if chest symptoms occur. Airway obstruction, as indicated by the presence of laryngeal stridor, may require the placement of an airway by trained emergency medical technicians. Chest symptoms may be treated by responding emergency medical technicians, by the administration of anticholinergic inhalants (such as Atrovent) and bronchodilators (such as albuterol). Prompt treatment by a physician is required, even if no symptoms of exposure are evident, since the symptoms may be delayed.

Eye Exposure. In case of contact of dichlorosilane with the eye, immediately flush with plenty of water and continue flushing for at least 15 minutes. Two or three drops of pontocaine (0.5% aqueous solution) may be instilled into the eye to facilitate irrigation. The eyelids should be held apart so that all parts of the eye and eyelids are thoroughly washed.

Prompt treatment by a physician, preferably an ophthalmologist, should be obtained immediately after the initial washing is completed. In some cases, irritation of the eye may not be evident immediately after exposure. The onset of symptoms may be delayed several hours.

Skin Exposure. In case of skin contact with liquid or vapor, immediately flush areas of exposure with large quantities of water while removing contaminated clothing and shoes. Continue to flush with water for at least 15 minutes or until medical attention is obtained.

Prompt treatment by a physician is required, even if no symptoms are present. Contaminated clothing should be washed before reuse.

See *Effects of Exposure to Toxic Gases—First Aid and Medical Treatment* for further information. [3]

METHODS OF SHIPMENT

Under the appropriate regulations, dichlorosilane is authorized for shipment as follows:

By Rail: Forbidden.

By Highway: On trucks utilizing cylinders with a water capacity not to exceed 250 lb (113.4 kg); also in larger portable containers (by DOT exemption).

By Water: In cylinders with a water capacity not to exceed 250 lb (113.4 kg); also in larger portable containers (by DOT exemption). On cargo vessels on deck only and away from living quarters; forbidden on passenger vessels.

By Air: Forbidden.

For specific requirements, see the "flammable liquids" listing in 49 CFR 172.101, "Hazardous Materials Table," or "dichlorosilane" in 49 CFR 172.102 as appropriate, or equivalent Canadian regulations. [4] and [5]

CONTAINERS

Dichlorosilane authorized for shipment must meet the requirements of 49 CFR 173.3(a) for packaging of certain poisonous materials.

Steel cylinders that meet TC/DOT Specifications 3A480 and 4BW240 and that are not over 250 lb (113.4 kg) water capacity are authorized for the shipment of dichlorosilane. Specification 3A and 3AA cylinders rated for 1800 psig (12410 kPa abs) minimum pressure may also be used.

DOT exemptions have also been authorized for stainless steel cylinders and portable containers with greater than 250 lb (113.4 kg) water capacities.

Valve Outlet Connections

The most common cylinder valve used for dichlorosilane is a stainless steel diaphragm valve with a Connection CGA 678 outlet. The cylinder valve body has a combination CG-4 type pressure relief device consisting of a fusible metal plug, melting about 165°F (74°C), which is protected from the cylinder contents by a frangible metal disk, i.e. a rupture disk, rated at 250 psig (1724 kPa).

Alternatively, dichlorosilane can be packaged in accordance with the requirements for Poison A materials, in which case the use of a pressure relief device is prohibited. The use of valves containing pressure relief devices requires approval from the DOT.

METHODS OF MANUFACTURE

Dichlorosilane is most commonly produced by the disproportionation of trichlorosilane in a catalytic redistribution reactor. The trichlorosilane is initially produced from metallurgical silicon which is reacted with hydrogen and silicon tetrachloride.

REFERENCES

[1] CGA P-1, *Safe Handling of Compressed Gases in Containers,* Compressed Gas Association, Inc., 1235 Jefferson Davis Highway, Arlington, VA 22202.

[2] *Uniform Fire Code,* Western Fire Chiefs Association and International Conference of Building Officials, 5360 South Workman Mill Road, Whittier, CA 90601.

[3] *Effects of Exposure to Toxic Gases—First Aid and Medical Treatment,* 3rd ed., Matheson Gas Products, Inc., Secaucus, NJ 07094. (1988)

[4] *Code of Federal Regulations,* Title 49 CFR Parts 100–199, (Transportation), Superintendent of Documents, U.S. Government Printing Office, Washington DC 20402.

[5] *Transportation of Dangerous Goods Regulations,* Canadian Government Publishing Centre, Supply and Services Canada, Ottawa, Ontario, Canada K1A 0S9.

ADDITIONAL REFERENCES

Dichlorosilane Product Information Sheet (F-4190), Union Carbide Corporation, Linde Division, 39 Old Ridgebury Road, Danbury, CT 06817. (1980)

Encylopedie Des Gaz, L'Air Liquide. Elsevier/North-Holland Inc., 52 Vanderbilt Avenue, New York, NY 10017.

Material Safety Data Sheet for Dichlorosilane (L-4587A), Union Carbide Corporation, Linde Division, 39 Old Ridgebury Road, Danbury, CT 06817. (1985)

Matheson Gas Data Book, 6th ed., Matheson Gas Products, Inc., Secaucus, NJ 07094. (1980)

Matheson Unabridged Gas Data Book, 5th ed., Matheson Gas Products, Inc., Secaucus NJ 07094. (1977)

Ethane

Chemical Symbol: C_2H_6 (or CH_3CH_3)
Synonyms: Bimethyl, dimethyl, ethyl hydride, methyl-methane
CAS Registry Number: 74-84-0
DOT Classification: Flammable gas
DOT Label: Flammable gas
Transport Canada Classification: 2.1
UN Number: UN 1035 (compressed gas); UN 1961 (refrigerated liquid)

PHYSICAL CONSTANTS

	U.S. Units	SI Units
Chemical formula	C_2H_6	C_2H_6
Molecular weight	30.068	30.068
Vapor pressure		
at 70°F (21.1°C)	544 psig	3753 kPa
Density of the gas		
at 70°F (21.1°C) and 1 atm	0.07990 lb/ft^3	1.2799 kg/m^3
Specific gravity of the gas		
at 60°F (15.6°C) and 1 atm		
(air = 1)	1.0469	1.0469
Specific volume of the gas		
at 60°F (15.6°C) and 1 atm	12.5151 ft^3/lb	0.7813 m^3/kg
Density of the liquid		
at saturation pressure		
at 60°F (15.6°C)	23.52 lb/ft^3	376.7 kg/m^3
at 70°F (21.1°C)	22.40 lb/ft^3	358.8 kg/m^3
Boiling point at 1 atm	−127.53°F	−88.630°C
Melting point at 1 atm	−297.76°F	−183.2°C
Critical temperature	86.96°F	32.20°C
Critical pressure	708.35 psia	4883.9 kPa abs
Critical density	12.67 lb/ft^3	203.0 kg/m^3
Triple point at 1 atm	−297.89°F	−183.27°C
Latent heat of vaporization		
at boiling point and 1 atm	210.41 Btu/lb	489.41 kJ/kg
Latent heat of fusion		
at triple point	40.9 Btu/lb	95.1 kJ/kg

	U.S. Units	SI Units
Specific heat of the gas at 60°F (15.6) and 1 atm		
C_p	0.4097 Btu/(lb)(°F)	1.7153 kJ/(kg)(°C)
C_v	0.3436 Btu/(lb)(°F)	1.4386 kJ/(kg)(°C)
Ratio of specific heats (C_p/C_v)	1.192	1.192
Solubility in water, vol/vol at 68°F (20°C) and 1 atm	0.047	0.047
Pressure in typical full cylinder (approximate) at 130°F (54.4°C)	2474 psig	17 058 kPa
Critical volume	0.0788 ft^3/lb	0.00492 m^3/kg
Specific gravity of the liquid at 60°F (15.6°C) at saturation pressure	0.3771	0.3771
Gross heat of combustion		
Gas at 60°F (15.6°C) and 1 atm (for the real gas)	1783.7 Btu/ft^3	66 453 kJ/m^3
Liquid at 77°F (25°C) and saturation pressure	22 169 Btu/lb	51 565 kJ/kg
Net heat of combustion		
Gas at 60°F (15.6°C) and 1 atm (for the real gas)	1631.5 Btu/ft^3	60 783 kJ/m^3
Liquid at 77°F (25°C) and saturation pressure	20 281 Btu/lb	47 173 kJ/kg
Air required for combustion at 60°F (15.6°C) and 1 atm (volume of air per volume of the real gas)	16.845	16.845
Air required for combustion (weight of air per weight of gas)	16.090	16.090
Flammable limits in air by volume	3.0–12.4%	3.0–12.4%
Flash point	−211°F	−135°C

DESCRIPTION

Ethane is a colorless, odorless, flammable gas that is relatively inactive chemically and is considered nontoxic. It is shipped as a liquefied compressed gas under its vapor pressure of 544 psig at 70°F (3751 kPa at 21.1°C).

GRADES AVAILABLE

Ethane is typically available for commercial and industrial purposes in a C.P. grade (minimum purity of 99.0 mole percent) or a technical grade (minimum purity of 95.0 mole percent). A typical analysis of a technical grade of ethane is as follows:

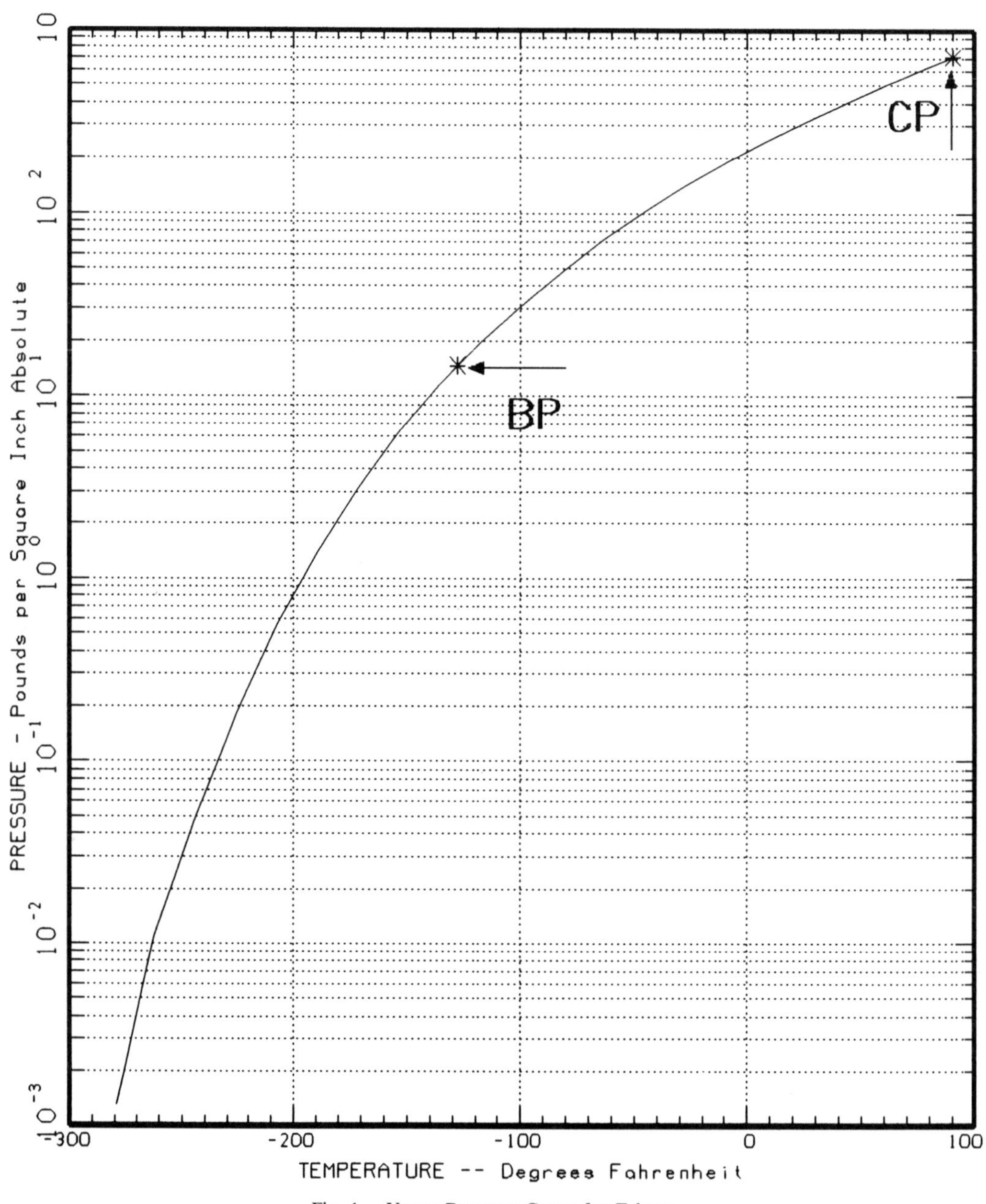

Fig. 1. Vapor Pressure Curve for Ethane.

Ethane	97.41 mole %
Methane	0.03 mole %
Ethylene	0.4 mole %
Hydrogen	2.15 mole %
n-Butane	0.01 mole %
Acetylene	not detectable

USES

Major uses of ethane include its employment as a fuel and in organic synthesis. For example, it can be chlorinated to give ethyl chloride, and can yield ethylene with a greater heat input than is required for obtaining ethylene by propane cracking. Ethane is also used as a refrigerant.

PHYSIOLOGICAL EFFECTS

Inhalation of ethane in concentrations in air up to 5 percent produces no definite symptoms, but inhalation of higher concentrations has an anesthetic effect. It can act as a simple asphyxiant by displacing the oxygen in the air. Contact between liquid ethane and skin can cause freezing of the tissue, and should be avoided.

MATERIALS OF CONSTRUCTION

Ethane is noncorrosive and may be contained in installations constructed of any common metals designed to withstand the pressure involved.

SAFE STORAGE, HANDLING, AND USE

All the precautions required for the safe handling of any flammable compressed gas must be observed with ethane. It is important that ignition sources be kept away from containers, including situations in which leakage could cause the gas to be ignited by such sources as a spark from a motor. All piping and equipment used with ethane should be grounded.

Ethane should not be stored with cylinders containing oxygen, chlorine, or other oxidizing or combustible materials.

See Chapter 5 for additional general recommendations concerning safe storage, handling, and use. Also see the monograph on liquefied petroleum gases.

DISPOSAL

Disposal of ethane, as with other gases, should be undertaken only by personnel familiar with the gas and the procedures for disposal. Contact the supplier for instructions. In general, the best procedure for disposal of flammable gases, including ethane, is to burn the gas, if a burning unit is available in the plant. This should be done in accordance with appropriate regulations.

HANDLING LEAKS AND EMERGENCIES

To detect leaks from containers, connections, or piping, use a soapy water solution. Leaks will be indicated by the formation of bubbles. Alternative means of detection involve the use of instrumental methods. Never use a flame for leak detection. See the liquefied petroleum gases monograph for steps to take regarding an emergency involving ethane.

First Aid

Inhalation of low concentrations can be remedied by promptly going to an uncontaminated area and inhaling fresh air. In the event of a massive exposure, wherein the victim has become unconscious or symptoms of asphyxiation may persist, the person should be removed promptly to an uncontaminated atmosphere, and given artificial respiration if breathing has stopped. This should be followed by oxygen, after breathing has been restored.

Contact of liquid ethane with the skin can result in frostbite. First aid treatment for frostbite consists of putting the frostbitten part in warm water of about 108°F (42°C). If warm water is not available, or is impractical to use, wrap the affected area gently in blankets. Encourage the victim to exercise the affected part while it is being warmed. This will aid circulation to reestablish itself naturally. Medical attention by a physician should be obtained.

In the event of eye contact with liquid ethane, flush with tap water for 15 minutes. If irritation persists, the patient should be referred to a physician.

METHODS OF SHIPMENT

Under the appropriate regulations, ethane is authorized for shipment as follows: [1] and [2]

By Rail: In cylinders.

By Highway: In cylinders on trucks and in tube trailers. Refrigerated liquid in cargo tanks.

By Water: In cylinders on cargo vessels either on deck or preferably under deck. Refrigerated liquid ethane must be stowed on deck. On passenger vessels, only compressed gas stowed either on deck or preferably under deck and subject to the requirements of 49 CFR 173.304 and 173.306. [1] In cylinders on barges of U.S. Coast Guard classes A, CA, and CB only.

By Air: In cylinders aboard cargo aircraft only up to 300 lb (136 kg) maximum net weight per cylinder. Refrigerated liquid forbidden.

CONTAINERS

Cylinders

Cylinders that comply with the following TC/DOT specifications are authorized for ethane service: 3A1800, 3AX1800, 3AA1800, 3AAX1800, 3AL1800, 3T1800, 3E1800, and 39. Containers of the same types with higher service pressures may also be used. DOT-3 cylinders may also be continued in service, but new construction is not authorized.

All types of cylinders authorized for ethane service must be requalified by periodic hydrostatic retest every five years under present regulations, except that no periodic retest is required for 3E cylinders. See 49 CFR Part 173 for further information. [1]

Authorized pressure relief devices for cylinders containing ethane are the Type CG-1 rupture disk. See CGA S-1.1, *Pressure Relief Device Standards—Part I—Cylinders for Compressed Gases,* for further information and requirements. [3]

Filling Limits

A maximum filling density of 35.8 percent (percent water capacity by weight) is authorized for ethane in cylinders meeting TC/DOT specifications 3A1800, 3AX1800, 3AA1800, 3AAX1800, 3AL1800, 3T1800, 3, 3E1800, and 39. A maximum filling density of 36.8 percent is authorized in cylinders meeting TC/DOT specifications 3A2000, 3AX2000, 3AA2000, 3AAX2000, 3T2000, 3AL2000, and 39. When used for shipment of flammable gas, the internal volume of a Specification 39 cylinder must not exceed 75 $in.^3$ (0.00123 m^3).

Valve Outlet Connections

The standard valve outlet connection in the United States and Canada for ethane cylinders is Connection CGA 350. [4]

Cargo Tanks

Authorized cargo tanks for ethane are those constructed in accordance with Specifications MC-331 and MC-338. [1] and [2] See 49 CFR 173.315.[1]

METHODS OF MANUFACTURE

Ethane is produced commercially from the cracking of light petroleum fractions, and also by fractionation from natural gas.

REFERENCES

[1] *Code of Federal Regulations,* Title 49 CFR Parts 100–199 (Transportation), Superintendent of Documents, U.S. Government Printing Office, Washington, DC 20402.

[2] *Transportation of Dangerous Goods Regulations,* Canadian Government Publishing Centre, Supply and Services Canada, Ottawa, Ontario, Canada K1A 0S9.

[3] CGA S-1.1, *Pressure Relief Device Standards—Part 1—Cylinders for Compressed Gases,* Compressed Gas Association, Inc., 1235 Jefferson Davis Highway, Arlington, VA 22202.

[4] CGA V-1, *American National, Canadian, and Compressed Gas Association Standard for Compressed Gas Cylinder Valve Outlet and Inlet Connections* (ANSI/CSA/CGA V-1), Compressed Gas Association, Inc., 1235 Jefferson Davis Highway, Arlington, VA 22202.

ADDITIONAL REFERENCE

Effects of Exposure to Toxic Gases—First Aid and Medical Treatment, 3rd ed., Matheson Gas Products, Inc., Secaucus, NJ 07094. (1988)

Ethylene

Chemical Symbol: C_2H_2 (or $H_2C:CH_2$)
Synonyms: Ethene (also olefiant gas, elayl, or etherin)
DOT Classification: Flammable gas
DOT Label: Flammable gas
Transport Canada Classification: 2.1
UN Number: UN 1078

PHYSICAL CONSTANTS

	U.S. Units	SI Units
Chemical formula	C_2H_4	C_2H_4
Molecular weight	28.05	28.05
Density of the gas at 32°F (0°C) and 1 atm	0.0787 lb/ft^3	1.261 kg/m^3
Specific gravity of the gas at 32°F and 1 atm (air = 1)	0.978	0.978
Specific volume of the gas at 70°F (21.1°C) and 1 atm	12.7 ft^3	0.793 m^3/kg
Density of the liquid at boiling point	35.42 lb/ft^3	567.47 kg/m^3
Boiling point at 1 atm	−154.8°F	−103.8°C
Melting point at 1 atm	−272.9°F	−169.4°C
Critical temperature	49.82°F	9.900°C
Crticial density	14.2 lb/ft^3	228 kg/m^3
Triple point	−272.47°F at 1.0146 psia	−169.15°C at 0.1014 kPa abs
Latent heat of vaporization at boiling point	208 Btu/lb	484 kJ/kg
Latent heat of fusion at melting point	51.2 Btu/lb	119 kJ/kg
Specific heat of the gas at 59°F (15°C) and 1 atm		
C_p	0.3622 Btu/(lb)(°F)	1.516 kJ/(kg)(°C)
C_v	0.2914 Btu/(lb)(°F)	1.220 kJ/(kg)(°C)
Ratio of specific heats (C_p/C_v)	1.243	1.243

	U.S. Units	SI Units
Solubility in water, vol/vol at 32°F (0°C)	0.26	0.26
Weight of the liquid at boiling point	4.735 lb/gal	567.4 kg/m^3
Gross heat of combustion	21 652 Btu/lb	50 300 kJ/kg
Flammable limits in air by volume	2.7–36%	2.7–36%
Flammable limits in oxygen by volume	2.9–79.9%	2.9–79.9%
Autoignition temperature		
in air	914°F[a]	490°C
in oxygen	905°F	485°C

[a]Also reported to be 1009°F (542.8°C).

DESCRIPTION

Ethylene is a colorless, flammable gas with a faint odor that is sweet and musty. It is nontoxic and has been used as an anesthetic (but is generally no longer used for this purpose in the United States and Canada). The hazardous properties of ethylene are its flammability and its potential to cause asphyxia by displacement of air with the resultant lowering of the oxygen content below that necessary to support life.

Chemically, ethylene reacts chiefly by addition to give saturated paraffins or derivatives of paraffin hydrocarbons. Ethylene is widely used as a raw material in the synthetic organic chemical industry. It is shipped as a gas at about 1250 psig at 70°F (8618 kPa at 21.1°C). Below 50°F (10°C) at such charging pressure, it is a liquefied gas in the cylinder.

GRADES AVAILABLE

Ethylene is typically available for commercial and industrial purposes in a C.P. grade (minimum purity of 99.5 mole percent), and a technical grade (minimum purity of 98.0 mole percent). A typical lot analysis of technical grade is as follows:

Ethylene	98.5 mole %
Methane	0.4 mole %
Ethane	1.0 mole %
Propane	0.1 mole %
Oil and foreign material	none
Dew point	−50°F (−45.6°C)

USES

Ethylene finds use in the manufacture of ethyl benzene, ethanol, ethylene oxide, ethylene glycol, and ethylene dichloride. About half of the ethylene produced in the United States is used for the production of high and low density polyethylene plastics. Other chemical raw materials made with ethylene include ethyl chloride, dichloroethane, vinyl chloride, ethyl ether, methyl acrylate, and styrene.

Ethylene is also employed as a refrigerant and a fuel for metal cutting and welding, and it has been used for anesthesia. It is also used to accelerate plant growth and fruit ripening.

PHYSIOLOGICAL EFFECTS

When used for anesthesia, ethylene is a nontoxic gas found pleasant and nonirritating by patients. Prolonged inhalation of substantial concentrations results in unconsciousness; light and moderate anesthesia is attained, and deep anesthesia seldom occurs. Inhalation is fatal only if the gas acts as a simple asphyxiant, depriving the body of

Vapor Pressure

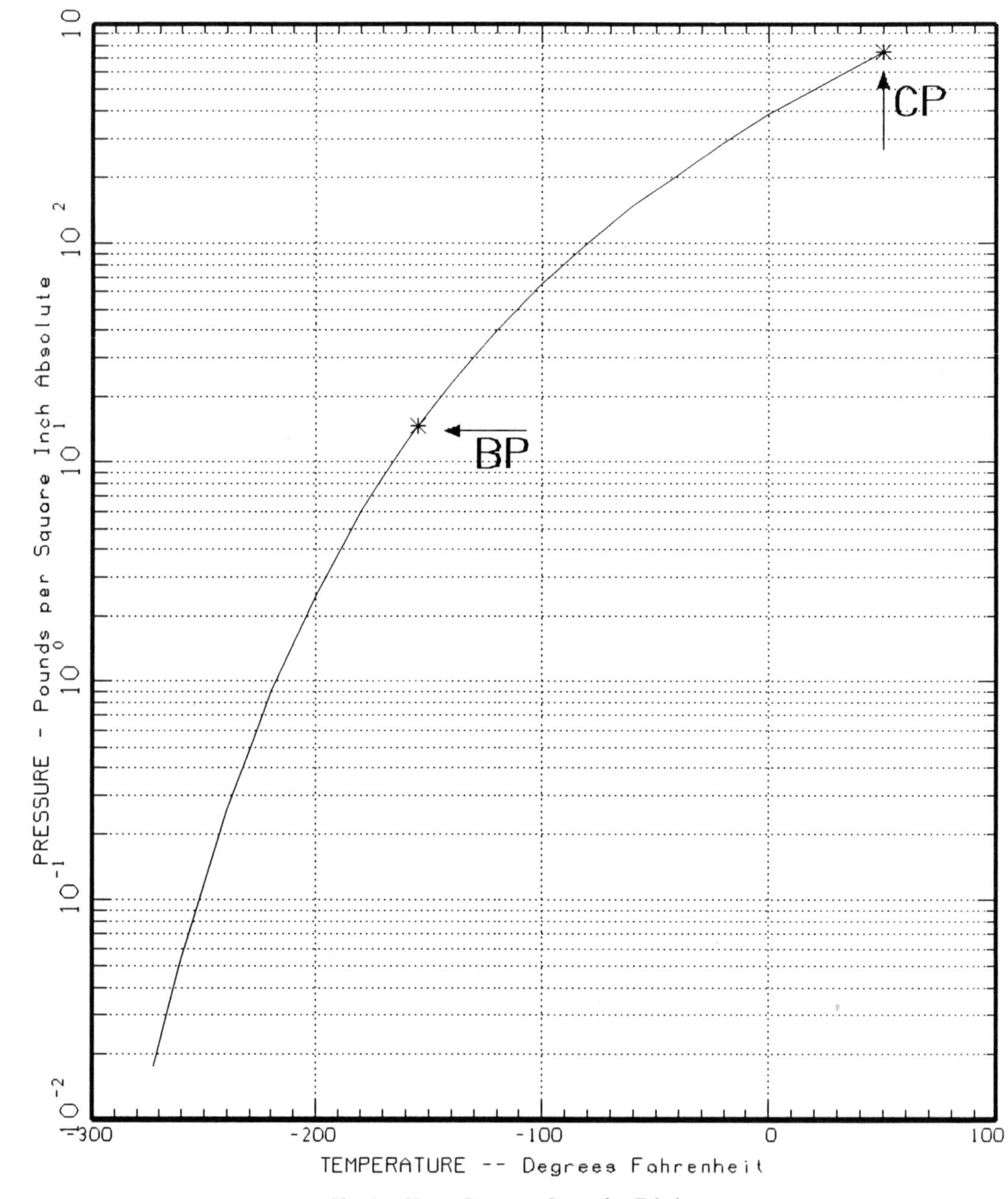

Fig. 1. Vapor Pressure Curve for Ethylene.

necessary oxygen. Because of its flammability, however, other agents have replaced ethylene for use in anesthesia in the United States and Canada.

No deleterious action by ethylene on circulatory, respiratory, or other systems or organs has been observed. Exhalation eliminates the major portion of ethylene within minutes, although complete desaturation from body fat takes several hours. Minute traces can be detected in the blood a number of hours after anesthesia has ended.

MATERIALS OF CONSTRUCTION

Installations must be designed to withstand the pressures involved and must comply with all applicable regulations. Because it is noncorrosive, any common commercially available metals may be used with ethylene.

SAFE STORAGE, HANDLING, AND USE

Ethylene poses hazards to personnel through its flammability, and the precautions necessary for the safe handling of any flammable gas must be observed in its use. It is important that ignition sources be kept away from containers, including situations in which leakage could cause the gas to be ignited by such sources as a spark from a motor. All piping and equipment used with ethylene should be grounded.

Ethylene should not be stored with cylinders containing oxygen, chlorine, or other oxidizing or combustible materials.

See Chapter 5 for additional general recommendations concerning safe storage, handling, and use of flammable gases.

DISPOSAL

Disposal of ethylene, as with other gases, should be undertaken only by personnel familiar with the gas and the procedures for disposal. Contact the supplier for instructions. In general, the best procedure for disposal of flammable gases, including ethylene, is to burn the gas if a burning unit is available in the plant. This should be done in accordance with appropriate regulations.

HANDLING LEAKS AND EMERGENCIES

To detect leaks from containers, connections, or piping, use a soapy water solution. Leaks will be indicated by the formation of bubbles. Alternative means of detection involve the use of instrumental methods. Never use a flame for leak detection.

All sources of ignition should be eliminated at once. If practical, a leaking ethylene cylinder should be moved to a safe area and plainly tagged as defective. Warnings should be posted in the area to prevent persons from approaching the cylinder with lit cigarettes or open flames.

Refer to the liquefied petroleum gases monograph for additional information on recommended procedures likewise relevant to an emergency involving ethylene.

First Aid

Inhalation of low concentrations can be remedied by promptly going to an uncontaminated area and inhaling fresh air. In the event of a massive exposure wherein the victim has become unconscious or symptoms of asphyxiation may persist, the person should be removed promptly to an uncontaminated atmosphere and given artificial respiration if breathing has stopped. This should be followed by oxygen after breathing has been restored.

Contact of liquid ethylene with the skin can result in frostbite. First aid treatment for frostbite consists of putting the frostbitten part in warm water of about 108°F (42°C). If warm water is not available or is impractical to use, wrap the affected area gently in blankets. Encourage the victim to exercise the affected part while it is being warmed. This will aid circulation to reestablish itself naturally. Medical attention by a physician should be obtained.

In the event of eye contact with liquid ethylene, the affected eyes should be flushed with tap water for 15 minutes. If irritation persists, the patient should be referred to a physician.

METHODS OF SHIPMENT

Under the appropriate regulations, ethylene is authorized for shipment as follows: [1] and [2]

By Rail: In cylinders, and by DOT exemption in DOT 113A tank cars in low temperature liquid form.

By Highway: In cylinders on trucks and in tube trailers; and by DOT exemption in insulated truck cargo tanks liquefied at low temperatures.

By Water: In cylinders via only cargo vessels. In cylinders on barges of U.S. Coast Guard classes A and C only, and by DOT exemption in DOT 113A tank cars.

By Air: Aboard cargo aircraft only in appropriate cylinders up to 300 lb (136 kg) maximum net weight per cylinder.

CONTAINERS

Ethylene is authorized for shipment in cylinders under U.S. Department of Transportation and Transport Canada regulations. It is also shipped in bulk quantities under DOT exemption.

Filling Limits

The maximum filling limits prescribed for ethylene in cylinders are as follows (percent water capacity by weight): for cylinders of 1800 psig (12 411 kPa) maximum service pressure, 31 percent; for cylinders of 2000 psig (13 790 kPa) maximum service pressure, 32.5 percent; and for cylinders of 2400 psig (16 547 kPa) maximum service pressure, 35.5 percent.

It is also shipped by highway in tube trailers. Under DOT exemption, ethylene is also shipped liquefied at low temperatures in insulated cargo tanks on trucks and truck trailers, and in insulated, low temperature tank cars of the DOT 113 type. Bulk shipment of ethylene may also be available by pipeline.

Cylinders

Cylinders with maximum service pressure ratings of 1800, 2000, and 2400 psig for the following TC/DOT specifications are authorized for ethylene service: 3A, 3AA, 3AX, 3AAX, 3AL, and 3T. Specification 3E1800, 3AA2265, and 39 cylinders are also authorized. Cylinders manufactured under the now obsolete specifications DOT-3 may be continued in service, but new construction is not authorized. When used for shipment of flammable gas, the internal volume of a Specification 39 cylinder must not exceed 75 in.^{3} (0.0123 m^3).

Cylinders of Specifications 3A and 3AA (as well as 3) used in ethylene service must be requalified by periodic hydrostatic retest every five years, or ten years under certain limitations, under present regulations. Cylinders of Specifications 3AX, 3AAX, 3AL, and 3T must be requalified by periodic hydrostatic retest every five years. For cylinders of Specification 3E, no periodic retest is required. See 49 CFR 173.34. [1]

Authorized pressure relief devices for ethylene cylinders are the Type CG-1 rupture disk. See CGA S-1.1, *Pressure Relief Device Standards—Part I—Cylinders for Compressed Gases,* for further information and requirements. [3]

Valve Outlet Connections

The standard valve outlet connection in the United States and Canada for ethylene cylinders is Connection CGA 350. The standard yoke connection in the United States and Canada (for medical use of the gas) is Connection CGA 900. [4]

METHODS OF MANUFACTURE

The most common method of producing ethylene commercially is high temperature coil cracking of propane or of a mixture of ethane and propane. Recovery of ethylene from the cracked gases is often accomplished by low temperature, high pressure straight fractionation. Another customary manufacturing method is by catalytic decomposition of ethyl alcohol.

REFERENCES

[1] *Code of Federal Regulations,* Title 49 CFR Parts 100–199 (Transportation), Superintendent of Documents, U.S. Government Printing Office, Washington, DC 20402.

[2] *Transportation of Dangerous Goods Regulations,* Canadian Government Publishing Centre, Supply and Services Canada, Ottawa, Ontario, Canada K1A 0S9.

[3] CGA S-1.1, *Pressure Relief Device Standards—Part 1—Cylinders for Compressed Gases,* Compressed Gas Association, Inc., 1235 Jefferson Davis Highway, Arlington, VA 22202.

[4] CGA V-1, *American National, Canadian, and Compressed Gas Association Standard for Compressed Gas Cylinder Valve Outlet and Inlet Connections* (ANSI/CSA/CGA V-1), Compressed Gas Association, Inc., 1235 Jefferson Davis Highway, Arlington, VA 22202.

Ethylene Oxide

Chemical Symbol: C_2H_4O or CH_2CH_2O
Synonyms: 1,2-Epoxyethane; oxirane; dimethylene oxide
CAS Registry Number: 75-21-8
DOT Classification: Flammable liquid
DOT Label: Flammable liquid
Transport Canada Classification: 2.1 (6.1)
UN Number: UN 1040

PHYSICAL CONSTANTS [1]

	U.S. Units	SI Units
Chemical formula	C_2H_4O	C_2H_4O
Molecular weight	44.05	44.05
Vapor pressure at 68°F (20°C)	21.1 psia	146.0 kPa
Density of the gas at 68°F (20°C) and 21.1 psia (146.0 kPa abs)	0.1751 lb/ft^3	2.804 kg/m^3
Specific gravity (air = 1)	1.52	1.52
Specific volume at 70°F (21.1°C) and 14.7 psia (101.3 kPa abs)	8.78 ft^3/lb	0.548 m^3/kg
Density of the liquid at 68°F (20°C) and 21.1 psia (146.0 kPa abs)	54.30 lb/ft^3	869.8 kg/m^3
Boiling point at 14.7 psia (101.3 kPa abs)	50.7°F	10.4°C
Freezing point	−170.7°F	−112.6°C
Critical temperature	385°F	196°C
Critical pressure	1043 psia	7191 kPa abs
Heat capacity of the gas C_p	0.262 Btu/(lb)(°F)	1.096 kJ/(kg)(°C)
Latent heat of vaporization at 14.7 psia (101.3 kPa abs)	249 Btu/lb	578.8 kJ/kg
Latent heat of fusion	50.5 Btu/lb	117.4 kJ/kg
Specific heat of the liquid at 68°F (20°C)	0.48 Btu/(lb)(°F)	2.01 kJ/(kg)(°C)
Solubility in water	completely miscible	completely miscible

DESCRIPTION

Ethylene oxide is a highly reactive colorless gas that condenses to a colorless liquid boiling at 50.7°F (10.4°C) and 14.7 psia (101.3 kPa abs). It is miscible in all proportions with water, alcohol, ether, and most organic solvents. The vapors of ethylene oxide are flammable and explosive. It is generally noncorrosive to metals and leaves no residual odor or taste.

GRADES AVAILABLE

Ethylene oxide is sold only in grades of 99.7 percent or higher purity. A typical specification is as follows:

Appearance	clear
Acetylene, weight percent	0–0.0005
Aldehydes, as acetaldehyde, weight percent maximum	0.0003–0.01
Water, weight percent maximum	0.01–0.05
Carbon dioxide weight percent maximum	0.005
Acidity, as acetic acid, weight percent maximum	0.001–0.005
Chlorides, weight percent maximum	
inorganic	0.0001–0.01
organic	0.005
Nonvolatile matter, g/100 ml, max.	0.005–0.01
Suspended matter	none to substantially free
Color, Pt-Co, max.	10–15
Odor	mild to nonresidual

Ranges in specification values indicate a difference in the limits given by different producers. [2]

USES

The major use of ethylene oxide is as a chemical intermediate for the manufacture of ethylene glycol and higher glycols. These glycols are used as drying agents, antifreezes, and raw materials for the manufacture of other chemical derivatives such as terphthalates, surfactants, ethanolamines, glycol ethers, nonionic surface-active agents, urethane intermediates, polyethylene oxides, hydroxyethyl cellulose, and crown ethers. Ethylene oxide, both pure and mixed with carbon dioxide or halocarbons, is also used as a sterilant and fumigant for heat-sensitive materials. [1]

PHYSIOLOGICAL EFFECTS

Ethylene oxide is a toxic liquid and gas. Contact of the eyes with liquid ethylene oxide can cause severe irritation and corneal injury. Eye contact with the vapor can cause moderate irritation. Skin contact with the liquid or vapor or water solutions can cause severe delayed chemical burns. Inhalation of vapor will cause irritation of the respiratory tract that may result in headache, nausea, and vomiting. All cases of inhalation or contact with ethylene oxide liquid or vapor must receive immediate first aid action followed by medical attention.

The U.S. Occupational Safety and Health Administration (OSHA) has set a permissible exposure limit (PEL) of 1 ppm for ethylene oxide. [3] Likewise, the American Conference of Governmental Industrial Hygienists (ACGIH) has adopted an 8-hour time-weighted average threshold limit value (TLV) of 1 ppm for exposure to ethylene oxide. [4] See 29 CFR 1910.1047 for additional OSHA regulations pertaining to workplace exposure to ethylene oxide. [3]

Ethylene oxide has been shown to produce mutagenic and cytogenic effects in laboratory tests and should be regarded as a suspect cancer hazard. OSHA has concluded that ethylene oxide "should be regarded as a potential human carcinogen". [5] The ACGIH lists ethylene oxide as a "suspected human carcinogen" [4]

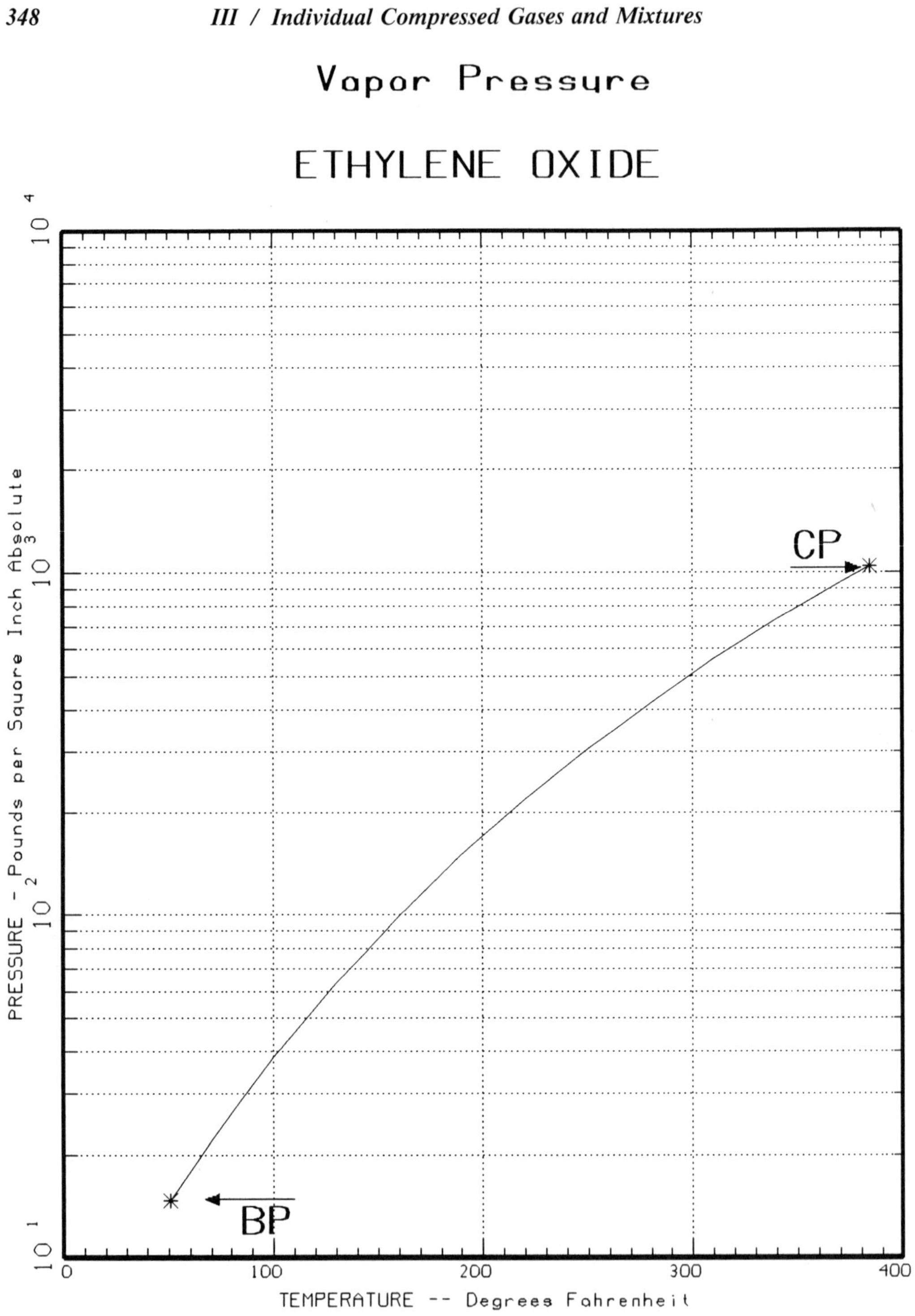

Fig. 1. Vapor Pressure Curve for Ethylene Oxide.

As health hazard information for ethylene oxide is currently an area for active research, users of ethylene oxide are urged to become familiar with the most current data, rules, and regulations. As an absolute minimum, the user should have and be conversant with the latest version of the Material Safety Data Sheet (MSDS) available from their product supplier, and the OSHA regulations in 29 CFR 1910.1047, "Occupation Exposure to Ethylene Oxide, Final Standard". [3]

MATERIALS OF CONSTRUCTION

Steel and stainless steel are suitable materials for equipment and piping in ethylene oxide service. Dangerous runaway reactions can result from contact with copper, silver, magnesium and their alloys; mercury and its salts; oxidizers of all types; alkalies and acids, alcohols, mercaptans, and alkali metals. Ethylene oxide will polymerize violently if contaminated with aqueous alkalies, amines, mineral acids, metal chlorides, or metal oxides.

SAFE STORAGE, HANDLING, AND USE

Ethylene oxide is toxic and flammable. All precautions necessary for the safe handling of any toxic, flammable liquid or gas must be taken. This includes keeping heat, flames, and spark-producing devices away from storage areas, pipelines, processing equipment, and cylinders. All cylinders and lines should be grounded and spark-proof tools used. Check valves, traps, or vacuum breaks in lines from cylinders will prevent reverse flow and avoid the possibility of undesirable and possibly violent reactions within the cylinder. Ethylene oxide may rearrange chemically and/or polymerize violently when in contact with highly catalytic surfaces such as anhydrous iron, tin, and aluminum chlorides; pure iron and aluminum oxides; and alkali metal hydroxides. [6]

Chapter 5 covers general safety guidelines for storage, handling, and use of compressed gases, and will direct readers to pertinent standards and additional sources of information. For recommendations on storage tank design and inert gas pressure padding requirements for ethylene oxide, consult references [1] and [7].

DISPOSAL

The preferred method of disposal of ethylene oxide is incineration.

HANDLING LEAKS AND EMERGENCIES

Systems designed to handle ethylene oxide must be pressure tested with nitrogen or other inert gas and be leakproof prior to being put into service. Such systems should be retested at routine intervals thereafter. Any leak or spill situation with ethylene oxide has the potential of fire and explosion coupled with exposure to a toxic material, and appropriate measures in recognition of these dangers are required, including evacuation of personnel.

First Aid

Users of ethylene oxide are urged to become familiar with the most current toxicity data. Material Safety Data Sheets will have the latest suggested first aid procedures. The following first aid recommendations are given in *Effects of Exposure to Toxic Gases—First Aid and Medical Treatment.* [8]

Inhalation. Workers who develop symptoms or signs of ethylene oxide toxicity should be removed immediately from the contaminated area. Oxygen can be administered to workers with persistent dyspnea.

Eye Contact. If liquid or vapor ethylene oxide contacts the eyes, they should be irrigated copiously with water for 15 minutes. An ophthalmic anesthetic drop may be used to decrease eyelid spasm and to facilitate irrigation. If eye irritation persists, the eye should be irrigated for a second 15-minute period, and a physician, preferably an eye specialist, should be consulted.

Skin Contact. Leather objects which become contaminated with ethylene oxide, such as shoes, belts, and watch bands, cannot be decontaminated and should be discarded. If they continue to be worn, skin burns or allergic rashes may result. Skin burns from exposure to aqueous solutions of ethylene oxide should receive copious irrigation with normal saline followed by the application of a topical antimicrobial agent, such as silver sulfadiazine cream, and a dressing. Signs of skin damage may not appear for up to 5 hours following skin exposure.

METHODS OF SHIPMENT

Under the appropriate regulations, ethylene oxide is authorized for shipment as follows: [9] and [10]

By Rail: In cylinders not exceeding 30 U.S. gallons (113.5 L) nominal water capacity; in containers not exceeding 61 gallons (231 L) nominal water capacity; in single-unit tank cars.

By Highway: In cylinders and containers on trucks and in portable tanks.

By Water: In cylinders, containers, and tanks on deck for passenger vessels and on deck and under deck for cargo vessels. Ethylene oxide cylinders are subject to the same segregation as flammable gases. See 49 CFR 172.100(h)(1), 172.101 and 176.63(b) for specific requirements. [9] In Canada, the regulations of the *International Maritime Dangerous Goods Code* apply. [11]

By Air: The maximum quantity in one package for cargo-only aircraft is limited to 300 lb (136 kg) in cylinders and 15 lb (6.8 kg) in other packagings.

CONTAINERS

Ethylene oxide is authorized for shipment in cylinders, special containers, portable tanks, and single-unit tank cars.

Filling Limits

The maximum filling densities authorized for ethylene oxide under present regulations are: in cylinders, such that the container will not be liquid full at 185°F (85°C); in 5P containers, such that the container will not be liquid full at 185°F (85°C); in single-unit tank cars (complying with TC/DOT 105A100W or 111A100W4 specifications), such that the tank will not be liquid full at 105°F (40.6°C); in portable tanks (complying with TC/DOT-51 specifications and 49 CFR 173.124), such that the tank will not be liquid full at 105°F (40.6°C).

Cylinders

Ethylene oxide is authorized for shipment in cylinders prescribed for any compressed gas, except acetylene, which meet the special requirements outlined in 49 CFR 173.124, which includes requirements for pressure relief devices. [9]

5P Containers

Ethylene oxide is authorized for shipment in lagged (insulated) steel drums complying to the TC/DOT 5P specifications and the special requirements in 49 CFR 173.124, or in equivalent sections of the *Transportation of Dangerous Goods Regulations* of Transport Canada. [9] and [10]

Tank Cars

Ethylene oxide is authorized for shipment in single-unit tank cars complying with DOT 105A100W or 111A100W4 specifications and special requirements in 49 CFR 173.124.

Valve Outlet Connections

The standard valve outlet connection in the United States and Canada for ethylene oxide cylinders is Connection CGA 510.

Portable Tanks

Ethylene oxide is authorized for shipment in portable tanks complying with TC/DOT-51 specifications and the special requirements in 49 CFR 173.124 or the *Transportation of Dangerous Goods Regulations* of Transport Canada. [9] and [10]

METHODS OF MANUFACTURE

Ethylene oxide is prepared commercially by two basic methods: direct oxidation processes and the process via ethylene chlorohydrin. The most commonly used process today involves the direct catalytic oxidation of ethylene with air or oxygen over a silver based catalyst to yield ethylene oxide. The major by-products of this process are carbon dioxide and water. The reaction product is purified by distillation. In 1979, this method accounted for over 96 percent of the nameplate capacity of ethylene oxide manufacture in the United States and Puerto Rico. [12]

An alternate process, rarely used for commercial production, involves the reaction of ethylene with hypochlorous acid to form ethylene chlorohydrin. This chlorohydrin is then dehydrochlorinated with lime to produce ethylene oxide and calcium chloride. Major by-products of this process are 1,2-dichloroethane, bis-(2-chloroethyl) ether, acetaldehyde, trace acetylenes, and other chlorinated hydrocarbons. The reaction product is purified by fractional distillation.

REFERENCES

[1] *Ethylene Oxide Product Information Bulletin F-7618,* Union Carbide Corporation, 39 Old Ridgebury Rd., Danbury, CT 06817 (1984).

[2] Anderson, R. L., ''Ethylene Oxide,'' *Encyclopedia of Industrial Chemical Analysis,* Vol. 12, Wiley, New York, 1971, p. 319.

[3] *Code of Federal Regulations,* Title 29 CFR Parts 1900–1910 (Labor), Superintendent of Documents, U.S. Government Printing Office, Washington, DC 20402.

[4] *Threshold Limit Values and Biological Exposure Indices,* 1989–90 ed., American Conference of Governmental Industrial Hygienists, 6500 Glenway Ave., Bldg D-5, Cincinnati, OH 45211.

[5] *Federal Register,* Vol. 49, No. 122, page 25740, June 22, 1984, Superintendent of Documents, U.S. Government Printing Office, Washington, DC 20402.

[6] *Matheson Gas Data Book,* 6th ed., Matheson Gas Products, Inc., Secaucus, NJ 07094. (1980)

[7] Cawse, J. N., et al., ''Ethylene Oxide,'' *Kirk-Othmer Encyclopedia of Chemical Technology,* 3rd ed., Vol. 9, Wiley, New York, p. 462.

[8] *Effects of Exposure to Toxic Gases—First Aid and Medical Treatment,* 3rd ed., Matheson Gas Products, Inc., Secaucus, NJ 07094. (1988)

[9] *Code of Federal Regulations,* Title 49 CFR Parts 100–199, (Transportation), Superintendent of Documents, U.S. Government Printing Office, Washington, DC 20402.

[10] *Transportation of Dangerous Goods Regulations,* Canadian Government Publishing Centre, Supply and Services Canada, Ottawa, Ontario, Canada K1A 0S9.

[11] *International Maritime Dangerous Goods Code,* International Maritime Organization, 4 Albert Embankment, London, United Kingdom SE1 7SR.

[12] Cawse, J. N., et al., ''Ethylene Oxide,'' *Kirk-Othmer Encyclopedia of Chemical Technology,* 3rd ed., Vol. 9, Wiley, New York, pp. 439–459.

ADDITIONAL REFERENCE

Material Safety Data Sheet for Ethylene Oxide (L-4798 B) (January 1985), Union Carbide Corporation, 39 Old Ridgebury Rd., Danbury, CT 06817.

Fluorine

Chemical Symbol: F_2
CAS Registry Number: 7782-41-4
DOT Classification: Nonflammable gas
DOT Label: Poison and Oxidizer
Transport Canada Classification: 2.3 (5.1)
UN Number: UN 1045

PHYSICAL CONSTANTS

	U.S. Units	SI Units
Chemical formula	F_2	F_2
Molecular weight	38.00	38.00
Density of the gas		
at 32°F (0.0°C) and 1 atm	0.106 lb/ft^3	1.70 kg/m^3
at 70°F (21.1°C) and 1 atm	0.098 lb/ft^3	1.57 kg/m^3
Specific gravity of the gas		
at 70°F (21.1°C) and 1 atm	1.312	1.312
Specific volume of the gas		
at 70°F (21.1°C) and 1 atm	10.17	0.635
Density of the liquid		
at −306.8°F (−188.2°C)	94.2 lb/ft^3	1509 kg/m^3
at −320.4°F (−195.8°C)	97.9 lb/ft^3	1568 kg/m^3
Boiling point at 1 atm	−306.8°F	−188.2°C
Melting point at 1 atm	−363.4°F	−219.6°C
Critical temperature	−199.9°F	−128.8°C
Critical pressure	756.4 psia	5215 kPa abs
Critical density	35.8 lb/ft^3	573.6 kg/m^3
Triple point	−363.4°F at 0.324 psia	−219.6°C at 0.223 kPa abs
Latent heat of vaporization		
at −306.8°F (−188.2°C)	74.8 Btu/lb	173.5 kJ/kg
Latent heat of fusion		
at −363.4°F (−219.6°C)	5.8 Btu/lb	13.4 kJ/kg

	U.S. Units	SI Units
Weight of liquid per gallon		
at −306.8°F (−188.1°C)	12.6 lb/gal	1509.8 kg/m^3
at −320.4°F −195.8°C)	13.1 lb/gal	1569.7 kg/m^3
Heat capacity of the gas, C_p		
at 32°F (0°C)	0.198 Btu/(lb)(°F)	0.828 kJ/(kg)(°C)
at 70°F (21.1°C)	0.197 Btu/(lb)(°F)	0.825 kJ/(kg)(°C)
Heat capacity of the gas, C_v		
at 70°F (21.1°C)	0.146 Btu/(lb)(°F)	0.610 kJ/(kg)(°C)
Ratio of specific heats, C_p/C_v		
at 70°F (21.1°C)	1.353	1.353
Thermal conductivity of gas		
at 32°F (0°C) and 1 atm	0.172 Btu·in./hr°F	0.248 W/m·K
Viscosity		
of gas at 32°F (0°C) and 1 atm	0.0218 cP	0.0218 mPa·s
of liquid at −314°F (−192.2°C)	0.275 cP	0.275 mPa·s

DESCRIPTION

Fluorine is a highly toxic, pale yellow gas about 1.3 times as heavy as air at atmospheric temperature and pressure. When cooled below its boiling point (−306.8°F or −188.2°C), it is a liquid about 1.5 times as dense as water.

Fluorine is the most powerful oxidizing agent known, reacting with practically all organic and inorganic substances. Exceptions are metal fluorides and a few completely fluorinated organic compounds in pure form. However, the latter may also react with fluorine if they are contaminated with a combustible material or in high flow situations.

Heats of reaction with fluorine are always high, and most reactions take place with ignition. Fluorine at low pressures and concentrations reacts slowly with many metals at room temperatures, however, and the reaction often results in formation of a metal fluoride film on the metal's surface. In the case of some metals, this film retards further reaction. This process is often denoted by the term *passivation*.

Fluorine readily displaces other halogens from their compounds, but such reactions are not always feasible for preparing fluorides. It reacts with water to form a mixture containing principally oxygen and hydrogen fluoride plus small amounts of ozone, hydrogen peroxide, and oxygen fluoride.

GRADES AVAILABLE

Fluorine is available at a minimum purity of 97%.

USES

Fluorine is used in producing uranium hexafluoride, sulfur hexafluoride, the halogen fluorides, and other fluorine compounds that require the high reactivity of elemental fluorine for their preparation. Fluorine is also used for the direct fluorination of organic materials when diluted with inert gas or at reduced pressures.

PHYSIOLOGICAL EFFECTS

Fluorine gas is a powerful corrosive irritant and is highly toxic. [1] In one series of animal experiments, inhalation of acute exposures of 10 000 ppm for 5 minutes, 1000 ppm for 30 minutes, and 500 ppm for 1 hour produced 100 percent mortality in rats, mice, guinea pigs, and rabbits. Inhalation of 100 ppm for 7 hours produced wide variation in species mortality, ranging from 0 percent in guinea pigs to 96 percent in mice.

Daily subacute exposures to a concentration of 2 ppm for periods of time varying from totals of 30 to 176 hours resulted in high mortality, rabbits appearing to be the most susceptible species and guinea pigs the least. Pulmonary irritation varying from severe at 16 ppm in some species to mild at 2 ppm represented the major pathological change, while similar subacute exposure at 0.5 ppm resulted in no significant pathology but some retention of fluorine, primarily in the bones and teeth.

Contact between the skin and high concentrations of fluorine gas under pressure will produce burns comparable to thermal burns along with chemical burn effects; contact with lower concentrations results in a chemical type of burn resembling that caused by acid. A TLV of 1 ppm (2 mg/m^3) for an 8-hour day for fluorine has been adopted by the American Conference of Governmental Industrial Hygienists. [2] The U.S. Occupational Safety and Health Administration's 8-hour time-weighted average exposure limit is 0.1 ppm (0.2 mg/m^3). [3]

MATERIALS OF CONSTRUCTION

Nickel, iron, aluminum, magnesium, copper, and certain of their alloys are quite satisfactory for handling fluorine at room temperature, for these are among the metals with which formation of a surface fluoride film retards further reaction.

Listed in Table 1 are various materials that have been used with satisfactory results in gaseous fluorine service at normal temperatures and liquid service at low temperatures. Nickel and Monel are generally considered by far the best materials for fluorine service at high temperatures and pressures, but selection of suitable materials for service at elevated temperatures and pressures must be based on the conditions of the specific application.

Equipment Preparation and Decontaminating

Equipment to be used for fluorine service must first be thoroughly cleaned, degreased, and dried, then treated with increasing concentrations of fluorine gas so that any impurities will be burned out without the simultaneous ignition of the equipment. The passive metal fluoride film thus formed will inhibit further reaction with fluorine.

SAFE STORAGE, HANDLING, AND USE

All precautions necessary for the safe handling of any strongly oxidizing gas must be observed with fluorine, in addition to the precautions outlined below. Accidental fluorine fires may be most simply extinguished by cutting off the fluorine supply at the fluorine source, then employing conventional fire-fighting methods. Dry type extinguishers are recommended.

Before introducing any application of fluorine, users should fully work out all details of first aid and treatment with the medical personnel who would be called to administer aid in case of an accident.

Only trained and competent personnel should be permitted to handle fluorine. It is recommended that personnel work in pairs and within sight and sound of each other, but not in the same working area. Supervisory personnel should make frequent checks of the operation.

Essential additional precautions in handling of fluorine cylinders are outlined in the subsequent section on cylinders.

Personal Protective Equipment

Loose leather gloves should be worn when handling fluorine cylinders and when operating fluorine gas-handling equipment. Leather gloves provide optimum protection against potential fluorine-fed fires. Care

TABLE 1. MATERIALS GIVING SATISFACTORY RESULTS IN FLUORINE SERVICE.[a]

Type of Equipment	Gaseous Service Normal Temperature	Liquid Service Low Temperature
Storage tanks	Stainless steel 300 series Aluminum 6061 Mild steel (low pressure)	Monel Stainless steel 300 series Aluminum 6061
Lines and fittings	Nickel Monel Copper Aluminum silicon bronze Stainless steel 300 series Aluminum 2017, 2024, 5052, 6061 Mild steel (low pressure)	Monel Stainless steel 300 series Copper Aluminum 2017, 2024, 2050
Valve bodies	Stainless steel 300 series Bronze Aluminum silicon bronze	Monel Stainless steel 300 series Bronze
Valve seats	Copper Aluminum 1100 Stainless steel 300 series Aluminum silicon bronze Monel	Copper Aluminum 1100 Monel
Valve plugs	Stainless steel 300 series Monel	Stainless steel 300 series Monel
Valve packing	Tetrafluoroethylene polymer	Tetrafluoroethylene polymer
Valve bellows	Stainless steel 300 series Monel Bronze	Stainless steel 300 series Monel Bronze
Gaskets	Aluminum 1100 Lead Copper Tin Tetrafluoroethylene polymer Red rubber (5 psig) Neoprene (5 psig)	Aluminum 1100 Copper

[a]Contact equipment manufacturers for recommendations of other materials that are satisfactory for fluorine service.

must be exercised to avoid contamination of the gloves in the event of previous fluorine leakage since fluorine will react with atmospheric moisture, forming hydrofluoric acid which can result in severe and painful tissue destruction. When working with acid-contaminated systems, all fluorine should be removed from the system, and clean, chemical-resistant gloves (such as neoprene) should be worn while decontaminating, repairing, or modifying the system.

Similarly, loose leather coats (such as a welder's jacket) may be worn for routine cylinder handling and equipment operation for fire protection. Chemical-resistant outerwear will provide protection against hydrofluoric acid burns when working on depressurized systems where hydrogen fluoride formation may have occurred.

All chemical-resistant protective clothing should be thoroughly cleaned after use to eliminate acidic contamination, and gloves should be carefully examined for holes, cracks, etc. before each use. All protective clothing must be designed and used so that it can be shed easily and quickly if necessary. Safety glasses should be worn at all times.

Face shields, preferably made of transparent highly fluorinated polymers such as Aclar, should be worn whenever operators must approach equipment containing fluorine under pressure. Face shields made of any conventional materials afford limited, though valuable, protection against air-diluted blasts of fluorine.

Safe Handling and Storage of Cylinders

Personnel working with fluorine cylinders must be protected by use of a cylinder enclosure or barricade and remote-control valves, preferably ones operated by manual extension handles passing through the barricade. The main function of a barricade is to dissipate and prevent the breakthrough of any flame or flow of molten metal which, in case of equipment failure, could issue from any part of a system containing fluorine under pressure. Barricades of $\frac{1}{4}$-in. (6.35-mm) steel plate, brick, or concrete provide satisfactory protection for fluorine in cylinder quantities. Adequate ventilation of enclosed working spaces is essential. Installation in a fume hood is recommended for laboratory use of fluorine cylinders.

Fluorine cylinders should be securely supported while in use to prevent movement or straining of connections.

Store full or empty fluorine cylinders in a well-ventilated area protected from excessive heat, located away from organic or flammable materials, and chained in place to prevent falling. Valve protection caps and valve outlet caps must be securely attached to cylinders not in use.

Always protect fluorine cylinders from mechanical shock or abuse, and never heat them with a torch or heat lamp.

DISPOSAL

Before opening or refilling equipment that has contained fluorine, thoroughly purge the system with a dry inert gas (such as nitrogen) and evacuate it if possible. Small quantities of fluorine may be disposed of by slowly venting through either a solid, or preferably a liquid, caustic scrubber. Large quantities should be disposed of via a liquid caustic scrubber. The fluorine flow rate must be controlled to avoid overheating the disposal unit.

Soda lime, a sodium hydroxide–calcium oxide mixture, is the preferred solid scrubbing medium, although alumina or molecular sieves may be used. Activated charcoal and carbon should be avoided because fluorine often first absorbs on these materials at room temperature, followed by an explosive reaction. For liquid scrubbing of fluorine, a potassium hydroxide solution is the preferred medium. The potassium hydroxide concentration should be held to between 5 and 15 percent by weight in water for optimum performance.

Should a purged fluorine system require evacuation, a soda lime tower followed by a drier should be included in the vacuum system to pick up trace amounts of fluorine in order to protect it.

HANDLING LEAKS AND EMERGENCIES

Leak Detection

All areas containing fluorine under pressure should be inspected for leaks at suitable intervals, and any leaks discovered should be repaired at once after fluorine has been removed from the system. Ammonia vapor expelled from a squeeze bottle of ammonium hydroxide at suspected points of leakage may be used to detect leaks. Filter paper impregnated with potassium iodide solution provides a very sensitive means for detecting fluorine leaks (effective down to about 25 ppm), changing from light brown to black. In using it, hold the paper with metal tongs or forceps about 18 to 24 inches (46 to 61 cm) long.

Liquid Fluorine Spills

In the event of a large spillage of liquid fluorine, the contaminated area can be neutralized with sodium carbonate. The dry powder can be sprayed on the spill area from a fluidized system similar in principle to that of dry-chemical fire extinguishers. If major spillages occur in areas where the formation of hydrofluoric acid liquid and vapor pose no undue danger, water in the form of a fine mist or fog is recommended.

First Aid

It is unlikely that persons not injured or trapped would continue to inhale highly toxic concentrations of fluorine because of its strong odor and its irritation of eyes, nose, and mucous membranes. Liquid fluorine will severely burn the skin and eyes. Speed in removing the patient from the contaminated atmosphere or removing the vapor or liquid from the skin or eyes is essential. First aid must be started immediately in all cases of contact. All affected persons should be referred to a physician, no matter how slight the injury. The physician should be given a detailed account of the accident. The following first aid recommendations are given in *Effects of Exposure to Toxic Gases—First Aid and Medical Treatment,* in which additional information on the effects of exposure and recommended medical treatment may be found. [4] Also refer to the current Material Safety Data Sheet (MSDS) provided by the manufacturer.

Inhalation: Remove the victim to an uncontaminated atmosphere and administer 100% oxygen immediately. It has been reported helpful to even borderline cases to give 100% oxygen at half-hour intervals for three to four hours. Bronchospasm may be treated by emergency response technicians with a bronchodilator, such as albuterol, and an anticholinergic inhalant, such as Atrovent.

Eye Exposure: Immediately irrigate eyes with water for 15 minutes. Anesthetic eye drops used after an initial one minute water irrigation will reduce spasm of the eyelids and facilitate irrigation.

Skin Exposure: Victims of skin contact with fluorine should receive a drenching shower of water for 15 minutes with clothing removed as rapidly as possible while the victim is in the shower. For burns of less than 2 in.^{2} or burns involving the face, 2.5% calcium gluconate gel (2.5 g calcium gluconate USP to 100 ml KY-Jelly prepared by a pharmacist) should then be rubbed onto the affected area continuously for $1\frac{1}{2}$ hours or until further medical care is available. Rubber gloves should be worn by those initially applying the gel.

For larger burns or burns involving the finger tips and nail beds, the affected part should be placed in an iced Zephiran (1:750) or Hyamine 1622 (1:1500) bath for 4 hours. If a bath cannot be used, iced compresses with one of these agents should be used, changing the compress every two minutes. Solutions of Zephiran or Hyamine 1622 should not be used on the face.

METHODS OF SHIPMENT

Under the appropriate regulations, fluorine is authorized for shipment as follows:

By Rail: In cylinders (via freight and express to a maximum quantity of 6 lb in one outside container).

By Highway: In cylinders on trucks, and, under special DOT permit, in trailer-mounted tank transports.

By Water: In cylinders on cargo vessels only stowed on deck in well ventilated space away from organic materials. On barges of U.S. Coast Guard classes A, CA, and CB only.

By Air: Forbidden.

CONTAINERS

Fluorine is authorized for shipment in cylinders as a compressed gas under U.S. Department of Transportation and Transport Canada regulations, and as a liquefied low temperature gas in liquid nitrogen refrigerated tanks mounted on truck trailers by special permit of the DOT. [5] and [6]

Cylinders

Cylinders that meet TC/DOT specifications 3A1000, 3AA1000, and 3BN400 are authorized for fluorine service. The cylinders must not be equipped with pressure relief devices and must be fitted with valve protection caps. Commonly available sizes of cylinders contain 0.5 lb, 4.9 lb, and 6 lb (0.2 kg, 2.2 kg, and 2.7 kg) net weight of fluorine. See 49 CFR 173.302 (d). [5]

All cylinders authorized for fluorine must be requalified by hydrostatic retest every five years under current regulations.

Valve Outlet Connections

The standard valve outlet connection for fluorine in the United States and Canada is Connection CGA 679. [7]

Trailer-Mounted Tanks

Fluorine is authorized for shipment as a liquid at low temperature and atmospheric pressure in tanks mounted on motor vehicle trailers under exemption of the DOT. These tanks commonly have a 5000-pound (2268-kg) capacity.

Such trailer-mounted units consist of three concentric tanks. The liquid fluorine is contained in an inner baffled tank made of Monel metal or stainless steel. The inner tank is enclosed by a stainless steel tank filled with liquid nitrogen. The third, outer tank is made of carbon steel, and the annular space between it and the liquid nitrogen cooling jacket is filled with insulation and evacuated.

Vaporization of the liquid nitrogen (at −320.4°F or −195.8°C and 1 atm) in the specially constructed units keeps the liquid fluorine below its boiling point (of −306.8°F or −188.2°C at 1 atm), and radiation heat loss is minimized by the outer insulation. The ullage or vacant space in the liquid fluorine tank is brought to atmospheric pressure with helium to prevent subsequent in-leakage of moist air in case of valve or piping failure.

The liquid fluorine tank has two connections, one to a vapor space line and the other to a dip line in the liquid. Both lines are double-valved. There is no rupture disk or pressure relief valve for relieving excess pressure, but a pressure gauge and high pressure alarm are installed in the vapor line.

More than 300 gallons (1.4 m^3) of liquid nitrogen are held by the inner cooling jacket, which has fill, vent, and drain lines, each protected as required with a pressure relief valve and rupture disk. Gauges showing the level and pressure of the liquid nitrogen are on the control panel. The outer jacket of insulation is protected from excess pressure and has an electronic vacuum gauge attached.

Safety equipment on the trailer body includes a water fire extinguisher, a dry-chemical fire extinguisher, and a special tool chest containing hand tools, safety clothing, tubing, and breathing apparatus for road emergency use by drivers only (not for unloading). Full information on unloading the units is available from fluorine producers operating them.

Storage and Piping Equipment

Stationary installations for storing and piping fluorine must be designed with due regard for the reactivity of flourine and must be made of materials proven in the type of fluorine service planned, such as high temperature, high pressure, or cold liquefied compressed gas. Extreme care must be taken to keep all lines, fittings, tanks, and other equipment clean. Pipe and fittings for service in lines not to be dismantled should be welded and, in general, the number of non-welded lines should be kept to a minimum. Valves should have seatings of dissimilar metals in order to prevent galling, and packless stem seals should be used if possible.

METHODS OF MANUFACTURE

Fluorine is produced commercially by electrolytic decomposition of an anhydrous hydrofluoric acid, potassium bifluoride (KF•HF) solution. The melt formed (approximately KF•2HF) is solid at normal ambient temperatures and liquefies at approximately 160°F (71.1°C). The commercial electrolytic cell uses carbon anodes and a metal cathode with a method for separate collection of the gas release at each electrode. (These cells are heated when not operating or operating at low current rates to prevent their freezing, and are cooled when operated at normal rates to prevent overheating). When voltage is applied to the cell, current passing through the melt decomposes the hydrofluoric acid with the release of fluorine at the anode and hydrogen at the cathode. Continuous addition of hydrofluoric acid replaces the acid decomposed.

The fluorine thus produced is purified and then either used directly in a production process, compressed for cylinder filling, or condensed to a liquid for charging into refrigerated containers.

REFERENCES

[1] Voegtlin, C., and Hodge, H. C., *Pharmacology and Toxicology of Uranium Compounds,* National Nuclear Energy Series, Division VI, Vol. 1, pp. 1021–1042.

[2] *Threshold Limit Values and Biological Exposure Indices,* 1989–90 ed., American Conference of Governmental Industrial Hygienists, 6500 Glenway Avenue, Bldg. D-7, Cincinnati, OH 45211-4438.

[3] *Code of Federal Regulations,* Title 29 CFR Parts 1900–1910 (Labor), Superintendent of Documents, U.S. Government Printing Office, Washington, DC 20402.

[4] *Effects of Exposure to Toxic Gases—First Aid and Medical Treatment,* 3rd ed., Matheson Gas Products. Inc., Secaucas, NJ 07094. (1988)

[5] *Code of Federal Regulations,* Title 49 CFR Parts 100–199 (Transportation), Superintendent of Documents, U.S. Government Printing Office, Washington, DC 20402.

[6] *Transportation of Dangerous Goods Regulations,* Canadian Government Publishing Centre, Supply and Services Canada, Ottawa, Ontario, Canada K1A 0S9.

[7] CGA V-1, *American National, Canadian, and Compressed Gas Association Standard for Compressed Gas Cylinder Valve Outlet and Inlet Connections* (ANSI/CSA/CGA V-1), Compressed Gas Association, Inc., 1235 Jefferson Davis Highway, Arlington, VA 22202.

Fluorocarbons

IDENTIFYING INFORMATION

The term *fluorocarbons* is defined here as carbon compounds containing fluorine. If they contain only fluorine, they are sometimes referred to as FCs. The compounds may also contain chlorine, bromine, or hydrogen. Other descriptive names include chlorofluorocarbons (CFCs), fluorinated compounds, and halogenated hydrocarbons. Hydrochlorofluorocarbons (HCFCs) and hydrofluorocarbons (HFCs) are compounds in this family which contain hydrogen. Unless otherwise specified, the term *fluorocarbons* in this monograph applies to FCs, CFCs, HCFCs, and HFCs.

A standard numbering system was developed for identifying these products and has been published as ASHRAE 34, *Number Designation of Refrigerants,* by the American Society of Heating, Refrigerating and Air-Conditioning Engineers. [1] Letters are often used with the number to indicate the intended use of the material. For example, R-12 refers to dichlorodifluoromethane used as a refrigerant. When used as an aerosol propellant it might be called P-12.

The chemical names of these products conform to practice in the United States. Halogens are listed in alphabetical order as, bromo, chloro, fluoro, iodo. For mixtures, the higher-pressure material is listed first. An exception to this rule is found in some aerosol propellant mixtures where products not containing fluorine (hydrocarbons, chlorinated compounds, etc.) are listed last. Higher-boiling fluorocarbons are usually placed with other fluorocarbons in order of decreasing pressure.

The information included here is intended only to illustrate the properties and regulations governing the shipment of the fluorocarbon gases. Current regulations and the latest editions of publications concerned with shipping and handling these products should be used for details. It should be noted that some of these gases are nonflammable while others are flammable. Some information on naming and registry numbering for compressed gas fluorocarbons is given in Table 1. Under the *Canadian Transportation of Dangerous Goods Act,* these products are classified either as a nonflammable compressed gas or as a flammable compressed gas.

TABLE 1. FLUOROCARBON GASES.[a,b]

Fluorocarbon Number	Chemical symbol	Shipping Name[c]		CAS Registry Number	UN Number[d]	
		United States[e]	International		U.S.	Internatl.
12	CCl_2F_2	Dichlorodifluoromethane	Dichlorodifluoromethane	75-71-8	1028	1028
13	$CClF_3$	Chlorotrifluoromethane	Chlorotrifluoromethane	75-72-9	1022	1022
13B1	$CBrF_3$	Bromotrifluoromethane	Bromotrifluoromethane	75-63-8	1009	1009
14	CF_4	Compressed gas N.O.S.	Tetrafluoromethane	75-73-0	1956	1982
22	$CHClF_2$	Chlorodifluoromethane	Chlorodifluoromethane	75-45-6	1018	1018
23	CHF_3	Compressed gas N.O.S.	Trifluoromethane	75-46-7	1956	1984
115	$CClF_2CF_3$	Chloropentafluoroethane	Chloropentafluoroethane	76-15-3	1020	1020
116	CF_3CF_3	Compressed gas N.O.S.	Hexafluoroethane	76-16-4	1956	2193
142b	CH_3CClF_2	Chlorodifluoroethane	Chlorodifluoroethane	75-68-3	2517	2517
152a	CH_3CHF_2	Difluoroethane	Difluoroethane	75-37-6	1030	1030
500	CCl_2F_2/ CH_3CHF_2 (73.8/26.2 % by wt)	Refrigerant gas N.O.S.	Dichlorodifluoromethane and difluoroethane, azeotropic mixture	75-71-8 75-37-6	1078	2602
502	$CHClF_2$/ $CClF_2CF_3$ (48.8/51.2 % by wt)	Refrigerant gas N.O.S.	Chlorodifluoromethane and chloropentafluoro-ethane mixture	75-45-6 76-15-3	1078	1973
503	CHF_3/$CClF_3$ (40.1/59.9 % by wt)	Refrigerant gas N.O.S.	Chlorotrifluoromethane and Trifluoromethane azeotropic mixture	75-46-7 75-72-9	1078	2599

[a]The products listed in Table 1 are classified by the U.S. Department of Transportation as nonflammable gas and shipped with a nonflammable gas green label with the exception of fluorocarbon 142b and fluorocarbon 152a. These two products are classified as flammable gas and shipped with a flammable gas red label.

[b]Transport Canada Hazard Classification Numbers are: 2.1, flammable compressed gas; 2.2, nonflammable compressed gas; 2.3, corrosive; and 2.4, poisonous. The products in Table 1 are classified as 2.2 nonflammable compressed gas except for fluorocarbon 142b and fluorocarbon 152a, classified as 2.1 flammable compressed gas.

[c]Transport Canada shipping names are the same as international.

[d]The UN number for compressed gas N.O.S. (nonflammable) is 1956 and for refrigerant gas N.O.S. (nonflammable) is 1078.

[e]Historically, the prefix *mono* was used when only one atom of bromine, chlorine, or sometimes fluorine was present in a compound. At present, the use of *mono* is optional, and it is omitted from the shipping names in this table. Under the Canadian *Transportation of Dangerous Goods Act,* use of "mono" on documents is not acceptable.

PHYSICAL CONSTANTS [2], [3], [4], and [5]

(12, Dichlorodifluoromethane)

	U.S. Units	SI Units
Chemical formula	CCl_2F_2	CCl_2F_2
Molecular weight	120.93	120.93
Vapor pressure		
at 0°F (−17.8°C)	23.86 psia	164.51 kPa abs
at 70°F (21.1°C)	84.90 psia	585.36 kPa abs
at 105°F (40.6°C)	141.07 psia	972.64 kPa abs
at 115°F (46.1°C)	161.19 psia	1111.36 kPa abs
at 130°F (54.4°C)	195.20 psia	1345.85 kPa abs
Density of the gas		
at 70°F (21.1°C) and 1 atm	0.319 lb/ft^3	5.110 kg/m^3
Specific gravity of the gas		
at 70°F and 1 atm (air = 1)	4.32	4.32
Specific volume of the gas		
at 70°F (21.1°C) and 1 atm	3.1348 ft^3/lb	0.1957 m^3/kg
Density of the liquid		
at 0°F (−17.8°C)	90.57 lb/ft^3	1450.8 kg/m^3
at 70°F (21.1°C)	82.70 lb/ft^3	1324.7 kg/m^3
at 105°F (40.6°C)	78.14 lb/ft^3	1251.7 kg/m^3
at 115°F (46.1°C)	76.72 lb/ft^3	1228.9 kg/m^3
at 130°F (54.4°C)	74.45 lb/ft^3	1192.6 kg/m^3
Boiling point at 1 atm	−21.2°F	−29.79°C
Freezing point at 1 atm	−252°F	−158°C
Critical temperature	233.2°F	111.8°C
Critical pressure	598.3 psia	4125 kPa abs
Critical density	34.83 lb/ft^3	557.9 kg/m^3
Critical volume	0.0287 ft^3/lb	0.001792 m^3/kg
Latent heat of vaporization		
at boiling point	71.42 Btu/lb	166.01 kJ/kg
Specific heat of the gas		
at 86°F (30°C) and 1 atm		
C_p	0.147 Btu/(lb)(°F)	0.615 kJ/(kg)(°C)
C_v	0.129 Btu/(lb)(°F)	0.540 kJ/(kg)(°C)
Ratio of specific heats		
(C_p/C_v)	1.139	1.139
Solubility in water,		
weight percent,		
at 77°F (25°C) and 1 atm	0.028 percent	0.028 percent
Density of the liquid		
at 70°F (21.1°C)	11.06 lb/gal	1325.3 kg/m^3
Density of saturated vapor		
at 70°F (21.1°C)	2.07 lb/ft^3	33.16 kg/m^3
Specific heat of the liquid		
at 86°F (30°C)	0.240 Btu/(lb)(°F)	1.004 kJ/(kg)(°C)

PHYSICAL CONSTANTS [2], [4], and [5]

(13, Chlorotrifluoromethane)

	U.S. Units	SI Units
Chemical formula	$CClF_3$	$CClF_3$
Molecular weight	104.47	104.47
Vapor pressure		
at 0°F (−17.8°C)	177.55 psia	1224.2 kPa abs
at 70°F (21.1°C)	471.54 psia	3251.1 kPa abs
Density of the gas		
at 70°F (21.1°C) and 1 atm	0.273 lb/ft^3	4.37 kg/m^3
Specific gravity of the gas		
at 70°F and 1 atm (air = 1)	3.69	3.69
Specific volume of the gas		
at 70°F (21.1°C) and 1 atm	3.66 ft^3/lb	0.228 m^3/kg
Density of the liquid		
at 0°F (−17.8°C)	76.96 lb/ft^3	1232.8 kg/m^3
at 70°F (21.1°C)	56.47 lb/ft^3	904.6 kg/m^3
Boiling point at 1 atm	−114.6°F	−81.44°C
Freezing point at 1 atm	−294°F	−181°C
Critical temperature	83.9°F	28.83°C
Critical pressure	561.3 psia	3870 kPa abs
Critical volume	0.02772 ft^3/lb	0.001731 m^3/kg
Critical density	36.07 lb/ft^3	577.8 kg/m^3
Latent heat of vaporization		
at boiling point	64.39 Btu/lb	149.7 kJ/kg
Specific heat of the gas		
at −30°F (−34.4°C) and 1 atm		
C_p	0.138 Btu/(lb)(°F)	0.577 kJ/(kg)(°C)
C_v	0.118 Btu/(lb)(°F)	0.494 kJ/(kg)(°C)
Ratio of specific heats		
(C_p/C_v)	1.17	1.17
Solubility in water,		
weight percent,		
at 77°F (25°C) and 1 atm	0.009%	0.009%
Density of the liquid		
at 70°F (21.1°C)	7.55 lb/gal	904.7 kg/m^3
Density of saturated vapor		
at 70°F (21.1°C)	17.28 lb/ft^3	276.8 kg/m^3
Specific heat of the liquid		
at −30°F (−34.4°C)	0.248 Btu/(lb)(°F)	1.038 kJ/(kg)(°C)

PHYSICAL CONSTANTS [2], [6], [7], and [8]

(13B1, Bromotrifluoromethane)

	U.S. Units	SI Units
Chemical formula	$CBrF_3$	$CBrF_3$
Molecular weight	148.93	148.93
Vapor pressure		
at 0°F (−17.8°C)	71.12 psia	490.4 kPa abs
at 70°F (21.1°C)	213.6 psia	1472.7 kPa abs
at 105°F (40.6°C)	335.0 psia	2309.7 kPa abs
at 115°F (46.1°C)	377.6 psia	2603.5 kPa abs
at 130°F (54.4°C)	448.8 psia	3094.4 kPa abs
Density of the gas at 70°F (21.1°C) and 1 atm	0.391 lb/ft³	6.26 kg/m³
Specific gravity of the gas at 70°F and 1 atm (air = 1)	5.29	5.29
Specific volume of the gas at 70°F (21.1°C) and 1 atm	2.56 ft³/lb	0.160 m³/kg
Density of the liquid		
at 0°F (−17.8°C)	112.49 lb/ft³	1801.9 kg/m³
at 70°F (21.1°C)	97.80 lb/ft³	1566.6 kg/m³
at 105°F (40.6°C)	87.80 lb/ft³	1406.4 kg/m³
at 115°F (46.1°C)	84.22 lb/ft³	1349.1 kg/m³
at 130°F (54.4°C)	77.63 lb/ft³	1243.5 kg/m³
Boiling point at 1 atm	−71.92°F	−57.73°C
Freezing point at 1 atm	−270°F	−168°C
Critical temperature	152.6°F	67.00°C
Critical pressure	574.9 psia	3963.8 kPa abs
Critical density	46.50 lb/ft³	744.86 kg/m³
Critical volume	0.0215 ft³/lb	0.00134 m³/kg
Latent heat of vaporization at boiling point	51.10 Btu/lb	118.78 kJ/kg
Specific heat of the gas at 86°F (30°C) and 1 atm		
C_p	0.113 Btu/(lb)(°F)	0.473 kJ/(kg)(°C)
C_v	0.099 Btu/(lb)(°F)	0.414 kJ/(kg)(°C)
Ratio of specific heats (C_p/C_v)	1.143	1.143
Solubility in water, weight percent, at 77°F (25°C) and 1 atm	0.03%	0.03%
Density of the liquid at 70°F (21.1°C)	13.07 lb/gal	1566.1 kg/m3
Density of saturated vapor at 70°F (21.1°C)	7.44 lb/ft³	119.2 kg/m³
Specific heat of the liquid at 86°F (30°C)	0.215 Btu/(lb)(°F)	0.900 kJ/(kg)(°C)

PHYSICAL CONSTANTS [2] and [5]

(14, Tetrafluoromethane)

	U.S. Units	SI Units
Chemical formula	CF_4	CF_4
Molecular weight	88.01	88.01
Vapor pressure		
at −100°F (−73.3°C)	223.7 psia	1542 kPa abs
at −60°F (−51.1°C)	460.5 psia	3175 kPa abs
Density of the gas		
at 70°F (21.1°C) and 1 atm	0.228 lb/ft³	3.65 kg/m³
Specific gravity of the gas		
at 70°F (21.1°C) and 1 atm		
(air = 1)	3.09	3.09
Specific volume of the gas		
at 70°F (21.1°C) and 1 atm	4.38 ft³/lb	0.273 m³/kg
Density of the liquid		
at −150°F (−101.1°C)	91.90 lb/ft³	1472 kg/m³
at −100°F (−73.3°C)	78.58 lb/ft³	1259 kg/m³
Boiling point at 1 atm	−198.3°F	−127.9°C
Freezing point at 1 atm	−299°F	−184°C
Critical temperature	−50.2°F	−45.7°C
Critical pressure	543.2 psia	3745 kPa abs
Critical density	39.06 lb/ft³	625.7 kg/m³
Critical volume	0.0256 ft³/lb	0.00160 m³/kg
Latent heat of vaporization		
at boiling point	57.74 Btu/lb	134.2 kJ/kg
Specific heat of the gas		
at −100°F (−73.3°C) and 1 atm		
C_p	0.136 Btu/(lb)(°F)	0.569 kJ/(kg)(°C)
C_v	0.112 Btu/(lb)(°F)	0.469 kJ/(kg)(°C)
Ratio of specific heats		
(C_p/C_v)	1.217	1.217
Specific heat of the liquid		
at −100°F (−73.3°C)	0.286 Btu/(lb)(°F)	1.197 kJ/(kg)(°C)
Solubility in water,		
weight percent,		
at 77°F (25°C) and 1 atm	0.0015	0.0015

PHYSICAL CONSTANTS [2], [5], and [6]

(22, Chlorodifluoromethane)

	U.S. Units	SI Units
Chemical formula	$CHClF_2$	$CHClF_2$
Molecular weight	86.48	86.48
at 0°F (−17.8°C)	38.73 psia	267.0 kPa abs
at 70°F (21.1°C)	136.2 psia	939.1 kPa abs
at 105°F (40.6°C)	225.5 psia	1554.8 kPa abs
at 115°F (46.1°C)	257.5 psia	1775.4 kPa abs
at 130°F (54.4°C)	311.7 psia	2149.1 kPa abs
Density of the gas at 70°F (21.1°C) and 1 atm	0.227 lb/ft^3	3.636 kg/m^3
Specific gravity of the gas at 70°F and 1 atm (air = 1)	3.07	3.07
Specific volume of the gas at 70°F (21.1°C) and 1 atm	4.41 ft^3/lb	0.275 m3/kg
Density of the liquid		
at 0°F (−17.8°C)	83.62 lb/ft^3	1339.5 kg/m^3
at 70°F (21.1°C)	75.27 lb/ft^3	1205.7 kg/m^3
at 105°F (40.6°C)	70.30 lb/ft^3	1126.1 kg/m^3
at 115°F (46.1°C)	68.72 lb/ft^3	1100.8 kg/m^3
at 130°F (54.4°C)	66.17 lb/ft^3	1059.9 kg/m^3
Boiling point at 1 atm	−41.47°F	−40.82°C
Freezing point at 1 atm	−256°F	−160°C
Critical temperature	205.1°F	96.17°C
Critical pressure	723.4 psia	4987.6 kPa abs
Critical density	32.03 lb/ft^3	513.1 kg/m^3
Latent heat of vaporization at boiling point	100.38 Btu/lb	233.32 kJ/kg
Specific heat of the gas at 86°F (30°C) and 1 atm		
C_p	0.158 Btu/(lb)(°F)	0.661 kJ/(kg)(°C)
C_v	0.134 Btu/(lb)(°F)	0.561 kJ/(kg)(°C)
Ratio of specific heats (C_p/C_v)	1.18	1.18
Solubility in water, weight percent, at 77°F (25°C) and 1 atm	0.3%	0.3%
Density of the liquid at 70°F (21.1°C)	10.06 lb/gal	1205.4 kg/m3
Density of saturated vapor at 70°F (21.1°C)	2.485 lb/ft^3	39.81 kg/m^3
Critical volume	0.0312 ft^3/lb	0.00195 m^3/kg
Specific heat of the liquid at 86°F (30°C)	0.305 Btu/(lb)(°F)	1.276 kJ/(kg)(°C)

PHYSICAL CONSTANTS [2],* [5], [6], and [8]

(115, Chloropentafluoroethane)

	U.S. Units	SI Units
Chemical formula	$CClF_2CF_3$	$CClF_2CF_3$
Molecular weight	154.48	154.48
Vapor pressure		
at 0°F (−17.8°C)	36.94 psia	254.70 kPa abs
at 70°F (21.1°C)	119.1 psia	821.2 kPa abs
at 105°F (40.6°C)	195.4 psia	1347 kPa abs
at 115°F (46.1°C)	222.7 psia	1534 kPa abs
at 130°F (54.4°C)	268.8 psia	1853 kPa abs
Density of the gas		
at 70°F (21.1°C) and 1 atm	0.405 lb/ft^3	6.487 kg/m^3
Specific gravity of the gas		
at 70°F and 1 atm (air = 1)	5.48	5.48
Specific volume of the gas		
at 70°F (21.1°C) and 1 atm	2.47 ft^3/lb	0.154 m^3/kg
Density of the liquid		
at 0°F (−17.8°C)	91.78 lb/ft^3	1470 kg/m^3
at 70°F (21.1°C)	81.38 lb/ft^3	1304 kg/m^3
at 105°F (40.6°C)	74.97 lb/ft^3	1201 kg/m^3
at 115°F (46.1°C)	72.89 lb/ft^3	1168 kg/m^3
at 130°F (54.4°C)	69.44 lb/ft^3	1112 kg/m^3
Boiling point at 1 atm	−38.4°F	−39.1°C
Freezing point at 1 atm	−159°F	−106.1°C
Critical temperature	175.9°F	79.94°C
Critical pressure	458 psia	3157.8 kPa abs
Critical density	38.3 lb/ft^3	613.5 kg/m^3
Latent heat of vaporization		
at boiling point	53.37 Btu/lb	124.1 kJ/kg
Specific heat of the gas		
at 86°F (30°C) and 1 atm		
C_p	0.181 Btu/(lb)(°F)	0.757 kJ/(kg)(°C)
C_v	0.168 Btu/(lb)(°F)	0.703 kJ/(kg)(°C)
Ratio of specific heats		
(C_p/C_v)	1.08	1.08
Solubility in water,		
weight percent,		
at 77°F (25°C) and 1 atm	0.0058%	0.0058%
Density of the liquid		
at 70°F (21.1°C)	10.88 lb/gal	1303.7 kg/m^3
Density of saturated vapor		
at 70°F (21.1°C)	4.107 lb/ft^3	65.788 kg/m^3
Critical volume	0.0261 ft^3/lb	0.00163 m^3/kg
Specific heat of the liquid		
at 86°F (30°C)	0.263 Btu/(lb)(°F)	1.10 kJ/(kg)(°C)

*See the 1969 edition of reference [2]. Data not available in 1987 edition of reference [2] as of 5-13-87.

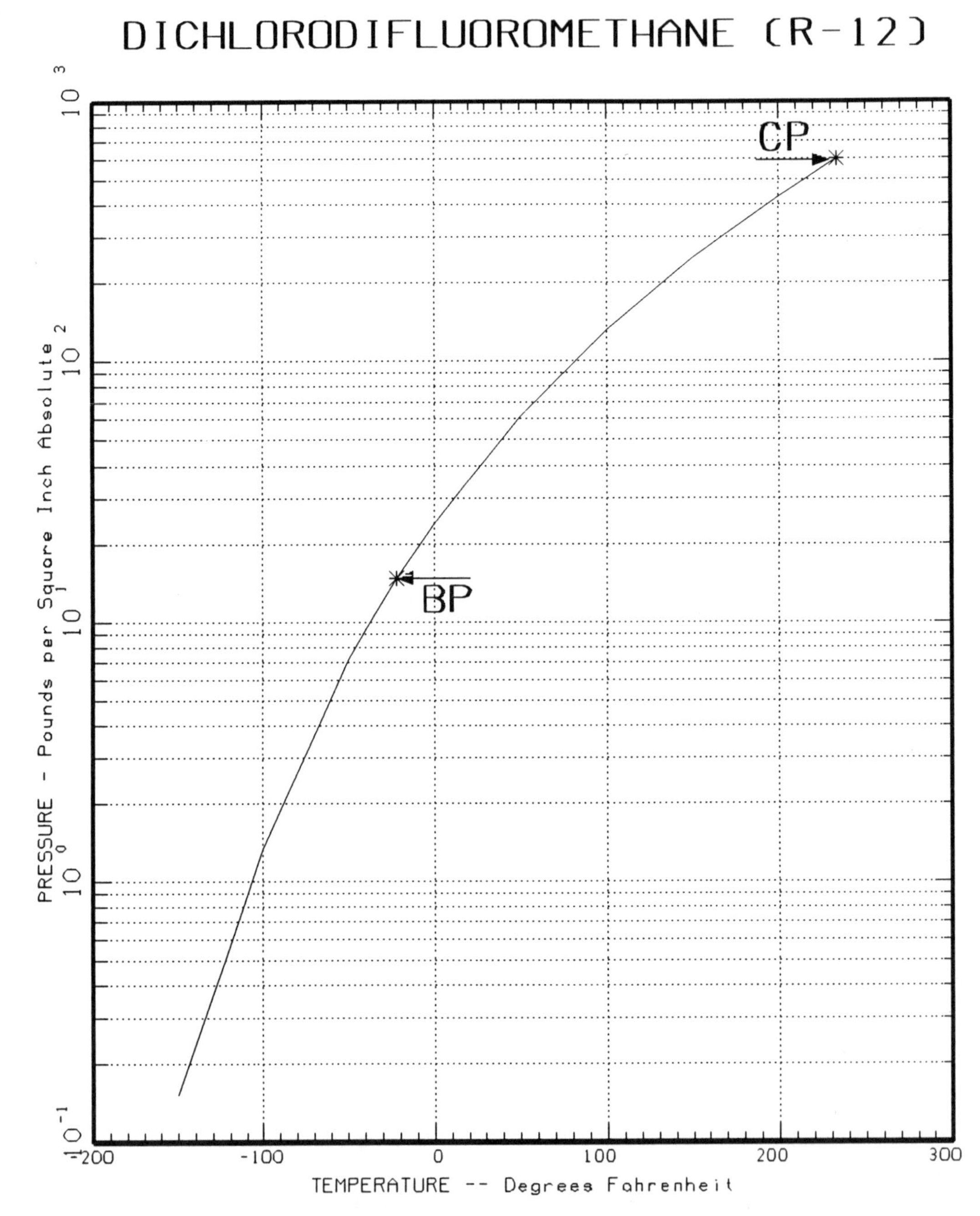

Fig. 1. Vapor Pressure Curve for Dichlorodifluoromethane (R-12).

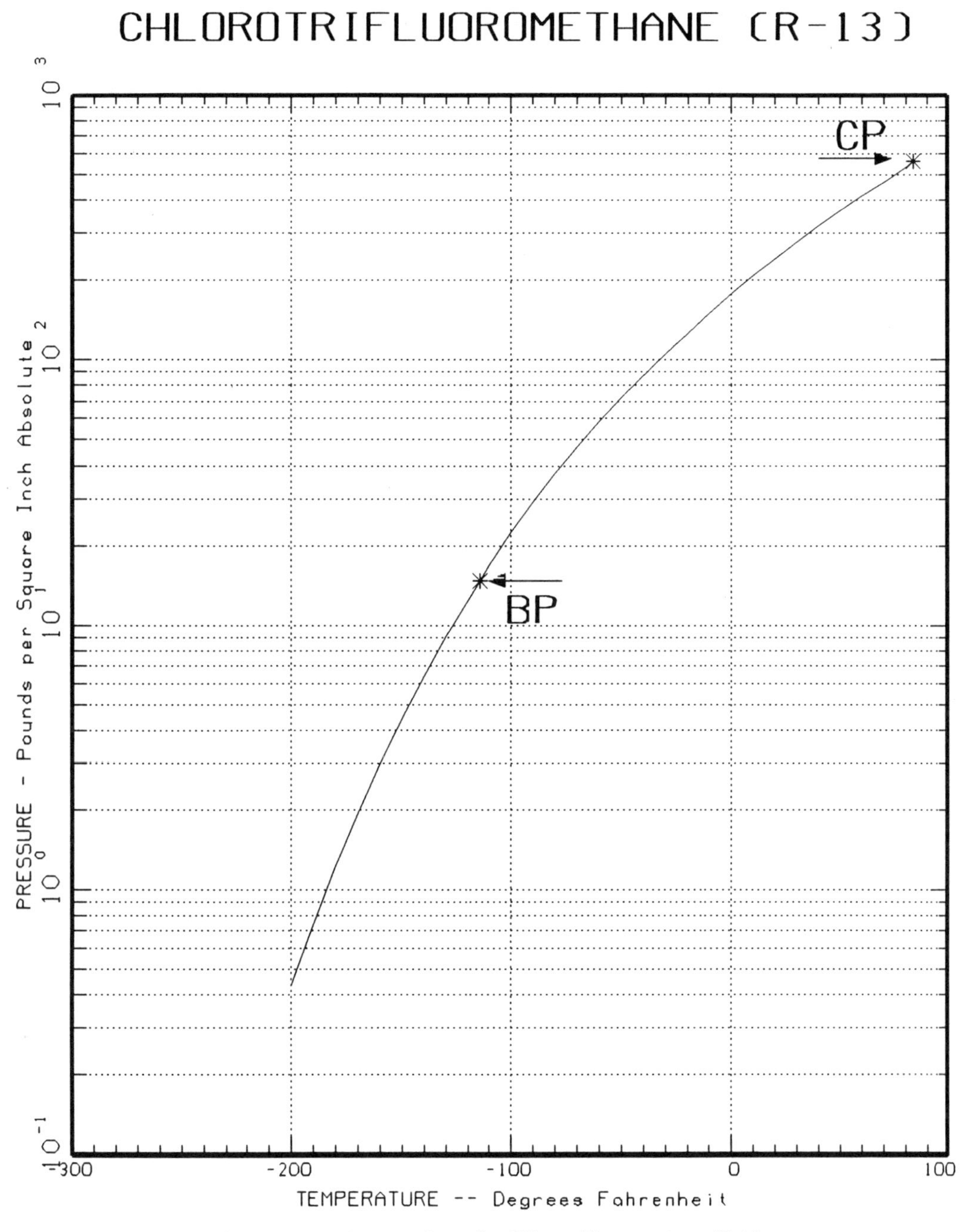

Fig. 2. Vapor Pressure Curve for Chlorotrifluoromethane (R-13).

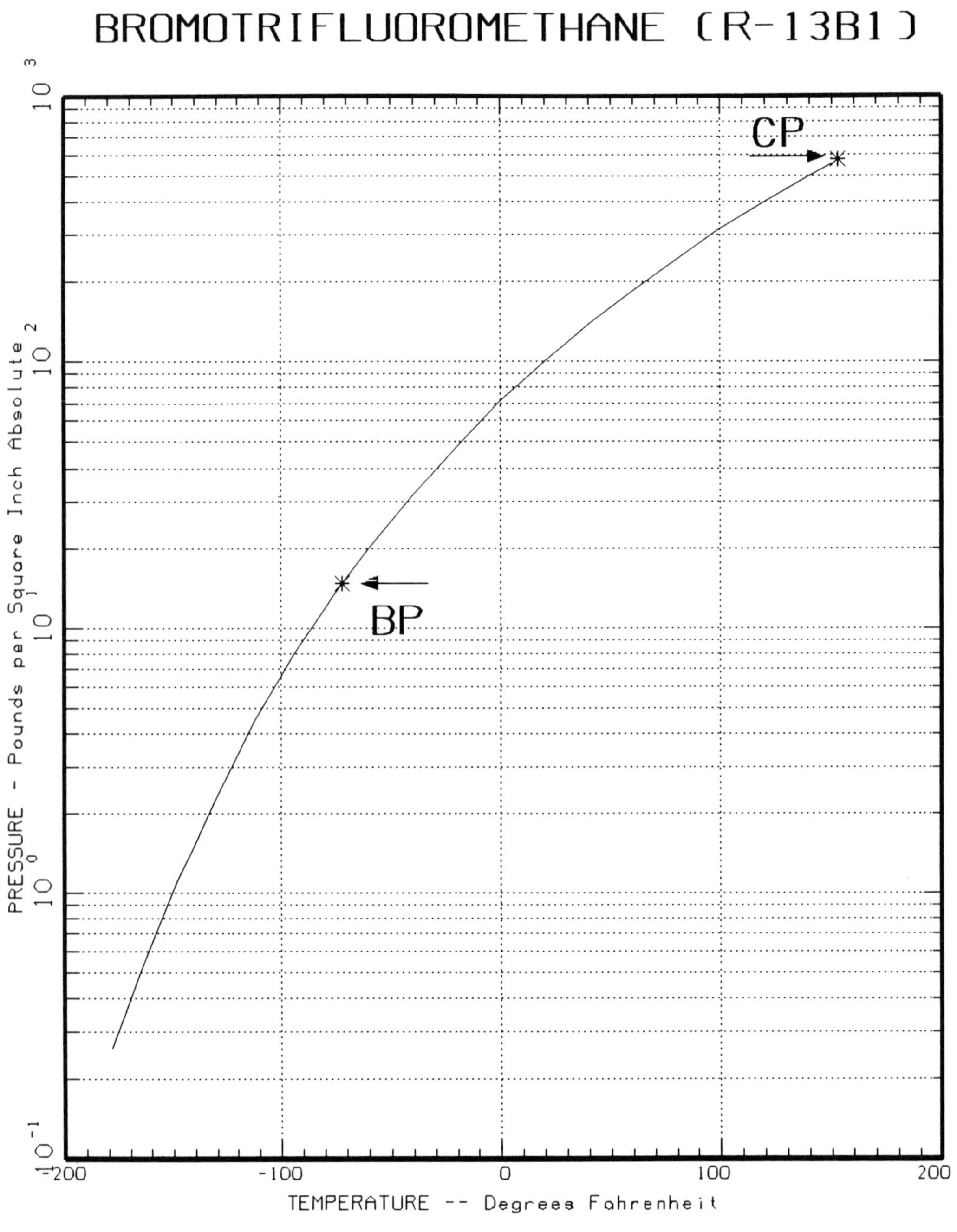

Fig. 3. Vapor Pressure Curve for Bromotrifluoromethane (R-13B1).

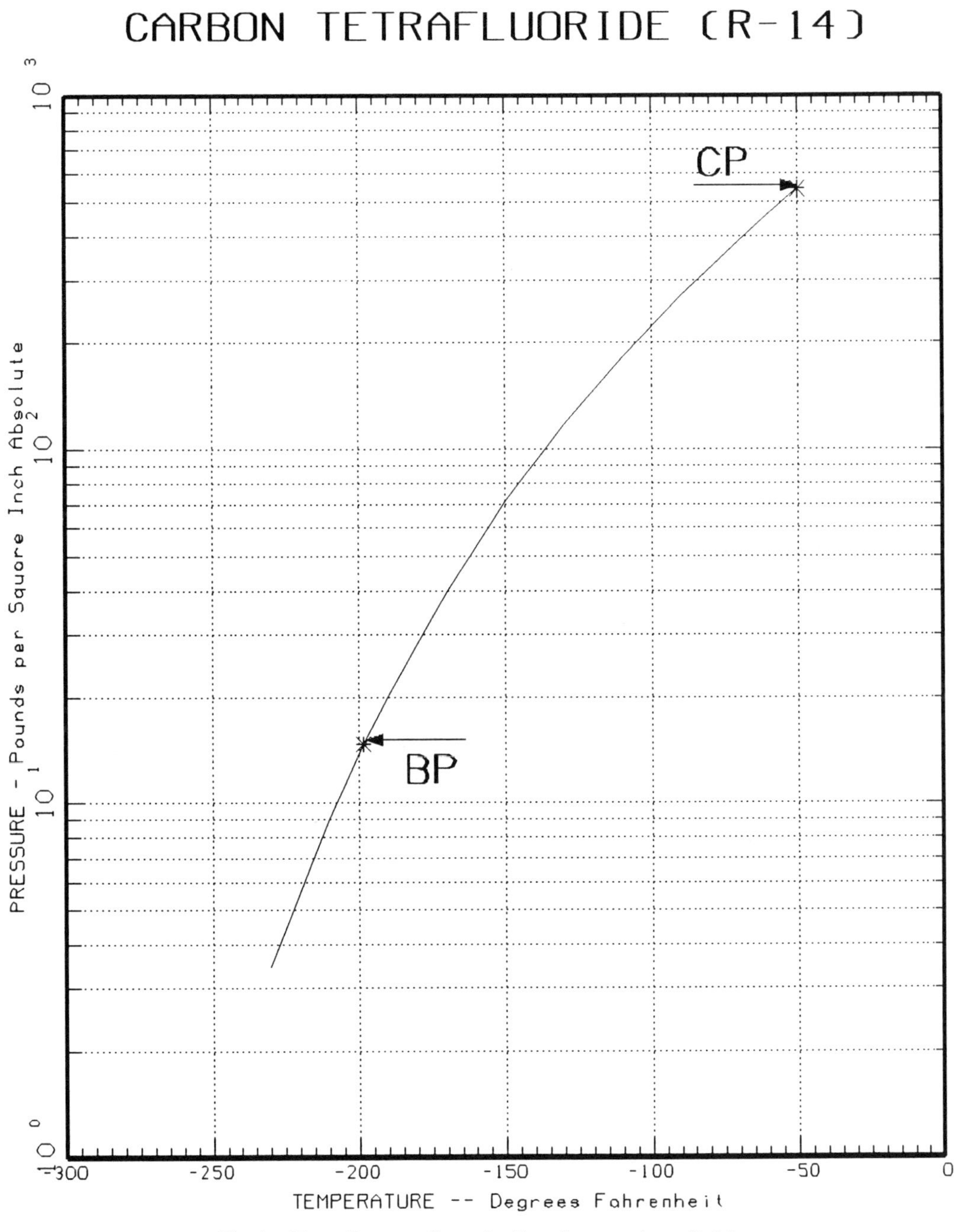

Fig. 4. Vapor Pressure Curve for Tetrafluoromethane (R-14).

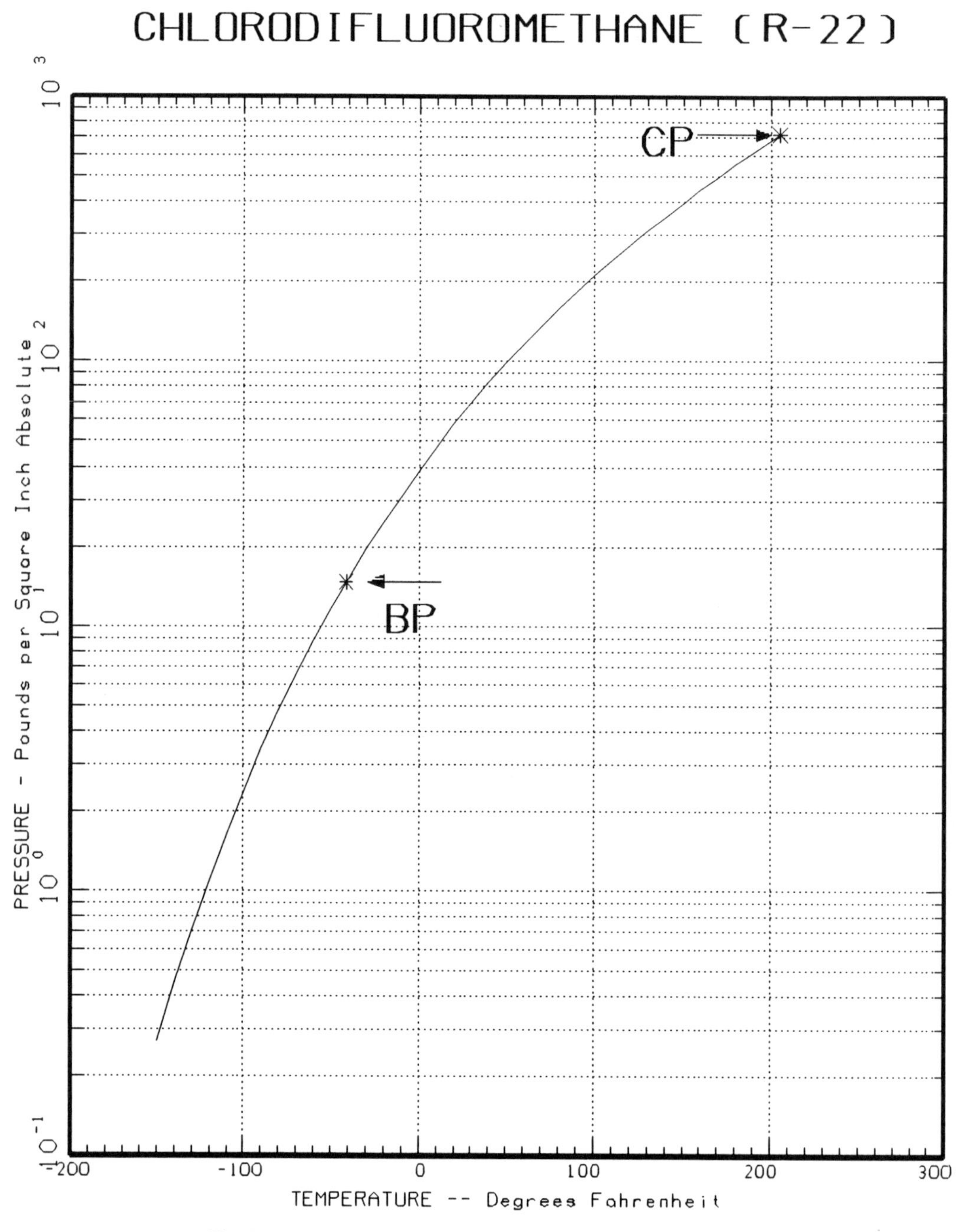

Fig. 5. Vapor Pressure Curve for Chlorodifluoromethane (R-22).

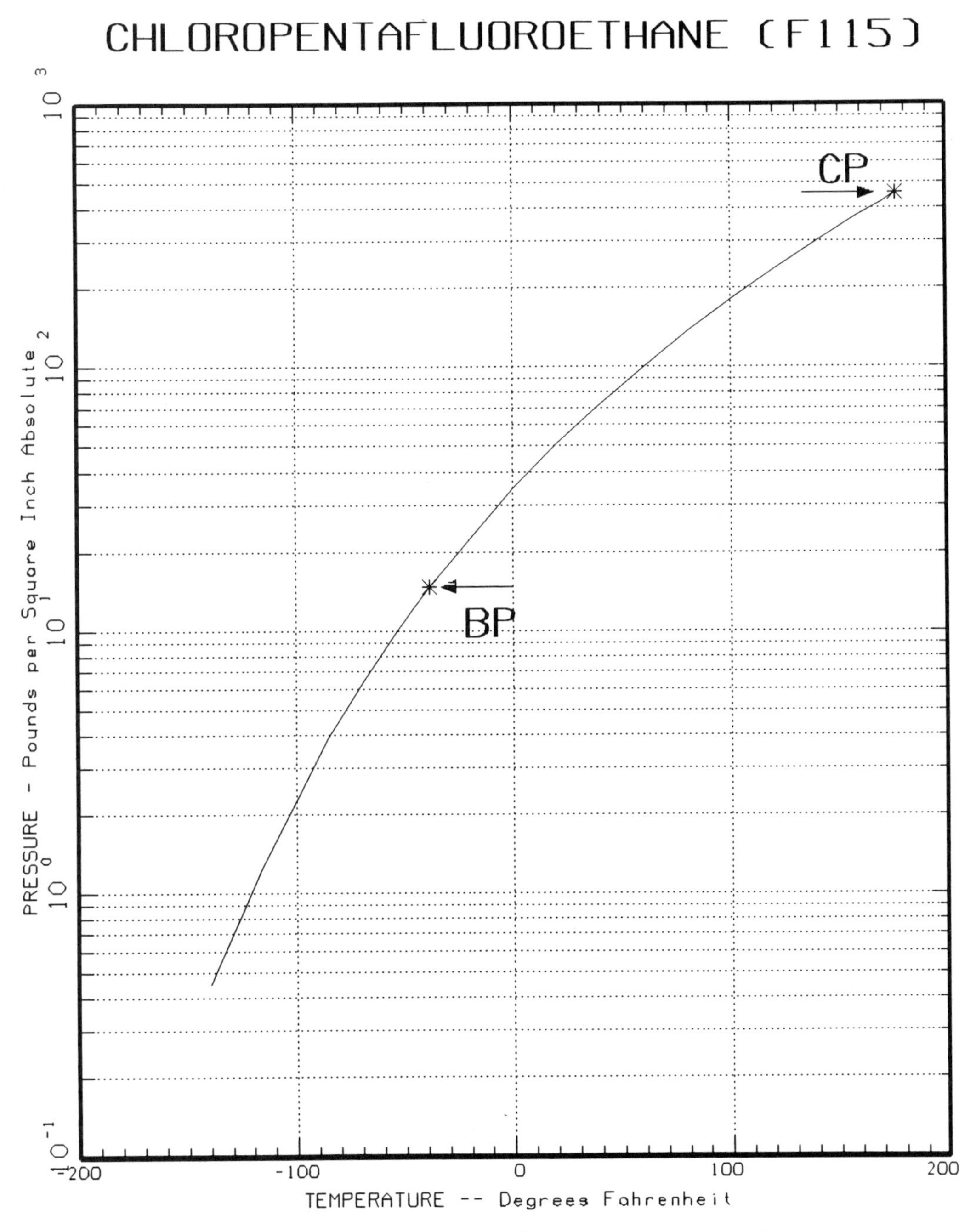

Fig. 6. Vapor Pressure Curve for Chloropentafluoroethane (R-115).

PHYSICAL CONSTANTS [2]

(142B, Chlorodifluoroethane)

	U.S. Units	SI Units
Chemical formula	$CClF_2CH_3$	$CClF_2CH_3$
Molecular weight	100.5	100.5
Vapor pressure		
at 0°F (−17.8°C)	10.61 psia	73.15 kPa abs
at 70°F (21.1°C)	43.52 psia	300.1 kPa abs
at 105°F (40.6°C)	77.30 psia	533.0 kPa abs
at 115°F (46.1°C)	89.93 psia	620.2 kPa abs
at 130°F (54.4°C)	111.75 psia	770.5 kPa abs
Density of the gas at 70°F (21.1°C) and 1 atm	0.268 lb/ft^3	4.29 kg/m^3
Specific gravity of the gas at 70°F and 1 atm (air = 1)	3.625	3.625
Specific volume of the gas at 70°F (21.1°C) and 1 atm	3.73 ft^3/lb	0.233 m^3/kg
Density of the liquid		
at 0°F (−17.8°C)	75.72 lb/ft^3	1212.9 kg/m^3
at 70°F (21.1°C)	69.93 lb/ft^3	1120.2 kg/m^3
at 105°F (40.6°C)	66.68 lb/ft^3	1068.1 kg/m^3
at 115°F (46.1°C)	65.68 lb/ft^3	1052.1 kg/m^3
at 130°F (54.4°C)	64.11 lb/ft^3	1026.9 kg/m^3
Boiling point at 1 atm	14.5°F	−9.72°C
Freezing point at 1 atm	−204°F	−131.11°C
Critical temperature	278.8°F	137.11°C
Critical pressure	598 psia	4123 kPa abs
Critical density	27.2 lb/ft^3	435.7 kg/m^3
Critical volume	0.0368 ft^3/lb	0.00230 m^3/kg
Latent heat of vaporization at boiling point	92.53 Btu/lb	215.1 kJ/kg
Specific heat of the gas at 86°F (30°C) and 1 atm		
C_p	0.200 Btu/(lb)(°F)	0.837 kJ/(kg)(°C)
C_v	0.180 Btu/(lb)(°F)	0.753 kJ/(kg)(°C)
Ratio of specific heats (C_p/C_v)	1.11	1.11
Solubility in water, weight percent, at 77°F (25°C) and 1 atm	slight	slight
Density of the liquid at 70°F (21.1°C)	9.35 lb/gal	1120.4 kg/m^3
Density of saturated vapor at 70°F (21.1°C)	0.849 lb/ft^3	13.60 kg/m^3
Specific heat of the liquid at 86°F (30°C)	0.286 Btu/(lb)(°F)	1.20 kJ/(kg)(°C)

PHYSICAL CONSTANTS [2]

(152A, Difluoroethane)

	U.S. Units	SI Units
Chemical formula	CHF_2CH_3	CHF_2CH_3
Molecular weight	66.05	66.05
Vapor pressure		
at 0°F (−17.8°C)	19.82 psia	136.7 kPa abs
at 70°F (21.1°C)	77.19 psia	532.2 kPa abs
at 105°F (40.6°C)	134.25 psia	925.6 kPa abs
at 115°F (46.1°C)	155.32 psia	1070.9 kPa abs
at 130°F (54.4°C)	191.46 psia	1320.1 kPa abs
Density of the gas		
at 70°F (21.1°C) and 1 atm	0.174 lb/ft^3	2.79 kg/m^3
Specific gravity of the gas		
at 70°F and 1 atm (air = 1)	0.236	0.236
Specific volume of the gas		
at 70°F (21.1°C) and 1 atm	5.74 ft^3/lb	0.358 m^3/kg
Density of the liquid		
at 0°F (−17.8°C)	62.19 lb/ft^3	996.2 kg/m^3
at 70°F (21.1°C)	56.75 lb/ft^3	909.0 kg/m^3
at 105°F (40.6°C)	53.54 lb/ft^3	857.6 kg/m^3
at 115°F (46.1°C)	52.54 lb/ft^3	841.6 kg/m^3
at 130°F (54.4°C)	50.94 lb/ft^3	816.0 kg/m^3
Boiling point at 1 atm	−12.99°F	−25.0°C
Freezing point at 1 atm	−178.6°F	−117.0°C
Critical temperature	236.3°F	113.50°C
Critical pressure	652 psia	4495 kPa abs
Critical density	22.79 lb/ft^3	365.1 kg/m^3
Critical volume	0.0439 ft^3/lb	0.00274 m^3/kg
Latent heat of vaporization		
at boiling point	136.99 Btu/lb	318.42 kJ/kg
Specific heat of the gas		
at 86°F (30°C) and 1 atm		
C_p	0.255 Btu/(lb)(°F)	1.07 kJ/(kg)(°C)
C_v	0.225 Btu/(lb)(°F)	0.942 kJ/(kg)(°C)
Ratio of specific heats		
(C_p/C_v)	1.133	1.133
Solubility in water,		
weight percent		
at 77°F (25°C) and 1 atm	slight	slight
Density of the liquid		
at 70°F (21.1°C)	7.59 lb/gal	909.48 kg/m^3
Density of saturated vapor		
at 70°F (21.1°C)	1.014 lb/ft^3	16.24 kg/m^3
Specific heat of the liquid		
at 86°F (30°C)	0.418 Btu/(lb)(°F)	1.75 kJ/(kg)(°C)

PHYSICAL CONSTANTS [2]

(500,(12)Dichlorodifluoromethane/(152A)Difluoroethane)

	U.S. Units	SI Units
Chemical formula	CCl_2F_2/CHF_2CH_3 (73.8:26.2)	CCl_2F_2/CHF_2CH_3 (73.8:26.2)
Molecular weight	99.31	99.31
Vapor pressure		
at 0°F (−17.8°C)	27.97 psia	192.8 kPa abs
at 70°F (21.1°C)	100.04 psia	689.7 kPa abs
at 105°F (40.6°C)	166.95 psia	1151.1 kPa abs
at 115°F (46.1°C)	190.98 psia	1316.8 kPa abs
at 130°F (54.4°C)	231.65 psia	1597.2 kPa abs
Density of the gas		
at 70°F (21.1°C) and 1 atm	0.262 lb/ft^3	4.20 kg/m^3
Specific gravity of the gas		
at 70°F and 1 atm (air = 1)	3.54	3.54
Specific volume of the gas		
at 70°F (21.1°C) and 1 atm	3.82 ft^3/lb	0.238 m^3/kg
Density of the liquid		
at 0°F (−17.8°C)	80.47 lb/ft^3	1289.0 kg/m^3
at 70°F (21.1°C)	73.00 lb/ft^3	1169.3 kg/m^3
at 105°F (40.6°C)	68.69 lb/ft^3	1100.3 kg/m^3
at 115°F (46.1°C)	67.27 lb/ft^3	1077.6 kg/m^3
at 130°F (54.4°C)	65.09 lb/ft^3	1042.6 kg/m^3
Boiling point at 1 atm	−28.31°F	−33.51°C
Freezing point at 1 atm	−254°F	−159°C
Critical temperature	222.0°F	105.6°C
Critical pressure	641.9 psia	4426 kPa abs
Critical density	31.10 lb/ft^3	498.2 kg/m^3
Critical volume	0.03215 ft^3/lb	0.00201 m^3/kg
Latent heat of vaporization		
at boiling point	86.39 Btu/lb	200.8 kJ/kg
Specific heat of the gas		
at 86°F (30°C) and 1 atm		
C_p	0.176 Btu/(lb)(°F)	0.737 kJ/(kg)(°C)
C_v	0.154 Btu/(lb)(°F)	0.645 kJ/(kg)(°C)
Ratio of specific heats (C_p/C_v)	1.14	1.14
Solubility in water,		
weight percent,		
at 77°F (25°C) and 1 atm	slight	slight
Density of the liquid		
at 70°F (21.1°C)	9.76 lb/gal	1169.5 kg/m
Density of saturated vapor		
at 70°F (21.1°C)	2.058 lb/ft^3	32.97 kg/m^3
Specific heat of the liquid		
at 86°F (30°C)	0.290 Btu/(lb)(°F)	1.214 kJ/(kg)(°C)

PHYSICAL CONSTANTS [2], [6], and [8]

(502,(22)Chlorodifluoromethane/(115)Chloropentafluoroethane)

	U.S. Units	SI Units
Chemical formula	$CHClF_2$/ $CClF_2CF_3$ (48.8:51.2)	$CHClF_2$/ $CClF_2CF_3$ (48.8:51.2)
Molecular weight	111.63	111.63
Vapor pressure		
at 0°F (−17.8°C)	45.78 psia	315.6 abs
at 70°F (21.1°C)	152.3 psia	1050 kPa abs
at 105°F (40.6°C)	246.4 psia	1699 kPa abs
at 115°F (46.1°C)	279.1 psia	1924 kPa abs
at 130°F (54.4°C)	335.5 psia	2313 kPa abs
Density of the gas at 70°F (21.1°C) and 1 atm	0.293 lb/ft^3	4.69 kg/m^3
Specific gravity of the gas at 70°F and 1 atm (air = 1)	3.97	3.97
Specific volume of the gas at 70°F (21.1°C) and 1 atm	3.41 ft^3/lb	0.213 m^3/kg
Density of the liquid		
at 0°F (−17.8°C)	86.68 lb/ft^3	1388 kg/m^3
at 70°F (21.1°C)	77.06 lb/ft^3	1234 kg/m^3
at 105°F (40.6°C)	71.02 lb/ft^3	1138 kg/m^3
at 115°F (46.1°C)	69.02 lb/ft^3	1106 kg/m^3
at 130°F (54.4°C)	65.66 lb/ft^3	1052 kg/m^3
Boiling point at 1 atm	−49.75°F	−45.42°C
Critical temperature	179.9°F	82.17°C
Critical pressure	591.0 psia	4075 kPa abs
Critical density	35.0 lb/ft^3	561 kg/m^3
Critical volume	0.0286 ft^3/lb	0.00179 m^3/kg
Latent heat of vaporization at boiling point	74.20 Btu/lb	172.4 kJ/kg
Specific heat of the gas at 86°F (30°C) and 1 atm		
C_p	0.168 Btu/(lb)(°F)	0.703 kJ/(kg)(°C)
C_v	0.147 Btu/(lb)(°F)	0.615 kJ/(kg)(°C)
Ratio of specific heats (C_p/C_v)	1.14	1.14
Solubility in water, weight percent, at 77°F (25°C) and 1 atm	slight	slight
Density of the liquid at 70°F (21.1°C)	10.30 lb/gal	1234.2 kg/m^3
Density of saturated vapor at 70°F (21.1°C)	3.736 lb/ft^3	59.84 kg/m^3
Specific heat of the liquid at 86°F (30°C)	0.304 Btu/(lb)(°F)	1.272 kJ/(kg)(°C)

PHYSICAL CONSTANTS [2], [7], and [8]

(503,(23)Trifluoromethane/(13)Chlorotrifluoromethane)

	U.S. Units	SI Units
Chemical formula	$CHF_3/CClF_3$ (40.1:59.9)	$CHF_3/CClF_3$ (40.1:59.9)
Molecular weight	87.5	87.5
Vapor pressure at 0°F (−17.8°C)	245.2 psia	1690.6 kPa abs
Density of the gas at 70°F (21.1°C) and 1 atm	0.227 lb/ft^3	3.636 kg/m^3
Specific gravity of the gas at 70°F (21.1°C) and 1 atm (air = 1)	3.07	3.07
Specific volume of the gas at 70°F (21.1°C) and 1 atm	4.40 ft^3/lb	0.275 m^3/kg
Density of the liquid at 0°F (−17.8°C)	72.32 lb/ft^3	1158
Boiling point at 1 atm	−126.15°F	−87.86°C
Critical temperature	67.0°F	19.4°C
Critical pressure	631.9 psia	4357 kPa abs
Critical density	35.21 lb/ft^3	564.0 kg/m^3
Critical volume	0.0284 ft^3/lb	0.00177 m^3/kg
Latent heat of vaporization at boiling point	77.17 Btu/lb	179.4 kJ/kg
Specific heat of the gas at −30°F (−34.4°C) and 1 atm		
C_p	0.143 Btu/(lb)(°F)	0.598 kJ/(kg)(°C)
C_v	0.121 Btu/(lb)(°F)	0.494 kJ/(kg)(°C)
Ratio of specific heats (C_p/C_v)	1.21	1.21
Solubility in water, weight percent, at 77°F (25°C) and 1 atm	slight	slight
Density of the liquid at 60°F (15.5°C)	6.86 lb/gal	822 kg/m^3
Specific heat of the liquid at −30°F (−34.4°C)	0.290 Btu/(lb)(°F)	1.213 kJ/(kg)(°C)

DESCRIPTION

Strictly speaking, fluorocarbon compounds contain only the elements carbon, fluorine, and sometimes hydrogen. However, in industrial applications such as refrigerants and aerosol propellants, the term fluorocarbon has been used to include compounds containing chlorine and/or bromine atoms as well. These industrial products have somewhat similar chemical and physical properties. Their relatively inert character and wide range of vapor pressures and boiling points make them especially well suited as refrigerants in a variety of applications, blowing agents for plastic foams, and aerosols.

Some of the fluorocarbon compounds are listed by name in the regulations of the U.S. Department of Transportation and Transport Canada, and others are shipped under DOT exemption. [9] and [10] At one time, the prefix *mono* was used in DOT listings to show the presence of one atom of the indicated element, but it is omitted in modern terminology. The prefix *mono* is not acceptable on documents in Canada.

FCs and CFCs are nonflammable in all concentrations and are low in toxicity. Some HCFCs and HFCs are flammable and others are not. They are colorless as liquids and freeze to white solids. The fluorocarbons are odorless in concentrations of less than 20 percent by volume in air, but some have a faint and ethereal odor in higher concentrations.

Chemically, fluorocarbons are analogs of hydrocarbons in which all or nearly all of the hydrogen has been replaced by fluorine and/or bromine or chlorine. The presence of fluorine atoms in the molecule accounts for their pronounced stability. They are more dense than corresponding hydrocarbons, have lower refractive indices, lower solubilities, and lower surface tension. The viscosity is comparable to that of hydrocarbons. They also have relatively high dielectric strength.

Fluorinated organic compounds are unusually stable. In general, resistance toward thermal decomposition is high but varies with each product. When decomposition does occur, toxic products such as hydrofluoric and hydrochloric acids may be formed, especially in intimate contact with flames. The toxic products are very irritating and usually give adequate warning of their presence even in very low concentrations in air. Resistance toward hydrolysis is also high, especially in neutral or acid solutions. However, some products containing hydrogen, such as HCFC 22, are quite susceptible to alkaline hydrolysis.

GRADES AVAILABLE

Fluorocarbons are available for commercial and industrial use in various grades with essentially the same composition from one producer to another. Purities of most fluorocarbon products are a minimum of 99 mole percent. For special applications, very high purity grades are available.

USES

The fluorocarbons covered in this monograph are widely used as refrigerants, polymer intermediates, and blowing agents in the manufacture of polymerized foams used in insulation and comfort cushioning. They are also used for making packaging foams as well as aerosol propellants for products applied in foam or spray form, and fire-extinguishing agents. Special mixtures of two or more fluorocarbons, or fluorocarbons and hydrocarbons, are often used to provide desired special properties in particular refrigeration or aerosol propellant applications.

As a result of recently recognized deleterious effects on the ozone layer of the earth's atmosphere from emissions of certain halogenated hydrocarbons, the production and use of certain of these compounds has been curtailed by international agreement and a search is under way to develop suitable substitute materials.

PHYSIOLOGICAL EFFECTS

The fluorocarbons used in industry generally have low levels of toxicity, and hazards related to their use are minimal. These compounds have been extensively studied by many investigators under a variety of conditions.

Acute Exposure

The Underwriters Laboratories studied the effect of acute exposure (rather large concentrations for short periods of time—intended to represent conditions if a major leak occurred in a large commercial refrigeration or air conditioning system. [11] They found some variation among the different compounds but in general little effect under practical conditions. Other investigators confirmed this result. However, the Underwriters Laboratories' tests were not completely definitive, and possible hazards from breathing fluorocarbons have been recognized.

With any inert gas, high concentrations may lead to asphyxiation if the concentration of oxygen is sufficiently reduced. The fluorocarbons are heavy gases and tend to collect in low places or where ventilation is not adequate. The resulting decrease in air concentration can be fatal.

Some volatile organic compounds may sensitize the heart, causing irregular rhythm (arrhythmia) and, in severe cases, cardiac arrest. Chlorinated products and, to a lesser degree, some chlorofluoro compounds may have this effect. Adrenalin produced during physical or emotional stress enhances the effect. If heart trouble from exposure to organic compounds is suspected, epinephrin or other heart stimulants should not be administered and medical attention should be obtained.

Chronic Exposure

Chronic exposure is defined as breathing relatively small concentrations of a gas for long periods of time. Fluorocarbon products have been studied under these conditions by several different investigators. The results have been evaluated by the American Conference of Governmental Industrial Hygienists (ACGIH). They have established threshold limit values (TLVs) to show relative hazards. [12]

The TLV represents the concentration under which it is believed that nearly all workers may be repeatedly exposed day after day without adverse effect, that is for 8 hours/day, 40 hours/week. Apart from carbon dioxide (threshold limit value 5000 ppm) and simple asphyxiants such as nitrogen, the highest threshold limit value (least toxic) is 1000 ppm. A threshold limit value of 1000 ppm has been assigned for some fluorocarbons, and comparative data suggest similar values are reasonable for the other fluorocarbons discussed in this section. In many cases, the threshold limit values as established by the ACGIH have been adopted by the U.S. Occupational Safety and Health Administration (OSHA) as workplace exposure limits. See 29 CFR 1910.1000. [13]

Due to their low boiling points, the fluorocarbons evaporate very quickly at ambient temperature, minimizing dermal, eye, and ingestion toxicity. The rapid evaporation and resultant chilling can cause tissue freezing and frostbite when contact is made with the boiling liquid. Should liquid splash into the eyes, wash them thoroughly with water for at least 15 minutes and call an eye specialist at once.

The critical mode of entry into the body is by inhalation. These products generally show inhalation effects similar to anesthetics, causing central nervous system (CNS) depression and some increase in activity with an initial feeling of intoxication and euphoria (psychological effects). Human exposure has shown that the onset of psychomotor effects becomes statistically detectable with fluorocarbon 12 at 1 percent (10 000 ppm) in a 2.5-hour exposure, that is, at ten times the threshold limit value. Under conditions of progressively greater exposure, there occurs loss of coordination, loss of consciousness, and eventually death. No adverse effects have been observed during or after long exposures at the threshold limit value.

More complete discussions of fluorocarbon toxicity can be found in the *Handbook of Aerosol Technology, Patty's Industrial Hygiene and Toxicology,* and *Effects of Exposure to Toxic Gases—First Aid and Medical Treatment.* [14], [15], and [16]

MATERIALS OF CONSTRUCTION

The fluorocarbons are generally compatible with most of the common metals except at high temperatures. At elevated temperatures, the following metals resist fluorocarbon corrosion (and are named in decreasing order of their corrosive resistance): Inconel, stainless steel, nickel, steel, and bronze. Water or water vapor in fluorocarbon systems will corrode magnesium alloys or aluminum containing over 2 percent magnesium. These metals are not recommended for use with fluorocarbon systems in which water may be present.

Systems using fluorocarbons as refrigerants should be dry in order to prevent the possibility of malfunctioning from the icing of components such as regulating valves, bellows, diaphragms, hermetically sealed coils, and so forth. Most fluorocarbon compounds can be used with elastomeric and plastic packing and gasketing materials, but there are some exceptions. Consult manufacturers for specific information or refer to the *ASHRAE Handbook and Product Directory,* "Fundamentals" volume. [8]

SAFE STORAGE, HANDLING, AND USE

All the precautions necessary for the safe handling of any nonflammable or flammable gas must be observed with the fluorocarbons. For further information on guidelines for safe storage, handling, and use, see Chapter 5. For flammable fluorocarbons, it should be noted that adherence to pertinent electrical standards is necessary. Personnel should not weld, solder, or braze in atmospheres containing fluorocarbons of any type.

DISPOSAL

If disposal is necessary, follow all federal, state, provincial, and local regulations. Because of their stability, and in the interest of conservation, fluorocarbons can often be reclaimed and processed for reuse. Contact your supplier for details on the proper means by which this can be done.

HANDLING LEAKS AND EMERGENCIES

Significant liquid leaks in fluorocarbon systems may be detected visually. As the material escapes, moisture in the air surrounding the leak condenses and then freezes around the leak due to the refrigerating effect of the vaporizing fluorocarbons. The frost thus formed is readily apparent.

Smaller leaks may be located through the use of:

(1) A solution of liquid detergent in water, applied directly to the area being tested. The formation of bubbles indicates a leak.

(2) Electronic leak detectors, which are capable of sensitivities far greater than the other methods—often in terms of fractions of an ounce of fluorocarbon per year. When the probe of the instrument is placed near a leak, positive identification of the leak is indicated by a flashing light, by meter deflection, or by audible means.

(3) A halide torch equipped with a hose through which air is sucked to the flame. The flame will change color when it comes in contact with a leak. This method should not be used with the few fluorocarbon products that are flammable.

It should be noted that the vapors of these fluorocarbons are all much heavier than air and in the absence of good ventilation will tend to collect in low areas. In handling major leaks, personnel should wear appropriate self-contained breathing apparatus and personal protective equipment. Furthermore, the use of the "buddy system" is recommended, especially when working in confined areas.

As noted previously, the vapors will undergo decomposition when drawn through a flame or if in contact with very hot surfaces. The products of decomposition include hydrogen fluoride and hydrogen chloride. The halogen acids are both toxic and intensely irritating to the nose and throat. The irritating action of these decomposition products is readily noticeable before hazardous levels are reached. If such a situation develops, the affected areas should be vacated, the heat source and leak eliminated, and the area well ventilated before resuming work. Also, it should be remembered that some HCFCs are flammable.

Monitoring Concentrations in Air

Instruments and analytical methods are available for monitoring the concentration of fluorocarbon gases in air. Consult your supplier for specific information.

First Aid

Accidental exposures to concentrations higher than the TLV should be treated by prompt removal to fresh air. Severe exposures requiring medical attention should not be treated with stimulants or adrenalin, since high concentrations of these fluorocarbons may result in a sensitization of the heart to adrenalin (a relatively common effect of many volatile organic compounds).

Fluorocarbons in the liquefied gas form which come in contact with the skin can cause severe freezing or frostbite because of their low boiling points. In cases of frostbite, place the frostbitten area in warm water, about 108°F (42°C). If warm water is not available, or it is impractical to use, wrap the affected part gently in blankets. Let the circulation reestablish itself naturally. Encourage the victim to exercise the affected part while it is being warmed. Consult a physician.

Should liquid splash into the eyes, wash them thoroughly with water for at least 15 minutes and call an eye specialist at once.

METHODS OF SHIPMENT

Fluorocarbons are authorized for shipment by rail, highway, water, and air in interstate commerce in Title 49 of the U.S. *Code of Federal Regulations,* Parts 100–199. [9] Regulations pertaining to shipment of fluorocarbons in Canada are found in *Transportation of Dangerous Goods Regulations.* [10] Due to frequency of change in these regulations, it is recommended that current editions be consulted. Tables 2 and 3 in the following section list the types of containers and maximum filling densities for fluorocarbons that are authorized in these regulations as of October 1988. Pressure relief devices and proper CGA valve outlet connections are shown in Table 4. Table 5 is a listing describing each type of pressure relief device with notations on conditions for how they are to be used as referenced in Table 4.

CONTAINERS

Those fluorocarbons which are liquefied compressed gases are stored and shipped under their own vapor pressures in cylinders, portable tanks, truck cargo tanks, TMU tanks, and single-unit tank cars as shown in Table 2.

Filling Limits

Table 3 lists maximum permitted filling densities (as given in current regulations) for each of the fluorocarbons. Filling density is expressed as percentage of the water capacity of the container, in pounds. For these calculations, the weight of water at 60°F (15.6°C) in air is 8.32828 lb/gal (997.94781 kg/m^3), 1 gallon (0.0038 m^3) occupying 231 $in.^3$ or 0.1336 ft^3. Table 4 lists authorized pressure relief devices for use on fluorocarbon cylinders as given in CGA S-1.1, *Pressure Relief*

TABLE 2. AUTHORIZED SPECIFICATION CONTAINERS FOR SHIPPING FLUOROCARBONS.

Fluorocarbon Number	Cylinders	Railcars (TMU, Single-Unit)	Motor Vehicle Tanks (Portable, Cargo)
12	3A225, 3AA225	106A500-X	DOT-51
	3B225, 4A225,	110A500-W	MC-330, MC-331
	4B225, 4BA225,	112A340-W	
	4BW225, 4B240ET,	114A340-W	
	4E225, 3AL225,	105A300-W	
	9, 39, 41, 3E1800		
13	3A1800, 3AA1800,		DOT-51
	3AL1800	105A300-W	MC-330, MC-331
	3, 3E1800, 39	114A340-W	
13B1	3A400, 3AA400,	110A800-W	DOT-51
	3B400,		MC-330, MC-331
	4A400, 4AA480,		
	4B400,		
	4BA400, 4BW400,		
	3E1800, 39, 3AL400		
22	3A240, 3AA240,	106A500-X	DOT-51
	3B240,	110A500-W	MC-330, MC-331
	4B240, 4BA240,	105A300-W	
	4BW240, 4B240ET,	112A400-W	
	4E240, 3AL240,	114A340-W	
	39, 41, 3E1800		
115	3A225, 3AA225,	106A500-X	DOT-51
	3B225, 3AL225,	110A500-W	MC-330, MC-331
	4A225, 4B225,	105A300-W	
	4BA225, 4BW225,	114A340-W	
	39, 3E1800		
142b	3A150, 3AA150,	106A500-X	DOT-51
	3B150,	110A500-W	MC-330, MC-331
	4B150, 4BA225,	105A100-W	
	4BW225,	114T340-W	
	39, 3E1800	114J340-W	
	3AL150		
152a	3A150, 3AA150,	106A500-X	DOT-51
	3B150,	110A500-W	MC-330, MC-331
	4B150, 4BA225,	105A300-W	
	4BW225,	112T400-W	
	3E1800, 3AL150	112J400-W	
		114T340-W	
		114J340-W	
500	3A240, 3AA240,	106A500-X	DOT-51
	3B240,	110A500-W	MC-330, MC-331
	4A240, 4B240	112A340-W	
	4BA240, 4BW240,	114A340-W	
	9, 39, 3E1800	105A300-W	
	4E240		

TABLE 3. MAXIMUM FILLING DENSITIES PERMITTED FOR AUTHORIZED SHIPPING CONTAINERS CONTAINING FLUOROCARBONS (PERCENT WATER CAPACITY BY WEIGHT).

		Tank Car[a] (TMU, Single-Unit)		Motor Vehicle Tanks[b] (Portable, Cargo)	
Fluorocarbon Number	Cylinders Filling Density (pecent)	Specification	Filling Density (percent)	Filling Density (percent)	Minimum Design Pressure (psig)
12	119	106A500-X	119	119	150
		110A500-W	119		
		105A300-W	125		
		112A340-W	123		
		114A340-W	123		
13	100	—	—	—	—
13B1	124	110A800-W	124	133	365
22	105	106A500-X	105	105	250
		110A500-W	105		
		105A300-W	110		
		112A400-W	108		
115	110	—	—	—	—
142b	100	106A500-X	100	100	100
		110A500-X			
		105A100-W			
		114T340-W			
		114J340-W			
152a	79	106A500-X	79	79	150
		110A500-W	79		
		112T400-W	79		
		112J400-W	79		
		114T340-W	79		
		114J340-W	79		
		105A300-W	84		
500	not liquid full at 130°F	112A340-W	[c]	[d]	250
		114A340-W			
		106A500-X			
		110A500-W			
		105A300-W			

[a]The gas pressure at 105°F (40.5°C) in any insulated tank car tank of the DOT-105A class, at 115°F (46.1°C) in any uninsulated tank car tank of the DOT-112A-W and 114A-W class, or at 130°F in any uninsulated tank car tank of the DOT-106A and 110A-W class must not exceed 3.4 times the prescribed retest pressure of the tank.

[b]The gas pressure at 115°F (46.1°C) must not exceed the design pressure of the portable tank or cargo tank container.

[c]The liquid portion of fluorocarbon 500 at 105°F (40.5°C) must not completely fill an insulated tank, nor at 130°F (54.4°C) must not completely fill an uninsulated tank with the exception that the liquid portion of fluorocarbon 500 at 115°F (46.1°C) must not completely fill an uninsulated tank car tank of the DOT-112A-W and 114A-W classes.

[d]The liquid portion of fluorocarbon 500 shall not fill the tank at 105°F (40.5°C) if the tank is lagged, or at 115°F (46.1°C) if the tank is unlagged.

TABLE 4. VALVE OUTLET CONNECTIONS AND PRESSURE RELIEF DEVICES FOR CYLINDERS CONTAINING FLUOROCARBONS.

Fluorocarbon Number	CGA Valve Outlet Connection[a]		Pressure Relief Device	
	Standard	Limited Standard	Type[b]	Notes[b]
12	CGA 660	CGA 165, CGA 182	CG-1	A
			CG-2	M
			CG-3	M
			CG-7	A
13	CGA 660	CGA 165, CGA 182	CG-1	A
		CGA 320	CG-4	P
13B1	CGA 660	CGA 165, CGA 182	CG-1	A
			CG-7	A
14	CGA 580	CGA 320	CG-1	A
			CG-4	B
			CG-5	B
			CG-7	K
22	CGA 660	CGA 165, CGA 182	CG-1	A
			CG-2	M
			CG-3	M
			CG-7	A
23	CGA 660	CGA 165, CGA 182	CG-1	A
			CG-4	E
115	CGA 660	CGA 165, CGA 182	CG-1	A
			CG-7	A
116	CGA 660	CGA 165, CGA 182	CG-1	A
		CGA 320	CG-4	B
142b	CGA 510		CG-2	M
			CG-3	M
			CG-7	A
152a	CGA 510		CG-2	M
			CG-3	M
			CG-7	A
500[c]	CGA 660	CGA 165, CGA 182	CG-1	A
			CG-2	M
			CG-3	M
			CG-7	A
502[c]	CGA 660	CGA 165, CGA 182	CG-1	A
			CG-2	M
			CG-3	M
			CG-7	A

[a]Alternate connections that are obsolete as of January 1, 1992, are Connection CGA 660 for fluorocarbons 142b and 152a and Connection CGA 668 for the other fluorocarbons in Table 4.

[b]See Table 5.

[c]Valve outlets for fluorocarbons 500 and 502 are not in current CGA standards. Entries in Table 4 are based on standards for other fluorocarbons with similar properties.

TABLE 5. TYPES OF PRESSURE RELIEF DEVICES.

CG-1	Rupture disk
CG-2	165°F (73.9°C) Fusible plug
CG-3	212°F (100°C) Fusible plug
CG-4	Combination rupture disk with 165°F (73.9°C) Fusible alloy backing
CG-5	Combination rupture disk with 212°F (100°C) Fusible alloy backing
CG-7	Pressure relief valve

Notes (see Table 4):

A. This device is required in one end of the cylinder only, regardless of length, exclusive of trailer tubes, in which this device is required in both ends.

B. When cylinders are over 65 inches (1651 mm) long, exclusive of neck, this device is required at both ends. For shorter cylinders, the device is required in one end only.

E. When cylinders are over 30 inches (762 mm) long, exclusive of neck, this device is required at both ends. For shorter cylinders, the device is required in one end only.

K. This device can be used up to 500 psig (3450 kPa) charging pressure.

M. May be used in addition to CG-7.

P. For use only on cylinders over 65 inches (1651 mm) long. This device is required on both ends.

Device Standards—Part 1—Cylinders for Compressed Gases. [17]

Valve Outlet Connections

Table 4 lists the standard valve outlet connections for fluorocarbon cylinders in the United States and Canada. Limited standard connections generally confined to use in the refrigerant industry are also listed. The information in Table 4 is based on CGA V-1, *American National, Canadian, and Compressed Gas Association Standard for Compressed Gas Cylinder Valve Outlet and Inlet Connections.* [18]

Bulk Storage and Handling

Shipments of fluorocarbons in single-unit tank cars and cargo or portable tanks can range from approximately 30 000 to 200 000 lb (13 607 to 90 718 kg) net weight subject to weight limit regulations at local, regional, and federal levels. Refer to Chapter 6 for information on unloading of bulk quantities of liquefied compressed gases.

Fluorocarbon storage vessels are commonly fabricated of steel and are built according to Section VIII of the *ASME Boiler and Pressure Vessel Code* for unfired pressure vessels. [19] All tanks and installations must also comply with the requirements of appropriate insurance and regulatory agencies having jurisdiction within the locality of the installations. Recommended design pressures (or working pressures) are based on the material to be stored; these all should be at least as great as the design pressures of the bulk container in which the commodities are normally transported, or as required in appropriate codes or regulations, whichever is greater.

Piping may be of copper, preferably Type K, with silver-soldered fittings. ASTM Schedule 40 steel pipe with welded joints may also be used. Screwed fittings, also widely used, exhibit a greater tendency to work loose, resulting in leaks, particularly in areas subject to vibration due to operation of machinery. Polytetrafluoroethylene luting and gaskets are often recommended for fluorocarbon service.

A recommended safe practice is to install hydrostatic pressure relief valves in all pipelines wherever the possibility exists that liquid fluorocarbons may become trapped (such as between two valves) to prevent piping rupture through thermal expansion of the fluorocarbon.

Facilities for the storage and handling of fluorocarbons should be designed, fabricated, and installed in consultation with experts who are thoroughly familiar with the fluorocarbons and their handling.

METHODS OF MANUFACTURE

In general, fluorocarbons are produced commercially by the reaction of hydrofluoric acid with chlorocarbons, or by the disproportionation of other fluorocarbons. Fluorocarbon 12, for example, is made by the reaction of carbon tetrachloride and hydrofluoric acid in the presence of antimony chloride as a catalyst.

Fluorocarbon 13 is produced by the disproportionation of fluorocarbon 12 in the vapor phase in the presence of aluminum chloride or bromide. Fluorocarbon 22 is made by treating chloroform with hydrofluoric acid in the presence of a small amount of antimony chloride at elevated temperatures and pressures.

Fluorocarbon 152a is a component of fluorocarbon 500 and is manufactured by the addition of hydrofluoric acid to acetylene with boron trifluoride as a catalyst.

REFERENCES

[1] ASHRAE 34, *Number Designation of Refrigerants,* American Society of Heating, Refrigerating and Air Conditioning Engineers, 1791 Tullie Circle, N.E., Atlanta, GA 30329.

[2] *Thermodynamic Properties of Refrigerants,* American Society of Heating, Refrigerating and Air Conditioning Engineers (ASHRAE), 1791 Tullie Circle N.E., Atlanta, GA 30329. (1987)

[3] *Properties of Commonly Used Refrigerants,* Air Conditioning and Refrigeration Institute, 1501 Wilson Blvd., Suite 600, Arlington, VA 22209 (1967).

[4] *Thermophysical Properties of Refrigerants,* American Society of Heating, Refrigerating and Air Conditioning Engineers (ASHRAE), 1791 Tullie Circle, N.E., Atlanta, GA 30329. (1969)

[5] Parmelee, H. M., "Water Solubility of Freon Refrigerants", *Refrigerating Engineering,* **61**, 1341 (December 1953).

[6] Bulletins C-30 (1973) and C-30A (1966), E. I. du Pont de Nemours & Co., Freon Products Division, Wilmington, DE 19898.

[7] Bulletin B-2 (1969), E. I. du Pont de Nemours & Co., Freon Products Division, Wilmington, DE 19898.

[8] *ASHRAE Handbook and Product Directory* (4 volumes: Fundamentals (1985), Systems (1985), Equipment (1983), Applications (1982), American Society of Heating, Refrigerating and Air Conditioning Engineers, 1791 Tullie Circle N.E., Atlanta, GA 30329.

[9] *Code of Federal Regulations,* Title 49 CFR Parts 100–199 (Transportation), Superintendent of Documents, U.S. Government Printing Office, Washington, DC 20402.

[10] *Transportation of Dangerous Goods Regulations,* Canadian Government Publishing Centre, Supply and Services Canada, Ottawa, Ontario, Canada K1A 0S9.

[11] Nuckolls, A. H. "The Comparative Life, Fire, and Explosion Hazard of Common Refrigerants," Underwriters Laboratories, Miscellaneous Hazard No. 2375, November (1933). Other UL reports: MH-3134 (R-22), MH-2256 (R-115), MH-3135 (R-124a), MH-2256 (R-502). Underwriters Laboratories, 333 Pfingsten Rd., Northbrook, IL 60062.

[12] *Threshold Limit Values and Biological Exposure Indices,* 1989-90 ed., American Conference of Governmental Industrial Hygienists, 6500 Glenway Ave., Bldg D-5, Cincinnati, OH 45211.

[13] *Code of Federal Regulations,* Title 29 CFR Parts 1900–1910 (Labor), Superintendent of Documents, U.S. Printing Office, Washington, DC 20402.

[14] Sanders, Paul A., *Handbook of Aerosol Technology,* 2nd ed., Van Nostrand Reinhold, New York, (1979).

[15] Clayton, G. D., and Clayton, F. E., eds., *Patty's Industrial Hygiene and Toxicology,* rev. 3rd ed., Wiley, New York, 1981.

[16] *Effects of Exposure to Toxic Gases—First Aid and Medical Treatment,* 3rd ed., Matheson Gas Products, Inc., Secaucus NJ 07094. (1988)

[17] CGA S-1.1, *Pressure Relief Device Standards—Part 1—Cylinders for Compressed Gases,* Compressed Gas Association, Inc., 1235 Jefferson Davis Highway, Arlington, VA 22209.

[18] CGA V-1, *American National, Canadian, and Compressed Gas Association Standard for Compressed Gas Cylinder Valve Outlet and Inlet Connections* (ANSI/CSA/CGA V-1), Compressed Gas Association, Inc., 1235 Jefferson Davis Highway, Arlington, VA 22209.

[19] *ASME Boiler and Pressure Vessel Code* (Section VIII), American Society of Mechanical Engineers, 345 E. 47th Street, New York, NY 10017.

Helium

Chemical Symbol: He
CAS Registry Number: 7440-59-7
DOT Classification: Nonflammable gas
DOT Label: Nonflammable gas
Transport Canada Classification: 2.2
UN Number: UN 1046 (compressed gas); UN 1963 (refrigerated liquid)

PHYSICAL CONSTANTS

	U.S. Units	SI Units
Chemical formula:	He	He
Molecular weight	4.00	4.00
Density of the gas at 70°F (21.1°C) and 1 atm	0.0103 lb/ft^3	0.165 kg/m^3
Specific gravity of the gas at 70°F (21.1°C) and 1 atm	0.138	0.138
Specific volume of the gas at 70°F (21.1°C) and 1 atm	97.09 ft^3/lb	6.061 m^3/kg
Density of the liquid at boiling point and 1 atm	7.802 lb/ft^3	124.98 kg/m^3
Boiling point at 1 atm	−452.1°F	−268.9°C
Melting point at 1 atm	None	None
Critical temperature	−450.3°F	−267.9°C
Critical pressure	33.0 psia	227 kPa abs
Critical density	4.347 lb/ft^3	69.64 kg/m^3
Triple point	None	None
Latent heat of vaporization at boiling point and 1 atm	8.72 Btu/lb	20.28 kJ/kg
Latent heat of fusion at triple point	No T.P.	No T.P.
Specific heat of the gas at 70°F (21.1°C) and 1 atm		
C_p	1.24 Btu/(lb)(°F)	5.19 kJ/(kg)(°C)
C_v	0.745 Btu/(lb)(°F)	3.121 kJ/(kg)(°C)
Ratio of specific heats (C_p/C_v)	1.66	1.66
Solubility in water, vol/vol at 32°F (0°C)	0.0094	0.0094
Weight of the liquid at boiling point	1.043 lb/gal	125.0 kg/m^3

DESCRIPTION

Helium is the second lightest element; only hydrogen is lighter. It is one-seventh as heavy as air. Helium is one of the rare gases of the atmosphere, in which it is present in a concentration of only 5 ppm. Natural gas containing up to 2 percent helium has been found in the American Southwest. Other helium-bearing natural gas fields have been discovered in Saskatchewan, Canada, and near the Black Sea.

Helium is chemically inert. It has no color, odor, or taste. Liquid helium is extremely important in cryogenic research since it is the only known substance to remain liquid at temperatures near absolute zero, and hence has a unique use as a refrigerant in cryogenics. It is also the only known nuclear reactor coolant that does not become radioactive. Helium is nonflammable and is only slightly soluble in water. It is shipped at high pressures—at or above 2400 psig at 70°F (16 547 kPa at 21.1°C) in cylinders and in bulk units. It is also shipped as a cryogenic liquid.

GRADES AVAILABLE

Table 1, from CGA G-9.1, *Commodity Specification for Helium,* presents the component maxima, in parts per million (mole/mole) unless otherwise shown, for specific grades of helium, also known as quality verification levels. [1] A blank indicates no maximum limiting characteristic. The absence of a value in a listed quality verification level does not mean to imply that the limiting characteristic is or is not present, but merely indicates that the test is not required for compliance with the specification.

USES

Helium is used as an inert gas shield in arc welding, as a lifting gas for lighter-than-air aircraft, as a gaseous cooling medium in nuclear reactors, to provide a protective atmosphere for growing germanium and silicon crystals for transistors, to provide a protective atmosphere in the production of such reactive metals as titanium and zirconium, to fill cold-weather fluorescent lamps, to trace leaks in refrigeration and other closed systems, and to fill neutron and gas thermometers. It is used in cryogenic research such as for superconductivity. In mixtures with oxygen, it has medical applications. Radioactive mixtures of helium with krypton are available to users licensed by the Nuclear Regulatory Commission. Liquid helium is also gaining wide use for cooling superconductive magnets used in magnetic resonance imaging.

PHYSIOLOGICAL EFFECTS

Helium is nontoxic and inert. It can act as a simple asphyxiant by diluting the concentration of oxygen in air below levels necessary to support life. Inhalation in excessive concentrations can result in dizziness, nausea, vomiting, loss of consciousness, and death. Death may result from errors in judgment, confusion, or loss of consciousness which prevents self-rescue. At low oxygen concentrations, unconsciousness and death may occur in seconds without warning.

Gaseous helium must be handled with all the precautions necessary for safety with any nonflammable, nontoxic compressed gas (see Chapter 5).

All precautions necessary for the safe handling of any gas liquefied at very low temperatures must be observed with liquid helium. Extensive tissue damage similar to burns can result from exposure to liquid helium or cold helium vapors. See CGA P-12, *Safe Handling of Cryogenic Liquids,* for recommended practices. [2]

MATERIALS OF CONSTRUCTION

Gaseous helium is noncorrosive and inert, and may consequently be contained in systems constructed of any common metals and designed to safely withstand the pressures involved. At the temperature of liquid helium, ordinary carbon steels and most alloy steels lose their ductility and are considered unsafe for liquid helium service. Satisfactory materials for use with liquid helium include Type

TABLE 1. HELIUM GRADES AVAILABLE. (Units in ppm (mole/mole) unless otherwise stated)

Limiting Characteristics	Maxima for Gaseous Helium[a]							
	H	J	K	L	M	N	P	G
Helium min. % (mole/mole)	97.5	99.0	99.99	99.995	99.995	99.997	99.999	99.9999
Water ppm v/v (vapor)				15	9	3	1.5	Sum of all these impurities less than 1 ppm
Dew Point °F							−100	
Total Hydrocarbon Content (as Methane)					5	1	0.5	
Oxygen				5	3	3	1	
Nitrogen + Argon					14	5[b]	5	
Neon					23	23	2	
Hydrogen					Sum = 37 ppm	1	1	
Carbon Dioxide							Sum = 0.5 ppm	
Carbon Monoxide		10						
Odor		None						
Buoyancy	Yes							
Identity		Yes						
USP		Yes						

[a]The quality verification levels (grades) apply only to gaseous helium. Impurity limits for liquid helium are not specified since sufficient technical data and analytical procedures are not available to warrant a definitive quantitative specification. The requirement for ensuring that the loaded fluid in a container is liquid helium can be satisfied by analyzing the shipping container vent gas or by demonstrating that the temperature of the loaded fluid is below the hydrogen triple point (13.8°K).

[b]Maximum 5 ppm nitrogen only.

Note. The 1972 edition of G-9.1 listed 7 quality verification levels of helium lettered A to G. Except for QVL G, these letter designations were dropped from the 1986 edition since they no longer represent major volume of usage by industry. To get the specifications of quality verification levels dropped, see CGA G-9.1-1972 or contact the Compressed Gas Association.

18-8 stainless steel and other austenitic nickel-chromium alloys, copper, Monel, brass, and aluminum.

SAFE STORAGE, HANDLING, AND USE

Gaseous helium is commonly stored in high pressure cylinders, hydril tubes, or tube trailers. Liquid helium is commonly stored at the consumer site in cryogenic liquid cylinders and specially designed insulated tanks. To minimize helium transfer losses, the shipping container for liquid helium is normally used for storage.

Users of liquid helium must also take special precautions in addition to those necessary for the safe handling of such inert liquefied gases as nitrogen and argon. The extremely low temperature of liquid helium makes these special precautions imperative; it can solidify all other gases and it causes air to condense on any uninsulated or inadequately insulated pipe through which is passes. This can result in a localized oxygen-enriched atmosphere and possible dripping of oxygen-enriched liquid air. Refer to the oxygen monograph for applicable precautions. Liquid helium must not be allowed to come in contact with air and must be equipped with pressure relief devices that prevent back-leakage of air into liquid helium equipment. Plugging by solidified air constitutes a serious safety hazard.

Similarly, if air enters and plugs the vent of a helium container, a serious hazard is created. Therefore, the vents of liquid helium containers must be tested on delivery and periodically checked to make sure they remain clear. The use of open-neck dewar flasks for liquid helium also increases the possibility of neck tube plugging from transfer or gauging of the contents. Users of liquid helium should obtain information on safe handling precautions and equipment from their suppliers and CGA P-12. [2]

Liquid and gaseous systems should be designed and installed only under the direction of personnel thoroughly familiar with liquid and gaseous helium equipment and in full compliance with all state, provincial, and local requirements. Additional information regarding helium may be found in CGA P-9, *The Inert Gases—Argon, Nitrogen, and Helium.* [3]

DISPOSAL

Disposal of helium gas may be accomplished by slowly venting to a well-ventilated outdoor location remote from personnel work areas. Do not attempt to dispose of any residual helium in compressed gas cylinders. Return cylinders to the supplier with residual pressure, the cylinder valve tightly closed, and with the valve protective cap in place.

Allow liquid helium to evaporate in well-ventilated outdoor locations which are remote from work areas.

HANDLING LEAKS AND EMERGENCIES

Ventilate enclosed areas to prevent the formation of oxygen-deficient atmospheres caused by the release of gaseous helium or by the evaporation of liquid helium. Personnel, including rescue workers, should not enter areas where the oxygen concentration is below 19 percent, unless provided with a self-contained breathing apparatus or air-line respirator.

Avoid contact of the skin with liquid helium or its cold boil-off gas. Flush liquid helium spills with water to disperse the spill.

METHODS OF SHIPMENT

Under the appropriate regulations, helium is authorized for shipment as follows (helium gas, except where liquid helium is indicated): [4] and [5]

By Rail: In cylinders and in tank cars. Liquid helium, in insulated cylinders.

By Highway: In cylinders on trucks, and in tube trailers. Liquid helium, in insulated cylinders, portable tanks, and cargo tanks.

By Water: In cylinders on cargo and passenger vessels, and on ferry and railroad car ferry vessels (passenger or vehicle). In authorized cargo tanks on cargo vessels only. Liquid helium, in insulated cylinders, portable tanks, and cargo tanks on cargo and passenger vessels and ferry and railroad car ferry vessels (passenger or vehicle).

By Air: In cylinders aboard passenger aircraft up to 150 lb (68 kg) and aboard cargo aircraft up to 300 lb (136 kg) maximum net weight per cylinder. Liquid helium aboard passenger aircraft up to 100 lb (45 kg) maximum net contents per container, aboard cargo aircraft up to 1100 lb (499 kg) maximum net weight per container. Shipment of nonpressurized (open dewars) is prohibited aboard aircraft.

CONTAINERS

Helium gas is authorized for shipment in cylinders, tank cars, and tube trailers in accordance with U.S. Department of Transportation and Transport Canada regulations. [4] and [5] Liquid helium is shipped as a cryogenic fluid in insulated cylinders, insulated portable and cargo tanks, and insulated intermodal containers.

Filling Limits

The maximum filling limits authorized for containers of gaseous helium are as follows:

(1) In cylinders and tube trailers up to the authorized service pressures marked on the cylinders or tube assemblies at 70°F (21.1°C). In the case of cylinders of TC/DOT specifications 3A, 3AA, 3AX, 3AAX, and 3T that meet special requirements, up to 10 percent in excess of their marked service pressures is authorized.

(2) In uninsulated tank cars of the DOT-107A type, a filling limit of not more than 10 percent in excess of the marked maximum gas pressure at 130°F (54.4°C) is authorized.

The maximum filling limits authorized for liquid helium are:

(1) Specification TC/DOT-4L cylinders are authorized for the transportation of liquid helium when carried in the vertical position. The maximum permitted filling density is 12.5 percent by weight.

(2) Liquid helium shipped at pressures below 25 psig (172 kPa) in insulated containers is not subject to DOT fill densities. However, in Canada, compressed gases and refrigerated liquids are regulated regardless of pressure.

Cylinders

Cylinders which comply with TC/DOT specifications 3A and 3AA are the types usually used to ship gaseous helium, but it is authorized for shipment in any cylinders approved for nonliquefied compressed gas. (These include cylinders meeting TC/DOT specifications 3A, 3AA, 3AX, and 3AAX; 3B, 3E, 3T, 3AL, 39, 4B, 4BA, and 4BW; in addition, continued use of cylinders complying with TC/DOT specifications 3C, 3D, 4, 4A, 4C, 3, 25, 26, 33, and 38 is authorized, but new construction is not authorized.)

Liquid helium is authorized for shipment in cylinders which meet TC/DOT specification 4L. At operating pressures under 25 psig (172 kPa), the container specification is not regulated by the U.S. Department of Transportation.

All cylinders authorized for gaseous helium service must be requalified by hydrostatic retest every 5 or 10 years under present regulations, with the following exceptions: DOT-4 cylinders, every 10 years; and no periodic retest is required for cylinders of Specification 3C, 3E, and 4C.

Also, for cylinders of the 4L type authorized for liquid helium service, no periodic retest is required for requalification.

Pressure relief devices authorized for use on helium cylinders include the CG-1, CG-4, CG-5, and CG-7 type devices. These include rupture disk, combination rupture disk/fusible plug, and pressure relief valve type devices. For further information and requirements, see CGA S-1.1, *Pressure Relief*

Device Stanadards—Part 1—Cylinders for Compressed Gases. [6]

Valve Outlet Connections

Standard connections in the United States and Canada for helium cylinders are as follows: for service pressure up to 3000 psig (20 684 kPa), Connection CGA 580; for 3001 to 5500 psig (20 691) to 37 921 kPa), Connection CGA 680; for 5501 to 7500 psig (37 928 to 51 711 kPa), Connection CGA 677; and for cryogenic liquid withdrawal, Connection CGA 792.

The standard pin-indexed yoke connection in the United States and Canada (for medical use of the gas) is Connection CGA 930 for service pressures up to 3000 psig (20 684 kPa). Further information on connections can be found in CGA V-1, *American National, Canadian, and Compressed Gas Association Standard for Compressed Gas Cylinder Valve Outlet and Inlet Connections.* [7]

Tank Cars

Gaseous helium is authorized for rail shipment in tank cars that comply with TC/DOT specification 107A. DOT regulations require that the pressure to which the containers are charged must not exceed 10 percent in excess of the marked maximum gas pressure at 130°F (54.4°C).

Tube Trailers

Gaseous helium is shipped in tube trailers with capacities ranging to more than 40 000 ft^3 (1100 m^3). These trailers are built to comply with DOT cylinder specifications 3A, 3AA, 3AX, 3AAX, or 3T. The trailers commonly serve as the storage supply for the user, with the supplier replacing trailers as they are emptied.

Tank Trailers (Cargo Tanks)

Liquid helium is shipped in bulk in special insulated tank trailers, with capacities in excess of 400 000 ft^3 (11 100 m^3). As previously noted, compressed gases and refrigerated liquids are regulated in Canada regardless of pressure.

Small Portable Containers (Dewars)

Liquid helium is shipped and stored in small insulated portable containers which hold quantities ranging from 1 to 25 gallons (4 to 90 L) or more.

METHOD OF MANUFACTURE

The principal source of helium is natural gas, from which it is recovered in essentially a stripping operation involving liquefaction and purification.

REFERENCES

[1] CGA G-9.1, *Commodity Specification for Helium,* Compressed Gas Association, Inc., 1235 Jefferson Davis Highway, Arlington, VA 22202.

[2] CGA P-12, *Safe Handling of Cryogenic Liquids,* Compressed Gas Association, Inc., 1235 Jefferson Davis Highway, Arlington, VA 22202.

[3] CGA P-9, *The Inert Gases—Argon, Nitrogen, and Helium,* Compressed Gas Association, Inc., 1235 Jefferson Davis Highway, Arlington, VA 22202.

[4] *Code of Federal Regulations,* Title 49 CFR Parts 100–199 (Transportation), Superintendent of Documents, U.S. Government Printing Office, Washington, DC 20402.

[5] *Transportation of Dangerous Goods Regulations,* Canadian Government Publishing Centre, Supply and Services Canada, Ottawa, Ontario, Canada K1A 0S9.

[6] CGA S-1.1, *Pressure Relief Device Standards—Part 1—Cylinders for Compressed Gases,* Compressed Gas Association, Inc., 1235 Jefferson Davis Highway, Arlington, VA 22202.

[7] CGA V-1, *American National, Canadian, and Compressed Gas Association Standard for Compressed Gas Cylinder Valve Outlet and Inlet Connections* (ANSI/CSA/CGA V-1), Compressed Gas Association, Inc., 1235 Jefferson Davis Highway, Arlington, VA 22202.

ADDITIONAL REFERENCES

CGA P-14, *Accident Prevention in Oxygen-Rich and Oxygen-Deficient Atmospheres,* Compressed Gas Association, Inc., 1235 Jefferson Davis Highway, Arlington, VA 22202.

McCarty, R. D., *Interactive Fortran IV Computer Programs for the Thermodynamic and Transport Properties of Selected Cryogens (Fluid Pack)* (National Bureau of Standards Technical Note 1025, October 1980), National Institute of Standards and Technology, Gaithersburg, MD 20899.

Younglove, B. A., *Interactive Fortran Program to Calculate Thermophysical Properties of Six Fluids* (National Bureau of Standards Technical Note 1048, July 1982), National Institute of Standards and Technology, Gaithersburg, MD 20899.

Hydrogen

Chemical Symbol: H_2
CAS Registry Number: 1333-74-0
DOT Classification: Flammable gas
DOT Label: Flammable gas
Transport Canada Classification: 2.1
UN Number: UN 1049 (compressed gas); UN 1966 (refrigerated liquid)

PHYSICAL CONSTANTS

Normal Hydrogen

	U.S. Units	SI Units
Chemical formula	H_2	H_2
Molecular weight	2.016	2.016
Density of the gas at 70°F (21.1°C) and 1 atm	0.00521 lb/ft^3	0.08342 kg/m^3
Specific gravity of the gas at 32°F (0°C) and 1 atm (air = 1)	0.06960	0.06960
Specific volume of the gas at 70°F (21.1°C) and 1 atm	192.0 ft^3/lb	11.99 m^3/kg
Boiling point at 1 atm	−423.0°F	−252.8°C
Melting point at 1 atm	−434.55°F	−259.2°C
Critical temperature	−399.93°F	−239.96°C
Critical pressure	190.8 psia	1315 kPa abs
Critical density	1.88 lb/ft^3	30.12 kg/m^3
Triple point	−434.55°F at 1.045 psia	−259.2°C at 7.205 kPa abs
Latent heat of vaporization at boiling point	191.7 Btu/lb	446.0 kJ/kg
Latent heat of fusion at triple point	24.97 Btu/lb	58.09 kJ/kg
Specific heat of the gas at 70°F (21.1°C) and 1 atm		
C_p	3.425 Btu/(lb)(°F)	14.34 kJ/(kg)(°C)
C_v	2.418 Btu/(lb)(°F)	10.12 kJ/(kg)(°C)
Ratio of specific heats (C_p/C_v)	1.42	1.42

	U.S. Units	SI Units
Solubility in water, vol/vol at 60°F (15.6°C)	0.019	0.019
Density of the gas at boiling point and 1 atm	0.083 lb/ft^3	1.331 kg/m^3
Density of the liquid at boiling point and 1 atm	4.43 lb/ft^3 (0.5922 lb/gal)	70.96 kg/m^3
Gas/liquid ratio (liquid at boiling point, gas at 70°F (21.1°C) and 1 atm), vol/vol	850.3	850.3
Heat of combustion at 70°F (21.1°C) and 1 atm		
Gross	318.1 Btu/ft^3	11 852 kJ/m^3
Net	268.6 Btu/ft^3	10 009 kJ/m^3

Para Hydrogen
(The following constants are different from normal hydrogen)

	U.S. Units	SI Units
Boiling point at 1 atm	−423.2°F	−252.9°C
Melting point	−434.8°F	−259.3°C
Critical temperature	−400.31°F	−240.17°C
Critical pressure	187.5 psia	1293 kPa abs
Critical density	1.96 lb/ft^3	31.43 kg/m^3
Triple point	−434.8°F at 1.021 psia	−259.3°C at 7.042 kPa abs
Latent heat of vaporization at boiling point	191.6 Btu/lb	445.6 kJ/kg
Latent heat of fusion at triple point	25.06 Btu/lb	58.29 kJ/kg
Specific heat of the gas at 70°F (21.1°C) and 1 atm		
C_p	3.555 Btu/(lb)(°F)	14.88 kJ/(kg)(°C)
C_v	2.570 Btu/(lb)(°F)	10.76 kJ/(kg)(°C)
Ratio of specific heats (C_p/C_v)	1.38	1.38
Density of the gas at boiling point and 1 atm	0.084 lb/ft^3	1.338 kg/m^3
Density of the liquid at boiling point and 1 atm	4.42 lb/ft^3 (0.5907 lb/gal)	70.78 kg/m^3
Gas/liquid ratio (liquid at boiling point, gas at 70°F (21.1°C) and 1 atm, vol/vol	848.3	848.3

DESCRIPTION

Hydrogen is colorless, odorless, tasteless, flammable, and nontoxic. It exists as a gas at ambient temperatures and atmospheric pressures. It is the lightest gas known, with a density approximately 0.07 that of air. Hydrogen is present in the atmosphere, occurring in concentrations of only about 0.5 ppm by volume at lower altitudes.

Hydrogen burns in air with a pale blue, almost invisible flame. Its ignition temperature will not vary greatly from the range 1050–1074°F (565.5–578.9°C) in mixtures with either air or oxygen at atmospheric pressure. The flammable limits of hydrogen in dry air at atmospheric pressure are 4.0–75.0 percent hydrogen by volume. In dry oxygen at atmospheric pressure, the flammable limits are 4.6–93.9 percent hydrogen by volume. Its flammable limits in air or oxygen vary somewhat with pressure, temperature, and water vapor content.

When cooled to its normal boiling point of −423°F (−252.8°C) and condensed, hydrogen becomes a colorless liquid only one-fourteenth as heavy as water. All gases except helium become solids at the temperature of liquid hydrogen. Because of its extremely low temperature, it can make ductile or pliable materials with which it comes in contact brittle and easily broken (an effect that must be considered whenever liquid hydrogen is handled). Liquid hydrogen has a relatively high thermal coefficient of expansion compared with other cryogenic liquids.

Ortho and Para Molecules

The hydrogen molecule can exist in two forms: ortho and para, named according to the relative spin directions of the nuclei of its two atoms. Ortho-hydrogen molecules have their nuclei spinning in the same (parallel) direction; para-hydrogen molecules have their two nuclei spinning in opposite (antiparallel) directions. There is no difference in the chemical properties of these forms, but there is a difference in physical properties. As a gas at room temperature and above, hydrogen consists of about three parts ortho and one part para. This 3 ortho/1 para ratio mixture is called normal hydrogen. The equilibrium concentration of the para form increases with decreasing temperature until, as a liquid, the para concentration is nearly 100 percent.

Conversion from the ortho form to the para form is accompanied by the release of heat (exothermic). Conversion in the para-to-ortho direction is endothermic. If normal hydrogen is cooled and liquefied rapidly without catalyst, the relative three-to-one ratio of ortho-to-para does not immediately change. In the liquid phase, ortho to para conversion occurs (even in the absence of catalyst) at a significant rate. For each pound (or 0.45 kg) of rapidly cooled liquid hydrogen that changes from the ortho to the para form, enough heat is liberated to vaporize 1.566 lb (0.71 kg) of liquid hydrogen. Consequently, liquid made from normal hydrogen has a high boil-off loss caused by the ortho-to-para conversion releasing heat into the liquid and that heat boiling away a large part of the liquid.

However, if a catalyst is used in the liquefaction cycle, para-hydrogen can be produced directly with minimal vaporization loss from self-generated heat of conversion in the liquid storage tank. For this reason, para-hydrogen is the preferred form for storage and transport of the liquid, with 95 percent para being the usual specification.

Throttled expansion from high to low pressure at ordinary temperatures cools most common gases (such as oxygen, nitrogen, and carbon dioxide). Hydrogen, though, is an exception. It becomes heated to a slight extent under these conditions (increasing by about 10°F (5.6°C)) in temperature, for example, when throttled from 2500 psig (17 232 kPa) to atmospheric pressure.

Hydrogen diffuses rapidly through porous materials and through some metals at red heat. It may leak out of a system which is gastight for air or common gases at equivalent pressures. It can diffuse into carbon steel and combine with carbon to form methane, which then causes lamination and loss of strength in the steel. A Nelson curve indicates the proper alloy to use depending on the temperature and pressure.

In its chemical properties, hydrogen is fundamentally a reducing agent and is frequently applied as such in chemical technology.

GRADES AVAILABLE

Table 1, from CGA G-5.3, *Commodity Specification for Hydrogen,* presents the

TABLE 1. HYDROGEN GRADES AVAILABLE. (Units in ppm (mole/mole) unless otherwise stated)

Limiting Characteristics	Maxima for Type I (Gaseous)								Maxima for Type II (Liquid)			
	Quality Verification Levels (Grades)											
	A	B	C	D	E	F	G	H	A	C	B	D
Hydrogen min. % (mole/mole)	99.8	99.95	99.95	99.99	99.995	99.995[1]	99.9991	99.997[a,b]	99.995[c]	99.9991[c]	99.9997[c]	99.9997[a,b]
Water ppm (v/v)	None condensed	32.0	7.8	3.5	3.5	1.5	3	0.2		3		0.2
Dew Point (°F)		−60	−80	−90	−90	−100	−92.5	−92.5		−92.5		
Oxygen		10	10	5	5	1	1	0.2		1	1	0.2
Argon												
Nitrogen		400	400	25	20	2	2	2	9[d]	2	2	2
Total Hydrocarbons (as methane)	10	10	10	5	1	0.5	1	0.2		1		0.2
Helium									39			
Carbon Dioxide	10	10	10	0.5			2	0.2	1	2		0.2
Carbon Monoxide	10	10	10	1								
Mercury Vapor (ppb)			4									
Para Content min. %									95	[e]	95	
Permanent Particulates								[e]	Filtering req.[f]			[e]

[a]A purifier is allowed in order to meet the specification.
[b]Percent purity includes trace quantities of rare gases.
[c]Includes up to 50 ppm neon + helium.
[d]Includes water.
[e]To be determined between supplier and user.
[f]By a 10–40 micron (10 microns nominal, 40 microns absolute) filter assembly installed in transfer system.

component maxima in parts per million (mole/mole), unless shown otherwise, for the types and grades of hydrogen. [1] These are also known as quality verification levels (QVLs). Gaseous hydrogen is denoted as Type I and liquefied hydrogen as Type II in the table. A blank indicates no maximum limiting characteristics are specified.

USES

Large quantities of hydrogen are produced on-site or pipelined for use by refineries, petrochemical, and bulk chemical facilities for hydrotreating, catalytic reforming, and hydrocracking. Smaller quantities of hydrogen are produced on-site or pipelined for use in industries such as the following: (1) chemical, (2) metallurgical, (3) fats and oils, (4) glass, and (5) electronic.

Some of these smaller users have hydrogen delivered to their manufacturing location as gaseous hydrogen in cylinders or tube trailers, or by cascade into on-site storage cylinders. Certain smaller users have liquid hydrogen delivered into an on-site liquid hydrogen storage system.

Hydrogen is used in the production of a wide variety of chemicals:

(1) Dyes: in the manufacture of derivatives of aniline produced from nitrobenzene and derivatives of toluidine produced from nitrotoluene.

(2) Catalysts: to produce aluminum alkyls, cobalt compounds, metal hydrides, and nickel as well as a wide variety of others.

(3) Flavors and fragrances: used to make many flavors and fragrances which are generally compounded from a few chemical intermediates.

(4) Pesticides: used in the manufacture of complex, proprietary chemicals several of which are known to require hydrogenation.

(5) Halogen organics: to manufacture a variety of chemicals used in organic synthesis such as dibromobutane, pentachlorobenzene, or fluorocarbons.

(6) Plastic and synthetic fibers: used for the production of high-density polyethylene and polypropylene and also used in the production of intermediates for polyurethane, nylon, and polyamide fibers.

(7) Specialty chemicals: used in the production of a variety of specialty chemicals including intermediates in the production of high-energy rocket fuels and other unique applications.

(8) Petroleum: used to hydrogenate various unsaturated petroleum products.

Metallurgical companies use hydrogen in the production of their products:

(1) Heat treating: Ferrous metals are treated under controlled atmospheres to change their physical properties. Protective atmospheres are used primarily to exclude oxygen and prevent oxidation of the metal at elevated temperatures in the treatment furnace. Protective atmospheres containing hydrogen will also remove any oxides that were present prior to heat treating.

Stainless steels are usually blanketed with enriched hydrogen atmospheres because they contain chromium, titanium, and other metals which are reactive with nitrogen at high temperatures.

(2) Metal production: Basic metal refiners of tungsten, molybdenum, and magnesium using hydrometallurgical processes require large volumes of hydrogen to reduce oxides and prevent oxidation of the metal. Hydrogen is also used as an atmosphere in some powder metallurgy applications. It is also used in processing nuclear fuels.

(3) Welding and cutting: Hydrogen is used with oxygen in oxyhydrogen welding and cutting, being employed in certain brazing operations. It is also used for welding aluminum and magnesium (especially in thin sections) and for welding lead. The oxyhydrogen flame has a temperature of about 4000°F (2204°C) and is well suited for such comparatively low-temperature welding and brazing. It is also used to some extent in cutting metals, particularly in underwater cutting because hydrogen can be safely compressed to the pressures necessary to overcome water pressures at the depths involved in salvage operations. The oxyhydro-

gen flame is also applied in the working and fabrication of quartz and glass.

Atomic hydrogen welding, another important application, is particularly suitable for thin stock and can be used with practically all nonferrous metals and alloys as well as with ferrous alloys; jobs performed with this welding process range from the fabrication of nickel and Monel tanks to the welding of aluminum aircraft parts and propellers and the repair of steel molds and dies. In the process, an arc with a temperature of about 11 000°F (6093°C) is maintained between two nonconsumable metal electrodes. Molecular hydrogen fed into the arc is transformed into atomic hydrogen, which transmits heat from the arc to the weld zone. At the relatively colder surface of the weld area, the atomic hydrogen recombines to molecular hydrogen with the release of heat. The hydrogen also provides a protective atmosphere around the weld area.

Food companies hydrogenate fats, oils, and fatty acids to control various physical and chemical properties. Both edible and inedible fats and oils are hydrogenated, resulting in an increase in melting temperature and an improvement in color, odor, and the stability of the material. Hydrogenated edible oils are used in oleomargarine, shortening, and other food products. Inedible oils and fatty acids that are hydrogenated are used in the production of soaps, industrial greases and oils, surfactants, and plasticizers. Edible oils are reacted with low pressure (20–50 psig; 138–345 kPa) hydrogen, while inedible oils are reacted under higher pressures (300–500 psig; 2068–3447 kPa). Both reactions involve bubbling hydrogen through oil in the presence of a nickel catalyst.

Glass manufacturers use hydrogen as a protective atmosphere in a process whereby molten glass is floated on a surface of molten tin. As the glass hardens, it forms a much smoother surface than is attainable by other production techniques. Hydrogen is used as an atmosphere to prevent the oxidation of molten tin. In the working of leaded glass, hydrogen content must be carefully controlled to prevent burning the lead out of the glass.

Electronics manufacturers use hydrogen at several steps in the complex processes for manufacturing semiconductors (integrated circuits). Hydrogen is used:

(1) As a protective atmosphere during the initial step of silicon crystal growth
(2) As a reactant and carrier gas during the deposition of a thin layer of silicon on the surface of a chip
(3) As a reactant with high-purity oxygen to produce steam during various oxidation steps
(4) As a reducing atmosphere during the sintering to ensure good, low-resistance electrical connections
(5) As a reducing atmosphere to protect the chips when wire leads are bonded to the contact points

In the electrical power industry, large electrical generators are run in a hydrogen atmosphere to reduce windage losses and remove heat.

Because of its high specific impulse which influences the resultant thrust, liquid hydrogen has assumed importance as a fuel for powering missiles and rockets. It was used as a fuel in the NASA Apollo Space Program and in the NASA Space Shuttle Program. It is employed in laboratory research on the properties of materials at cryogenic temperatures, among them the superconductivity of metals (a state in which they have extremely low electrical resistance). Liquid hydrogen is also used in the study of high-energy physics as the liquid medium in bubble chambers.

PHYSIOLOGICAL EFFECTS

Hydrogen is nontoxic, but it can act as a simple asphyxiant by displacing or diluting atmospheric air to the point where the oxygen content cannot support life. Unconsciousness can occur without any warning symptoms from inhaling air which contains a sufficiently large amount of hydrogen.

However, the lower flammable limit of only 4 percent by volume of hydrogen in air is reached significantly before the asphyxiation hazard level is achieved. The flammable gas hazards of hydrogen far exceed the asphyxiant hazards. Safe handling of hydrogen requires close attention to its flammability hazards.

Liquid hydrogen and the cold gas evolving from the liquid can produce severe cryogenic burns similar to thermal burns upon contact with the skin and other tissues. The eyes can be injured by exposure to the cold gas or splashed liquid that would otherwise be too brief to affect the skin of the hands or face. Contact between unprotected parts of the body with uninsulated piping or vessels containing liquid hydrogen can cause the flesh to stick and tear when an attempt is made to withdraw.

MATERIALS OF CONSTRUCTION

Hydrogen gas is noncorrosive and may be contained at ambient temperatures by most common metals used in installations designed to have sufficient strength for the working pressures involved. Equipment and piping built to utilize hydrogen should be selected with consideration of the possibility of embrittlement, particularly at elevated pressures and temperatures above 450°F (232°C). A Nelson curve should be consulted to select the proper alloys.

Metals used for liquid hydrogen equipment must have satisfactory properties at very low operating temperatures. Ordinary carbon steels lose their ductility at liquid hydrogen temperatures and are considered too brittle for this service. Suitable materials include austenitic chromium-nickel steels (stainless steels), copper, copper silicon alloys, aluminum, Monel, and some brasses and bronzes.

Hydrogen Embrittlement

When absorbed into steel, hydrogen can cause several undesirable effects such as flaking, hydrogen embrittlement, or delayed brittle fracture, and at high temperatures, it can cause decarburization.

Hydrogen embrittlement is probably the most prevalent of the three and causes lowered ductility of the steel at certain strain rates and temperatures. The steps involved are hydrogen adsorption on the surface of the steel, migration of the hydrogen, and its retention at high-stress regions and discontinuities in the steel. Hydrogen may be introduced into the steel as a result of exposure of the metal to hydrogen at high temperatures and pressures or as a result of corrosion of the metal in aqueous solutions. [2]

Hydrogen sulfide, cyanide, and arsenic, even in trace amounts, are examples of materials that greatly increase the amount of hydrogen that becomes absorbed by steel. Therefore, under corrosive conditions, particularly those environments that also contain these substances, hydrogen damage can become severe.

SAFE STORAGE, HANDLING, AND USE

Regulators and Control Valves

Contrary to general practice with other gas cylinders, it is inadvisable to partially open, that is, "crack," hydrogen cylinder valves before connecting them to a regulator or manifold since "self-ignition" of the issuing hydrogen may occur.

Cylinders

All the general rules for compressed gas cylinders apply, but extra precautions are necessary in the handling of hydrogen gas. The cylinder valve should not be cracked before fitting the regulator (the issuing gas may catch fire). After attaching a regulator and before the cylinder valve is opened, see that the regulator is closed and the adjusting screw is turned out. Then open the cylinder valve slowly with the valve outlet pointed away from you. Fully open the valve (to minimize self-ignition) when the cylinder is in use (connected into a piping system).

Storage

Specific requirements for storage of hydrogen are contained in NFPA 50A, *Standard for Gaseous Hydrogen Systems at Consumer Sites,* and NFPA 50B, *Standard for Liquefied Hydrogen Systems at Consumer Sites.* [3] and [4]

Storage banks, both portable and stationary, are to be inspected periodically for: (1) corrosion of vessels and supports, (2) condition of pressure relief devices, (3) proper operations of shutoff valves, and (4) condition of manifold piping. A leak test with soap solution or other appropriate leak detection equipment should be performed periodically at maximum operating pressure.

Where the storage banks consist of TC/DOT cylinders, they should be checked for the last hydrostatic test date to ascertain that the cylinders are within the allowable time period between tests. Cylinders in these banks must be retested every five years. TC/DOT cylinders must be filled and periodically retested in accordance with appropriate regulations. Retesting is not required for ASME storage tubes.

The condition of pigtails should be checked and worn or damaged ones replaced. The wheels on portable banks should be checked for proper operation and lubricated if required. Drainage holes in bottom supports should be unplugged to prevent cylinder bottoms from rusting.

Safe Handling of Cryogenic Hydrogen

As with other cryogenic liquids, cryogenic hydrogen is not to be used or handled by any personnel unless they are familiar with its properties and skilled in the procedures necessary for its safe use. Cryogenic liquids other than oxygen which vaporize in enclosed areas may reduce the oxygen content in the atmosphere below the level necessary to sustain life. Adequate ventilation must be provided in areas where these materials are being vaporized. The atmosphere should be checked at regular intervals for oxygen content.

All items pertaining to transfer of cryogenic liquids also pertain to liquid hydrogen. See CGA P-12, *Safe Handling of Cryogenic Liquids.* [5] However, because of its wide flammable range and ease of ignition, extra precautions are necessary. Measures to ensure safe handling are described in CGA G-5, *Hydrogen,* [6], and NFPA 50B, *Standard for Liquefied Hydrogen Systems at Consumer Sites.* [4]

The following are several of the precautions that need to be observed during transfer operations:

(1) Only personnel thoroughly acquainted with liquid hydrogen properties, equipment, and operating procedures should be permitted to perform transfer operations.

(2) Transfer hoses in liquid hydrogen service must be purged with helium or gaseous hydrogen before using.

(3) Hydrogen tankers must be adequately grounded during loading and unloading operations.

(4) Hydrogen transfer operations should be discontinued during thunderstorms.

Containers for cryogenic liquids must be kept clean and restricted to this service. These containers must be made of Type 300 series stainless steel, copper, brass, aluminum, etc., except for special laboratory equipment of glass. Several materials, such as carbon steel, become brittle at cryogenic temperatures and cannot be used for handling these liquids.

Dewars and Liquid Cylinders

Dewars are generally the open type, nonpressurized vacuum-jacketed vessels used to contain cryogenic liquids. Cryogenic liquid cylinders are closed, pressurized vessels.

Bulk Liquid Storage Systems

Bulk liquid storage systems usually consist of a storage tank, vaporizer, and associated valves and piping. Storage system capacities may vary from several hundred to thousands of gallons. The tanks are usually either cylindrical or spherical, and normal pressures may be from atmospheric to several hundred pounds per square inch. Persons required to work with these systems should be specifically trained in the storage of flammable cryogenic liquids.

Recommended standards for liquefied hydrogen systems are given in NFPA 50B, *Standard for Liquefied Hydrogen Systems at Consumer Sites.* [4] Extensive information on equipment for handling liquid hydrogen appears in *Handbook for Hydrogen Handling Equipment* (PB161835) issued by the U.S. Department of Commerce. [7] Electrical equipment in areas for the storage of large quantities of liquid hydrogen should be installed in accordance with Article 500 of NFPA 70, *National Electrical Code.* [8] Separate rooms or buildings devoted to the handling of liquid hydrogen in substantial quantities should conform with the recommendations of NFPA 68, *Guide for Explosion Venting.* [9]

DISPOSAL

The most common method of disposing of hydrogen is by venting to the atmosphere. Venting should be done through a vent stack that discharges at an elevated point. It should be done at a controlled rate, in an isolated area remote from sources of ignition and away from air intakes.

HANDLING LEAKS AND EMERGENCIES

Leaking hydrogen cylinders should be handled with special care. If hydrogen leaks from the cylinder valve even when the valve is closed, or if a leak occurs at the pressure relief device, carefully remove the cylinder to an outdoor open space and well away from any possible source of ignition. Plainly tag the cylinder as having an unserviceable valve or pressure relief device, and immediately notify the cylinder supplier and ask for instructions.

Extreme care is recommended in protecting access to the defective cylinder because the leaking hydrogen may ignite in the absence of any normally apparent source of ignition and if so, will burn with an almost invisible flame that can instantly injure anyone coming into contact with it. The presence of a hydrogen flame can be detected by approaching with a straw broom outstretched in front to make the flame visible.

For leaks on liquid hydrogen systems, observe the same precautions noted above. Isolate and restrict unauthorized access to the area. Eliminate any ignition sources in the immediate area of the leak. Immediately notify your hydrogen supplier for assistance and/or instructions. Proper personal protective equipment should be used before attempting to handle or repair any leak.

Should hydrogen ignite, let it burn, keeping the surroundings cool with water spray to prevent the spread of fire, but do not spray water on the vent stack. In this manner, an accumulation of the gas that might lead to an explosion can be avoided.

Caution: Water should never be directed onto the vent system of a bulk liquid facility due to the potential of plugging as a result of ice formation.

If unignited and uncontrolled leakage is encountered which cannot be stopped by shutting off the closest appropriate valve or main supply valve (without risk), implement a plan for evacuation and quickly contact the local fire department.

First Aid

If the person is conscious and becomes aware of symptoms of asphyxia, he or she should go to an uncontaminated area and inhale fresh air or oxygen. An unconscious person must be carried to an uncontaminated area and given artificial respiration with simultaneous administration of oxygen as soon as possible.

Cryogenic liquids will cause burns similar to thermal burns when they come in contact with the skin and must be treated immediately. Thoroughly flush the affected skin area with tepid water (between 105°F and 115°F) until the skin temperature returns to normal. Get medical attention as soon as possible.

METHODS OF SHIPMENT

Under the appropriate regulations, gaseous and liquid hydrogen (minimum 95 percent para hydrogen) is authorized for shipment as follows: [10] and [11]

By Rail: Gaseous hydrogen is authorized to be shipped in TC/DOT specification cylinders and in TC/DOT Specification 107A seamless tank cars. Liquid hydrogen is authorized to be shipped in TC/DOT Specification 113 cryogenic liquid tank cars and in insulated portable tanks under exemption.

By Highway: Gaseous hydrogen is authorized to be shipped in TC/DOT specification cylinders on trucks and in tube trailers. Liquid hydrogen is authorized to be shipped in TC/DOT specification insulated welded cylinders, in TC/DOT specification insulated cargo tanks, and in insulated portable tanks under exemption.

By Water: Gaseous hydrogen is authorized to be shipped by cargo vessel in TC/DOT specification cylinders and in tube trailers. Liquid hydrogen is currently not authorized to be shipped by water except with the approval of the appropriate authority (DOT or TC). Liquid hydrogen is shipped under exemption in TC/DOT specification insulated cargo tanks and in insulated portable tanks.

By Air: Gaseous hydrogen is authorized for shipment in TC/DOT specification cylinders aboard cargo aircraft only up to 300 lb (136 kg) maximum net weight per container. Liquid hydrogen is forbidden for air shipment.

CONTAINERS

Gaseous hydrogen is authorized for shipment in TC/DOT specification cylinders, tube trailers, and tank cars.

Liquid hydrogen is authorized for shipment in TC/DOT specification insulated welded cylinders, TC/DOT specification insulated cargo tanks, TC/DOT specification cryogenic liquid tank cars, and insulated portable tanks under exemption.

Filling Limits

Gaseous hydrogen in TC/DOT specification cylinders may be charged to the marked service pressure on the cylinder at 70°F (21.1°C) and not in excess of 5/4 times the service pressure at 130°F (54.4°C). Gaseous hydrogen is usually shipped in high pressure cylinders at pressures of 2200–2400 psig (15 168–16 547 kPa). Gaseous hydrogen is also shipped in tube trailers. Cylinders and tube trailers may be shipped under exemption at pressures (at 70°F) up to 110 percent of the marked service pressure.

Liquid hydrogen is shipped in TC/DOT specification insulated welded cylinders at a fill density of 6.7 percent. The pressure in these cylinders must be limited by a pressure-controlling valve set to limit pressure to not more than 17 psig (117 kPa).

Gaseous hydrogen in Specification 107A tank cars must be shipped at pressures which do not exceed 0.7 of the marked test pressure of the tank.

Liquid hydrogen in cryogenic liquid tank cars must be shipped with a maximum fill density of 6.6 percent, and each tank car must have a pressure-controlling valve with a maximum set to discharge pressure of 17 psig (117 kPa).

Liquid hydrogen in insulated cargo tanks must be shipped with fill densities shown in the applicable regulations for the maximum set-to-discharge pressure of the pressure control valve. Liquid hydrogen, when shipped in portable tanks, is subject to fill densities shown in the authorizing exemptions. See 49 CFR 173.316 (c)(3). [10]

Cylinders

Cylinders built in accordance with TC/DOT specifications 3A, 3AA, 3AX, and 3AAX are the types of cylinder most commonly used, but cylinders built to TC/DOT specifications 3B, 3E, 4B, 4BA, 4BW, 39, and 3AL are also authorized. Cylinders built to TC/DOT specifications 3, 3C, 3D, 4, 4A, 4C, 25, 26, 33, or 38 may be continued in gaseous hydrogen service, but new construction is not authorized. TC/DOT 4L insulated cylinders are authorized for liquid hydrogen.

Pressure relief devices authorized for use on hydrogen cylinders include the CG-1 rupture disk, CG-4 and CG-5 combination rupture disk/fusible plug devices, and the CG-7 pressure relief valve, in accordance with requirements given in CGA S-1.1, *Pressure Relief Device Standards—Part 1—Cylinders for Compressed Gases.* [12]

All cylinders authorized for gaseous hydrogen service must be requalified by hydrostatic retest every 5 or 10 years under present regulations, with the following exceptions: TC/DOT 4 cylinders, every 10 years; no periodic retest is required for cylinders of types 3C, 3E, and 4C. No periodic retest is required for TC/DOT 4L insulated cylinders. Refer to 49 CFR 173.34. [10]

Valve Outlet Connections

The standard valve outlet connections in the United States and Canada for hydrogen cylinders are: Connection CGA 350 for pressures up to 3000 psig (20 684 kPa), Connection CGA 695 for pressures of 3001–5500 psig (29 691–37 921 kPa), Connection CGA 703 for pressures of 5501–7500 psig (37 928–51 711 kPa), and Connection CGA 795 for cryogenic liquid withdrawal. [13]

Cargo Tanks and Portable Tanks

Liquefied hydrogen is shipped in specially designed portable tanks under DOT exemption and in TC/DOT MC-338 insulated cargo tanks. Approximate capacities for the cargo tanks and portable tanks are 13 000 gallon (49.2 m^3) and 10 000 gallon (37.8 m^3), respectively. The tank consists of an inner vessel with an outer jacket. The annular space is filled with a highly efficient insulating material and evacuated. Should the insulating vacuum deteriorate in transit, drivers are required to leave the highway and go to a safe area to reduce the pressure with a manual blowdown valve.

Liquid hydrogen is also shipped internationally in intermodal, or ISO type, containers under DOT exemption. These specially designed 11 000 gallon (41.6 m^3) vessels can be transported by highway, rail, or sea for periods exceeding 30 days without venting. Their evacuated superinsulation system employs the use of liquid nitrogen shielding.

Portable tanks for long-term shipment are provided with liquid nitrogen shields to intercept the heat leak into liquid hydrogen. Though often equipped with an initial pressure relief device, the tanks do not normally vent any hydrogen in transit since they are ordinarily filled at a pressure of 2 psig (14 kPa) at the plant and will rise in pressure to only about 6–8 psig (41–55 kPa) during transits of up to 3000 miles (4828 km). Liquid nitrogen shielded tanks can hold liquid hydrogen for more than six months without hydrogen loss.

Tube Trailers

Hydrogen gas is authorized for bulk highway shipment in tube trailers at pressures to approximately 2600 psig (17 926 kPa). These trailers are made up of a number of large horizontally mounted tubes constructed to TC/DOT specifications 3A, 3AA, 3AX, or 3AAX and manifolded together to a common header. Valves must be installed on each tube and closed in transit. Such trailers are usually transported to the point of utilization and replace an empty unit, which is then transported to the supplier's plant for refilling. Capacities of the tube trailers range up to 200 000 ft^3 (5663 m^3).

Tank Cars

Gaseous hydrogen is authorized for rail shipment in tank cars that comply with TC/DOT specification 107A. These cars consist of a number of seamless vessels permanently mounted on a special car frame. Regulations require that on such cars the pressure relief device header outlet must be equipped with an approved ignition device which will instantly ignite any hydrogen discharged through the relief devices. Other requirements are also set forth in the regulations.

Liquefied hydrogen containing a minimum of 95 percent para hydrogen is authorized for shipment in tank cars meeting TC/DOT specifications 113A175W and 113A60W (no new construction is authorized for 113A175W) which must be outfitted with pressure-controlling valves set at a pressure not exceeding 17 psig (117 kPa). A common size for these cars is 68 ft (20.7 m) length with 28 300 gallon (107 m^3) capacity. The cars are so well insulated for internal cooling that they usually do not vent any hydrogen while being shipped distances up to 3000 miles (4828 km). Should the pressure approach the relief valve setting, excess hydrogen is mixed with air in a special device to a concentration of less than half its lower flammable limit before being vented.

METHODS OF MANUFACTURE

The majority of the hydrogen produced in the United States is consumed at or within pipeline distances of the production site. Hydrogen is produced directly by: (1) steam reforming of natural gas, and (2) electrolysis of water. It may also be produced directly by (1) steam reforming of heavy hydrocarbons, and (2) ammonia dissociation.

As a by-product, hydrogen is also produced by: (1) electrolytic processes such as chlorine–sodium hydroxide production, (2) catalytic reforming processes in refineries, (3) olefin production, and (4) recovery from ammonia plant purge gas.

Depending on the purity required, hydrogen can be further processed by: (1) cryogenic adsorption, (2) solvent absorption, (3) pressure swing adsorption (PSA), (4) cryogenic absorption using liquid nitrogen, (5) membrane separation, and (6) metal hydride adsorption.

Hydrogen can be liquefied provided the liquefier feed is very-high-purity hydrogen. Liquefied hydrogen has a minimum purity of 99.995 percent. Para hydrogen liquefies at −423.2°F (−252.9°C), and at that temperature, any contaminant (other than helium), if present in the liquefication feedstock, can solidify and plug the liquefier.

Liquid hydrogen is produced by first precooling compressed, virtually contaminant-free hydrogen to liquid nitrogen temperature (−320°F; −195°C), then passing it through a "guard adsorber" at that temperature to remove all impurities to concentrations below their solubilities in liquid hydrogen. It is then liquefied by further cooling the hydrogen stream to −400 to −410°F (−240 to −246°C). This high pressure liquid stream is reduced in pressure by means of a Joule-Thompson valve and the liquid sent to storage.

Since hydrogen at ambient temperature is 25 percent para and hydrogen at −420°F (−251°C) is 99.8 percent para, about 75 percent of the hydrogen changes form, either during the liquefaction process or in the storage tank. Since the conversion is exothermic, a number of catalyst beds at successively lower temperatures are incorporated into the design of the liquefaction plant, so that very little conversion occurs in the storage tank. This minimizes the loss of liquid hydrogen from the storage tank.

Refrigeration to −320°F (−195°C) for liquid hydrogen plants is provided by liquid nitrogen or by expanders using hydrogen gas. The approximate ratio of liquid nitrogen to hydrogen required to liquefy hydrogen is 10:1 by weight. Refrigeration at liquid hydrogen temperature is provided by hydrogen expanders. Some processes use high pressure inlets to the expanders (1500 psi; 10 342 kPa) and use reciprocating machines which are very efficient but relatively expensive. Other processes utilize lower pressures and several turbo expanders in series. Their expanders are considerably less efficient. All current existing commercial facilities making tonnage quantities of the liquid use the high pressure process and reciprocating machinery.

REFERENCES

[1] CGA G-5.3, *Commodity Specification for Hydrogen,* Compressed Gas Association, Inc., 1235 Jefferson Davis Highway, Arlington, VA 22202.

[2] LaQue and Copson, *Corrosion Resistance of Metals and Alloys,* 2nd ed., Van Nostrand Reinhold, New York, 1963.

[3] NFPA 50A, *Standard for Gaseous Hydrogen Systems at Consumer Sites,* National Fire Protection Association, Batterymarch Park, Quincy, MA 02269.

[4] NFPA 50B, *Standard for Liquefied Hydrogen Systems at Consumer Sites,* National Fire Protection Association, Batterymarch Park, Quincy, MA 02269.

[5] CGA P-12, *Safe Handling of Cryogenic Liquids,* Compressed Gas Association, Inc., 1235 Jefferson Davis Highway, Arlington, VA 22202.

[6] CGA G-5, *Hydrogen,* Compressed Gas Association, Inc., 1235 Jefferson Davis Highway, Arlington, VA 22202.

[7] *Handbook for Hydrogen Handling Equipment* (PB161835), U.S. Department of Commerce, Office of Technical Services, Washington, DC 20025 (deals with liquid hydrogen only).

[8] NFPA 70, *National Electrical Code,* National Fire Protection Association, Batterymarch Park, Quincy, MA 02269.

[9] NFPA 68, *Guide for Explosion Venting,* National Fire Protection Association, Batterymarch Park, Quincy, MA 02269.

[10] *Code of Federal Regulations,* Title 49 CFR Parts 100–199, (Transportation), Superintendent of Documents, U.S. Government Printing Office, Washington, DC 20402.

[11] *Transportation of Dangerous Goods Regulations,* Canadian Government Publishing Centre, Supply and Services Canada, Ottawa, Ontario, Canada K1A 0S9.

[12] CGA S-1.1, *Pressure Relief Device Standards—Part 1—Cylinders for Compressed Gases,* Compressed Gas Association, Inc., 1235 Jefferson Davis Highway, Arlington, VA 22202.

[13] CGA V-1, *American National, Canadian, and Compressed Gas Association Standard for Compressed Gas Cylinder Valve Outlet and Inlet Connections* (ANSI/CSA/CGA V-1), Compressed Gas Association, Inc., 1235 Jefferson Davis Highway, Arlington, VA 22202.

ADDITIONAL REFERENCES

CGA P-6, *Standard Density Data, Atmospheric Gases and Hydrogen,* Compressed Gas Association, Inc., 1235 Jefferson Davis Highway, Arlington, VA 22202.

Engineering Data Book, 10th ed., Gas Processors Suppliers Association, 6526 E. 60th Street, Tulsa OK 74145. (1987)

NBS Monograph 168, *Selected Properties of Hydrogen,* National Institute of Standards and Technology, Gaithersburg, MD 20899. (1981)

NBS Technical Note 1079, *Tables of Industrial Gas Container Contents and Density for Oxygen, Argon, Nitrogen, Helium and Hydrogen,* National Institute of Standards and Technology, Gaithersburg, MD 20899. (1985)

NFPA 51, *Standard for the Design and Installation of Oxygen-Fuel Gas Systems for Welding and Cutting,* National Fire Protection Association, Batterymarch Park, Quincy, MA 02269.

Hydrogen Bromide

Chemical Symbol: HBr
Synonyms: Anhydrous hydrobromic acid
CAS Registry Number: 10035-10-6
DOT Classification: Nonflammable gas
DOT Label: Nonflammable gas
Transport Canada Classification: 2.4
UN Number: UN 1048

PHYSICAL CONSTANTS

	U.S. Units	SI Units
Chemical formula	HBr	HBr
Molecular weight	80.912	80.912
Vapor pressure		
at −87.7°F (−66.5°C)	14.7 psia	101.3 kPa abs
at −60.7°F (−51.5°C)	29.4 psia	202.7 kPa abs
at 62.2°F (16.8°C)	294 psia	2027 kPa abs
at 93.0°F (33.9°C)	441 psia	3039 kPa abs
at 194.0°F (90.0°C)	1240 psia	8548 kPa abs
Density of the gas (1 atm)		
at −4°F (−20°C)	0.24 lb/ft^3	3.923 kg/m^3
at 32°F (0°C)	0.23 lb/ft^3	3.636 kg/m^3
at 77°F (25°C)	0.21 lb/ft^3	3.33 kg/m^3
at 122°F (50°C)	0.19 lb/ft^3	2.98 kg/m^3
at 212°F (100°C)	0.17 lb/ft^3	2.662 kg/m^3
Density of the liquid		
at −88.8°F (−67°C)	169.3 lb/ft^3	2717 kg/m^3
at −76.0°F (−60°C)	139.4 lb/ft^3	2238 kg/m^3
at −50.8°F (−47°C)	135.5 lb/ft^3	2174 kg/m^3
Specific gravity of the gas at 77°F (25°C) (air = 1)	2.812	2.812
Specific volume of the gas at 1 atm, 77°F (25°C)	4.76 ft^3/lb	0.297 m^3/kg
Freezing point at 1 atm	−124.4°F	−86.9°C
Boiling point at 1 atm	−88.1°F	−66.7°C

	U.S. Units	SI Units
Critical volume	0.0218 ft^3/lb	0.0013609 m^3/kg
Critical temperature	193.6°F	89.8°C
Critical pressure	1234 psia	8508 kPa abs
Critical density	45.81 lb/ft^3	735 kg/m^3
Critical compressibility factor	0.310	0.310
Triple point	−124.3°F	−86.8°C
Latent heat of vaporization at boiling point	93.8 Btu/lb	217.7 kJ/kg
Latent heat of fusion at triple point	12.8 Btu/lb	29.74 kJ/kg
Heat capacity at constant pressure (77°F; 25°C)	0.0881 Btu/(lb)(°F)	0.368 kJ/(kg)(°C)
Refractive index (gas, 77°F; 25°C)	1.0005591	1.0005591
Viscosity (gas, 1 atm, 68°F; 20°C)	1.17×10^{-5} lb/ft sec	1.75×10^{-5} Pa•s
Thermal conductivity (gas, 1 atm, 60°F; 15.5°C)	1.33×10^{-6} Btu/(ft)(sec)(°F)	83×10^{-6} J/(cm)(sec)(°K)
Surface tension at −51°F −46.2°C)	3.08×10^{-6} lb/ft	22.7×10^{-7} N/m
Dielectric constant (liquid, −121°F; −85°C)	7	7
Dielectric constant (gas, 1 atm, 32°F; 0°C)	1.00313	1.00313
Solubility in water (1 atm, 68°F; 20°C), wt/wt solution	0.49	0.49
Heat capacity at constant volume (77°F; 25°C)	0.062 Btu/(lb)(°F)	0.259 kJ/(kg)(°C)
Specific heat ratio (86°F; 30°C, 0.3–1.5 atm)	1.42	1.42
Entropy (gas, 1 atm 77°F; 25°C)	0.586 Btu/(lb)(°F)	2.45 kJ/(kg)(°C)
Heat of formation (gas, 77°F; 25°C)	−192 Btu/lb	−447 kJ/kg

DESCRIPTION

Hydrogen bromide is a colorless, toxic, and corrosive gas at room temperature and pressure. The liquid is yellowish in color. The gas is heavier than air and fumes with the formation of hydrobromic acid on contact with moist air. It is shipped in cylinders under its own vapor pressure of 320 psig (2206 kPa) at 70°F (21.1°C). Hydrogen bromide is thermally stable and is resistant to oxidation. It is very soluble in water and is soluble in nonpolar solvents such as benzene. It reacts vigorously with bases, fluorine, ammonia, and unsaturated organics. In steel storage vessels, hydrogen bromide may decompose into bromine and hydrogen due to surface catalytic action. The rate of decomposition increases with temperature.

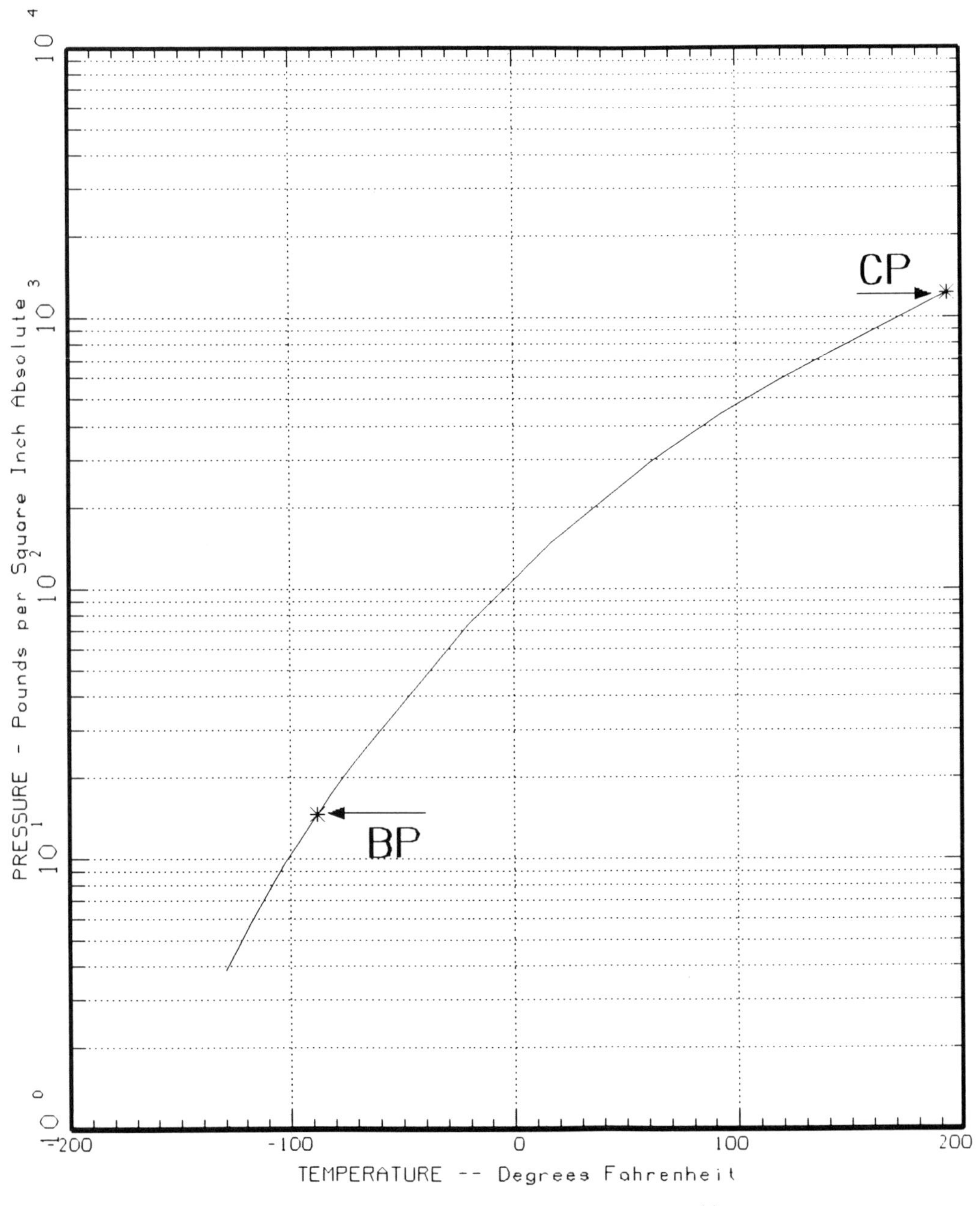

Fig. 1. Vapor Pressure Curve for Hydrogen Bromide.

GRADES AVAILABLE

Hydrogen bromide is typically available in 99.8 percent purity.

USES

Hydrogen bromide is used as a reagent and catalyst in several types of organic reactions, such as the formation of alkyl bromides from alcohols.

It is also used as a source material in the preparation of inorganic bromides. Hydrogen bromide serves as a catalyst in alkylation reactions. It has also been reportedly used in the controlled oxidation of aliphatic and alicyclic hydrocarbons to peroxides, ketones, and acids. In organic synthesis, hydrogen bromide is used to substitute bromine for aliphatic chlorine in the presence of aluminum catalyst.

PHYSIOLOGICAL EFFECTS

Hydrogen bromide is extremely irritating to the eyes, mucous membranes of the respiratory tract, and skin. Contact may cause burns. Repeated or prolonged short exposures to concentrations of about 35 ppm can cause irritation to the nose and throat with mucous production and indigestion. Inhalation of higher concentrations can cause pulmonary edema and laryngeal spasm, and can be fatal. Skin contact with the vapor or liquid causes severe tissue irritation and necrosis. [1]

Concentrations of 1300–2000 ppm are lethal on brief exposure, of about one minute. Concentrations in the range of 1000–1300 ppm are dangerous if breathed for 30 minutes. The maximum tolerable concentration for exposure of 1 hour is in the range of 50 ppm. The maximum tolerable exposure for several hours' duration is approximately 10 ppm.

The U.S. Occupational Safety and Health Administration has established a ceiling limit for exposure to hydrogen bromide at 3 ppm (10 mg/m^3). [2] Similarly, the American Conference of Governmental Industrial Hygienists has adopted a threshold limit value ceiling limit of 3 ppm (10 mg/m^3) for exposure to hydrogen bromide. [3]

MATERIALS OF CONSTRUCTION

Hydrogen bromide does not aggressively attack common metals of construction while in the anhydrous state. However, in the presence of moisture, hydrogen bromide will attack most metals except platinum and silver. Galvanized pipe, brass, and bronze should be avoided. Steel, Monel, and aluminum-silicon-bronze have proven satisfactory in anhydrous hydrogen bromide service.

SAFE STORAGE, HANDLING, AND USE

Persons handling hydrogen bromide should be trained in its properties and physiological effects.

Hydrogen bromide should be used in a well-ventilated area. Electrical equipment should be weather tight to resist the corrosive effects of the gas when wet. Personnel should be equipped with chemically resistant gloves, full face shield and chemical goggles, and chemically resistant aprons. Arms and legs should be fully covered, and open or fabric shoes should not be permitted. [4]

Emergency showers and eyewash stations should be accessible as well as NIOSH approved respirators.

No more hydrogen bromide should be inventoried than can be used in a six month period. Cylinders should be moved with valve protector caps in place using carts designed for the purpose.

DISPOSAL

Hydrogen bromide is an acid gas and as such can be neutralized with caustic solution. When possible, the discharge from a valve leak should be controlled by a flow-limiting valve and piped into a 15 percent solution of sodium hydroxide. Use a trap to prevent reverse flow of the solution to the cylinder valve.

HANDLING LEAKS AND EMERGENCIES

Leaks of hydrogen bromide will be evident as a white smoke when in contact with moist air. Small leaks are made visible as a white mist by holding an open bottle of ammonium hydroxide in the vicinity of the leak. Also, moistened, blue litmus paper will turn red in the presence of hydrogen bromide.

Cylinder valve stem leaks may be controlled by tightening of the valve packing nut by turning clockwise as viewed from above. Persistent leaks should be called to the attention of the gas supplier.

First Aid

In case of inhalation, the victim should be moved to fresh air. If not breathing, administer artificial respiration. If breathing is difficult, administer oxygen. Airway obstruction, as indicated by the presence of laryngeal stridor, may require the placement of an airway by trained emergency medical technicians.

In the case of skin contact, wash the affected area for 10-30 minutes with water. After the initial washing, treat the affected areas with ice water poultices for 30 minutes.

Any contact of hydrogen bromide with the eyes should be regarded as serious and prolonged contact may result in impairment or loss of vision. Eyes should be irrigated without delay with copious amounts of running water. Irrigation should continue for at least 15 minutes and the eyelids should be spread apart to ensure contact with all accessible tissues.

See *Effects of Exposure to Toxic Gases—First Aid and Medical Treatment,* for additional information and recommendations for medical treatment. [1]

METHODS OF SHIPMENT

Under the appropriate regulations, hydrogen bromide is authorized for shipment as follows:

By Rail: Freight, in cylinders.

By Highway: On trucks, in cylinders.

By Water: Cargo vessel, in cylinders, on deck; passenger vessel, in cylinders, on deck.

By Air: Cargo only, in cylinders, limited to 300 lb (136 kg) net weight per package.

Hydrogen bromide has a hazard class of nonflammable gas and carries a nonflammable gas label in domestic U.S. shipments. [5] In Canada, hydrogen bromide is classified as a 2.4 corrosive gas, and the appropriate labels should be used. [6]

Hydrogen bromide is shipped as compressed liquefied gas under its own vapor pressure in cylinders. The maximum filling density for hydrogen bromide is 136 percent.

CONTAINERS

Cylinders with the following minimum TC/DOT specifications can be used for hydrogen bromide service: 3A1800, 3AA1800, 3AX1800, 3AAX1800, and 3E1800. [5] and [6] Specification 3A and 3AA cylinders require five-year periodic retest when in hydrogen bromide service. Specification 3E cylinders do not require retests.

Cylinders must be equipped with a CG-4 type pressure relief device. This device must be in both ends of the cylinder when the cylinder is over 30 inches (76 cm) in length. This type of device is a combination rupture disk/fusible plug, utilizing a fusible alloy with a yield temperature not over 170°F (76.7°C). [7]

Valve Outlet Connections

The standard valve outlet connection in the United States and Canada for hydrogen bromide cylinders is Connection CGA 330. [8]

METHODS OF MANUFACTURE

Hydrogen bromide can be synthesized by direct combination of hydrogen and bromine over platinized silica at 662°F (350°C). Alternately, heating anhydrous methyl bromide with a nonoxidizing acid will produce hydrogen bromide, as will reaction of bromine with a covalent hydride.

REFERENCES

[1] *Effects of Exposure to Toxic Gases—First Aid and Medical Treatment,* 3rd ed., Matheson Gas Products, Inc., Secaucus, NJ 07094. (1988)

[2] *Code of Federal Regulations,* Title 29 CFR Parts 1900–1910 (Labor), Superintendent of Documents, U.S. Government Printing Office, Washington, DC 20402.

[3] *Threshold Limit Values and Biological Exposure Indices,* 1989–90 ed., American Conference of Governmental Industrial Hygienists, 6500 Glenway Avenue, Bldg. D-7, Cincinnati, OH 45211-4438.

[4] *Guide to Safe Handling of Compressed Gases,* Matheson Gas Products, Inc., Secaucus, NJ 07094. (1982)

[5] *Code of Federal Regulations,* Title 49 CFR Parts 100–199 (Transportation), Superintendent of Documents, U.S. Government Printing Office, Washington, DC 20402.

[6] *Transportation of Dangerous Goods Regulations,* Canadian Government Publishing Centre, Supply and Services Canada, Ottawa, Ontario, Canada K1A 0S9.

[7] CGA S-1.1, *Pressure Relief Device Standards—Part 1—Cylinders for Compressed Gases,* Compressed Gas Association, Inc., 1235 Jefferson Davis Highway, Arlington, VA 22202.

[8] CGA V-1, *American National, Canadian, and Compressed Gas Association Standard for Compressed Gas Cylinder Valve Outlet and Inlet Connections* (ANSI/CSA/CGA V-1), Compressed Gas Association, Inc., 1235 Jefferson Davis Highway, Arlington, VA 22202.

ADDITIONAL REFERENCES

Matheson Gas Data Book, 6th ed., Matheson Gas Products, Inc., Secaucus, NJ 07094. (1980)

"Matheson Material Safety Data Sheet for Hydrogen Bromide," Matheson Gas Products, Inc., Secaucus, NJ 07094.

The Merck Index, 9th ed., Merck & Co., Inc., Rahway, NJ 07065. (1976)

Hydrogen Chloride, Anhydrous

Chemical Symbol: HCl
Synonyms: Hydrochloric acid (anhydrous)
CAS Registry Number: 7647-01-0
DOT Classification: Nonflammable gas
DOT Label: Nonflammable gas
Transport Canada Classification: 2.4
UN Number: UN 1050 (compressed gas); UN 2186 (refrigerated liquid)

PHYSICAL CONSTANTS

	U.S. Units	SI Units
Chemical formula	HCl	HCl
Molecular weight	36.465	36.465
Vapor pressure		
at 70°F (21.1°C)	613 psig	4227 kPa
at 77°F (25°C)	676 psig	4661 kPa
at 105°F (40.6°C)	950 psig	6550 kPa
at 115°F (46.1°C)	1075 psig	7412 kPa
at 124.5°F (51.4°C)	1185 psig	8170 kPa
Density of the gas (1 atm)		
at −50°C (−45.6°C)	0.7370 lb/ft^3	11.81 kg/m^3
at 32°F (0°C)	0.102 lb/ft^3	1.634 kg/m^3
at 70°F (21.1°C)	0.0950 lb/ft^3	1.522 kg/m^3
Density of the liquid		
at −121°F (−85°C)	74.3 lb/ft^3	1190 kg/m^3
at −50°F (−45.6°C)	67.8 lb/ft^3	1086 kg/m^3
at 2°F (−16.7°C)	62.9 lb/ft^3	1005 kg/m^3
at 70°F (21.1°C)	52.7 lb/ft^3	842 kg/m^3
at 105°F (40.6°C)	43.8 lb/ft^3	700 kg/m^3
at 115°F (46.1°C)	39.7 lb/ft^3	634 kg/m^3
at 124.5°F (51.4°C)	26.2 lb/ft^3	420 kg/m^3
Specific gravity of the gas at 32°F (0°C) and 1 atm (air = 1)	1.266	1.266
Specific volume of the gas at 70°F (21.1°C) and 1 atm	10.6 ft^3/lb	0.6617 m^3/kg
Boiling point at 1 atm	−121°F	−85°C

	U.S. Units	SI Units
Melting point at 1 atm	−173.6°F	−114.2°C
Critical temperature	124.5°F	51.4°C
Critical pressure	1198 psia	8260 kPa abs
Critical density	26.2 lb/ft^3	420 kg/m^3
Triple point at 2.61 psia (18 kPa abs)	−167.8°F	−111°C
Latent heat of vaporization at boiling point	190.5 Btu/lb	443.1 kJ/kg
Latent heat of fusion at melting point	23.49 Btu/lb	54.64 kJ/kg
Specific heat of the gas at 312.8°F (156°C) and 1 atm		
C_p	0.1939 Btu/(lb)(°F)	0.8118 kJ/(kg)(°C)
C_v	0.1375 Btu/(lb)(°F)	0.5757 kJ/(kg)(°C)
Ratio of specific heats (C_p/C_v)	1.41	1.41
Solubility in water, wt/wt of water, at 32°F (0°C)	0.823	0.823
Weight of the liquid at 2°F (−16.7°C)	8.346 lb/gal	1 kg/L

DESCRIPTION

Anhydrous hydrogen chloride is a colorless gas which fumes strongly in moist air and has a highly irritating effect on body tissues. It has a sharp, suffocating odor and is shipped in cylinders as a liquefied compressed gas under its vapor pressure of 613 psig at 70°F (4226 kPa at 21.1°C).

Chemically, hydrogen chloride is relatively inactive and noncorrosive in the anhydrous state. However, it is readily absorbed by water to yield the highly corrosive hydrochloric (muriatic) acid. It also dissolves readily in alcohol and ether and reacts rapidly (violently, in some cases) with many organic substances. At high temperatures (3240°F or 1782°C and above), hydrogen chloride tends to dissociate into its constituent elements.

GRADES AVAILABLE

Anhydrous hydrogen chloride is typically available for commercial and industrial purposes in a technical grade (minimum purity of 99.0 percent). A typical technical grade analysis is as follows:

Hydrogen chloride (liquid phase)	99.5 weight %
Hydrocarbons (ethylene, 1, 1-dichloroethane, and ethyl chloride)	0.04 weight %
Water	0.01 weight %
Carbon dioxide	0.01 weight %
Inert materials	0.1 weight %

For semiconductor applications, anhydrous hydrogen chloride is available in an electronic grade specified by the Semiconductor Equipment and Materials International (SEMI) with a minimum purity of 99.9944 percent with the following specifications for impurities: [1]

Impurity	Gas Phase
Nitrogen	16 ppm max.
Oxygen and argon	5 ppm max.
Hydrocarbons (C_1 and C_2)	5 ppm max.
Water	10 ppm max.
Carbon dioxide	10 ppm max.
Hydrogen	10 ppm max.
Iron	50 ppb by wt. in the liquid

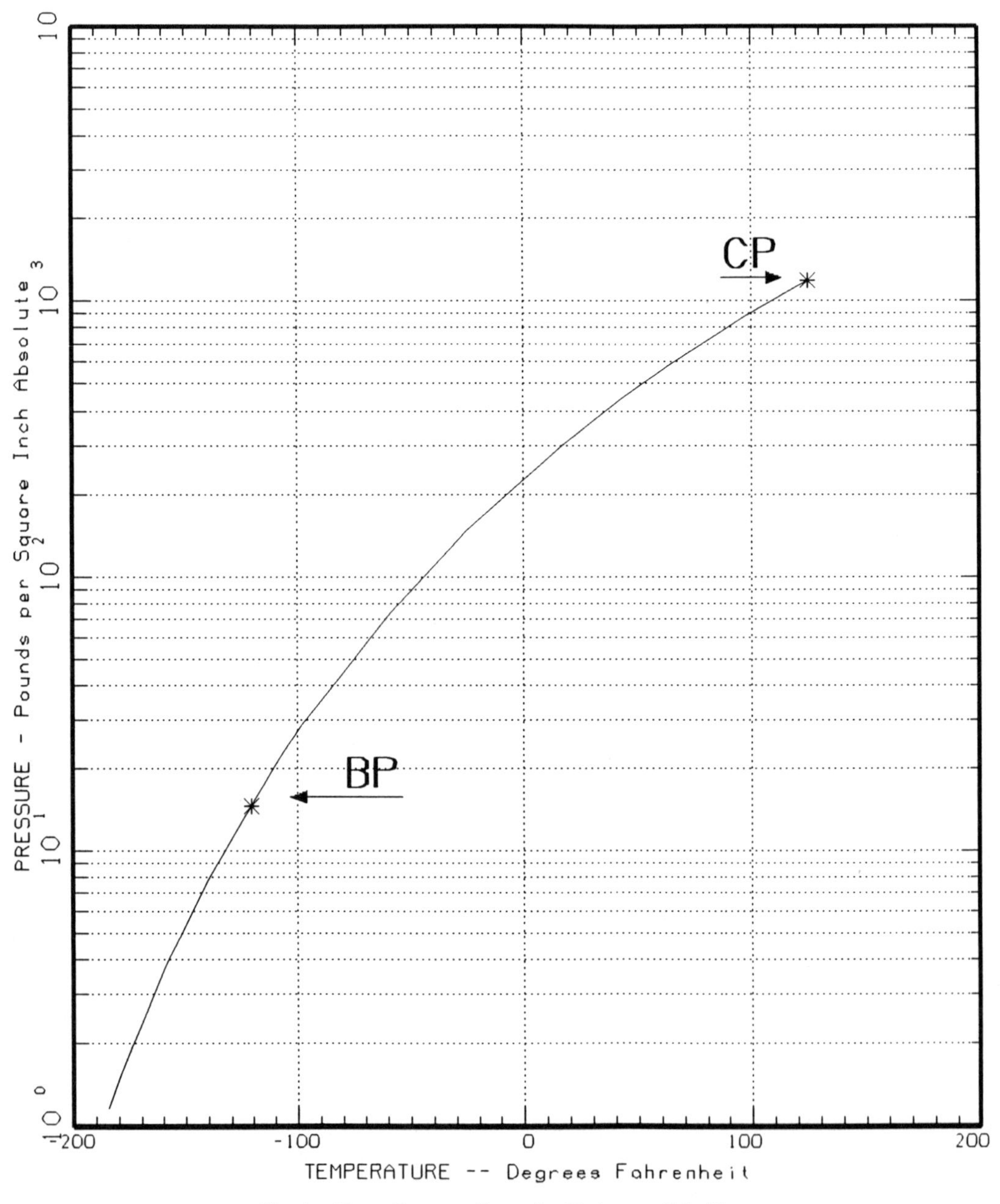

Fig. 1. Vapor Pressure Curve for Hydrogen Chloride.

USES

An important industrial chemical, hydrogen chloride is widely used in the manufacture of rubber, pharmaceuticals, and both inorganic and organic chemicals, as well as in gasoline refining, metals processing, and wool reclaiming. Rubber hydrochloride, which results from the treatment of natural rubber with hydrogen chloride, can be cast in film from solutions; such rubber hydrochloride films provide a strong, water-resistant packaging material for meats and other foods, paper products, and textiles.

The chemicals industry uses hydrogen chloride to produce a large variety of chlorinated derivatives through both addition and substitution reactions with organic compounds. Examples include the manufacture of ethyl chloride from ethylene in making vinyl plastics, the manufacture of chloromethanes and monochlorobenzene, alkyl chlorides, and methyl chloride from methyl alcohol. Hydrogen chloride is utilized in the gasoline industry as a promoter for the aluminum chloride catalyst used for converting *n*-butane to isobutane.

In the metals industry, uses of gaseous hydrogen chloride include application as a flux in bonding Babbitt metal to steel strip in manufacturing insert bearing blanks, and for treating steel strip at elevated temperatures to improve the bond in later hot galvanizing. The textiles industry uses hydrogen chloride to decompose vegetable fibers with which wool has been woven in reclaiming wool for fabrics. Cotton seeds are also delinted and disinfected by being tumbled in an atmosphere of gaseous hydrogen chloride.

In the electronics industry, a high-purity grade of hydrogen chloride is used as an etchant for semiconductor materials.

PHYSIOLOGICAL EFFECTS

A threshold limit value (TLV) ceiling limit of 5 ppm (7 mg/m^3) for hydrogen chloride has been adopted by the U.S. Occupational Safety and Health Administration and the American Conference of Governmental Industrial Hygienists. [2] and [3] This concentration should not be exceeded during any part of the working day.

Inhalation of excess quantities of hydrogen chloride produces coughing, burning of the throat, and a choking sensation. Occasionally, ulceration of the nose, throat, and larynx, or edema of the lungs has resulted. Prolonged inhalation of high concentrations may cause death. The irritating character of the vapors provides warning of dangerous concentrations well before injury can result, unless personnel are trapped or disabled.

A concentration of 50 ppm in air cannot be tolerated for more than one hour, and concentrations of around 1500–2000 ppm are fatal for humans within a few minutes.

Repeated exposure of the skin to concentrated anhydrous hydrogen chloride vapor may result in burns or dermatitis.

MATERIALS OF CONSTRUCTION

Piping, valves, and other equipment used in direct contact with anhydrous hydrogen chloride should be of stainless steel or of cast or mild steel. Carbon steel may be employed in some components, but only if their temperature is controlled to remain below about 265°F (129°C). In the presence of moisture, however, hydrogen chloride will corrode most metals other than silver, platinum, or tantalum.

Smaller-sized valves, such as those used on cylinders, constructed of aluminum-silicon-bronze with Monel stems, have had satisfactory service experience due to frequent maintenance. The satisfactory extension of these materials to other applications should be confirmed by testing prior to use.

SAFE STORAGE, HANDLING, AND USE

Safety precautions include adequate ventilation of working areas and readily accessible safety shower and eye bath facilities located in areas not likely to become contaminated. Rubber or plastic aprons and gloves should be used as protective clothing with wool or other hydrochloric acid resistant outer garments and chemical safety goggles. Other important precautions include the availability of proper full-face respirators and/or self-contained breathing apparatus and the thorough familiarity with safety measures and equipment operation on the part of personnel. Should eye contact, skin contact, or inhalation occur, immediately give first aid and call a physician.

Hydrogen chloride is often unloaded from special tank cars and tank trucks directly to process, thus using the transport container for on-site storage. Unloading piping and valves must be of stainless steel or of cast or mild steel and capable of withstanding the high pressures of the unloading process. Galvanized or copper-alloyed metals such as brass or bronze are corroded by hydrogen chloride and must not be used in pipe or fittings. Design and specification of components for unloading systems should be made with the help of suppliers or engineers thoroughly familiar with anhydrous hydrogen chloride.

DISPOSAL

Disposal of hydrogen chloride may be accomplished by slowly discharging the gas into a scrubber or other suitable vessel containing approximately 15 percent sodium hydroxide or other alkali and water. It is necessary to place a reverse flow check valve or trap in the discharge line to prevent the caustic solution from drawing back into the cylinder.

HANDLING LEAKS AND EMERGENCIES

Leaks in anhydrous hydrogen chloride system connections may be detected by the white fumes which form when the gas comes into contact with the moisture of the atmosphere. Small leaks may be found with an open bottle of concentrated ammonium hydroxide solution (which forms dense white fumes in the presence of hydrogen chloride) or with wet blue litmus paper (which is turned pink by hydrogen chloride).

First Aid

Hydrogen chloride is toxic, causing severe irritation of the eyes and the skin on contact, and severe irritation of the upper respiratory tract on inhalation. Should contact occur, the eyes and eyelids must be washed immediately and thoroughly with large quantities of flowing water in order to avoid impairment of vision or even loss of sight. Irrigation should continue for at least 15 minutes. The eyelids should be spread apart with the fingers to ensure irrigation of all accessible tissues. See a physician, preferably an ophthalmologist, immediately. Prompt washing of the skin after removing contaminated clothing and shoes is also required after contact to prevent severe burns. Discard contaminated clothing and shoes.

If hydrogen chloride is inhaled, remove the victim to fresh air. Give artificial respiration if not breathing. Give oxygen if breathing is difficult. Keep the victim warm. Seek professional medical attention promptly.

METHODS OF SHIPMENT

Under the appropriate regulations, anhydrous hydrogen chloride is authorized for shipment as follows: [4] and [5]

By Rail: In cylinders and in insulated tank cars.

By Highway: In cylinders on trucks; also in insulated tank trucks and tube trailers.

By Water: In cylinders via cargo vessels, passenger vessels, and ferry and railroad car ferry vessels (passenger or vehicle).

By Air: Forbidden on passenger aircraft. Limited to 300 lb (136 kg) maximum net weight per container on cargo aircraft.

CONTAINERS

Anhydrous hydrogen chloride is shipped as a compressed liquefied gas under its own vapor pressure in cylinders and in tube trailers.

It is also shipped in liquefied form in insulated tank cars and cargo tanks.

Filling Limit

The maximum filling density authorized for hydrogen chloride in cylinders by the DOT is 65 percent (percent water capacity by weight).

Cylinders

Cylinders that meet the following TC/DOT specifications are authorized for hydrogen chloride service: 3A1800, 3AA1800, 3AX1800, 3AAX1800, and 3E1800 (cylinders manufactured under the obsolete specification DOT-3 may be continued in service, but new construction is not authorized). DOT-3A2015 cylinders are also often used in anhydrous hydrogen chloride service. DOT-3T tubes may also be used. The Type CG-4 combination rupture disk/fusible plug is the pressure relief device authorized for use on hydrogen chloride cylinders. When cylinders are over 65 inches (1.65 m) long, exclusive of the neck, this device is required at both ends. For shorter cylinders, the device is required in one end only. [6]

Specification 3A and 3AA (as well as 3) cylinders used in hydrogen chloride service must be requalified by hydrostatic retest every five years under present regulations. For Specification 3E cylinders, no periodic retest is required.

Because of the tendency for buildup of particulates to occur, cylinder valves on hydrogen chloride cylinders are normally wrench operated to facilitate application of torques in excess of that which can be applied manually with a handwheel.

Valve Outlet Connections

The standard valve outlet connection in the United States and Canada for hydrogen chloride cylinders is Connection CGA 330. [7]

Cargo Tanks

Hydrogen chloride is authorized for shipment in insulated cargo tanks (tank trucks) of special design. Constructed of stainless steel, the tanks conform to TC/DOT specification MC-331 or MC-338 and must be insulated. Whenever the normal travel time is 24 hours or less, the tank holding time must be at least twice the travel time. When the travel time exceeds 24 hours, the tank's holding time must be at least 24 hours greater than the normal travel time.

Tube Trailers

Shipment of hydrogen chloride is also made in tube trailers. Manifolding is authorized for specification cylinders containing hydrogen chloride provided that each cylinder is equipped with an individual shut-off valve. Each cylinder must be equipped with a Type CG-4 combination pressure relief device on each end.

Special Single-Unit Insulated Tank Cars

Shipment of hydrogen chloride as a refrigerated liquid in DOT 105A600W tank cars is authorized provided:

(1) The shipper notifies the Bureau of Explosives whenever a car is not received by the consignee within 20 days from the date of shipment.

(2) Prior to the release of an "empty" car for transportation, the pressure in the car must be reduced to below 70 psig (483 kPa). Current regulations should be checked for possible change.

(3) The tank car containing hydrogen chloride refrigerated liquid must have the auxiliary valve on the pressure relief device closed during transportation.

METHODS OF MANUFACTURE

Hydrogen chloride is produced as a byproduct from the chlorination of benzene and other hydrocarbons, and by burning hydrogen, methane, or water gas in a chlorine atmosphere.

REFERENCES

[1] *Book of SEMI Standards 1986,* Semiconductor Equipment and Materials International, 805 East Middlefield Road, Mountain View, CA 94043.

[2] *Code of Federal Regulations,* Title 29 CFR Parts 1900–1910, (Labor), Superintendent of Documents, U.S. Government Printing Office, Washington, DC 20402.

[3] *Threshold Limit Values and Biological Exposure Indices,* 1989-90 ed., American Conference of Governmental Industrial Hygienists, 6500 Glenway Avenue, Bldg. D-7, Cincinnati, OH 45211-4438.

[4] *Code of Federal Regulations,* Title 49 CFR Parts 100–199, (Transportation), Superintendent of Documents, U.S. Government Printing Office, Washington, DC 20402.

[5] *Transportation of Dangerous Goods Regulations,* Canadian Government Publishing Centre, Supply and Services Canada, Ottawa, Ontario, Canada K1A 0S9.

[6] CGA S-1.1, *Pressure Relief Device Standards—Part 1—Cylinders for Compressed Gases,* Compressed Gas Association, Inc., 1235 Jefferson Davis Highway, Arlington, VA 22202.

[7] *CGA V-1, American National, Canadian, and Compressed Gas Association Standard for Compressed Gas Cylinder Valve Outlet and Inlet Connections* (ANSI/CSA/CGA V-1), Compressed Gas Association, Inc., 1235 Jefferson Davis Highway, Arlington, VA 22202.

ADDITIONAL REFERENCES

AGA Gas Handbook, AGA AB, Lidingö, Sweden, 1985.

Matheson Gas Data Book, 6th ed., Matheson Gas Products, Inc. Secaucus, NJ 07094. (1980)

Hydrogen Fluoride

Chemical Symbol: HF
Synonyms: Anhydrous hydrofluoric acid
CAS Registry Number: 7664-39-3
DOT Classification: Corrosive material
DOT Label: Corrosive
Transport Canada Classification: 2.4 (6.1) (9.2)
UN Number: UN 1052

PHYSICAL CONSTANTS

	U.S. Units	SI Units
Chemical formula	HF	HF
Molecular weight	20.01	20.01
Boiling point	67.14°F	19.52°C
Vapor pressure at 70°F (21.1°C)	15.54 psia	107 kPa abs
Specific volume at 70°F (21.1°C)	19.3 ft^3/lb	1.2 m^3/kg
Critical temperature	370.4°F	188°C
Critical pressure	940.5 psia	6470 kPa abs
Critical density	18.2 lb/ft^3	290 kg/m^3
Melting point	−118.43°F	−83.6°C
Latent heat of vaporization at boiling point	161 Btu/lb	373 kJ/kg
Vapor density at boiling point	0.1985 lb/ft^3	3.2 kg/m^3
Liquid density at boiling point	59.9 lb/ft^3	959 kg/m^3
Specific heat of the liquid at 67.2°F (19.54°C)	0.603 Btu/(lb)(°F)	2.523 kJ/(kg)(°C)
Triple point	−118.1°F	−83.4°C
Specific gravity of the gas at 77°F (25°C) (air = 1)	1.858	1.858

DESCRIPTION

Hydrogen fluoride is a colorless, corrosive, inorganic acid. It is shipped as a liquid under its own vapor pressure. It has a normal boiling point of 67.1°F (19.5°C) and therefore has a vapor pressure which is close to atmospheric pressure under most shipping conditions. When exposed to air, anhydrous hydrogen fluoride and concentrated solutions of hydrofluoric acid produce pungent fumes. Hydrogen fluoride dissolves in water with a high heat of solution.

GRADES AVAILABLE

Anhydrous hydrogen fluoride is available from a number of suppliers with grades ranging from 99.0 to 99.96 percent. The major impurities are water (H_2O) and sulfur dioxide (SO_2). A normal specification for 99.90 mole percent hydrogen fluoride is as follows:

	Mole Percent
Hydrofluoric acid	99.9
Hydrofluosilicic acid	0.05
Water	0.025
Nonvolatile acidity	0.005
Sulfur dioxide	0.01

USES

Hydrogen fluoride has a wide variety of uses in a number of different industries. It is used in the chemical industry as an electrolyte in the manufacture of pure fluorine, as a fluorinating agent used to produce inorganic fluorides and fluorocarbons, and in the production of aqueous hydrofluoric acid.

The aluminum industry uses hydrogen fluoride as a fluoride ion source in the production of aluminum fluoride and cryolite. It is also used as a catalyst in petroleum refining and in the production of uranium hexafluoride used in the production of atomic energy fuels.

PHYSIOLOGICAL EFFECTS

Hydrogen fluoride is highly corrosive to all living tissue. Contact with liquid anhydrous hydrogen fluoride, its vapor, or hydrogen fluoride solutions can cause severe burns to skin, eyes, or respiratory tract. A threshold limit value (TLV) ceiling limit of 3 ppm (2.5 mg/m^3) for hydrogen fluoride has been adopted by the American Conference of Governmental Industrial Hygienists. [1] The U.S. Occupational Safety and Health Administration has adopted an 8-hour time-weighted average exposure limit of 3 ppm for hydrogen fluoride (as F) and a 15-minute short term exposure limit (STEL) of 6 ppm. [2]

As with most acids, the initial extent of the burn depends on concentration and duration of contact. Unlike other acids, hydrofluoric acid penetrates the skin, causing destruction of deep tissue layers including bone. This process may continue for days.

Another problem unique to hydrofluoric acid burns is delayed symptoms. Concentrations less than 20 percent may not produce symptoms for 24 hours. Symptoms of exposure to concentrations between 20 and 50 percent may be delayed for one to eight hours. Concentrations above 50 percent produce immediate burning, pain, and tissue destruction.

Severe exposure to hydrogen fluoride by any route will cause the lowering of serum calcium (hypocalcemia), which can be fatal if not treated. Hypocalcemia is possible in all instances of inhalation or ingestion and whenever skin burns exceed 25 in.^{2} (160 cm^2).

Inhalation of hydrofluoric acid vapors leads to excessive irritation of the lungs, causing acute pneumonitis (deep lung inflammation) and pulmonary edema (abnormal fluid buildup in the lungs). Symptoms include coughing, tearing of eyes, labored breathing, chest tightness, and excessive saliva and sputum formation.

MATERIALS OF CONSTRUCTION

Carbon steel (without nonmetallic inclusions) is acceptable for handling hydrogen fluoride up to approximately 150°F (65°C). Aluminum-silicon-bronze, stainless steel, or nickel are suitable for cylinder valves. For

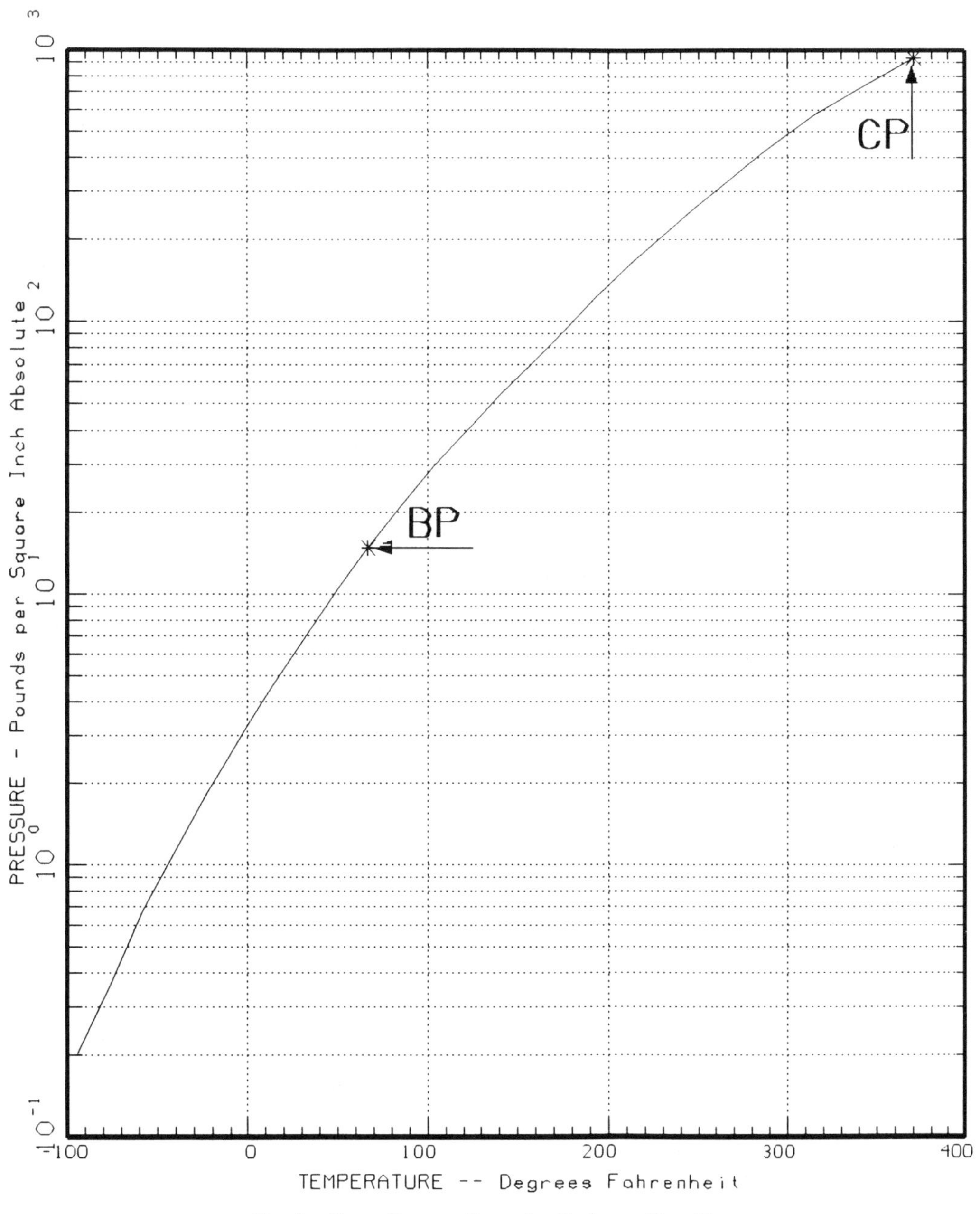

Fig. 1. Vapor Pressure Curve for Hydrogen Fluoride.

higher temperatures, Monel, Inconel, nickel, or copper should be used. Cast iron or malleable fittings should be avoided. Polyethylene, lead, soft copper, Kel-F, and Teflon are acceptable gasket materials. Polyethylene, Kel-F, and Teflon are acceptable packing materials.

SAFE STORAGE, HANDLING, AND USE

Cylinders should be protected from physical damage and stored in a cool, dry, well-ventilated area. The temperature in the storage area should not exceed 125°F (52°C). The valve protection cap should remain in place unless the cylinder is secured and has the valve outlet connected through appropriate piping to the use point. Do not heat the cylinder to increase the discharge rate. Use a check valve or trap in the discharge line to prevent hazardous backflow into the cylinder.

When liquid product is used, care must be taken to avoid trapping liquid in the piping system. Hydrostatic pressure formed by trapped liquid may cause catastrophic failure of the process line. Areas where possible liquid trapping may occur must be protected by a pressure relief device.

On-site bulk storage tanks should be sized one and one-half times the size of proposed deliveries, or ten days' requirements, whichever is greater. Storage tanks should be designed for a minimum working pressure of 60 psig (414 kPa) or a pressure to suit the particular installation. Storage tanks should be equipped with pressure relief devices safely vented to a scrubber.

All anhydrous hydrogen fluoride systems must be kept scrupulously dry. Moisture in the system increases the rate of corrosion of hydrogen fluoride, thus reducing the working life of the system. Piping should be purged with dry inert gas prior to introducing hydrogen fluoride. Systems should also be purged before opening to the atmosphere. If a system is to be shut down for a period of time, it should be sealed and left under a slight positive pressure of inert gas.

Personnel working with hydrogen fluoride should wear safety glasses plus a face shield, acid-resistant gauntlet-type gloves, and acid-resistant jacket, pants, and boots. For emergency use, a positive-pressure air line with mask or self-contained breathing apparatus and acid suits should be available.

Safety equipment, especially gloves, should be inspected frequently for tears or pinholes. All contaminated protective clothing should be washed before removal to prevent accidental exposure. Contaminated tools should be washed after use.

When removing a cylinder from service, either temporarily or for return to the supplier, close the valve and tightly reinstall the outlet cap. Compressed gas cylinders should not be refilled except by qualified suppliers of compressed gases.

DISPOSAL

Routine disposal of hydrogen fluoride may be accomplished by discharging the gas directly into a counterflow caustic (KOH) scrubber. The system should contain an "anti-suckback" device to prevent reverse flow into the cylinder. The scrubber vent should discharge into a well-ventilated area away from personnel and building ventilation air intakes, and be monitored for acid breakthrough.

HANDLING LEAKS AND EMERGENCIES

Leaks of anhydrous hydrogen fluoride will appear as white fumes or vapors. This is due to the reaction of the hydrogen fluoride with moisture in the air to form hydrofluoric acid. The "fume cloud" is very acidic and dangerous. Proper protective equipment is required for all personnel entering the area.

If the leak is in the process system, shut the gas off at the source. Vent process gas to a scrubber and purge lines with dry inert gas before making permanent repairs.

If the leak is in a container or container valve, isolate the container in a remote area away from personnel and notify the supplier.

First Aid

Speed in removing the victim from the contaminated atmosphere or removing the vappor or liquid from the skin or eyes is essential. First aid must start immediately in all cases of contact with the particular gas in any form. All affected persons should be referred to a physician, no matter how slight the injury, and the physician given a detailed account of the accident. First aid for various routes of exposure follow. [3] As soon as possible, more thorough medical treatment should be administered by a physician, preferably one familiar with hydrogen flouride exposure treatment.

Inhalation

Remove the victim to an uncontaminated atmosphere and administer 100 percent oxygen immediately. It has been found helpful to treat even borderline cases with 100 percent oxygen at half-hour intervals for three or four hours. Bronchospasm may be treated by emergency response technicians with a bronchodilator, such as albuterol, and an anticholinergic inhalant, such as Atrovent.

Eye Contact

In the event of eye contact, immediately flush with water for 15 minutes. Anesthetic eye drops should be used after an initial one-minute water irrigation to reduce muscle spasm of the eyelids and to facilitate irrigation.

Skin Contact

Workers who have had skin contact should receive a drenching shower for 15 minutes. The clothing should be removed as rapidly as possible while the victim is in the shower. For small burns (2 in.^{2}; 6.45 cm^2) or burns involving the face, 2.5 percent calcium gluconate gel (made up by a pharmacist by adding 2.5 g calcium gluconate USP to 100 ml KY-Jelly) should then be rubbed onto the affected area continuously for $1\frac{1}{2}$ hours or until further medical care is available. Those initially applying the gel should wear rubber gloves. For larger burns or burns involving the fingertips and nail beds, the affected part should be placed in an iced Zephiran (1:750) or Hyamine 1622 (1:500) bath for four hours. If a bath cannot be used, iced compresses with one of these agents should be used, changing the compress every two minutes. Solutions of Zephiran or Hyamine 1622 should not be used on the face. [3]

See *Effects of Exposure to Toxic Gases—First Aid and Medical Treatment* for additional information. [3]

METHODS OF SHIPMENT

Under the appropriate regulations, anhydrous hydrogen fluoride is authorized for shipment as follows: [4] and [5]

By Rail: In cylinders (freight cars) and in insulated tank cars. Forbidden on passenger cars.

By Highway: In cylinders and tank transports on trucks.

By Water: In cylinders on cargo vessels stowed above deck. Forbidden on passenger vessels.

By Air: Cargo aircraft only with a 110 lb (49.9 kg) maximum net weight per container limit. Forbidden on passenger-carrying aircraft.

CONTAINERS

Anhydrous hydrogen fluoride is shipped in cylinders, tank trucks, and insulated tank cars.

Filling Limit

The maximum filling density authorized for hydrogen fluoride by DOT is 85 percent (percent water capacity by weight).

Cylinders

Cylinders that meet the following TC/DOT specifications are authorized for hydrogen fluoride service: 3A, 3AA, 3B, 3C,

3E, 4, or 4A; also 4B, 4BA, 4BW, or 4C if not brazed. Cylinders manufactured under DOT-3, DOT-25, or DOT-38 may be continued in service, but new construction is not authorized. Pressure relief devices are not required on hydrogen fluoride cylinders. [6]

Valve Outlet Connections

The standard valve outlet connection in the United States and Canada for hydrogen fluoride cylinders is Connection CGA 670. A limited standard connection for hydrogen fluoride cylinders is Connection CGA 660. [7]

Tanks

DOT specification 106A500W or 110A500W tanks (ton containers) are approved for hydrogen fluoride service provided they do ***not*** have pressure relief devices of any type and the container valves are protected by metal caps. Cargo tanks under specification MC-310, MC-311, or MC-312 are also authorized as are TC/DOT Specification 51 portable tanks.

Tank Cars

Tank cars of TC/DOT specification 105A300W, 112A400W, or 114A400W equipped with special valves and fittings compatible with hydrogen fluoride are approved for service. Specification 114A400W tanks must have valves and fittings located on top of the tank. Bottom openings in tanks are prohibited.

METHODS OF MANUFACTURE

Hydrogen fluoride is produced by the reaction with sulfuric acid and fluorospar (CaF_2) or by heating potassium fluoride acid (KHF_2).

REFERENCES

[1] *Threshold Limit Values and Biological Exposure Indices,* 1989–90 ed., American Conference of Governmental Industrial Hygienists, 6500 Glenway Avenue, Bldg. D-7, Cincinnati, OH 45211-4438.

[2] *Code of Federal Regulations,* Title 29 CFR Parts 1900–1910 (Labor). Superintendent of Documents, U.S. Government Printing Office, Washington, DC 20402.

[3] *Effects of Exposure to Toxic Gases—First Aid and Medical Treatment,* 3rd ed., Matheson Gas Products, Inc., Secaucus, NJ 07094. (1988)

[4] *Code of Federal Regulations,* Title 49 CFR Parts 100–199, (Transportation). Superintendent of Documents, U.S. Government Printing Office, Washington, DC 20402.

[5] *Transportation of Dangerous Goods Regulations,* Canadian Government Publishing Centre, Supply and Services Canada, Ottawa, Ontario, Canada K1A 0S9.

[6] CGA S-1.1, *Pressure Relief Device Standards—Part 1—Cylinders for Compressed Gases,* Compressed Gas Association, Inc., 1235 Jefferson Davis Highway, Arlington, VA 22202.

[7] CGA V-1, *American National, Canadian, and Compressed Gas Association Standard for Compressed Gas Cylinder Valve Outlet and Inlet Connections* (ANSI/CSA/CGA V-1), Compressed Gas Association, Inc., 1235 Jefferson Davis Highway, Arlington, VA 22202.

ADDITIONAL REFERENCES

Material Safety Data Sheet for Hydrogen Fluoride, Air Products and Chemicals, Inc., Allentown, PA 18195.

Matheson Gas Data Book, 6th ed., Matheson Gas Products, Inc., Secaucus, NJ 07094. (1980)

"Recommended Medical Treatment for Hydrofluoric Acid Exposure," Allied Signal Inc., P.O. Box 1139R, Morristown, NJ 07960.

Storage and Handling Hydrofluoric Acid Anhydrous, E. I. du Pont de Nemours & Co., Wilmington, DE 19898.

Hydrogen Sulfide

Chemical Symbol: H_2S
Synonyms: Sulfuretted hydrogen, hydrogen sulphide, hydrosulfuric acid
CAS Registry Number: 7783-06-4
DOT Classification: Flammable gas
DOT Label: Flammable gas and Poison
Transport Canada Classification: 2.1
UN Number: UN 1053

PHYSICAL CONSTANTS

	U.S. Units	SI Units
Chemical formula	H_2S	H_2S
Freezing point at 1 atm	−117.2°F	−82.9°C
Boiling point at 1 atm	−76.4°F	−60.2°C
Specific gravity of the liquid at 60°F (15.5°C)	0.79	0.79
Vapor pressure at 60°F (15.5°C)	229 psia	1579 kPa abs
Density of saturated gas		
at 59°F (15°C)	1.658 lb/ft^3	26.55 kg/m^3
at 70°F (21.1°C)	1.938 lb/ft^3	31.04 kg/m^3
Density of saturated liquid		
at −76°F (−60.3°C)	57.1 lb/ft^3	915.3 kg/m^3
at 59°F (15°C)	49.2 lb/ft^3	787.8 kg/m^3
at 70°F (21.1°C)	48.3 lb/ft^3	774.2 kg/m^3
at 105°F (40.6°C)	45.5 lb/ft^3	728.1 kg/m^3
at 115°F (46.1°C)	44.5 lb/ft^3	712.5 kg/m^3
at 130°F (54.4°C)	43.0 lb/ft^3	688.7 kg/m^3
Specific gravity of the vapor at 59°F (15°C) (air = 1)	1.189	1.189
Solubility in water at 68°F (20°C) and 1 atm	0.317 lb/gal	38 kg/m^3
Percent volatile	100	100
Upper flammable limit (approximate)	46%	46%
Lower flammable limit (approximate)	4.3%	4.3%
Molecular weight	34.08	34.08

	U.S. Units	SI Units
Autoignition temperature (approximate)	500°F	260°C
Critical temperature	212.9°F	100.5°C
Critical pressure	1306.5 psia	9008 kPa abs
Specific heat of the liquid at 60°F (15.5°C)	0.49 Btu/(lb)(°F)	0.931 kJ/(kg)(°C)
Specific heat of the vapor at 60°F (15.5°C)	0.035 Btu/(lb)(°F)	0.1465 kJ/(kg)(°C)
Latent heat of the vapor at 60°F (15.5°C)	191.1 Btu/lb	444.5 kJ/kg
Latent heat of fusion at 117.2°F (47.33°C)	29.99 Btu/lb	697.6 kJ/kg
Vapor pressure of the saturated gas		
at 59°F (15°C)	211.2 psig	1456.2 kPa
at 70°F (21.1°C)	249.8 psig	1722.3 kPa

DESCRIPTION

Hydrogen sulfide is a colorless, flammable, poisonous gas or liquid, with an offensive odor and irritant properties. Hydrogen sulfide is slightly heavier than air, and the liquid phase is somewhat less dense than water. Combustion of hydrogen sulfide in air forms sulfur dioxide and water.

Hydrogen sulfide can act as a reducing agent. It reacts readily with all metals in the electromotive series down to and including silver. However, in some cases the resultant sulfide coating prevents further reaction. Hydrogen sulfide is somewhat soluble in water, alcohol, petroleum solvents, and crude petroleum.

In commerce, hydrogen sulfide is transported as a flammable liquefied compressed gas. It is transported and stored under its own vapor pressure in authorized cylinders, tank cars, and cargo tanks (tank trucks). It is used in research, in metals refining, and in the production of a number of basic and specialty chemicals.

GRADES AVAILABLE

Hydrogen sulfide is available in a technical or commercial grade which is 99 mole percent minimum hydrogen sulfide. It is also available in higher purities, up to 99.99 mole percent minimum hydrogen sulfide.

USES

Hydrogen sulfide is used commercially to purify hydrochloric and sulfuric acid, to precipitate sulfides of metals, and to manufacture elemental sulfur and organosulfur compounds. Chemical production processes using hydrogen sulfide include the manufacture of mercaptans; pharmaceuticals; plastics; adhesives; television, CRT, and fluorescent tube phosphors; dyes; pigments; biodegradable pesticides; ethylene; nylon; soda ash; sodium hydrosulfide; heavy water; and others.

Hydrogen sulfide is used as a reducing agent in cresylic acid recovery. It is also employed as a reagent in analytical chemistry. In waste water cleanup and groundwater restoration projects, hydrogen sulfide is used to immobilize heavy metals.

PHYSIOLOGICAL EFFECTS

Hydrogen sulfide is a toxic, irritating, and asphyxiant gas. The substance is known to be produced and metabolized naturally in the human body at low concentrations, but

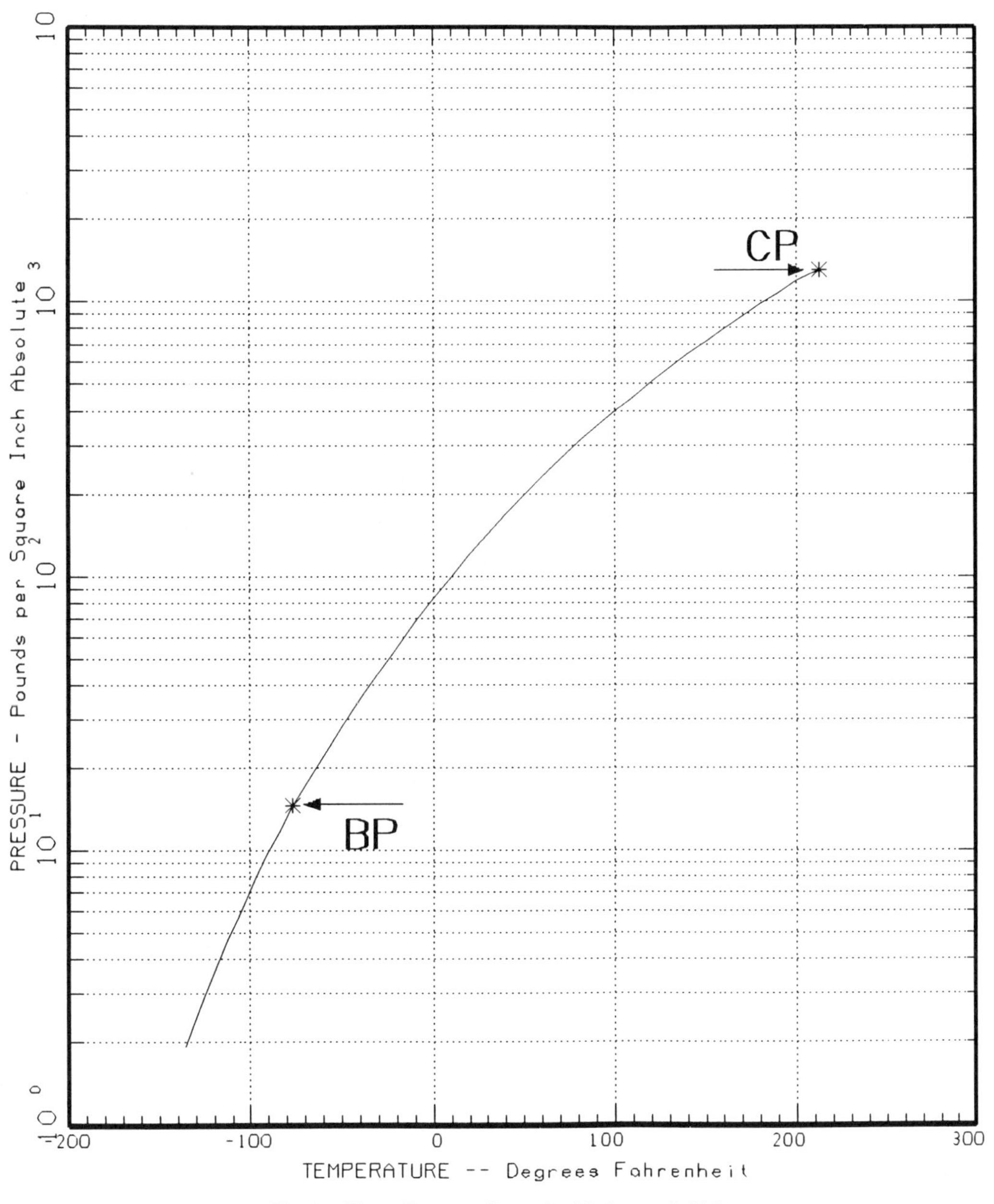

Fig. 1. Vapor Pressure Curve for Hydrogen Sulfide.

can be quickly fatal once the natural bodily defenses are overwhelmed.

In the United States, the Occupational Safety and Health Administration (OSHA) has promulgated an acceptable time-weighted average (TWA) concentration of 10 ppm (14 mg/m^3) during an eight-hour shift with a permissible short term exposure limit (STEL) of 15 ppm (21 mg/m^3) for 15-minute time-weighted average exposures. [1] The American Conference of Governmental Industrial Hygienists has adopted an eight-hour time-weighted average threshold limit value (TLV-TWA) of 10 ppm (14 mg/m^3) and a 15-minute short-term exposure limit (STEL) of 15 ppm (21 mg/m^3). [2] To avoid eye irritation, levels below 10 ppm on a time-weighted average basis have been recommended by manufacturers. The NIOSH criteria document recommends exposure be limited to 10 ppm for 10 minutes, based on the potential for eye irritation in some individuals. [3]

The primary routes of entry to be addressed in occupational use are inhalation and eye exposure, both of which are easily prevented by the use of chemical goggles or faceshields and approved respirators. Frostbite may occur on contact with escaping liquid hydrogen sulfide or with equipment containing evaporating liquid.

Effects of overexposure to hydrogen sulfide include irritation of the eye and throat at concentrations of about 20 ppm, pain in these areas at concentrations of about 100 ppm, deadening of the sense of smell, and coughing. Concentrations of 250–450 ppm produce weariness, headache, and dizziness. Higher exposures in the range of 500–800 ppm lead to rapid loss of consciousness, respiratory paralysis, coma, and death. The above concentrations are for illustration of short-term effects of one hour or less; the severity of the effects of exposure vary with both length of exposure and concentration, and also among individuals.

Respiratory paralysis can be almost immediate at concentrations above 600–800 ppm, and a single breath at slightly higher concentrations can cause unconsciousness. Death due to respiratory paralysis will follow unless the victim is removed to fresh air and resuscitation quickly administered. If breathing is weak or has ceased, give artificial respiration and oxygen at once.

The hazards of various concentrations of hydrogen sulfide in air are as given in Table 1.

There are no reports in the literature associating hydrogen sulfide in air with carcinogenesis, mutagenesis, or teratogenesis.

Odor and the Sense of Smell

Hydrogen sulfide has a strong odor of "rotten eggs" at concentrations as low as 0.1 ppm changing to a "sickening sweet odor" as levels rise above 50–200 ppm. Exposure above 100 ppm may rapidly deaden the sense of smell, reportedly in as little as 2–15 minutes, particularly at the higher concentrations. Above 200 ppm, loss of smell is very rapid.

Olfactory fatigue may also, reportedly, re-

TABLE 1. PHYSIOLOGICAL EFFECTS OF VARIOUS CONCENTRATIONS OF HYDROGEN SULFIDE IN AIR. [4]

Concentration (ppm)	Response
10–30	Keratitis after prolonged exposures
100–150	Rapid olfactory fatigue
150–250	Mucous membrane and eye irritation within one hour
250–600	Pulmonary edema after prolonged exposures
600–1000	Apnea and death after 30–60 min.
1800	Immediate collapse and respiratory paralysis

sult from prolonged exposure to levels somewhat below 100 ppm. This latter effect has not been observed in normal outdoor operations and the normally associated chronic intermittent exposures which result from handling routine small leakages from closed piping systems. Users should beware, however, that in enclosed spaces, buildings, and so on, where concentrations may be relatively constant, olfactory acclimatization or fatigue will occur, and rising concentrations may not be noticed in such cases.

Related Physiological Considerations

Consideration should be given to limiting exposures that are mixed with other toxic materials, such as carbon disulfide, because synergistic effects may occur and are suggested in the literature. The use of central nervous system depressants, including alcohol, during or prior to work should be discouraged for this reason.

Skin absorption from exposure to high concentrations (100 percent gas) has been demonstrated in the laboratory on shaved animals. However, no evidence exists to suggest that skin absorption is an ordinary route of entry in humans, even at the exposure concentrations and durations encountered in handling relatively large leaks outdoors. Such leaks have been routinely handled without rubber suits and in ordinary work clothes and gloves. Prudence suggests that exposure should be minimized by working upwind of leaks. Respiratory and eye protection are advised in leak situations.

MATERIALS OF CONSTRUCTION

Dry hydrogen sulfide is satisfactorily handled under pressure at normal ambient temperatures in carbon steel or black iron piping. Carbon steels in wet applications are known to be subject to sulfide stress cracking and low-temperature brittle fracture under some conditions of temperature, stress, and pressure. While hydrogen sulfide itself is relatively noncorrosive to steel in many uses, factors such as impurities, pH, erosive conditions, and high thermal or mechanical stresses in the metal can cause severe corrosion problems. High-strength steels are subject to crack formation when exposed to hydrogen sulfide.

Aluminum is considered to be a Class A material of construction for wet or dry hydrogen sulfide even at elevated temperatures. Carpenter-20 stainless steel is also considered a Class A material for dry or moist hydrogen sulfide at 100–170°F (37.7–76.7°C). [5] At high temperatures, for example, 900°F (482°C), high nickel/chromium alloys (made of 15–18 percent Cr, 35–75 percent Ni) are rapidly attacked. [6] Other materials resistant to both dry and moist hydrogen sulfide are Hastelloy C (up to 170°F; 76.7°C), Inconel (70°F; 21.1°C), polyethylene, epoxy, Teflon, and PVC. [7] and [8]

Hydrogen sulfide, or possibly the sulfide ion, is thought to act as a catalyst in the absorption of hydrogen into the metal surface, which leads to embrittlement. As a result, mixtures of hydrogen sulfide with hydrogen have been shown to cause accelerated stress corrosion cracking of high-strength steel under moist conditions. [9] The action of these mixtures has also resulted in premature failure of combination pressure relief devices (fuse metal-backed rupture disks). It is important to ensure dryness with mixtures of hydrogen sulfide with hydrogen. Furthermore, cylinder suppliers are encouraged to routinely replace combination pressure relief devices with new ones for cylinders in hydrogen/hydrogen sulfide mixture service.

Brass valves, though tarnished by dry hydrogen sulfide, have been found to withstand years of cylinder service without appreciable malfunction. Fasteners are generally of annealed (softened) carbon steel or stainless steel. Good results in resisting environmental and sulfide attack in air have been had with B-8M stainless bolting.

SAFE STORAGE, HANDLING, AND USE

All the precautions required for the safe handling of any pressurized, flammable, toxic gas must be observed with hydrogen sulfide. Personnel should be trained in the handling of this gas prior to its use. For more information on hydrogen sulfide, and procedures and equipment for its use, see CGA G-12, *Hydrogen Sulfide.* [10]

Containers and piping should be protected from physical damage, backflow, fire, and abuse. Procedures and equipment should be in place to prevent exposure of workers and/or the public to released gas and to deal with potential releases. Do not store hydrogen sulfide near oxidizing materials, acids, or other corrosive materials. Use isolated storage areas clear of large accumulations of flammable materials, which, if ignited, might expose containers to high temperatures. Protect from unauthorized access. Adequate water and hoses should be nearby to cool tanks in case of fire.

Positive-pressure full-face breathing apparatus should be on site for use in handling potential leaks. Persons working where hydrogen sulfide is present should be trained in the appropriate actions to take during an emergency, and should also be trained in providing artificial respiration. Areas not continuously manned should have emergency telephone numbers posted conspicuously. During transfer operations, investigation of leaks, or handling of emergencies, the buddy system should be used.

Maintenance Precautions

Before entry into vessels, enclosed spaces, and trenches which may have contained hydrogen sulfide, purge thoroughly with air and test for oxygen and hydrogen sulfide concentrations. Where applicable, steaming followed by forced airing is most readily effective to drive residual material from vessels. Before purging and entry, isolate vessels completely and positively from other process lines and equipment (line blinds/disconnection/locked double block and bleed). Workers should be equipped with approved respiratory protection. Keep workers who must be inside vessels and enclosed spaces under constant surveillance and have rescue equipment at hand. Personnel, while inside vessels, should not use steam, hot water, or cleaning agents likely to react with remaining sulfide residues (acids, for example) and liberate hydrogen sulfide.

Dissolved hydrogen sulfide in liquids such as water, alcohol, oils, or sludge may be released upon warming. Retest the air in vessels frequently and whenever vessel temperature rises (as from sunlight or other forms of heating).

For maximum safety, avoid use or storage indoors. Indoor areas should have positive ventilation with at least six volumes of air changes per hour. Use of a fume hood in laboratories is advisable. Automatic air-monitoring equipment is advisable in indoor applications. Isolated outdoor areas provide maximum handling safety. Natural ventilation helps to dissipate leaks rapidly. Gas dispersion analysis should be performed for a specific location to determine potential geographic exposure.

Detection Equipment

Portable chemical detectors, such as the "detector tube" variety or the electronic variety, should be made available to workers. Leak location can be assisted with lead acetate tape and soap solutions. Automatic fixed-point monitors may also be used outdoors, but may be of limited value.

DISPOSAL

Depending on local conditions and regulations, waste gas may be safely disposed of to flare, to process, or by wet scrubbing. Absorption techniques may generate a liquid hazardous material which must be sold, treated, recycled, or properly disposed. A large quantity of material in need of emergency disposal is hazardous and may require disposal by flaring, which should be done with the consent of local air quality officials. Notify the shipper of any leaking or damaged shipping containers, and of any unintentional discharges of gas from shipping containers to the atmosphere. Shipping containers must be returned to the shipper when empty.

Emissions, spills, inventory, and disposal of hydrogen sulfide are also regulated under the Superfund Act and under the Resource Recovery and Conservation Act in the United States. The reportable quantity (RQ) for hydrogen sulfide is 100 lb (45.4 kg).

HANDLING LEAKS AND EMERGENCIES

No one should enter a leak area without proper respiratory protection, even to attempt a rescue! Without proper protection, there is a great risk of the rescue personnel themselves being overcome by gas, which will only compound the rescue problem.

Keep unprotected persons well way and upwind from an area of hazardous concentrations of hydrogen sulfide. As an initial precaution with sizable leaks, advise such persons to keep several hundred yards (meters) away from the leak.

Notify the supplier immediately to obtain advice and/or assistance. Notify appropriate public agencies. Evacuation of downwind areas may be necessary.

Work in the buddy system whenever working with a known or suspected leak. Wear proper protective equipment, such as positive-pressure full-face air masks in handling leaks. Prevent ignition of the leak by eliminating sources of ignition such as sparks, smoking materials, and open flames. Shut off the source of the leaking gas. Isolate the leak from large reservoirs of hydrogen sulfide liquid or gas such as delivery vehicles, cylinders, and storage tanks.

Mitigation

If the leak cannot be stopped by closing valves, it may be possible to transfer the material to another suitable vessel. The magnitude of any leak can also be greatly diminished by removing vapor (to lower the pressure) from the leaking vessel and disposing of it by safe means as quickly as practical. If the leaking vessel is movable, and not seriously damaged, it is advisable to move it to a nearby, lower-risk area.

Small Leaks

The presence of small leaks is generally indicated by odor alone, and such leaks may be found with indicating detectors and/or soap solution (bubble checking). Occasionally, leaks may become manifest as "frosted" areas of drips of liquid and may have a hissing sound associated with them. Even small leaks can be dangerous in areas of poor ventilation, particularly indoors. Leaks should always be repaired rapidly.

First Aid

Inhalation: Remove the exposed person to fresh air immediately, without endangering rescuers. If victim's breathing has stopped, give artificial respiration immediately and continue until breathing is restored. If breathing is difficult, administer oxygen. If more than one victim is involved, enlist the aid of others. Do not delay or cease administration of resuscitation to any victim whose breathing has stopped as seconds are important. After natural breathing is restored, treat as for shock; keep patient warm and quiet. Reassure the victim. Recovery is normally rapid and complete with prompt treatment. As soon as practical, seek medical aid.

For additional information concerning the treatment of severe exposure, see *Effects of Exposure to Toxic Gases—First Aid and Medical Treatment.* [4]

Skin and eyes: If irritation occurs, flush eyes and/or body for 15 minutes with clean, preferably tepid water. Remove contaminated clothing. Seek medical attention.

Frostbite: In case of frostbite, place the frostbitten part in warm water, about 108°F (42°C). If warm water is not available or is impractical to use, wrap the affected part gently in blankets. Let the circulation reestablish itself naturally. Encourage the victim to exercise the affected part while it is being warmed. Consult a physician. [4]

METHODS OF SHIPMENT

Under the appropriate regulations, hydrogen sulfide is authorized for shipment as follows:

By Rail: In cylinders and multi-unit tank cars (TC/DOT-106A800X) and tank cars under exemption.

By Highway: In cylinders, tube trailers; and cargo and portable tanks by exemption.

By Water: On deck on cargo vessels; forbidden on passenger vessels.

By Air: On cargo aircraft only, 300 lb (136 kg) maximum net weight per container. Forbidden on passenger aircraft.

CONTAINERS

Cylinders

A number of cylinders are authorized for hydrogen sulfide service, at a filling density of 62.5 percent, including TC/DOT specification 3A480, 3AA480, 3B480, 4A480, 4B480, 4BA480, 4BW480, 3AL480, 3E1800, and 26-480. Pressure relief devices authorized for use on hydrogen sulfide cylinders include the Type CG-2 fusible plug and the CG-4 combination rupture disk/fusible plug device. For further information and additional requirements, see CGA S-1.1, *Pressure Relief Device Standards—Part I—Cylinders for Compressed Gases.* [11]

Valve Outlet Connections

The standard valve outlet connection in the United States and Canada for hydrogen sulfide cylinders is Connection CGA 330. [12]

Tank Cars

Bulk shipments of liquid hydrogen sulfide are authorized by specific exemption by the United States Department of Transportation and Transport Canada. Bulk transportation, at a filling density of 72 percent, is in specially insulated and equipped MC-331 trailers with a working pressure of 500 psig (3447 kPa), and specially insulated and equipped specification 105J600W, 120A600W, and 120J600W tank cars. Multi-unit tank cars of TC/DOT specification 106A800X are authorized for use at a filling density of 68 percent and requiring fusible plug pressure relief devices having a maximum yield temperature of 170°F (76.6°C) and not less than 157°F (69.4°C). Valve outlets must be sealed with a threaded outlet cap or plug, and valves protected by a metal cover.

Other containers and fill densities may be authorized by exemption.

METHODS OF MANUFACTURE

Most hydrogen sulfide produced today is made as a by-product of other processes. The largest tonnage currently made originates as a by-product from the natural gas industry.

Many methods have been used in the past to produce hydrogen sulfide. The classical method consisted of iron sulfide plus an acid (usually hydrochloric). It also can be made by combining sulfur with hydrogen in a noncatalytic reactor at elevated temperature.

REFERENCES

[1] *Code of Federal Regulations,* Title 29 CFR Parts 1900–1910 (Labor), Superintendent of Documents, U.S. Government Printing Office, Washington, DC 20402.

[2] *Threshold Limit Values and Biological Exposure Indices,* 1989–90 ed., American Conference of Governmental Industrial Hygienists, 6500 Glenway Avenue, Bldg. D-7, Cincinnati, OH 45211.

[3] *Criteria for Recommended Standard for Occupational Exposure to Hydrogen Sulfide* (May 1977), National Institute for Occupational Safety and Health (NIOSH), U.S. Government Printing Office, Washington, DC 20402.

[4] *Effects of Exposure to Toxic Gases—First Aid and Medical Treatment,* 3rd ed., Matheson Gas Products, Inc., Secaucus, NJ 07094. (1988)

[5] *A Guide to Corrosion Resistance,* Climax Molybdenum Company, a Division of American Metal Climax, Inc.

[6] Uhlig, *Corrosion Handbook,* Wiley, New York, 1948.

[7] *Corrosion of Metals,* American Society of Metals, Cleveland, OH, 1958.

[8] *Compass Corrosion Guide,* La Mesa, CA, 1980.

[9] LaQue and Copson, *Corrosion Resistance of Met-*

als and Alloys, 2nd ed., Van Nostrand Reinhold, New York. (1963)

[10] CGA G-12, *Hydrogen Sulfide,* Compressed Gas Association, Inc., 1235 Jefferson Davis Highway, Arlington, VA 22202.

[11] CGA S-1.1, *Pressure Relief Device Standards—Part 1—Cylinders for Compressed Gases,* Compressed Gas Association, Inc., 1235 Jefferson Davis Highway, Arlington, VA 22202.

[12] CGA V-1, *American National, Canadian, and Compressed Gas Association Standard for Compressed Gas Cylinder Valve Outlet and Inlet Connections* (ANSI/CSA/CGA V-1), Compressed Gas Association, Inc., 1235 Jefferson Davis Highway, Arlington, VA 22202.

ADDITIONAL REFERENCES

Clayton & Clayton, eds., *Patty's Industrial Hygiene & Toxicology,* Wiley, New York, 1978.

Code of Federal Regulations, Title 49 CFR Parts 100–199 (Transportation), Superintendent of Documents, U.S. Government Printing Office, Washington, DC 20402.

Material Safety Data Sheet for Hydrogen Sulfide, Montana Sulphur & Chemical Co., Billings, MT 59107-1118, 1988.

Review of Emergency Systems, Report to Congress, Final Report, June 1988, U.S. Environmental Protection Agency, Office of Solid Waste and Emergency Response, Washington, DC 20460.

Sax, Irving, et al., *Dangerous Properties of Industrial Materials,* 5th ed., Van Nostrand Reinhold, New York, 1979.

Transportation of Dangerous Goods Regulations, Canadian Government Publishing Centre, Supply and Services Canada, Ottawa, Ontario, Canada K1A 0S9.

Liquefied Petroleum Gases

Note: The tables of Physical Constants that follow are for research grade products.

PHYSICAL CONSTANTS
Butane

	U.S. Units	SI Units
Chemical formula	C_4H_{10}	C_4H_{10}
Molecular weight	58.124	58.124
Vapor pressure		
at 70°F (21.1°C)	16.54 psig	114.04 kPa
at 100°F (37.8°C)	36.92 psig	254.55 kPa
at 115°F (46.1°C)	50.26 psig	346.53 kPa
at 130°F (54.4°C)	66.03 psig	455.26 kPa
Density of the gas		
at 70°F (21.1°C) and 1 atm	0.15537 lb/ft^3	2.489 kg/m^3
Specific gravity of the gas		
at 70°F and 1 atm (air = 1)	2.0064	2.0064
Specific volume of the gas		
at 60°F (15.6°C) and 1 atm	6.3356 ft^3/lb	0.3955 m^3/kg
Density of the liquid		
at saturation pressure		
at 60°F (15.6°C)	36.39 lb/ft^3	582.91 kg/m^3
at 70°F (21.1°C)	35.95 lb/ft^3	575.86 kg/m^3
at 105°F (40.6°C)	34.38 lb/ft^3	550.71 kg/m^3
at 115°F (46.1°C)	34.01 lb/ft^3	544.79 kg/m^3
at 130°F (54.4°C)	33.38 lb/ft^3	534.70 kg/m^3
Boiling point at 1 atm	31.10°F	−0.51°C
Freezing point at 1 atm	−217.05°F	−138.36°C
Critical temperature	305.65°F	152.03°C
Critical pressure	550.7 psia	3796.94 kPa abs
Critical density	14.2 lb/ft^3	227.46 kg/m^3
Latent heat of vaporization		
at 31.10°F (0.51°C)	153.59 Btu/lb	357.25 kJ/kg
Latent heat of fusion		
at 217.05°F (138.36°C)	10.64 Btu/lb	25.75 kJ/kg
Specific heat of the gas		
at 60°F (15.6°C) and 1 atm		
C_p	0.3991 Btu/(lb)(°F)	1.671 kJ/(kg)(°C)
C_v	0.3649 Btu/(lb)(°F)	1.528 kJ/(kg)(°C)

	U.S. Units	SI Units
Ratio of specific heats, C_p/C_v	1.094	1.094
Solubility in water, vol/vol at 100°F (37.8°C)	0.000 061	0.000 061
Weight of the liquid at saturation pressure and 60°F (15.6°C)	4.865 lb/gal	582.955 kg/m^3
Specific heat of the liquid at 1 atm	0.5636 Btu/(lb)(°F)	2.3597 kJ/(kg)(°C)
Gross heat of combustion		
Ideal gas at 60°F (15.6°C) and 1 atm	3262.1 Btu/ft^3	121 542 kJ/m^3
Liquid at 60°F (15.6°C) and saturation pressure	21 139 Btu/lb	49 169.3 kJ/kg
Liquid at 60°F (15.6°C) and saturation pressure	102 989 Btu/gal	28 704 713 kJ/m^3
Net heat of combustion		
Ideal gas at 60°F (15.6°C) and 1 atm	3010.4 Btu/ft^3	112 164.3 kJ/m^3
Liquid at 77°F (25.0°C) and saturation pressure	19 494 Btu/lb	45 343 kJ/kg
Net heat of combustion		
Liquid at 77°F (25.0°C) and saturation pressure	93 201 Btu/gal	25 976 637 kJ/m^3
Air required for combustion		
Volume of air per 1-unit volume of the ideal gas	31.02 ft^3 air	0.8784 m^3 air
Weight of air per 1-unit weight of the ideal gas	15.459 lb air	7.0121 kg air
Flammable limits in air, percent by volume	1.8–8.4%	1.8–8.4%
Flash point	−101°F	−73.9°C

IDENTIFYING INFORMATION

	Butane	Isobutane	Propane	Propylene
Chemical Symbol:	C_4H_{10}	C_4H_{10} [or $(CH_3)_2CHCH_3$]	C_3H_8	C_3H_6 [or $CH_3CH{:}CH_2$]
Synonyms:	Normal butane N-butane butyl hydride	2-methylpropane trimethylmethane	Dimethylmethane	Propene
CAS Registry: Number	106-97-8	75-28-5	74-98-6	115-07-1
DOT Classification:	Flammable gas	Flammable gas	Flammable gas	Flammable gas
DOT Label:	Flammable gas	Flammable gas	Flammable gas	Flammable gas
Transport Canada Classification:	2.1	2.1	2.1	2.1
UN Number:	UN 1075	UN 1969	UN 1978	UN 1077

	1-Butene	cis-2-Butene	trans-2-Butene	Isobutylene
Chemical Symbol:	C_4H_8 [or $CH_2{:}CHCH_2CH_3$]	C_4H_8 [or $CH_3CH{:}CHCH_3$]	C_4H_8 [or $CH_3CH{:}CHCH_3$]	C_4H_8 [or $(CH_3)_2C{:}CH_2$]
Synonyms:	Ethylethylene alpha-butene	Dimethylethylene beta-butylene "high-boiling" butene-2	Dimethylethylene beta-butylene "low-boiling" butene-2	2-methylpropene isobutene
CAS Registry Number:	106-98-9	590-18-1	624-64-6	115-11-7
DOT Classification:	Flammable gas	Flammable gas	Flammable gas	Flammable gas
DOT Label:	Flammable gas	Flammable gas	Flammable gas	Flammable gas
Transport Canada Classification:	2.1	2.1	2.1	2.1
UN Number:	UN 1012	UN 1012	UN 1012	UN 1055

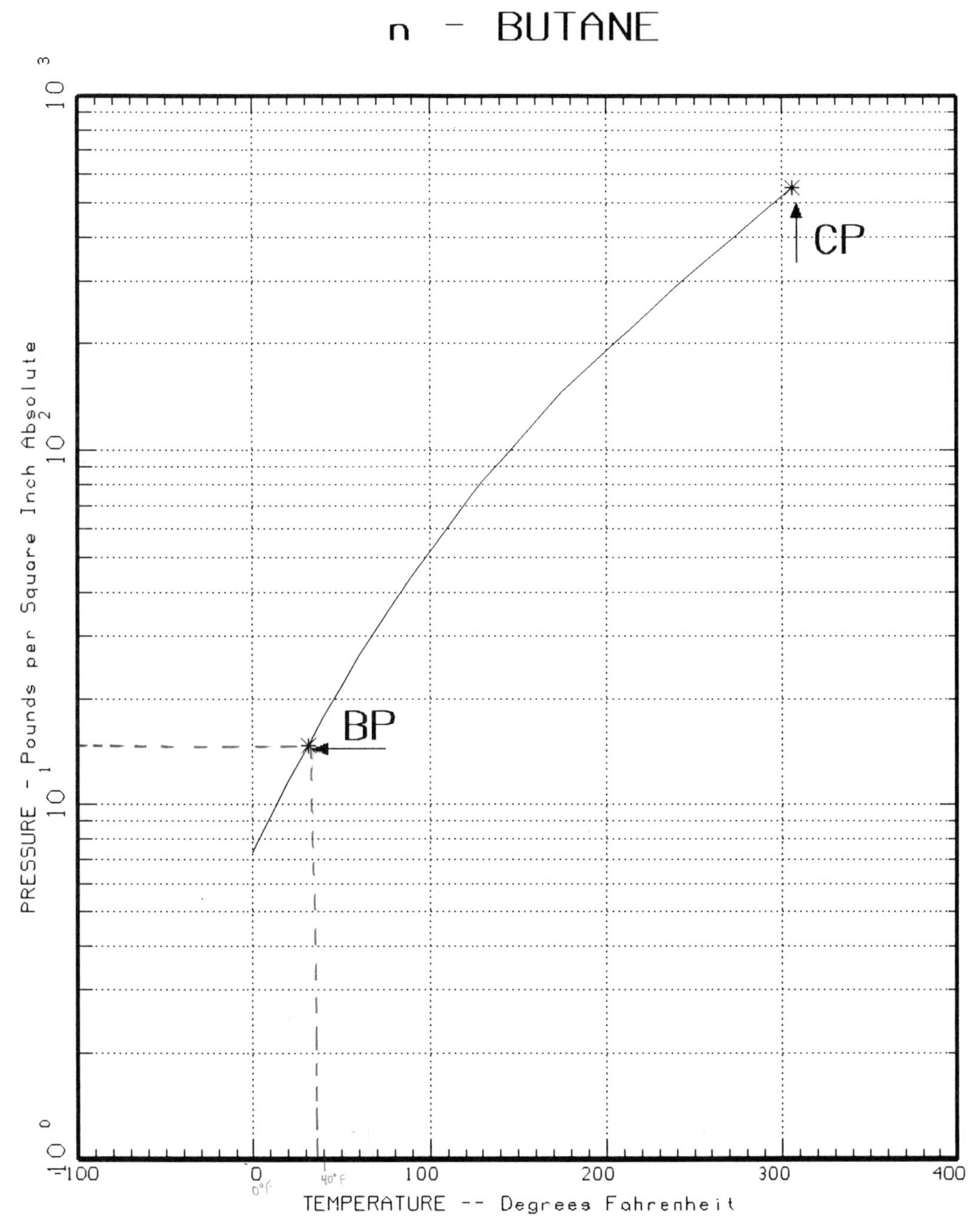

Fig. 1. Vapor Pressure Curve for Butane.

PHYSICAL CONSTANTS

1-Butene

	U.S. Units	SI Units
Chemical formula	C_4H_8 [or CH_2: $CHCH_2CH_3$]	C_4H_8 [or CH_2: $CHCH_2CH_3$]
Molecular weight	56.108	56.108
Vapor pressure		
at 70°F (21.1°C)	23.45 psig	161.68 kPa
at 100°F (37.8°C)	47.59 psig	328.12 kPa
at 115°F (46.1°C)	63.27 psig	436.23 kPa
at 130°F (54.4°C)	81.72 psig	563.44 kPa
Density of the gas at 70°F (21.1°C) and 1 atm	0.149 49 lb/ft^3	2.395 kg/m^3
Specific gravity of the gas at 70°F and 1 atm (air = 1)	1.9368	1.9368
Specific volume of the gas at 60°F (15.6°C) and 1 atm	6.551 ft^3/lb	0.4090 m^3/kg
Density of the liquid at saturation pressure		
at 60°F (15.6°C)	37.43 lb/ft^3	599.57 kg/m^3
at 70°F (21.1°C)	36.82 lb/ft^3	589.80 kg/m^3
at 105°F (40.6°C)	35.26 lb/ft^3	564.81 kg/m^3
at 115°F (46.1°C)	34.69 lb/ft^3	555.68 kg/m^3
at 130°F (54.4°C)	33.90 lb/ft^3	543.03 kg/m^3
Boiling point at 1 atm	20.75°F	−6.25°C
Freezing point at 1 atm	−301.63°F	−185.35°C
Critical temperature	295.5°F	146.39°C
Critical pressure	583.0 psia	4019.64 kPa abs
Critical density	14.6 lb/ft^3	233.87 kg/m^3
Latent heat of vaporization at boiling point	167.94 Btu/lb	390.63 kJ/kg
Latent heat of fusion	0.065 01 Btu/lb	0.151 21 kJ/kg
Flammable limits in air	1.6–10%	1.6–10%
Specific heat of the gas at 60°F (15.6°C) and 1 atm		
C_p	0.3543 Btu/(lb)(°F)	1.483 kJ/(kg)(°C)
C_v	0.3189 Btu/(lb)(°F)	1.335 kJ/(kg)(°C)
Ratio of specific heats, C_p/C_v	1.11	1.11
Solubility in water, vol/vol at 100°F (37.8°C)	0.000 22	0.000 22
Weight of liquid at saturation pressure at 60°F (15.6°C)	5.004 lb/gal	599.61 kg/m3
Heat of combustion		
gross	3177.9 Btu/ft^3	118 405.2 kJ/m^3
net	2970.3 Btu/ft^3	110 670.2 kJ/m^3

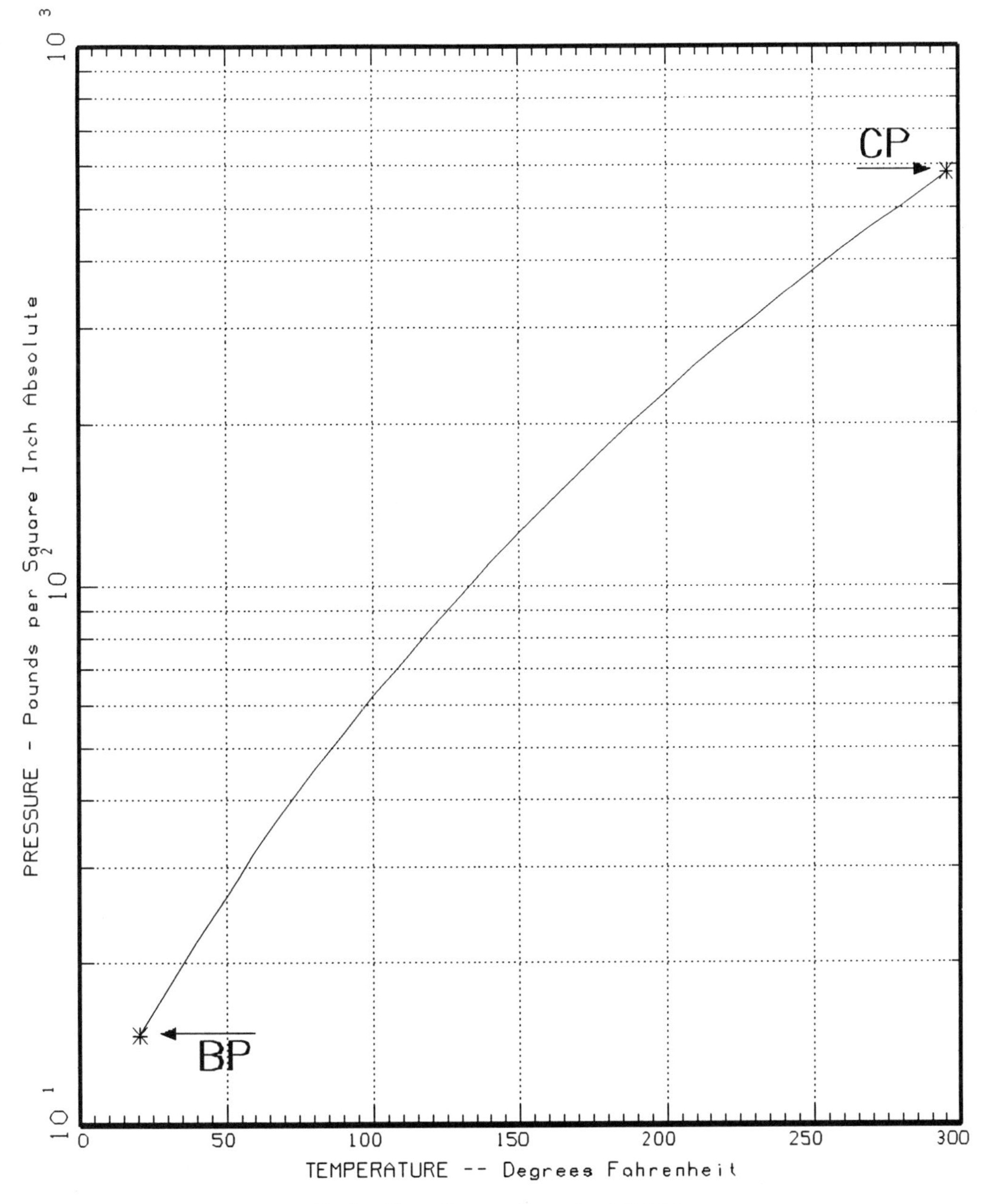

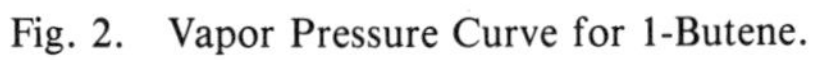
Fig. 2. Vapor Pressure Curve for 1-Butene.

PHYSICAL CONSTANTS

cis-2-Butene

	U.S. Units	SI Units
Chemical formula	C_4H_8 [or $CH_3CH{:}CHCH_3$]	C_4H_8 [or $CH_3CH{:}CHCH_3$]
Molecular weight	56.108	56.108
Vapor pressure		
at 70°F (21.1°C)	12.67 psig	87.36 kPa
at 100°F (37.8°C)	31.37 psig	216.29 kPa
at 115°F (46.1°C)	43.73 psig	301.51 kPa
at 130°F (54.4°C)	58.44 psig	402.93 kPa
Density of the gas		
at 70°F (21.1°C) and 1 atm	0.150 04 lb/ft^3	2.4034 kg/m^3
Specific gravity of the gas		
at 70°F and 1 atm (air = 1)	1.9368	1.9368
Specific volume of the gas		
at 60°F (15.6°C) and 1 atm	6.523 ft^3/lb	0.4072 m^3/kg
Density of the liquid		
at saturation pressure		
at 60°F (15.6°C)	39.04 lb/ft^3	625.36 kg/m^3
at 70°F (21.1°C)	38.50 lb/ft^3	616.71 kg/m^3
at 105°F (40.6°C)	37.01 lb/ft^3	592.84 kg/m^3
at 115°F (46.1°C)	36.55 lb/ft^3	585.47 kg/m^3
at 130°F (54.4°C)	35.86 lb/ft^3	574.42 kg/m^3
Boiling point at 1 atm	38.69°F	3.72°C
Freezing point at 1 atm	−218.06°F	−138.92°C
Critical temperature	324.32°F	162.4°C
Critical pressure	610 psia	4206 kPa abs
Critical density	15.0 lb/ft^3	240.28 kg/m^3
Latent heat of vaporization		
at boiling point	178.91 Btu/lb	416.14 kJ/kg
Latent heat of fusion		
at −218.06°F (−138.92°C)	0.123 47 Btu/lb	0.287 19 kJ/kg
Specific heat of the gas		
at 60°F (15.6°C) and 1 atm C_p	0.3222 Btu/(lb)(°F)	1.349 kJ/(kg)(°C)
C_v	0.2868 Btu/(lb)(°F)	1.201 kJ/(kg)(°C)
Ratio of specific heats, C_p/C_v	1.123	1.123
Flammable limits in air	1.6–9.7%	1.6–9.7%
Weight of liquid		
at saturation pressure		
at 60°F (15.6°C)	5.219 lb/gal	625.374 kg/m^3
Heat of combustion		
gross	3168 Btu/ft^3	118 036 kJ/m^3
net	2960.5 Btu/ft^3	110 305 kJ/m^3

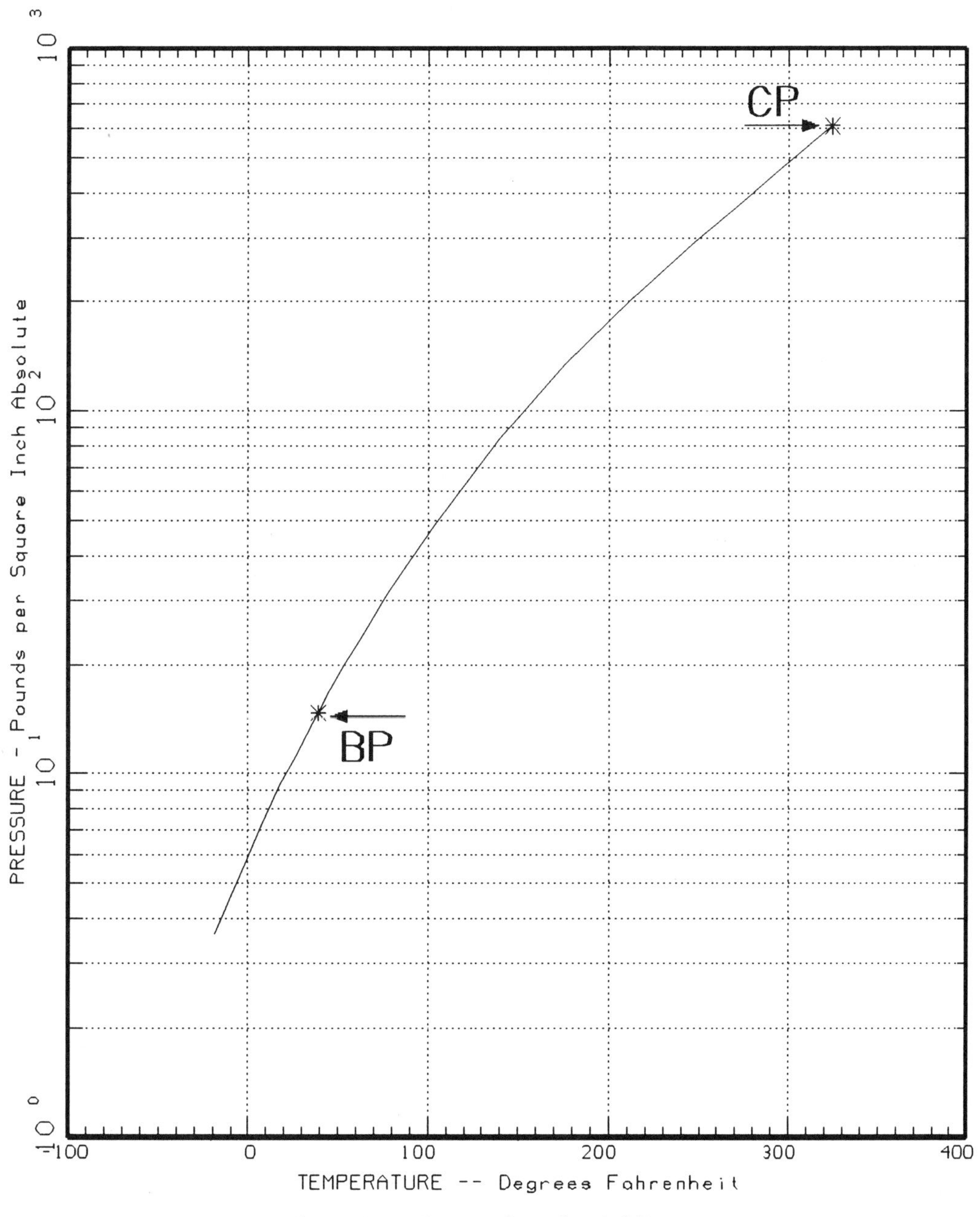

Fig. 3. Vapor Pressure Curve for cis-2-Butene.

PHYSICAL CONSTANTS
trans-2-Butene

	U.S. Units	SI Units
Chemical formula	C_4H_8 [or $CH_3CH{:}CHCH_3$]	C_4H_8 [or $CH_3CH{:}CHCH_3$]
Molecular weight	56.108	56.108
Vapor pressure		
at 70°F (21.1°C)	15.20 psig	104.80 kPa
at 100°F (37.8°C)	35.05 psig	241.66 kPa
at 115°F (46.1°C)	48.10 psig	331.64 kPa
at 130°F (54.4°C)	63.61 psig	438.58 kPa
Density of the gas		
at 70°F (21.1°C) and 1 atm	0.150 00 lb/ft^3	2.4028 kg/m^3
Specific gravity of the gas		
at 70°F and 1 atm (air = 1)	1.9368	1.9368
Specific volume of the gas		
at 60°F (15.6°C) and 1 atm	6.5245 ft^3/lb	0.4073 m^3/kg
Density of the liquid		
at saturation pressure		
at 60°F (15.6°C)	37.97 lb/ft^3	608.22 kg/m^3
at 70°F (21.1°C)	37.47 lb/ft^3	600.21 kg/m^3
at 105°F (40.6°C)	35.89 lb/ft^3	574.90 kg/m^3
at 115°F (46.1°C)	36.43 lb/ft^3	567.53 kg/m^3
at 130°F (54.4°C)	34.71 lb/ft^3	556.00 kg/m^3
Boiling point at 1 atm	33.58°F	0.8777°C
Freezing point at 1 atm	−157.96°F	−105.53°C
Critical temperature	311.83°F	155.46°C
Critical pressure	595 psia	4102 kPa abs
Critical density	14.7 lb/ft^3	234.47 kg/m^3
Latent heat of vaporization		
at 33.58°F (0.8777°C)	174.39 Btu/lb	405.63 kJ/kg
Latent heat of fusion		
at −157.96°F (−105.53°C)	0.164 83 Btu/lb	0.383 39 kJ/kg
Specific heat of gas		
at 60°F (15.6°C) and 1 atm		
C_p	0.3618 Btu/(lb)(°F)	1.5147 kJ/(kg)(°C)
C_v	0.3264 Btu/(lb)(°F)	1.3666 kJ/(kg)(°C)
Ratio of specific heats, C_p/C_v	1.108	1.108
Flammable limits in air	1.6–9.7%	1.6–9.7%
Weight of the liquid		
at saturation pressure		
at 60°F (15.6°C)	5.076 lb/gal	608.239 kg/m^3
Heat of combustion		
gross	3163 Btu/ft^3	117 850 kJ/m^3
net	2957 Btu/ft^3	110 175 kJ/m^3

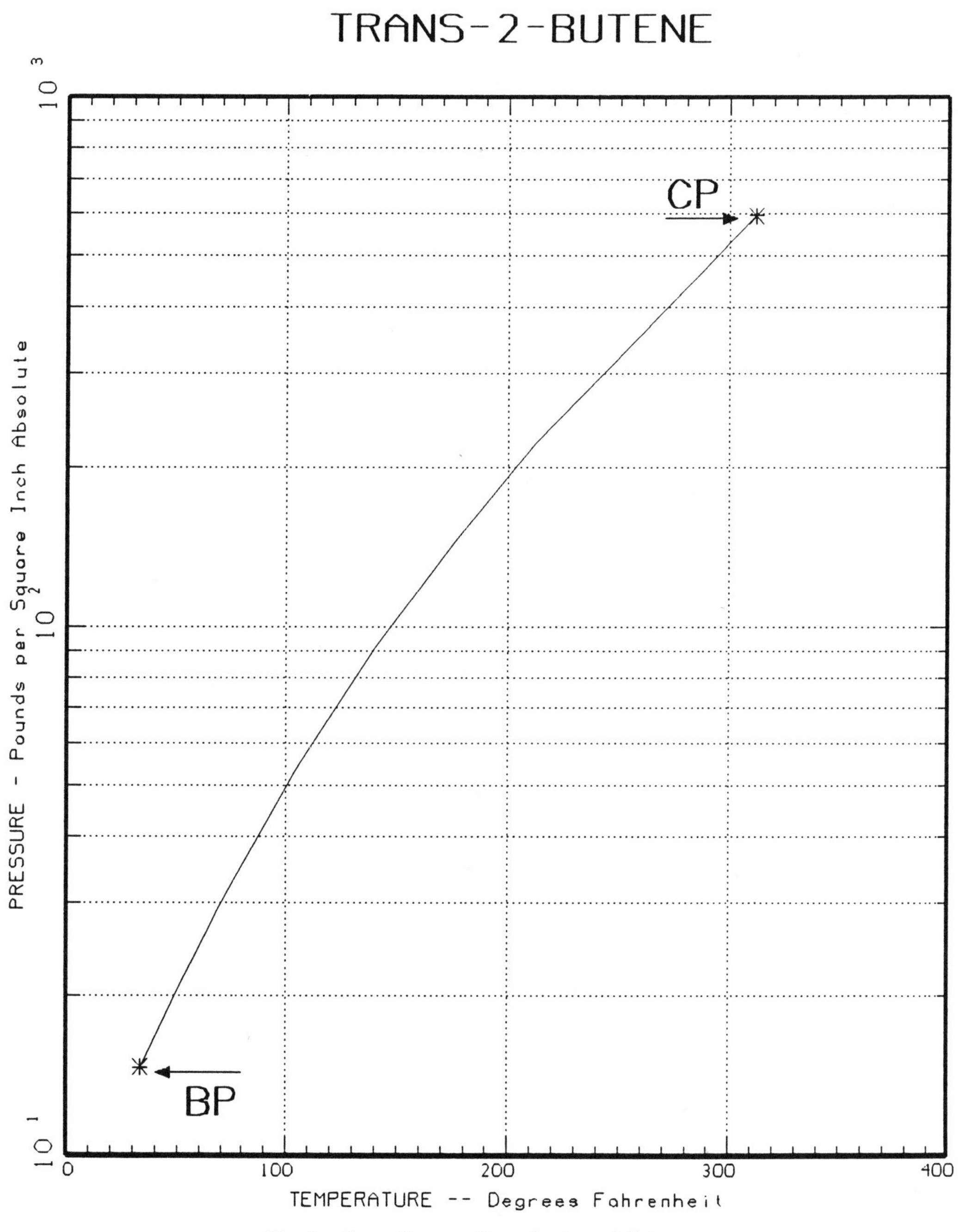

Fig. 4. Vapor Pressure Curve for trans-2-Butene.

PHYSICAL CONSTANTS

Isobutylene

	U.S. Units	SI Units
Chemical formula	C_4H_8 [or $(CH_3)_2C{:}CH_2$]	C_4H_8 [or $(CH_3)_2C{:}CH_2$]
Molecular weight	56.108	56.108
Vapor pressure		
at 70°F (21.1°C)	23.85 psig	164.44 kPa
at 100°F (37.8°C)	48.04 psig	331.22 kPa
at 115°F (46.1°C)	63.78 psig	439.75 kPa
at 130°F (54.4°C)	82.33 psig	567.65 kPa
Density of the gas		
at 70°F (21.1°C) and 1 atm	0.14957 lb/ft^3	2.3959 kg/m^3
Specific gravity of the gas		
at 70°F and 1 atm (air = 1)	1.997	1.997
Specific volume of the gas		
at 60°F (15.6°C) and 1 atm	6.545 ft^3/lb	0.4086 m^3/kg
Density of the liquid		
at saturation pressure		
at 60°F (15.6°C)	37.37 lb/ft^3	598.61 kg/m^3
at 70°F (21.1°C)	36.90 lb/ft^3	591.08 kg/m^3
at 105°F (40.6°C)	35.25 lb/ft^3	564.65 kg/m^3
at 115°F (46.1°C)	34.76 lb/ft^3	556.80 kg/m^3
at 130°F (54.4°C)	34.01 lb/ft^3	544.79 kg/m^3
Boiling point at 1 atm	19.59°F	−6.8944°C
Freezing point at 1 atm	−220.61°F	−140.34°C
Critical temperature	292.51°F	144.73°C
Critical pressure	580.2 psia	4000.34 kPa abs
Critical density	14.7 lb/ft^3	235.47 kg/m^3
Latent heat of vaporization		
at boiling point	169.48 Btu/lb	394.21 kJ/kg
Latent heat of fusion		
at −220.61°F (−140.34°C)	0.100 20 Btu/lb	0.233 07 kJ/kg
Specific heat of gas		
at 60°F (15.6°C) and 1 atm C_p	0.3701 Btu/(lb)(°F)	1.550 kJ/(kg)(°C)
C_v	0.3347 Btu/(lb)(°F)	1.401 kJ/(kg)(°C)
Ratio of specific heats, C_p/C_v	1.106	1.106
Flammable limits in air	1.8–9.6%	1.8–9.6%
Weight of the liquid		
at saturation pressure		
at 60°F (15.6°C)	4.996 lb/gal	598.65 kg/m^3
Heat of combustion		
gross	3156 Btu/ft^3	117 589.24 kJ/m^3
net	2949 Btu/ft^3	109 876.64 kJ/m^3

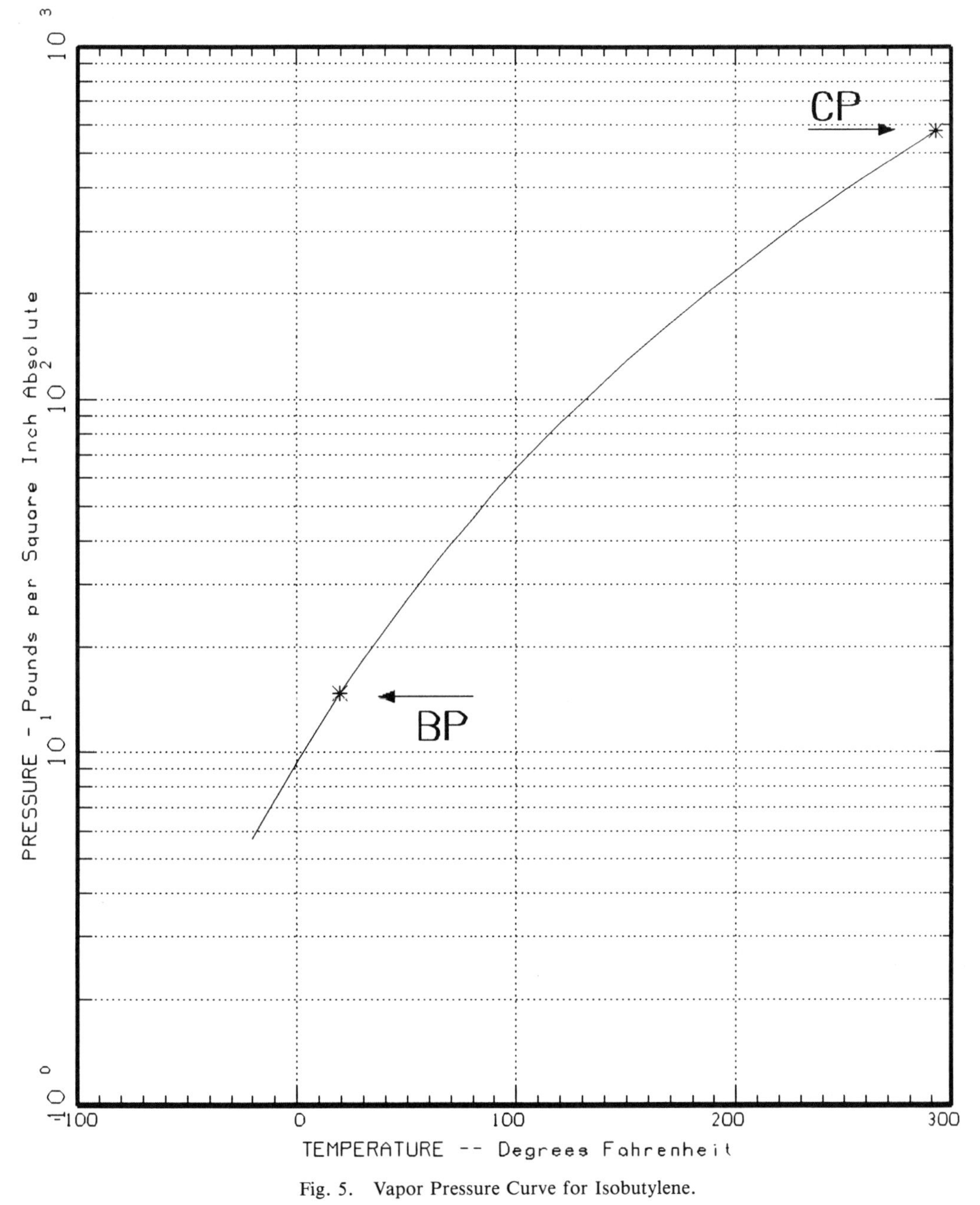

Fig. 5. Vapor Pressure Curve for Isobutylene.

PHYSICAL CONSTANTS

Isobutane

	U.S. Units	SI Units
Chemical formula	C_4H_{10} [or $(CH_3)_2CHCH_3$]	C_4H_{10} [or $(CH_3)_2CHCH_3$]
Molecular weight	58.124	58.124
Vapor pressure		
at 70°F (21.1°C)	30.58 psig	210.84 kPa
at 100°F (37.8°C)	57.87 psig	399.00 kPa
at 115°F (46.1°C)	75.38 psig	519.73 kPa
at 130°F (54.4°C)	95.86 psig	660.93 kPa
Density of the gas		
at 70°F (21.1°C) and 1 atm	0.154 74 lb/ft^3	2.4787 kg/m^3
Specific gravity of the gas		
at 70°F and 1 atm (air = 1)	2.006 36	2.006 36
Specific volume of the gas		
at 60°F (15.6°C) and 1 atm	6.3355 ft^3/lb	0.3955 m^3/kg
Density of the liquid		
at saturation pressure		
at 60°F (15.6°C)	35.05 lb/ft^3	561.45 kg/m^3
at 70°F (21.1°C)	34.82 lb/ft^3	557.76 kg/m^3
at 105°F (40.6°C)	33.26 lb/ft^3	532.77 kg/m^3
at 115°F (46.1°C)	32.82 lb/ft^3	525.73 kg/m^3
at 130°F (54.4°C)	32.01 lb/ft^3	512.75 kg/m^3
Boiling point at 1 atm	10.90°F	−11.72°C
Freezing point at 1 atm	−255.29°F	−159.61°C
Critical temperature	274.96°F	134.98°C
Critical pressure	529.1 psia	3648.02 kPa abs
Critical density	13.8 lb/ft^3	221.05 kg/m^3
Latent heat of vaporization		
at boiling point	157.53 Btu/lb	366.41 kJ/kg
Specific heat of the gas		
at 60°F (15.6°C) and 1 atm		
C_p	0.3906 Btu/(lb)(°F)	1.6354 kJ/(kg)(°C)
C_v	0.3564 Btu/(lb)(°F)	1.4922 kJ/(kg)(°C)
Ratio of specific heats, C_p/C_v	1.096	1.096
Flammable limits in air	1.8–8.4%	1.8–8.4%
Solubility in water, vol/vol		
at 100°F (37.8°C)	0.000 052	0.000 052
Weight of the liquid		
at saturation pressure		
at 60°F (15.6°C)	4.686 lb/gal	561.51 kg/m^3
Heat of combustion		
gross	3352.7 Btu/ft^3	124 918.08 kJ/m^3
net	3001.1 Btu/ft^3	111 817.83 kJ/m^3

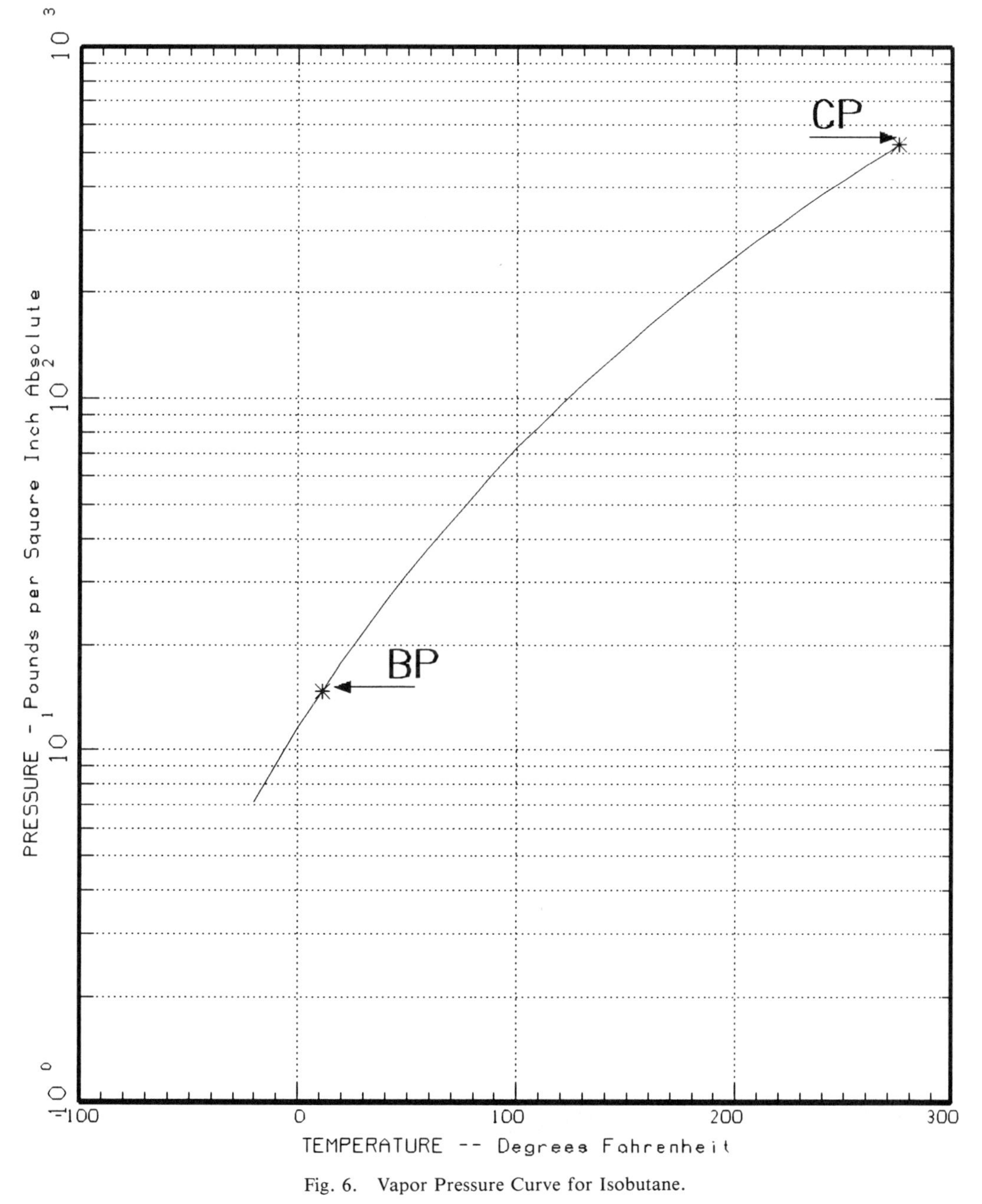

Fig. 6. Vapor Pressure Curve for Isobutane.

PHYSICAL CONSTANTS

Propane

	U.S. Units	SI Units
Chemical formula	C_3H_8	C_3H_8
Molecular weight	44.097	44.097
Vapor pressure		
at 70°F (21.1°C)	109.73 psig	756.56 kPa
at 100°F (37.8°C)	173.38 psig	1195.41 kPa
at 115°F (46.1°C)	212.95 psig	1468.24 kPa
at 130°F (54.4°C)	258.37 psig	1781.40 kPa
Density of the gas		
at 70°F (21.1°C) and 1 atm	0.115 99 lb/ft^3	1.8580 kg/m^3
Specific gravity of the gas		
at 70°F and 1 atm (air = 1)	1.5223	1.5223
Specific volume of the gas		
at 60°F (15.6°C) and 1 atm	8.4515 ft^3/lb	0.5276 m^3/kg
Density of the liquid		
at saturation pressure		
at 60°F (15.6°C)	31.59 lb/ft^3	506.02 kg/m^3
at 70°F (21.1°C)	31.20 lb/ft^3	499.78 kg/m^3
at 105°F (40.6°C)	29.33 lb/ft^3	469.82 kg/m^3
at 115°F (46.1°C)	28.70 lb/ft^3	459.73 kg/m^3
at 130°F (54.4°C)	27.77 lb/ft^3	444.83 kg/m^3
Boiling point at 1 atm	−43.67°F	−42.04°C
Freezing point at 1 atm	−305.84°F	−187.69°C
Critical temperature	206.01°F	96.672°C
Critical pressure	616.3 psia	4249.24 kPa abs
Critical density	13.5 lb/ft^3	216.25 kg/m^3
Latent heat of vaporization		
at at boiling point	183.05 Btu/lb	425.77 kJ/kg
Specific heat of gas		
at 60°F (15.6°C) and 1 atm		
C_p	0.3881 Btu/(lb)(°F)	1.625 kJ/(kg)(°C)
C_v	0.3430 Btu/(lb)(°F)	1.436 kJ/(kg)(°C)
Ratio of specific heats, C_p/C_v	1.131	1.131
Flammable limits in air	2.2–9.5%	2.2–9.5%
Solubility in water, vol/vol		
at 100°F (37.8°C)	0.065	0.065
Weight of the liquid		
at saturation pressure		
at 60°F (15.6°C)	4.223 lb/gal	506.03 kg/m^3
Heat of combustion		
gross	2517.5 Btu/ft^3	93 799.41 kJ/m^3
net	2316.1 Btu/ft^3	86 295.45 kJ/m^3

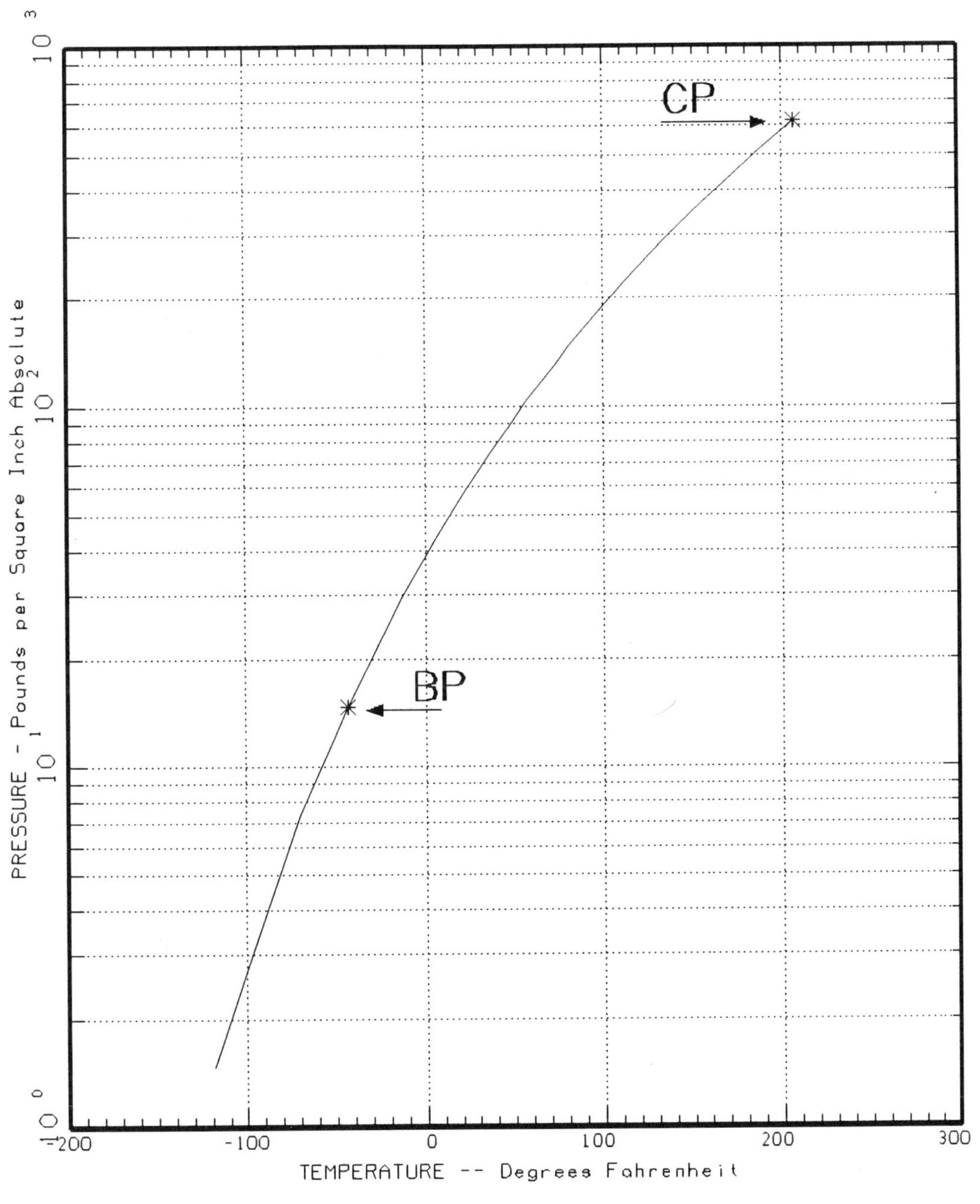

Fig. 7. Vapor Pressure Curve for Propane.

PHYSICAL CONSTANTS

Propylene

	U.S. Units	SI Units
Chemical formula	C_3H_6	C_3H_6
Molecular weight	42.081	42.081
Vapor pressure		
at 70°F (21.1°C)	132.81 psig	915.69 kPa
at 100°F (37.8°C)	206.92 psig	1426.66 kPa
at 115°F (46.1°C)	253.03 psig	1744.58 kPa
at 130°F (54.4°C)	306.05 psig	2113.24 kPa
Density of the gas		
at 70°F (21.1°C) and 1 atm	0.110 447 lb/ft^3	1.7692 kg/m^3
Specific gravity of the gas		
at 70°F and 1 atm (air = 1)	1.4529	1.4529
Specific volume of the gas		
at 60°F (15.6°C) and 1 atm	8.875 ft^3/lb	0.554 m^3/kg
Density of the liquid		
at saturation pressure		
at 60°F (15.6°C)	37.43 lb/ft^3	599.57 kg/m^3
at 70°F (21.1°C)	32.07 lb/ft^3	513.71 kg/m^3
at 105°F (40.6°C)	29.89 lb/ft^3	478.79 kg/m^3
at 115°F (46.1°C)	29.20 lb/ft^3	467.74 kg/m^3
at 130°F (54.4°C)	28.08 lb/ft^3	449.80 kg/m^3
Boiling point at 1 atm	−53.90°F	−47.72°C
Freezing point at 1 atm	−301.45°F	−185.25°C
Critical temperature	197.2°F	91.77°C
Critical pressure	670.0 psia	4619.49 kPa abs
Critical density	14.5 lb/ft^3	232.27 kg/m^3
Latent heat of vaporization		
at boiling point	188.18 Btu/lb	437.71 kJ/kg
Latent heat of fusion		
at −301.45.°F (−185.25°C)	0.0673 Btu/lb	0.1565 kJ/kg
Specific heat of gas		
at 60°F (15.6°C) and 1 atm C_p	0.3549 Btu/(lb)(°F)	1.4859 kJ/(kg)(°C)
C_v	0.3077 Btu/(lb)(°F)	1.2883 kJ/(kg)(°C)
Ratio of specific heats, C_p/C_v	1.153	1.153
Solubility in water, vol/vol		
at 100°F (37.8°C)	0.0009	0.0009
Weight of the liquid		
at saturation pressure		
at 60°F (15.6°C)	4.343 lb/gal	520.41 kg/m^3
Heat of combustion		
gross	2371.7 Btu/ft^3	883 67.05 kJ/m^3
net	2218.3 Btu/ft^3	826 51.53 kJ/m^3
Flammable limits in air	1.9–11.1%	1.9–11.1%

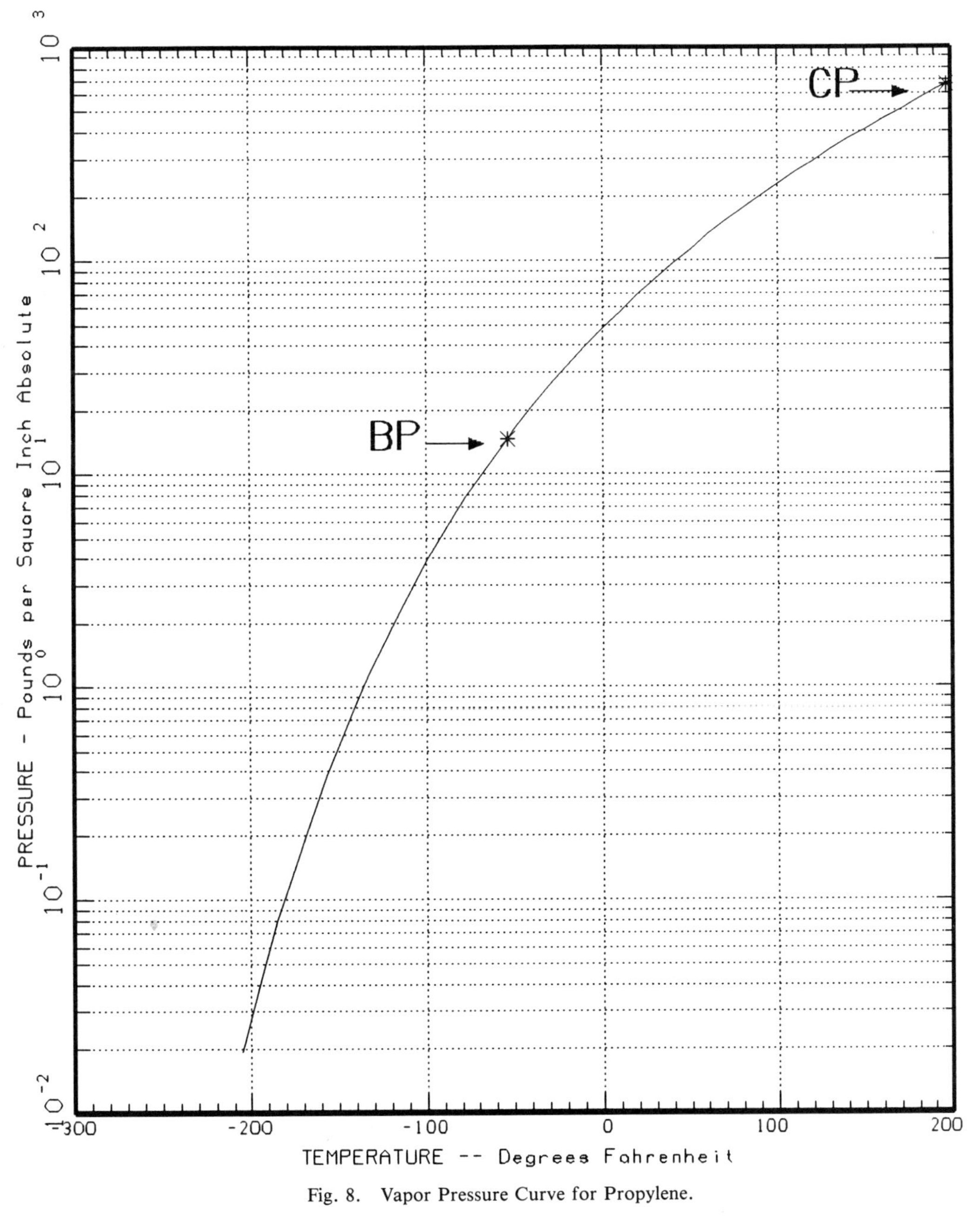

Fig. 8. Vapor Pressure Curve for Propylene.

DESCRIPTION

The liquefied petroleum gases are butane, isobutane, propane, propylene (propene), butylenes (butenes) and any mixtures of these hydrocarbons in the generally accepted definition of the National Fire Protection Association. [1] The gases are also denoted by the terms LP-gas or LPG. They are flammable, colorless, noncorrosive, and nontoxic. These gases are easily liquefied under pressure at ambient temperature, and are shipped and stored as liquids. They are largely used in gaseous and liquid form as fuels in many diverse applications.

Propane, isobutane, and butane are among the lightest hydrocarbons in the liquid phase in the paraffin series, occurring between ethane and pentane (the lightest natural gasoline fraction). Propylene, isobutylene, 1-butene, and 2-butene are among the lightest hydrocarbons in the monoolefin series, occurring between ethylene and pentene. (The ending *-ane* indicates a member of the paraffin series, while *-ene* indicates a member of the monoolefin series.)

The liquefied petroleum gases in the paraffin series are chemically stable and odorless, and the DOT and other regulating bodies require artificial odorization of propane and butane (except in technical uses where the odorant would harm further processing and the odorant warning action would not be important). Propylene and the butenes (also denoted as butylenes) have an unpleasant odor characteristic of petroleum refinery gas or coal gas.

All the liquefied petroleum gases are soluble to varying degrees in alcohol and ether. Propane and propylene are slightly soluble in water.

GRADES AVAILABLE

The liquefied petroleum gases are supplied in various scientific and commercial grades. Their properties differ according to the grade being used, and their compositions vary depending upon whether the LPG is extracted from refinery streams or from natural gas. Properties are given for the research grades of these gases in the preceding Physical Constants sections. [2], [3], [4], and [5]

As an example of the way properties of the various commercial grades of the liquefied petroleum gases generally differ from the properties for research grades that are shown in the Physical Constants tables, Table 1 compares vapor pressure properties for research grades and average commercial grades of propane and butane from one producer.

Moreover, these gases are often used in mixtures designed to have certain desired properties; in particular, propane and butane are frequently ordered as mixtures to meet certain boiling point and other requirements of individual applications. Suppliers furnish data on physical constants for the various grades and mixtures they make available.

USES

Propane and butane (known most extensively in commercial and popular terms as LP-gas or LPG) have an extremely wide range of domestic, industrial, commercial, agricultural, and internal combustion engine uses. It is estimated that the two gases, unmixed and in mixtures, have several thousand industrial applications and many more in other fields. Their very broad application stems from their occurrence as hydrocarbons between natural gas and natural gasoline, and from their corresponding properties.

The liquefied petroleum gases are used:

(1) As appliance fuel for space heating, water heating, boiler heating, cooking, baking, air conditioning, and refrigeration in rural or urban areas beyond the reach of gas mains.

(2) In bulk by utilities and industries (especially industries using kilns or furnaces which must be maintained continuously at given temperatures), as standby fuel to protect against failure or interruption of natural or artificial gas supply. By utilities to bridge peak load demands for natural or synthetic

TABLE 1. COMPARISON OF VAPOR PRESSURE PROPERTIES FOR RESEARCH GRADES AND AVERAGE COMMERCIAL GRADES OF PROPANE AND BUTANE.[a]

	Vapor Pressure (psig)			
Temperature	Research Grade Propane	Commercial Propane	Research Grade Butane	Commercial Butane
70°F	109.73	124	16.54	31
100°F	173.38	192	36.92	59
105°F	—	206	—	65
130°F	258.37	286	66.03	97

[a]One must be cautious in using "average" properties for commercial liquefied petroleum gases, since the industry has not been able to arrive at average figures because they vary from season to season, year to year, and supplier to supplier.

or substitute natural gas, and for gas enrichment. By utilities serving rural communities from central plants.

(3) For space heating during the erection of buildings.

(4) As fuel for the entire range of industrial heating processes, especially those where the heating must be accurately controlled. Industrial heating process uses include heat treating, stress relieving, annealing, enamel baking and firing ceramic kilns and furnaces, brazing, metal cutting, and soldering.

(5) As fuel for such operations as poultry brooding, cotton and grain drying, tobacco curing, crop dehydration, weed burning, and orchard heating.

(6) As fuel for vehicles such as trucks, buses, taxicabs, and forklift trucks, and for mobile farm machinery such as tractors and harvesters. As fuel for stationary engines powering well pumps, electric generators, and so on. Engines especially designed for LP-gas are available, and gasoline engines may readily be converted for LP-gas operation.

(7) Isobutane and the gases in the monoolefin series are used less extensively than propane and butane. Isobutane and isobutylene are used in manufacturing alkylate for increasing gasoline octane ratings. Isobutylene is employed in the manufacture of synthetic rubber. Propylene is used substantially in the chemical industry for synthesis in the production of a wide range of products. The butenes are used in preparing a large number of organic compounds.

PHYSIOLOGICAL EFFECTS

The liquefied petroleum gases are nontoxic. Prolonged inhalation of high concentrations has an anesthetic effect; also, since they can displace oxygen in the air, they can act as simple asphyxiants. Contact of the skin by these gases in liquid form can cause freezing of tissue and may result in injury similar to a thermal burn.

For butane, the American Conference of Governmental Industrial Hygienists has adopted a threshold limit value of 800 ppm (1900 mg/m^3) which is also the eight-hour time-weighted average permissible exposure limit established by the U.S. Occupational Safety and Health Administration. For propane, the TLV adopted by ACGIH and the PEL established by OSHA is 1000 ppm (1800 mg/m^3).

MATERIALS OF CONSTRUCTION

Any common, commercially available metals may be used with commercial (or higher) grades of liquefied petroleum gases because they are noncorrosive, though installations must be designed to withstand the pressures involved and must comply with all state and local regulations. Widely accepted recommendations on storage systems and safe usage are given in NFPA 58, *Standard for the Storage and Handling of Liquefied Petroleum Gases.* [1] Similar recommendations for larger storage systems appear in NFPA 59, *Standard for the Storage and Handling of Liquefied Petroleum Gases at Utility Gas Plants.* [6]

SAFE STORAGE, HANDLING, AND USE

Steel tanks and large underground chambers are used for the storage of liquefied petroleum gas. Steel tanks above or below ground range up to 120 000 gallons (450 m^3) or more in capacity, while below-ground caverns or pits have been found to offer safe and economical storage where a suitable geological formation exists at a site of large-volume handling, such as at pipelines or marine terminals and at natural gasoline plants or refineries. Widely recognized recommendations for LP-gas installations are presented in NFPA 58 and NFPA 59. [1] and [6]

DISPOSAL

Disposal of liquefied petroleum gases, as with other gases, should be undertaken only by personnel familiar with the gas and the procedures for disposal. Contact the supplier for instructions. In general, should it become necessary to dispose of liquefied petroleum gases, the best procedure, as for other flammable gases, is to burn them in any suitable burning unit available in the plant. This should be done in accordance with appropriate regulations.

HANDLING LEAKS AND EMERGENCIES

Never use a flame to test for an LP-gas leak; instead, apply a soapy water solution to areas suspected of leaking. Alternative means of detection involve the use of instrumental methods. Frost around valve stems, at piping joints, oı at other points may indicate a liquid leak.

In an emergency, keep unauthorized persons away and eliminate possible sources of ignition.

If LP-gas is escaping and not on fire, the following steps can be taken provided they can be done without risk to personnel:

(1) An attempt should be made to close a valve which will stop the flow of gas.

(2) Small lines, such as copper tubing, can be flattened or crimped to stop the flow of gas.

(3) If possible, the container should be moved to a safe place.

(4) Water spray is effective in dispersing LP-gas vapor.

(5) If a valve is not available to stop the flow, an expert emergency responder may consider igniting the escaping gas from a safe distance and upwind of the leak. Be certain all personnel are sufficiently clear of the area. After ignition, allow the fire to burn, applying sufficient water to cool the container and any exposed pipe.

If LP-gas is escaping and on fire:

(1) Fire should not be extinguished unless the leakage can be stopped immediately.

(2) The container vapor space and appurtenances should be kept cool with water spray applied from the upwind direction.

(3) Dry chemical or CO_2 extinguishers are suitable for putting out small LP-gas fires. The extinguishing agent should be directed at the base of the flame.

(4) Shooting holes in an LP-gas tank that is involved in a fire does not serve any useful purpose and should not be permitted. It may cause the fire to become more serious.

(5) Ordinarily, no attempt should be made to move any tank involved in a fire because usually little will be gained in reducing the hazard.

(6) A tank involved in a fire should never

be dragged because valves or appurtenances on the tank might be damaged.

First Aid

Inhalation of low concentrations can be remedied by promptly going to an uncontaminated area and inhaling fresh air. In the event of massive exposure wherein the victim has become unconscious or symptoms of asphyxiation may persist, the person should be removed promptly to an uncontaminated area and given artificial respiration if breathing has stopped. This should be followed by the administration of oxygen after breathing has been restored.

Contact with the liquid phase of liquefied petroleum gases with the skin can result in frostbite. First aid treatment for frostbite consists of putting the frostbitten part in warm water of about 108°F (42°C). If warm water is not available or is impractical to use, wrap the affected area gently in blankets. Encourage the victim to exercise the affected part while it is being warmed. This will aid circulation to reestablish itself naturally. Medical attention by a physician should be obtained.

In the event of eye contact with the liquid phase, the affected eyes should be flushed with tap water for 15 minutes. If irritation persists, the patient should be referred to a physician.

METHODS OF SHIPMENT

Under the appropriate regulations, liquefied petroleum gases are authorized for shipment as follows: [7] and [8]

By Rail: In cylinders, in insulated and uninsulated single-unit tank cars, and in TMU tank cars.

By Highway: In cylinders on trucks, in portable tanks, and in cargo tanks on trucks or on semi and full trailers.

By Water: In cylinders on passenger vessels and on ferry or railroad car ferry vessels (passenger or vehicle); on cargo vessels in cylinders and portable tanks (meeting TC/DOT 51 specifications and not over 20 000 lb (9072 kg) gross weight, with vapor pressure at 115°F (46.1°C) not exceeding the container service pressure); in authorized tank cars on trainships only, and in authorized tank trucks on trailerships and trainships only. In cylinders on barges of U.S. Coast Guard classes A, CA, and CB only. In cargo tanks aboard tankships and tank barges (to maximum filling densities by specific gravity as prescribed in Coast Guard regulations).

By Air: In cylinders on cargo aircraft only up to 300 lb (136 kg) maximum net weight per cylinder.

By Pipeline: Propane and butane are also transported by pipeline from points of production to distant bulk storage facilities.

CONTAINERS

Propane and butane are classified as flammable gases and are authorized for shipment as liquefied compressed gases in cylinders, portable tanks and cargo tanks, and insulated or uninsulated single-unit tank cars. TC/DOT regulations also provide for their shipment in TMU (ton multi-unit) tanks and tank cars, but they are not generally shipped in TMU containers. [7] and [8] The two gases are stored in large tanks above and below ground, and also in very large underground chambers such as in natural salt deposits.

In addition to such storage in liquefied form under their vapor pressure at ambient temperatures, refrigerated storage in cold liquid form under atmospheric pressures is used for propane and butane. Refrigerated storage systems are closed and insulated, and in them the LP-gas vapor is circulated through pumps and compressors to serve as the systems' refrigerant. Propane and butane are stored in pits in the earth capped by metal domes as well as in underground chambers; one of the largest storage pits of this kind, located in Utah, holds propane in the millions of gallons in refrigerated liquid form.

Caution: Refrigerated compressed gases must be heated to ambient temperatures prior to transfer to single-unit tank cars, tank trucks, or other containers constructed of steels that are intended for ambient-temperature service.

Isobutane and the monoolefins are authorized for shipment in single-unit tank cars and truck cargo tanks, and are usually shipped in bulk units because they are generally used in large quantities. They are also authorized for shipment in cylinders, portable tanks, and TMU tanks.

Filling Densities

The maximum filling densities authorized by TC/DOT regulations for liquefied petroleum gases are prescribed according to the specific gravity of the liquid material at 60°F (15.6°C) in detailed tables that are part of the regulations. Producers and suppliers who charge LP-gas containers should consult these tables in the current regulations for the maximum densities to which to fill containers with the particular grades and mixtures they are handling. The lower and upper limits of the maximum filling densities authorized in the present regulations are as follows (percent water capacity by weight):

In cylinders: from 26 percent for 0.271–0.289 specific gravity, to 57 percent for 0.627–0.634 specific gravity. See 49 CFR 173.304(d). [7]

In single-unit tank cars and TMU tanks: from 45.50 percent (insulated tanks, April through October) for 0.500 specific gravity, to 61.57 percent (uninsulated tanks, November through March with no storage in transit) for 0.635 specific gravity. Moreover, filling must not exceed various specified limits of pressure and liquid content at temperatures of 105°F (40.6°C) or 130°F (54.4°C) as given in U.S. Department of Transportation and Transport Canada regulations. See 49 CFR 173.314(f). [7]

In cargo tanks and portable tanks: from 38 percent (tanks of 1200-gallon or 4542-L capacity or less) for 0.473–0.480 specific gravity, to 60 percent (tanks of over 1200-gal capacity) for 0.627 and over specific gravity (except when using fixed-length dip tube or other fixed maximum liquid level indicators). Moreover, the tank must not be liquid full at 105°F (40.6°C) if insulated or at 115°F (46.1°C) if uninsulated, and the gauge vapor pressure at 115°F must not exceed the tank's design pressure. See 49 CFR 173.315(b). [7]

Cylinders

Cylinders authorized for liquefied petroleum gas service include those which comply with the following TC/DOT specifications: 3A, 3AA, 3B, 3E, 3AL, 4B, 4BA, 4B240ET, 4B240FLW, 4BW, and 39. Cylinders meeting the following TC/DOT specifications may be continued in use but new construction is not authorized: 3, 4, 4A, 9, 4B240X, 25, 26, 38, and 41. TC/DOT regulations also authorize shipment in several special types of small containers. Shipments of flammable gases in Specification 3AL cylinders are authorized only when transported by highway, rail, and cargo-only aircraft. The internal volume of Specification 39 cylinders must not exceed 75 in.^{3} (0.00123 m^3) when used for shipment of flammable gas.

All of these types of cylinders must be requalified by hydrostatic retest every five years under present TC/DOT regulations except as follows: (1) no periodic retest is required for 3E cylinders; and (2) ten years is the required retest interval for type 4 cylinders; and (3) external visual inspection every five years may be used in lieu of hydrostatic retest for cylinders that are used exclusively for liquefied petroleum gas which is commercially free from corroding components and that are of the following types (including cylinders of these types with higher service pressures): 3A480, 3AA480, 3A480X, 3B, 4B, 4BA, 4E, 4BW, 26-240, and 260-300.

Authorized pressure relief devices for use on cylinders of liquefied petroleum gases are the Type CG-7 pressure relief valve. For butane cylinders, the Type CG-4 combination rupture disk/fusible plug device is also authorized. For propane cylinders, the Type CG-3 fusible plug is also authorized. See CGA S-1.1, *Pressure Relief Device Standards—Part I—Cylinders for Compressed Gases,* for additional information and further requirements. [9]

Valve Outlet Connections

The standard valve outlet connection in the United States and Canada for cylinders of butane, isobutane, propane, propylene, or the various butenes is Connection CGA 510. Connection CGA 555 is used for butane and propane liquid withdrawal. Connection CGA 600 is the limited standard valve outlet connection for gaseous withdrawal of propane and propylene. [10]

Cargo Tanks

Liquefied petroleum gas is authorized for shipment in truck or truck-trailer cargo tanks that comply with TC/DOT specifications MC-330 or MC-331. Various design pressures may be used, but the gauge vapor pressure at 115°F (46.1°C) of a shipment must not exceed the tank's design pressure.

Portable Tanks

Liquefied petroleum gases may be shipped in portable tanks meeting specifiations TC/DOT 51. Design pressures are the same as for cargo tanks. See 49 CFR 173.315. [7]

TMU Tanks

TMU tank car shipment, though little used, is also authorized for LP-gas with pressure not exceeding 375 psi at 130°F (2590 kPa at 54.4°C).

Tank Cars

Bulk shipment of liquefied petroleum gas is authorized for single-unit tank cars in TC/DOT 105J, 112J, 112T, 114J, and 114T series tank cars with the required service pressure of the tank specified with respect to the maximum allowable vapor pressure at 105°F (40.6°C) for 105J series tank cars and at 115°F (46.1°C) for 112J, 112T, 114J, and 114T tank cars. See 49 CFR 173.314. [7] TC/DOT specification 106A500X tank cars are also authorized for pressure not exceeding 375 psig (2586 kPa) at 130°F (54.4°C).

METHODS OF MANUFACTURE

Butane and propane (with other hydrocarbons in the paraffin series) are recovered from "wet" natural gas, from natural gas associated with or dissolved in crude oil, and from petroleum refinery gases. They may be separated from wet natural gas or crude oil through absorption in light "mineral seal" oil, through adsorption on surfaces such as activated charcoal, or by refrigeration, followed in each case by fractionation. Propylene and other gases in the monoolefin series are recovered from petroleum gases by fractionation.

REFERENCES

[1] NFPA 58, *Standard for the Storage and Handling of Liquefied Petroleum Gases,* National Fire Protection Association, Batterymarch Park, Quincy, MA 02269.

[2] NGPA 2145, *Physical Constants for Paraffin Hydrocarbons and Other Components of Natural Gas,* Gas Processors Association, 6526 E. 60th Street, Tulsa, OK 74145.

[3] *Engineering Data Book,* 10th ed., Gas Processors Suppliers Association, 6526 E. 60th Street, Tulsa, OK 74145. (1987)

[4] API Pub. 999, *Technical Data Book—Petroleum Refining,* 4th ed., American Petroleum Institute, 1220 L Street N.W., Washington, DC 20005. (1983)

[5] ASTM DS 4A, *Physical Constants of Hydrocarbons* C_1 to C_{10} (Data Series 4A), ASTM, 1916 Race Street, Philadelaphia, PA 19103.

[6] NFPA 59, *Standard for the Storage and Handling of Liquefied Petroleum Gases at Utility Gas Plants,* National Fire Protection Association, Batterymarch Park, Quincy, MA 02269.

[7] *Code of Federal Regulations,* Title 49 CFR Parts 100–199 (Transportation), Superintendent of Doc-

uments, U.S. Government Printing Office, Washington, DC 20402.

[8] *Transportation of Dangerous Goods Regulations,* Canadian Government Publishing Centre, Supply and Services Canada, Ottawa, Ontario, Canada K1A 0S9.

[9] CGA S-1.1, *Pressure Relief Device Standards—Part 1—Cylinders for Compressed Gases,* Compressed Gas Association, Inc., 1235 Jefferson Davis Highway, Arlington, VA 22202.

[10] CGA V-1, *American National, Canadian, and Compressed Gas Association Standard for Compressed Gas Cylinder Valve Outlet and Inlet Connections* (ANSI/CSA/CGA V-1), Compressed Gas Association, Inc., 1235 Jefferson Davis Highway, Arlington, VA 22202.

Methane

Chemical Symbol: CH_4
Synonyms: Marsh gas, methyl hydride
CAS Registry Number: 74-82-8
DOT Classification: Flammable gas
DOT Label: Flammable gas
Transport Canada Classification: 2.1
UN Number: UN 1971 (compressed gas); UN 1972 (refrigerated liquid)

PHYSICAL CONSTANTS

	U.S. Units	SI Units
Chemical formula	CH_4	CH_4
Molecular weight	16.042	16.042
Density of the gas at 60°F (15.6°C) and 1 atm	0.042 35 lb/ft^3	0.6784 kg/m^3
Specific gravity of the gas at 60°F (15.6°C) and 1 atm (air = 1)	0.554 91	0.554 91
Specific volume of the gas at 60°F (15.6°C) and 1 atm	23.6113 ft^3/lb	1.474 00 m^3/kg
Density of the liquid at boiling point	26.57 lb/ft^3	425.61 kg/m^3
Boiling point at 1 atm	−258.68°F	−161.49°C
Freezing point at 1 atm	−296.7°F	−182.61°C
Critical temperature	−115.78°F	−82.100°C
Critical pressure	673.1 psia	4640.86 kPa abs
Critical density	10.09 lb/ft^3	161.63 kg/m^3
Triple point	−296.5°F at 1.69 psia	−182.5°C at 11.65 kPa abs
Latent heat of vaporization at boiling point	219.22 Btu/lb	509.91 kJ/kg
Latent heat of fusion at −296.5°F (−182.5°C)	0.055 62 Btu/lb	0.1294 kJ/kg
Specific heat of the ideal gas at 60°F (15.6°C) and 1 atm		
C_p	0.5271 Btu/(lb)(°F)	2.207 kJ/(kg)(°C)
C_v	0.4032 Btu/(lb)(°F)	1.688 kJ/(kg)(°C)

	U.S. Units	SI Units
Ratio of specific heats, C_p/C_v	1.307	1.307
Weight of the liquid at boiling point	3.552 lb/gal	425.6 kg/m³
Heat of combustion of the real gas at 60°F (15.6°C) and 1 atm	1011.6 Btu/ft³	37 691.15 kJ/m³
Net heat of combustion of the real gas at 60°F (15.6°C) and 1 atm	910.77 Btu/ft³	33 934.33 kJ/m³
Air required for combustion at 60°F (15.6°C) and 1 atm		
per ft³ of the real gas	9.563 ft³ air	0.2708 m³ air
per lb of the real gas	17.233 lb air	7.8167 kg air
Flammable limits in air	5.0–15%	5.0–15%
Flash point	−306°F	−187.7°C

DESCRIPTION

Methane is a colorless, odorless, tasteless, flammable gas. It is the first member of the paraffin (aliphatic or saturated) series of hydrocarbons. It is the major constituent of natural gas. Methane is soluble in alcohol or ether, and slightly soluble in water. Methane is shipped as a nonliquefied compressed gas in cylinders at pressures up to 6000 psig at 70°F (41 368 kPa at 21.1°C). Liquefied methane (LNG) is shipped in bulk quantities as a cryogenic liquid on barges and tankers.

Methane is a major constituent of coal gas and is present to an extent in air in coal mines.

GRADES AVAILABLE

Methane is typically available for commercial and industrial purposes in a C.P. Grade (minimum purity of 99 mole percent), a technical grade (minimum purity of 98.0 mole percent), and a commercial grade which is actually natural gas as it is received from the pipeline (there is no guaranteed purity, but methane content usually runs about 93 percent or better). A typical analysis for commercial grade methane is as follows:

Methane	93.63 mole %
Carbon Dioxide	0.70 mole %
Nitrogen	0.47 mole %
Ethane	3.58 mole %
Propane	1.02 mole %
Isobutane	0.21 mole %
n-Butane	0.19 mole %
Isopentane	0.06 mole %
n-Pentane	0.06 mole %
Hexane	0.02 mole %
Heptanes plus	0.06 mole %

Tertiary-butyl mercaptan is added in trace amounts as an odorant. This grade of gas has a sulfur content of 0.002 grains/100 ft³ (0.0457 mg/m³) and a typical gross heating value of 1044 Btu/ft³ (38 898 kJ/m³).

USES

Natural gas, which is mostly methane, is widely used as a fuel. In the chemical industry, methane is used heavily as a raw material for making important products that include acetylene, ammonia, ethanol, and methanol; its chlorination also yields carbon tetrachloride, chloroform, methyl chloride, and methylene chloride. It is used to produce carbon black for use in the manufacture of rubber products and printing inks. The burning of high-purity methane is used to make carbon black of special quality for electronic devices. Natural gas has seen limited use as a motor fuel handled as a compressed gas in high pressure cylinders or liquid dewars.

PHYSIOLOGICAL EFFECTS

Methane is generally considered nontoxic. Exposures to concentrations of up to 9 percent methane have been reported without apparent ill effects; inhalation of higher concentrations eventually causes a feeling of pressure on the forehead and eyes, but the sensation ends after returning to fresh air. Methane is a simple asphyxiant.

MATERIALS OF CONSTRUCTION

Methane is noncorrosive and may be contained by any common, commercially available metals, with the exception of cryogenic liquid applications. Handling equipment must, however, be designed to safely withstand the temperatures and pressures to be encountered.

At the temperature of liquid methane, ordinary carbon steels and most alloy steels lose their ductility and are considered unsafe for liquid methane service. Satisfactory materials for use with liquid methane include Type 18-8 stainless steel and other austenitic nickel-chromium alloys, copper, Monel, brass, and aluminum.

SAFE STORAGE, HANDLING, AND USE

Methane poses hazards to personnel through its flammability. All the precautions necessary for the safe handling of any flammable compressed gas must be observed in working with methane. Any shipping mode or regulation applicable for methane may also apply for natural gas.

It is important that ignition sources be kept away from containers, including situations in which leakage could cause the gas to be ignited by such sources as a spark from a motor. All piping and equipment used with methane should be grounded.

Methane should not be stored with cylinders containing oxygen, chlorine, or other oxidizing or combustible materials. See Chapter 5 for additional general recommendations concerning safe storage, handling, and use of flammable gases.

Precautions for the safe handling of liquefied methane are found in NFPA 59A, *Standard for the Production, Storage, and Handling of Liquefied Natural Gas [LNG].* [1]

DISPOSAL

Disposal of methane, as with other gases, should be undertaken only by personnel familiar with the gas and the procedures for disposal. Contact the supplier for instructions. In general, the best procedure for disposal of flammable gases, including methane, is to burn the gas if a burning unit is available in the plant. This should be done in accordance with appropriate regulations.

HANDLING LEAKS AND EMERGENCIES

Leaks

Storage and use areas should be monitored for leakage of methane, since methane (unless odorized) may not give adequate warning of its dangerous presence in the atmosphere. It is lighter than air and presents hazards similar to hydrogen with regard to its ignitability.

In the absence of appropriate monitors, localized leaks may be difficult to detect unless the gas is odorized. Therefore, periodic checks should be made of connections and joints, flanges, and components subject to leakage. Sensitive instrumental detectors are available. Alternatively, leak detection solutions will indicate leakage through bubble formation.

Eliminate all sources of ignition until leaks have been repaired.

Once a leak is found, shut off the source of gas and lower the pressure in the system to minimize the leak. After the leak has been repaired, the system should be leak tested prior to use.

If uncontrolled leakage is encountered which cannot be stopped by shutting off the closest appropriate valve or main supply valve (without risk), implement a plan for evacuation and quickly contact a local fire department.

Fires

If a methane fire should occur, shut off the gas source if this can be done without risk. If the source cannot be shut off, let the fire burn itself out while making sure that gas storage vessels and piping containing gas in close proximity, as well as combustible materials in the area, are kept cool by spraying with water. The local fire department should be called.

Only personnel specifically trained and wearing appropriate personal protective equipment should be permitted to work at the fire scene.

See Chapter 5 for additional recommendations that may assist in training personnel in firefighting.

First Aid

Inhalation of low concentrations can be remedied by promptly going to an uncontaminated area and inhaling fresh air or oxygen. In the event of a massive exposure wherein the victim has become unconscious or symptoms of asphyxiation are present, the person should be removed promptly to an uncontaminated atmosphere and given artificial respiration if breathing has stopped. This should be followed by administering oxygen after breathing has been restored.

Skin contact of liquid methane can result in frostbite. First aid treatment for frostbite consists of putting the frostbitten part in warm water of about 108°F (42°C). If warm water is not available or is impractical to use, wrap the affected area gently in blankets. Encourage the victim to exercise the affected part while it is being warmed. This will aid circulation to reestablish itself naturally. Medical attention by a physician should be obtained.

In the event of eye contact with liquid methane, flush with tap water for 15 minutes. If irritation persists, the patient should be referred to a physician.

METHODS OF SHIPMENT

Under the appropriate regulations, methane or natural gas is authorized for shipment as follows: [2] and [3]

By Rail: In cylinders. As a liquid in tank cars designed for cryogenic gases, shipped under TC or DOT exemption.

By Highway: In cylinders on trucks and in tube trailers. As a liquid in tanks (specification MC-338) designed for cryogenic liquids.

By Water: In cylinders aboard cargo vessels only. As a liquid in cargo tankers or in barge tanks designed for cryogenic liquids.

By Air: In cylinders aboard cargo aircraft only up to 300 lb (136 kg) maximum net weight per cylinder.

By Pipeline: As gas in high pressure pipelines or as a liquid in lines designed for cryogenic liquids.

CONTAINERS

Methane may be shipped in any cylinders of the types authorized for nonliquefied compressed gases. Bulk industrial users of methane usually receive it as natural gas by pipeline and purify it if necessary for further processing. Ocean shipment of methane is made in specially designed ships carrying it as a cryogenic liquid.

Filling Limits

The maximum fill pressure permitted for methane or natural gas in cylinders at 70°F (21.1°C) is the authorized service pressure of the cylinder into which it is filled.

Cylinders

Any cylinders authorized for the shipment of a compressed gas may be used in methane gas service under TC/DOT regulations, but cylinder types of Specifications 3A and 3AA are probably those most commonly used for methane. Authorized cylinders for methane service include those meeting TC/DOT specifications 3A, 3AA, 3B, 3E, 4B, 4BA, 4BW, 3AX, 3AAX, 3T, 3AL, and 39. When used for shipment of flammable gas, the internal volume of a Specification 39 cylinder must not exceed 75 in.^{3} (0.00123 m^3). Cylinders meeting specifications, 3, 3C, 3D, 4, 4A, 4C, 25, 26, 33, and 38 may also be continued in methane service, but new construction is not authorized. All types of cylinders authorized for methane shipment must be requalified by periodic hydrostatic retest every five years under present regulations, with the following exceptions: TC/DOT-4 cylinders, every ten years; and no periodic retest is required for cylinders of specifications 3C, 3E, 4C, and 7.

Pressure relief devices authorized for use with methane cylinders are the Type CG-1 rupture disk, the Type CG-4 and Type CG-5 combination rupture disk/fusible plug devices, and the Type CG-7 pressure relief valve. See CGA S-1.1, *Pressure Relief Device Standards—Part 1—Cylinders for Compressed Gases,* for additional information and requirements. [4]

Valve Outlet Connections

Standard connections in the United States and Canada for methane (and natural gas) cylinders are as follows: for service pressure up to 3000 psig (20 684 kPa), Connection CGA 350; for 3001 to 5500 psig (20 691 to 37 921 kPa), Connection CGA 695; for 5501 to 7500 psig (37 928 to 51 711 kPa), Connection CGA 703; and for cryogenic liquid withdrawal, Connection CGA 450. [5]

METHODS OF MANUFACTURE

Methane is produced commercially from natural gas by absorption or adsorption methods of purification; supercooling and distillation are sometimes employed to secure methane of very high purity. Some California natural gas wells produce methane of high purity. It can also be obtained by cracking petroleum fractions.

REFERENCES

[1] NFPA 59A, *Standard for the Production, Storage, and Handling of Liquefied Natural Gas [LNG],* National Fire Protection Association, Batterymarch Park, Quincy, MA 02269.

[2] *Code of Federal Regulations,* Title 49 CFR Parts 100–199 (Transportation), Superintendent of Documents, U.S. Government Printing Office, Washington, DC 20402.

[3] *Transportation of Dangerous Goods Regulations,* Canadian Government Publishing Centre, Supply and Services Canada, Ottawa, Ontario, Canada K1A 0S9.

[4] CGA S-1.1, *Pressure Relief Device Standards—Part 1—Cylinders for Compressed Gases,* Compressed Gas Association, Inc., 1235 Jefferson Davis Highway, Arlington, VA 22202.

[5] CGA V-1, *American National, Canadian, and Compressed Gas Association Standard for Compressed Gas Cylinder Valve Outlet and Inlet Connections* (ANSI/CSA/CGA V-1), Compressed Gas Association, Inc., 1235 Jefferson Davis Highway, Arlington, VA 22202.

MAPP[R] Gas

Chemical Symbol: C_3H_4
Synonyms: MAPP[R] Gas; liquefied petroleum gas with methylacetylene-propadiene; methylacetylene-propadiene mixture; methylacetylene-propadiene stabilized
CAS Registry Number: none
DOT Classification: Flammable gas
DOT Label: Flammable gas
Transport Canada Classification: 2.1
UN Number: UN 1075

PHYSICAL CONSTANTS

	U.S. Units	SI Units
Chemical formula (base)	C_3H_4	C_3H_4
Molecular weight (avg.)	41.72	41.72
Density of the gas at 60°F (15.6°C)	0.113 lb/ft³	1.82 kg/m³
Specific gravity of the gas at 60°F (15.6°C)	1.48	1.48
Specific volume of the gas at 70°F (21.1°C)	9.1 ft³/lb	0.57 m³/kg
Density of the liquid at 60°F (15.6°C)	4.68 lb/ft³	551 kg/m³
Specific gravity of the liquid at 60°F (15.6°C)	0.571	0.571
Boiling range	−54°F to −10°F	−47°C to −23°C
Vapor pressure at 70°F (21.1°C)	107 psig	738 kPa
Critical temperature	245°F	118°C
Critical pressure	752 psia	5185 kPa abs
Latent heat of vaporization	227 Btu/lb	529.7 kJ/kg
Specific heat of the liquid at 70°F (21.1°C) and 1 atm C_p	0.362 Btu/(lb)(°F)	1.515 kJ/(kg)(°C)
Specific heat of the liquid at 70°F (21.1°C) and 1 atm C_v	0.227 Btu/(lb)(°F)	0.950 kJ/(kg)(°C)

	U.S. Units	SI Units
Ratio of specific heats, C_p/C_v	1.59	1.59
Gross heat of combustion	21 000 Btu/ft^3	49 000 kJ/kg
Gross heat after vaporization	2404 Btu/ft^3	90 MJ/m^3
Flammable limits in oxygen	2.5–60%	2.5–60%
Flammable limits in air	3.5–11.0%	3.5–11.0%

DESCRIPTION

MAPP Gas is a stabilized mixture of methylacetylene and propadiene. Alkane and alkene hydrocarbons are added as stabilizers. The stabilizers serve to render the methylacetylene and propadiene shock insensitive and to ensure that their concentration remains nearly uniform at all times during vaporization of the mixture. [1]

MAPP Gas is transported and stored as a liquid under its own vapor pressure. Its transportation, storage, and handling characteristics are the same as those of liquefied petroleum gas. It has a characteristic odor detectable at concentrations as low as 100 ppm in air. Stabilized liquefied MAPP Gas and MAPP Gas vapor are insensitive to shock. [1], [2], and [3]

Its vapor is not subject to exothermic decomposition from a 100-J energy source to 419°F (215°C) and 285 psig (1965 kPa). With a probe temperature of 825°F (440°C), the gas is stable up to 600°F (315°C) and 1100 psig (7584 kPa). Thus, it can be safely used up to full cylinder pressure.

GRADES AVAILABLE

MAPP Gas is available in only one grade.

USES

MAPP Gas is primarily used as an industrial fuel gas for oxygen-fuel gas cutting, welding, heating, wire metallizing, flame hardening, soldering, and brazing. Small amounts are used for other diverse purposes such as mold smoking, simulation of weapons firing, flame polishing of plastics, special effects, and furnace atmospheres.

PHYSIOLOGICAL EFFECTS

The toxicity of MAPP Gas is very slight, but high concentrations (5000 ppm) can have an anesthetic effect or cause nausea. Contact with MAPP Gas vapor causes no adverse effects on the eyes or skin. Contact with MAPP Gas liquid may cause frostbite. [4]

The strong odor of MAPP Gas (detectable at 100 ppm and objectionable at 1000 ppm) should provide sufficient warning to prevent exposure. The strong odor also serves as a means of avoiding exposures above the lower detectable odor limit in air.

MATERIALS OF CONSTRUCTION

Piping systems for MAPP Gas must be made of steel. [5] Schedule 40 pipe is suitable if joints are welded. Schedule 80 pipe is recommended for threaded joints. In no case shall threaded joints be welded after assembly. Fittings may be rolled, forged, or cast steel, malleable iron, or modular iron. Gray or white cast iron fittings must not be used.

SAFE STORAGE, HANDLING, AND USE

MAPP Gas should be treated with the same care and caution as any other flammable gas. Cylinders in storage or use should always be secured in an upright position. Do not transport cylinders in a closed vehicle.

MAPP Gas cylinder storage inside buildings, except those in actual use, is limited to 368 lb (167 kg). [5] Larger quantities must be stored outside, in a special building or in a separate room in accordance with NFPA 51, *Standard for the Design and Installation of Oxygen-Fuel Gas Systems for Welding, Cutting, and Allied Processes.* [5]

MAPP Gas can be used with most commonly available torches, hoses, and regulators. Acetylene regulators can be used for working pressures up to 15 psig (103 kPa). If a working pressure of more than 15 psig (103 kPa) is required, a suitable LP-gas regulator should be used. If cylinders are manifolded, do not use acetylene pigtails with dry flash arrestors. The flash arrestors are not necessary and will impede the flow of MAPP Gas.

Torch tips should be those designed for MAPP Gas since the use of acetylene torch tips will often result in the flame lifting off before a sufficient flame size can be achieved. When MAPP Gas is used with some propane-natural gas tips, backfire can occur at low flow rates.

Commonly accepted safe practice for use of MAPP Gas is available in ANSI Z49.1, *Safety in Welding and Cutting,* and AWS C4.2, *Operator's Manual for Oxyfuel Gas Cutting.* [6] and [7]

DISPOSAL

Disposal of MAPP Gas, as with other gases, should be undertaken only by personnel familiar with the gas and the procedures for disposal. In general, the best procedure for disposal of flammable gases, including MAPP Gas, is to burn the gas if a burning unit is available in the plant. This should be done in accordance with appropriate regulations.

HANDLING LEAKS AND EMERGENCIES

MAPP Gas cylinders have a pressure relief device built into the cylinder valve. If the cylinder is exposed to high temperature, the pressure will increase to a level which may be sufficient to open the relief device. In such an event, the rapid vaporization may cause enough cooling of the product to reduce the vapor pressure and cause the device to close. If an ignition source is present, the escaping gas may ignite. In any event, cool the cylinder and remove it from high temperature locations.

If a cylinder valve is leaking, attempt to stop the leak. Possible sources for valve leakage are through the stem packing, through the valve outlet, from the pressure relief device, and from the neck threads. The supplier should be consulted in the event of a leak. If a leak persists, remove the cylinder to an outside location away from any ignition source and tag the cylinder as having a defective valve. Immediately notify the supplier and follow the supplier's instructions.

First Aid

A person who is exposed to high concentrations and exhibits anesthetic effects should be removed to fresh air. Apply artificial respiration if necessary and consult a physician.

METHODS OF SHIPMENT

MAPP Gas is authorized for shipment under the appropriate regulations for liquefied petroleum gas.

By Rail: As a liquid under its own pressure in tank cars approved for LP-gas service and in cylinders.

By Water: In cylinders as deck cargo on cargo vessels. In DOT 51 tanks as deck cargo on cargo vessels.

By Highway: In cylinders on open trucks. In tank trucks or trailers that comply with TC/DOT specifications MC-330 and MC-331.

By Air: In limited quantities on cargo aircraft only.

CONTAINERS

MAPP Gas is shipped in steel cylinders conforming to DOT 4BW or in aluminum cylinders conforming to TC/DOT specification 4E. Service pressure must be at least 240 psig (1654 kPa). It is also shipped in one-pound disposable containers under special DOT permits. Cylinder valves are equipped with a CG-7 type pressure relief valve.

Valve Outlet Connections

In the United States and Canada, cylinders are equipped with valves having Connection CGA 510.

Filling Limits

Maximum permissible filling density of portable containers is 50 percent of water capacity in pounds. [8]

REFERENCES

[1] Huston, R. F., Barrios, C. A., and Holleman, R. A., "Weathering and Stability of Methylacetylene-Propadiene-Hydrocarbon Mixtures," *Journal of Chemical and Engineering Data,* **15,** no. 1 (1970). Library, American Chemical Society, 1155 16th Street, N.W., Washington, DC 20036.

[2] "Flammability and Shock Sensitivity Characteristics of Methylacetylene, Propadiene and Propylene Mixtures," Explosives Research Laboratory, Bureau of Mines Report No. 3849, Cooperative Agreement with Dow Chemical Company 14-09-0050-2152. Bureau of Mines, Pittsburgh Research Center, Cochrans Mill Road, P.O. Box 18070, Pittsburgh, PA 15236-0070. *Journal of Chemical & Engineering Data,* **9,** no. 3, 467–472 (July 1964). American Chemical Society, 1155 16th Street N.W., Washington, DC 20036.

[3] Factory Mutual Report No. 18093, "MAPP Industrial Gas," Factory Mutual Research Corp., 1151 Boston Providence Turnpike, Norwood, MA 02062.

[4] Torkleson, T. K., and Rower, U. K., "Results of Repeated Inhalation by Laboratory Animals and a Limited Human Sensory Study of a Mixture of Saturated and Unsaturated C_3 and C_4 Hydrocarbons (MAPP Industrial Gas)," *American Industrial Hygiene Association Journal,* **25,** 554–559 (November-December 1964). American Industrial Hygiene Association, 475 Wolf Ledges Parkway, Akron, OH 44311-1087.

[5] NFPA 51, *Standard for the Design and Installation of Oxygen-Fuel Gas Systems for Welding, Cutting, and Allied Processes,* National Fire Protection Association, Batterymarch Park, Quincy, MA 02269.

[6] ANSI Z49.1, *Safety in Welding and Cutting,* American National Standards Institute, 1430 Broadway, New York, NY 10018.

[7] AWS C4.2, *Operator's Manual for Oxyfuel Gas Cutting,* American Welding Society, Box 351040, Miami, FL 33135.

[8] NFPA 58, *Standard for the Storage and Handling of Liquefied Petroleum Gases,* National Fire Protection Association, Batterymarch Park, Quincy, MA 02269.

Methylamines (Anhydrous)

	Monomethylamine	Dimethylamine	Trimethylamine
Chemical Symbol:	CH_3NH_2	$(CH_3)_2NH$	$(CH_3)_3N$
Synonyms:	Methylamine, aminomethane		
CAS Registry Number:	74-89-5	124-40-3	75-50-3
DOT Classification:	Flammable gas	Flammable gas	Flammable gas
DOT Label:	Flammable gas	Flammable gas	Flammable gas
Transport Canada Classification:	Class 2.1 (9.2)	Class 2.1 (9.2)	Class 2.1 (9.2)
UN Number:	1061	1032	1083

PHYSICAL CONSTANTS

Monomethylamine

	U.S. Units	SI Units
Chemical formula	CH_3NH_2	CH_3NH_2
Molecular weight	31.058	31.058
Vapor pressure		
at 68°F (20°C)	43.5 psia	300 kPa abs
Specific gravity of the gas		
at 68°F (20°C) and 1 atm		
(air = 1)	1.08	1.08
Specific volume of the gas		
at 70°F (21.1°C) and 1 atm	12.1 ft^3/lb	0.755 m^3/kg
Density of the liquid		
at 70°F (21.1°C)	41.4 lb/ft^3	663.2 kg/m^3
at 105°F (40.5°C)	40.1 lb/ft^3	642.3 kg/m^3
at 115°F (46.1°C)	39.8 lb/ft^3	637.5 kg/m^3
at 130°F (54.4°C)	39.2 lb/ft^3	627.9 kg/m^3
Boiling point at 1 atm	20.6°F	−6.33°C
Freezing point at 1 atm	−136.3°F	−93.5°C
Critical temperature	314.4°F	156.9°C
Critical pressure	1081.9 psia	7459.4 kPa abs
Critical density	13.47 lb/ft^3	215.8 kg/m^3
Latent heat of vaporization		
at boiling point	357.5 Btu/lb	831.5 kJ/kg
Latent heat of fusion		
at freezing point	84.95 Btu/lb	197.6 kJ/kg
Specific heat of gas		
at 68°F (20°C) and 1 atm		
C_p	0.396 Btu/(lb)(°F)	1.657 kJ/(kg)(°C)
C_v	0.329 Btu/(lb)(°F)	1.378 kJ/(kg)(°C)
Ratio of specific heats, C_p/C_v	1.202	1.202
Solubility in water, wt/wt		
at 77°F (25°C), 1 atm	108%	108%
Weight of the liquid		
at 68°F (20°C)	5.55 lb/gal	665 kg/m^3
Autoignition temperature in		
air	806°F	430°C
Flammable limits in air, by		
volume	4.9–20%	4.9–20%

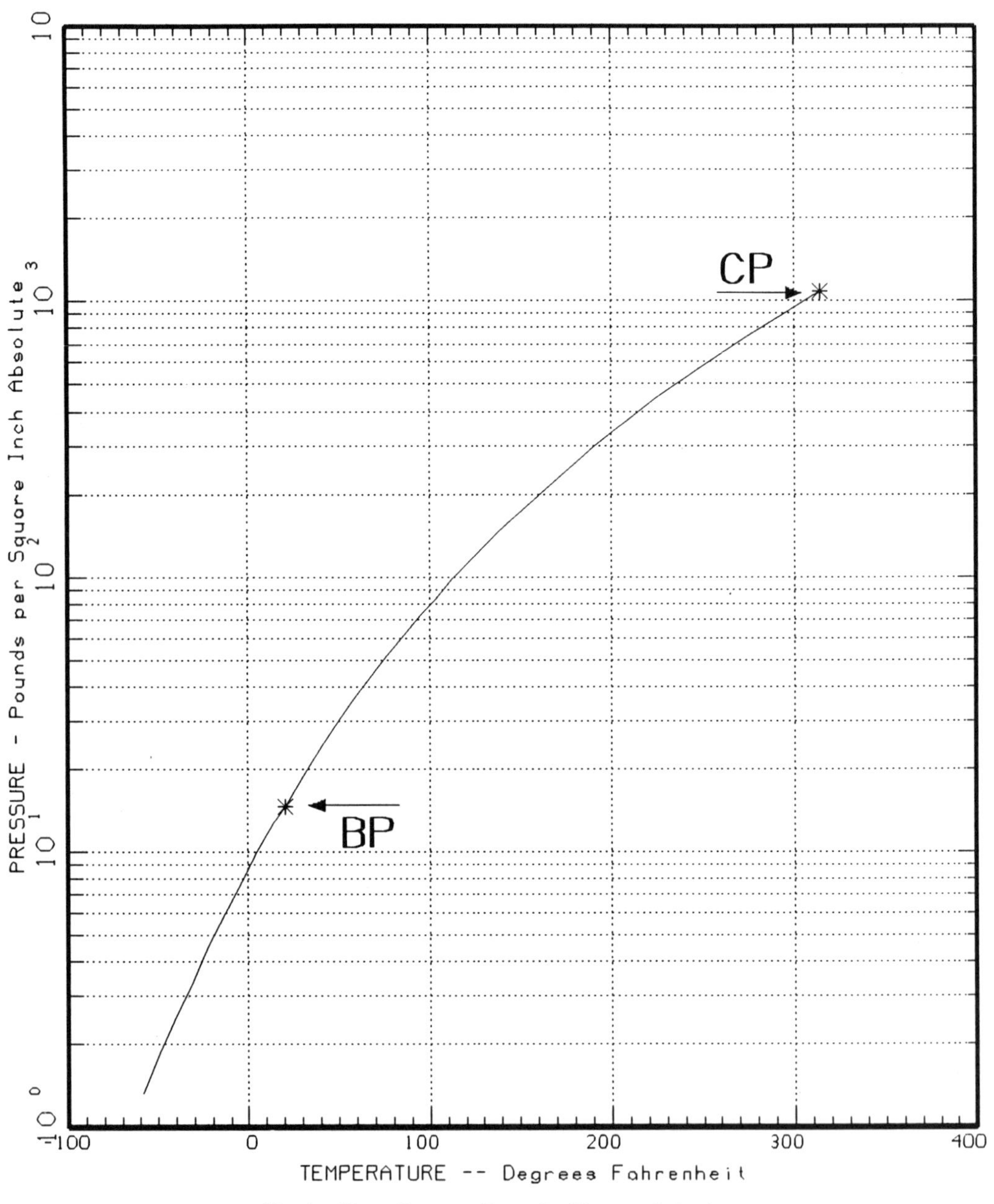

Fig. 1. Vapor Pressure Curve for Monomethylamine.

PHYSICAL CONSTANTS

Dimethylamine

	U.S. Units	SI Units
Chemical formula	$(CH_3)_2NH$	$(CH_3)_2NH$
Molecular weight	45.08	45.08
Vapor pressure at 70°F (21.1°C)	26.0 psig	179.3 kPa
Specific gravity of the gas at 59°F and 1 atm (air = 1)	1.55	1.55
Specific volume of the gas at 70°F (21.1°C) and 1 atm	8.6 ft^3/lb	0.54 m^3/kg
Density of liquid		
at 70°F (21.1°C)	40.8 lb/ft^3	653.6 kg/m^3
at 105°F (40.5°C)	39.4 lb/ft^3	631.1 kg/m^3
at 115°F (46.1°C)	39.0 lb/ft^3	624.7 kg/m^3
at 130°F (54.4°C)	38.5 lb/ft^3	616.7 kg/m^3
Boiling point at 1 atm	44.4°F	6.89°C
Freezing point at 1 atm	−134°F	−92.22°C
Critical temperature	328.3°F	164.6°C
Critical pressure	770.1 psia	5311 kPa abs
Critical density	15.98 lb/ft^3	255.97 kg/m^3
Latent heat of vaporization at boiling point	252.8 Btu/lb	588.01 kJ/kg
Latent heat of fusion at freezing point	56.7 Btu/lb	131.88 kJ/kg
Specific heat of the gas at 77°F (25°C) and 1 atm		
C_p	0.3819 Btu/(lb)(°F)	1.60 kJ/(kg)(°C)
C_v	0.3324 Btu/(lb)(°F)	1.40 kJ/(kg)(°C)
Ratio of specific heats, C_p/C_v	1.149	1.149
Solubility in water, weight percent at 140°F (60°C) and 1 atm	23.7%	23.7%
Weight of the liquid at 68°F (20°C)	5.48 lb/gal	657 kg/m^3
Autoignition temperature in air	756°F	402°C
Flammable limits in air, by volume	2.8–14.4%	2.8–14.4%

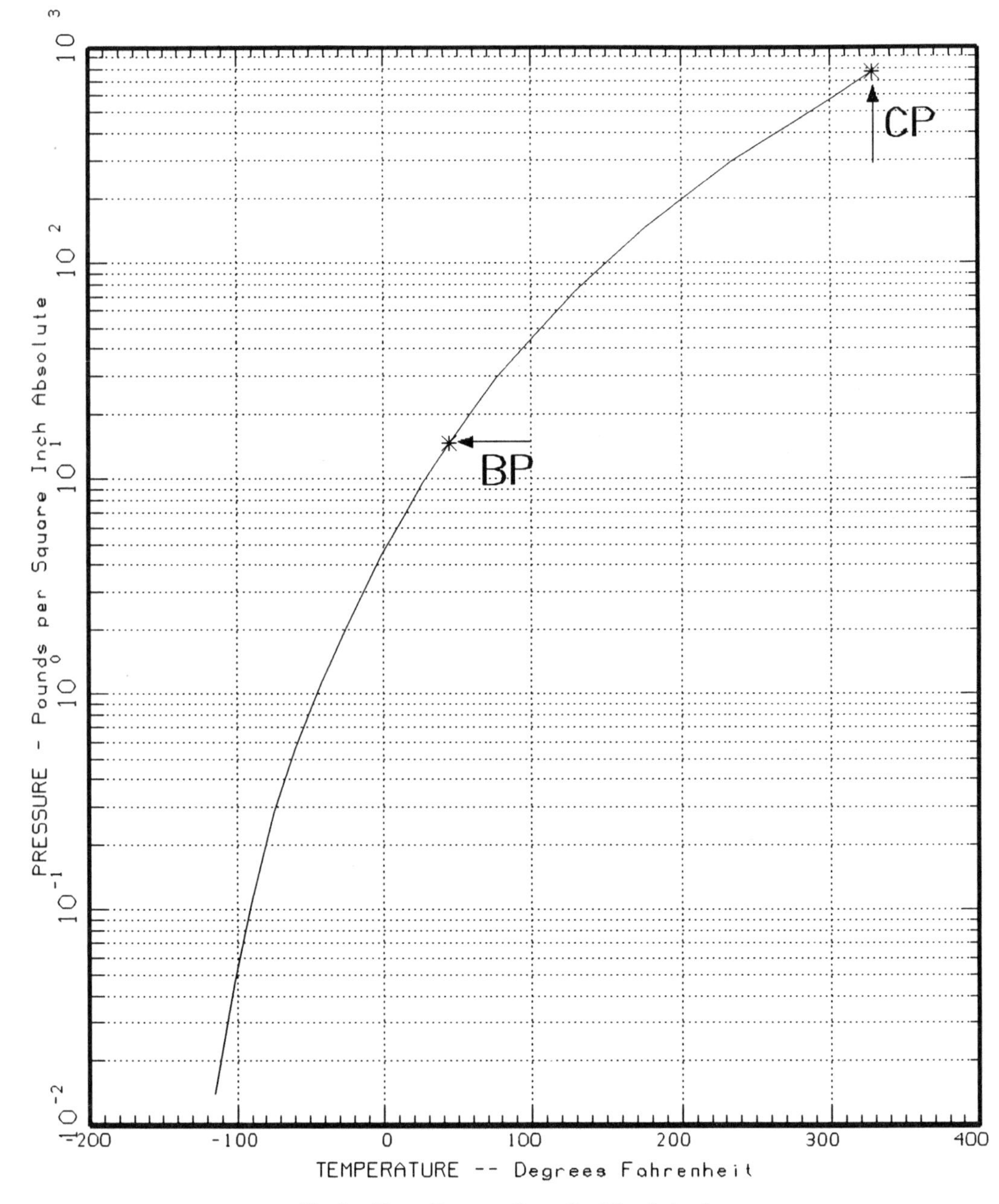

Fig. 2. Vapor Pressure Curve for Dimethylamine.

PHYSICAL CONSTANTS

Trimethylamine

	U.S. Units	SI Units
Chemical formula	$(CH_3)_3NH$	$(CH_3)_3NH$
Molecular weight	59.11	59.11
Vapor pressure		
at 68°F (20°C)	28.0 psig	193.1 kPa
Specific volume of the gas		
at 70°F (21.1°C) and 1 atm	6.4 ft^3/lb	0.40 m^3/kg
Density of the liquid		
at 70°F (21.1°C)	39.4 lb/ft^3	631.1 kg/m^3
at 105°F (40.5°C)	38.1 lb/ft^3	610.3 kg/m^3
at 115°F (46.1°C)	37.7 lb/ft^3	603.9 kg/m^3
at 130°F (54.4°C)	37.1 lb/ft^3	594.3 kg/m^3
Boiling point at 1 atm	37.2°F	2.89°C
Freezing point at 1 atm	−178.8°F	−117.1°C
Critical temperature	320.2°F	160.1°C
Critical pressure	590.9 psia	4074.11 kPa abs
Critical density	14.55 lb/ft^3	233.07 kg/m^3
Latent heat of vaporization		
at boiling point	166.9 Btu/lb	388.2 kJ/kg
Latent heat of fusion		
at freezing point	47.6 Btu/lb	110.7 kJ/kg
Specific heat of the gas		
at 77°F (25°C) and 1 atm		
C_p	0.3717 Btu/(lb)(°F)	1.56 kJ/(kg)(°C)
C_v	0.3139 Btu/(lb)(°F)	1.31 kJ/(kg)(°C)
Ratio of specific heats, C_p/C_v	1.184	1.184
Solubility in water, weight percent		
at 86°F (30°C) and 1 atm	47.5%	47.5%
Weight of the liquid		
at 68°F (20°C)	5.31 lb/gal	636.3 kg/m^3
Autoignition temperature in air	374°F	190°C
Flammable limits in air, by volume	2.0–11.6%	2.0–11.6%

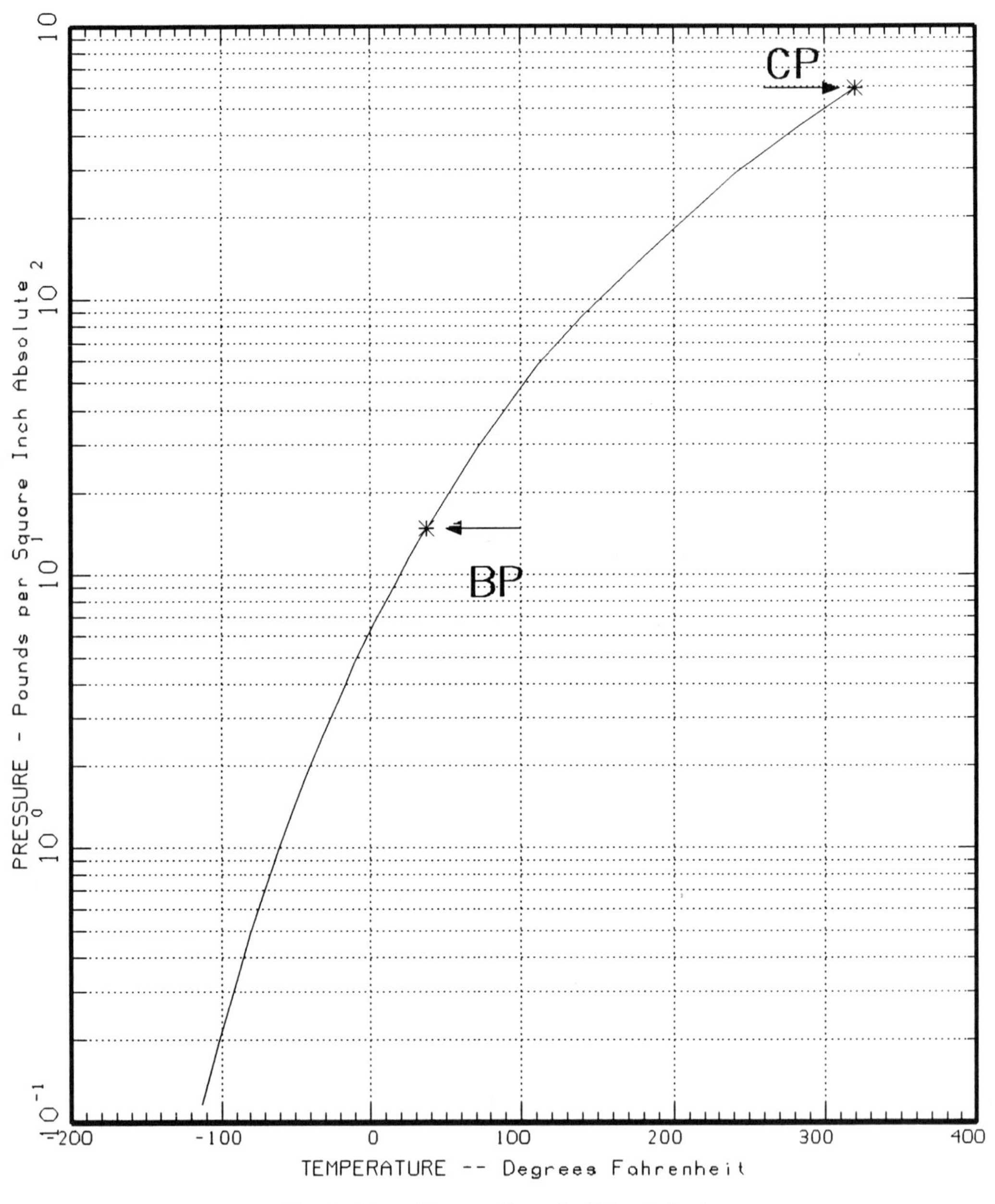

Fig. 3. Vapor Pressure Curve for Trimethylamine.

DESCRIPTION

The methylamines (monomethylamine, dimethylamine, and trimethylamine) are colorless, flammable, and toxic gases at normal room temperatures and atmospheric pressure in their anhydrous form. They have a distinct and disagreeable fishy odor in concentrations up to 100 ppm. In higher concentrations they have an odor like ammonia, which they resemble and from which they are derived. They are easily liquefied and are shipped in their anhydrous form as liquefied compressed gases. They are highly soluble in water and in alcohol, ether, and various other organic solvents.

Vapors of methylamines in air can burn within their respective flammable ranges. Gaseous methylamines and their solutions are alkaline materials and in sufficient concentrations can irritate and burn the skin, eyes, and respiratory system.

Chemically, the methylamines are slightly stronger than ammonia as bases. They hydrate in water solutions and neutralize acids to form methylammonium salts. They do not corrode iron and steel, but do attack copper and its alloys, zinc, and aluminum. The methylamines can form explosive compounds with mercury, so they must never be brought into contact with mercury.

Methylamines are used and shipped both in the form of anhydrous gases and in aqueous solutions. Only the anhydrous form is treated here.

GRADES AVAILABLE

Methylamines (anhydrous) are available for commercial and industrial use in various grades ranging from 98 percent purity and above.

USES

As sources of reactive organic nitrogen, the methylamines serve as intermediates in the synthesis of pharmaceuticals, agricultural chemicals, dyes, rubber chemicals, and explosives. Derivatives serve in agriculture as fungicides, insecticides, and feed supplements. Derivatives have also been employed in producing antihistamines, tranquilizers, and other drugs; in making dyestuffs, explosives, and rocket fuel; and in curing resins.

Rubber industry applications of derivatives include use as accelerators, vulcanizing agents, and chain terminators in synthetic rubber production. Derivatives are also solvents for various organic plastics, resins, gums, dyes, and pharmaceuticals.

PHYSIOLOGICAL EFFECTS

The methylamines are toxic. They are irritating to the nose, throat, and eyes in low concentrations, and require suitable gas masks and eye-protective devices for safe handling by exposed personnel. Severe exposure of the eyes may lead to loss of sight. Dermatitis results from contact of the methylamines with the skin. Inhalation of sufficiently high concentrations is followed by violent sneezing, a burning sensation of the throat with constriction of the larynx, and difficulty in breathing with congestion of the chest and inflammation of the eyes.

The eight-hour time weighted average exposure limits adopted by the U.S. Occupational Safety and Health Administration and the threshold limit values (TLVs) adopted by the American Conference of Governmental Industrial Hygienists for the methylamines are: [1] and [2]

Monomethylamine	10 ppm (12 mg/m^3)
Dimethylamine	10 ppm (18 mg/m^3)
Trimethylamine	10 ppm (24 mg/m^3)

OSHA has also adopted a 15-minute short term exposure limit (STEL) of 15 ppm (36 mg/m^3) for trimethylamine allowable during an eight-hour workday. [1]

MATERIALS OF CONSTRUCTION

Iron, steel, stainless steel, Monel, and some plastics have proven satisfactory in methylamine service. Copper, copper alloys (including brass and bronze), zinc (together with zinc alloys and galvanized surfaces), and aluminum are corroded by the methylamines and should not be used in direct contact with them. Mercury and the methylamines can explode on contact, and instruments containing mercury must never be used with the methylamines. Among gasket and packing materials satisfactory for use with them are compressed asbestos, polyethylene, Teflon, and carbon steel or stainless steel wound asbestos.

SAFE STORAGE, HANDLING, AND USE

Storage tanks for the methylamines should be made of steel and designed to comply with Section VIII of the *ASME Boiler and Pressure Vessel Code* as well as with all state and local regulations. [3] Safe working pressures vary with the vapor pressure–temperature relationship of the particular methylamine being stored, and with the high-temperature ranges at the plant location.

Important parts of well-designed tanks include proper dual pressure relief valves, a vapor absorption system, liquid level and pressure gauges, liquid and vapor transfer valves, and an adequate electrical ground. Pipes, fittings, pumps, gauges, and other equipment should be of steel, iron, or other material not subject to corrosion by the methylamines. Storage and handling installations should be designed with the help of technically competent personnel thoroughly familiar with the gases.

All the precautions necessary for the safe handling of any flammable, toxic gas must be observed with the anhydrous methylamines. See Chapter 5 for precautions and recommendations on the safe handling of flammable compressed gases.

DISPOSAL

Methylamines may be disposed of by slowly discharging the gas into a scrubber or other suitable vessel containing 10–20 percent aqueous acid solution (i.e., sulfuric, hydrochloric, etc.). It is necessary to place a check valve or reverse flow trap in line to prevent the acid solution from drawing back into the cylinder. The resulting solution should be disposed of in accordance with federal, state, provincial, and local regulations.

HANDLING LEAKS AND EMERGENCIES

Leaks in systems using methylamines may be detected by a number of methods. Leaks may be found with an open bottle of muriatic acid, which will fume in the presence of methylamines, or with moist red litmus paper, which turns blue upon contact. Never use flames to detect leaks. Leaks which cannot be immediately stopped should be covered with wet rags in order to prevent contamination of the atmosphere until repairs are made.

First Aid

Exposed personnel should be removed to fresh air. Artificial respiration and oxygen should be administered simultaneously in the event breathing has stopped.

For contact with the skin, contaminated clothing should be removed and the affected area of the body should be immediately flushed with large quantities of water. Since the reaction is alkaline, the skin should be washed with a mild acidic solution such as vinegar or 1–2 percent acetic acid. Use copious amounts gently and try to avoid abrading the skin. For exposure by the nose or throat, irrigate continuously for 15 minutes and, if the patient can swallow, encourage drinking large quantities of 1–2 percent citric acid solution or lemonade to neutralize alkali.

In the event of eye contact, immediately irrigate thoroughly with water or saline, holding the eyelids wide open. Continue irrigation for 30–60 minutes. Contact lenses must be removed prior to irrigation. An evaluation by an ophthalmologist should be made once irrigation has been completed.

See *Effects of Exposure to Toxic Gases—First Aid and Medical Treatment* for additional information. [4]

METHODS OF SHIPMENT

Under the appropriate regulations, the anhydrous methylamines are authorized for shipment as follows: [5] and [6]

By Rail: In cylinders and by insulated single-unit tank cars and TMU (multi-unit) tank cars.

By Highway: In cylinders on trucks, in tank trucks, and in portable tanks on trucks.

By Water: In cylinders and portable tanks (not over 20 000 lb or 9072 kg gross weight) via cargo vessels and in railroad tank cars complying with DOT provisions on trainships only. Dimethylamine and trimethylamine are not permitted on passenger vessels, ferry vessels, and railroad car ferry vessels. Monomethylamine may be shipped in cylinders on passenger vessels, ferries, and railroad car ferries (including passenger or vehicle ferry vessels). The methylamines are also authorized for shipment in cylinders on barges of U.S. Coast Guard classes A and C only and in cargo tanks aboard tankships and tank barges (with maximum filling densities by specific gravity as prescribed in Coast Guard regulations). See 46 CFR Parts 0-199. [7]

By Air: Aboard cargo aircraft only in appropriate cylinders up to 300 lb (136 kg) maximum net weight per cylinder.

CONTAINERS

The anhydrous methylamines are shipped as flammable gases under their own vapor pressures in cylinders, tank cars, and cargo tank trucks. They are also authorized by the DOT for shipment in TMU (ton multi-unit) tank cars and portable tanks.

Filling Limits

The maximum filling densities allowed under Transport Canada and U.S. Department of Transportation regulations for cylinders, TMU tank car tanks (DOT 106A type), and truck cargo tanks and portable tanks in percent water capacity by weight are as follows: monomethylamine, 60 percent; dimethylamine, 59 percent; and trimethylamine, 57 percent. Corresponding maximum filling densities for single-unit tank cars (DOT 105J300W with properly fitted loading and unloading valves) are: monomethylamine, 62 percent; dimethylamine, 62 percent; and trimethylamine, 59 percent.

For DOT 112T340W and 112J340W single-unit tank cars, the maximum filling densities are: monomethylamine, 61 percent; dimethylamine, 61 percent; and trimethylamine, 58 percent. See 49 CFR 173.314. [5]

Cylinders

Cylinders meeting the following TC/DOT specifications are authorized for methylamine service: 3A150, 3AA150, 3B150, 4B150, 4BA225, 4BW225 and 3E1800. Pressure relief devices are not required on cylinders charged with the methylamines.

Specification 3A, 3AA, and 3B cylinders used in methylamine service must be requalified by hydrostatic retest every five years under present regulations. Specification 4B, 4BA, and 4BW must be retested after expiration of the first 12-year period and each 7 years thereafter. Cylinders in compliance with Specifications 3A, 3AA, 3B, 4B, 4BA, or 4BW which are used specifically for monomethylamine, dimethylamine, and trimethylamine service and are free from corroding components may also be qualified by an external visual inspection as described in CGA C-6, *Standards for Visual Inspection of Steel Compressed Gas Cylinders.* [8] Periodic hydrostatic retest is not required for Specification 3E cylinders.

Valve Outlet Connections

The standard valve outlet connection in the United States and Canada for cylinders containing monomethylamine, dimethylamine, or trimethylamine is Connection CGA 705. [9]

Cargo Tanks

Anhydrous methylamines are authorized for shipment in motor vehicle cargo tanks complying with DOT specification MC-330 and MC-331 with a minimum design pressure of 150 psig (1034 kPa).

Portable Tanks

Anhydrous methylamines are also authorized for shipment via motor vehicle in steel portable tanks built to DOT-51 specification and with a minimum design pressure of 150 psig (1034 kPa).

TMU Tanks

Authorized rail shipment of the methylamines may also be made in TMU tank cars of DOT specification 106A500X.

Tank Cars

Bulk quantities of the anhydrous methylamines are commonly shipped by rail in tank cars of DOT specification 105J300W (an insulated single-unit tank car with properly fitted loading and unloading valves suitable for methylamine service). Other authorized tank cars are specification 112T340W and 112J340W. Also, Specification 105A300W is authorized for anhydrous methylamines if the tank car was built before September 1, 1981, and has a water capacity not exceeding 18 500 U.S. gallons. See 49 CFR 173.314. [5]

METHOD OF MANUFACTURE

The methylamines are produced by the reaction of methyl alcohol and ammonia over a catalyst at high temperature.

REFERENCES

[1] *Code of Federal Regulations,* Title 29 CFR Parts 1900–1910 (Labor). Superintendent of Documents, U.S. Government Printing Office, Washington, DC 20402.

[2] *Threshold Limit Values and Biological Exposure Indices,* 1989–90 ed., American Conference of Governmental Industrial Hygienists, 6500 Glenway Avenue, Bldg. D-7, Cincinnati, OH 45211-4438.

[3] *ASME Boiler and Pressure Vessel Code* (Section VIII), American Society of Mechanical Engineers, 345 E. 47th St., New York, NY 10017.

[4] *Effects of Exposure to Toxic Gases—First Aid and Medical Treatment,* 3rd ed., Matheson Gas Products, Inc., Secaucus, NJ 07094. (1988)

[5] *Code of Federal Regulations,* Title 49 CFR Parts 100–199 (Transportation), Superintendent of Documents, U.S. Government Printing Office, Washington, DC 20402.

[6] *Transportation of Dangerous Goods Regulations,* Canadian Government Publishing Centre, Supply and Services Canada, Ottawa, Ontario, Canada K1A 0S9.

[7] *Code of Federal Regulations,* 46 CFR Parts 0-199 (Shipping), Superintendent of Documents, U.S. Government Printing Office, Washington, DC 20402.

[8] CGA C-6, *Standards for Visual Inspection of Steel Compressed Gas Cylinders,* Compressed Gas Association, Inc., 1235 Jefferson Davis Highway, Arlington, VA 22202.

[9] CGA V-1, *American National, Canadian, and Compressed Gas Association Standard for Compressed Gas Cylinder Valve Outlet and Inlet Connections* (ANSI/CSA/CGA V-1), Compressed Gas Association, Inc., 1235 Jefferson Davis Highway, Arlington, VA 22202.

Methyl Bromide

Chemical Symbol: CH_3Br
Synonyms: Bromomethane
CAS Registry Number: 74-83-9
DOT Classification: Poison B
DOT Label: Poison
Transport Canada Classification: 2.3
UN Number: UN 1062

PHYSICAL CONSTANTS

	U.S. Units	SI Units
Chemical formula	CH_3Br	CH_3Br
Molecular weight	99.944	99.944
Vapor pressure		
at 70°F (21.1°C)	27.5 psig	189.6 kPa
at 105°F (40.6°C)	50.0 psig	344.7 kPa
at 115°F (46.1°C)	58.5 psig	405.3 kPa
at 130°F (54.4°C)	74.0 psig	510.2 kPa
Density of the gas at 77°F (25°C) and 1 atm	0.248 lb/ft^3	3.974 kg/m^3
Specific gravity of the gas at 77°F (25°C) and 1 atm (air = 1)	3.355	3.355
Specific volume of the gas at 77°F (25°C) and 1 atm	4.031 ft^3/lb	0.252 m^3/kg
Density of the liquid		
at 68°F (20°C)	104.63 lb/ft^3	1676 kg/m^3
at 77°F (25°C)	103.75 lb/ft^3	1662 kg/m^3
at 104°F (40°C)	103.63 lb/ft^3	1660 kg/m^3
at 140°F (60°C)	99.88 lb/ft^3	1600 kg/m^3
at 176°F (80°C)	97.07 lb/ft^3	1555 kg/m^3
Boiling point at 1 atm	38.4°F	3.56°C
Melting point at 1 atm	−137.2°F	−94°C
Critical temperature	381.2°F	194°C
Critical pressure	757.1 psia	5220 kPa abs
Critical density	38.1 lb/ft^3	610 kg/m^3
Latent heat of vaporization at 38.4°F (3.56°C)	108.41 Btu/lb	252.11 kJ/kg

	U.S. Units	SI Units
Heat capacity of the gas at 77°F (25°C) and 14.7 psia		
C_p	0.1106 Btu/(lb)(°F)	0.4630 kJ/(kg)(°C)
Solubility in water	1.75 lb/100 lb	1.75 g/100 g

DESCRIPTION

Methyl bromide is a colorless liquid or gas with practically no odor. It is a poisonous gas at room temperature and atmospheric pressure. At high concentrations, it has a chloroformlike odor. Detection of lower concentrations is often facilitated by a warning odorant, chloropicrin, which is added by the manufacturer.

Methyl bromide is flammable only in the range of 10 to 16 percent by volume in air. Its physical properties do not meet the definition of a flammable gas as defined by the U.S. Department of Transportation. The onset of thermal decomposition occurs at approximately 752°F (400°C).

GRADES AVAILABLE

Methyl bromide is sold either as a pure compound, 99.8 percent minimum, or with an odorant, chloropicrin (typically 2 percent).

USES

Methyl bromide is used primarily as a fumigant to control insects infesting various grains and non-food material. It is also used in small quantities in organic synthesis for methylations.

PHYSIOLOGICAL EFFECTS

Methyl bromide is toxic. The vapor is odorless at low concentrations and nonirritating to the skin or eyes during exposure; therefore, detection of overexposure is not possible until the onset of symptoms. Early symptoms of overexposure are dizziness, headache, nausea and vomiting, weakness, and collapse. Lung edema may develop 2 to 48 hours after exposure, accompanied by cardiac irregularities. These effects may cause death. Repeated overexposures can result in blurred vision, staggering gait, and mental imbalance, with probable recovery after a period of no exposure. Blood bromide levels suggest the occurrence, but not the degree, of exposure. Treatment is symptomatic. [1]

A threshold limit value (TLV) of 5 ppm has been adopted by the American Conference of Governmental Industrial Hygienists. [2] The U.S. Occupational Safety and Health Administration has adopted an 8-hour time weighted average (TWA) permissible exposure limit of 5 ppm (20 mg/m^3) for methyl bromide. OSHA also requires the use of gloves, coveralls, goggles, or other appropriate personal protective equipment, engineering controls or work practices to prevent or reduce to the extent necessary in the circumstances skin exposure to methyl bromide. [3]

Liquid methyl bromide in contact with the skin may cause superficial burns with blistering.

MATERIALS OF CONSTRUCTION

Dry methyl bromide is inert and noncorrosive in the presence of most structural metals. However, in the presence of impurities such as alcohols and water, reactions will take place on zinc, tin, and iron surfaces. [4] Aluminum and its alloys should ***not*** be used for methyl bromide service because of the formation of trimethyl aluminum, which is pyrophoric. [5]

SAFE STORAGE, HANDLING, AND USE

The specific precautions that should be observed in storage and handling of methyl bromide are as follows: [1] and [5]

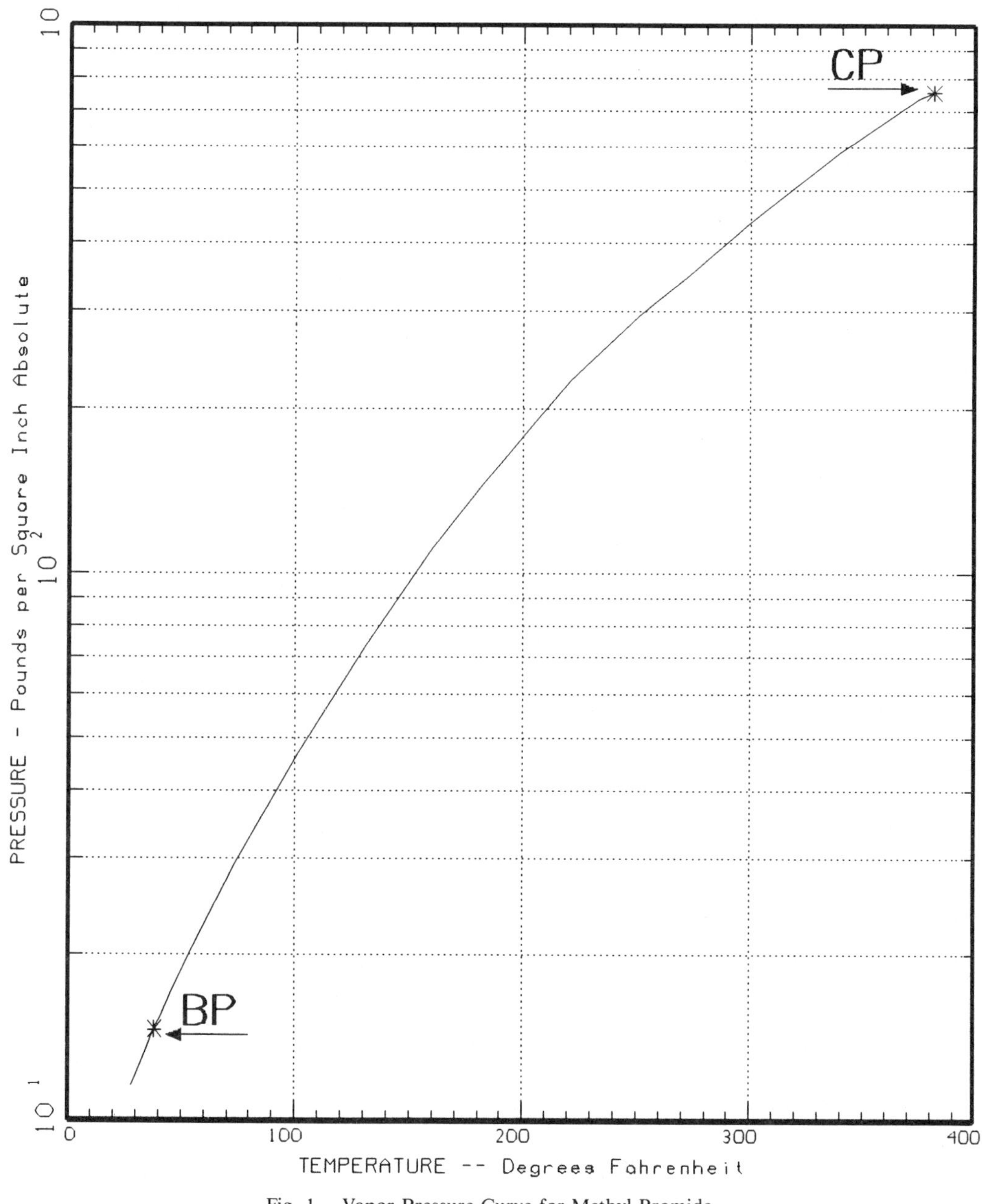

Fig. 1. Vapor Pressure Curve for Methyl Bromide.

(1) Methyl bromide should be stored in a secured, dry, cool, well-ventilated area posted as a pesticide storage area. Avoid possible contamination of water, food, or feed by associated storage.

(2) Methyl bromide should be isolated from flammable materials during storage.

(3) Cylinders should be stored upright and secured to a rack or wall to prevent them from falling.

(4) A NIOSH/MSHA-approved self-contained breathing apparatus or combination air-supplied/SCBA respirator should be provided for use in the event of an accidental leak or when atmospheric concentration is greater than 5 ppm (20 mg/m^3).

(5) Jewelry and tight clothing should be avoided when handling methyl bromide, since skin irritation will result from trapped vapors. Wear a full-face shield or goggles for eye protection, plastic gloves and apron (*do not use rubber or leather*), high-top safety shoes, and wool clothes. Do not reuse contaminated clothing.

(6) A safety shower and eyewash station should be provided in an area close to where methyl bromide is being used but which is not subject to contamination.

(7) Cylinders containing methyl bromide should not be subjected to rough handling or mechanical shock. A cylinder hand truck is recommended for transport.

(8) All labels, Material Safety Data Sheets, and manufacturer recommendations should be read prior to use.

(9) The cylinder valve protection cap and valve outlet connection cap should not be removed until immediately before use and should be replaced when not in use.

(10) When applied as a pesticide, methyl bromide should be used only by a certified applicator or persons under his or her direct supervision, and only for those uses covered by the certified applicator's certification.

(11) Manual controls directly attached to the cylinder outlet for flow control are recommended.

(12) Methyl bromide sold in 1- and $1\frac{1}{2}$-lb cans should be treated in the same manner as cylinders. The contents of the cans are under pressure and should only be opened in the manner recommended by the manufacturer.

(13) Federal, state or provincial, and local regulations must be observed should it be necessary to vent methyl bromide from process piping or vessels.

General safety guidelines for storage and handling of compressed gas containers may be found in Chapter 5.

DISPOSAL

Methyl bromide can be absorbed by organic solvents such as toluene, carbon tetrachloride, chloroform, and ethyl alcohol. The resulting solution should be disposed of in a manner recommended by the State Pesticide Agency, Environmental Control Agency, or regional EPA office.

HANDLING LEAKS AND EMERGENCIES

In the event of a spill or leak, evacuation of the immediate area is essential. Use a self-contained breathing apparatus or combination air-supplied/SCBA respirator for entry into the affected area to correct the problem. If the leak is in a cylinder or can, and cannot be shut off, remove the container to an isolated area outdoors if possible. Allow the leak or spill to evaporate. Personnel working with the container should position themselves upwind at all times.

First Aid

If inhaled, remove the victim to fresh air and consult a doctor. Keep the person warm, and make sure that he or she is breathing freely. If the victim is not breathing, give artificial respiration. Give oxygen if breathing is labored. If the person is conscious, rinse his or her mouth with water. Consult a doctor immediately in all cases of methyl bromide exposure.

If eyes have been exposed, hold open the eyelids and flush them with a steady stream of water for at least 15 minutes. Obtain medical attention promptly.

If methyl bromide is spilled on the skin, remove clothing immediately, flush the affected area with water, then wash with soap and water. Contaminated clothing and shoes should be discarded. Skin lesions should be bathed in sodium bicarbonate solution (2 percent) and blisters treated like second-degree thermal burns.

For further information, see *Effects of Exposure to Toxic Gases—First Aid and Medical Treatment.* [6]

METHODS OF SHIPMENT

Under the appropriate regulations, methyl bromide is authorized for shipment as follows: [7] and [8]

By Rail: In tank cars and portable tanks.

By Highway: In cylinders and cargo tanks.

By Water: On deck only on cargo vessels; forbidden on passenger vessels.

By Air: On cargo aircraft up to 55 gallons; forbidden on passenger aircraft.

Methyl bromide is a DOT hazardous Class B poison, which requires a *POISON* label.

CONTAINERS

Cylinders

Methyl bromide is shipped in low pressure (225 psi minimum) steel cylinders. Specification 3A225, 3AA225, 3B225, 3E1800, 4B225, 4BA225, or 4BW225 cylinders are authorized. Specification 4A225 cylinders may also be used, but new construction is not authorized. Pressure relief devices are not required. Methyl bromide is also shipped in 1- and $1\frac{1}{2}$-lb cans. For further information, see 49 CFR 173.353. [7]

Valve Outlet Connections

The standard valve outlet connection in the United States and Canada for methyl bromide cylinders is Connection CGA 330. The limited standard valve outlet connection is Connection CGA 320. [9]

Tank Cars

Specification 105A100W, 111A100W4, and 106A500X tank cars are authorized for shipping. Specification 105A100 tank cars may also be used, but new construction is not authorized. See 49 CFR 173.353. [7]

Portable Tanks

Specification 51 steel portable tanks having a design pressure of not less than 250 psi (1724 kPa) and equipped with a spring-loaded pressure relief device are authorized for shipment of methyl bromide. See 49 CFR 173.353. [7]

Cargo Tanks

Specifications MC330 and MC331 cargo tanks having a design pressure not less than 250 psi (1724 kPa) and equipped with an approved spring-loaded pressure relief valve are authorized.

METHODS OF MANUFACTURE

The manufacture of methyl bromide generally involves the reaction of hydrobromic acid with methanol. The acid is generated by adding sulfuric acid to a strong solution of sodium bromide and methanol in a reactor. The methyl bromide generated is distilled in a reflux column, which returns methanol and hydrobromic acid to the reactor for further reaction. The product is then dried by passing it through sulfuric acid before condensation, then further fractionated to remove impurities. [4]

More direct routes, such as the reaction of hydrogen bromide with excess methyl chloride at 400–500°C, have been proposed.

REFERENCES

[1] *Directions for Use of the Products Meth-O-Gas and Terr-O-Gas 100,* Great Lakes Chemical Corporation, Highway 52 Northwest, West Lafayette, IN 47906. (1986)

[2] *Threshold Limit Values and Biological Exposure Indices,* 1989–90 ed., American Conference of Gov-

ernmental Industrial Hygienists, 6500 Glenway Avenue, Bldg. D-7, Cincinnati, OH 45211-4438.

[3] *Code of Federal Regulations,* Title 29 CFR Parts 1900–1910 (Labor). Superintendent of Documents, U.S. Government Printing Office, Washington, DC 20402.

[4] *Kirk-Othmer Encyclopedia of Chemical Technology,* 3rd ed., Vol. 4, "Methyl Bromide," 1980, p. 251.

[5] Braker, W., and Mossman, A. L., *Methyl Bromide,* Matheson Gas Data Book, 6th ed., Matheson Gas Products, Inc., Secaucus, NJ, 1980, p. 457.

[6] *Effects of Exposure to Toxic Gases—First Aid and Medical Treatment,* 3rd ed., Matheson Gas Products, Inc. Secaucus, NJ 07094. (1988)

[7] *Code of Federal Regulations,* Title 49 CFR Parts 100–199 (Transportation), Superintendent of Documents, U.S. Government Printing Office, Washington, DC 20402.

[8] *Transportation of Dangerous Goods Regulations,* Canadian Government Publishing Centre, Supply and Services Canada, Ottawa, Ontario, Canada K1A 0S9.

[9] CGA V-1, *American National, Canadian, and Compressed Gas Association Standard for Compressed Gas Cylinder Valve Outlet and Inlet Connections,* (ANSI/CSA/CGA V-1), Compressed Gas Association, Inc., 1235 Jefferson Davis Highway, Arlington, VA 22202.

Methyl Chloride

Chemical Symbol: CH_3Cl
Synonyms: Chloromethane, monochloromethane
CAS Registry Number: 74-87-3
DOT Classification: Flammable gas
DOT Label: Flammable gas
Transport Canada Classification: 2.1 (6.1)
UN Number: UN 1063

PHYSICAL CONSTANTS

	U.S. Units	SI Units
Chemical formula	CH_3Cl	CH_3Cl
Molecular weight	50.488	50.488
Vapor pressure		
at 70°F (21.1°C)	73.4 psig	506.1 kPa abs
at 105°F (40.6°C)	123.7 psig	852.9 kPa abs
at 115°F (46.1°C)	143.7 psig	990.3 kPa abs
at 130°F (54.4°C)	173.7 psig	1197.6 kPa abs
Density of the gas		
at 70°F (21.1°C) and 1 atm	0.1330 lb/ft^3	2.130 kg/m^3
at −11.6°F (−24.2°C) and 1 atm	0.159 lb/ft^3	2.547 kg/m^3
Specific gravity of the gas at 32°F (0°C) and 1 atm (air = 1)	1.74	1.74
Specific volume of the gas at 70°F (21.1°C) and 1 atm	7.5 ft^3/lb	0.47 m^3/kg
Density of the liquid		
at boiling point	62.2 lb/ft^3	996.3 kg/m^3
at 70°F (21.1°C)	57.3 lb/ft^3	917.9 kg/m^3
at 105°F (40.6°C)	55.6 lb/ft^3	890.6 kg/m^3
at 115°F (46.1°C)	53.4 lb/ft^3	855.4 kg/m^3
at 130°F (54.4°C)	53.0 lb/ft^3	849.0 kg/m^3
Boiling point at 1 atm	−11.6°F	−24.2°C
Melting point at 1 atm	−143.7°F	−97.6°C
Critical temperature	289.4°F	143.0°C

	U.S. Units	SI Units
Critical pressure	968.7 psia	6678.9 kPa abs
Critical density	23.1 lb/ft^3	370.0 kg/m^3
Triple point	−144°F at 1.27 psia	−97.8°C at 8.76 kPa abs
Latent heat of vaporization at −11.6°F (−24.2°C) and 1 atm	183.42 Btu/lb	426.6 kJ/kg
Latent heat of fusion at −143.7°F (−97.6°C) and 1 atm	55.8 Btu/lb	129.8 kJ/kg
Specific heat of the gas at 77°F (25°C) and 1 atm		
C_p	0.199 Btu/(lb)(°F)	0.833 kJ/(kg)(°C)
C_v	0.155 Btu/(lb)(°F)	0.649 kJ/(kg)(°C)
Ratio of specific heats, C_p/C_v	1.284	1.284
Weight of the liquid at 70°F (21.1°C)	7.68 lb/gal	920.27 kg/m^3
Specific gravity of the liquid (water = 1)		
at −11.11°F (−23.95°C)	1.000	1.000
at 70°F (21.1°C)	0.919	0.919
Specific heat of the liquid average from 5°F to 86°F (−15°C to 30°C)	0.376 Btu/(lb)(°F)	1.574 kJ/(kg)(°C)
Solubility of gas in water at 1 atm, vol/vol		
at 32°F (0°C)	3.4	3.4
at 68°F (20°C)	2.2	2.2
at 86°F (30°C)	1.7	1.7
at 104°F (40°C)	1.3	1.3
Flammable limits in air	8.1–17.2%	8.1–17.2%
Autoignition temperature	1170°F	632.2°C
Flash point (open cup)	below 32°F	below 0°C

DESCRIPTION

Methyl chloride is a colorless, flammable gas with a faintly sweet, nonirritating odor at room temperatures. It is shipped as a transparent liquid under its vapor pressure of about 59 psig at 70°F (407 kPa at 21.1°C).

Methyl chloride burns feebly in air but forms mixtures with air that can be explosive within its flammability range.

Dry methyl chloride is very stable at normal temperatures and in contact with air. In the presence of moisture, it hydrolyzes slowly, which results in the formation of corrosive hydrochloric acid. At temperatures above 700°F (371°C), methyl chloride may decompose into toxic end products (hydrochloric acid, phosgene, chlorine, and carbon monoxide). It is slightly soluble in water and very soluble in alcohol, mineral oils, chloroform, and most organic liquids.

GRADES AVAILABLE

Methyl chloride is available for commercial and industrial use in various grades having much the same component proportions

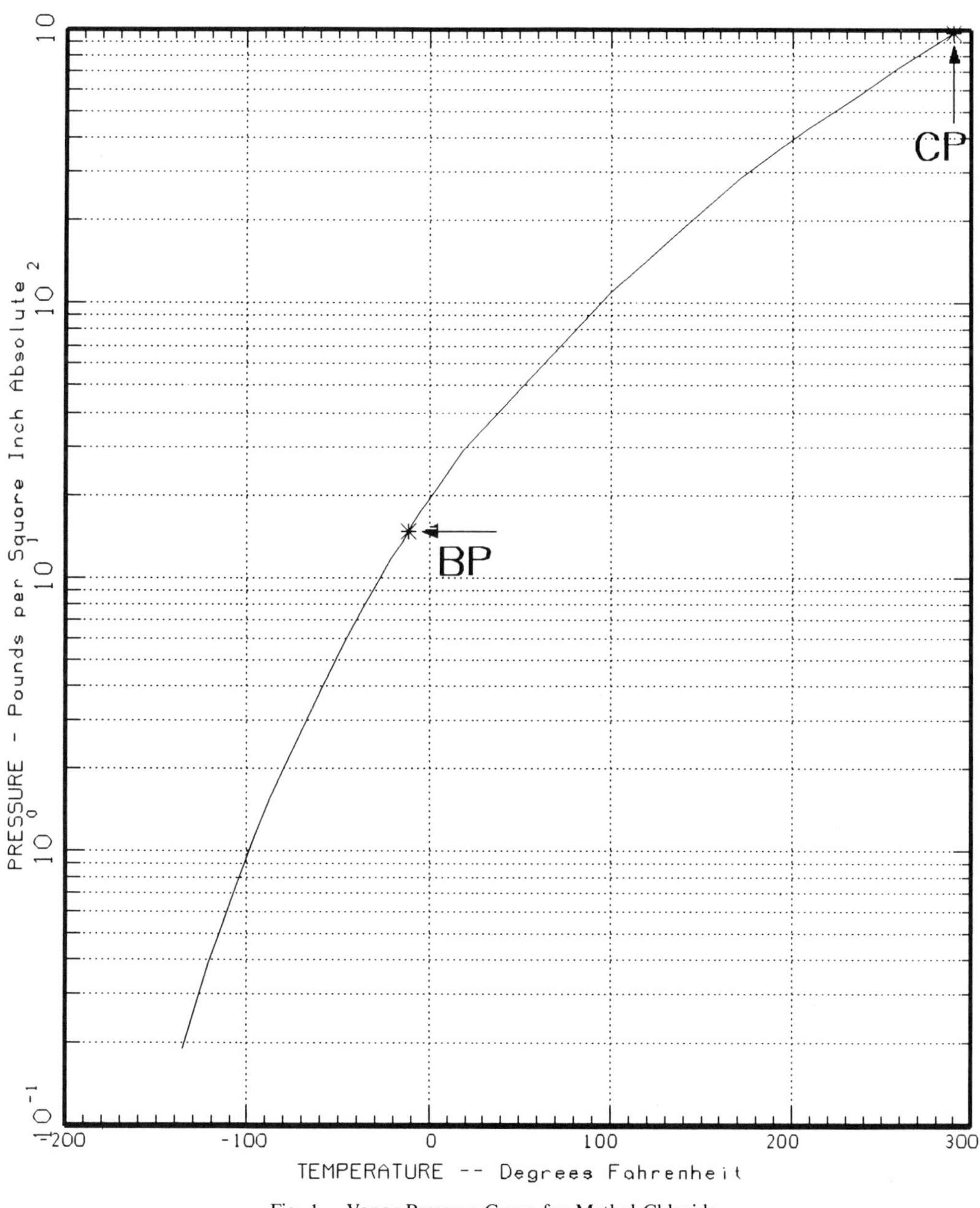

Fig. 1. Vapor Pressure Curve for Methyl Chloride.

from one producer to another. Purities generally range from a minimum of 99.5 mole percent.

USES

Methyl chloride is used as a catalyst carrier in the low-temperature polymerization of such products as the silicones and butyl and other types of synthetic rubber; as a refrigerant gas; and as a methylating agent in organic synthesis of such compounds as Grignard reagents, methyl ethers, and quaternary ammonium compounds. Examples of products synthesized from methyl chloride are methyl mercaptan, methylene chloride, chloroform, carbon tetrachloride, and various bromochloro- and chlorofluoromethanes. It is also used as a chlorinating agent; as an extractant for greases, waxes, essential oils, and resins; as a low-temperature solvent; and as a fluid for thermometric and thermostatic equipment.

PHYSIOLOGICAL EFFECTS

Methyl chloride is toxic, and areas in which it is handled must be adequately ventilated. It is particularly dangerous in that it has no pronounced odor that can serve as a warning.

It acts as an anesthetic about one-fourth as potent as chloroform, and also acts as a narcotic. Inhalation must be avoided. Mild cases of methyl chloride poisoning usually suffer from ataxia, lightheadedness, confusion, tremors, nausea and vomiting, and frequently from anorexia after a latent period of one-half to several hours. Hiccough and constricting pain in the neck may also be experienced. Visual disturbances such as double vision are frequently reported. [1]

Severe nonfatal poisonings are also characterized by a latent period of several hours between exposure and the onset of the first signs or symptoms. This varies with individual susceptibility and the intensity of exposure. Exposure to high concentrations of several hundred ppm or more leads successively to dizziness, headache, vertigo, loss of coordination, nausea and vomiting, abdominal pain, tremors, extreme nervousness, mental confusion, convulsion, unconsciousness, coma, and eventually death. Rapid pulse, lowered blood pressure, elevated body temperature, and rapid respiration are among additional signs of exposure that may be present. Some victims may show signs of liver injury associated with jaundice and porphyrinuria, and renal disturbances characterized by albuminuria and oliguria which may pass into anuria. [1] Complete recovery from severe methyl chloride poisoning may take weeks or months.

Fatal methyl chloride poisoning can have symptoms similar to those of severe nonfatal poisoning. Apparent recovery from what seems a mild exposure through inhalation may be followed by serious or prolonged or even fatal aftereffects within a few days or weeks as a result of cerebral and pulmonary edema and circulatory failure. Repeated exposures are dangerous because methyl chloride is eliminated slowly from the body, in which it is converted into hydrochloric acid and methyl alcohol (wood alcohol).

A time-weighted average threshold limit value (TLV-TWA) of 50 ppm (105 mg/m^3) during an eight-hour workday for exposure to methyl chloride has been adopted by the American Conference of Governmental Industrial Hygienists. [2] The U.S. Occupational Safety and Health Administration has adopted a 50 ppm (105 mg/m^3) eight-hour time-weighted average (TWA) exposure limit for methyl chloride, and a 100 ppm (205 mg/m^3) 15-minute short term exposure limit (STEL) during the workday. [3]

Contact of methyl chloride liquid (or vapor in a concentrated stream) with the skin or the eyes must also be avoided, for such contact can result in a condition resembling frostbite of the tissues.

MATERIALS OF CONSTRUCTION

Dry methyl chloride may be contained in such common metals as steel, iron, copper, and bronze, but it has a corrosive action on zinc, aluminum, die castings, and possibly magnesium alloys. Methyl chloride must not be used with aluminum, since in the presence of moisture it forms spontaneously flammable methyl aluminum compounds upon contact with that metal. No reaction occurs, however, with the drying agent, activated alumina.

Gaskets made of natural rubber and many neoprene compositions should be avoided because methyl chloride dissolves many organic materials. Pressed-fiber gaskets, including those made of asbestos, may be used with methyl chloride. Polyvinyl alcohol is unaffected by methyl chloride, and its use is also recommended. Medium-soft metal gaskets may be used for applications where alternating stresses such as those resulting from large temperature changes do not lead to "ironing out" and consequent leakage.

SAFE STORAGE, HANDLING, AND USE

All personnel handling methyl chloride cylinders should be fully informed about the dangers that can arise from improper handling of methyl chloride. The cylinder and system should be grounded before use. Before introducing methyl chloride into any apparatus or equipment, this apparatus should be tested for leaks, all leaks repaired, and the apparatus thoroughly dried. Only nonsparking tools should be used with methyl chloride. Chemical safety goggles and/or a full face shield should be used when handling liquid methyl chloride. [4]

The operator should wear an approved (NIOSH) air- or oxygen-supplied mask with a full face piece where high concentrations of methyl chloride may develop, and have it ready for emergencies. Unless very large, spills usually evaporate rapidly and cause little damage. However, ample ventilation should be provided to prevent the formation of toxic or explosive mixtures. [4]

Leaking cylinders should be moved outdoors or to an isolated, well-ventilated area where the contents may be released. Each employee should know the location of safety showers, eye fountains, and fire-extinguishing equipment. [4]

Use methyl chloride in a well-ventilated area only, preferably a hood with forced ventilation. The cylinder should never be directly connected to a container of liquid, since reverse flow can occur, causing a reaction within the cylinder. To prevent reverse flow, a trap, check valve, or vacuum break should be inserted into the line. Cylinders of methyl chloride should be stored and used only in well-ventilated areas away from heat and all sources of ignition such as flames and sparks in order to avoid fire and explosion hazards. Flames should never be used to detect methyl chloride leaks; use a soapy water solution. Do not use methyl chloride around sparking motors or other non-explosion-proof equipment. Reserve stocks of methyl chloride cylinders should not be stored with cylinders containing oxygen, chlorine, or other highly oxidizing or flammable materials. [4]

DISPOSAL

Methyl chloride may be absorbed from leaking cylinders into a suitable organic solvent such as xylene, toluene, benzene, or chloroform after attachment of the appropriate regulator and flexible tubing. It may subsequently be recovered for use, or it may be disposed of in a manner suitable for disposal of a flammable gas and in accordance with federal, state or provincial, and local regulations. [4]

HANDLING LEAKS AND EMERGENCIES

In detecting leaks, it is advisable to transfer the methyl chloride vapor or liquid from refrigerating units into gastight containers before testing the units for suspected leaks. The units may then be placed under carbon dioxide, air, or nitrogen pressure and tested by the application of soapy water (or glycerine, in freezing weather) to the suspected points. Soapy water and glycerine are also recommended for testing possible leaks in cylinders and other containers. A leak will be shown by the formation of bubbles. An open flame or a halide torch should not be used to detect leaks.

First Aid

Speed is of primary importance in mitigating methyl chloride exposures. The victim should be promptly moved from a contaminated atmosphere, and contaminated clothing removed. First aid must be started at once in all cases of contact with methyl chloride as serious injury may result. If the victim is not breathing, give artificial respiration. If breathing is difficult, give oxygen. Remove methyl chloride from the skin or eyes by washing with tap water for at least 15 minutes. All affected persons should be referred to a physician, even when immediate injury is not apparent. The physician should be given a detailed account of the accident. [1]

METHODS OF SHIPMENT

Under the appropriate regulations, methyl chloride is authorized for shipment as follows: [5] and [6]

By Rail: In cylinders, in TMU tank cars, and in single-unit tank cars.

By Highway: In cylinders on trucks, in TMU tanks on trucks, and in portable tanks and cargo tanks.

By Water: In cylinders on cargo vessels, passenger vessels, and ferry and railroad car ferry vessels (passenger or vehicle). On cargo vessels only in portable tanks (complying with DOT-51 specification) not over 20 000 pounds (9072 kg) gross weight. In authorized tank cars on trainships only, and in authorized tank trucks on trailerships and trainships only. In cylinders on barges of U.S. Coast Guard classes A, CA, and CB only. In cargo tanks on tankships and tank barges (to maximum filling densities as prescribed in Coast Guard regulations). See 46 CFR 151.50-30(e). [7]

By Air: In cylinders aboard cargo aircraft only up to 300 lb (136 kg) maximum net weight per cylinder. Forbidden on passenger aircraft.

CONTAINERS

Methyl chloride is authorized for shipment in cylinders, insulated single-unit tank cars, and TMU tanks and tank cars. It is also shipped in tank barges. Methyl chloride is also authorized for shipment in portable tanks and cargo tanks.

Filling Limits

The maximum filling densities authorized for methyl chloride under present regulations are (percent water capacity by weight):

In cylinders—84 percent.

In TMU tanks—84 percent.

In single-unit tank cars complying with DOT 105J300W specifications—86 percent; in single-unit tank cars complying with DOT specification 112T340W or 112J340W—85 percent.

In cargo tanks and portable tanks—84 percent by weight; 88.5 percent by volume.

Tank car loading and unloading interior piping must be equipped with excess flow valves.

Cylinders

Methyl chloride is authorized for shipment in cylinders meeting TC/DOT specifications as follows: 3A225, 3AA225, 3B225, 4A225, 4B225, 4BA225, 4BW225, 4, 3E1800, and 4B240ET. Cylinders which comply with DOT specifications 3, 25, 26-300, and 38 may also be continued in methyl chloride service, but new construction is not authorized. Cylinders complying with DOT specifications 3A150, 3B150, 4A150, or 4B150, manufactured before December 7, 1936, are also authorized.

Cylinders authorized for methyl chloride service must be requalified by hydrostatic retest every 5 years under present regulations with the following exceptions: (a) DOT-4 cylinders require hydrostatic retest every 10 years; (b) no retest is required for Specification 3E cylinders; and (c) Specification 4B, 4BA, and 26-300 cylinders require retest every 12 years if they are used exclusively for methyl chloride that is free from corroding components and if they are protected externally by suitable corrosion-resistant coatings such as galvanizing and painting. As an alternative, Specification 4B, 4BA, and 26-300 cylinders may also be requalified by being retested every 7 years to an internal hydrostatic pressure at least two times the marked service pressure and without determination of expansions. See 49 CFR 173.34(e)(9). [5]

Authorized pressure relief devices for methyl chloride cylinders are Type CG-7 pressure relief valves. [8]

Valve Outlet Connections

The standard valve outlet connection in the United States and Canada for methyl chloride cylinders is Connection CGA 510. The limited standard connection for methyl chloride cylinders is Connection CGA 660. [9]

Cargo Tanks

Cargo tanks that comply with TC/DOT specifications MC-330 or MC-331 are authorized for the shipment of methyl chloride via motor vehicle. Minimum design pressure for cargo tanks is 150 psig (1034 kPa).

Portable Tanks

Portable tanks meeting DOT-51 specifications and with a minimum design pressure of 150 psig (1034 kPa) are authorized for shipment of methyl chloride via motor vehicle.

TMU Tanks

TMU tanks complying with DOT specification 106A500X are authorized for methyl chloride shipment via motor vehicle and rail.

Tank Cars

Methyl chloride is authorized for shipment in insulated, single-unit tank cars which comply with TC/DOT specifications 105J300W, 112T340W, or 112J340W provided that interior pipes of loading and unloading valves are equipped with excess flow valves of approved design. See 49 CFR 173.314. [5]

METHODS OF MANUFACTURE

Methyl chloride is made commercially mainly by two methods: the reaction of hydrogen chloride gas or hydrochloric acid with methyl alcohol (in the presence of a catalyst to accelerate the reaction); and the chlorination of methane.

In the first process, the products are gaseous methyl chloride with unreacted hydrogen chloride and methyl alcohol and several by-products. These are removed in a series of chemical purification steps, and the methyl chloride gas is compressed and dried. A small amount of air remaining in the condensate is distilled off before the liquid is charged into the shipping container.

Natural gas is the source of the methane used in the second process, in which the methane is removed by fractional distillation and reacted with chlorine. Undesired reaction products (including methylene chloride, chloroform, carbon tetrachloride, hydrochloric acid, and some chlorinated hydrocarbons) are similarly removed from the methyl chloride in subsequent chemical purification steps.

REFERENCES

[1] *Effects of Exposure to Toxic Gases—First Aid and Medical Treatment,* 3rd ed., Matheson Gas Products, Inc., Secaucus, NJ 07094. (1988)

[2] *Threshold Limit Values and Biological Exposure Indices,* 1989–90 ed., American Conference of Governmental Industrial Hygienists, 6500 Glenway Avenue, Bldg. D-7, Cincinnati, OH 44211-4438.

[3] *Code of Federal Regulations,* Title 29 CFR Parts 1900–1910 (Labor), Superintendent of Documents, U.S. Government Printing Office, Washington, DC 20402.

[4] *Matheson Gas Data Book,* 6th ed., Matheson Gas Products, Inc., Secaucus, NJ 07094. (1980).
[5] *Code of Federal Regulations,* Title 49 CFR Parts 100–199 (Transportation), Superintendent of Documents, U.S. Government Printing Office, Washington, DC 20402.
[6] *Transportation of Dangerous Goods Regulations,* Canadian Government Publishing Centre, Supply and Services Canada, Ottawa, Ontario, Canada K1A 0S9.
[7] *Code of Federal Regulations,* Title 46 CFR Parts 0-199, (Shipping), Superintendent of Documents, U.S. Government Printing Office, Washington, DC 20402.
[8] CGA S-1.1, *Pressure Relief Device Standards—Part 1—Cylinders for Compressed Gases,* Compressed Gas Association, Inc., 1235 Jefferson Davis Highway, Arlington, VA 22202.
[9] CGA V-1, *American National, Canadian, and Compressed Gas Association Standard for Compressed Gas Cylinder Valve Outlet and Inlet Connections* (ANSI/CSA/CGA V-1), Compressed Gas Association, Inc., 1235 Jefferson Davis Highway, Arlington, VA 22202.

Nitric Oxide

Chemical Symbol: NO
Synonyms: Nitrogen (II) oxide
CAS Registry Number: 10102-43-9
DOT Classification: Poison A
DOT Label: Poison gas
Transport Canada Classification: 2.3 (5.1)
UN Number: UN 1660

PHYSICAL CONSTANTS

	U.S. Units	SI Units
Chemical formula	NO	NO
Molecular weight	30.006	30.006
Density of the gas at 70°F (21.1°C) and 1 atm	0.0777 lb/ft^3	1.245 kg/m^3
Specific gravity of the gas at 70°F (21.1°C) and 1 atm (air = 1)	1.04	1.04
Specific volume of the gas at 70°F (21.1°C) and 1 atm	13 ft^3/lb	0.81 m^3/kg
Density of the liquid at −241.2°F (−151.8°C)	79.3 lb/ft^3	1270.26 kg/m^3
Boiling point at 1 atm	−241.2°F	−151.8°C
Melting point at 1 atm	−262.6°F	−163.6°C
Critical temperature	−135.2°F	−92.9°C
Critical pressure	949.4	6544.0 kPa
Critical density	32.375 lb/ft^3	518.6 kg/m^3
Latent heat of vaporization at −241.2°F (−151.7°C)	197.4 Btu/lb	459.1 kJ/kg
Heat capacity gas at 14.7 psia C_p	0.2376 Btu/(lb)(°F)	0.9946 kJ/(kg)(°C)
Solubility in water at 32°F (0°C) and 1 atm, vol/vol	.0734	.0734

References for above: [1], [2], [3], and [4].

DESCRIPTION

Nitric oxide at room temperature is a colorless, nonflammable, toxic gas. Nitric oxide in the presence of air forms brown fumes of nitrogen dioxide, which is extremely reactive and a strong oxidizing agent. The conversion of nitric oxide to nitrogen dioxide is rate dependent on the concentration of oxygen and the square of the concentration of nitric oxide. It also reacts vigorously with fluorine oxides and, when moisture is present, chlorine. Nitric oxide is shipped in a nonliquefied form at a cylinder pressure of 514.7 psia (3549 kPa abs) at 70°F (21.1°C).

Nitric oxide is an oxidizer and will support combustion. It will react with oxygen, oxidizing agents, halides, and hydrocarbons. Nitric oxide is noncorrosive and most structural materials are unaffected; however, in the presence of moisture, corrosion can develop with the formation of nitrous and nitric acids.

Nitric oxide is unstable at higher pressures and temperatures and has been known to cause violent rupture of a container with an adequate energy input.

GRADES AVAILABLE

Nitric oxide is available in a minimum purity of 98.5 percent or 99.0 percent.

USES

Nitric oxide is an important intermediate in the production of nitric acid. Fluorine, chlorine, and bromine react with nitric oxide to form the corresponding nitrosyl halide. It is also used in making mixtures for calibration standards for stationary and mobile exhaust emission measurements.

PHYSIOLOGICAL EFFECTS

Nitric oxide, with the attendant formation of nitrogen dioxide, results in a strong respiratory irritant which may be fatal. Symptoms may be moderate at first, and include tightness in the chest, headaches, irritation of the eyes, nausea, and a slow loss of strength. Delayed symptoms may be severe and cause increased difficulty in breathing, chemical pneumonitis, and pulmonary edema. Untreated cases could lead to eventual death. Exposure to 100–150 ppm of nitrogen oxides for 30–60 minutes could lead to delayed pulmonary edema, and a few breaths of nitrogen oxides in 200–700 ppm concentrations may result in fatal pulmonary edema after 5–8 hours have passed. [2], [3], and [4]

Acute exposure to nitric oxide alone at high levels results in the rapid formation of methemoglobin. Nitric oxide itself does not have the marked irritant effect of nitrogen dioxide. But due to the rapid formation of methemoglobin, greater lethality can be associated with high acute levels of exposure to nitric oxide than with nitrogen dioxide exposures. Initial symptoms such as muscular tremors, drowsiness, a brownish-blue hue to the mucous membranes, increased heart and respiratory rate, vertigo, and vomiting can occur at methemoglobin levels of 30-40%. Coma and death can ensue when methemoglobin levels reach 70-90%. [5]

A time-weighted threshold limit value of 25 ppm (30 mg/m^3) for an eight-hour period has been adopted by the American Conference of Governmental Industrial Hygienists. [6] The U.S. Occupational Safety and Health Administration similarly has established 25 ppm (30 mg/m^3) as the eight-hour time weighted average (TWA) exposure limit for nitric oxide. [7] At normal ambient temperatures, nitric oxide combines with atmospheric oxygen to form nitrogen dioxide at a rate dependent on the concentration of oxygen and the square of the concentration of nitric oxide.

MATERIALS OF CONSTRUCTION

Nitric oxide is noncorrosive and most common structural materials may be used. However, in the presence of moisture and oxygen, corrosive conditions will develop as a result of the formation of nitric and nitrous acids. Prior to use, systems to contain nitric oxide must first be purged with an inert gas. Where air contamination cannot be eliminated, stainless steel should be used.

SAFE STORAGE, HANDLING, AND USE

Nitric oxide should only be used in well-ventilated areas. Valve protection caps and valve outlet threaded plugs must remain in place unless the container is secured and the valve outlet piped to the point of use. Do not drag, slide, or roll cylinders. Use a suitable hand truck to move cylinders. Use a pressure-reducing regulator when connecting a cylinder to lower-pressure (1000 psig; 6895 kPa) piping systems. Do not heat a cylinder of nitric oxide by any means to increase the discharge rate from the cylinder. Use a check valve or trap in the discharge line to prevent hazardous reverse flow into the cylinder.

Protect cylinders from physical damage. Store them in a cool, dry, well-ventilated area away from heavily trafficked areas and emergency exits. Do not allow the temperature where cylinders are stored to exceed 125°F (51.7°C). Cylinders should be stored upright and firmly secured to prevent them from falling or being knocked over. Full and empty cylinders should be segregated. Use a first in/first out inventory system to prevent full cylinders from being stored for excessive periods of time.

Personnel should have NIOSH-approved gas masks available for immediate use in emergencies, and areas in which nitric oxide is being handled should have enough exits to permit personnel to leave quickly enough in the event of an emergency. See Chapter 5 for additional guidelines.

DISPOSAL

Cylinders should be moved outside or to a ventilated hood by persons wearing self-contained breathing apparatus. Gas should be vented into an adequate solution of alkaline potassium permanganate. The resulting reaction is:

$$NO + KMnO_4 \rightarrow KNO_3 + MnO_2$$

An alternate method of disposing of nitric oxide is to mix it with air. The resulting mixture of NO and NO_2 is then introduced into a caustic solution. A mixture of sodium nitrate and sodium nitrite will result. [1]

HANDLING LEAKS AND EMERGENCIES

Evacuate all personnel from the affected area. Use appropriate personal protection and breathing apparatus. If the leak is in the user's equipment, shut the cylinder valve and be certain to purge the piping with an inert gas prior to attempting repairs.

For leaks, discharge the gas at a moderate rate into a solution of caustic soda and slaked lime. Keep in a fume hood until cylinder has emptied.

If in a fire situation, keep cylinders cool with a water spray if possible. Wear self-contained breathing apparatus with a full facepiece operated in pressure-demand or other positive-pressure mode and full-body protective clothing. [8]

First Aid

If nitric oxide is inhaled, remove the individual to fresh air and get immediate medical attention. If breathing is difficult, administer oxygen. If breathing has stopped, give artificial respiration. Keep the person warm and quiet. Keep victims under competent medical observation for 48 to 72 hours or until the hazard of delayed pulmonary edema has passed. First aid in treating methemoglobin formation involves the administration of oxygen.

For eye exposure, immediately flush eyes with large amounts of water for at least 15 minutes, keeping eyelids apart to assure contact of water with all exposed tissues. If pain is still present, continue irrigation for an additional 15 minutes. Get immediate medical attention.

In case of skin exposure, remove all contaminated clothing under an emergency shower. Wash the affected skin areas under running water with soap for at least 15 minutes. Subsequent medical treatment is as for thermal burns.

Refer to *Effects of Exposure to Toxic Gases—First Aid and Medical Treatment* for further information. [5]

METHODS OF SHIPMENT

Under the appropriate regulations, nitric oxide is authorized for shipment as follows: [9] and [10]

By Rail: In tank cars and cylinders.

By Highway: In cylinders on trucks.

By Water: In cylinders on cargo vessels only, stored above deck only.

By Air: Forbidden.

CONTAINERS

Nitric oxide is authorized for shipment in cylinders and tank car tanks (ton containers) as a nonliquefied gas. The container must not be equipped with a pressure relief device. Valve outlets must be sealed with a solid threaded cap or plug. See 49 CFR 173.337. [9]

Cylinders

TC/DOT specification cylinders 3A1800, 3AA1800, 3E1800, and 3AL1800 are approved for nitric oxide service. Cylinders may not be pressurized higher than 750 psig (5171 kPa) at 70°F (21.1°C). The use of pressure relief devices is prohibited. [11]

Valve Outlet Connections

The standard valve outlet connection in the United States and Canada for nitric oxide cylinders is Connection CGA 660. [12]

Tank Cars

Specification 106A500X tank cars are approved for nitric oxide service. Tanks may not be pressurized higher than 200 psig (1379 kPa) at 70°F (21.1°C). Each tank must be equipped with a gas-tight valve protection cap. See 49 CFR 173.337. [9]

METHODS OF MANUFACTURE

Nitric oxide is prepared as an intermediate for the manufacture of nitric acid by oxidation of ammonia above 932°F (500°C) with a platinum catalyst.

$$4NH_3 + 5O_2 \rightarrow 4NO + 6H_2O$$

Nitric oxide is also produced when acid solutions of nitrates or nitrites are reduced by metals.

$$8HNO_3 + 3Cu \rightarrow 3Cu(NO_3)_2 + 2NO + 4H_2O$$

$$2KNO_3 + 6Hg + 4H_2SO_4 \rightarrow 2NO + 3Hg_2SO_4 + K_2SO_4 + 4H_2O$$

Even though equilibrium is unfavorable at high temperatures, direct combination of nitrogen and oxygen can occur under energy-rich conditions with 5 percent by volume of nitric oxide being formed at 5792°F (3200°C). [1] and [13]

$$N_2 + O_2 \rightarrow 2NO$$

REFERENCES

[1] *Matheson Unabridged Gas Data Book,* Matheson Gas Products, Inc., Secaucus, NJ 07094. (1974)

[2] Material Safety Data Sheet for Nitric Oxide (4/78), Air Products and Chemicals, Inc., Allentown, PA 18195.

[3] Material Safety Data Sheet for Nitric Oxide (10/85), Alphagaz Division, Liquid Air Corp., 2121 N. California Blvd., Walnut Creek, CA 94596.

[4] Material Safety Data Sheet for Nitric Oxide (11/85), Scientific Gas Products, 2330 Hamilton Blvd., Box 648, South Plainfield, NJ 07080.

[5] *Effects of Exposure to Toxic Gases—First Aid and Medical Treatment,* 3rd ed., Matheson Gas Products, Inc., Secaucus, NJ 07094. (1988)

[6] *Threshold Limit Values and Biological Exposure Indices,* 1989–90 ed., American Conference of Governmental Industrial Hygienists, 6500 Glenway Avenue, Bldg. D-7, Cincinnati, OH 45211-4438.

[7] *Code of Federal Regulations,* Title 29 CFR Parts 1900–1910 (Labor), Superintendent of Documents, U.S. Government Printing Office, Washington, DC 20402.

[8] Sax, N. I., *Dangerous Properties of Industrial Materials,* 6th ed., Van Nostrand Reinhold, New York, 1984.

[9] *Code of Federal Regulations,* Title 49 CFR Parts 100–199, (Transportation), Superintendent of

Documents, U.S. Government Printing Office, Washington, DC 20402.

[10] *Transportation of Dangerous Goods Regulations,* Canadian Government Publishing Centre, Supply and Services Canada, Ottawa, Ontario, Canada K1A 0S9.

[11] CGA S-1.1, *Pressure Relief Device Standards—Part 1—Cylinders for Compressed Gases,* Compressed Gas Association, Inc., 1235 Jefferson Davis Highway, Arlington, VA 22202.

[12] CGA V-1, *American National, Canadian, and Compressed Gas Association Standard for Compressed Gas Cylinder Valve Outlet and Inlet Connections,* (ANSI/CSA/CGA V-1), Compressed Gas Association, Inc., 1235 Jefferson Davis Highway, Arlington, VA 22202.

[13] *Matheson Gas Data Book,* 6th ed., Matheson Gas Products, Inc., Secaucus, NJ 07094. (1980)

Nitrogen

Chemical Symbol: N_2
Synonyms: LIN (liquid only)
CAS Registry Number: 7727-37-9
DOT Classification: Nonflammable gas
DOT Label: Nonflammable gas
Transport Canada Classification: 2.2
UN Number: UN 1066 (compressed gas); UN 1977 (refrigerated liquid)

PHYSICAL CONSTANTS

	U.S. Units	SI Units
Chemical formula	N_2	N_2
Molecular weight	28.01	28.01
Density of the gas at 70°F (21.1°C) and 1 atm	0.072 lb/ft^3	1.153 kg/m^3
Specific gravity of the gas at 70°F (21.1°C) and 1 atm (air = 1)	0.967	0.967
Specific volume of the gas at 70°F (21.1°C) and 1 atm	13.89 ft^3/lb	0.867 m^3/kg
Density of the liquid at boiling point and 1 atm	50.47 lb/ft^3	808.5 kg/m^3
Boiling point at 1 atm	−320.4°F	−195.8°C
Melting point at 1 atm	−345.8°F	−209.9°C
Critical temperature	−232.4°F	−146.9°C
Critical pressure	493 psia	3399 kPa abs
Critical density	19.60 lb/ft^3	314.9 kg/m^3
Triple point at 1.81 psia (12.5 kPa abs)	−346.0°F	−210.0°C
Latent heat of vaporization at boiling point	85.6 Btu/lb	199.1 kJ/kg
Latent heat of fusion at melting point	11.1 Btu/lb	25.1 kJ/kg
Specific heat of the gas at 70°F (21.1°C) and 1 atm		
C_p	0.249 Btu/(lb)(°F)	1.04 kJ/(kg)(°C)
C_v	0.177 Btu/(lb)(°F)	0.741 kJ/(kg)(°C)

	U.S. Units	SI Units
Ratio of specific heats, C_p/C_v	1.41	1.41
Solubility in water vol/vol at 32°F (0°C)	0.023	0.023
Weight of liquid at boiling point	6.747 lb/gal	808.5 kg/m^3
Gas/liquid ratio (gas at 70°F (21.1 °C) and 1 atm, liquid at boiling point, vol/vol)	696.5	696.5

DESCRIPTION

Nitrogen makes up the major portion of the atmosphere (78.03 percent by volume, 75.5 percent by weight). It is a colorless, odorless, tasteless, nontoxic, and almost totally inert gas, and is colorless as a liquid. Nitrogen is nonflammable, will not support combustion, and is not life supporting. It combines with some of the more active metals, such as lithium and magnesium, to form nitrides, and at high temperatures it will also combine with hydrogen, oxygen, and other elements. It is employed as an inert protection against atmospheric contamination in many nonwelding applications. Nitrogen is only slightly soluble in water and most other liquids, and is a poor conductor of heat and electricity. As a liquid at cryogenic temperatures, it is nonmagnetic. It is shipped as a nonliquefied gas at pressures of 2000 psig (13 790 kPa) or above, and also as a cryogenic fluid at pressures below 200 psig (1379 kPa).

GRADES AVAILABLE

Table 1, from CGA G-10.1, *Commodity Specification for Nitrogen,* presents the component maxima in ppm (mole/mole), unless shown otherwise, for the grades of nitrogen. [1] These are also known as quality verification levels. A blank indicates no maximum limiting characteristic. The absence of a value in a listed grade does not mean to imply that the limiting characteristic is or is not present, but merely indicates that the test is not required for compliance with the specification. Type II notations in the table refer to liquid nitrogen.

A table which provides a means of converting moisture data into the particular units of interest appears in the introduction to Part III.

USES

Nitrogen has many commercial and technical applications. As a gas, it is used in: heat treating of primary metals; blanketing of oxygen-sensitive liquids and of volatile liquid chemicals; the production of semiconductor electronic components, as a blanketing atmosphere; the blowing of foam-type plastics; the deaeration of oxygen-sensitive liquids; the degassing of nonferrous metals; food processing and packing; inhibition of aerobic bacteria growth; magnesium reduction of aluminum scrap; and the propulsion of liquids through pipelines.

Gaseous nitrogen is also used in: pressurizing aircraft tires and emergency bottles to operate landing gear; purging, in the brazing of copper tubing for air-conditioning and refrigeration systems; the purging and filling of electronic devices; the purging, filling, and testing of high-voltage compression cables; the purging and testing of pipelines and related instruments; and the treatment of alkyd resins in the paint industry.

Liquid nitrogen also has a great many uses, among them: the freezing of highly perishable foods, such as shrimp, hamburgers, and chicken; deflashing of rubber tires; cooling of concrete; and the cold-trapping of materials such as carbon dioxide from gas

TABLE 1. NITROGEN GRADES AVAILABLE.[a]
(Units in ppm (mole/mole) unless otherwise shown)

Limiting Characteristics	Maxima for Gaseous and Liquid Nitrogen								
	B	E	F	G	H	K	L	M	Q
Nitrogen[b] Min. % (mole/mole)	99.0	99.5	99.9	99.95	99.99	99.995	99.998	99.999	99.999
Oxygen		0.5%	0.1%	500	50	20	10	5	1
Water ppm (v/v)		26	32	26	11	16	4	2	2
Dew Point, °F		−63	−60	−63	−75	−70	−90	−100	−100
Odor	None			None					
Total Hydrocarbon Content		58			5			5	1
Hydrogen									
Argon, Neon, Helium									5
Carbon Monoxide	10								5
National Formulary	Yes								
Permanent Particulates[c]		Type II less than 1.0 mg/L			Type II less than 1.0 mg/L				

[a]The 1976 edition of CGA G-10.1 listed 14 grades of nitrogen lettered A to P. Some of those grades were dropped from the 1985 edition since they no longer represent major volume of usage by industry. Also, one new grade, Q, has been added to reflect a current specification. To get a listing of grades dropped, see CGA G-10.1-1976 or contact the Compressed Gas Association.

[b]Unless otherwise shown, percent nitrogen includes trace quantities of neon, helium, and argon.

[c]The permanent particulate content of liquid nitrogen is determined by passing the liquid nitrogen sample through a specified low-micron-rated, tared, analytical filter disk contained in a suitable holder. The filtered liquid is collected and measured in an open containter. The assembly is warmed and the disk reweighed. The size of particulate matter can be evaluated by the examination of the filter disk with a suitable optical magnifier.

streams (and it is commonly employed in this way in systems which produce high vacuums). It is used as a coolant for electronic equipment, for pulverizing plastics, and for simulating the conditions of outer space.

Other ways in which liquid nitrogen is employed include: creating a very-high-pressure gaseous nitrogen (15 000 psig; 103 421 kPa) through liquid nitrogen pumping; in food and chemical pulverization; for the freezing of liquids in pipelines for emergency repairs; for low temperature stabilization and hardening of metals; for low temperature research; for low temperature stress relieving of aluminum alloys; for the preservation of whole blood, livestock sperm, and other biologicals; for refrigerating foods in long-distance hauling as well as local delivery; for refrigeration shielding of liquid hydrogen, helium, and neon; for the removal of skin blemishes in dermatology; and for shrink fitting of metal parts.

Liquid nitrogen also has a number of classified applications in the missile and space programs of the United States, in which it is used in large quantities.

PHYSIOLOGICAL EFFECTS

Nitrogen is nontoxic and largely inert. It can act as a simple asphyxiant by diluting the concentration of oxygen in air below levels necessary to support life. Inhalation of nitrogen in excessive concentrations can result in dizziness, nausea, vomiting, loss of consciousness, and death. Death may result from errors in judgment, confusion, or loss of consciousness which prevents self-rescue. At low oxygen concentrations, unconsciousness and death may occur in seconds without warning.

Gaseous nitrogen must be handled with all the precautions necessary for safety with any nonflammable, nontoxic compressed gas.

All precautions necessary for the safe handling of any gas liquefied at very low temperatures must be observed with liquid nitrogen. Extensive tissue damage or burns can result from exposure to liquid nitrogen or cold nitrogen vapors. See Chapter 5 and CGA P-12, *Safe Handling of Cryogenic Liquids,* for further information. [2]

MATERIALS OF CONSTRUCTION

Gaseous nitrogen is noncorrosive and inert, and may consequently be contained in systems constructed of any common metals and designed to safely withstand the pressures involved. At the temperature of liquid nitrogen, ordinary carbon steels and most alloy steels lose their ductility and are considered unsafe for liquid nitrogen service. Satisfactory materials for use with liquid nitrogen include Type 18-8 stainless steel and other austenitic nickel-chromium alloys, copper, Monel, brass, and aluminum.

SAFE STORAGE, HANDLING, AND USE

Gaseous nitrogen is commonly stored in high pressure cylinders, tubes, or tube trailers. Liquid nitrogen is commonly stored at the consumer site in cryogenic liquid cylinders and specially designed vacuum-insulated storage tanks.

All of the precautions necessary for the handling of any nonflammable gas or cryogenic liquid must be taken. For additional details, see Chapter 5 and the compressed gas manufacturer's Material Safety Data Sheet. Also see CGA P-9, *The Inert Gases, Argon, Nitrogen and Helium.* [3] As previously noted, nitrogen can act as a simple asphyxiant by diluting the concentration of oxygen in air below levels necessary to support life. Recommendations for means by which to prevent such situations are given in CGA P-14, *Accident Prevention in Oxygen-Rich and Oxygen-Deficient Atmospheres* [4], and in CGA Safety Bulletin SB-2, "Oxygen Deficient Atmospheres."

Liquid and gaseous systems should be designed and installed only under the direction of personnel thoroughly familiar with liquid and gaseous nitrogen equipment, and in compliance with state, provincial, and local requirements.

DISPOSAL

Vent nitrogen gas slowly to a well-ventilated outdoor location remote from personnel work areas and building air intakes. Return cylinders to the supplier with residual pressure, and with the cylinder valve tightly closed and the valve caps in place.

Allow liquid nitrogen to evaporate in well-ventilated outdoor locations which are remote from work areas and building air intakes.

HANDLING LEAKS AND EMERGENCIES

Ventilate adjacent enclosed areas to prevent the formation of oxygen-deficient atmospheres caused by the release of gaseous nitrogen or by the evaporation of liquid nitrogen. Personnel, including rescue workers, should not enter areas where the oxygen concentration is below 19 percent, unless provided with a self-contained breathing apparatus or air-line respirator.

Avoid contact of the skin with liquid nitrogen or its cold boil-off gas. Flush liquid nitrogen spills with water to accelerate evaporation.

First Aid

Nitrogen is generally inert and can cause asphyxiation due to displacement of oxygen in the atmosphere. Rescuers wearing a self-contained breathing apparatus or air-line respirator should move the affected person from the hazardous exposure to fresh air at once. If supplemental oxygen is available, administer by nasal canula or mask. Perform artificial respiration if the person is not breathing. Persons who have been unconscious should be taken to a hospital for evaluation and care.

In case of frostbite from exposure to liquid nitrogen, the frostbitten part should be placed in warm water of about 108°F (42°C). If warm water is not available, or it is impractical to use, wrap the affected part gently in blankets. Let circulation re-establish itself naturally. Encourage the victim to exercise the affected part while it is being warmed. Consult a physician. [5]

METHODS OF SHIPMENT

Under the appropriate regulations, nitrogen is authorized for shipment as follows (nitrogen gas, except where liquid nitrogen is indicated): [6] and [7]

By Rail: In cylinders and in tube tank cars. Liquid nitrogen, in vacuum-insulated cylinders and tank cars.

By Highway: In cylinders on trucks, and in tube trailers. Liquid nitrogen, in vacuum-insulated cylinders and cargo tanks.

By Water: In cylinders on cargo and passenger vessels, and on ferry and railroad car ferry vessels (passenger or vehicle). In authorized cargo tanks on cargo vessels only. Liquid nitrogen, in pressurized cylinders on cargo and passenger vessels and ferry and railroad car ferry vessels (passenger or vehicle).

By Air: In gaseous cylinders aboard passenger aircraft up to 150 lb (68 kg), and aboard cargo aircraft up to 300 lb (136 kg), maximum net weight per cylinder. Liquid nitrogen, aboard passenger aircraft up to 100 lb (45 kg) maximum net weight per container; liquid nitrogen aboard cargo aircraft up to 1100 lb (499 kg) maximum net weight per container.

CONTAINERS

Nitrogen gas is authorized for shipment in cylinders, tube tank cars, and tube trailers. Liquid nitrogen is shipped as a cryogenic fluid in vacuum-insulated cylinders, and in insulated portable tanks, tank trucks, and tank cars.

Filling Limits

The maximum filling limits authorized for gaseous nitrogen are as follows:

In cylinders and tube trailers—the authorized service pressures marked on the cylinders or tube assemblies at 70°F (21.1°C). In the case of cylinders of Specifications 3A, 3AA, 3AX, 3AAX, and 3T that meet special requirements, up to 10 percent in excess of their marked service pressures. See 49 CFR 173.302(c). [6]

In tube tank cars—in uninsulated cars of the TC/DOT-107A type, not more than seven-tenths of the marked test pressure at 130°F (54.4°C).

The maximum filling limits authorized for liquid nitrogen are: Specification TC/DOT-4L cylinders are authorized for the transportation of liquid nitrogen when carried in the vertical position. The filling density must be in accordance with Table 2. See 49 CFR 173.316. [6]

Cylinders

Cylinders which comply with TC/DOT specifications 3A and 3AA are the types usually used to ship gaseous nitrogen, but it is authorized for shipment in any cylinders approved for nonliquefied compressed gas. (These include cylinders meeting Specifications 3A, 3AA, 3AX, 3AAX; 3B, 3E, 3T, 3AL, 39, 4B, 4BA, and 4BW. In addition, continued use of cylinders complying with Specifications 3C, 3D, 4, 4A, 4C, 3, 25, 26, 33, and 38 is authorized, but new construction is not authorized.

TABLE 2. LIQUID NITROGEN IN TC/DOT-4L CYLINDERS.

Pressure Control Valve Setting (Maximum start-to-discharge pressure) psig	(kPa)	Maximum Permitted Filling Density (Percent by weight)
45	(310)	76
75	(517)	74
105	(724)	72
170	(1172)	70
230	(1586)	69
295	(2034)	68
360	(2482)	65
450	(3103)	61
540	(3723)	58
625	(4309)	55

Liquid nitrogen is authorized for shipment in cylinders which meet TC/DOT Specification 4L. Liquid nitrogen at pressures under 25.3 psig (174 kPa) is not regulated by the DOT. However, in Canada compressed gases and refrigerated liquids are regulated regardless of pressure.

All cylinders authorized for gaseous nitrogen service must be requalified by hydrostatic retest every five or ten years under present regulations, with the following exceptions: DOT-4 cylinders, every ten years; and no periodic retest is required for cylinders of Specifications 3C, 3E, and 4C. See 49 CFR 173.34. [6]

Also, for cylinders of Specification 4L authorized for liquid nitrogen service, no periodic retest is required for requalification.

Pressure relief devices authorized for use on nitrogen cylinders include the CG-1, CG-3, CG-4, CG-5, and CG-7 type devices. These include rupture disk, combination rupture disk/fusible plug, and pressure relief valve type devices. See CGA S-1.1, *Pressure Relief Device Standards—Part 1—Cylinders for Compressed Gases,* for further information and requirements. [8]

Valve Outlet Connections

Standard connections in the United States and Canada are as follows: for service pressure up to 3000 psig (20 684 kPa), Connection CGA 580; for 3001 to 5500 psig (20 684 to 37 921 kPa), Connection CGA 680; for 5501 to 7500 psig (37 928 to 51 711 kPa), Connection CGA 677; and for cryogenic liquid withdrawal, Connection CGA 295. Connections CGA 555 and 590 are limited standard connections suitable for use at service pressures up to 3000 psig (20 684 kPa) in other than medical gas applications.

The standard pin-indexed yoke connection in the United States and Canada (for medical use of the gas) is Connection CGA 960 for service pressures up to 3000 psig (20 684 kPa). Further information on connections can be found in CGA V-1, *American National, Canadian, and Compressed Gas Association Standard for Compressed Gas Cylinder Valve Outlet and Inlet Connections.* [9]

Tank Cars

Gaseous nitrogen is authorized for rail shipment in tank cars that comply with TC/DOT specification 107A. Regulations require that the pressure to which the containers are charged must not exceed seven-tenths of the marked test pressure at 130°F (54.4°C). See 49 CFR 173.314. [6]

Liquid nitrogen is also shipped in vacuum-insulated tank cars (AAR204W) at pressures less than 25.3 psig (174.4 kPa).

Tube Trailers

Gaseous nitrogen is shipped in tube trailers with capacities ranging to more than 40 000 ft^3 (1100 m^3). These trailers are built to comply with DOT cylinder specifications 3A, 3AA, 3AX, 3AAX, or 3T. The trailers commonly serve as the storage supply for the user, with the supplier replacing trailers as they are emptied.

Tank Trailers (Cargo Tanks)

Liquid nitrogen is shipped in bulk at pressures below 25.3 psig (174.4 kPa) in special insulated tank trailers, with capacities in excess of 400 000 ft^3 (11 100 m^3). See 49 CFR 173.318. [6]

As previously noted, compressed gases and refrigerated liquids are regulated in Canada regardless of pressure.

Small Portable Containers

Liquid nitrogen is shipped and stored in small portable containers (dewar flasks) which hold quantities ranging from 1 to 25 gallons (4 to 90 L) or more. These containers are encased in shells and are heavily insulated; they maintain the liquid at atmospheric pressure.

METHODS OF MANUFACTURE

Nitrogen is produced commercially at air separation plants by liquefaction of atmospheric air and separation of the nitrogen from it by fractionation.

Pressure swing adsorption and membrane systems are also used to produce low-purity nitrogen gas.

REFERENCES

[1] CGA G-10.1, *Commodity Specification for Nitrogen,* Compressed Gas Association, Inc., 1235 Jefferson Davis Highway, Arlington, VA 22202.

[2] CGA P-12, *Safe Handling of Cryogenic Liquids,* Compressed Gas Association, Inc., 1235 Jefferson Davis Highway, Arlington, VA 22202.

[3] CGA P-9, *The Inert Gases, Argon, Nitrogen and Helium,* Compressed Gas Association, Inc., 1235 Jefferson Davis Highway, Arlington, VA 22202.

[4] CGA P-14, *Accident Prevention in Oxygen-Rich and Oxygen-Deficient Atmospheres,* Compressed Gas Association, Inc., 1235 Jefferson Davis Highway, Arlington, VA 22202.

[5] *Effects of Exposure to Toxic Gases—First Aid and Medical Treatment,* 3rd ed., Matheson Gas Products, Inc., Secaucus, NJ 07094. (1988)

[6] *Code of Federal Regulations,* Title 49 CFR Parts 100–199 (Transportation), Superintendent of Documents, U.S. Government Printing Office, Washington, DC 20402.

[7] *Transportation of Dangerous Goods Regulations,* Canadian Government Publishing Centre, Supply and Services Canada, Ottawa, Ontario, Canada K1A 0S9.

[8] CGA S-1.1, *Pressure Relief Device Standards—Part 1—Cylinders for Compressed Gases,* Compressed Gas Association, Inc., 1235 Jefferson Davis Highway, Arlington, VA 22202.

[9] CGA V-1, *American National, Canadian, and Compressed Gas Association Standard for Compressed Gas Cylinder Valve Outlet and Inlet Connections,* Compressed Gas Association, Inc., 1235 Jefferson Davis Highway, Arlington, VA 22202.

ADDITIONAL REFERENCES

McCarty, R. D., *Interactive Fortran IV Computer Programs for the Thermodynamic and Transport Properties of Selected Cryogens (Fluid Pack)* (National Bureau of Standards Technical Note 1025, October 1980), National Institute of Standards and Technology, Gaithersburg, MD 20899.

Saturated Liquid Densities of Oxygen, Nitrogen, Argon and Para-Hydrogen, (National Bureau of Standards Technical Note 361), National Institute of Standards and Technology, Gaithersburg, MD 20899.

Younglove, B. A., *Interactive Fortran Program to Calculate Thermophysical Properties of Six Fluids* (National Bureau of Standards Technical Note 1048, July 1982), National Institute of Standards and Technology, Gaithersburg, MD 20899.

Nitrogen Dioxide

Chemical Symbol: NO_2 or N_2O_4
Synonyms: Dinitrogen tetroxide, nitrogen peroxide, and nitrogen tetroxide
CAS Registry Number: 10102-44-0
DOT Classification: Poison A
DOT Label: Poison gas and Oxidizer
Transport Canada Classification: 2.3 (5.1) (9.2)
UN Number: UN 1067

PHYSICAL CONSTANTS

	U.S. Units	SI Units
Chemical formula	NO_2/N_2O_4	NO_2/N_2O_4
Molecular weight	46.0055/92.011	46.0055/92.011
Vapor pressure		
at 70°F (21.1°C)	14.66 psia	101.08 kPa abs
at 105°F (40.6°C)	34.44 psia	237.46 kPa abs
at 115°F (46.1°C)	43.01 psia	296.54 kPa abs
at 130°F (54.4°C)	59.32 psia	409.00 kPa abs
Density of the gas		
at 70°F (21.1°C) and 1 atm	0.212 lb/ft^3	3.394 kg/m^3
Specific gravity of the gas		
at 70°F (21.1°C) and 1 atm		
(air = 1)	2.62	2.62
Specific volume of the gas		
at 70°F (21.1°C) and 1 atm	4.7 ft^3/lb	0.2934 m^3/kg
Density of the liquid		
at 70°F (21.1°C)	89.85 lb/ft^3	1439 kg/m^3
at 100°F (37.8°C)	87.41 lb/ft^3	1400 kg/m^3
at 130°F (54.4°C)	84.82 lb/ft^3	1359 kg/m^3
Boiling point at 1 atm	70.1°F	21.2°C
Melting point at 1 atm	11.8°F	−11.2°C
Critical temperature	316.8°F	158.2°C
Critical pressure	1469.6 psia	10 132.5 kPa abs
Critical density	34.772 lb/ft^3	557 kg/m^3
Latent heat of vaporization		
at 70.2°F (21.2°C)	9.110 kcal/mol	38.116 kJ/mol

	U.S. Units	SI Units
Heat capacity at 76.73°F (24.85°C) (gas at 14.7 psia) C_p	8.837 cal/mol·K	36.974 J/mol·K
Viscosity of the liquid at 68°F (20°C)	0.416 cP	0.416 mPa·s
Solubility in water	Decomposes to nitric and nitrous acids	

References for above: [1], [2], [3], [4], and [5]

DESCRIPTION

Nitrogen dioxide is a nonflammable, toxic, reddish brown gas at room temperature. In the liquid state it is brown, and in solid form it is colorless. Nitrogen dioxide exists as an equilibrium mixture of NO_2 and N_2O_4. At 100°C, the liquid state contains less than 1 percent NO_2 and the vapor state 90 percent NO_2. The solid form is entirely N_2O_4. Nitrogen dioxide is a strong oxidizing agent and extremely reactive. It may form explosive mixtures with alcohols, hydrocarbons, organic materials, and fuels. Nitrogen dioxide also decomposes on contact with water to nitric and nitrous acids. It also acts as a catalyst for the explosive reaction of oxygen and hydrogen. [1] and [3]

GRADES AVAILABLE

Nitrogen dioxide is available in grades of 99.5 percent or 99.995 percent.

USES

Nitrogen dioxide has been used as a catalyst in certain oxidation reactions, as an inhibitor to prevent polymerization of acrylates during distillation, as a nitrating agent for organic compounds, as an oxidizing agent, and as a rocket fuel. It is also used as a flour bleaching agent, in the manufacture of liquid explosives, and for increasing the wet strength of paper. [2]

PHYSIOLOGICAL EFFECTS

A major hazard regarding exposure to nitrogen dioxide is that serious effects are not felt until several hours after the exposure. Exposure to nitrogen dioxide at levels of 90 ppm and higher has resulted in delayed pulmonary edema occuring anywhere from a few hours to 72 hours after exposure ceases. Symptoms include cyanosis, shortness of breath, restlessness, headache, and the production of a frothy yellow or brown sputum. With appropriate treatment, symptoms usually resolve rapidly but can persist for several weeks. [6]

Exposures to nitrogen dioxide of 10 minutes or less at a level of about 150 ppm produces cough, nose and throat irritation, tearing, headache, nausea, and vomiting. Exposures ranging from 50-150 ppm have been associated with moderate irritation to the eyes and mucous membranes. Permanent eye damage can occur, however, if exposures at these levels are prolonged. [6]

Delayed pulmonary edema may follow exposure to 100-150 ppm for only 30-60 minutes, while a few breaths at a concentration of 200-700 ppm will produce severe pulmonary damage which may result in fatal pulmonary edema after 5-8 hours have elapsed. [2]

Liquid nitrogen dioxide or strong vapor concentrations can cause severe injury to the skin and eyes.

The American Conference of Governmental Industrial Hygienists has adopted a time weighted average threshold limit value (TLV-TWA) of 3 ppm for an eight-hour period. [7] The U.S. Occupational Safety and Health Administration has adopted a short-term exposure limit (STEL) of 1 ppm (1.8 mg/m^3) for a 15-minute time weighted average for exposure to nitrogen dioxide. [8] Nitrogen dioxide in 10–20 ppm concentrations in air is slightly irritating to mucous membranes and

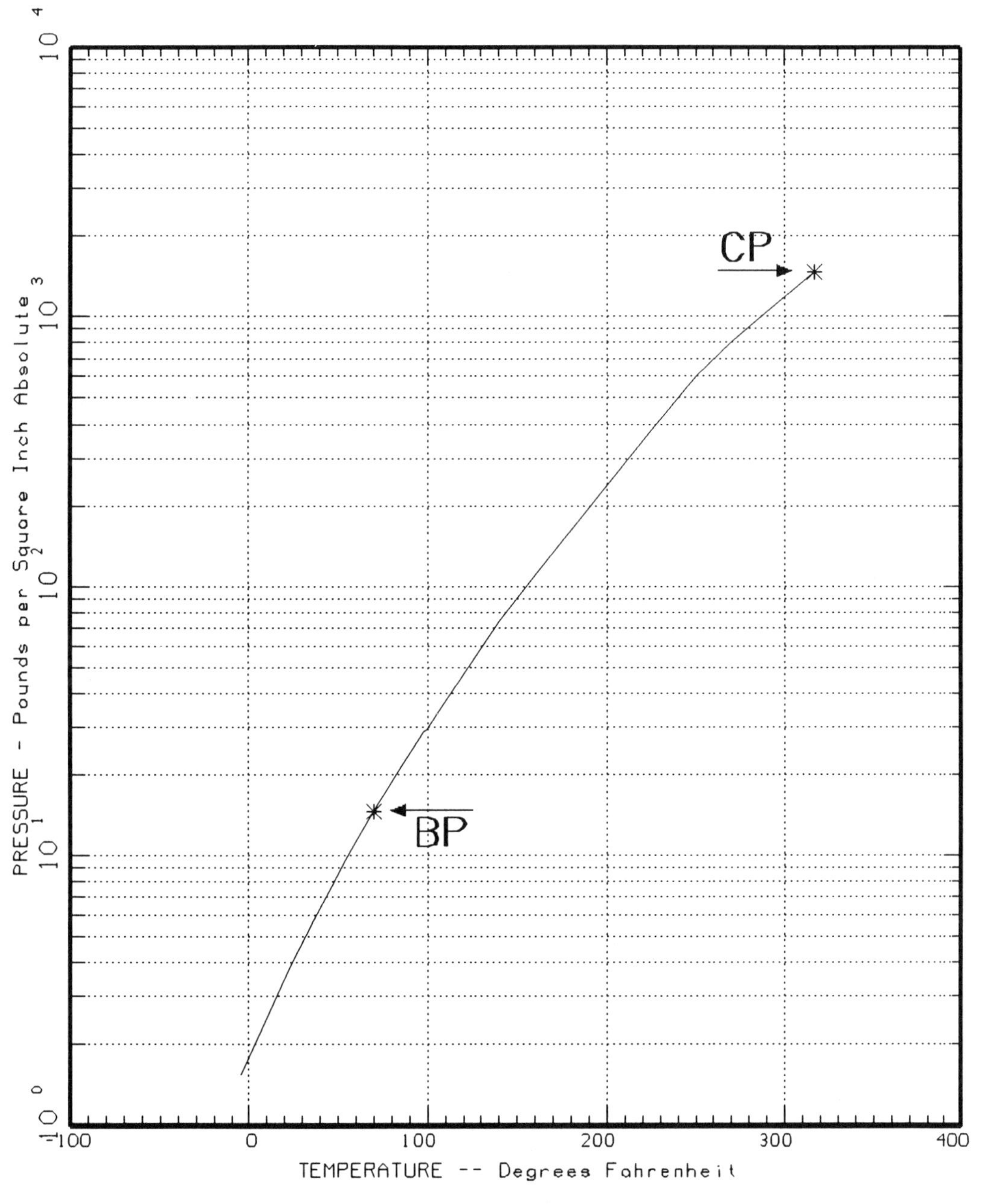

Fig. 1. Vapor Pressure Curve for Nitrogen Dioxide.

the upper respiratory tract. The odor is distinct in concentrations of 5 ppm. Concentrations above 100 ppm in air cause immediate distress. Exposure of the skin to liquid nitrogen dioxide can cause severe burns.

MATERIALS OF CONSTRUCTION

When dry (0.1 percent moisture or less), nitrogen dioxide is not corrosive to mild steel at ordinary temperatures and pressures. Numerous metals and alloys such as carbon steel, stainless steel, aluminum, nickel, and Inconel are satisfactory for handling and storage. Under wet conditions, stainless steels resistant to about 60 percent nitric acid serve best.

Equipment parts, such as valve stems, which are partly in contact with the atmosphere should be stainless steel with sufficient chromium content to resist corrosion caused by leaks through stuffing boxes. Good-quality ceramic bodies and Pyrex are satisfactory for handling wet or dry nitrogen dioxide.

Among the plastics, Teflon and Kel-F films are most satisfactory. Koroseal and Saran are useful but have a limited service life. In general, the vinyl plastics do not hold up well with nitrogen dioxide. Asbestos and asbestos-graphite are satisfactory for valve stuffing boxes. Koroseal has given reasonably good service in this use. For use on pipe threads, graphite-disodium silicate (waterglass) is recommended and hydrocarbon lubricants should be avoided. [2]

SAFE STORAGE, HANDLING, AND USE

Protect cylinders of nitrogen dioxide from physical damage. Store in a cool, dry, well-ventilated area away from heavily traveled areas and emergency exits. Do not allow the temperature where cylinders are stored to exceed 125°F (51.7°C). Cylinders should be stored upright and firmly secured to prevent falling or being knocked over. Full and empty cylinders should be segregated. Use a first in/first out inventory system to prevent full cylinders from being stored for excessive periods of time. [2] and [4]

Precautions in Handling

- Nitrogen dioxide should be handled only in well-ventilated areas, preferably a hood with forced ventilation.
- Self-contained breathing apparatus should be available for emergencies and stored in convenient locations not likely to be contaminated.
- Areas in which nitrogen dioxide is being handled should be provided with enough exits to permit personnel to leave quickly in emergencies.
- Personnel should have available, for immediate use, gas masks with Universal canisters for emergencies. These canisters are satisfactory only for short exposures (about 5 minutes for a 2 percent concentration of nitrogen dioxide, 20 minutes for a concentration of 0.5 percent) and should be changed on an exact time schedule and after each use.
- Instant-acting safety showers and eyewashing facilities should be conveniently located in an area not likely to become contaminated.
- Personnel handling nitrogen dioxide should wear chemical safety goggles and/or a full shield and rubber gloves.

DISPOSAL

Dispose of liquid nitrogen dioxide by running the liquid into excess 5–10 percent aqueous sodium hydroxide solution at a moderate rate. Transfer the resulting solution to a plant disposal unit for neutralization and disposal. [2] and [4]

HANDLING LEAKS AND EMERGENCIES

Should it become necessary to dispose of nitrogen dioxide from a leaking cylinder, the following procedure may be used. Move the cylinder to a hood or safe out-of-doors area.

Attach an appropriate needle valve with a long piece of flexible tubing to the cylinder valve, and run the liquid nitrogen dioxide into excess 5–10 percent aqueous sodium hydroxide solution at a moderate rate. Transfer the resulting solution to the plant disposal unit for neutralizing and disposal. [2], [3], and [4]

First Aid

Prompt medical attention is mandatory in all cases of exposure to nitrogen dioxide. Rescue personnel should be equipped with self-contained breathing apparatus and wear protective clothing and boots if the liquid has spilled or is being discharged.

Skin exposure: Immediately flush exposed area with water for at least 15 minutes; get medical attention. Remove contaminated clothing. Discard contaminated clothing and shoes.

Eye exposure: Immediately flush with large amounts of water for at least 15 minutes. The eyelids must be held open and away from the eyeball to ensure that all surfaces are flushed thoroughly. Contact a physician immediately, preferably an ophthalmologist. [9]

Swallowing: Due to the nature of the material, it is very unlikely that it could be taken internally.

Inhalation: Conscious persons should be carried (not assisted) to an uncontaminated area and breathe fresh air supplemented with oxygen. Keep the patient warm, quiet, and under competent medical observation until the danger of delayed pulmonary edema has passed (at least 72 hours). Any physical exertion during this period should be discouraged as it may increase the severity of the pulmonary edema or chemical pneumonitis. Bed rest is indicated.

METHODS OF SHIPMENT

Under the appropriate regulations, nitrogen dioxide is authorized for shipment as follows: [10] and [11]

By Rail: In tank cars and cylinders.

By Highway: In cylinders on trucks.

By Water: In cylinders on cargo vessels stowed above deck.

By Air: Forbidden.

CONTAINERS

Nitrogen dioxide is authorized for shipment in TC/DOT approved cylinders and tank cars (ton containers) as a liquefied gas under its own vapor pressure of 14.66 psia (101.1 kPa abs) at 70°F (21.1°C). Approved cylinders include those of Specifications 3A480, 3AA480, and 3AL1800. Cylinders and ton containers must not be equipped with a pressure relief device. Valve outlets must be sealed with a solid threaded cap or plug. Authorized containers and specific requirements for shipment of nitrogen dioxide are given in 49 CFR 173.336. [10]

Valve Outlet Connections

The standard valve outlet connection in the United States and Canada for nitrogen dioxide cylinders is Connection CGA 660.

Tank Cars

Nitrogen dioxide is also authorized for shipment in Specification 106A500X ton containers and 105A500W tank cars, the latter of which are equipped with a combination pressure relief valve including a stainless steel or platinum frangible disk. Specification 110A500W tank cars are also authorized. See 49 CFR 173.336. [10]

METHODS OF MANUFACTURE

High-quality nitrogen dioxide is obtained in connection with the production of sodium nitrate from sodium chloride and nitric acid. It is also obtained when sodium nitrate is treated with nitric acid and the evolved nitric oxide (NO) is combined with oxygen; by treating nitrosyl chloride with oxygen to give nitrogen dioxide and chlorine; by passing ammonia and air heated to 1112°F (600°C) over a platinum catalyst and treating the NO formed with oxygen to give nitrogen dioxide. Nitrogen dioxide is also formed when concentrated nitric acid reacts with copper. [2]

REFERENCES

[1] *Matheson Unabridged Gas Data Book,* Matheson Gas Products, Inc., Secaucus, NJ 07094. (1974)

[2] *Matheson Gas Data Book,* 6th ed., Matheson Gas Products, Inc., Secaucus, NJ 07094. (1980)

[3] Material Safety Data Sheet for Nitrogen Dioxide (10/85), Alphagaz Division, Liquid Air Corp., 2121 N. California Blvd., Walnut Creek, CA 94596.

[4] Material Safety Data Sheet for Nitrogen Dioxide (11/85), Scientific Gas Products, 2330 Hamilton Blvd., Box 648, South Plainfield, NJ 07080.

[5] Material Safety Data Sheet for Nitrogen Dioxide (12/78), Material Bureau of Standards (under license from Genium Publishing Company).

[6] *Effects of Exposure to Toxic Gases—First Aid and Medical Treatment,* 3rd ed., Matheson Gas Products, Inc. Secaucus, NJ 07094. (1988)

[7] *Threshold Limit Values and Biological Exposure Indices,* 1989-90 ed., American Conference of Governmental Industrial Hygienists, 6500 Glenway Avenue, Bldg. D-7, Cincinnati, OH 45211-4438.

[8] *Code of Federal Regulations,* Title 29 CFR Parts 1900–1910 (Labor), Superintendent of Documents, U.S. Government Printing Office, Washington, DC 20402.

[9] Material Safety Data Sheet for Nitrogen Dioxide (12/85), Union Carbide Corp., Linde Division, 39 Old Ridgebury Road, Danbury, CT 06817-0001.

[10] *Code of Federal Regulations,* Title 49 CFR Parts 100–199 (Transportation), Superintendent of Documents, U.S. Government Printing Office, Washington, DC 20402.

[11] *Transportation of Dangerous Goods Regulations,* Canadian Government Publishing Centre, Supply and Services Canada, Ottawa, Ontario, Canada K1A 0S9.

Nitrogen Trifluoride

Chemical Symbol: NF_3
CAS Registry Number: 7883-54-2
DOT Classification: Nonflammable gas
DOT Label: Nonflammable gas
Transport Canada Classification: 2.3
UN Number: UN 2451

PHYSICAL CONSTANTS

	U.S. Units	SI Units
Chemical formula	NF_3	NF_3
Molecular weight	71.00	71.00
Density of the gas at 70°F (21.1°C) and 1 atm	0.1843 lb/ft^3	2.95 kg/m^3
Specific gravity of the gas at 70°F (21.1°C) and 1 atm (air = 1)	2.46	2.46
Specific volume of the gas at 70°F (21.1°C) and 1 atm	5.43 ft^3/lb	0.337 m^3/kg
Density of the liquid at −200.3°F (−129°C)	96.01 lb/ft^3	1538 kg/m^3
Boiling point at 1 atm	−200.3°F	−129.1°C
Melting point at 1 atm	−340.2°F	−206.8°C
Critical temperature	−38.5°F	−39.2°C
Critical pressure	646.9 psia	4460 kPa abs
Critical density	35.1 lb/ft^3	562.3 kg/m^3
Latent heat of vaporization at −200.3°F (−129°C)	70.1 Btu/lb	162.6 kJ/kg

DESCRIPTION

Nitrogen trifluoride is a colorless gas with little odor. However, commercial grades contaminated with trace levels of active fluorides have a pungent, musty odor.

Nitrogen trifluoride is an oxidizer that is thermodynamically stable except at elevated temperatures. At temperatures up to about 662°F (350°C), its reactivity is comparable to oxygen. At higher temperatures, its reactivity is similar to fluorine owing to appreciable dissociation into NF_2 and F^-. The thermal dissociation of nitrogen trifluoride has been studied by a number of investigators and has been found to peak in the temperature range of 1100 to 1500 K.

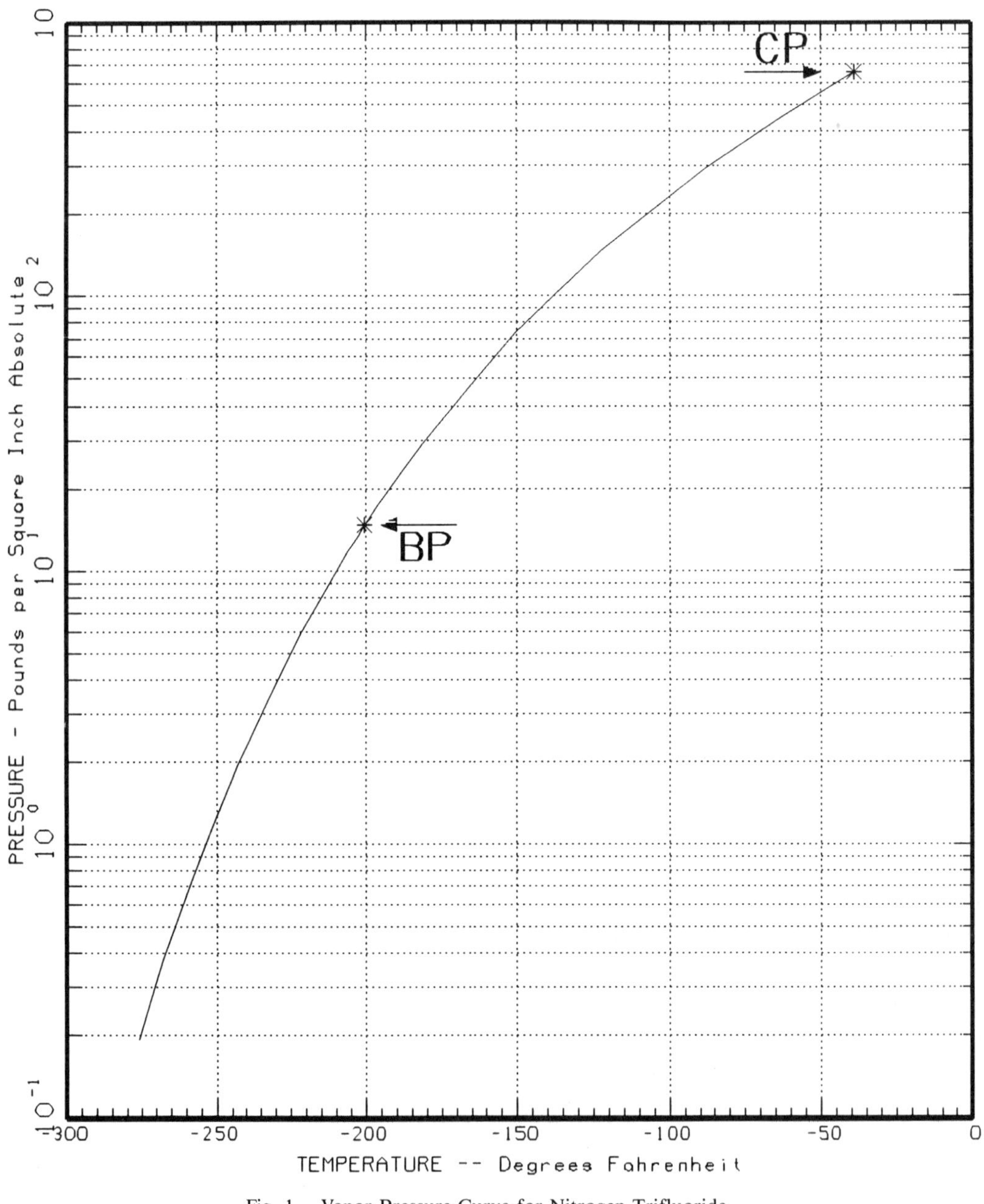

Fig. 1. Vapor Pressure Curve for Nitrogen Trifluoride.

Nitrogen trifluoride acts primarily upon the elements as a fluorinating agent, but not a very active one at lower temperatures. At elevated temperatures, nitrogen trifluoride pyrolyzes with many of the elements to produce nitrogen tetrafluoride (N_2F_4) and the corresponding fluoride. The pyrolysis of nitrogen trifluoride over copper turnings produces N_2F_4 in a 62–71 percent yield at 707°F (375°C). Pyrolysis over carbon is more favorable.

Hydrogen reacts with nitrogen trifluoride with the rapid liberation of large amounts of heat and is the basis for the use of nitrogen trifluoride in high-energy chemical lasers. The flammability range for NF_3/H_2 mixtures is 9.4–95 mole percent NF_3. Nitrogen trifluoride reacts with organic compounds, but generally an elevated temperature is required to initiate the reaction. Under these conditions, the reaction will often proceed explosively, and great care must be exercised when exposing nitrogen trifluoride to organic compounds. Therefore, nitrogen trifluoride has found little use as a fluorinating agent for organic compounds.

GRADES AVAILABLE

Nitrogen trifluoride is available in grades ranging from 98 percent minimum purity.

USES

Nitrogen trifluoride has been used successfully in large quantities as a fluorine source for high-energy chemical lasers. It is preferred over fluorine because of its comparative ease of handling at ambient conditions.

Recently, an increasing amount of nitrogen trifluoride is being used in the semiconductor industry as a dry etchant, showing significantly higher etch rates and selectivities when compared to carbon tetrafluoride (CF_4) and mixtures of CF_4 and oxygen.

Minor amounts of nitrogen trifluoride are used as chemical intermediates in the production of tetrafluorohydrazine and a series of perfluoroammonium salts. Nitrogen trifluoride was also used as an oxidizer in rocketry in the early 1960s, but this application was not commercialized.

PHYSIOLOGICAL EFFECTS

Nitrogen trifluoride is a toxic substance and is hazardous by inhalation, with an eight-hour time-weighted average threshold limit value (TLV-TWA) of 10 ppm (29 mg/m^3) as recommended by the American Conference of Governmental Industrial Hygienists (ACGIH). [1] This is also the exposure limit established by the U.S. Occupational Safety and Health Administration (OSHA). [2]

The toxicity of nitrogen trifluoride is related to its capacity to form methemoglobin, a modified form of hemoglobin incapable of oxygen transport, and to destroy red blood cells (hemolysis). Upon cessation of exposure, methemoglobin spontaneously reverts to hemoglobin. However, at high levels of exposure, therapeutic intervention may be necessary (oxygen, methylene blue, exchange transfusion). The occurrence of hemolysis requires careful monitoring for degree of anemia and the potential for impaired kidney function.

The inhalation toxicity of nitrogen trifluoride on rats, dogs, and monkeys has been studied extensively. Experiments were done with concentrations from 100 to 400 000 ppm and exposure periods of 10 to 40 000 minutes. Typical LC_{50} values reported are 38 800 ppm for dogs and 19 300 ppm for mice at 15-minute exposure. For 60-minute exposure, these values are 9600 ppm for dogs and 7500 ppm for mice. The National Academy of Sciences, National Research Council Committee on Toxicology, recommends the following emergency exposure limits for nitrogen trifluoride: 10 minutes, 2250 ppm; 30 minutes, 750 ppm; 60 minutes, 375 ppm. Gaseous nitrogen trifluoride is considered innocuous to the skin and a minor irritant to the eyes and mucous membranes.

MATERIALS OF CONSTRUCTION

At temperatures less than 662°F (350°C), nitrogen trifluoride has a reactivity similar to that of oxygen and is relatively inert to most materials of construction. At ambient temperatures, brass, aluminum, copper, steel, and stainless steels can be used, corrosion rates of less than 0.1 mil/yr. at 160°F (71°C) having been determined for these materials. Nitrogen trifluoride is also compatible with fluorinated materials such as Teflon at ambient temperatures.

At increased temperatures and pressures, nitrogen trifluoride's reactivity increases, becoming more like that of fluorine, with nickel and Monel being the preferred materials of construction.

SAFE STORAGE, HANDLING, AND USE

All the precautions necessary for the safe handling of any compressed gas must also be observed with nitrogen trifluoride. In addition, all combustible materials, especially hydrocarbon oils and greases, must be kept from contact with high nitrogen trifluoride concentrations.

Since nitrogen trifluoride has little odor, it cannot be detected at concentrations within the TLV. Therefore, adequate ventilation must be provided, and detection devices should be used to monitor personnel exposure.

While most materials will not spontaneously react with nitrogen trifluoride except at elevated temperatures, safe handling demands that care be taken to avoid materials of questionable compatibility and to minimize possible sources of ignition. All lines, fittings, and other equipment should be cleaned, degreased, and dried before assembly to remove organic materials such as oils, solvents, and other combustibles. After assembly, the gas-handling system should be purged and evacuated. Following this, nitrogen trifluoride should be slowly introduced into the system to a few atmospheres of pressure at room temperature.

For systems which will be operated at elevated temperatures, fluorine should be used to passivate the system prior to the introduction of nitrogen trifluoride. (See the monograph on fluorine for further discussion of passivation.) Passivation results in the removal of oxidizable material and forms a protective metal fluoride film which prevents further reaction. Passivated systems or components that have been exposed to moisture should be repassivated before use. Materials of construction and valve designs normally acceptable for oxygen service can also be used in a nitrogen trifluoride system. Generally acceptable for use are valves employing stainless-steel bodies, metal-to-metal or metal-to-Teflon seats, and Teflon stem packings, metal diaphragms, or metal bellows.

Safe Handling and Storage of Cylinders

Nitrogen trifluoride cylinders must be securely supported while in use to prevent movement and straining of connections. Full cylinders must be stored in a well-ventilated area, protected from excessive heat (125°F; 52°C), located away from organic or flammable materials, and secured. Valve protection caps and valve outlet caps must be securely in place at all times when the cylinder is not in use.

Nitrogen trifluoride cylinders must also always be protected from mechanical shock or abuse and must never be heated with a torch or heat lamp.

DISPOSAL

Disposal of nitrogen trifluoride can be accomplished by combustion with a fuel such as activated charcoal, hydrocarbons, or metals at elevated temperatures. Reactions with activated charcoal produce nitrogen and carbon tetrafluoride. Carbon bed temperatures must be maintained above 1000°F (538°C) to assure reaction and to prevent gas adsorption. Similarly, nitrogen trifluoride streams can be scrubbed by introducing the gas through a packed bed of nickel or Monel turnings maintained above 1000°F (538°C), followed by caustic scrubbing.

Nitrogen trifluoride effluents have also been disposed of by introducing the gas stream below the flame of a burner operating on a hydrocarbon fuel. While nitrogen trifluoride is destroyed by this method, both hydrogen fluoride (HF) and oxides of nitrogen (NO_x) may be formed which require subsequent downstream scrubbing techniques for their removal.

HANDLING LEAKS AND EMERGENCIES

Leaks of nitrogen trifluoride in process lines or equipment may be detected by applying soap solution to the suspected area. Leaks will be indicated by bubble formation. Leaks on the cylinder or cylinder valve area can be detected in the same manner. If a leaking cylinder is found, move the cylinder to a well-ventilated area and contact the manufacturer. Proper safety precautions should be followed, especially for other than minor leaks, which will require the use of personal protective equipment, including self-contained breathing apparatus.

First Aid

Treatment is usually only required in the event of cyanosis, indicating 20% or greater methemoglobin formation, or if there is underlying hypoxemia or coronary artery disease. First aid is limited to treatment with oxygen. Methemoglobinemia resolves rapidly (in hours) with this treatment alone. [3]

METHODS OF SHIPMENT

Under the appropriate regulations, nitrogen trifluoride is authorized for shipment as follows: [4] and [5]

By Rail: In cylinders.

By Highway: In cylinders on trucks and in tube trailers.

By Water: In cylinders on cargo vessels only, stowed above deck only.

By Air: Aboard cargo aircraft only up to 300 lb (136 kg) maximum net weight per container.

CONTAINERS

Nitrogen trifluoride is authorized for shipment as a nonflammable compressed gas in cylinders, including DOT approved trailer tubes. Nitrogen trifluoride is normally compressed to 1450 psig (9997 kPa) in cylinders.

Cylinders

Nitrogen trifluoride is authorized for shipment in DOT specified cylinders as identified in 49 CFR 173.302. [4] Also see 49 CFR 173.301(d)(2). [4] Among the cylinders authorized in 49 CFR 173.302 are Specifications 3A, 3AA, 3B, 3E, 4B, 4BA, and 4BW. Cylinders must be equipped with pressure relief devices composed of either a fusible plug designed to relieve at 212°F (used on low-pressure—500 psig—cylinders only) or a frangible rupture disk backed by a fusible plug. See CGA S-1.1, *Pressure Relief Device Standards—Part I—Cylinders for Compressed Gases,* for further information and requirements concerning the Type CG-3, CG-4, and CG-5 pressure relief devices authorized for use with nitrogen trifluoride. [6] Cylinders must all be fitted with valve protection caps, boxes or crates, or recessed valve(s).

All cylinders authorized for nitrogen trifluoride must be requalified every five years in accordance with the requirements of 49 CFR 173.34. [4]

Valve Outlet Connections

The standard valve outlet connection for nitrogen trifluoride cylinders in the United States and Canada is Connection CGA 330. [7]

Tube Trailers

Nitrogen trifluoride is authorized for bulk highway shipment at high pressure (usually around 1450 psig) or 9997 kPa in tube trailers. These trailers are made up of a number of large tubes, usually constructed to Specification 3A or 3AA, which are manifolded together to a common header. Approved pressure relief devices and valves must be installed on each tube, and valves must be closed during transit.

METHODS OF MANUFACTURE

Nitrogen trifluoride can be formed from a wide variety of chemical reactions. The commercial process for production involves direct fluorination of ammonia with fluorine gas in the presence of ammonium fluoride.

REFERENCES

[1] *Threshold Limit Values and Biological Exposure Indices,* 1989-90 ed., American Conference of Governmental Industrial Hygienists, 6500 Glenway Avenue, Bldg. D-7, Cincinnati, OH 45211-4438.

[2] *Code of Federal Regulations,* Title 29 CFR Parts 1900-1910 (Labor), Superintendent of Documents, U.S. Government Printing Office, Washington, DC 20402.

[3] *Effects of Exposure to Toxic Gases—First Aid and Medical Treatment,* 3rd ed., Matheson Gas Products, Inc., Secaucus, NJ 07094. (1988)

[4] *Code of Federal Regulations,* Title 49 CFR Parts 100-199 (Transportation), Superintendent of Documents, U.S. Government Printing Office, Washington, DC 20402.

[5] *Transportation of Dangerous Goods Regulations,* Canadian Government Printing Centre, Supply and Services Canada, Ottawa, Ontario, Canada K1A 0S9.

[6] CGA S-1.1, *Pressure Relief Device Standards—Part I—Cylinders for Compressed Gases,* Compressed Gas Association, Inc., 1235 Jefferson Davis Highway, Arlington, VA 22202.

[7] CGA V-1, *American National, Canadian, and Compressed Gas Association Standard for Compressed Gas Cylinder Valve Outlet and Inlet Connections* (ANSI/CSA/CGA V-1), Compressed Gas Association, Inc., 1235 Jefferson Davis Highway, Arlington, VA 22202.

ADDITIONAL REFERENCES

Kirk-Othmer, *Encyclopedia of Chemical Technology,* 3rd ed., 10, Wiley, New York, 1980.

Woytek, A. J., Lileck, J. T., and Barbanic, J. A., *Solid State Technology,* Vol. 27, No. 3 (March 1984).

Nitrous Oxide

Chemical Symbol: N_2O
Synonyms: Nitrogen monoxide, dinitrogen monoxide, laughing gas
CAS Registry Number: 10024-97-2
DOT Classification: Nonflammable gas
DOT Label: Nonflammable gas
Transport Canada Classification: 2.2
UN Number: UN 1070 (liquefied gas); UN 2201 (refrigerated liquid)

PHYSICAL CONSTANTS

	U.S. Units	SI Units
Chemical formula	N_2O	N_2O
Molecular weight	44.0128	44.0128
Specific volume		
at 59°F (15°C) and 1 atm	8.538 ft^3/lb	0.533 m^3/kg
at 70°F (21.1°C) and 1 atm	8726 ft^3/lb	0.5447 m^3/kg
Specific gravity of the gas (air = 1)	1.5297	1.5297
Density of the gas		
at 59°F (15°C) and 1 atm	0.1172 lb/ft^3	1.877 kg/m^3
at 70°F (21.1°C) and 1 atm	0.1146 lb/ft^3	1.947 kg/m^3
Critical density	28.15 lb/ft^3	450.4 kg/m^3
Boiling point at 1 atm	−127.4°F	−88.5°C
Latent heat of vaporization		
at boiling point	161.8 Btu/lb	376.1 kJ/kg
at 32°F (0°C)	107.5 Btu/lb	249.9 kJ/kg
at 70°F (21.1°C)	77.7 Btu/lb	180.6 kJ/kg
Latent heat of fusion at triple point	63.9 Btu/lb	148.5 kJ/kg
Specific heat of the gas		
at 59°F (15°C) and 1 atm C_p	0.207 Btu/(lb)(°F)	0.866 kJ/(kg)(°C)
at 59°F (15°C) and 1 atm C_v	0.158 Btu/(lb)(°F)	0.665 kJ/(kg)(°C)

	U.S. Units	SI Units
Ratio of specific heats		
at 59°F (15°C) and 1 atm		
C_p/C_v	1.303	1.303
Solubility in water, vol/vol, and 1 atm		
at 32°F (0°C)	1.3	1.3
at 68°F (20°C)	0.68	0.68
at 77°F (25°C)	0.59	0.59
Viscosity of the gas		
at 32°F (0°C)	0.326 lb/(ft)(hr)	0.135×10^{-3} Pa(s)
at 80°F (27°C)	0.359 lb/(ft)(hr)	0.149×10^{-3} Pa(s)
Thermal conductivity of the gas		
at 32°F (0°C)	0.0083 Btu (ft)/ h(ft^2)(°F)	0.0144 W/m(°C)
at 212°F (100°C)	0.0135 Btu (ft)/ h(ft^2)(°F)	0.0233 W/m(°C)

DESCRIPTION

Nitrous oxide at room temperature and atmospheric pressure is a colorless gas with a barely perceptible sweet odor and taste. It is nonflammable but will support combustion. At elevated temperatures, nitrous oxide decomposes into nitrogen and oxygen. Decomposition in the absence of catalysts is negligible at temperatures below 1200°F (649°C). Nitrous oxide is moderately soluble in water, alcohol, and oils. Unlike some higher oxides of nitrogen, nitrous oxide does not affect the acidity of water solutions.

GRADES AVAILABLE

Nitrous oxide is available in medical, commercial, and high-purity grades. The medical (USP) grade is the most widely used. Manufacturers typically produce nitrous oxide for this use to the specification published in the *United States Pharmacopeia/National Formulary.* [1] CGA G-8.2, *Commodity Specification for Nitrous Oxide,* describes the requirements for particular grades of nitrous oxide. [2] Other specifications to meet particular requirements are available from suppliers. Table 1, from CGA G-8.2, presents the component maxima, in parts per million (mole/mole) unless otherwise shown, for specific grades of nitrous oxide, also known as quality verification levels. The absence of a value in a listed quality verification level does not mean to imply that the limiting characteristic is or is not present, but merely indicates that the test is not required for compliance with the specification.

USES

The major use of nitrous oxide is for anesthesia and analgesia. It is also used for cryosurgery. Nitrous oxide also finds use as an oxidizing gas for atomic absorption spectrophotometry, a propellant for pressure or aerosol products (whipped cream is most prevalent), and a fuel oxidant for racing vehicles. Some product for this latter application is supplied with an additive to deter abuse of the product. Nitrous oxide is also used in the manufacture of semiconductors.

PHYSIOLOGICAL EFFECTS

Nitrous oxide's primary physiological effect is central nervous system (CNS) depression. At high concentrations, anesthetic levels can be obtained, but the low potency of nitrous oxide necessitates concomitant administration of other depressant drugs. Ni-

Vapor Pressure

NITROUS OXIDE

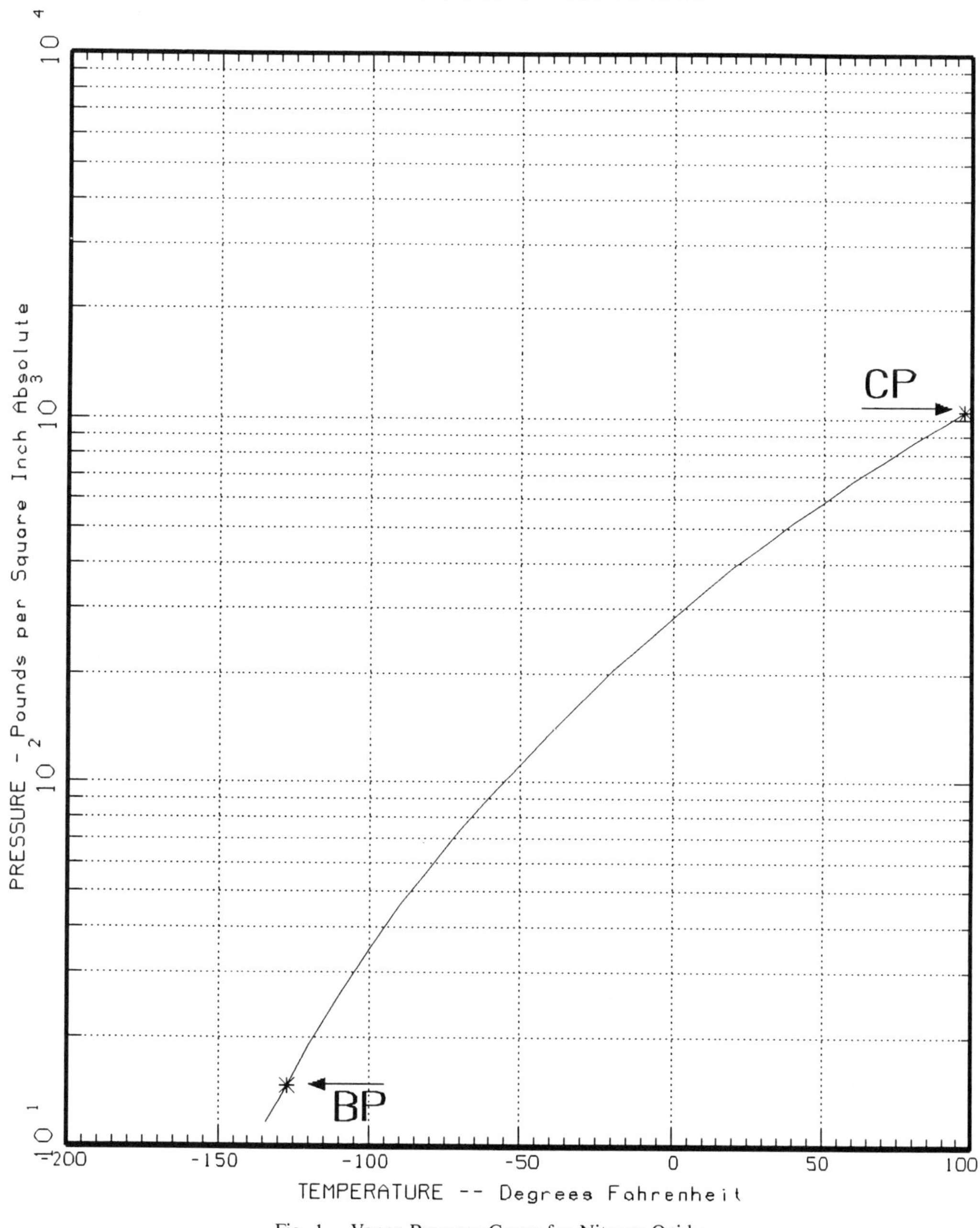

Fig. 1. Vapor Pressure Curve for Nitrous Oxide.

TABLE 1. NITROUS OXIDE GRADES AVAILABLE.
(Units in ppm (mole/mole) unless otherwise shown)

Limiting Characteristics [a,b]	Maxima for Gaseous and Liquid Nitrous Oxide		
	A	B	C
Nitrous Oxide Min. % (mole/mole)[c]	99.0 (liquid)	99.0	99.9974
Carbon Monoxide	10 (vapor)		1
Nitric Oxide	1 (vapor)		1
Nitrogen Dioxide	1 (liquid)		1
Halogens (as chlorine)	1 (vapor)		
Carbon Dioxide	300 (vapor)		2
Ammonia	25 (vapor)		5
Water, ppm (v/v)	200 (vapor)		3
Dew Point, °F	−33 (vapor)		−92
United States Pharmacopeia[c,d]	Yes		
Oxygen		10 000	2
Nitrogen			10
Hydrocarbons (C_1-C_5)			1

[a]Below the critical temperature, nitrous oxide exists in two phases. Contaminants such as carbon monoxide, nitrogen, oxygen, and nitric oxide are concentrated in the gas phase; whereas contaminants such as nitrogen dioxide, ammonia, and chlorine are concentrated in the liquid phase. It is therefore important that a two-phase sample be analyzed in an analytical sequence which will not materially reduce contaminants to be determined in subsequent procedures. The analytical procedures are listed in a sequence which will minimize concentration changes. Analytical sequence may be altered as agreed upon between the supplier and the customer for non-USP nitrous oxide.

[b]Unless otherwise specified, the sample shall be obtained from the vapor phase of the container, or as agreed upon between the supplier and the customer.

[c]For *United States Pharmacopeia* (USP) quality verification, the percent nitrous oxide shall be defined as the difference between 100% and the percent air. This test is performed last in the sequence of USP tests, after all other trace constituents have been determined.

[d]For *United States Pharmacopeia* (USP) quality verification, perform the verification tests in the sequence listed (after the filled container has warmed to ambient temperature); a gas/liquid equilibrium must be established in the container. For a "vapor" sample, test nitrous oxide gas from the vapor phase of the container; for a "liquid" sample, a representative vaporized sample of the liquid phase must be obtained. The sequence listed minimizes concentration changes and allows for maximum detection of toxic contaminants on the assumption that the usage is inhalation.

trous oxide has been associated with several side effects from long-term exposure. The most strongly substantiated effect is neuropathy. Epidemiological studies also suggest feto-toxic effects and higher incidents of spontaneous abortion in exposed personnel. Although no cause-and-effect relationship has been firmly established, exposure to the gas should be minimized.

Inhalation of nitrous oxide without the provision of a sufficient oxygen supply may be fatal or cause brain damage. Due to the concern over long-term exposure effects, release of the product into general work areas should be minimized. The National Institute for Occupational Safety and Health (NIOSH) has recommended limits of 25 ppm for hospital operating rooms and 50 ppm for dental offices measured on an eight-hour time-weighted average basis. Contact with liquid nitrous oxide can freeze skin tissue and must be avoided. The American Conference of Governmental Industrial Hygienists has adopted a time-weighted average threshold limit value of 50 ppm (90 mg/m^3) for an eight-hour workday exposure to nitrous oxide. [3]

Warning: The misuse of nitrous oxide can cause death by reducing the oxygen necessary to support life. Nitrous oxide abuse can impair an individual's ability to make and implement life-sustaining decisions.

MATERIALS OF CONSTRUCTION

Nitrous oxide is noncorrosive and may therefore be used with any of the common, commercially available metals. Because of its oxidizing action, however, all equipment being prepared to handle nitrous oxide, particularly at high pressures, must be free of oil, grease, and other readily combustible materials. Nitrous oxide may cause swelling of some elastomers.

SAFE STORAGE, HANDLING, AND USE

Nitrous oxide will support combustion. It must be kept away from oil, grease, and other combustible materials. Never permit oil, grease or any other readily combustible substance to come in contact with cylinders or other equipment containing nitrous oxide.

Store and use nitrous oxide with adequate ventilation. Containers that become exposed to fire, including bulk storage tanks, could rupture violently if subject to localized heating.

See Chapter 5 for general guidelines concerning the safe storage and handling of compressed gases, including nitrous oxide. Detailed recommendations for nitrous oxide storage locations are given in CGA G-8.1, *Standard for Nitrous Oxide Systems at Consumer Sites.* [4] Requirements for nitrous oxide systems in hospitals and other health care facilities are found in NFPA 99, *Standard for Health Care Facilities.* [5]

DISPOSAL

Nitrous oxide is not harmful to the environment and can be released to the atmosphere, provided ventilation is adequate for protection of personnel in the immediate vicinity of the release point. Do not release in the vicinity of building air intakes.

HANDLING LEAKS AND EMERGENCIES

Turn off ignition sources in the general area of the leak if possible. Nitrous oxide is nonflammable but is an oxidizer which can cause or intensify fires. Evacuate the area to prevent asphyxiation. Use self-contained breathing apparatus to enter the area. Provide as much ventilation as possible. Avoid contact with liquid spills. Contact supplier for assistance.

First Aid

People acutely exposed to high levels of nitrous oxide should be immediately removed from exposure, after which symptoms should rapidly reverse. If obtundation (dulling or blunting of sensitivity, such as to pain) is present, oxygen should be administered. Mechanical ventilation should be instituted in the event of respiratory arrest. Vomiting may occur as the person awakes. In order to prevent aspiration, exposed individuals should be placed on their side with their head at the level of or slightly lower than their body. [6]

METHODS OF SHIPMENT

Under the appropriate regulations, nitrous oxide is shipped as a refrigerated liquid, typically 0°F and 268 psig (−17°C and 1848 kPa), and as a liquefied compressed gas in high-pressure cylinders under ambient conditions. [7] and [8] See Table 2 concerning the vapor pressure and temperature properties of nitrous oxide.

By Rail: In cylinders, portable tanks, and tank cars.

By Highway: In cylinders, cargo tanks, and portable tanks.

By Water: For cargo and passenger vessels, as a compressed gas stored on deck or below deck in a well ventilated area. As a refrigerated liquid stored on deck on cargo and passenger vessels.

By Air: On cargo aircraft as a compressed gas up to 300 pounds (136 kg) maximum net weight per container. On passenger aircraft up to 150 pounds (68 kg) maximum net weight per container. Refrigerated liquid nitrous oxide forbidden on passenger and cargo aircraft.

TABLE 2. PROPERTIES OF SATURATED LIQUID NITROUS OXIDE.

Temperature		Vapor Pressure		Density	
°F	°C	psig	kPa	lb/gal	kg/m³
−127	−88	0	0	10.70	1280
−120	−84	4	30	10.56	1270
−100	−73	20	140	10.16	1220
−80	−62	43	300	9.89	1190
−60	−51	76	525	9.43	1130
−40	−40	122	845	9.09	1090
−20	−29	188	1300	8.76	1050
−10	−23	225	1550	8.40	1010
−5	−21	240	1655	8.34	999
0	−18	260	1790	8.23	986
5	−15	290	2000	8.13	974
10	−12	315	2170	8.02	961
15	−9	345	2380	7.90	947
20	−7	365	2515	7.83	938
30	−1	430	2965	7.58	909
40	4	490	3380	7.37	883
50	10	565	3895	7.09	849
60	16	655	4515	6.77	810
70	21	735	5070	6.45	773
80	27	840	5790	5.98	717
90	32	945	6515	5.44	652
98	37	1040	7170	3.78	453

CONTAINERS

Cylinders

Cylinders meeting the following TC/DOT specifications are authorized for nitrous oxide service: 3A1800, 3AX1800, 3AA1800, 3AAX1800, 3, 3E1800, 3T1800, 3HT2000, 39, and 3AL1800. Insulated containers of the TC/DOT-4L type may be permitted by exemption.

Authorized pressure relief devices for cylinders of nitrous oxide are of the Type CG-1 rupture disk. [9] Type CG-7 devices can be used up to 500 psig charging pressure for DOT-4L cylinders by DOT exemption. The Type CG-7 device is a pressure relief valve. [9]

Valve Outlet Connections

The standard cylinder valve outlet connection in the United States and Canada for nitrous oxide cylinders is Connection CGA 326. The standard medical cylinder valve yoke connection is Connection CGA 910. [10]

Tank Cars

TC/DOT specification 105A500W and 105A600W tank cars with approved insulation are authorized for shipment of nitrous oxide. See 49 CFR 173.314. [7]

Cargo Tanks and Portable Tank Containers

Specification DOT-51 portable tanks and Specification MC-330 and MC-331 cargo tanks are authorized for shipment of nitrous oxide in the United States. See 49 CFR 173.315. [7]

In Canada, highway and portable tanks are specified by the Canadian Standards Association. See CSA B620, *Highway Tanks and Portable Tanks for the Transportation of Dangerous Goods,* and CSA B622, *CSA Standard for the Selection and Use of Tank Trucks, Tank Trailers and Portable Tanks for the Transportation of Dangerous Goods for Class 2 By Road.* [11] and [12]

METHODS OF MANUFACTURE

Nitrous oxide is obtained in commercial quantities by the thermal decomposition of ammonium nitrate and by recovery from a by-product stream from adipic acid manufacturing processes.

REFERENCES

[1] *United States Pharmacopeia/National Formulary,* current ed., United States Pharmacopeial Convention, Inc., 12601 Twinbrook Parkway, Rockville, MD 20852.

[2] CGA 6-8.2, *Commodity Specification for Nitrous Oxide,* Compressed Gas Association, Inc., 1235 Jefferson Davis Highway, Arlington, VA 22202.

[3] *Threshold Limit Values and Biological Exposure Indices,* 1989–90 ed., American Conference of Governmental Industrial Hygienists, 6500 Glenway Avenue, Bldg. D-7, Cincinnati, OH 45211-4438.

[4] CGA G-8.1, *Standard for Nitrous Oxide Systems at Consumer Sites,* Compressed Gas Association, Inc., 1235 Jefferson Davis Highway, Arlington, VA 22202.

[5] NFPA 99, *Standard for Health Care Facilities,* National Fire Protection Association, Batterymarch Park, Quincy MA 02269.

[6] *Effects of Exposure to Toxic Gases—First Aid and Medical Treatment,* 3rd ed., Matheson Gas Products, Inc., Secaucus, NJ 07094. (1988)

[7] *Code of Federal Regulations,* Title 49 CFR Parts 100–199 (Transportation), Superintendent of Documents, U.S. Government Printing Office, Washington, DC 20402.

[8] *Transportation of Dangerous Goods Regulations,* Canadian Government Publishing Centre, Supply and Services Canada, Ottawa, Ontario, Canada K1A 0S9.

[9] CGA S-1.1, *Pressure Relief Device Standards—Part I—Cylinders for Compressed Gases,* Compressed Gas Association, Inc., 1235 Jefferson Davis Highway, Arlington, VA 22202.

[10] CGA V-1, *American National, Canadian and Compressed Gas Association Standard for Compressed Gas Cylinder Valve Outlet and Inlet Connections,* (ANSI/CSA/CGA V-1) Compressed Gas Association, Inc., 1235 Jefferson Davis Highway, Arlington, VA 22202.

[11] CSA B620, *Highway Tanks and Portable Tanks for the Transportation of Dangerous Goods,* Canadian Standards Association, 178 Rexdale Blvd., Rexdale (Toronto), Ontario, Canada M9W 1R3.

[12] CSA B622, *CSA Standard for the Selection and Use of Tank Trucks, Tank Trailers and Portable Tanks for the Transportation of Dangerous Goods for Class 2 By Road,* Canadian Standards Association, 178 Rexdale Blvd., Rexdale (Toronto), Ontario, Canada M9W 1R3.

Oxygen

Chemical Symbol: O_2
CAS Registry Number: 7782-44-7
DOT Classification: Nonflammable gas
DOT Label: Oxidizer
Transport Canada Classification: 2.2 (5.1)
UN Number: UN 1072 (compressed gas); UN 1073 (refrigerated liquid)

PHYSICAL CONSTANTS

	U.S. Units	SI Units
Chemical formula	O_2	O_2
Molecular weight	31.9988	31.9988
Density of the gas at 70°F (21.1°C) and 1 atm	0.08279 lb/ft³	1.326 kg/m³
Specific gravity of the gas at 70°F (21.1°C) and 1 atm (air = 1)	1.105	1.105
Specific volume of the gas at 70°F (21.1°C) and 1 atm	12.08 ft³/lb	0.7541 m³/kg
Boiling point at 1 atm	−297.33°F	−182.96°C
Freezing point at 1 atm	−361.80°F	−218.78°C
Critical temperature	−181.43°F	−118.57°C
Critical pressure	731.4 psia	5043 kPa abs
Critical density	27.22 lb/ft³	436.1 kg/m³
Triple point	−361.82°F at 0.02147 psia	−218.79°C at 0.1480 kPa abs
Latent heat of vaporization at boiling point	91.7 Btu/lb	213 kJ/kg
Latent heat of fusion at −361.1°F (−218.4°C) melting point	5.959 Btu/lb	13.86 kJ/kg
Specific heat of the gas at 70°F (21.1°C) and 1 atm		
C_p	0.2197 Btu/(lb)(°F)	0.9191 kJ/(kg)(°C)
C_v	0.1572 Btu/(lb)(°F)	0.6578 kJ/(kg)(°C)

	U.S. Units	SI Units
Ratio of specific heats (C_p/C_v)	1.40	1.40
Solubility in water, vol/vol at 32°F (0°C)	0.0491	0.0491
Density of the liquid at boiling point	9.52 lb/gal or 71.23 lb/ft^3	1141 kg/m^3
Density of the gas at boiling point	0.2799 lb/ft^3	4.483 kg/m^3
Gas/liquid ratio (gas at 70°F (21.1°C) and 1 atm, liquid at boiling point, vol/vol)	860.5	860.5

DESCRIPTION

Oxygen, the colorless, odorless, tasteless elemental gas that supports life and combustion, constitutes about a fifth of the atmosphere (20.99 percent by volume and 23.2 percent by weight). At temperatures ranging below −300°F (−184°C), it is a transparent, pale blue liquid that is slightly heavier than water. All elements except the inert gases combine directly with oxygen to form oxides. However, oxidation of different elements occurs over a wide range of temperatures. Phosphorus ignites spontaneously in air at ambient temperatures, and the noble metals oxidize only at very high temperatures.

Oxygen is nonflammable but it readily supports combustion. All materials that are flammable in air burn much more vigorously in oxygen. Some combustibles, such as oil and grease, burn with nearly explosive violence in oxygen if ignited.

Oxygen is shipped as a nonliquefied gas at pressures of 2000 psig (13 790 kPa) or above, and also as a cryogenic liquid at pressures and temperatures below 200 psig (1379 kPa) and −232°F (−146.5°C).

GRADES AVAILABLE

Table 1, from CGA G-4.3, *Commodity Specification for Oxygen,* presents the component maxima in parts per million, ppm (mole/mole), unless otherwise shown, for generally available grades of oxygen. [1] These are sometimes denoted as quality verification levels. A blank indicates no maximum limiting characteristic. The absence of a value in a listed grade does not mean to imply that the limiting characteristic is or is not present, but merely indicates that the test is not required for compliance with the specification.

In the United States, oxygen USP must comply with the Federal Food, Drug, and Cosmetic Act as administered by the Food and Drug Administration in Title 21 CFR Parts 200–211, and the *Compressed Medical Gases Guideline.* [2] and [3]

In Canada, oxygen USP must be manufactured, filled and handled in accordance with the Food and Drug Act and Regulations of Health and Welfare Canada and the *Canadian General Standards Board Standard for Identification of Medical Gas Containers, Pipelines and Valves.* [4]

USES

The major uses of oxygen stem from its life-sustaining and combustion-supporting properties. It is used extensively in medicine for therapeutic purposes, for resuscitation in asphyxia, and with other gases in anesthesia. It is also used in high-altitude flying, deep-sea diving, and as both an inhalant and a power source in the United States space program.

Industrial applications include its very wide utilization with acetylene, propane, hydrogen, and other fuel gases for such purposes as metal cutting, welding, hardening,

TABLE 1. OXYGEN GRADES AVAILABLE. (Units in ppm (mole/mole) unless otherwise stated

Limiting Characteristics	Maxima for Type I (Gaseous)							Maxima for Type II (Liquid)				
	A	B	C	G	D	E	F	A	B	G	C	D
Oxygen Min. % (mole/mole)	99.0	99.5	99.5	99.5	99.5	99.6	99.995	99.0	99.5	99.5	99.5	99.5
Nitrogen				100						100		
Odor	None				None			None				None
Water ppm (v/v)			50	2	6.6	8	1.0		6.6	2	26.3	6.6
Dew Point °F			−54.5	−97	−82	−80	−105		−82	−97	−63.5	−82
Water (condensed)		5 ml/ container										
Total Hydrocarbon Content (as methane)				25		50	1.0			25	67.7	
Methane					50							25
Ethane and Other Hydrocarbons (as Ethane)					6							3

Ethylene					0.4							0.2
Acetylene					0.1		0.05				0.5	0.05
Carbon Dioxide	300[a]			5	10		1.0	300[a]		5		5
Carbon Monoxide	10[a]						1.0	10[a]				
Nitrous Oxide				2	4		0.1			2		2
Halogenated Refrigerants					2							1
Solvents					0.2							0.1
Other Components[b]					0.2							0.1
USP	Yes							Yes				
Permanent Particulates											1 mg/L 1mm	

[a]Test is not required when oxygen produced by air liquefacation

[b]Infrared absorbing components not specifically named may be determined from a survey by a dispersive infrared analyzer.

and scarfing. In the steel industry, oxygen helps increase the capacity and efficiency of production in steel and iron furnaces. One of its major uses is in the production of synthesis gas (a hydrogen–carbon monoxide mixture) from coal, natural gas or liquid fuels; synthesis gas is in turn used to make gasoline, methanol, and ammonia. Oxygen is similarly employed in manufacturing some acetylene through partial oxidation of the hydrocarbons in methane-rich feedstocks such as natural gas. It is also used in the production of nitric acid, ethylene, and other compounds.

PHYSIOLOGICAL EFFECTS

The inhalation of gaseous oxygen is used appropriately in many medical emergencies as well as for long-term therapy. Such use, except for emergencies, should be with the advice of a physician.

Inhalation of high concentrations of oxygen for a few hours has not been found to produce harmful effects except for some special classes of patients. Premature infants may suffer permanent visual impairment or blindness from inhalation of oxygen at high concentrations, and their oxygen therapy must be carefully controlled. Patients with chronic obstructive pulmonary disease retain carbon dioxide abnormally. If oxygen is administered to them, raising the oxygen concentration in the blood depresses their breathing and raises their retained carbon dioxide to a dangerous level. [5]

The two systems in adults most likely to be damaged by high concentrations of oxygen are the respiratory and central nervous systems. Sea divers or tunnel makers are the working groups most commonly affected by high pressure oxygen environments.

One hundred percent oxygen at atmospheric pressure can cause pulmonary irritation and edema after 24 hours of exposure. The earliest symptoms are pleuritic substernal pain and dry cough, occurring after only 6 hours. Adult respiratory distress syndrome, which involves interstitial and intra-alveolar fluid extravasation in the lung tissue, follows after 24 to 48 hours. Other known toxic effects include retrolental fibroplasia which has occurred in premature infants exposed to high concentrations of oxygen at birth, retinal circulatory injury, and erythrocyte hemolysis in adults. [6]

When pure oxygen is inhaled at two or more atmospheres, central nervous system (CNS) toxicity supervenes. Symptoms include nausea, vomiting, dizziness or vertigo, muscle twitching, vision changes, and loss of consciousness and generalized seizures. At three atmospheres, CNS toxicity occurs in less than two hours, and at six atmospheres in only a few minutes. Physical exertion shortens the period before toxic symptoms and signs appear. [6]

The toxic effects of oxygen can be attributed to its free radical activity. Biological oxidations and auto-oxidations can convert molecular oxygen to the free radical form, superoxide anion (O_2^-). Other reactive intermediates include hydrogen peroxide (H_2O_2) and hydroxyl free radical ($OH^\cdot$). When these forms react with lipids, potent lipoperoxides form which damage cell membranes and other vital cellular and subcellular structures. [6]

In general, pure oxygen is a local irritant to mucous membranes and, with extended continued exposure, can be destructive to lung tissue. Thus, when oxygen treatment is used to correct hypoxia (low oxygen concentration in the arterial blood), it should be with the minimum concentration of oxygen that will overcome the hypoxia. It should be continued only as long as necessary.

MATERIALS OF CONSTRUCTION

Gaseous oxygen is noncorrosive and may consequently be contained in systems constructed of any common metal and designed to safely withstand the pressures involved. At the temperature of liquid oxygen, ordinary carbon steels and most alloy steels lose their ductility and are therefore considered unsatisfactory for liquid oxygen service. Metals and alloys that have satisfactory ductility at the temperature of liquid oxygen include austenitic stainless steel (i.e., Types 304 and 316), and nickel-chromium alloys, nickel, Monel 400, copper, brasses, bronzes, and aluminum alloys. At various temperatures and pressures, there are substantial differences in the compatibilities of these various metals and alloys with oxygen. These compatibilities must be considered in their selection for service with either gaseous or liquid oxygen.

Care must be taken to remove all oil, grease, and other combustible material from piping systems and containers before putting them into oxygen service. Cleaning methods employed by manufacturers of oxygen equipment are described in CGA G-4.1, *Cleaning Equipment for Oxygen Service,* as well as in Chapter 10. [7]

SAFE STORAGE, HANDLING, AND USE

General precautions for safe handling of gaseous oxygen are contained in CGA G-4, *Oxygen.* [8] For liquid oxygen, a thorough discussion of necessary precautions can be found in CGA P-12, *Safe Handling of Cryogenic Liquids.* [9] Also see Chapter 5.

When liquid oxygen is held in any closed vessel or space, there must be an appropriate pressure relief device because of the very large pressure increases that can occur as the liquid oxygen is vaporized by heat flowing into the container. Liquid oxygen must also be handled with all the precautions required for safety with any cryogenic fluid. In addition, liquid oxygen exhibits strong oxidizing properties. Contact between the skin and liquid oxygen, or uninsulated piping or vessels containing it, can cause severe burnlike injuries.

The ability of oxygen to react with most materials is a hazard. All easily combustible materials, especially hydrocarbon oils and greases, must be kept from contact with high oxygen concentrations. Sources of ignition should be eliminated to the extent possible. Sudden opening of valves, and particles carried by oxygen flowing at high velocity, are to be avoided since these actions can result in ignition. Valves, therefore, should be opened slowly. When this cannot be done, precautions must be taken to prevent severe damage or injury to personnel should an ignition occur. Such safety measures include isolation and the provision of barriers as discussed in ASTM G-88, *Guide for Designing Systems for Oxygen Service.* [10]

High pressure cylinders, tube trailers, and tank cars often serve as the storage supply for gaseous oxygen. Larger amounts of oxygen are frequently stored more economically in liquid form at the user's site. Standards to ensure safety with oxygen systems at user sites are set forth in publications of the National Fire Protection Association. [11], [12], and [13]

Pressure Relief Devices

At least one pressure relief device is required on any compressed gas cylinder or trailer tube offered for transport of oxygen. A pressure relief device must be selected in accordance with CGA S-1.1, *Pressure Relief Device Standards—Part 1—Cylinders for Compressed Gases.* [14]

If the device selected is a CG-1 rupture disk, only one is required regardless of cylinder length. If a CG-4 combination rupture disk/fusible plug (nominal yield temperature 165°F; 73.9°C) is used, one is required at each end if the cylinder exceeds 65 inches in length, exclusive of neck. This requirement also applies if a CG-5 combination rupture disk/fusible plug (nominal yield temperature 212°F; 100°C) is used. The CG-7 pressure relief valve is also authorized for oxygen service up to 500 psig (3447 kPa) charging pressure.

A cargo tank or portable tank for liquid oxygen under pressure must be equipped with a primary system of one or more pressure relief valves and a secondary system of one or more rupture disks or pressure relief valves. Considerations for settings and capacity requirements are to be found in Title 49 of the U.S. *Code of Federal Regulations* and equivalent Canadian regulations. [15] and [16]. Also see CGA S-1.2, *Pressure Re-*

lief Device Standards—Part 2—Cargo and Portable Tanks for Compressed Gases. [17]

Medical Gas Requirements

A manufacturer or filler of medical gases, is required to be registered with the Food and Drug Administration (FDA) as a drug manufacturer. In addition, all medical gas related operations and procedures which are carried out in the plant must meet FDA requirements for Current Good Manufacturing Practices (CGMP) in compliance with the U.S. *Code of Federal Regulations,* Title 21, Parts 210 and 211. [2] A useful publication for guidance in meeting CGMP requirements available from the FDA is entitled *Compressed Medical Gases Guideline.* [3]

DISPOSAL

Gaseous oxygen for which disposal becomes necessary should be vented in a manner that does not create an oxygen-rich atmosphere in a confined space.

When disposing of liquid oxygen, be sure that the liquid and vapors do not come in contact with combustible materials, especially hydrocarbon materials such as oil, grease, and asphalt.

HANDLING LEAKS AND EMERGENCIES

Turn off ignition sources in general area if possible. The source of the leakage should be shut off if possible and the area adequately ventilated. Avoid contact with liquid spills. In the event liquid oxygen is spilled over asphalt or other surfaces contaminated with combustibles, such as oil-soaked concrete or gravel, do not walk on or roll equipment over the area for at least one-half hour after the frost has disappeared. A violent reaction may occur simply by impact or shock. Contact the supplier for assistance.

First Aid

In case of frostbite from contact with liquid oxygen, place the frostbitten part in warm water, about 108°F (42°C). If warm water is not available, or is impractical to use, wrap the affected part gently in blankets. Let the circulation re-establish itself naturally. Encourage the victim to exercise the affected part while it is being warmed. Consult a physician. [6]

METHODS OF SHIPMENT

Under the appropriate regulations, oxygen is authorized for shipment as follows (gaseous oxygen, except where liquid oxygen is noted):

By Rail: In cylinders and tube tank cars, for both gaseous and liquid oxygen (special insulated cylinders and tank cars for liquid oxygen are not subject to DOT regulations at pressures below 25 psig (172 kPa)). In Canada, transportation of oxygen is regulated regardless of pressure.

By Highway: In cylinders on trucks and in tube trailers. Liquid oxygen in insulated cylinders on trucks and cargo tanks.

By Water: In cylinders on cargo and passenger vessels and on ferry and railroad car ferry vessels (passenger or vehicle). In authorized tank cars on cargo vessels only. Pressurized liquid oxygen, in insulated cylinders on cargo and passenger vessels and on ferry and railroad car ferry vessels (passenger or vehicle).

By Air: A maximum net weight of oxygen per cylinder of 150 lb (68 kg) aboard passenger aircraft and of 300 lb (136 kg) aboard cargo aircraft. Liquid oxygen, either pressurized or nonpressurized, is not accepted for air shipment under present regulations.

CONTAINERS

The shipment of liquid or pressurized gaseous oxygen is under the jurisdiction of the U.S. Department of Transportation or Transport Canada and must be done in containers meeting specifications of the applicable regulations. [15] and [16] Gaseous oxygen is authorized for shipment in approved cylinders, tank cars, and tube trailers. Oxygen is shipped as a cryogenic liquid in insulated cylinders, tank trucks, and tank cars.

Oxygen service requires certain restrictions for container design and materials, and applicable regulations should be consulted.

Cylinders

Cylinders meeting TC/DOT specifications 3A, 3AA, or 3 AL are the types usually used to ship gaseous oxygen, but oxygen is authorized for shipment in any cylinders designated for nonliquefied compressed gases. These include cylinders that comply with DOT specifications 3A, 3AA, 3AX, 3AAX, 3AL, 3B, 3E, 3T, 4B, 4BA, 4BW, and 39. In addition, cylinders meeting DOT specifications 3, 3C, 3D, 4, 4A, 4C, 7, 25, 26, 33, and 38 may be continued in gaseous oxygen service, but new construction is not authorized. Some new higher-strength steel cylinders and fiber-wrapped aluminum cylinders are authorized under DOT exemptions.

Liquid oxygen is authorized for shipment in cylinders which meet TC/DOT specification 4L.

Most cylinders authorized for gaseous oxygen service must be requalified by hydrostatic retest every five or, under more restricted service requirements, ten years. No periodic retest is required for cylinders of Specifications 3C, 3E, 4C, and 7. Cylinders of Specification 4L authorized for liquid oxygen service are also exempt from periodic retest requirements. See 49 CFR 173.34. [15]

Types of pressure relief devices used on oxygen cylinders include the CG-1 rupture disk, the CG-4 and CG-5 combination rupture disk/fusible plug devices, and the CG-7 pressure relief valve. [14] See CGA S-1.1, *Pressure Relief Device Standards—Part 1—Cylinders for Compressed Gases,* for further information and requirements.

Valve Outlet Connections

Standard connections in the United States and Canada are as follows: for service pressure up to 3000 psig (20 684 kPa), Connection CGA 540; for 3001 to 4000 psig (20 691 to 27 579 kPa), Connection CGA 577; for 4001 to 5500 psig (27 586 to 37 921 kPa), Connection CGA 701; and for cryogenic liquid withdrawal, Connection CGA 440. The standard pin-indexed yoke connection in the United States and Canada (for medical use of the gas) is Connection CGA 870. Further information on connections can be found in CGA V-1, *American National, Canadian, and Compressed Gas Association Standard for Compressed Gas Cylinder Valve Outlet and Inlet Connections.* [18]

Small Portable Liquid Containers

Liquid oxygen is packaged in small portable containers that hold quantities ranging from 1 to more than 25 gallons (4 to 90 L). These containers, generally termed dewars, are encased in steel shells and are vacuum insulated; they maintain the liquid at atmospheric pressure and are consequently not subject to DOT regulations.

Tube Trailers and Cargo Tanks on Trucks or Trailers

Gaseous oxygen at pressures up to 2640 psig (18 202 kPa) is shipped in tube trailers with capacities that may exceed 40 000 ft^3 (1133 m^3). These trailers commonly serve as the storage supply for the user, with empty trailers being replaced periodically by the supplier.

Most liquid oxygen is shipped in bulk at pressures below 25 psig (172 kPa) in special insulated cargo tanks on trucks and truck trailers with capacities up to or in excess of 400 000 ft^3 (11 327 m^3).

Tank Cars

Gaseous oxygen is authorized for rail shipment in tank cars meeting DOT specification 107A.

Liquid oxygen is also shipped in special insulated tank cars at pressures below 25 psig (172 kPa) meeting Specification AAR 204W. [19]

Filling Limits

The maximum filling limits authorized for gaseous oxygen in shipment are as follows:

In cylinders and tube trailers, the authorized service pressures marked on the cylinders or tube assemblies at 70°F (21.1°C). In the case of cylinders of Specifications 3A, 3AA, 3AX, and 3AAX that meet special requirements, up to 10 percent in excess of their marked service pressures. See 49 CFR 173.302 (c). [15]

In uninsulated tank cars of the Specification 107A type, the gas pressure at 130°F (54.4°C) must not exceed seven-tenths of the marked test pressure of the tank.

For TC/DOT 4L cylinders, filling limits for liquid oxygen as expressed in percent water capacity by weight are given in terms of the pressure control valve start-to-discharge setting. For TC/DOT 4L cylinders with a design service temperature of −320°F (−195°C), filling limits range from 86 percent for a start-to-discharge setting of 625 psig (4309 kPa) to 108 percent for a start-to-discharge setting of 45 psig (310 kPa). See 49 CFR 173.317 for further information and requirements. [15]

Liquid oxygen shipped in insulated truck cargo tanks or in other insulated containers at pressures below 25 psig (172 kPa) is not subject to DOT regulations except placarding, shipping papers, and incident reporting. See 49 CFR 173.320. [15]

METHODS OF MANUFACTURE

Commercial oxygen is produced at air separation plants by liquefaction of atmospheric air and separation of the oxygen by fractionation. Very small quantities are produced by the electrolysis of water.

Oxygen ranging in purity between 90 and 97 percent is produced by pressure swing absorption (PSA) techniques. PSA oxygen is used chiefly in steel making, but there are also many small units in homes producing therapeutic oxygen. Recent innovations in membrane technology are also being used for production of oxygen from the air.

REFERENCES

[1] CGA G-4.3, *Commodity Specification for Oxygen,* Compressed Gas Association, Inc., 1235 Jefferson Davis Highway, Arlington, VA 22202.

[2] *Code of Federal Regulations,* Title 21 CFR Parts 200, 201, 207, and 211, (Food and Drugs), Superintendent of Documents, U.S. Government Printing Office, Washington, DC 20402.

[3] *Compressed Medical Gases Guideline,* Division of Drug Quality Compliance (HFN-320), U.S. Food and Drug Administration, Room 9B-09, 5600 Fishers Lane, Rockville, MD 20857.

[4] *Canadian General Standards Board Standard for Identification of Medical Gas Containers, Pipelines and Valves,* (Health and Welfare Canada), Canadian Government Publishing Centre, Supply and Services Canada, Ottawa, Ontario, Canada K1A 0S9.

[5] Osol, Arthur, Pratt, Robertson, and Altschule, Mark, *United States Dispensatory and Physicians Pharmacology,* 26th ed., Lippincott, Philadelphia.

[6] *Effects of Exposure to Toxic Gases—First Aid and Medical Treatment,* 3rd ed., Matheson Gas Products, Inc., Secaucus, NJ 07094. (1988)

[7] CGA G-4.1, *Cleaning Equipment for Oxygen Service,* Compressed Gas Association, Inc., 1235 Jefferson Davis Highway, Arlington, VA 22202.

[8] CGA G-4, *Oxygen,* Compressed Gas Association, Inc., 1235 Jefferson Davis Highway, Arlington, VA 22202.

[9] CGA P-12, *Safe Handling of Cryogenic Liquids,* Compressed Gas Association, Inc., 1235 Jefferson Davis Highway, Arlington, VA 22202.

[10] ASTM G-88, *Guide for Designing Systems for Oxygen Service,* ASTM, 1916 Race Street, Philadelphia, PA 19103.

[11] NFPA 51, *Standard for the Design and Installation of Oxygen-Fuel Gas Systems for Welding and Cutting,* National Fire Protection Association, Batterymarch Park, Quincy, MA 02269.

[12] NFPA 99, *Standard for Health Care Facilities,* National Fire Protection Association, Batterymarch Park, Quincy, MA 02269.

[13] NFPA 50, *Standard for Bulk Oxygen Systems at Consumer Sites,* National Fire Protection Association, Batterymarch Park, Quincy, MA 02269.

[14] CGA S-1.1, *Pressure Relief Device Standards—Part 1—Cylinders, for Compressed Gases,* Compressed Gas Association, Inc., 1235 Jefferson Davis Highway, Arlington, VA 22202.

[15] *Code of Federal Regulations,* Title 49 CFR Parts 100–199 (Transportation), Superintendent of Documents, U.S. Government Printing Office, Washington, DC 20402.

[16] *Transportation of Dangerous Goods Regulations,* Canadian Government Publishing Centre, Supply and Services Canada, Ottawa, Ontario, Canada KIA 0S9.

[17] CGA S-1.2, *Pressure Relief Device Standards—Part 2—Cargo and Portable Tanks for Compressed Gases,* Compressed Gas Association, Inc., 1235 Jefferson Davis Highway, Arlington, VA 22202.

[18] CGA V-1, *American National, Canadian and Compressed Gas Association Standard for Com-*

pressed Gas Cylinder Valve Outlet and Inlet Connections, (ANSI/CSA/CGA V-1) Compressed Gas Association, Inc., 1235 Jefferson Davis Highway, Arlington, VA 22202.

[19] *AAR Specifications for Tank Cars* (M-1002), Association of American Railroads, 50 F Street, Washington, DC 20001.

ADDITIONAL REFERENCES

Batino, R., ed., *IUPAC Solubility Data Series,* Vol. 7, "Oxygen and Ozone," Pergamon Press, New York, 1981.

CGA G-4.4, *Industrial Practices for Gaseous Oxygen Transmission and Distribution Piping Systems,* Compressed Gas Association, Inc., 1235 Jefferson Davis Highway, Arlington, VA 22202.

CGA P-6, *Standard Density Data, Atmospheric Gases and Hydrogen,* Compressed Gas Association, Inc., 1235 Jefferson Davis Highway, Arlington, VA 22202.

CGA P-14, *Accident Prevention in Oxygen-Rich and Oxygen-Deficient Atmospheres,* Compressed Gas Association, Inc., 1235 Jefferson Davis Highway, Arlington, VA 22202.

Daubert, T. E., and R. P. Danner, *Data Compilation Tables of Properties of Pure Compounds,* Design Institute for Physical Property Data, American Institute of Chemical Engineers, 345 East 47th Street, New York, NY 10017, 1985.

Younglove, B. A., *Interactive Fortran Program to Calculate Thermophysical Properties of Six Fluids,* National Bureau of Standards Technical Note 1048 (July 1982).

Phosgene

Chemical Symbol: $COCl_2$
Synonyms: Carbonyl chloride, carbon oxychloride
CAS Registry Number: 75-44-5
DOT Classification: Poison A
DOT Label: Poison gas
Transport Canada Classification: 2.3 (8)
UN Number: UN 1076

PHYSICAL CONSTANTS

	U.S. Units	SI Units
Chemical formula	$COCl_2$	$COCl_2$
Molecular weight	98.92	98.92
Vapor pressure		
at 70°F (21.1°C)	9.1 psig	63 kPa
at 105°F (40.6°C)	29.9 psig	206 kPa
at 115°F (46.1°C)	38.1 psig	263 kPa
at 130°F (54.4°C)	52.1 psig	359 kPa
Density of the gas		
at 70°F (21.1°C) and 1 atm	0.26 lb/ft^3	4.16 kg/m^3
Specific gravity of the gas		
at 68°F (20°C) and 1 atm		
(air = 1)	3.5	3.5
Specific volume of the gas		
at 70°F (21.1°C) and 1 atm	3.9 ft^3/lb	0.24 m^3/kg
Density of the liquid		
at 68°F (20.0°C)	86.65 lb/ft^3	1388 kg/m^3
at 70°F (21.1°C)	86.47 lb/ft^3	1385 kg/m^3
at 105°F (40.6°C)	83.47 lb/ft^3	1337 kg/m^3
at 115°F (46.1°C)	82.66 lb/ft^3	1324 kg/m^3
at 130°F (54.4°C)	81.41 lb/ft^3	1304 kg/m^3
Boiling point at 1 atm	46.76°F	8.200°C
Freezing point at 1 atm	−198.0°F	−127.8°C
Critical temperature	359.4°F	181.9°C
Critical pressure	823 psia	5674 kPa abs
Critical density	32.5 lb/ft^3	521 kg/m^3
Latent heat of vaporization		
at boiling point	106.2 Btu/lb	247.0 kJ/kg

	U.S. Units	SI Units
Latent heat of fusion at melting point	24.95 Btu/lb	58.03 kJ/kg
Specific heat of the gas at 77°F (25°C) and 1 atm		
C_p	0.1393 Btu/(lb)(°F)	0.5832 kJ/(kg)(°C)
C_v theoretical	0.1189 Btu/(lb)(°F)	0.4978 kJ/(kg)(°C)
Ratio of specific heats, C_p/C_v	1.171	1.171
Solubility in water, vol/vol	hydrolyzes	hydrolyzes
Weight of the liquid at 68°F (20°C)	11.57 lb/gal	1386 kg/m^3
Specific gravity of the liquid at 68°F (20°C) (water = 1)	1.388	1.388
Heat capacity of the liquid at boiling point	0.244 Btu/(lb)(°F)	1.02 kJ/(kg)(°C)
Coefficient of expansion at 32°F (0°C)	0.002207 per °F	0.003973 per °C

DESCRIPTION

Phosgene is a nonflammable, colorless gas more than three times as heavy as air, and is designated a Class A poison or of the "extremely dangerous" group by the U.S. Department of Transportation. Phosgene under pressure is a colorless to light yellow liquid. It has a characteristic odor which is often stifling or suffocating and strong, but sometimes not unpleasant, depending on the concentration; it has been said to smell like sour green corn or moldy hay when diluted with air. Phosgene vapors strongly irritate the eyes.

Completely dry, pure phosgene is stable at ordinary temperatures. It dissociates into its component parts, carbon monoxide and chlorine, at elevated temperatures, to an extent ranging from 0.45 percent dissociation at 214°F (101.1°C) to 100 percent at 1472°F (800°C).

Phosgene is slightly soluble in water and is slowly hydrolyzed by water to form corrosive hydrochloric acid and carbon dioxide.

In other solvents, phosgene dissolves as follows (parts per 100 parts of solvent by weight, at 1 atm and 68°F (20°C), or as indicated): carbon tetrachloride, 28; chloroform, 59; glacial acetic acid, 62; Russian mineral oil, 35; chlorinated paraffin, 81; ethyl acetate, 98; benzene, 99; toluene (at 63°F or 17.2°C), 244; xylene (at 54°F or 12.2°C), 457; and chlorobenzene (at 54°F or 12.2°C), 422.

GRADES AVAILABLE

Phosgene is available for commercial and industrial use from various suppliers at a typical minimum purity of 99 mole percent.

USES

Phosgene is used mainly as an intermediate in the manufacture of many types of compounds, including: barbiturates; chloroformates and thiochloroformates; carbamoyl chlorides, acid chlorides, and acid anhydrides; carbamates; carbonates and pyrocarbonates; urethanes; ureas; azo-urea dyes, triphenylmethane dyes, and substituted benzophenones; isocyanates and isothiocyanates; carbazates and carbohydrazides; malonates; carbodiimides; and oxazolidinediones. It is also used in bleaching sand for glass manufacture and as a chlorinating agent.

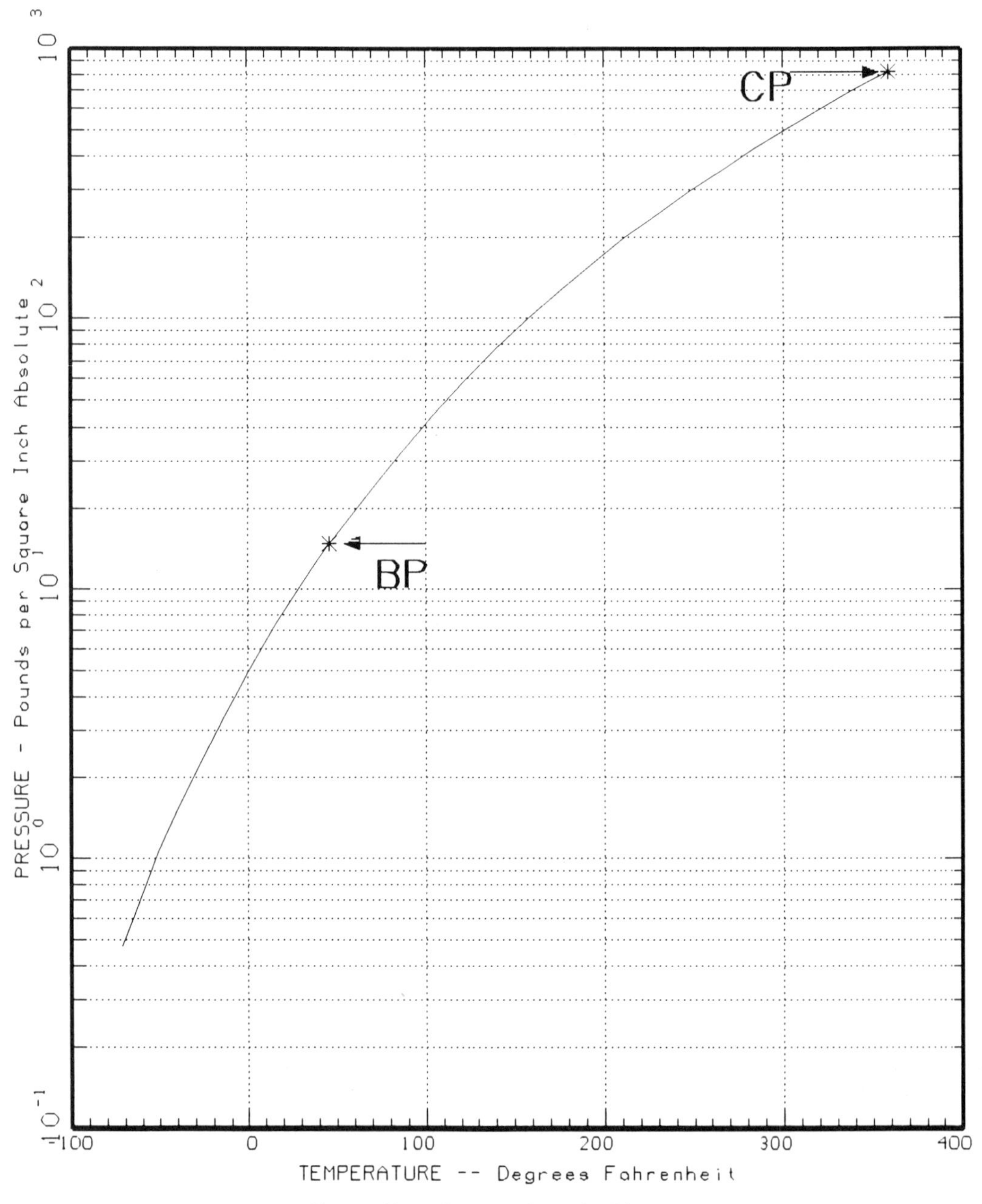

Fig. 1. Vapor Pressure Curve for Phosgene.

PHYSIOLOGICAL EFFECTS

Phosgene is a strong lung irritant and also attacks other parts of the respiratory system. Low concentrations in air cause watering of the eyes and coughing which may result in a thin, frothy expectoration. It will also result in upper respiratory tract irritation and bronchitis. High concentrations cause greater distress, such as shortness of breath, choking, coughing, chest tightness, and painful breathing. Physiological responses to various concentrations of phosgene are shown in Table 1.

A time-weighted average threshold limit value (TLV-TWA) for phosgene of 0.1 ppm (0.4 mg/m^3) for an 8-hour workday has been adopted by the U.S. Occupational Safety and Health Administration (OSHA) and the American Conference of Governmental Industrial Hygienists. [1] and [2]

One serious difficulty with the treatment of persons exposed to phosgene is that more serious symptoms may not appear until 2–24 hours after the exposure. These include bloody sputum, increasing shortness of breath, pulmonary edema, and respiratory failure. The delayed action of phosgene can be particularly injurious if the victim performs heavy exercise after having been exposed. Persistent effects after acute lung injury from phosgene can include bronchiolitis obliterans.

Phosgene is hydrolyzed to hydrogen chloride and chlorine in the alveoli of the lungs which in turn cause irritation and result in pulmonary edema and subsequently leads to respiratory and cardiac failure.

All persons who have been exposed to phosgene must be examined by a physician as soon as possible, because serious symptoms may develop subsequently.

TABLE 1. PHYSIOLOGICAL RESPONSE TO PHOSGENE GAS.

Response	Concentration (ppm)
Cough, eye, and throat irritation within 1 minute	3–5
Irritation of eyes and respiratory tract in less than 1 minute	10
Severe lung injury within 1–2 minutes	20
Rapidly fatal (30 minutes or less)	50–90

Adapted from *Patty's Industrial Hygiene and Toxicology,* 3rd rev. ed., Wiley, New York, 1983, p. 4128.

MATERIALS OF CONSTRUCTION

Anhydrous phosgene in the liquid state is compatible with a variety of common metals, including aluminum (of 99.5 percent purity), copper, pure iron or cast iron, steel (including cast steel and chrome-nickel steels), lead (up to 250°F or 121.1°C), nickel, and silver; it is also compatible with platinum and platinum alloys in instruments. Nonmetallic materials with which liquid anhydrous phosgene is also compatible include acid-resistant linings (ceramic plates and carbon blocks), enamel on cast iron or glass-lined steel, Jena special glass (as well as Pyrex or Kimax), porcelain, quartzware, granite or basalt natural stone, stoneware, and Teflon.

In the presence of moisture, phosgene is not compatible with copper, steel, or pure or cast iron. Detailed data on the corrosion resistance of various materials to phosgene under a range of conditions are given in *Corrosion Data Survey—Metals Section* of the National Association of Corrosion Engineers. [3]

For commercial (nonlaboratory) applications, steel piping with seamless fittings is recommended for handling phosgene and pipe no smaller than $\frac{3}{4}$-inch (1.9-cm) nominal size should be used to ensure rigidity and minimize possible leaks. For pipe size up to 4 inches (10.2 cm), Schedule 80 seamless (or alloy steel to ASTM A333 GR3) piping is recommended; 6-inch (15.2-cm) diameter Schedule 40 seamless may be used as a larger pipe size.

Screwed or flanged joints should be kept to a minimum, and cast iron or malleable iron fittings and valves should not be used; nonarmored porcelain valves must not be used, regardless of the pressure with either liquid or gaseous phosgene. Only outside yoke or rising stem valves are recommended to reduce the possibility of accident; nonindicating valves should not be used. Monel is the material generally used in manually operated valves for the disk, seat, and stem components.

A pipe joint compound composed of litharge and glycerin is recommended, or Teflon envelope for flat gaskets, depending on the temperature. Detailed recommendations on these and other materials for various purposes in phosgene service may be obtained from phosgene suppliers.

SAFE STORAGE, HANDLING, AND USE

Since phosgene is a highly toxic gas, appropriate precautions must be taken in its storage and handling as with any poison gas. It is imperative to prevent moisture from entering any closed phosgene container because of the formation of hydrochloric acid and carbon dioxide resulting in possible corrosion of the container. Reaction products, such as hydrogen and carbon dioxide, in combination with a weakened container, may result in its rupture. Refer to Chapter 5 for more comprehensive general recommendations on storage and handling of compressed gases.

DISPOSAL

Phosgene may be disposed of by neutralization with alkalis such as agricultural lime (CaO), crushed limestone ($CaCO_3$), or sodium bicarbonate ($NaHCO_3$), or alkali solutions. The resulting products should be discarded in conformance with all federal, state, provincial, and local environmental regulations.

HANDLING LEAKS AND EMERGENCIES

Field neutralization of phosgene in emergencies is accomplished with alkali or alkali solutions. The reaction of phosgene with an ammonia solution (which forms urea) is particularly effective.

Leak Detection

Suspected leaks should be investigated only by personnel who are wearing positive-pressure, self-contained or air-supplied full face masks, of a design approved by the U.S. National Institute for Occupational Safety and Health.

Personnel wearing masks can detect phosgene leaks with ammonia vapor devices, as phosgene produces white fumes in the presence of ammonia. Other detection devices, such as detector tubes, and specialized leak detection instruments may also be used.

First Aid

Transport the victim immediately to an uncontaminated atmosphere. Victims should not be allowed to talk or otherwise exert themselves. Summon a physician to examine any person exposed to phosgene. Then take the following steps:

(1) Remove contaminated clothing.

(2) Administer oxygen by mask, if person is breathing. If breathing has ceased, give artificial respiration, preferably with simultaneous administration of oxygen.

(3) Have the patient lie in a flat position. The head may be elevated. Warm patient with blankets.

(4) Keep the patient at rest. An occasional change of position from lying to sitting may be beneficial. Keep the patient calm and have him or her try to suppress desires to cough.

Seek immediate professional medical assistance for further treatment. [4] and [5]

METHODS OF SHIPMENT

Under the appropriate regulations, phosgene is authorized for shipment as follows: [6] and [7]

By Rail: In cylinders and in TMU tank cars.

By Highway: In cylinders on trucks, and in TMU tanks on trucks and on full or semi-trailers (provided that tanks are securely chocked or clamped to prevent movement).

By Water: In cylinders and TC/DOT specification 106A500X TMU tank cars authorized by the DOT on cargo vessels only.

By Air: Forbidden.

CONTAINERS

Phosgene is authorized for shipment by the TC/DOT in cylinders and in TMU tanks of TC/DOT specification 106A500X, which usually have net capacities of 150 lb (or less) and 2000 lb (68 kg or less and 907 kg), respectively. See 49 CFR 173.328 and 173.333. [6]

Filling Limits

The maximum filling densities permitted for phosgene are as follows:

In cylinders—125 percent (percent of water capacity by weight) plus the requirement that the cylinder must not contain more than 150 lb (68 kg) of phosgene.

In TMU tanks (Specification 106A500X)—not liquid full at 130°F (54.4°C).

Cylinders

Cylinders that meet TC/DOT specifications 33 or 3D and that are of not over 125 lb (56.7 kg) water capacity (nominal) are authorized for the shipment of phosgene. Use of existing 33 or 3D cylinders is authorized, but no new construction is authorized. TC/DOT 3A, 3AA, 3AL, or 3E cylinders with a minimum pressure rating of 1800 psig (12 410 kPa) may also be used in phosgene service. Unless Specification 3D or 33 cylinders have valve protection extension rings, the cylinders must be packed in wooden boxes as prescribed; if gaskets are used between the caps and cylinder necks, they must be renewed for each shipment even though they may appear to be in good condition. Specification 3E1800 cylinders must be packaged in wooden or metal boxes.

Each filled cylinder must show absolutely no leakage in an immersion test made before shipment. Cylinders must be tested without their valve protection caps. For the test, the cylinder and valve must be kept submerged in a bath of water heated to approximately 150°F (65.6°C) for at least 30 minutes, and frequent examinations must be made during that time to note any escape of gas. The cylinder valve must not be loosened after the test and before shipment.

Whether boxed or protected with caps or cap protection rings, the unit must be able to pass a drop test (6 ft (1.8 m) drop onto a concrete floor impacting at the weakest point). See 49 CFR 173.327. [6]

Valves must be either a nonperforated diaphram type (with outlets sealed with a solid metal cap or plug) or a packed type provided the assembly is made gas-tight by means of a seal cap to the valve body or to the cylinder.

Pressure relief devices are prohibited for use on phosgene cylinders.

All the precautions necessary for the safe handling, shipping and storage of any compressed gas cylinder must be observed with phosgene cylinders.

The 3D, 3A1800, 3AA1800, 3AL1800, and 33 types of cylinders authorized for phosgene service must be requalified by hydrostatic retest every five years under present regulations. For 3E1800 cylinders no hydrostatic retest is required.

Warming Cylinders to Help Remove Contents. Phosgene cylinders may be heated by warm air or warm water to facilitate removal of their contents. ***Never use steam, boiling water, or direct flame to remove contents.*** Never under any circumstances allow the outside of a cylinder to reach temperatures above 125°F (51.7°C).

Valve Outlet Connections

The standard valve outlet connection in the United States and Canada for phosgene cylinders is Connection CGA 660. The limited standard connection is Connection CGA 160. [8]

TMU Tanks

Only TMU tanks that meet TC/DOT specification 106A500X are authorized for phosgene service. These tanks may be shipped either by rail or by motor vehicle. The tanks must be equipped with gas-tight valve protection caps approved by the Bureau of Explosives, and they must not be equipped with pressure relief devices of any type. See 49 CFR 179.302(a). [6]

METHOD OF MANUFACTURE

Phosgene is produced commercially by passing chlorine and carbon monoxide (in excess) over activated carbon as a catalyst under carefully controlled conditions.

REFERENCES

[1] *Code of Federal Regulations,* Title 29 CFR Parts 1900–1910 (Labor), Superintendent of Documents, U.S. Government Printing Office, Washington, DC 20402.

[2] *Threshold Limit Values and Biological Exposure Indices,* 1989–90 ed., American Conference of Governmental Industrial Hygienists, 6500 Glenway Avenue, Bldg. D-7, Cincinnati, OH 45211-4438.

[3] *Corrosion Data Survey—Metals Section,* 6th ed., National Association of Corrosion Engineers, P.O. Box 218340, Houston, TX 77218.

[4] *Matheson Gas Data Book,* 6th ed., Matheson Gas Products, Inc., Secaucus, NJ 07094. (1980)

[5] *Effects of Exposure to Toxic Gases—First Aid and Medical Treatment,* 3rd ed., Matheson Gas Products, Inc., Secaucus, NJ 07094. (1988)

[6] *Code of Federal Regulations,* Title 49 CFR Parts 100–199, (Transportation), Superintendent of Documents, U.S. Government Printing Office, Washington, DC 20402.

[7] *Transportation of Dangerous Goods Regulations,* Canadian Government Publishing Centre, Supply and Services Canada, Ottawa, Ontario, Canada K1A 0S9.

[8] CGA V-1, *American National, Canadian, and Compressed Gas Association Standard for Compressed Gas Cylinder Valve Outlet and Inlet Connections* (ANSI/CSA/CGA V-1), Compressed Gas Association, Inc., 1235 Jefferson Davis Highway, Arlington, VA 22202.

Phosphine

Chemical Symbol: PH_3
Synonyms: Hydrogen phosphide, phosphorous hydride, phosphorated hydrogen
CAS Registry Number: 7803-51-2
DOT Classification: Poison A
DOT Label: Poison gas and Flammable gas
Transport Canada Classification: 2.3 (2.1)
UN Number: UN 2199

PHYSICAL CONSTANTS

	U.S. Units	SI Units
Chemical formula	PH_3	PH_3
Molecular weight	33.998	33.998
Vapor pressure		
at 70°F (21.1°C)	583 psig	4020 kPa
at 105°F (40.6°C)	914 psig	6302 kPa
at 115°F (46.1°C)	1015 psig	6998 kPa
Density of the gas at 70°F (21.1°C) and 1 atm	0.0877 lb/ft^3	1.405 kg/m^3
Specific gravity of the gas at 77°F (25°C) and 1 atm (air = 1)	1.184	1.184
Density of the liquid		
at 32°F (0°C)	48.4 lb/ft^3	623 kg/m^3
at 59°F (15°C)	45.2 lb/ft^3	581 kg/m^3
at 122°F (50°C)	30.4 lb/ft^3	391 kg/m^3
Boiling point at 1 atm	−125.9°F	−87.7°C
Melting point at 1 atm	−207.4°F	−133°C
Critical temperature	124.9°F	51.6°C
Critical pressure	947.9 psia	6535.8 kPa abs
Critical density	18.76 lb/ft^3	300.5 kg/m^3
Latent heat of vaporization at 77°F (25°C)	88 Btu/lb	204.7 kJ/kg
Heat capacity (gas at 1 atm) C_p	0.261 Btu/(lb)(°F)	1.093 kJ/(kg)(°C)
Solubility in water, vol gas/vol water at 62.6°F (17°C)	0.26	0.26

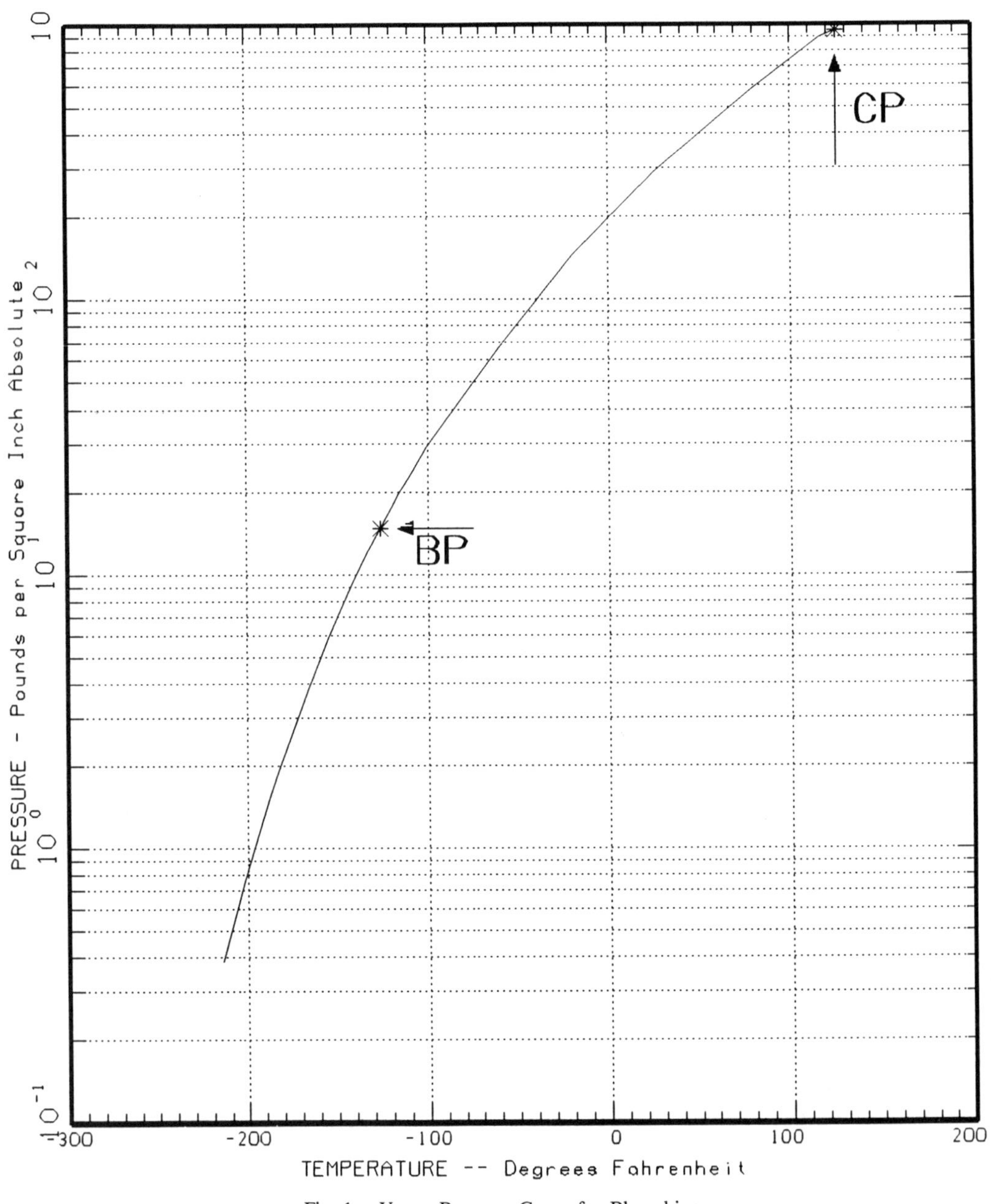

Fig. 1. Vapor Pressure Curve for Phosphine.

DESCRIPTION

Phosphine is a colorless, flammable gas that is heavier than air and has a characteristic odor described as being similar to decaying fish. Pure phosphine is claimed to be odorless, even at a level of 200 ppm. The odor threshold for commercially available phosphine ranges from 0.02 to 3 ppm. It has an autoignition temperature of 100°F (37.8°C) and ignites spontaneously when traces of other phosphorous hydrides, such as diphosphine, are present. For all practical purposes, phosphine should be handled both as a pyrophoric and highly toxic gas.

Phosphine is stable at room temperature and begins to decompose at about 707°F (375°C), with complete decomposition at about 1100°F (600°C). Phosphine is readily oxidized by common oxidizers such as potassium permanganate and sodium hypochlorite. Unlike arsine, it will have some reaction with the alkalis. Phosphine is a strong reducing agent and can precipitate a number of heavy metals from solutions of their salts. It will react violently with oxidizers such as oxygen, chlorine, fluorine, and nitric oxide.

Phosphine is shipped in the pure form as a liquefied gas, and is also commonly available as a mixture when blended with hydrogen or inert gases.

GRADES AVAILABLE

Phosphine is supplied in a number of grades, primarily as electronic grade, with a purity of 99.999 percent on a hydrogen-free basis. Typical maximum impurities by volume are:

Arsine	2 ppm
Carbon Dioxide	1 ppm
Carbon Monoxide	0.5 ppm
Total Hydrocarbons	1 ppm
Nitrogen	3 ppm
Oxygen	1 ppm
Water	1 ppm

An MOCVD grade is also offered, with a purity of 99.9998 percent. Typical maximum impurities by volume are:

Arsine	0.1 ppm
Carbon Dioxide	0.1 ppm
Carbon Monoxide	0.1 ppm
Total Hydrocarbons	0.1 ppm
Nitrogen	0.1 ppm
Water	0.1 ppm

USES

Phosphine is used in a variety of organic preparations and in the preparation of phosphonium halides. It is commonly used (in gas mixtures) as a doping agent for *n*-type semiconductors, and in the manufacture of light-emitting diodes.

PHYSIOLOGICAL EFFECTS

Phosphine is a highly toxic gas that can cause death from delayed pulmonary edema or from tissue anoxia secondary to interference with tissue respiration. Phosphine is both an irritant and a general systemic poison. Its action is similar to that of hydrogen sulfide.

Symptoms of irritation include lacrimation, substernal chest pain and chest tightness, shortness of breath, a slight cough, and cyanosis. Nonlethal exposures can result in symptoms referable to the gastrointestinal tract and the nervous system. Abdominal symptoms include nausea, vomiting, severe epigastric pain, and diarrhea. Neurologic symptoms include vertigo, headache, restlessness, intentional tremor, lack of muscular coordination, double vision, drowsiness, and a decreased sensation in the extremities. Death in humans has occurred after exposures as low as 8 ppm for 1–2 hours. [1]

Additional acute toxic symptoms involve cardiac abnormalities, liver dysfunction, and kidney inflammation. Agitated psychotic behavior can occur. For additional information, refer to *Effects of Exposure to Toxic Gases—First Aid and Medical Treatment.* [1]

The American Conference of Governmental Industrial Hygienists has adopted a time-weighted average threshold limit value (TLV-TWA) for phosphine of 0.3 ppm for an eight-hour workday, with a 15-minute short-term exposure limit (STEL) of 1 ppm. [2] The U.S. Occupational Safety and Health Administration similarly has adopted a time-weighted average exposure limit of 0.3 ppm (0.4 mg/m^3) for phosphine for an eight-hour period and a 1 ppm (1 mg/m^3) short-term exposure limit for a 15-minute period. [3]

MATERIALS OF CONSTRUCTION

Pure phosphine appears to be noncorrosive to common metals. Steel is usually recommended for use with phosphine; however, Type 316 and 316L stainless steel is the metal of choice in the semiconductor industry. In all cases, systems should be adequately designed to withstand the pressures to be encountered.

SAFE STORAGE, HANDLING, AND USE

Phosphine should be stored in a cool, well-ventilated, dry area equipped to handle flammable and toxic gases. Storage locations should be built in conformance with applicable building codes and local regulations. Phosphine cylinders should be isolated from other noncompatible gases, such as oxidants and corrosive gases, either by adequate distance (20 ft minimum) or fire-resistant separators having at least a 1-hour fire rating.

DISPOSAL

Phosphine is usually treated in scrubbers for disposal by oxidation with potassium permanganate or sodium hypochlorite solutions in accordance with the following reactions:

$$PH_3 + 2KMnO_4 \rightarrow Mn_2O_3 + K_2HPO_4 + H_2O$$

$$PH_3 + 4NaOCl \rightarrow H_3PO_4 + 4NaCl$$

Phosphine may also be adsorbed on charcoal containing special catalysts, where it is subsequently oxidized. Other dry-type scrubbers containing proprietary adsorbents are supplied by specialty gas and equipment suppliers.

In all cases the waste reaction products must be disposed of in accordance with federal, state, provincial and local regulations.

HANDLING LEAKS AND EMERGENCIES

Leaks

It is recommended that storage areas be monitored for leakage and that emergency procedures be developed for handling leaks. Employees should be specially trained in handling phosphine cylinders to minimize chances of leakage due to unwarranted abuse and mishandling. Suspected leaks should be investigated by trained personnel who are wearing positive pressure, self-contained or air-supplied full face masks, of a design approved by the National Institute for Occupational Safety and Health.

Leaks may be detected by special hydride detectors, chemically treated paper tapes that will change color when exposed to phosphine, detector tubes, and other detection devices. If leaks are detected, the main source of gas supply should be shut off. Evacuate or purge the system with inert gas to ensure that those repairing the leak cannot be exposed to phosphine. After repairing the leak, pressurize the system with an inert gas such as helium, and check to see that leaks have been corrected before introducing phosphine into the system.

Emergencies

If a phosphine fire should occur, shut off the source of gas if this can be done without risk. If the source cannot be shut off, let the fire burn itself out while making sure that cylinders and combustible materials in the area are kept cool by spraying with water. Keep the area well ventilated and monitor for the presence of phosphine to ensure that those handling the emergency are not exposed to dangerous concentrations of phosphine.

First Aid

Individuals exposed to phosphine at concentrations exceeding the TLV, or those exhibiting symptoms of phosphine exposure, should receive professional medical care immediately. First aid consists of moving the exposed person (with minimum exertion) to an uncontaminated area and administering pure oxygen, bronchodilators (such as albuterol) and inhalant anticholinergics (such as Atrovent) if chest symptoms occur. [1]

METHODS OF SHIPMENT

Under the appropriate regulations, phosphine is authorized for shipment as follows: [4] and [5]

By Rail: In TC/DOT approved cylinders.

By Highway: In TC/DOT approved cylinders.

By Water: On-deck stowage on cargo vessels only; shipment aboard passenger vessels is prohibited. Water shipments require segregation, the same as for flammable gases.

By Air: Forbidden.

Specification 3AL cylinders containing phosphine may only be shipped by highway and rail.

CONTAINERS

Cylinders

Pure phosphine and its mixtures are shipped in TC/DOT approved cylinders in accordance with regulations governing the shipment of Poison A materials. Shipment is permitted in specification 3A1800, 3AA1800, 3AL1800, or 3E1800 cylinders. Use of existing specification 33 or 3D cylinders is also authorized, but new construction is not authorized.

Specification 3A, 3AA, and 3AL cylinders must not exceed 125 lb water capacity. Cylinders must have valve protection or be packed in strong wooden or metal boxes as described in 49 CFR 173.327(a)(2). [4] Specification 3E1800 cylinders must be packed in strong wooden or metal boxes.

The use of pressure relief devices on phosphine cylinders is prohibited. [6]

Filling Limits

Cylinders must be filled so that they are not liquid full at 130°F (54.4°C). Valves utilized for cylinders of phosphine are usually of the diaphragm type. When shipped, the outlet must be sealed with a threaded cap or threaded solid plug. If shipped with a packed type valve, the valve must be fitted with a seal cap to contain leakage around the stem, or the complete valve must be sealed to the cylinder with a gasketed gas-tight cap. See 49 CFR 173.327 and 173.328 for general packaging requirements for Poison A materials. [4]

Valve Outlet Connections

The standard valve outlet in the United States and Canada for phosphine cylinders is Connection CGA 350. The limited standard valve outlet is Connection CGA 660. [7]

METHODS OF MANUFACTURE

Phosphine can be prepared by the reaction of concentrated (30–40 percent) sodium hydroxide solution on yellow phosphorous. When heated, it reacts to form sodium hypophosphite and phosphine:

$$3NaOH + 4P + 3H_2O \rightarrow 3NaH_2PO_2 + PH_3$$

It can also be prepared by disproportionation of phosphorous acid or hypophosphorous acid, or by direct combination of the elements under pressure.

REFERENCES

[1] *Effects of Exposure to Toxic Gases—First Aid and Medical Treatment,* 3rd ed., Matheson Gas Products, Inc., Secaucus, NJ 07094. (1988)

[2] *Threshold Limit Values and Biological Exposure Indices,* 1989–90 ed., American Conference of Governmental Industrial Hygienists, 6500 Glenway Avenue, Bldg. D-7, Cincinnati, OH 45211-4438.

[3] *Code of Federal Regulations,* Title 29 CFR Parts 1900–1910, (Labor), Superintendent of Documents, U.S. Government Printing Office, Washington, DC 20402.

[4] *Code of Federal Regulations,* Title 49 CFR Parts 100–199, (Transportation), Superintendent of Documents, U.S. Government Printing Office, Washington, DC 20402.

[5] *Transportation of Dangerous Goods Regulations,* Canadian Government Publishing Centre, Supply and Services Canada, Ottawa, Ontario, Canada K1A 0S9.

[6] CGA S-1.1, *Pressure Relief Device Standards—Part 1—Cylinders for Compressed Gases,* Compressed Gas Association, Inc., 1235 Jefferson Davis Highway, Arlington, VA 22202.

[7] CGA V-1, *American National, Canadian, and Compressed Gas Association Standard for Compressed Gas Cylinder Valve Outlet and Inlet Connections,* (ANSI/CSA/CGA V-1), Compressed Gas Association, Inc., 1235 Jefferson Davis Highway, Arlington, VA 22202.

ADDITIONAL REFERENCES

Gas Encyclopaedia, L'Air Liquide, Elsevier/North Holland Inc., 52 Vanderbilt Avenue, New York, NY 10017. (1976)

Matheson Gas Data Book, 6th ed., Matheson Gas Products, Inc., Secaucus, NJ 07094. (1980)

Matheson Unabridged Gas Data Book, Matheson Gas Products, Inc., Secaucus, NJ 07094.

Rare Gases: Krypton, Neon, Xenon

	Krypton	Neon	Xenon
Chemical Symbol:	Kr	Ne	Xe
CAS Registry Number:	7439-90-9	7440-01-9	7440-63-3
DOT Classification:	Nonflammable gas	Nonflammable gas	Nonflammable gas
DOT Label:	Nonflammable gas	Nonflammable gas	Nonflammable gas
Transport Canada Classification:	2.2	2.2	2.2
UN Number:			
(compressed gas):	UN 1056	UN 1065	UN 2036
(refrigerated liquid):	UN 1970	UN 1913	UN 2591

PHYSICAL CONSTANTS

Krypton

	U.S. Units	SI Units
Chemical formula	Kr	Kr
Molecular weight	83.80	83.80
Density of the gas at 70°F (21.1°C) and 1 atm	0.2172 lb/ft^3	3.479 kg/m^3
Specific gravity of the gas at 70°F (21.1°C) and 1 atm (air = 1)	2.899	2.899
Specific volume of the gas at 70°F (21.1°C) and 1 atm	4.604 ft^3/lb	0.287 m^3/kg
Boiling point at 1 atm	−244.0°F	−153.4°C
Melting point at 1 atm	−251°F	−157°C
Critical temperature	−82.8°F	−63.8°C
Critical pressure	798.0 psia	5502 kPa abs
Critical density	56.7 lb/ft^3	908 kg/m^3
Triple point	−251.3°F at 10.6 psia	−157.4°C at 73.2 kPa abs
Latent heat of vaporization at boiling point	46.2 Btu/lb	107.5 kJ/kg
Latent heat of fusion at triple point	8.41 Btu/lb	19.57 kJ/kg
Specific heat of the gas at 70°F (21.1°C) and 1 atm		

	U.S. Units	SI Units
C_p	0.060 Btu/(lb)(°F)	0.251 kJ/(kg)(°C)
C_v	0.035 Btu/(lb)(°F)	0.146 kJ/(kg)(°C)
Ratio of specific heats, C_p/C_v	1.69	1.69
Solubility in water, vol/vol at 68°F (20°C)	0.0594	0.0594
Weight of the liquid at boiling point	20.15 lb/gal	2415 kg/m^3
Density of the liquid at boiling point	150.6 lb/ft^3	2412.38 kg/m^3
Gas/liquid volume ratio (liquid at boiling point, gas at 70°F and 1 atm, vol/vol)	693.4	693.4

PHYSICAL CONSTANTS

Neon

	U.S. Units	SI Units
Chemical formula	Ne	Ne
Molecular weight	20.183	20.183
Density of the gas at 70°F (21.1°C) and 1 atm	0.05215 lb/ft^3	0.83536 kg/m^3
Specific gravity of the gas at 70°F (21.1°C) and 1 atm (air = 1)	0.696	0.696
Specific volume of the gas at 70°F (21.1°C) and 1 atm	19.18 ft^3/lb	1.197 m^3/kg
Boiling point at 1 atm	−410.9°F	−246.0°C
Melting point at 1 atm	−415.6°F	−248.7°C
Critical temperature	−379.8°F	−228.8°C
Critical pressure	384.9 psia	2654 kPa abs
Critical density	30.15 lb/ft^3	483 kg/m^3
Triple point	−415.4°F at 6.29 psia	−248.6°C at 43.4 kPa abs
Latent heat of vaporization at boiling point	37.08 Btu/lb	86.3 kJ/kg
Latent heat of fusion at triple point	7.14 Btu/lb	16.6 kJ/kg
Specific heat of the gas at 70°F (21.1°C) and 1 atm		
C_p	0.25 Btu/(lb)(°F)	1.05 kJ/(kg)(°C)
C_v	0.152 Btu/(lb)(°F)	0.636 kJ/(kg)(°C)
Ratio of specific heats, C_p/C_v	1.64	1.64
Solubility in water, vol/vol at 68°F (20°C)	0.0105	0.0105
Weight of the liquid at boiling point	10.07 lb/gal	1207 kg/m^3
Density of saturated vapor at 1 atm	0.5862 lb/ft^3	9.390 kg/m^3

	U.S. Units	SI Units
Density of the gas at boiling point	0.6068 lb/ft^3	9.7200 kg/m^3
Density of the liquid at boiling point	75.35 lb/ft^3	1207 kg/m^3
Gas/liquid volume ratio (liquid at boiling point, gas at 70°F and 1 atm, vol/vol)	1445	1445

PHYSICAL CONSTANTS

Xenon

	U.S. Units	SI Units
Chemical formula	Xe	Xe
Molecular weight	131.3	131.3
Density of the gas at 70°F (21.1°C) and 1 atm	0.3416 lb/ft^3	5.472 kg/m^3
Specific gravity of the gas at 70°F and 1 atm (air = 1)	4.560	4.560
Specific volume of the gas at 70°F (21.1°C) and 1 atm	2.927 ft^3/lb	0.183 m^3/kg
Boiling point at 1 atm	−162.6°F	−108.2°C
Melting point at 1 atm	−168°F	−111°C
Critical temperature	61.9°F	16.6°C
Critical pressure	847.0 psia	5840 kPa abs
Critical density	68.67 lb/ft^3	1100 kg/m^3
Triple point	−169.2°F at 11.84 psia	−111.8°C at 81.6 kPa abs
Latent heat of vaporization at boiling point	41.4 Btu/lb	96.3 kJ/kg
Latent heat of fusion at triple point	7.57 Btu/lb	17.6 kJ/kg
Specific heat of the gas at 70°F (21.1°C) and 1 atm		
C_p	0.038 Btu/(lb)(°F)	0.269 kJ/(kg)(°C)
C_v	0.023 Btu/(lb)(°F)	0.096 kJ/(kg)(°C)
Ratio of specific heats, C_p/C_v	1.667	1.667
Solubility in water, vol/vol at 68°F (20°C)	0.108	0.108
Weight of the liquid at boiling point	25.51 lb/gal	3057 kg/m^3
Density of the liquid at boiling point	190.8 lb/ft^3	3057 kg/m^3
Gas/liquid volume ratio (liquid at boiling point, gas at 70°F and 1 atm, vol/vol)	558.5	558.5

DESCRIPTION

Krypton, neon, and xenon are rare atmospheric gases. Each is odorless, colorless, tasteless, nontoxic, monatomic, and chemically inert. All three together constitute less than 0.002 percent of the atmosphere, with approximate concentrations in the atmosphere of 18 ppm for neon, 1.1 ppm for krypton, and 0.09 ppm for xenon. Few users of the three gases need them in bulk quantities, and the three are shipped most often in single cylinders and glass liter flasks.

Radon, a radioactive rare gas, is not treated in this book because it has little or no practical application at present. It is the heaviest gas known (density at 70°F and 1 atm, 0.61 lb/ft^3; at 21.1°C and 1 atm, 9.8 kg/m^3).

Among the rare gases, neon, krypton, and xenon in particular ionize at lower voltages than other gases, and the brilliant, distinctive light they emit while conducting electricity in the ionized state accounts for one of their primary uses. Their characteristic colors as ionized conductors are: red for neon, yellow-green for krypton, and blue to green for xenon. Similarly, argon and helium are also used for this purpose and emit red or blue for argon and yellow for helium. These latter two gases are treated in separate monographs.

GRADES AVAILABLE

Neon, krypton, and xenon are available in various grades for industrial, medical, and advanced-technology use. Neon is available with a minimum purity ranging from 75 to 99.999 mole percent. Krypton and xenon are each available with a minimum purity ranging from 99.95 to 99.997 mole percent.

USES

Neon, krypton, and xenon are used principally to fill lamp bulbs and tubes. The electronics industry uses them singly or in mixtures in many types of gas-filled electron tubes (among them, voltage regulator tubes, starter tubes, phototubes, counter tubes, T.R. tubes, xenon thryatron tubes, half-wave xenon rectifier tubes, and Geiger-Muller tubes). Large quantities of neon (as well as of atmospheric helium and specially purified argon) are employed as fill gases in illuminated signs. Small quantities of krypton and xenon are used for special effects.

In the lamp industry, the three gases serve as fill gas in specialty lamps, neon glow lamps, 100-watt fluorescent lamps, ultraviolet sterilizing lamps, and very-high-output lamps. The three gases have additional applications in the atomic energy field as fill gas for ionization chambers, bubble chambers, gaseous scintillation counters, and other detection and measurement devices.

PHYSIOLOGICAL EFFECTS

Neon, krypton, and xenon are nontoxic and largely inert. They can act as simple asphyxiants by displacing air, thereby diluting the concentration of oxygen below levels necessary to support life. Inhalation in excessive concentrations can result in dizziness, nausea, vomiting, loss of consciousness, and death. Death may result from errors in judgment, confusion, or loss of consciousness which prevents self-rescue. At low oxygen concentrations, unconsciousness and death may occur in seconds without warning.

MATERIALS OF CONSTRUCTION

Gaseous neon, krypton, and xenon are noncorrosive and inert, so they may be contained in systems constructed of any common metals and designed to withstand safely the pressures involved. At the temperatures encountered with liquid neon, krypton, and xenon, ordinary carbon steels and most alloy steels lose their ductility and are considered unsafe for use with these cryogenic liquids. Satisfactory materials for use with liquid neon, krypton, and xenon include Type 18-8 stainless steel and other austenitic nickel-chromium alloys, copper, Monel, brass, and aluminum.

SAFE STORAGE, HANDLING, AND USE

Gaseous neon, krypton, and xenon must be handled with all the precautions necessary for safety with any nonflammable, nontoxic compressed gas.

All precautions necessary for the safe handling of any gas liquefied at very low temperatures must be observed with liquid neon, krypton, and xenon. Extensive tissue damage or burns can result from exposure to liquid neon, krypton, or xenon or their cold vapors.

CGA P-1, *Safe Handling of Compressed Gases in Containers,* provides basic guidelines and requirements for the safe handling and storage of compressed gas cylinders. [1] Refer also to Chapter 5 for general recommendations. Also refer to CGA P-12, *Safe Handling of Cryogenic Liquids,* for information concerning safe handling of neon, krypton, and xenon in liquid form. [2] Another useful reference concerning inert gases is CGA P-14, *Accident Prevention in Oxygen-Rich and Oxygen-Deficient Atmospheres.* [3]

DISPOSAL

When disposal becomes necessary, vent neon, krypton, and xenon gas slowly to a well-ventilated outdoor location remote from personnel work areas and building air intakes. Do not dispose of any residual neon, krypton, and xenon in compressed gas cylinders. Return cylinders to the supplier with residual pressure, the cylinder valve tightly closed, and the valve caps in place.

Allow liquid neon, krypton, and xenon to evaporate in well-ventilated outdoor locations which are remote from work areas.

HANDLING LEAKS AND EMERGENCIES

Leaks should be quickly repaired, particularly in confined spaces. Leakage of cryogenic liquids is of particular concern because of the large quantity of gas resulting from vaporization of the liquid. In such cases, rapid asphyxiation can take place. Areas in which cryogenic liquids will be handled should be well ventilated and leaks quickly repaired. Cryogenic liquid or pressure in lines should be eliminated before leaks are repaired.

Avoid contact of the skin with liquid neon, krypton, and xenon or their cold boil-off vapor.

First Aid

Since the rare gases are inert, they can cause asphyxiation due to displacement of oxygen in the atmosphere. Rescuers wearing a self-contained breathing apparatus or air-line respirator should move the affected person from the hazardous exposure to fresh air at once. If supplemental oxygen is available, administer by nasal canula or mask. Perform artificial respiration if the person is not breathing. Persons who have been unconscious should be taken to a hospital for evaluation and care. [4]

In case of frostbite from exposure to liquid neon, krypton, or xenon, the frostbitten part should be placed in warm water of about 108°F (42°C). If warm water is not available, or it is impractical to use, wrap the affected part gently in blankets. Let circulation reestablish itself naturally. Encourage the victim to exercise the affected part while it is being warmed. Consult a physician. [4]

METHODS OF SHIPMENT

Under the appropriate regulations, neon, krypton, and xenon are authorized for shipment as follows: [5] and [6]

By Rail: In cylinders.

By Highway: In cylinders.

By Water: In cylinders via cargo and passenger vessels, and in ferry and railroad car ferry vessels (either passenger or vehicle).

By Air: For gaseous neon, krypton, and xenon, aboard passenger aircraft in appropriate cylinders up to 150 lb (68 kg) maximum net weight per cylinder, and aboard cargo aircraft in appropriate cylinders up to 300 lb (136 kg) maximum net weight per cylinder. Liquefied neon, krypton, or xenon, either pressurized or low pressure, is authorized for shipment by passenger aircraft up to 100 lb (45.4 kg) and by cargo aircraft only up to 1100 lb (499 kg) maximum net weight per container.

CONTAINERS

Neon, krypton, and xenon are shipped most often in individual cylinders or in liter quantities in glass flasks, since they are seldom used in bulk quantities. Neon is shipped as a liquefied, cryogenic gas in special insulated containers.

Filling Limits

The maximum authorized filling limit for neon, krypton, and xenon in approved types of cylinders is the marked service pressure of the cylinder at 70°F (21.1°C) (or, in the case of Specification 3A and 3AA cylinders meeting additional specified requirements, 10 percent in excess of the marked service pressure). Because commercial grades of xenon deviate markedly in physical properties from the ideal gas at elevated pressures, maximum filling limits for the specific commercial grade of xenon being used must be determined experimentally.

Cylinders

Neon, krypton, and xenon are authorized for shipment in cylinders of any type approved by Transport Canada or DOT for nonliquefied compressed gases. See Title 49 of the U.S. *Code of Federal Regulations,* Parts 100–199. [5] In Canada, the requirements of the *Transportation of Dangerous Goods Regulations* of Transport Canada pertain. [6] Cylinders meeting TC/DOT specifications, 3A, 3AA, 3AL, 3B, 3E, 4B, 4BA, 4BW, and 39 are authorized; also, cylinders meeting Specifications 3, 3C, 3D, 4, 4A, 4C, 25, 26, 33, and 38 may be continued in service with these gases, but new construction is not authorized.

Neon and xenon are shipped in cylinders by rail, highway, water, or air as "Nonflammable compressed gas." Krypton is shipped as "Nonflammable compressed gas, n.o.s."

Under present regulations, cylinders of all types authorized for service with neon, krypton, and xenon must be requalified by hydrostatic test every five years with the following exceptions: DOT-3A and 3AA used exclusively for krypton, neon, and xenon may be retested every ten years instead of every five years under special requirements as given in Title 49 of the U.S. *Code of Federal Regulations,* 49 CFR 173.34(e)(15), or equivalent Transport Canada regulations. [5] and [6] DOT-4 may be retested every ten years; and DOT-3C, 3E, and 4C require no periodic retest.

Pressure relief devices authorized for use on containers of krypton and neon are the Type CG-1 rupture disk, the Type CG-4 and CG-5 combination rupture disk/fusible plugs, and the Type CG-7 pressure relief valve. [7] For xenon, the Type CG-1, CG-4, and CG-7 pressure relief devices are authorized. See CGA S-1.1, *Pressure Relief Device Standards—Part 1—Cylinders for Compressed Gases,* for further information and requirements. [7]

Valve Outlet Connections

Standard valve outlet connections in the United States and Canada for cylinders of krypton, neon, or xenon are Connection CGA 580 for pressures up to 3000 psi (20 684 kPa); Connection CGA 680 for pressures between 3001 and 5500 psig (20 691 and 37 921 kPa); and Connection CGA 677 for pressure between 5501 and 7500 psig (37 928 and 51 711 kPa). Connection CGA 792 is the standard connection for cryogenic liquid withdrawal of neon. [8]

METHOD OF MANUFACTURE

Neon, krypton, and xenon are produced commercially at air separation plants in two stages—an initial stage of partial separation by liquefaction and fractional distillation, and a final purification stage requiring complex processing.

REFERENCES

[1] CGA P-1, *Safe Handling of Compressed Gases in Containers,* Compressed Gas Association, Inc., 1235 Jefferson Davis Highway, Arlington, VA 22202.

[2] CGA P-12, *Safe Handling of Cryogenic Liquids,* Compressed Gas Association, Inc., 1235 Jefferson Davis Highway, Arlington, VA 22202.

[3] CGA P-14, *Accident Prevention in Oxygen-Rich and Oxygen-Deficient Atmospheres,* Compressed Gas Association, Inc., 1235 Jefferson Davis Highway, Arlington, VA 22202.

[4] *Effects of Exposure to Toxic Gases—First Aid and Medical Treatment,* 3rd ed., Matheson Gas Products, Inc., Secaucus, NJ 07094. (1988)

[5] *Code of Federal Regulations,* Title 49 CFR Parts 100–199 (Transportation), Superintendent of Documents, U.S. Government Printing Office, Washington, DC 20402.

[6] *Transportation of Dangerous Goods Regulations,* Canadian Government Publishing Centre, Supply and Services Canada, Ottawa, Ontario, Canada K1A 0S9.

[7] CGA S-1.1, *Pressure Relief Device Standards—Part 1—Cylinders for Compressed Gases,* Compressed Gas Association, Inc., 1235 Jefferson Davis Highway, Arlington, VA 22202.

[8] CGA V-1, *American National, Canadian, and Compressed Gas Association Standard for Compressed Gas Cylinder Valve Outlet and Inlet Connections* (ANSI/CSA/CGA V-1), Compressed Gas Association, Inc., 1235 Jefferson Davis Highway, Arlington, VA 22202.

ADDITIONAL REFERENCES

Cook, G. A., ed., *Argon, Helium, and the Rare Gases,* 2 vols., Wiley, New York, 1961, pp. 435–437.

CGA P-6, *Standard Density Data, Atmospheric Gases and Hydrogen,* Compressed Gas Association, Inc., 1235 Jefferson Davis Highway, Arlington, VA 22202.

Silane

Chemical Symbol: SiH_4
Synonyms: Silicon tetrahydride, monosilane, silicane
CAS Registry Number: 7803-62-5
DOT Classification: Flammable gas
DOT Label: Flammable gas
Transport Canada Classification: 2.3 (2.1)
UN Number: UN 1954 (domestic U.S. shipments, compressed flammable gas n.o.s.)
UN 2203 (international shipments)

PHYSICAL CONSTANTS

	U.S. Units	SI Units
Chemical formula	SiH_4	SiH_4
Molecular weight	32.112	32.112
Vapor pressure at −188°F (−122.2°C)	7.73 psia	53.3 kPa abs
Density of the gas at 32°F (0°C) and 14.7 psia (101.3 kPa abs)	0.0899 lb/ft^3	1.44 kg/m^3
Specific gravity at 70°F (21.1°C) and 14.7 psia (101.3 kPa abs) (air = 1)	1.2	1.2
Specific volume of the gas at 70°F (21.1°C) and 14.7 psia (101.3 kPa abs)	12.0 ft^3/lb	0.749 m^3/kg
Density of the liquid at −301°F (−185°C)	44.4 lb/ft^3	711 kg/m^3
Boiling point at 1 atm	−169°F	−112°C
Melting point at 1 atm	−300.5°F	−184.7°C
Critical temperature	25.8°F	−3.4°C
Critical pressure	703 psia	4842 kPa abs
Critical density	15.4 lb/ft^3	247 kg/m^3
Latent heat of vaporization at −188°F (−122.2°C)	166 Btu/lb	386 kJ/kg

	U.S. Units	SI Units
Heat capacity of the gas at 77°F (25°C) and 14.7 psia (101.3 kPa) C_p	0.5738 Btu/(lb)(°F)	1.334 kJ/(kg)(°C)
Solubility in water	negligible	negligible

DESCRIPTION

Silane is a colorless, spontaneously flammable (pyrophoric) gas. It has a choking odor and may form explosive mixtures with air. Silane will react violently with heavy metal halides and free halogens other than hydrogen chloride.

GRADES AVAILABLE

Silane is sold only in very-high-purity grades primarily for use in the electronics industry. Table 1 gives specification requirements for silane that have been developed and published by Semiconductor Equipment and Materials International. [1]

USES

In the electronics industry, silane is widely used for epitaxial deposition of single-crystal films and for the production of polycrystalline silicon. Important features of silane are that deposition temperatures are relatively low, that is, 1472–1832°F (800–1000°C), and no corrosive by-products are formed.

Other applications of silane include low-temperature chemical vapor deposition of silicon dioxide films by controlled oxidation, chemical vapor deposition of silicon nitride films by controlled reaction with ammonia, and deposition of amorphous silicon films in the production of solar cells and copier drums.

TABLE 1. SEMI STANDARDS FOR SILANE. [1]

	Maximum Acceptable Level (ppm)		
Impurities	Silane Epitaxial Grade (% purity, 98.8897)	Silane Polysilicon and/or Silicon Dioxide Grade (% purity, 98.8937)	Silane Silicon Nitrade Grade (% purity, 99.9417)
Argon and Helium	—	—	40
Carbon monoxide and Carbon dioxide ($CO + CO_2$)	10	10	10
Chlorosilane (ionizable chlorides including HC1 reported as chloride	1000	1000	10
Heavy metals	—	—	—
Hydrocarbons (C_1-C_3)	40	40	10
Hydrogen (H_2)	10 000	10 000	500
Nitrogen (N_2)	40		
Oxygen (O_2)	10	10	10
Particulates	—	—	—
Water (H_2O)	3	3	3
Total Impurities	11 103	11 063	583

Note: The electrical resistivity specification for all three SEMI grades is greater than 100 ohm-cm (n-type).

Silane is well suited for the chemical vapor deposition described in the above paragraphs as it is a gas and can be precisely and reproducibly introduced into reactors. [2] and [3]

PHYSIOLOGICAL EFFECTS

The major hazards of silane stem from its ability to combust spontaneously and its irritating properties. Silane reacts with water to form silicic acid and can therefore cause irritation of the eyes, mucous membranes, and respiratory tract. Inhalation may result in headache, nausea, and irritation of the upper respiratory tract. The offensive odors of silane should be taken as a warning signal for the presence of dangerous concentrations. [4] and [5]

An eight-hour time-weighted average threshold limit value (TLV-TWA) of 5 ppm (7 mg/m^3) for exposure to silane in occupational environments has been adopted by the American Conference of Governmental Industrial Hygienists. [6] Similarly, the U.S. Occupational Safety and Health Administration has adopted an eight-hour time-weighted average exposure limit of 5 ppm (7 mg/m^3). [7]

MATERIALS OF CONSTRUCTION

Piping and equipment for silane service may be of steel or stainless steel construction. Piping and equipment must be designed to withstand the pressures involved. Extreme care must be taken to avoid the contact of silane with materials containing heavy-metal halides or free halogens. Silane will react violently or explosively with these compounds.

Silane systems should be purged of air to prevent silane ignition and contamination with silicon dioxide.

All systems to handle silane should be designed with the following factors in mind:

(1) The prevention of leakage, both in and out, under vacuum and pressure.

(2) The minimum necessary internal volume.

(3) The elimination of dead spaces.

(4) The isolation of system components in case of a leak, rupture, or other failure.

(5) The ability to easily evacuate and purge the system and components with inert gas.

(6) That silane should never be purged through a vacuum pump.

(7) All fittings preferably of the welded face seal gasketed type to minimize the likelihood of leaks.

(8) The use of diaphragm packless valves with resilient seats such as Kel-F.

(9) Removal of backplates from gauges and rotometers where gases may collect upon leakage to allow pressure venting away from personnel if an explosion takes place.

(10) The use of metal diaphragm regulators to minimize air diffusion leakage.

(11) Consideration of the use of flow restrictions in cylinder valves.

SAFE STORAGE, HANDLING, AND USE

With the proper precautions, silane and silane mixtures can be handled safely to avoid contact with air at all times. Almost all recommendations are directed at not violating this one rule. Specific precautions that should be observed in handling silane are as follows: [3] and [8]

(1) Before using silane, read all equipment instructions, cylinder labels, data sheets, and other associated information pertaining to silane and its use.

(2) Handle silane in a well-ventilated area while avoiding the presence of combustible materials.

(3) Store silane at positive pressures.

(4) Do not condense silane by avoiding temperatures of −148°F (−100°C) or less. Condensation of silane runs the risk of leaks developing with subsequent sucking back of air into the system, forming explosive mixtures.

(5) Do not use silane in conjunction with heavy-metal halides or free halogens, with which it will react violently. Care should be taken that all components of any silane-han-

dling system are purged of free halogens that might exist from degreasing agents or chlorinated hydrocarbons.

(6) Evacuate and thoroughly pressure check all systems, preferably with helium, for leaks at pressures two to three times the anticipated working pressure. In addition, a regular leak test procedure and testing schedule should be instituted and followed as part of normal preventive maintenance.

(7) Ground all equipment and lines using silane.

(8) Use an alternate vacuum and inert gas purge of the system to purge all air out of the system after it has been leak checked or opened for any reason.

(9) When pressurizing equipment with silane or a silane mixture, open the cylinder valve slowly. All other equipment adjustments of regulators, needle valves, and so on, should also be made slowly.

(10) Before disconnecting any system that has had silane in it, thoroughly purge the system of silane with an inert gas. Any portion of a system that is dead-ended or allows "pocketing" of silane should be treated by considerable purging, on the order of ten times the trapped volume.

(11) Vent silane or silane mixtures through small-diameter pipe or tubing ending under a shallow water seal to prevent back diffusion of air. Venting should be to an area designed for silane disposal, preferably by burning. Concentrations even in the low percentage range are dangerous and should not be exposed directly to air. Silane can also be vented by diluting with inert gas to prevent ignition upon discharge to the atmosphere.

In addition, all precautions necessary for the safe handling of any toxic flammable gas must be observed. See Chapter 5 for further information on general guidelines.

DISPOSAL

The preferred method of disposal for silane is incineration, accomplished by slowly bleeding silane-containing gases into a continuously burning pilot flame.

HANDLING LEAKS AND EMERGENCIES

Silane will ignite spontaneously on contact with air. Any leaks will cause a fire and may form explosive mixtures with air. If a leak occurs, immediately evacuate all personnel from the danger area. Use a self-contained breathing apparatus and protective clothing to shut off the leak if it can be done without risk. Immediately cool containers with water spray from the maximum possible distance. If the flow of gas cannot be shut off, allow the fire to burn out.

Combustion products can be reduced with a water spray or fog. Ventilate the area of a leak prior to the return of personnel. In some cases, leaks of silane will not ignite immediately. A delay of ignition can take place with devastating force. Storage and use locations should be constructed to prevent worker exposure to such possible detonations. The use of silane mixtures with inert gas or hydrogen at 2 percent or lower silane concentrations will eliminate the pyrophoric hazard in the use of silane.

First Aid

If silane is inhaled, remove the victim to fresh air. If breathing has stopped, give artificial respiration. If breathing is difficult or weak, give oxygen. Call a physician.

Exposure to a high concentration of silane may result in delayed pulmonary edema. Therefore, individuals should be observed for delayed respiratory impairment and treated appropriately if pulmonary edema occurs.

In the event of skin or eye contact, the affected areas should be washed with tap water for 15 minutes. If irritation persists, a physician should be consulted. [5]

METHODS OF SHIPMENT

Under the appropriate regulations, silane is authorized for shipment as follows: [9] and [10]

By Rail: In cylinders.

By Highway: In cylinders on trucks. Also,

in high pressure tube trailers in the United States by DOT exemption; not permitted in Canada.

By Water: In cylinders.

By Air: Consult Title 49 CFR (or equivalent Canadian) regulations.

CONTAINERS

Silane is shipped as a compressed gas in cylinders and, in the United States, also in tube trailers under DOT exemption.

Cylinders

Any cylinders authorized for the shipment of a high-pressure nonliquefied compressed gas may be used in silane service under federal regulations, but cylinders made in accordance with TC/DOT specifications 3A, 3AA, and 3E are probably those most commonly used for silane. See 49 CFR 173.302. [9]

Authorized pressure relief devices for use on silane cylinders are the Type CG-4 combination rupture disk/fusible plug device wherein the fusible metal has a nominal yield temperature of 165°F (74°C). Consult CGA S-1.1, *Pressure Relief Device Standards—Part 1—Cylinders for Compressed Gases,* for further information and requirements. [11]

Valve Outlet Connections

The standard valve outlet connections in the United States and Canada for silane cylinders are Connection CGA 510 for pressures up to 500 psig (3447 kPa) and Connection CGA 350 for pressures up to 3000 psig (20 684 kPa). [12]

METHODS OF MANUFACTURE

There are two basic routes for the commercial preparation of silane. The first involves the acid hydrolysis of magnesium silicide (Mg_2S:) to give a mixture of silicon hydrides which are then separated by fractional distillation. The second method involves the direct or indirect reduction of chlorosilane compounds with hydrogen to give silane. Further purification is generally necessary. [13]

REFERENCES

[1] *1988 SEMI International Standards,* Semiconductor Equipment and Materials International, 805 East Middlefield Road, Mountainview, CA 94043.

[2] Blanchard, R., and Gise, P., *Semiconductor and Integrated Circuit Fabrication Techniques,* Reston Publishing, 1979, pp. 38–41.

[3] Silane Product Information Sheet (F-3492), Union Carbide Corporation, Linde Division, 39 Old Ridgebury Road, Danbury, CT 06817. (1980)

[4] Material Safety Data Sheet for Silane (F-4649), Union Carbide Corporation, 39 Old Ridgebury Road, Danbury, CT 06817. (1980)

[5] *Effects of Exposure to Toxic Gases—First Aid and Medical Treatment,* 3rd ed., Matheson Gas Products, Inc., Secaucus, NJ 07094. (1988)

[6] *Threshold Limit Values and Biological Exposure Indices,* 1989–90 ed., American Conference of Governmental Industrial Hygienists, 6500 Glenway Avenue, Bldg. D-7, Cincinnati, OH 45211-4438.

[7] *Code of Federal Regulations,* Title 29 CFR Parts 1900–1910, (Labor), Superintendent of Documents, U.S. Government Printing Office, Washington, DC 20402.

[8] *Matheson Gas Data Book,* 6th ed., Matheson Gas Products, Inc., Secaucus, NJ 07094. (1980)

[9] *Code of Federal Regulations,* Title 49 CFR Parts 100–199, (Transportation), Superintendent of Documents, U.S. Government Printing Office, Washington, DC 20402.

[10] *Transportation of Dangerous Goods Regulations,* Canadian Government Publishing Centre, Supply and Services Canada, Ottawa, Ontario, Canada K1A 0S9.

[11] CGA S-1.1, *Pressure Relief Device Standards—Part 1—Cylinders for Compressed Gases,* Compressed Gas Association, Inc., 1235 Jefferson Davis Highway, Arlington, VA 22202.

[12] CGA V-1, *American National, Canadian, and Compressed Gas Association Standard for Compressed Gas Cylinder Valve Outlet and Inlet Connections* (ANSI/CSA/CGA V-1), Compressed Gas Association, Inc., 1235 Jefferson Davis Highway, Arlington, VA 22202.

[13] Arkles, B., and Peterson, W. R., "Silicon Compounds (Silanes)," *Kirk-Othmer Encyclopedia of Chemical Technology,* 3rd ed., vol. 20, 1980, pp. 887–909.

ADDITIONAL REFERENCES

Borreson, R. W., and Yaws, C. L., "Physical and Thermodynamic Properties of Silane," *Solid State Technology,* 43–46, 69 (January 1978).

Gases for the Electronics Industry (L-5526), Union Carbide Corporation, Linde Division, 39 Old Ridgebury Road, Danbury, CT 06817. (1983)

Sulfur Dioxide

Chemical Symbol: SO_2
Synonym: Sulfurous acid anhydride
CAS Registry Number: 7446-09-5
DOT Classification: Nonflammable gas
DOT Label: Nonflammable gas
Transport Canada Classification: 2.3
UN Number: UN 1079

PHYSICAL CONSTANTS

	U.S. Units	SI Units
Chemical formula	SO_2	SO_2
Molecular weight	64.06	64.06
Boiling point at 1 atm	14.0°F	−10.0°C
Critical density	32.6 lb/ft^3	520 kg/m^3
Critical pressure	1143 psia	7866.9 kPa abs
Critical temperature	315.4°F	157.4°C
Dielectric constant of the liquid at 32°F (°C)	15.6	15.6
Freezing point at 1 atm	−104.6°F	−75.9°C
Latent heat of fusion at −103.9°F (−75.5°C)	49.70 Btu/lb	115.5 kJ/kg
Latent heat of vaporization at 70°F (21.1°C)	155.5 Btu/lb	361.5 kJ/kg
Liquid density at 70°F (21.1°C) and 49.1 psia (338.5 kPa abs)	86.06 lb/ft^3	1379 kg/m^3
Refractive index of the liquid at 68°F (20°C)	1.410	1.410
Refractive index of the gas at 59°F (15°C)	1.000686	1.000686
Specific conductance of the liquid at 95°F (35°C)	1.5×10^{-8} mhos/cm	1.5×10^{-8} mhos/cm
Specific gravity of the vapor compared to dry air at 32°F (0°C) and 1 atm	2.2638	2.2638

Note: In this monograph, as throughout the Handbook, gauge pressure is denoted by the terms *psig* and *kPa,* whereas absolute pressure is denoted by the terms *psia* and *kPa abs.*

	U.S. Units	SI Units
Specific gravity of the liquid at 32°F (0°C) and 23.7 psia (163.4 kPa abs) compared to water at 39.2°F (4°C)	1.436	1.436
Specific heat of the gas at 77°F (25°C) and 1 atm		
C_p	0.149 Btu/(lb)(°F)	0.622 kJ/(kg)(°C)
C_v	0.116 Btu/(lb)(°F)	0.485 kJ/(kg)(°C)
Ratio of specific heats, C_p/C_v	1.28	1.28
Specific volume of the gas at 70°F (21.1°C) and 1 atm	5.99 ft³/lb	0.374 m³/kg
Surface tension of the liquid at 59°F (15°C) and 39.7 psia (237.7 kPa abs)	1.30×10^{-4} lb/in.	23.64 dynes/cm
Triple point	−103.9°F at 0.2429 psia	−75.5°C at 1.675 kPa abs
Vapor density at 32°F (0°C) and 1 atm	0.1810 lb/ft³	2.927 kg/m³
Viscosity of the liquid at 32°F (0°C) and 28.45 psia (196.16 kPa abs)	2.64×10^{-4} lb/(ft)(sec)	3.929×10^{-4} Pa·s
Viscosity of the vapor at 32°F (0°C) and 14.22 psia (98.05 kPa abs)	7.79×10^{-6} lb/(ft)(sec)	11.593×10^{-6} Pa·s

TABLE 1. PROPERTIES OF SULFUR DIOXIDE AT VARIOUS TEMPERATURES.[a]

Temperature (°F)	Temperature (°C)	Vapor Pressure (psia)	Vapor Pressure (psig)	Volume (ft³/lb) Liquid	Volume (ft³/lb) Vapor	Density (lb/ft³) Liquid	Density (lb/ft³) Vapor	Latent Heat (Btu/lb)
−40	−40.0	3.12	−11.57	0.01044	22.2	95.79	0.04505	178.4
−20	−28.9	5.88	−8.81	0.01062	12.5	94.16	0.08000	174.4
0	−17.8	10.26	−4.43	0.01082	7.35	94.42	0.13605	170.3
10	−12.2	13.3	−1.39	0.01092	5.77	91.58	0.17331	168.3
20	−6.7	16.9	2.2	0.01103	4.59	90.66	0.21786	166.3
30	−1.1	21.3	6.6	0.01114	3.70	89.77	0.27027	164.2
40	4.4	26.6	11.9	0.01125	3.02	88.89	0.33113	162.2
50	10.0	32.9	18.2	0.01137	2.48	87.95	0.40323	160.0
60	15.6	40.3	25.6	0.01149	2.05	87.03	0.48780	157.8
70	21.1	49.1	34.4	0.01162	1.70	86.06	0.58824	155.5
80	26.7	59.8	44.6	0.01175	1.42	85.11	0.70423	153.1
90	32.2	71.0	56.3	0.01189	1.20	84.10	0.83333	150.7
100	37.8	84.1	69.4	0.01204	1.02	83.06	0.98039	148.2
110	43.3	99.1	84.4	0.01219	0.868	82.03	1.15207	145.7
120	48.9	116.3	101.6	0.01235	0.746	80.97	1.34048	143.0
130	54.4	135.8	121.1	0.01251	0.646	79.94	1.54799	140.0
140	60.0	157.7	143.0	0.01269	0.554	78.80	1.80505	137.1

[a]Taken from "Thermodynamic Properties of Sulfur Dioxide," by D. F. Rynning and C. O. Hurd, *Transactions of the American Institute of Chemical Engineers,* **XLI.**

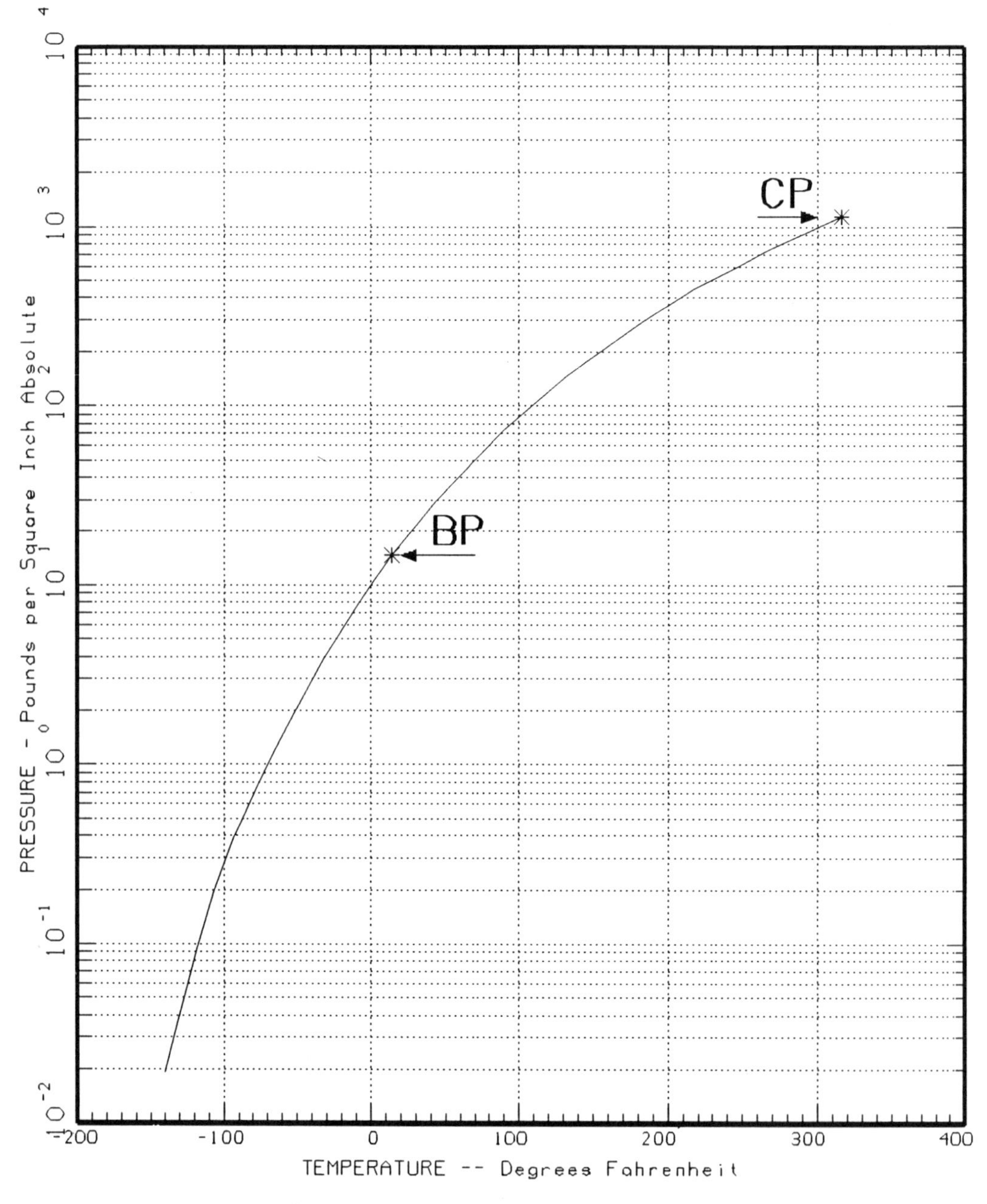

Fig. 1. Vapor Pressure Curve for Sulfur Dioxide.

DESCRIPTION

Sulfur dioxide is a compound formed by the combination of the elements sulfur and oxygen. On a weight basis, the proportion of the elements is about one part sulfur to one part oxygen, or more exactly, 50.05 percent to 49.95 percent, respectively. At standard conditions of temperature and pressure, sulfur dioxide is a colorless gas with a characteristic pungent odor. It may be cooled and compressed to a colorless liquid, which, at one atmosphere pressure, boils at 14°F (−10°C) and freezes at −104.6°F (−75.9°C). Sulfur dioxide liquid is heavier than water, having a specific gravity of 1.436 at 32°F (0°C). As a gas, it is more than twice as heavy as air, its relative density being 2.2638 at atmospheric pressure and 32°F (0°C).

Sulfur dioxide is not flammable or explosive in either the gaseous or liquid state. It is a relatively stable chemical. Temperatures above 3632°F (2000°C) are required to bring about detectable decomposition of sulfur dioxide. Dry sulfur dioxide (less than 100 ppm water) is not corrosive to ordinary metals. However, in the presence of even small amounts of water, sulfur dioxide becomes corrosive to most metals, with exceptions including lead, Type 316 stainless steel, and certain alloys. Glass and certain plastics are also resistant to moist sulfur dioxide.

GRADES AVAILABLE

Sulfur dioxide is available in technical and food grades for use in both commercial and industrial applications.

For food grade sulfur dioxide, specification requirements are as follows:

Sulfur Dioxide	99.9 weight percent minimum*
Arsenic (as As)	3 ppm maximum
Heavy metals (as Pb)	30 ppm maximum
Lead (as Pb)	10 ppm maximum
Nonvolatile residue	0.05 weight percent maximum
Selenium (as Se)	20 ppm maximum
Water	0.05 weight percent maximum

*Note: Percent SO_2 = 100 − (percent non-volatile residue + percent water)

USES

Sulfur dioxide's desirable chemical and physical properties result in its use in a diversity of applications. In addition to its oxidizing and reducing properties, it is an excellent solvent for many chemicals, and it acts as a preservative in numerous food applications. The following industries make use of sulfur dioxide: petroleum, pulp and paper, glass, sweetener, beverage, mining, water and wastewater treatment, clay, electric power, textiles, farming, and chemical. For more specific information, see the current edition of CGA G-3, *Sulfur Dioxide.* [1]

PHYSIOLOGICAL EFFECTS

Exposure to sulfur dioxide gas in low concentrations produces an irritating effect on the mucous membranes of the eyes, nose, throat, and lungs due to the formation of sulfurous acid as the gas comes in contact with the moisture on these surfaces. The effects of sulfur dioxide according to exposure pathway are as follows.

Inhalation

Acute exposure through inhalation may result in dryness and irritation of the nose and throat, choking, sneezing, coughing, and bronchospasm. Severe overexposure may cause death through a systemic acidosis, from pulmonary edema, or from respiratory arrest. Prolonged or repeated exposure may cause impaired lung function, bronchitis, hacking cough, nasal irritation and discharge, increased fatigue, alteration in the senses of taste and smell, and longer duration of common colds. In extreme cases, dental caries, loss of fillings, gum disorders, and the rapid and painless destruction of teeth may result from repeated overexposure.

Skin Contact

Liquid sulfur dioxide can cause frostbite and skin burns, and it converts to sulfurous acid in moist environments, which may cause skin irritation.

Eye Contact

Corneal burns, opacification of the cornea, and blindness may result if liquid sulfur dioxide is splashed in the eyes. Sulfur dioxide can penetrate the intact cornea and cause iritis.

Ingestion

Severe burns to the mouth, throat, and gastrointestinal system may occur.

Exposure Limits

The maximum exposure limits/odor threshold for sulfur dioxide are as follows: The U.S. Occupational Safety and Health Administration (OSHA) has adopted an eight-hour time-weighted average (TWA) exposure limit for sulfur dioxide of 2 ppm (5 mg/m^3) at 77°F (25°C) and 1 atm, which is also the threshold limit value time-weighted average (TLV-TWA) adopted by the American Conference of Governmental Industrial Hygienists (ACGIH). [2] and [3] OSHA has also adopted a short term exposure limit (STEL) of 5 ppm (10 mg/m^3) for a 15-minute period. [2]

Immediately Dangerous to Life and Health (IDLH): 100 ppm (262 mg/m^3) [4]
Odor Threshold: 0.5 ppm (1 mg/m^3) [5]

MATERIALS OF CONSTRUCTION

Service conditions must be defined to properly specify materials of construction for handling sulfur dioxide. It is customary, however, to use carbon steel for dry sulfur dioxide at ambient temperatures. Reference should be made to the ASME Code Section VIII Division 1, latest edition, for vessels, and to ANSI/ASME B31.3, latest edition, for piping specifications. [6] and [7]

Moist sulfur dioxide is corrosive to carbon steel; therefore, other materials of construction have to be considered in this case. A source of data on the corrosivity of sulfur dioxide to various materials is the *Corrosion Data Survey,* published by the National Association of Corrosion Engineers; suppliers of liquid sulfur dioxide are another source. [8]

SAFE STORAGE, HANDLING, AND USE

Prior to handling sulfur dioxide, users should become thoroughly familiar with information contained on the Material Safety Data Sheet (MSDS) provided from the supplier and the current edition of CGA G-3, *Sulfur Dioxide.* [1] In order to prevent injury to personnel or damage to property, only properly trained personnel should be allowed to handle sulfur dioxide.

Personal protective equipment, including

TABLE 2. NATIONAL RESEARCH COUNCIL RECOMMENDED EELs AND CEL FOR SULFUR DIOXIDE.[a]

		ppm	mg/m^3[b]
10-minute	Emergency Exposure Limit	30	79
30-minute	Emergency Exposure Limit	20	52
60-minute	Emergency Exposure Limit	10	26
24-hour	Emergency Exposure Limit	5	13
90-day	Continuous Exposure Limit	1	3

[a]Taken from the National Research Council Committee on Toxicology, "Emergency and Continuous Exposure Limits for Selected Airborne Contaminants," vol. 2, 1984, p. 99.
[b]At 77°F (25°C) and 1 atm.

respiratory equipment and protective clothing, must be provided, used, and maintained in accordance with the applicable provisions of Title 29 of the U.S. *Code of Federal Regulations,* Part 1910, Subpart I, "Occupational Safety and Health Standards." [2]

The Occupational Safety and Health Administration (OSHA) has promulgated a hazard communication standard which is located at 29 CFR 1910.1200. [2] Employees must be given information regarding the adverse health effects of sulfur dioxide; the proper working procedures required to minimize risk to themselves, their co-workers, and the public; and first aid procedures required in the event of overexposure.

For more specific details, including protective equipment and first aid procedures, refer to the supplier's current Material Safety Data Sheet (MSDS) and current edition of CGA G-3, *Sulfur Dioxide.* [1]

DISPOSAL

Due to the complexity and scope of sulfur dioxide disposal procedures, care must be taken to ensure that all existing regulations are complied with. For more detailed information or guidance, a local waste disposal firm or a sulfur dioxide manufacturer should be consulted.

If disposal of sulfur dioxide becomes necessary, such as from a leaking container or vessel, it can be vented to a lime or caustic soda solution. The resulting salt solution should be taken to a plant treating unit for neutralization and disposal.

HANDLING LEAKS AND EMERGENCIES

Emergency Action Plan

It is essential that every facility handling sulfur dioxide have an emergency plan outlining the actions that employees should take in case of specific emergencies. These actions should include alerting fellow employees and area emergency control groups of the nature and extent of the emergency. The plan should also include coordination procedures with area emergency control groups in the event of a major release. The area likely to be affected by various weather conditions and release quantities should be predetermined as accurately as possible. See Chapter 5 for more information on emergency response plans.

Leak Detection

A leak in a sulfur dioxide system can be detected by odor, or sulfur dioxide sensors may be used as an alarm in case of leaks in areas where there are no personnel. A small leak can be located by the dense white fumes resulting when a 10 percent ammonia solution is dispensed from an aspirator or squeeze bottle in the region where the leak is suspected. A leak can also be located by using an ammonia swab prepared by securing a small piece of cloth or sponge to a wire and soaking it in a 10 percent solution of ammonia. When ammonia vapor is passed over points of leaks, dense white fumes will form where the sulfur dioxide and ammonia come in contact.

Action If a Leak Occurs

Only personnel trained for and designated to handle emergencies should attempt to stop a leak. Respiratory equipment of a type suitable for sulfur dioxide must be worn. All persons not so equipped must leave the affected area until the leak has been stopped.

If sulfur dioxide vapor is released, the irritating effect of the vapor will force personnel to leave the area long before they have been exposed to dangerous concentrations. To facilitate their rapid evacuation, there should be sufficient well-marked and easily accessible exits. If, despite all precautions, persons should become trapped in a sulfur dioxide atmosphere, they should breathe as little as possible and open their eyes only when necessary. Partial protection may be gained by holding a wet cloth over the nose and mouth. Since sulfur dioxide vapor is heavier than air, the upper floors of buildings would normally have lower concentrations, but personnel may also get trapped there if assistance is not quickly forthcoming.

Sulfur dioxide is fairly soluble in cool water (less than 100°F or 37.8°C), and therefore the vapor concentration can be reduced by the use of spray or fog nozzles. If water is added to liquid sulfur dioxide, it will cause an increase in evaporation rate (unless the water is very cold). The evaporation will slow after sufficient water is added.

Water should never be sprayed at or into a tank or system which is leaking sulfur dioxide. The presence of water causes sulfur dioxide to be very corrosive, and water directed into a tank would also increase the venting rate.

When possible, leaking containers or vessels can be vented to a lime or caustic soda solution. The reduction in pressure that results from venting will slow the leak and may permit stopping it.

When sulfur dioxide is released to the environment, the appropriate regulatory agency should be notified. Sulfur dioxide does not have a reportable quantity listed in 49 CFR 172.101. [9] In the event of a release however, all state, provincial, municipal, and/or local reporting regulations must be complied with. It is most important that the response groups in the area affected be notified as quickly as possible. If the producer or supplier cannot be reached, the following action should be taken.

In the United States, ask for advice through CHEMTREC, the Chemical Transportation Emergency Center at the Chemical Manufacturers Association in Washington, DC. In the 48 contiguous states, Puerto Rico, Virgin Islands, Alaska, Hawaii, and for Canadian products being transported in the United States, call (toll free) (800) 424-9300. In the District of Columbia and foreign locations (exclusive of Canada), call (202) 483-7616.

In Canada, ask for advice through CANUTEC, Transport of Dangerous Goods Branch, Transport Canada, Ottawa, Ontario, Canada, at (613) 996-6666.

Fire Exposure

A sulfur dioxide container exposed to a fire should be removed. If for any reason it cannot be removed, the container should be kept cool with a water spray until well after the fire is out. Fire-fighting personnel should be equipped with protective clothing and respiratory equipment. For further information regarding handling leaks and emergencies, see DOT P5800, *Emergency Response Guidebook,* published by the U.S. Department of Transportation. [10]

First Aid

Move victims of sulfur dioxide inhalation to fresh air. If breathing has ceased, begin artificial respiration immediately. Administer oxygen if exposure has been severe and breathing is difficult.

Skin exposure first aid treatment includes flushing the contaminated skin with copious amounts of water and continuing as required in order to control burning sensation. Medical attention should be sought if irritation persists or if skin is broken or blistered.

In the event of eye contact, flush eyes immediately with copious amounts of water for at least 15 minutes. Eyelids should be held apart to ensure complete irrigation. Seek medical attention immediately.

Refer to CGA G-3, *Sulfur Dioxide,* for additional information on first aid procedures and medical treatment. [1]

METHODS OF SHIPMENT

Since the transportation of sulfur dioxide as a vapor is not commercially economical, it is shipped and stored as a liquefied gas. Under the appropriate regulations, it is authorized for shipment in approved containers by rail, highway, water, and air. [9] and [11]

By Rail: In single-unit tank cars, multi-unit tank car tanks (ton containers), and portable tanks.

By Highway: In cylinders, cargo tanks, and multi-unit tank car tanks.

By Water: On cargo vessels, preferably below deck stowed away from living quarters. On passenger vessels, in accordance with the requirements of 49 CFR 173.306, 173.304, 173.314, and 173.315. [9]

By Air: On cargo aircraft only, up to 300 lb (136 kg) maximum net weight per container; forbidden on passenger aircraft.

CONTAINERS

Sulfur dioxide must be shipped only in containers which have been manufactured to DOT specifications or in certain containers built in accordance with the American Society of Mechanical Engineers' *ASME Boiler and Pressure Vessel Code* [6], and which are authorized by the DOT and permitted by local jurisdictions. Sulfur dioxide is authorized for shipment in insulated pressure tank cars, cargo tanks (tank trucks), portable tanks, ton containers (multi-unit tank car tanks), and cylinders.

Filling Limits

Because liquid sulfur dioxide expands as its temperature rises, it is possible, with sufficient increase in temperature, for a container to become liquid full. The resulting hydrostatic pressure could cause a container to rupture or to discharge sulfur dioxide through a pressure relief device. Therefore, DOT regulations limit the amount of sulfur dioxide which may be put into containers of various types. These regulations provide that the liquid portion of the commodity must not completely fill a tank car, portable tank, cargo tank, ton container, or cylinder at 130°F (54.4°C). This filling limitation is expressed in terms of maximum filling density for each type of container.

The term *filling density* for liquefied gases is defined as the percent ratio of the weight of gas in a container to the weight of water at 60°F (15.6°C) that the container will hold. Maximum allowable filling density for all sulfur dioxide shipping containers is 125 percent (1.25 kg sulfur dioxide per dm^3). [9]

Cylinders

Sulfur dioxide is authorized for shipment in cylinders conforming to the following minimum TC/DOT specifications: 3A225, 3AA225, 3B225, 3AL225, 4A225, 4B225, 4BA225, 4B240ET, 3, 4, 25, 26–150, 38, 39 and 3E1800. Cylinders that meet DOT specifications 3, 4, 4A225, 25, 26, and 38 may be continued in sulfur dioxide service, but new construction is not authorized. [9] The 150-lb (68-kg) cylinder is the most practical and economical for use. It has a tare weight of about 55 lb (25 kg) and is approximately 10.25 inches (250 mm) outside diameter and 49.5 inches (1260 mm) high, with a water capacity of about 14.72 gallons.

Authorized pressure relief devices for use on sulfur dioxide cylinders are Type CG-2 fusible plugs.

Valve Outlet Connections

The standard valve outlet connection in the United States and Canada for sulfur dioxide cylinders is Connection CGA 660.

Tank Cars

Sulfur dioxide is authorized for shipment in insulated pressure tank cars conforming to a minimum TC/DOT specification 105A200-W. The two common sizes of tank cars used to transport sulfur dioxide have nominal capacities of either 55 tons (50 metric tons) or 90 tons (82 metric tons).

Cargo Tanks (Tank Trucks)

A cargo tank is a container permanently attached to, or forming a part of, any motor vehicle such as a tank truck, trailer, or semitrailer. A sulfur dioxide cargo tank must comply with TC/DOT specifications MC-330 (before May 15, 1967) or MC-331 (later construction). The DOT does not limit the size of cargo tanks, except if shipped on vessels under the jurisdiction of the U.S. Coast

Guard, where the gross weight must not exceed 55 000 lb (24 948 kg). [12] State regulations limit the gross vehicle weight on public roads.

Portable Tanks

Portable tanks used for the transportation of sulfur dioxide must comply with DOT specification 51. They are designed to be temporarily attached to a motor vehicle, vessel, or railroad car other than a tank car, and are equipped with skids, mountings, or accessories to facilitate handling of the tank by mechanical means. If a portable tank is used as a cargo tank, it must comply with all the requirements prescribed for cargo tanks. [9] Compliance with DOT specification 51 means they must have a capacity in excess of 1000 lb (454 kg) of water. If shipped on vessels under the jurisdiction of the U.S. Coast Guard, the gross weight must not exceed 55 000 lb (24 948 kg). [12]

Ton Containers (Multi-unit Tank Car Tanks)

Ton containers for sulfur dioxide service are tanks meeting the requirements of the following specifications: DOT-106A500X, DOT-110A500W, ICCWGA500, ICC-27, and BE-27. These tanks are authorized for transportation by rail freight, highway, and cargo vessel. They have a maximum capacity of 2000 lb (908 kg) of sulfur dioxide. The tare weights range from 1300 to 1650 lb (590 to 748 kg).

METHODS OF MANUFACTURE

Sulfur dioxide is produced commercially by the liquefaction of gases from the combustion of elemental sulfur, by the roasting or smelting of sulfide ores, and from the reaction of liquid sulfur with sulfur trioxide. The gas strength produced varies greatly from case to case, which results in differing energy inputs being required to achieve liquefaction. In the case of weaker gases, a concentration stage is usually necessary prior to liquefaction in order to reduce the energy input.

The gases from typical metallurgical roasting or smelting operations are unsuitable for direct liquefaction, and the cleaned gas is contacted with an absorbing liquid which selectively absorbs the sulfur dioxide. The liquid is then subjected to appropriate temperature and pressure conditions to release the sulfur dioxide gas, which is dried and liquefied. This approach is also used for gases resulting from the combustion of elemental sulfur, although it is possible to produce gas which is sufficiently high in sulfur dioxide content to make direct liquefaction economically feasible.

With the advent of oxygen-enriched flash smelting processes came a resultant increase in sulfur dioxide content of the metallurgical gases produced. In these processes, the sulfur dioxide is generated at a concentration of 70–80 percent, following which it undergoes a series of gas cleaning stages prior to liquefaction.

The reaction of liquid sulfur with sulfur trioxide produces sulfur dioxide gas, which is then liquefied.

REFERENCES

[1] CGA G-3, *Sulfur Dioxide,* Compressed Gas Association, Inc., 1235 Jefferson Davis Highway, Arlington, VA 22202.

[2] *Code of Federal Regulations,* Title 29 CFR Parts 1900–1910 (Labor), Superintendent of Documents, U.S. Government Printing Office, Washington, DC 20402.

[3] *Threshold Limit Values and Biological Exposure Indices,* 1989–90 ed., American Conference of Governmental Industrial Hugienists (ACGIH), 6500 Glenway Ave., Bldg. D-7, Cincinnati, OH 45211.

[4] *NIOSH/OSHA Pocket Guide to Chemical Hazards,* U.S. Dept. of Health, Public Health Service, Center for Disease Control, National Institute for Occupational Safety and Health, Publication No. 780210. NIOSH (Publications), 4676 Columbia Parkway, Cincinnati, OH 45226.

[5] Leonardus, G., Kendall, N., and Barnard, N., "Odor Determinations of 53 Odorant Chemicals," *Journal of the Air Pollution Control Association,* **19**(2) (February 1969), Air Pollution Control Association, Box 2861, Pittsburgh, PA 15230.

[6] *ASME Boiler and Pressure Vessel Code* (Section VIII), American Society of Mechanical Engineers, 345 E. 47th St., New York, NY 10017.

[7] ANSI/ASME B31.3, *Chemical Plant and Petroleum Refinery Piping,* American National Standards Institute, Inc., 1430 Broadway, New York, NY 10018.

[8] *Corrosion Data Survey, Metals Section,* 6th ed., National Association of Corrosion Engineers, P.O. Box 218340, Houston, TX 77218.

[9] *Code of Federal Regulations,* 49 CFR Parts 100–199 (Transportation), Superintendent of Documents, U.S. Government Printing Office, Washington, DC 20402.

[10] DOT P5800, *Emergency Response Guidebook,* U.S. Department of Transportation, Superintendent of Documents, U.S. Government Printing Office, Washington, DC 20402.

[11] *Transportation of Dangerous Goods Regulations,* Canadian Government Publishing Centre, Supply and Services Canada, Ottawa, Ontario, Canada K1A 0S9.

[12] *Code of Federal Regulations,* 46 CFR Parts 0–199 (Shipping), Superintendent of Documents, U.S. Government Printing Office, Washington, DC 20402.

Sulfur Hexafluoride

Chemical Symbol: SF_6
CAS Registry Number: 2551-62-4
DOT Classification: Nonflammable gas
DOT Label: Nonflammable gas
Transport Canada Classification: 2.2
UN Number: UN 1080

PHYSICAL CONSTANTS

	U.S. Units	SI Units
Chemical formula	SF_6	SF_6
Molecular weight	146.05	146.05
Vapor pressure		
at 70°F (21.1°C)	312.7 psia	2156 kPa abs
at 114°F (45.6°C)	544.3 psia	3753 kPa abs
Density of the gas		
at 68°F (20°C) and 1 atm	0.385 lb/ft^3	6.17 kg/m^3
Specific gravity of the gas		
at 68°F and 1 atm (air = 1)	5.11	5.11
Specific gravity of the liquid at 130°F (54.4°C) (water = 1)	0.728	0.728
Specific volume of the gas		
at 70°F (21.1°C) and 1 atm	2.5 ft^3/lb	0.16 m^3/kg
Density of the liquid under its own approximate vapor pressure		
at 70°F (21.1°C)	73.9 lb/ft^3	1183.76 kg/m^3
at 105°F (40.6°C)	84.9 lb/ft^3	1359.97 kg/m^3
Sublimation point at 1 atm	−82.7°F	−63.8°C
Melting point at 325 psia (2241 kPa abs)	−59.4°F	−50.8°C
Critical temperature	114.0°F	45.55°C
Critical pressure	544.3 psia	3753 kPa abs
Critical density	45.8 lb/ft^3	734 kg/m^3

	U.S. Units	SI Units
Latent heat of vaporization at 68°F (20°C)	27.1 Btu/lb	63.1 kJ/kg
Latent heat of fusion at −59.4°F (−50.8°C) and 32.5 psia (224 kPa abs)	14.8 Btu/lb	34.4 kJ/kg
Latent heat of sublimation at −82.7°F (−63.8°C) and 1 atm	69.6 Btu/lb	161.9 kJ/kg
Specific heat of the gas at 70°F (21.1°C) and 1 atm C_p	0.16 Btu/(lb)(°F)	0.67 kJ/(kg)(°C)
Solubility in water, vol/vol of water, at 77°F (25°C)	0.001	0.001
Weight of the liquid at 68°F (20°C) and 312.7 psia (2156 kPa abs)	11.4 lb/gal	1366 kg/m^3

DESCRIPTION

Sulfur hexafluoride is a colorless, odorless, nontoxic, nonflammable gas that has a high dielectric strength and serves widely as an insulating gas in electrical equipment. At atmospheric pressures it sublimes directly from the solid to the gas phase and does not have a stable liquid phase unless under a pressure of more than 32 psia (221 kPa abs). It is shipped as a liquefied compressed gas at its vapor pressure of 298 psig at 70°F (2055 kPa at 21.1°C).

One of the most chemically inert gases known, it is completely stable in the presence of most materials to temperatures of about 400°F (204°C) and has shown no breakdown or reaction in quartz at 900°F (482°C). Sulfur hexafluoride is slightly soluble in water and oil. No change in pH occurs when distilled water is saturated with sulfur hexafluoride.

GRADES AVAILABLE

Sulfur hexafluoride is available for commercial and industrial use in various grades (minimum 99.8 mole percent) having much the same component proportions from one producer to another.

USES

Sulfur hexafluoride is employed extensively as a gaseous dielectric in various kinds of electrical power equipment, such as switchgear, transformers, condensors, and circuit breakers. It has also been used as a dielectric at microwave frequencies and as an insulating medium for the power supplies of high-voltage machines.

Sulfur hexafluoride is also gaining use in nonelectrical applications, including blanketing of molten magnesium, leak detection, and plasma etching in the semiconductor industry. Sulfur hexafluoride also has some limited medical applications.

PHYSIOLOGICAL EFFECTS

Sulfur hexafluoride is completely nontoxic, and in fact has been used medically with humans in cases involving pneumoperitoneum, the introduction of gas into the abdominal cavity. It can act as a simple asphyxiant by displacing the amount of oxygen in the air necessary to support life.

Lower fluorides of sulfur, some of which are toxic, may be produced if sulfur hexafluoride is subjected to electrical discharge, and inhalation of the gas after electrical discharge must be guarded against.

An eight-hour time-weighted average threshold limit value (TLV-TWA) of 1000 ppm (6000 mg/m^3) for sulfur hexafluoride has been adopted by the American Conference of Governmental Industrial Hygienists. [1] This is also the time-weighted average eight-hour exposure limit adopted by the U.S. Occupational Safety and Health Administration. [2]

MATERIALS OF CONSTRUCTION

Sulfur hexafluoride is noncorrosive to all metals. It may be partially decomposed if subjected to an electrical discharge. Some of the breakdown products are corrosive, this corrosion being enhanced by the presence of moisture or at high temperature. Sulfur hexafluoride decomposes very slightly in the presence of certain metals at temperatures in excess of 400°F (204°C), and this effect is most pronounced with silicon and carbon steels. Such breakdown, presumably catalyzed by the metals, is of the order of only several tenths of one percent over one year. Decomposition at elevated temperatures does not occur with aluminum, copper, brass, and silver.

Most common gasket materials, including Teflon, neoprene, and natural rubber, are suitable for sulfur hexafluoride service.

SAFE STORAGE, HANDLING AND USE

All of the precautions necessary for the handling of any nonflammable gas must be taken. Basic guidelines and requirements can be found in CGA P-1, *Safe Handling of Compressed Gases in Containers*. [3] See also Chapter 5 for general recommendations.

DISPOSAL

Return all unused quantities to the supplier. In the event that small quantities must be disposed of on site, all federal, state, provincial, and local regulations regarding waste disposal should be followed.

Controlled venting of this product to the atmosphere is a potential disposal option. Prior to venting, approval of state and local agencies governing air emissions should be obtained. Provisions should also be made to avoid possible asphyxiating conditions upon venting this product.

HANDLING LEAKS AND EMERGENCIES

Standard halogen-detecting devices can be employed to find sulfur hexafluoride leaks and can help locate extremely small leaks. This equipment will detect concentrations in the parts per billion (ppb) range.

METHODS OF SHIPMENT

Under the appropriate regulations, sulfur hexafluoride is authorized for shipment as follows:

By Rail: In cylinders.

By Highway: In cylinders on trucks, or in tube trailers.

By Water: In cylinders on cargo vessels, passenger vessels, ferry vessels (passenger or vehicle ferry vessels). In cylinders on barges.

By Air: In cylinders aboard passenger aircraft up to 150 lb (68 kg), and aboard cargo aircraft up to 300 lb (136 kg), maximum net weight per cylinder.

CONTAINERS

Sulfur hexafluoride is authorized for shipment in cylinders under TC/DOT regulations. [4] and [5]

Filling Limits

The maximum filling density permitted for sulfur hexafluoride in cylinders is 120 percent (percent water capacity by weight).

Cylinders

Cylinders that meet the following TC/DOT specifications are authorized for sulfur hexafluoride service: 3A1000, 3AA1000,

3AAX2400, 3AL1000, 3T1800 and 3E1800 (DOT-3 cylinders may also be continued in use, but new construction is not authorized).

Cylinders authorized for sulfur hexafluoride service must be requalified every five years with the exception that no retest is required for Specification 3E cylinders. Furthermore, Specification 3A and 3AA cylinders may be requalified every ten years instead of every five if they meet the requirements of 49 CFR 173.34(e)(15). [4]

Pressure relief devices authorized for use on sulfur hexafluoride cylinders are the Type CG-1 rupture disk, CG-5 combination rupture disk/fusible plug, and CG-7 pressure relief valve. [6] When cylinders are over 65 inches (1651 mm) long exclusive of the neck and a Type CG-5 device is used, one is required at each end of the cylinder. [6]

Valve Outlet Connections

The standard connection in the United States and Canada for sulfur hexafluoride cylinders is Connection CGA 590. [7]

METHOD OF MANUFACTURE

Sulfur hexafluoride is made commercially by the direct fluorination of molten sulfur. Some higher and lower toxic fluorides formed in the process are removed, and the commercial product is more than 99.5 mole percent pure. A high-purity etchant grade is also available for the electronics industry. Common impurities include small amounts of carbon tetrafluoride, nitrogen, and water vapor.

REFERENCES

[1] *Threshold Limit Values and Biological Exposure Indices,* 1989–90 ed., American Conference of Governmental Industrial Hygienists, 6500 Glenway Avenue, Bldg. D-7, Cincinnati, OH 45211-4438.

[2] Code of Federal Regulations, Title 29 CFR Parts 1900–1910 (Labor), Superintendent of Documents, U.S. Government Printing Office, Washington, DC 20402.

[3] CGA P-1, *Safe Handling of Compressed Gases in Containers,* Compressed Gas Association, Inc., 1235 Jefferson Davis Highway, Arlington, VA 22202.

[4] *Code of Federal Regulations,* Title 49 CFR Parts 100–199 (Transportation), Superintendent of Documents, U.S. Government Printing Office, Washington, DC 20402.

[5] *Transportation of Dangerous Goods Regulations,* Canadian Government Publishing Centre, Supply and Services Canada, Ottawa, Ontario, Canada K1A 0S9.

[6] CGA S-1.1, *Pressure Relief Device Standards—Part 1—Cylinders for Compressed Gases,* Compressed Gas Association, Inc., 1235 Jefferson Davis Highway, Arlington, VA 22202.

[7] CGA V-1, *American National, Canadian, and Compressed Gas Association Standard for Compressed Gas Cylinder Valve Outlet and Inlet Connections* (ANSI/CSA/CGA V-1), Compressed Gas Association, Inc., 1235 Jefferson Davis Highway, Arlington, VA, 22202.

ADDITIONAL REFERENCES

Material Safety Data Sheet for Sulfur Hexafluoride, Matheson Gas Products, Inc., Secaucus, NJ 07094.

Material Safety Data Sheet for Sulfur Hexafluoride, Union Carbide Corporation, Linde Division, 39 Old Ridgebury Road, Danbury, CT 06817.

Matheson Gas Data Book, 6th ed., Matheson Gas Products, Inc., Secaucus, NJ 07094. (1980)

Vinyl Chloride

Chemical Symbol: CH_2CHCl (or C_2H_3Cl)
Synonyms: Chloroethylene, chloroethene, VCM, VCl
CAS Registry Number: 75-01-4
DOT Classification: Flammable gas
DOT Label: Flammable gas
Transport Canada Classification: 2.1
UN Number: UN 1086

PHYSICAL CONSTANTS

	U.S. Units	SI Units
Chemical formula	C_2H_3Cl	C_2H_3Cl
Molecular weight	62.50	62.50
Vapor pressure		
at 70°F (21.1°C)	35.3 psig	243.38 kPa
at 105°F (40.6°C)	75.3 psig	519.18 kPa
at 115°F (46.1°C)	90.3 psig	622.60 kPa
at 130°F (54.4°C)	114.3 psig	788.07 kPa
Density of the gas at 70°F (21.1°C) and 1 atm	0.160 lb/ft^3	2.56 kg/m^3
Specific gravity of the gas at 59°F (15°C) and 1 atm (air = 1)	2.15	2.15
Specific volume of the gas at 70°F (21.1°C) and 1 atm	6.25 ft^3/lb	0.390 m^3/kg
Density of the liquid		
at 70°F (21.1°C)	56.71 lb/ft^3	908.41 kg/m^3
at 105°F (40.6°C)	54.38 lb/ft^3	871.08 kg/m^3
at 115°F (46.1°C)	53.69 lb/ft^3	860.03 kg/m^3
at 130°F (54.4°C)	52.61 lb/ft^3	842.73 kg/m^3
Boiling point at 1 atm	7.93°F	−13.4°C
Melting point at 1 atm	−245°F	−153.9°C
Critical temperature	317.1°F	158.4°C
Critical pressure	774.7 psia	5341.37 kPa abs
Critical density	23.1 lb/ft^3	370.03 kg/m^3
Triple point (estimated)	−240.75°F at 0.00018 psia	−151.5°C at 0.00124 kPa abs

	U.S. Units	SI Units
Latent heat of vaporization at boiling point	143.7 Btu/lb	334.25 kJ/kg
Latent heat of fusion at melting point	32.65 Btu/lb	75.94 kJ/kg
Specific heat of the gas at 77°F (25°C) and 1 atm C_p	0.205 Btu/(lb)(°F)	0.858 kJ/(kg)(°C)
Solubility in water, wt/wt, at 77°F (25°C) and 1 atm	0.0011	0.0011
Weight of the liquid at 70°F (21.1°C)	7.58 lb/gal	908.28 kg/m^3
Specific gravity of the liquid at 68°F (20°C) (water = 1)	0.9121	0.9121
Flammable limits in air by volume	4.0–22.0%	4.0–22.0%
Autoignition temperature	881.6°F	472.0°C
Flash point (open cup)	−108°F	−77.8°C

DESCRIPTION

Vinyl chloride is a colorless, flammable gas with a sweet ethereal odor. It is shipped as a liquefied compressed gas. Contact of the liquid with the skin can result in freezing or frostbite. Vinyl chloride has been established as a human carcinogen. In addition, acute effects of vinyl chloride exposure include irritation of the skin and eyes on contact. Inhalation of concentrations of more than 500 ppm produces mild anesthesia.

Anhydrous vinyl chloride does not corrode metals at normal temperatures and pressures, but in the presence of moisture, vinyl chloride accelerates the corrosion of iron and steel at elevated temperatures.

Vinyl chloride polymerizes readily when exposed to air, sunlight, heat, or oxygen, although it is chemically stable as shipped with an inhibitor (phenol).

GRADES AVAILABLE

Vinyl chloride is available for commercial and industrial use in various grades having much the same composition from one producer to another. It typically has a minimum purity of 99.9 mole percent in the liquid phase.

USES

Vinyl resins produced from vinyl chloride monomer (VCM) are used in making polyvinyl chloride and copolymer plastics. VCM is also used in organic synthesis and in the manufacture of adhesives and other chemicals.

PHYSIOLOGICAL EFFECTS

Vinyl chloride is toxic and carcinogenic. The U.S. Occupational Safety and Health Administration (OSHA) has established an exposure limit of 1 ppm for an eight-hour time-weighted average workday and a ceiling limit of 5 ppm for a 15-minute exposure. A complete standard describing control of employee exposure to vinyl chloride as required by OSHA is given in 29 CFR 1910.1017. [1]

Vinyl chloride acts as a general anesthetic in concentrations over 500 ppm. It has been reported that acute exposures to vinyl chloride concentrations above 1000 ppm slowly produce mild disturbances such as drowsiness, blurred vision, staggering gait, and tingling and numbness in the feet and hands.

The occurrence of acro-osteolysis and hepatic angiosarcoma have been associated with vinyl chloride exposure. Liver changes

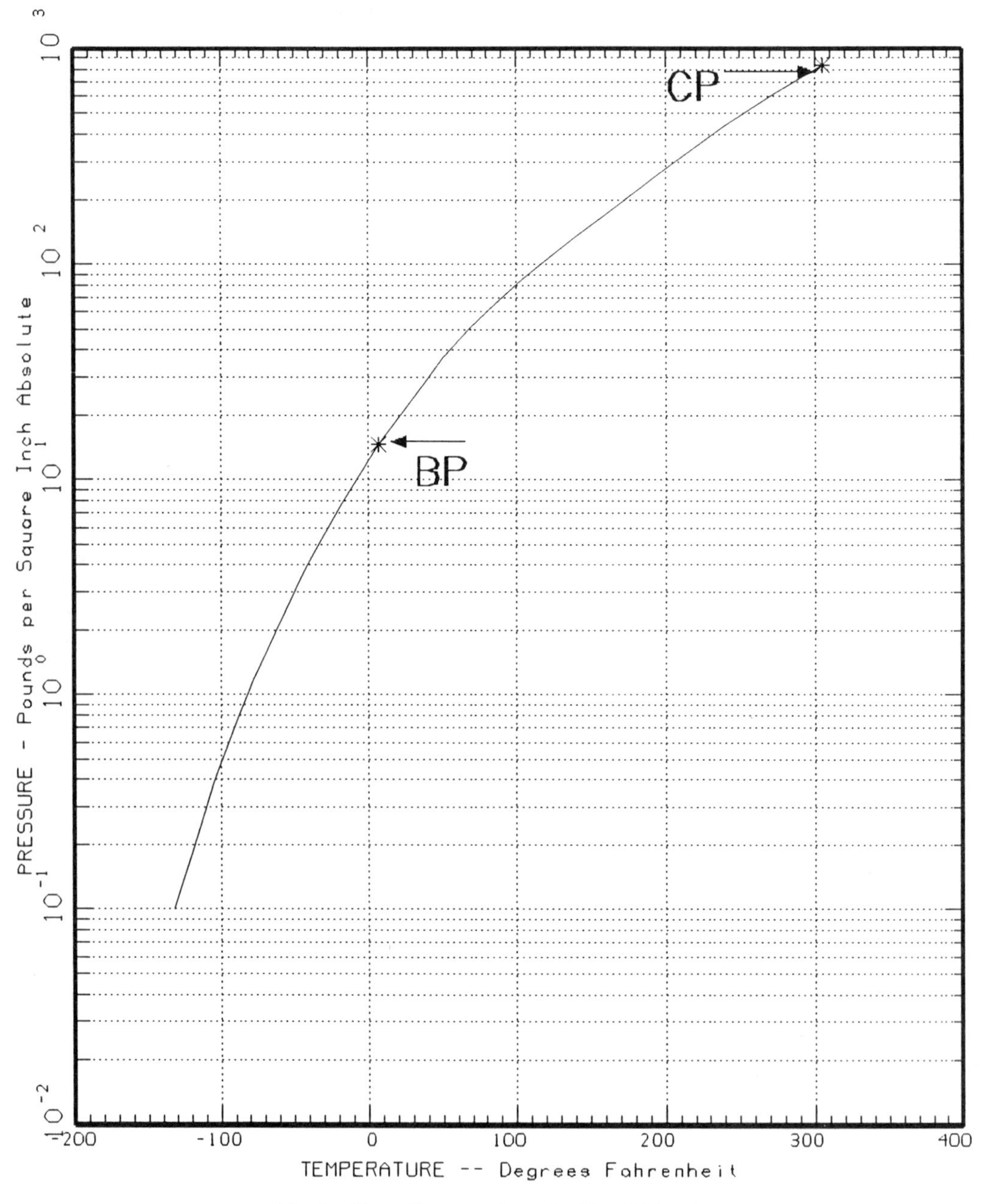

Fig. 1. Vapor Pressure Curve for Vinyl Chloride.

including hepatomegaly, liver function abnormalities, and parenchymal damage have been reported. [2]

Vinyl chloride can irritate or damage the eyes on contact. Liquid vinyl chloride also irritates the skin and can freeze the skin on prolonged contact.

MATERIALS OF CONSTRUCTION

Steel is recommended for all piping, storage tanks, and equipment used with vinyl chloride. However, at elevated temperatures, vinyl chloride in the presence of moisture speeds its corrosion. Stainless steel is also an acceptable material to use with vinyl chloride. Copper and copper alloys must not be used. Valves in vinyl chloride service must not contain copper or copper alloys. Acetylene may be present as an impurity in vinyl chloride and can form an explosive acetylide when exposed to copper.

Asbestos, Teflon, lead, and carbon are satisfactory gasket materials for fittings and connections.

SAFE STORAGE, HANDLING, AND USE

Vinyl chloride should be used in a well-ventilated area, preferably a hood with forced ventilation. Some authorities believe that the odor of vinyl chloride does not provide adequate warning of its presence in concentrations sufficient to produce dizziness and unconsciousness, so special caution is urged against leaks and poor ventilation.

Precautions required for the safe handling of all flammable gas must be observed with vinyl chloride. Adequate electrical grounding of all lines and equipment, and ditching or diking in storage tank areas to control the liquid in the event of vessel rupture, are among recommended precautions. Ditching is preferable because the material should not be retained at a location directly beneath or surrounding the storage tanks. Installations must be designed to comply with requirements for unfired pressure vessels and all state, provincial, and local regulations.

Personnel handling vinyl chloride should wear safety shoes, chemical safety goggles and/or a full face shield, and rubber gloves. An effective educational and training program must be instituted to inform the workers of the hazards involved in handling and using vinyl chloride and the first aid measures to be followed in the event of an emergency. For specific OSHA requirements, refer to 29 CFR 1910.1017. [1]

For respiratory protection, self-contained breathing apparatus, air-line gas masks, and U.S. Bureau of Mines or National Institute for Occupational Safety and Health approved canister-type gas masks should be available in emergencies. Instant-acting safety showers and eyewash fountains should be conveniently located near the site of the operation.

Store and use cylinders of vinyl chloride in well-ventilated areas away from heat and all sources of ignition such as flames and sparks. Do not use vinyl chloride around sparking motors or other non-explosion-proof equipment. Do not store reserve cylinder stocks of vinyl chloride with cylinders containing oxygen, chlorine, or other highly oxidizing or combustible materials. [3]

DISPOSAL

Discard any product, residue, disposable container, or liner in an environmentally acceptable manner in full compliance with federal, state, provincial, and local regulations.

HANDLING LEAKS AND EMERGENCIES

Should a leak occur, evacuate personnel from the area, shut off all ignition sources, ventilate the area, and reduce vapors with fog or fine water spray. Shut off the main source of gas supply. Flammable and toxic vapors may spread. The atmosphere should be checked with an appropriate detector while using self-contained breathing apparatus. For large leaks, personnel must wear special personal protective suits for fire/chemicals and positive-pressure self-contained breathing apparatus.

Vinyl chloride forms explosive mixtures with air and oxidizing agents. It is easily ignited. Should a fire occur, evacuate all personnel from the danger area. Do not extinguish flames due to the possibility of explosive reignition. Shut off the source of supply if without risk. Immediately cool containers with water spray from a maximum distance, taking care not to extinguish flames, and allow the fire to burn out. Remove ignition sources if possible without risk to personnel. If flames are accidentally extinguished, keep personnel out of the area as explosive reignition may occur. Use self-contained breathing apparatus.

First Aid

Vinyl chloride is toxic and a known carcinogen. Any person exposed to vinyl chloride vapors should be promptly removed to an uncontaminated area and inhale fresh air or be administered oxygen by trained personnel. In the event of overexposure, the victim may become unconscious. If breathing has ceased, administer artificial respiration and then oxygen. A physician should be called.

If contact with liquid vinyl chloride occurs, flush affected areas with copious amounts of water, then wash with soapy water. Irrigate eyes immediately with copious amounts of water for at least 15 minutes, holding eyelids apart to ensure contact of the water with all tissues of the eyes and lids. Obtain professional medical care, preferably by an ophthalmologist, as soon as possible.

METHODS OF SHIPMENT

Under the appropriate regulations, vinyl chloride is authorized for shipment, when inhibited against polymerization, as follows: [4] and [5]

By Rail: In cylinders, and in single-unit tank cars and TMU tank cars.

By Highway: In cylinders on trucks, and in tank trucks and TMU tanks on trucks.

By Water: In cylinders on cargo vessels or passenger vessels stowed away from living quarters and in accordance with packaging requirements referenced in 49 CFR 172.101. [4] In authorized tank cars aboard trainships only. In cylinders on barges of U.S. Coast Guard classes A and C only. In cargo tanks aboard ships and barges (with special warning signs) and aboard tankships (to maximum filling densities by specific gravity as specified in Coast Guard regulations).

By Air: On cargo aircraft only up to 300 lb (136 kg); forbidden on passenger aircraft.

CONTAINERS

Vinyl chloride is shipped as a liquefied compressed gas in cylinders, single-unit tank cars and TMU (ton multi-unit) tank cars, and in tank trucks and TMU tanks on trucks. For all these types of containers, it is required that all parts of valves and pressure relief devices in contact with the contents of the container must be of a metal or other material (suitably treated if necessary) which will not cause formation of any acetylides.

Filling Limits

The maximum filling densities authorized by the DOT for vinyl chloride are as follows (percent water capacity by weight):

In cylinders—84 percent.

In single-unit tank cars of specification TC/DOT 105J200W—87 percent; in single-unit tank cars of specifications TC/DOT-112T340-W, 112J340W, 114T340W, and 114J340W—86 percent. Interior pipes of loading and unloading valves must be equipped with excess flow valves of approved design.

In TMU tanks of TC/DOT specification 106A500X and in MC-330 or MC-331 cargo tanks—84 percent. See 49 CFR 173.314 and 49 CFR 173.315. [4]

Cylinders

Cylinders that meet the following TC/DOT specifications are authorized for vinyl chloride service: 3A150, 3AA150, 3AL150, 4BW225, 3E1800, 4B150 (without brazed seams), and 4BA225 (without brazed seams).

Cylinders meeting specification DOT-25 may be continued in service, but new construction is not authorized.

Under present regulations, cylinders of all types authorized for vinyl chloride service must be requalified by hydrostatic test every 5 years with the exception of Specification 3E (for which periodic hydrostatic retest is not required).

Authorized pressure relief devices for vinyl chloride cylinders include Type CG-2 fusible plugs which have a nominal yield temperature of 165°F (73.9°C) and Type CG-7 pressure relief valves. See CGA S-1.1, *Pressure Relief Device Standards—Part 1—Cylinders for Compressed Gases,* for specific requirements. [6]

Valve Outlet Connections

The standard valve outlet connection in the United States and Canada for vinyl chloride cylinders is Connection CGA 510. The limited standard connection is Connection CGA 290. [7]

Cargo Tanks

Shipment of vinyl chloride is authorized by the DOT in cargo tanks meeting TC/DOT specifications MC-330 or MC-331 and having a minimum design pressure of 150 psig (1034 kPa).

TMU Tank Cars

Shipment of vinyl chloride is authorized in TMU tanks of TC/DOT specification 106A500X on trucks or by rail.

Tank Cars

Vinyl chloride is authorized for shipment in single-unit tank cars meeting DOT specifications 105J200-W, 112T340W, 114T340W, or 114J340W (provided that they have loading and unloading valves with interior piping fitted with excess flow valves of approved design). Also, for cars of specification 105A200W built before January 1, 1975, openings in tank heads to facilitate application of nickel linings are authorized. These openings must be closed in an approved manner. Refer to 49 CFR 173.314. [4]

CGA P-10, *Standard for Vinyl Chloride Monomer Tank Car Manway Cover and Protective Housing Arrangement and Emergency Safety Kit,* describes and provides dimensional drawings for a standard manway cover and protective housing arrangement for vinyl chloride monomer (VCM) tank cars. It also describes an emergency safety kit for use in capping leaking fittings. [8]

METHODS OF MANUFACTURE

Most vinyl chloride monomer today is made via a three-step process employing ethylene oxyhydrochlorination. A small amount is made by the reaction of acetylene and hydrogen chloride, either as liquids or gases, with a copper chloride catalyst in the liquid process and a mercury catalyst in the gas process. Vinyl chloride is also made by the heating of ethylene chloride with alcoholic alkali.

REFERENCES

[1] *Code of Federal Regulations,* Title 29 CFR Parts 1900–1910 (Labor), Superintendent of Documents, U.S. Government Printing Office, Washington, DC 20402.

[2] *Effects of Exposure to Toxic Gases—First Aid and Medical Treatment,* 3rd ed., Matheson Gas Products, Inc., Secaucus, NJ 07094. (1988)

[3] *Matheson Gas Data Book,* 6th ed., Matheson Gas Products, Inc., Secaucus, NJ 07094. (1980)

[4] *Code of Federal Regulations,* Title 49 CFR Parts 100–199 (Transportation), Superintendent of Documents, U.S. Government Printing Office, Washington, DC 20402.

[5] *Transportation of Dangerous Goods Regulations,* Canadian Government Publishing Centre, Supply and Services Canada, Ottawa, Ontario, Canada KIA 0S9.

[6] CGA S-1.1, *Pressure Relief Device Standards—Part 1—Cylinders for Compressed Gases,* Compressed Gas Association, Inc., 1235 Jefferson Davis Highway, Arlington, VA 22202.

[7] CGA V-1, *American National, Canadian, and Compressed Gas Association Standard for Compressed Gas Cylinder Valve Outlet and Inlet Connections* (ANSI/CSA/CGA V-1), Compressed Gas Association, Inc., 1235 Jefferson Davis Highway, Arlington, VA 22202.

[8] CGA P-10, *Standard for Vinyl Chloride Monomer Tank Car Manway Cover and Protective Housing Arrangement and Emergency Safety Kit,* Compressed Gas Association, Inc., 1235 Jefferson Davis Highway, Arlington, VA 22202.

Gas Mixtures

INTRODUCTION

Throughout this presentation, the word *mixture* is used when referring to compressed gases that have been combined or blended into a single container. In actuality, these are solutions, either liquefied or nonliquefied.

USES

The uses for gas mixtures in today's technologically advanced society are virtually limitless. Following is a brief list of applications of some routinely used mixtures.

Analytical Instruments: In addition to pure gases, gas mixtures are used for the operation and/or calibration of analytical instruments. Zero gases (which are usually pure materials certified to contain negligible or known concentrations of a component of interest) are used for flame ionization detector instruments that require gases of low hydrocarbon content to achieve maximum sensitivity. "Zero air" may be used both to zero the instrument (calibrate the instrument to the low end of a concentration range) and to provide an oxidant of low hydrocarbon content for the operation of the analyzer.

Span gases, which usually contain a minor component of interest at a known concentration level, permit the analytical device to be calibrated at a value corresponding to that concentration. Fuel gas is used in instruments such as those with flame ionization detectors which require that the supply of fuel be enhanced by certain physical properties of the diluent. A mixture of 40 percent hydrogen in nitrogen is common.

Pollution Control: Small amounts of nitric oxide, nitrogen dioxide, sulfur dioxide, carbon monoxide, vinyl chloride, various hydrocarbons, and other airborne pollutants are added to air or other gases to calibrate analytical instruments for emission measurements.

X-Ray Fluorescence Spectroscopy: Mixtures containing small amounts of butane in a mixture of helium and neon are used in X-ray fluorescence spectroscopy.

Electron Capture: A mixture of 5 percent methane in argon is used in electron capture.

Special Carrier Gases: These include mixtures such as 8.5 percent hydrogen in helium.

Sterilizing: Mixtures of ethylene oxide, such as 12 percent with halocarbon-12, or 10 to 20 percent ethylene oxide with carbon dioxide, are commonly used in sterilization. Such mixtures require registration with the U.S. Environmental Protection Agency.

Medical Applications: A low concentration of carbon monoxide in air is used for diagnosing lung efficiency. Various percentage mixtures of carbon dioxide in oxygen are used for blood gas analysis. Cyclopropane, nitrous oxide, and certain other chemically active gases are used separately and as mixtures for anesthesia. Additionally, carbon dioxide in combination with other gases is used to produce atmospheres for biological study.

Nuclear Counter Gases: Mixtures of 0.95 percent isobutane in helium and 1.3 percent butane in helium are used as quenching gases in Geiger counters. Mixtures of 4 percent isobutane in helium and 10 percent methane in argon are used in proportional counters.

Electronic Component Manufacture: Gases which include low levels of dopants such as arsine, phosphine, diborane, and others in hydrogen, helium, argon, or nitrogen are used for making semiconductor devices. Highly purified ammonia and its mixtures are used for forming insulating nitride barriers in integrated circuits. Silane, either pure or in mixtures, is used for silicon deposition. Gas mixtures similar to the ones above but with higher percentage levels of arsine and phosphine are common in the manufacture of compound semiconductors.

Welding: Mixtures of oxygen or other gases with argon, carbon dioxide, helium, and nitrogen are used for welding.

Metallurgical Applications: Hydrogen-nitrogen mixtures (forming gases) are used to produce reducing atmospheres in heat-treating applications.

Leak Detection: A low percentage of halocarbons, such as halocarbon-12, or helium in air or nitrogen, is used for leak detection.

Illumination: Mixtures of rare gases with helium, nitrogen, or argon are used in lamps, signs, electronic tubes, and other devices.

Radioactive Gas Mixtures: Tritium and carbon-14, usually mixed with inert gases, are commonly used as tracer gases. Krypton-85 mixed with air is used for diagnosing heart and brain disorders.

Spark Chamber Gases: Generally, a mixture of 70–95 percent neon in helium is used in spark chambers.

Laser Mixtures: Mixtures of rare gases with nitrogen, and carbon dioxide with helium, are used for lasers. Excimer lasers use the rare gases mixed with fluorine or hydrogen chloride.

Diving: Mixtures of oxygen with helium and possibly other diluents are used for deep diving applications.

Propellants: Various mixtures of liquefied gases are commonly used in aerosol formulations as propellants.

GRADES AVAILABLE

Gas mixtures are routinely available from most suppliers in several grades, namely: primary standard, certified standard, and unanalyzed mixtures each having its own preparation tolerance and certification accuracy depending upon the minor component concentration.

When choosing a particular grade of gas mixture, the application must be considered in order to achieve the desired result with a minimum of cost. For example, gas mixtures used for welding, sterilizing, propellants, curing agents, leak detection, and certain process atmospheres need not be blended to the costly high tolerance of primary standard mixtures. However, other applications, such as mixtures used as standard reference materials (SRM's), and certified reference materials (CRM's) used for emission control testing, certain medical mixtures, and some standard blends used for calibration of monitoring instrumentation, require the highest accuracy available. Table 1 gives a typical specification chart for stable gas mixtures.

Table 1 is a representative sample of what is typically available. Different suppliers may have capabilities greater or less than those indicated. Additionally, there is no generally established definition of primary or certified standards, and these terms can be used differently by different suppliers. See also Fig. 1.

Mixtures of gases with known stability problems may be offered by some suppliers, but usually require longer lead times. The additional time is needed to permit special techniques such as passivation or other cylinder preparation to reduce or eliminate the degradation of the unstable component. For this reason, these mixtures are typically offered at lower tolerances.

PHYSIOLOGICAL EFFECTS

The physiological effects a gas mixture will have depends on the properties of the individual gases and the concentration of those gases in the final mixture. For example, if a

TABLE 1. EXAMPLES OF TYPICAL SPECIFICATIONS FOR GAS MIXTURE GRADES.

Type	Concentration Range	Preparation Tolerance	Analytical Accuracy
Primary Standards[a]	1000 ppm to 50%	1% of component	1% of component or 0.02% absolute, whichever is smaller
	20 to 1000 ppm	5% of component	1% of component or 0.02% absolute, whichever is smaller
	2 to 20 ppm	10% of component	0.2 ppm absolute
Certified Standards	10% to 50%	5% of component	2% of component
	50 ppm to less than 10%	10% of component	2% of component
	6 to less than 50 ppm	20% of component	5% of component
	2 to 6 ppm	20% of component	0.3 ppm absolute
Unanalyzed	10% to 50%	5% of component	Not analyzed, hence accuracy cannot be guaranteed
	50 ppm to less than 10%	10% of component	
	2 ppm to 49 ppm	20% of component	

10-ppm hydrogen sulfide in nitrogen mixture were to leak into a confined space and expose personnel, the primary potential hazard would be from oxygen displacement by the nitrogen rather than hydrogen sulfide poisoning, since the hydrogen sulfide is at its threshold limit value (TLV) in the mixture. [1] However, if 100 ft^3 of a 4000-ppm hydrogen cyanide mixture were suddenly released into a room with dimensions 20 ft × 20 ft × 8 ft and with poor to average ventilation, it could have severe toxicological effects on anyone in the room, including the possibility of death.

Fig. 1. High purity gases and gas mixtures require a state-of-the-art package. Here, specially prepared cylinders are internally examined using a boroscope.

Know the potential hazards of the gases in any mixture to which you may become exposed. Personnel using gas mixtures must familiarize themselves with the appropriate sections of the applicable Material Safety Data Sheets.

MATERIALS OF CONSTRUCTION

When determining what materials should be used with a gas mixture, it is not only important to use materials that are compatible with each of the components of the mixture separately but also with all of the components when they are combined. Conditions during use, such as temperature, pressure, and humidity, must also be taken into account. For assistance in making these determinations, consult the gas supplier.

SAFE STORAGE, HANDLING, AND USE

Gas mixtures should be stored in accordance with their physical and chemical properties. For example, flammable mixtures should not be stored within 20 ft (6.1 m) of oxidizer gases or mixtures. Pyrophoric mixtures should not be stored with toxic mixtures, and so on. Instead, pyrophoric mixtures should be stored in a special gas storage area specifically designed for those gases. Storage areas should be fire resistant, well ventilated, dry, and located away from sources of ignition or excessive heat.

Where cylinders are stored outside, they should be protected from the direct rays of the sun, particularly in areas where high ambient temperatures are common. Only cylinders actually in use should be permitted in the laboratory or work areas. All storage areas should be designed, constructed, and maintained in accordance with federal, state, provincial, and local regulations.

The storage area should be maintained at a temperature that will prevent one or more of the components of a mixture from condensing, thereby losing its homogeneity. Mixtures that contain a condensable component should be maintained at 60°F (15.6°C) or greater. Never allow a cylinder to attain a temperature over 125°F (51.7°C). If a mixture with a condensable component has been subjected to temperatures at or below its saturation temperature, it should be rehomogenized prior to the withdrawal of any of the gas by using one or more of the techniques described in the section on stability later in this presentation.

Before using any mixture, read the cylinder label to determine that the cylinder contains the gas mixture desired. Also, read the Material Safety Data Sheet (MSDS) on the mixture or mixture components, and become thoroughly familiar with the properties of the mixture before using it. Refer to the Methods of Manufacture section for additional information on mixture properties and to Chapter 5 for general safety guidelines.

DISPOSAL

No attempt should be made to dispose of any gas mixture before determining the following:

- What gases are in the mix?
- At what concentrations are they present?
- What is the total quantity for disposal?
- Is the mixture subject to environmental regulations?

Consult the appropriate sections of the Material Safety Data Sheets and the gas supplier for assistance. In many cases, sophisticated and expensive scrubbing equipment is necessary to destroy residual gases. It is best to return the unused portion of any gas or mixture to the supplier for disposition.

HANDLING LEAKS AND EMERGENCIES

Upon receipt of any gas or gas mixture, the user must become familiar with emergency procedures to be taken by reading the Material Safety Data Sheets provided. Gas mixtures that can pose serious hazards to personnel or property if they were to be released should be isolated from the area of use as much as possible. Where this is impractical, monitoring, or leak-detecting equipment should be installed, calibrated, and maintained on a routine schedule. In the event of a leak or release, thorough knowledge of the properties of the gases involved is essential. If there is any doubt as to what should be done, evacuate the area and consult the supplier. Other organizations that may be of assistance are CHEMTREC and the local hazardous materials emergency response committee.

After determining that it is safe to do so, remove the leaking container to an environment that will minimize exposure to personnel.

Many users of toxic or pyrophoric mixtures have chosen to keep these mixtures in specially designed ventilated cabinets during use. The use of these gas cabinets has received wide acceptance and in some geographical areas has become required by code. Such cabinets are available from specialty gas equipment suppliers.

METHODS OF SHIPMENT

Domestic Shipment

Before a mixture can be offered for transportation, it must be properly packaged, which may require special crating. It also needs to be properly classified according to its properties. Nearly all mixtures are classified by the following UN numbers:

- UN 1953 Compressed or liquefied gases (flammable, toxic) N.O.S.
- UN 1954 Compressed or liquefied gases (flammable, nontoxic) N.O.S.
- UN 1955 Compressed or liquefied gases (non-flammable, toxic) N.O.S.
- UN 1956 Compressed or liquefied gases (non-flammable, nontoxic) N.O.S.

Once classified and assigned the appropriate UN number, the mixture needs to be labeled with the proper DOT labels which are specified by Title 49 of the U.S. *Code of Federal Regulations,* Parts 172.101 and 172.102., or equivalent Canadian regulations. [2] and [3] See also 49 CFR 173.300–173.315. [2] Additional labeling should follow the guidelines in CGA C-4, *Method of Marking Portable Compressed Gas Containers to Identify the Material Contained,* and CGA C-7, *Guide for the Preparation of Precautionary Labeling and Marking of Compressed Gas Containers.* [4] and [5]

International Shipment

Labeling requirements are more strict with international shipments of compressed gas mixtures from the United States, as many of them move by sea or air. Since the vessels used in both of these modes of transportation are difficult to evacuate quickly in the event of an emergency, many gases and gas mixtures may have to be classified as poisons that would not be classified as such for domestic land transportation.

Sea shipments of hazardous materials are regulated by the International Maritime Organization (IMO) using the *International Maritime Dangerous Goods Code* (IMDG Code) which sets the documentation requirements for all shipment cargo. [6] International shipments may require a number of additional documents, including but not limited to:

(1) Hazardous goods declaration
(2) Import license
(3) Export license
(4) Container and valve manufacturer's certification

Air Shipments

Air shipments are governed by the International Civil Aviation Organization's (ICAO) technical instructions together with the International Air Transport Association's (IATA) Dangerous Goods Regulations. [7] and [8] Some gas mixtures are permitted aboard passenger aircraft, others are permitted aboard cargo aircraft only, while others are not permitted at all.

CONTAINERS

Gas mixtures can be supplied in many different types of containers. These containers can range in size from 0.25-L glass flasks to jumbo tube trailers. The overriding considerations in choosing a particular container for a given mixture are material compatibility, volume required, and minimum or maximum desired pressure.

Cylinders require the installation of a valve that will permit the withdrawal of the mixture. The valve used on a cylinder for a given mixture must be selected so that the materials of construction are compatible with all mixture components. The outlet connection is selected based on the physical and chemical characteristics of the components. CGA V-7, *Standard Method of Determining Cylinder Valve Outlet Connections for Industrial Gas Mixtures,* is a guide for the selection of outlet connections for mixtures. [9]

It includes numerous examples on how to apply the selection procedures.

Depending on the contents of a mixture, one or more pressure relief devices may be required on the container. Most pressure relief devices are an integral part of the valve. However, some devices are installed directly into the bottom or top of the container. For an in-depth discussion on pressure relief devices for mixtures, see CGA S-7, *Method for Selecting Pressure Relief Devices for Compressed Gas Mixtures in Cylinders.* [10] Also consult Chapter 7 of this *Handbook* for a presentation on pressure relief devices. The selection of the proper pressure relief device for a mixture involves a knowledge of the contents of the mixture, the final pressure, its physical state, and the type of container which will be used. To simplify the process of selecting the proper pressure relief device for a specific mixture, refer to the mixture safety selection algorithm in Fig. 2. Consult CGA S-7 for complete information on the use of this algorithm. [10]

METHODS OF MANUFACTURE

Gas Mixture Safety

The importance of safety cannot be overemphasized when discussing the combining of various gases into mixtures. It is a violation of DOT regulations to transport gases capable of combining chemically or capable of impairing the integrity of the container in which they are transported. ***Only persons knowledgeable of the potential hazards inherent in combining different gases and persons skilled in their manufacture should attempt to prepare them.***

The walls of certain compressed gas cylinders may catalyze some reactions.

Blends of flammable and oxidizing gases must only be mixed in proportions that are safe, not only after blending but also during the blending process. There are two sets of limits that must be considered when blending fuel gases with oxidizing gases. First, mixtures must not be made that will result in the final concentration falling between the lower explosive limit (L.E.L.) and the upper explosive limit (U.E.L.) at the final mixture pressure and temperature. The gases should be added in such a manner that if during the blending process the mixture were to go through the flammable range and ignite, the total energy released would be sufficiently low enough that if it were to all react the resultant instantaneous pressure rise would be safely contained within the cylinder.

Second, lower and upper detonation limits must be taken into consideration when determining the order in which gases are to be introduced. Detonation limits for fuel gases are within their L.E.Ls and U.E.Ls making this a narrower region but nonetheless a very important one. See Table 2.

Detonation reactions contain energies otherwise thought to be safe by nonrate-related energy calculations and produce shock waves that can rupture cylinders.

Gas Mixture Stability

Some gas mixtures degrade with time in one type of cylinder, whereas they may remain quite stable in one made of or coated with a more compatible material. Mixtures with low levels of hydrogen sulfide, sulfur dioxide, nitric oxide, nitrogen dioxide, carbon monoxide, and other substances used to be unstable over time. Extensive work has been done in this area, some of which has been published and some which is considered proprietary. Various internal treatments and coatings were used in an effort to stabilize these mixtures. Some, such as nickel plating for corrosive gases, enjoyed success. Others, such as wax coating and fluorocarbon polymer coating, were less successful. The introduction of cylinders manufactured of aluminum as well as other metals has helped solve this problem.

Impurities

Even the purest gases contain some impurities. The final mixture will contain the impurities of the "pure" gases used, to the extent of their component concentrations. Most

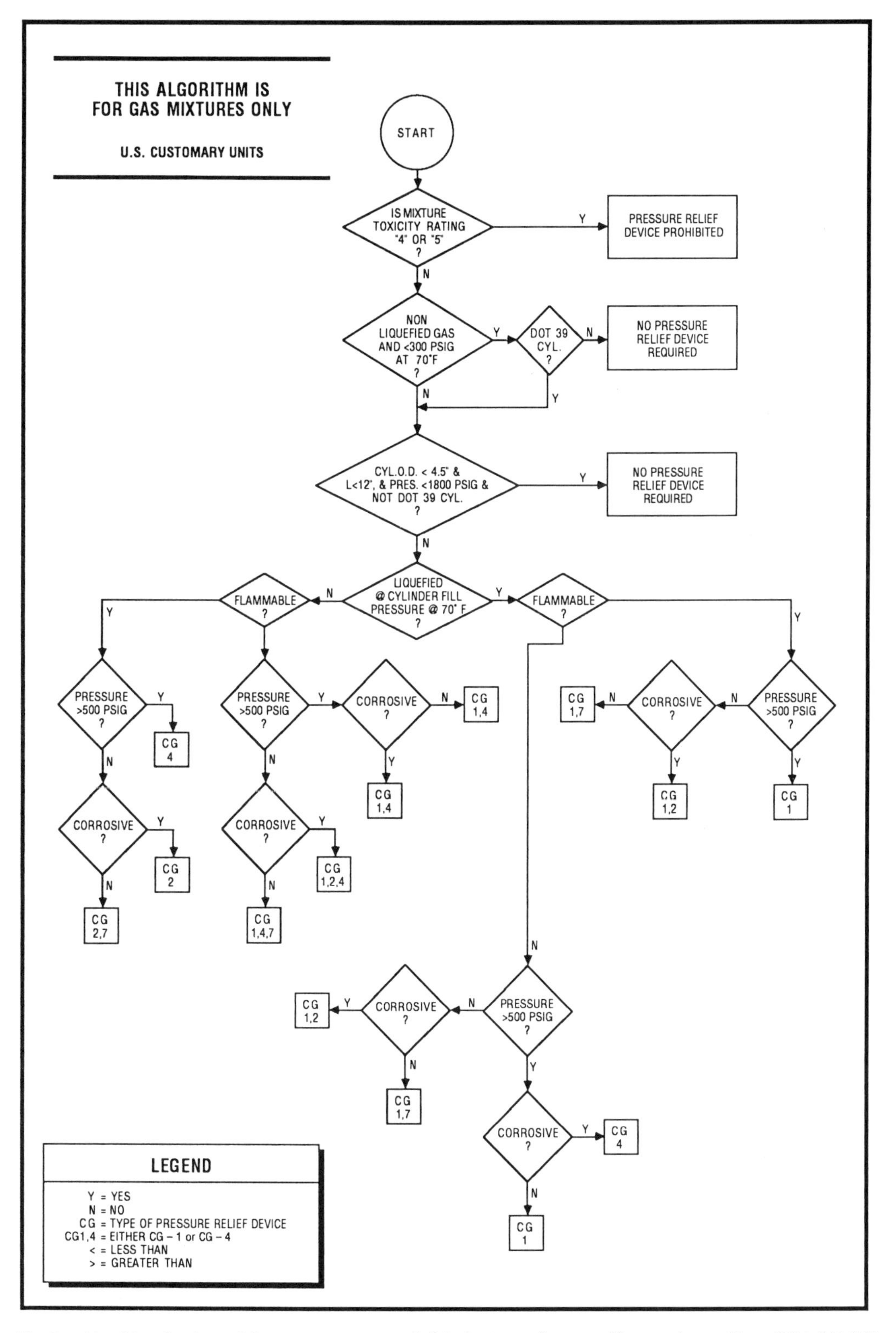

Fig. 2. Algorithm for determining proper pressure relief device to use for a specific gas mixture. From CGA S-7. [10]

TABLE 2. LIMITS OF FLAMMABILITY AND DETONABILITY.[a]

Gas	Limits of Flammability in Oxygen (Percent)	Limits of Detonability in Oxygen (Percent)
Hydrogen	4–94	15–90
Carbon monoxide	15.5–94	38.3–90
Ammonia	15–79	25.2–75.5
Acetylene	no data	3.5–92
Propane	2.3–55	3.1–27

[a]At atmospheric pressure; ranges widen with increasing pressure.

suppliers use high-purity raw material gases in their certified and unanalyzed grade mixtures, which is quite adequate for most applications. However, in cases where certain impurities or impurities in general must be at a minimum, more expensive research grade or ultrahigh-purity raw materials are used.

For custom specifications, special purification of the various component gases may be required. Some of these purifications are easily performed, while others are quite difficult. Overall, gas purities have improved dramatically over the last couple of decades. This has been spurred by the demanding requirements of industries such as air pollution monitoring, light bulb manufacturing, lasers, and semiconductor manufacturing.

In the case of hydrocarbon mixtures, the impurities in the starting materials may sometimes be compensated for, if the contaminants in the starting materials are also required in the final mix. The result can be high-quality mixtures blended from relatively lower purity raw materials. It is extremely important to have good, thorough analyses of the raw materials when using this technique.

Blending Techniques

To prepare a gas mixture of a specific composition, it is necessary to measure the correct amount of each component. Pressure and mass are the two fundamental properties of gases and vapors used in mixture preparation. Depending upon the degree of accuracy required in the final mix, either or both of these properties may be used by the mixture technician to achieve the desired result. For mixtures only requiring 5 to 10 percent preparation tolerances, the use of calibrated pressure gauges or transducers may be quite adequate.

When higher accuracies are required, mixtures are usually prepared on a high-load, high-accuracy, high-precision balance such as the one shown in Fig. 3, using weights calibrated with referenced masses of the National Institute of Standards and Technology (NIST) (formerly the National Bureau of Standards). The imprecision of a mixture made using this technique is a function of the imprecision of the balance and the calibrated masses as they relate to the amount of mass to be measured. Generally speaking, the greater the mass of each component, the greater the precision.

Mass can also be measured with electronic mass flowmeters. A mass flowmeter can be used in conjunction with a control device to control the amount of gas added to a system. This technique is used to blend gases at the final desired concentration while being compressed into cylinders for future use or while being consumed in a process. This dynamic method of blending gases has permitted users of high volumes of certain gas mixtures to obtain them essentially on demand. The

Fig. 3. High precision balance for preparing gas mixtures of highly accurate composition.

blended gases are monitored throughout the entire process by calibrated instrumentation. This technique is commonly used in vehicle emission calibration gas production and diving gas mixtures. The advantage of the dynamic blending system is its flexibility to produce a wide range of mixtures on demand, fully mixed and ready for use.

Homogenizing Mixtures

A concern some users have expressed over the years is whether gases can unmix or stratify due to gravitational forces. It has been demonstrated conclusively that once a gas mixture is homogeneous, it remains so and does not separate due to gravity. On the other hand, mixtures containing vapors of condensable components, if subjected to a temperature below their condensation temperature, will experience condensation of the condensable component.

No matter how painstakingly the measurements are made, if a mixture is used before its ingredients have been fully mixed, the gas withdrawn from the cylinder will not be the desired concentration and subsequent efforts to homogenize the mix will not produce the mixture at its original concentration.

There are a number of ways to homogenize gas mixtures. The simplest is to blend them prior to compression into the cylinders, as with the dynamic blending technique just described. Another way is to adjust the order in which the gases are added to the mix so that they are added in their increasing order of density (where physically possible). This yields a mix that will be homogeneous after standing for a relatively short period of time.

Alternatively, if vapor pressure limitations do not permit the gases or vapors to be introduced in an order that would produce a homogeneous mixture, then three options are available. After introducing all components into the mix, the container(s) can be placed in a horizontal position on a device that will rotate each container about its longitudinal axis. This is the most common method of homogenizing gas mixtures. Usually an hour of "rolling" is sufficient. Prior to the advent of "rolling" cylinders, the most common and effective way to homogenize a mixture was by heating the bottom of the cylinder, producing convection currents within the cylinder. Extreme caution must be exercised when using this method so as not to heat the cylinder above 125°F (51.7°C).

Some liquefied gas mixtures present an additional problem, especially where components have widely varying vapor pressures. Mixtures such as 12% dimethylethylamine/88% carbon dioxide and 20% ethylene oxide/80% carbon dioxide would not remain homogeneous during withdrawal from the container because they fractionate as they are drawn from the cylinder. For these mixtures, special eductor tubes have been designed to permit liquid withdrawal and thus minimize fractionation.

Some hydrocarbon mixtures which suffer from the same problem are provided in cylinders with a floating piston which can be pressurized on one side with an inert gas that pushes the piston towards the liquefied hydrocarbon blend, forcing the vaporized components back into liquid solution.

Quality Assurance

Quality assurance of mixed gases begins with analysis of the raw materials. Knowing the purity of the raw materials allows the technician to properly choose the components for the grade of mixture desired. Some of the instrumentation used for routine quality assurance of gases and vaporized liquids in the modern gas laboratory includes gas chromatographs equipped with detectors of the following types:

- Thermal conductivity
- Flame ionization
- Helium ionization
- Photoionization
- Thermistor
- Ultrasonic

In addition, infrared spectroscopy, atomic absorption spectroscopy, mass spectroscopy, and other dedicated analytical equipment are frequently used.

Most specialty gas mixture laboratories use "NIST traceable standards" for instrument calibration requiring high levels of accuracy such as Environmental Protection Agency (EPA) protocol mixtures. Where such standards are not available, suppliers typically make their own primary standards using gravimetric techniques. This can provide highly accurate standards of stable components. The grade of the desired mixture will determine the degree of analytical effort required.

Governmentally Regulated Mixtures

Mixed gases for medical applications such as pulmonary therapy and diagnostic gas mixtures require that specific quality assurance procedures be done as part of Good Manufacturing Practices under the authority, and subject to audit by, the U.S. Food and Drug Administration (FDA). Noncompliance may result in a warning or fine. Suppliers of these mixtures are required to be registered with the FDA.

EPA protocol mixtures for calibration of analytical instruments for air pollution emission measurements are required to pass quality assurance tests carried out in accordance with well-defined guidelines as to instrument calibration procedures and frequency (Protocols 1 and 2). [11] They must also pass stability testing prior to acceptance.

Suppliers of sterilant gas mixtures containing ethylene oxide must register with the EPA. Each cylinder must be labeled with an EPA registration number.

Suppliers of mixtures of radioactive gases are subject to regulation and must be licensed by the Nuclear Regulatory Commission (NRC).

REFERENCES

[1] *Threshold Limit Values and Biological Exposure Indices,* 1989–90 ed., American Conference of Governmental Industrial Hygienists, 6500 Glenway Avenue, Building D-7, Cincinnati, OH 45211-4438.

[2] *Code of Federal Regulations,* Title 49 CFR Parts 100–199 (Transportation), Superintendent of Documents, U.S. Government Printing Office, Washington, DC 20402.

[3] *Transportation of Dangerous Goods Regulations,* Canadian Government Publishing Centre, Supply and Services Canada, Ottawa, Ontario, Canada K1A 0S9.

[4] CGA C-4, *Method of Marking Portable Compressed Gas Containers to Identify the Material Contained,* Compressed Gas Association, Inc., 1235 Jefferson Davis Highway, Arlington, VA 22202.

[5] CGA C-7, *Guide for the Preparation of Precautionary Labeling and Marking of Compressed Gas Containers,* Compressed Gas Association, Inc., 1235 Jefferson Davis Highway, Arlington, VA 22202.

[6] *International Maritime Dangerous Goods Code,* International Maritime Organization, 4 Albert Embankment, London, England SE1 7SR.

[7] *Technical Instructions for the Safe Transport of Dangerous Goods by Air,* Intereg Group Inc., 5724 N. Pulaski Road, Chicago, IL 60646.

[8] *Dangerous Goods Regulations,* International Air Transport Association (IATA), 2000 Peel Street, Montreal, Quebec, Canada H3A 2R4.

[9] CGA V-7, *Standard Method of Determining Cylinder Valve Outlet Connections for Industrial Gas Mixtures,* Compressed Gas Association, Inc., 1235 Jefferson Davis Highway, Arlington, VA 22202.

[10] CGA S-7, *Method for Selecting Pressure Relief Devices for Compressed Gas Mixtures in Cylinders,* Compressed Gas Association, Inc., 1235 Jefferson Davis Highway, Arlington, VA 22202.

[11] *Traceability Protocol for Establishing True Concentrations of Gases Used for Calibration and Audits of Continuous Source Emission Monitors* (Protocol No. 1), and *Traceability Protocol for Establishing True Concentrations of Gases Used for Calibration and Audits of Air Pollution Analyzers* (Protocol No. 2), U.S. Environmental Protection Agency, Research Triangle Park, NC 27711.

ADDITIONAL REFERENCES

Coward, H. F., and Jones, G. W., *Limits of Flammability of Gases and Vapors,* Bureau of Mines, Bulletin 503, U.S. Department of the Interior, Washington, DC.

Gas Mixtures—Facts and Fables, Matheson Gas Products, Inc., Secaucus, NJ 07094.

International Hazardous Materials Transport Manual, The Bureau of National Affairs, Inc., 1231 25th Street, N.W., Washington, DC 22037.

Medard, Louis, "Collection Techniques et Documentation," Les Explosifs Occasionnels, 11 rue Lavoisier, Paris, France.

Reid, R. C., Prausnitz, J. M., and Sherwood, T. K., *The Properties of Gases and Liquids,* 3rd ed., McGraw-Hill, New York.

United States Pharmacopeia/The National Formulary, United States Pharmacopeial Convention, Inc., 12601 Twinbrook Parkway, Rockville, MD 20852.

Weast, Robert C., Ph.D., *Handbook of Chemistry and Physics,* The Chemical Rubber Co., 18901 Cranwood Parkway, Cleveland, OH 44128.

PART IV

Appendices

Part IV contains appendices which the reader will find helpful in clarifying terms and abbreviations encountered within the text of the *Handbook*.

Also, the appendix of State Regulatory Agencies and Codes lists some contact points, including addresses, for state authorities which may have jurisdictional authority over activities which pertain to compressed gases and compressed gas installations. State agencies listed are those with jurisdictional authority for the following:

- Environmental affairs
- Hazardous materials
- Highway safety
- Motor vehicles
- Occupational safety and health
- Public utilities
- Transportation
- Weights and measures

Those engaged in the manufacture, shipping, distribution, or use of compressed gases are responsible for compliance with all federal, state, provincial, and local governmental requirements. The addresses listed were the most current available at the time of publication and are provided as an aid to the reader. However, the reader is responsible for ascertaining the requirements of all current jurisdictional authorities regarding a given activity involving compressed gases or cryogenic liquids.

Some relevant codes developed by other organizations and pertaining to compressed gases are also listed.

Lastly, a listing providing descriptions of all publications and audiovisuals available from the Compressed Gas Association, Inc., is provided.

APPENDIX 1

Glossary of Terms

Absolute pressure Pressure measured with respect to zero pressure.

Absorption Penetration of a substance into the body of another.

Activity A thermodynamic quantity that denotes the effective concentration or intensity of a given substance in a given chemical system.

Adsorption The condensation of gases, liquids, or dissolved substances on the surfaces of solids.

Anhydrous A substance free of water.

Asphyxia A high degree of respiratory distress or suffocation due to lack of oxygen.

Atmospheric pressure The pressure exerted by the weight of the atmosphere at sea level. Equivalent to 14.696 pounds per square inch or 101.325 kPa. The pressure exerted by a column of mercury (Hg) 760 mm high.

Autoignition temperature The lowest temperature at which a flammable gas or vapor-air mixture ignites from its own heat source or a contacted heated surface without spark or flame. Vapors and gases spontaneously ignite at a lower temperature in oxygen than in air, and their autoignition temperature may be influenced by the presence of catalytic substances.

Bonnet A threaded member tightened onto or into the valve stem so as to clamp and compress the packing, diaphragm(s), or some other sealing members.

Cargo tank Any container designed to be permanently attached to any motor vehicle or other highway vehicle in which is to be transported any authorized compressed gas. A tank truck.

Carrier gases High-purity gases, primarily helium and nitrogen, which are used in gas chromatography or other processes to sweep another gas or vapor into or through a system. Argon and hydrogen are also commonly used.

Catalyst A substance that changes the speed of a chemical reaction but undergoes no permanent change itself.

Chromatography A method used to separate mixtures based on selective absorption or adsorption. It is used widely in analytical technology.

Compressed gas Any material or mixture having in the container an absolute pressure exceeding 40 psi at 70°F or, regardless of pressure at 70°F(21.1°C), having an absolute pressure exceeding 104 psi at 130°F, or any liquid material having a vapor pressure exceeding 40 psi absolute at 100°F as determined by ASTM Test D-323.

Compressibility factor The ratio between the real property (pressure, volume, temperature) and its value in the ideal state according to the ideal gas law. The value $z = 1$ indicates that the gas follows the ideal gas law without deviation.

Condensation point See *dew point*.

Container capacity For liquefied compressed gas containers, a container's capacity refers to the water capacity at 60° F which when multiplied by the authorized filling density for a particular gas equals the contents capacity for that gas. For example, a container with a water capacity of 277.8 pounds filled to the authorized filling density of 54% with ammonia would contain 150 pounds (68 kg) of ammonia. For nonliquefied compressed gas containers, refer to the authorized service pressures in U.S. Department of Transportation and Transport Canada regulations, and to NBS Technical Note

1079, *Tables of Industrial Gas Container Contents and Density for Oxygen, Argon, Nitrogen, Helium, and Hydrogen* (available from the Compressed Gas Association).

Cracking The act of slightly opening a valve to permit a very small amount of flow.

Critical pressure The pressure required to liquify a gas at the critical temperature.

Critical temperature The temperature above which a gas cannot be liquefied by pressure alone.

Critical volume The volume of one mole of a gas at critical temperature and pressure.

Cryogenic liquid A refrigerated liquefied gas having a normal boiling point below −130°F (−90°C).

Cryogenics The field of science dealing with the behavior of matter at very low temperatures.

Density The ratio of the mass of a specimen of a substance to its volume.

Dew point The temperature at which the partial pressure of a vapor in a gas is equal to the saturation pressure. Condensation will occur if the temperature continues to decrease. Dew point is also commonly expressed as ppm by volume.

Diaphragm valve A packless valve that utilizes clamped, flexible disk(s) to seal the opening through which the internal parts are installed. (See *packed valve.*)

DISS Diameter Index Safety System. Used in the medical gases and medical equipment segment of the compressed gas industry to help in the avoidance of erroneous connections. See CGA V-5 for details.

Dopant gas/semiconductor A gas or gas mixture used to incorporate a metallic impurity into a semiconductor substrate to impart particular electrical properties.

Edema A local or generalized condition in which the body tissues contain an excessive amount of tissue fluid.

Eductor tube (dip tube) A tube attached to the base of a cylinder valve, either straight or curved toward the wall of a cylinder, permitting the withdrawal of the liquid portion of the cylinder contents. The use of other terms, such as "gooseneck" tube, is discouraged.

Endothermic Characterized by heat absorption. A reaction in which heat is absorbed is called an endothermic reaction. An endothermic compound is formed from its basic constituents with the absorption of heat.

Enthalpy Heat function at constant pressure. Enthalpy is sometimes also called the heat content of the system. Symbol: H; unit: kJ or Btu.

Entropy A measure of the degree of disorder in a system, wherein every change that occurs and results in an increase of disorder is said to be a positive change in entropy. All spontaneous processes are accompanied by an increase in entropy according to the second law of thermodynamics. Symbol: S; unit: kJ/K or Btu/K.

Excess flow valve A device designed to close when the fluid passing through it exceeds a prescribed flow rate as determined by pressure drop.

Exothermic Characterized by heat evolution. A reaction in which heat is given off is called an exothermic reaction. An exothermic compound is formed from its elements with the evolution of heat.

Flammable range The difference between the lower and upper flammable limits, expressed in terms of percentage of vapor or gas in air or oxygen by volume; often referred to as the "explosive range." Oxygen-enriched concentrations will broaden the flammable range.

Freons Registered trade name for hydrocarbons where most of the hydrogen atoms have been replaced by fluorine and chlorine and/or bromine.

Fuel gas A combustible gas or gas mixture.

Fusible plug A nonreclosing pressure relief device designed to function by yielding or melting of a plug of material at a predetermined temperature.

Gas A state of matter in which the material has a very low density and viscosity, can expand and contract greatly in response to changes in temperature and pressure, easily diffuses into other gases, and readily and uniformly distributes itself throughout any container. A gas can be changed to the liquid or solid state only by the combined effect of increased pressure and decreased temperature (below the critical temperature).

Gas pressure The force exerted by a gas in its surroundings. In the United States, gas pressure is commonly designated in pounds per square inch (psi). The analogous SI unit is the kilopascal (kPa). One psi equals 6.894757 kPa. The term *psia* refers to absolute pressure. Absolute pressure is based on a zero reference point, the perfect vacuum. Measured from this reference, the standard atmospheric pressure at sea level is 14.696 psi; however, local atmospheric pressure may deviate from this standard value because of weather conditions and distance above or below sea level. Gauge pressure, designated by the term *psig,* is that pressure above local atmospheric pressure. Therefore, psia minus local atmospheric pressure equals psig.

GMP Good Manufacturing Practice. Used extensively in the medical gases service, usually in the context of U.S. Food and Drug Administration requirements for drug manufacture.

Handwheel A manually operable device attached to the valve stem through which the opening or closing of the valve is effected.

Heat capacity The quantity of heat required to raise the temperature of a given mass of material one degree. Heat capacity is always a positive quantity. The quantity of heat added to the material during heating is dependent on whether heating has taken place at constant volume or constant pressure. Symbol: C_p or C_v; unit: kJ/K (also kJ/°C) or Btu/°F.

Humidity Absolute humidity is the weight of water vapor per unit volume—pounds per cubic foot, or grams per cubic centimeter. Relative humidity is the ratio of the actual partial vapor pressure of the water vapor in a space to the saturation pressure of pure water at the same temperature.

Hydrocarbons Organic compounds composed solely of carbon and hydrogen. Several hundred thousand molecular combinations of C and H are known to exist. Basic hydrocarbons are: C_3H_8 (propane), CH_4 (methane), C_6H_{14} (hexane), and C_3H_6 (propylene).

Hydrogenation A reduction in which hydrogen is added to a substance by the use of gaseous hydrogen. The process is accomplished by means of a catalyst and proceeds more rapidly at high pressure.

Inert gas A nonreactive gas such as argon, helium, neon, and krypton.

Kelvin scale The fundamental temperature scale, also called the absolute or thermodynamic scale, in which the temperature measure is based on the average kinetic energy per molecule of a perfect gas. The zero of the Kelvin scale is −273.15°C (−459.67°F). A unit kelvin change is equivalent to a degree Celsius (°C) change. To convert from degree Fahrenheit (°F) to kelvin (K) use the formula: $T_K = (t°_F + 459.67)/1.8$. Symbol: K.

kPa Kilopascal. One pound per square inch (psi) of pressure is equivalent to 6.894757 kPa. As used in this *Handbook,* the term *kPa* refers to gauge pressure whereas the term *kPa abs* refers to absolute pressure.

Latent heat of fusion The heat required to convert a given mass of solid to liquid at the melting point.

Latent heat of vaporization The heat required to convert a unit mass of substance from the liquid state to the gaseous state at a given pressure and temperature.

LNG Liquid natural gas.

Lower explosive limit (L.E.L.) The lower limit of flammability of a gas or vapor at ordinary ambient temperatures and pressures expressed in percent of the gas or vapor in air by volume. The L.E.L. will vary with temperature and pressure.

LPG Liquefied petroleum gas. Refer to the monograph on liquefied petroleum gases in Part III.

Minimum pressure (minimum pressure rating) With respect to compressed gas containers, the term *minimum pressure* refers to the lowest service pressure rating authorized for cylinders of that specification in a particular gas service. For example, if Specification 3A480 cylinders were authorized, Specification 3A1000 or 3A2200 cylinders could also be used.

MOCVD Metal organic chemical vapor deposition.

Olefins A class of unsaturated hydrocarbons characterized by relatively great chemical activity. Obtained from petroleum and natural gas. Examples: butene, ethylene, and propylene. Generalized formula: C_nH_{2n}

Oxidizer A gas or liquid that accelerates combustion and that on contact with combustible material may cause fire or explosion.

Packed valve A valve that utilizes compressed packings to seal the openings through which the valve internal parts are installed, e.g. an O-ring valve. (See *diaphragm valve*)

Paraffins Paraffin series. (From *parum affinis,* "small affinity.") Those straight- or branched-chain hydrocarbon components of crude oil and natural gas whose molecules are saturated (i.e., carbon atoms attached to each other by single bonds) and therefore very stable. Examples: methane and ethane. Generalized formula: C_nH_{2n+2}

Permanent gas A gas which cannot be liquefied at normal ambient temperatures; a nonliquefied compressed gas.

Polymerization gas Gas used in the process of forming polymers. Vinyl chloride and butadiene are common examples.

Portable tank Any container designed primarily to be temporarily attached to a motor vehicle, other vehicle, railroad car other than tank car, or marine vessel, and equipped with skids, mountings, or accessories to facilitate handling of the container by mechanical means, in which is to be transported any authorized compressed gas.

ppm Parts per million. This can be expressed as ppm by volume or as ppm by weight.

Pressure Force applied to, or distributed over a surface; measured as force per unit area. See *absolute pressure, atmospheric pressure, gas pressure, static pressure,* and *vapor pressure.*

Pressure relief device A pressure- or temperature-actuated device used to prevent the pressure from rising above a predetermined maximum and thereby prevent the rupture of a normally charged cylinder when subjected to a standard fire test as required by 49 CFR 173.34(d) or equivalent Canadian regulations.

Pressure relief valve A type of pressure relief device designed to relieve excessive pressure and to reclose and reseal to prevent further flow of fluid from the cylinder after reseating pressure has been achieved.

psia Pounds per square inch absolute. (See *gas pressure.*)

psig Pounds per square inch gauge. (See *gas pressure.*)

Rare gas Gases such as argon, helium, neon, krypton, and xenon.

Relative humidity The ratio of the partial pressure of water vapor present in the air to the saturated vapor pressure of water at any specific temperature.

Rupture disk The operating part of a pressure relief device which, when installed in the device, is designed to rupture at a predetermined pressure to permit discharge of the cylinder contents.

Set pressure The pressure at which a pressure relief valve is set to discharge. Set pressure is normally marked on the pressure relief valve.

Specific gravity The ratio of the mass of a unit volume of a substance to the mass of the same volume of a standard substance at a standard temperature. Water at 4°C (39.2°F) is the standard usually referred to for liquids; for gases, dry air (at the same temperature and pressure as the gas) is often taken as the standard substance. See *density.*

Specific heat The heat required to raise the temperature of a given mass of a material one degree. See *heat capacity.*

Specific volume Volume per unit mass; material or material quantity unit. Unit: m^3/kg or ft^3/lb.

Static pressure The pressure exerted against the wall of a containing vessel in all directions by a fluid at rest. For a fluid in motion, it is measured in a direction normal (at right angles) to the direction of flow; thus it shows the tendency to burst or collapse the pipe. When added to velocity pressure, it gives total pressure.

Stridor A harsh sound during respiration; high pitched and resembling the blowing of wind, due to obstruction of air passages.

TC/DOT Transport Canada (TC) and/or the U.S. Department of Transportation, which are the government agencies of the respective countries with jurisdiction over the packaging and shipment of compressed gases.

Temperature The condition of a body that determines the transfer of heat to or from other bodies. Specifically, it is a manifestation of the average translational kinetic energy of the molecules of a substance due to heat agitation. See *Kelvin scale.*

TLV Threshold limit value. Also, TLV-TWA, time-weighted average threshold limit value. Time-weighted exposure level under which most people can work consistently for eight hours a day, day after day, with no harmful effects. A table of these values and accompanying precautions is published annually by the American Conference of Governmental Industrial Hygienists.

Triple point The only temperature and pressure at which three phases (gas, liquid, and solid) in a one-component system can exist in equilibrium.

Valve inlet The portion of the valve body that connects to the cylinder.

Valve outlet The portion of the valve body through which product is introduced into or discharged from the cylinder.

Vapor The gaseous form of substances that are normally in the solid or liquid state (at room temperature and pressure). The vapor can be changed back to the solid or liquid state either by increasing the pressure or decreasing the temperature alone. Vapors also diffuse. Evaporation is the process by which a liquid is changed into the vapor state and mixed with the surrounding air. Solvents with low boiling points will volatilize readily.

Vapor pressure Pressure exerted by a vapor. If a vapor is kept in confinement over its liquid so the vapor can accumulate above the liquid (the temperature being held constant), the vapor pressure approaches a fixed limit called the maximum (or saturated) vapor pressure, dependent only on the temperature and the liquid.

APPENDIX 2

List of Abbreviations

abs	absolute
ACGIH	American Conference of Governmental Industrial Hygienists
ASME	American Society of Mechanical Engineers
ASTM	American Society for Testing and Materials
atm	atmosphere
Btu	British thermal unit
°C	degrees Celsius
C_p	specific heat at constant pressure
C_v	specific heat of constant volume
cal	calorie(s)
CANUTEC	Canadian Transport Emergency Centre
CAS	Chemical Abstracts Service (American Chemical Society)
CFR	U.S. Code of Federal Regulations
CGA	Compressed Gas Association
CHEMTREC	Chemical Transportation Emergency Center
cm	centimeter(s)
cP	centipoise (viscosity)
CSA	Canadian Standards Association
DOT	U.S. Department of Transportation
EPA	U.S. Environmental Protection Agency
°F	degrees Fahrenheit
FDA	U.S. Food and Drug Administration
ft	foot, feet
g	gram(s)
gal	gallon(s)
IMO	International Maritime Organization
K	kelvin
kcal	kilocalorie(s)
kg	kilogram(s)
kJ	kilojoule(s)
kPa	kilopascal(s) gauge (gauge pressure)
kPa abs	kilopascal(s) absolute (absolute pressure)
L	liter(s)
lb	pound(s)
LP	liquefied petroleum
m	meter(s)
mg	milligram(s)
ml	milliliter(s)
mm	millimeter(s)
MPa	megapascal(s)
MSDS	material safety data sheet
NFPA	National Fire Protection Association
NPT	national pipe thread
OSHA	U.S. Occupational Safety and Health Administration
Pa•s	pascal-second(s) (viscosity)
ppm	parts per million
psia	pound-force per square inch absolute
psig	pound-force per square inch gauge
s	second(s)
TC	Transport Canada
TLV	threshold limit value (TWA, time-weighted average)
TMU	ton multi-unit tank
UN	United Nations
USP/NF	United States Pharmacopeia/ National Formulary
vol	volume
wt	weight

APPENDIX 3

State Regulatory Agencies and Codes

State Offices for Environmental Affairs

State	Agency and Location
Alabama	Dept. of Environmental Management, 1751 Federal Drive, Montgomery, AL 36130
Alaska	Dept. of Environmental Conservation, 3220 Hospital Drive, PO Box O, Juneau, AK 99811–1800
Arizona	Dept. of Environmental Quality, 2005 North Central Avenue, Phoenix, AZ 85004
Arkansas	Dept. of Pollution Control & Ecology, 8001 National Drive, PO Box 9583, Little Rock, AR 72219
California	Secretary of Environmental Affairs, 1102 Q Street, Sacramento, CA 95814
Colorado	Dept. of Natural Resources, 1313 Sherman Street, Denver, CO 80203
Connecticut	Dept. of Environmental Protection, 165 Capitol Avenue, Hartford, CT 06106
Delaware	Dept. of Natural Resources & Environmental Control, 715 Grantham Lane, New Castle, DE 19720
Dist. of Columbia	Dept. of Consumer & Regulatory Affairs, 5010 Overlook Drive S.W., Washington, DC 20032
Florida	Dept. of Environmental Regulation, 2600 Blair Stone Road, Tallahassee, FL 32399–2400
Georgia	Dept. of Natural Resources, 205 Butler Street, Suite 1252, Atlanta, GA 30334
Hawaii	Office of Environmental Quality Control, Dept. of Health, 465 King Street, Rm. 104, Honolulu, HI 96813
Idaho	Dept. of Health & Welfare, Towers Bldg, 450 West State Street, Boise, ID 83720
Illinois	Environmental Protection Agency, 2200 Churchill Road, Springfield, IL 62706
Indiana	Dept. of Environmental Management, 105 South Meridian Street, Indianapolis, IN 46225
Iowa	Dept. of Natural Resources, Wallace State Office Bldg., East 9th & Grand Avenue, Des Moines, IA 50319–0034
Kansas	Dept. of Health & Environment, Bldg. 740, Forbes Field, Topeka, KS 66620
Kentucky	Dept. of Environmental Protection, Fort Boone Plaza, 18 Reilly Road, Frankfort, KY 40601
Louisiana	Dept. of Environmental Quality, 625 North 4th Street, PO Box 44066, Baton Rouge, LA 70804
Maine	Dept. of Environmental Protection, Hospital Street, State House, Station 17, Augusta, ME 04333
Maryland	Dept. of Natural Resources, 2020 Industrial Drive, Annapolis, MD 21401
Massachusetts	Executive Office of Environmental Affairs, Leverett Saltonstall State Office Bldg., Rm. 2000, 100 Cambridge Street, Boston, MA 02202
Michigan	Dept. of Natural Resources, Stevens T. Mason Bldg., 7th Floor, PO Box 30028, Lansing, MI 48909
Minnesota	Environmental Quality Board, Centennial Office Bldg., Rm. 300, 658 Cedar Street, St. Paul, MN 55155
Mississippi	Dept. of Natural Resources, Highway 80 West at Ellis Avenue, PO Box 10385, Jackson, MS 39209

Missouri	Dept. of Natural Resources, 205 Jefferson Street, PO Box 176, Jefferson City, MO 65102
Montana	Dept. of Health & Environmental Sciences, W. F. Cogswell Bldg., Capitol Station, Helena, MT 59620
Nebraska	Dept. of Environmental Control, 301 Centennial Mall South, PO Box 98922, Lincoln, NE 68509–8922
Nevada	Dept. of Conservation & Natural Resources, 201 South Fall Street, Capitol Complex, Carson City, NV 89710
New Hampshire	Environmental Protection Bureau, State House Annex, 25 Capitol Street, Concord, NH 03301
New Jersey	Dept. of Environmental Protection, 401 East State Street, CN 402, Trenton, NJ 08625
New Mexico	Dept. of Health & Environment, 1190 St. Francis Drive, PO Box 968, Santa Fe, NM 87504–0968
New York	Dept. of Environmental Conservation, 50 Wolf Road, Albany, NY 12233–0001
North Carolina	Dept. of Natural Resources & Community Development, Archdale Bldg., Rm. 942, 512 North Salisbury Street, PO Box 27687, Raleigh, NC 27611
North Dakota	Dept. of Health, 1200 Missouri Avenue, PO Box 5520, Bismarck, ND 58502–5520
Ohio	Environmental Protection Agency, 1800 Watermark Drive, PO Box 1049, Columbus, OH 43266–0149
Oklahoma	Dept. of Pollution Control, 1000 N.E. 10th Street, PO Box 53504, Oklahoma City, OK 73152
Oregon	Dept. of Environmental Quality, Executive Bldg., 811 S.W. 6th Street, Portland, OR 97204
Pennsylvania	Dept. of Environmental Resources, 3rd & Locust Streets, PO Box 2063, Harrisburg, PA 17120
Rhode Island	Dept. of Environmental Management, 9 Hayes Street, Providence, RI 02908
South Carolina	Dept. of Health & Environmental Control, J. Marion Sims Bldg., Rm. 415, 2600 Bull Street, Columbia, SC 29201
South Dakota	Dept. of Water & Natural Resources, Joe Foss Bldg., Rm 209, 523 East Capitol Avenue, Pierre, SD 57501
Tennessee	Dept. of Health & Environment, Cordell Hull Bldg., Rm. 344, 436 6th Avenue North, Nashville, TN 37219–5404
Texas	Environmental Protection Division, 1124 South Interregional Hwy. 35, PO Box 12548, Capitol Station, Austin, TX 78711–2548
Utah	Dept. of Health, 288 North 1460 West, PO Box 16690, Salt Lake City, UT 84116–0690
Vermont	Agency of Natural Resources, Waterbury Office Complex, 103 South Main Street, Waterbury, VT 05676
Virginia	Council on the Environment, Ninth Street Office Bldg., Rm. 903, 202 North 9th Street, Richmond, VA 23219
Washington	Dept. of Ecology, St. Martins College, Mail Stop PV-11, Olympia, WA 98504
West Virginia	Dept. of Natural Resources, State Office Bldg. 3, Rm. 669, 1800 Washington Street East, Charleston, WV 25305
Wisconsin	Dept. of Natural Resources, Natural Resources Bldg., 101 South Webster Street, PO Box 7921, Madison, WI 53707
Wyoming	Dept. of Environmental Quality, Herschler Bldg., 4th Floor, 122 West 25th Street, Cheyenne, WY 82002
Puerto Rico	Environmental Quality Board, 204 Del Parque Street, PO Box 11488, Santurce PR 00910
Virgin Islands	Dept. of Planning & Natural Resources, 179 Altona & Welgunst, St. Thomas, VI 00802

State Offices for Hazardous Materials

State	Agency and Location
Alabama	Dept. of Environmental Management, 1751 Federal Drive, Montgomery, AL 36130
Alaska	Dept. of Environmental Conservation, 3220 Hospital Drive, PO Box O, Juneau, AK 99811–1800
Arizona	Radiation Regulatory Agency, 4814 South 40th Street, Phoenix, AZ 85040
Arkansas	Dept. of Pollution Control & Ecology, PO Box 9583, Little Rock, AR 72219
California	Waste Management Board, Environmental Affairs Agency, 1020 9th Street, Suite 300, Sacramento, CA 95814
Colorado	Office of Environmental Protection, Dept. of Health, 4210 East 11th Avenue, Denver, CO 80220
Connecticut	Dept. of Environmental Protection, 165 Capitol Avenue, Hartford, CT 06106
Delaware	Dept. of Natural Resources & Environmental Control, 89 Kings Hwy., Dover, DE 19903
Dist. of Columbia	Dept. of Consumer & Regulatory Affairs, 5010 Overlook Drive, S.W., Washington, DC 20032
Florida	Dept. of Environmental Regulation, Twin Towers Bldg., Rm. 626, 2600 Blair Stone Road, Tallahassee, FL 32399–2400
Georgia	Dept. of Natural Resources, 205 Butler Street, Suite 1152, Atlanta, GA 30334
Hawaii	Dept. of Health, Kinau Hale, 1250 Punchbowl Street, PO Box 3378, Honolulu, HI 96801
Idaho	Dept. of Health & Welfare, Towers Bldg., 5th Floor, 450 West State Street, Boise, ID 83720
Illinois	Dept. of Energy & Natural Resources, 1808 Woodfield Drive, Savoy, IL 61874
Indiana	Dept. of Environmental Management, 105 South Meridian Street, Indianapolis, IN 46225
Iowa	Dept. of Natural Resources, Wallace State Office Bldg., East 9th & Grand Avenue, Des Moines, IA 50319–0034
Kansas	Dept. of Health & Environment, Bldg. 740, Forbes Field, Topeka, KS 66620
Kentucky	Dept. of Environmental Protection, Fort Boone Plaza, 18 Reilly Road, Frankfort, KY 40601
Louisiana	Dept. of Environmental Quality, State Land & Natural Resources Bldg., 625 North 4th Street, PO Box 44066, Baton Rouge, LA 70804
Maine	Dept. of Environmental Protection, Ray Bldg., AMHI Complex, Hospital Street, Augusta, ME 04333
Maryland	Waste Management Administration, Dept. of the Environment, 201 West Preston Street, Rm. 2–11, Baltimore, MD 21201
Massachusetts	Division of Hazardous Waste, Dept. of Environmental Affairs, 80 Boylston Street, Suite 955, Boston, MA 02116
Michigan	Dept. of Natural Resources, Stevens T. Mason Bldg., 7th Floor, PO Box 30028, Lansing, MI 48909
Minnesota	Hazardous Waste Division, Pollution Control Agency, 520 Lafayette Rd., St. Paul, MN 55155
Mississippi	Dept. of Natural Resources, Southport Center, Highway 80 West at Ellis Avenue, PO Box 10385, Jackson, MS 39209

Missouri	Dept. of Natural Resources, Jefferson State Office Bldg., 205 Jefferson Street, PO Box 176, Jefferson City, MO 65102
Montana	Dept. of Health & Environmental Sciences, Cogswell Bldg., Rm. C108, Helena, MT 59620
Nebraska	Dept. of Environmental Control, State Office Bldg., 301 Centennial Mall South, PO Box 98922, Lincoln, NE 68509–8922
Nevada	Dept. of Conservation & Natural Resources, Nye Bldg., Rm. 221, 201 South Fall Street, Capitol Complex, Carson City, NV 89710
New Hampshire	Dept. of Environmental Services, Health & Human Services Bldg., 6 Hazen Drive, Concord, NH 03301
New Jersey	Dept. of Environmental Protection, 401 East State Street, CN 402, Trenton, NJ 08625
New Mexico	Bureau of Hazardous Waste, Dept. of Health & Environment, PO Box 968, Santa Fe, NM 87504–0968
New York	Hazardous Waste Program Office, 50 Wolf Road, Albany, NY 12205
North Carolina	Dept. of Human Resources, Bath Bldg., Rm. 213, 306 North Wilmington Street, PO Box 2091, Raleigh, NC 27602–2091
North Dakota	Dept. of Health, Missouri Office Bldg., 1200 Missouri Avenue, PO Box 5520, Bismarck, ND 58502–5520
Ohio	Environmental Protection Agency, 1800 Watermark Drive, PO Box 1049, Columbus, OH 43216
Oklahoma	Dept. of Health, 1000 N.E. 10th Street, PO Box 53551, Oklahoma City, OK 73152
Oregon	Dept. of Environmental Quality, Executive Bldg., 8th Floor, 811 SW 6th Ave., Portland, OR 97204
Pennsylvania	Emergency Management Agency, Transportation & Safety Bldg., Rm. B151, Commonwealth Avenue & Forster Street, P.O. Box 3321, Harrisburg, PA 17105
Rhode Island	Dept. of Environmental Management, Cannon Bldg., Rm. 204, 75 Davis Street, Providence, RI 02908
South Carolina	Dept. of Health & Environmental Control, J. Marion Sims Bldg., Rm. 415, 2600 Bull Street, Columbia, SC 29201
South Dakota	Dept. of Water & Natural Resources, Joe Foss Bldg., 523 East Capitol Avenue, Pierre, SD 57501–3181
Tennessee	Dept. of Health & Environment, Customs House, 4th Floor, 701 Broadway, Nashville, TN 37219–5403
Texas	Hazardous & Solid Waste Div., Stephen F. Austin State Office Bldg., Rm. 1150, 1700 North Congress Avenue, PO Box 13087, Capitol Station, Austin, TX 78711–3087
Utah	Division of Environmental Health, Dept. of Health, 288 North 1460 West, PO Box 16690, Salt Lake City, UT 84116–0690
Vermont	Dept. of Environmental Conservation, Waterbury Office Complex, West Bldg., 103 South Main Street, Waterbury, VT 05676
Virginia	Division of Health Hazards Control, Dept. of Health, Madison Bldg., 109 Governor Street, Richmond, VA 23219
Washington	Dept. of Ecology, 4224 6th Avenue S.E., Mail Stop PV-11, Olympia, WA 98503
West Virginia	Dept. of Health, 1800 Washington Street East, Charleston, WV 25305
Wisconsin	Dept. of Natural Resources, General Executive Facility II, 101 South Webster, PO Box 7921, Madison, WI 53707
Wyoming	Dept. of Environmental Quality, Herschler Bldg., 4th Floor, 122 West 25th Street, Cheyenne, WY 82002
Virgin Islands	Dept. of Planning & Natural Resources, 179 Altona & Welgunst, St. Thomas, VI 00802

State Offices for Highway Safety

State	Agency and Location
Alabama	Dept. of Ecomonic & Community Affairs, 3465 Norman Bridge Road, PO Box 2939, Montgomery, AL 36105–0939
Alaska	Dept. of Public Safety, 450 Whittier Street, Rm. 201, PO Box N, Juneau, AK 99811
Arizona	Dept. of Transportation, 3010 North 2nd Street, Suite 105, Phoenix, AZ 85012
Arkansas	Transportation Safety Agency, Justice Bldg., Rm 100, Little Rock, AR 72201
California	California Highway Patrol, 2555 1st Avenue, PO Box 942898, Sacramento, CA 94298–0001
Colorado	Dept. of Highways, West Annex, Rm. A-120, 4201 East Arkansas Avenue, Denver, CO 80222
Connecticut	Dept. of Transportation, 24 Wolcott Hill Road, Wethersfield, CT 06109
Delaware	Dept. of Public Safety, Thomas Collins Bldg., Rm. 363, 540 South DuPont Hwy., Dover, DE 19901
Dist. of Columbia	Dept. of Public Works, Frank Reeves Municipal Ctr., 2000 14th Street, N.W., Washington, DC 20009
Florida	Dept. of Highway Safety & Motor Vehicles, Neil Kirkman Bldg., 2900 Apalachee Pkwy., Tallahassee, FL 32399
Georgia	Governor's Office of Highway Safety, 100 Peachtree Street, Suite 2000, Atlanta, GA 30303
Hawaii	Dept. of Transportation, 79 South Nimitz Hwy., Honolulu, HI 96813
Idaho	Dept. of Transportation, 3311 West State Street, PO Box 7129, Boise, ID 83707
Illinois	Dept. of Transportation, IDOT Admin. Bldg., Rm. 319, 2300 South Dirksen Pkwy., Springfield, IL 62764
Indiana	Dept. of Highways, State Office Bldg., Rm. 801, 100 North Senate Avenue, Indianapolis, IN 46204
Iowa	Dept. of Public Safety, Wallace State Office Bldg., 3rd Floor, East 9th & Grand Avenue, Des Moines, IA 50319
Kansas	Dept. of Transportation, Docking State Office Bldg., 8th Floor, 915 Harrison Street, Topeka, KS 66612
Kentucky	Dept. of State Police, State Police Headquarters, 919 Versailles Road, Frankfort, KY 40601
Louisiana	Dept. of Public Safety & Corrections, 265 South Foster Drive, PO Box 66336, Baton Rouge, LA 70896
Maine	Dept. of Public Safety, Public Safety Bldg., 36 Hospital Street, State House, Station 42, Augusta, ME 04333
Maryland	Office of Transportation Planning, Dept. of Transportation, PO Box 8755, BWI Airport, MD 21240–0755
Massachusetts	Governor's Highway Safety Bureau, Leverett Saltonstall State Office Bldg., Rm. 2104, 100 Cambridge Street, Boston, MA 02202
Michigan	Office of Highway Safety Planning, Dept. of State Police, 300 South Washington Square, Suite 300, Lansing, MI 48913
Minnesota	Dept. of Public Safety, Transportation Bldg., Rm. 207, John Ireland Blvd., St. Paul, MN 55155
Mississippi	Criminal-Justice Planning, Federal-State Bldg., 301 West Pearl Street, Jackson, MS 39203–3088

Missouri	Dept. of Public Safety, 311 Ellis Blvd, PO Box 1406, Jefferson City, MO 65102
Montana	Highway Traffic Safety Division, Dept. of Justice, Scott Hart Bldg., Rm. 165, 303 North Roberts Street, Helena, MT 59620
Nebraska	Dept. of Motor Vehicles, State Office Bldg., 301 Centennial Mall South, PO Box 94612, Lincoln, NE 68509
Nevada	Dept. of Motor Vehicles & Public Safety, 555 Wright Way, Carson City, NV 89711–0999
New Hampshire	Highway Safety Agency, Pine Inn Plaza, 117 Manchester Street, Concord, NH 03301
New Jersey	Office of Highway/Traffic Safety, Dept. of Law & Public Safety, Bldg. No 5, Quakerbridge Plaza, CN 048, Trenton, NJ 08625
New Mexico	State Highway & Transportation Dept., Joseph Montoya Bldg., Rm. 3102, 1100 St. Francis Drive, PO Box 1149, Santa Fe, NM 87504–1149
New York	Governor's Traffic Safety Committee, Swan Street Bldg., 5th Floor, Empire State Plaza, Albany, NY 12228
North Carolina	Governor's Highway Safety Program, 215 East Lane Street, Raleigh, NC 27601
North Dakota	Highway Dept., Highway Bldg., 600 East Boulevard Avenue, Bismarck, ND 58505–0700
Ohio	Dept. of Highway Safety, 240 Parsons Avenue, PO Box 7167, Columbus, OH 43266–0563
Oklahoma	Dept. of Transportation Bldg., 200 N.E. 21st Street, Oklahoma City, OK 73105
Oregon	Traffic Safety Commission, State Library Bldg., Rm. 400, West Summer & Court Streets, Salem, OR 97310
Pennsylvania	Dept. of Transportation, Transportation & Safety Bldg., Rm. 1200, Commonwealth Avenue & Forster Street, Harrisburg, PA 17120
Rhode Island	Governor's Office on Highway Safety, 345 Harris Avenue, Providence, RI 02909
South Carolina	Office of Highway Safety Program, Edgar A. Brown Bldg., Rm. 412, 1205 Pendleton Street, Columbia, SC 29201
South Dakota	Dept. of Commerce & Regulation, Public Safety Bldg., 118 West Capitol Avenue, Pierre, SD 57501–2080
Tennessee	Dept. of Transportation, James K. Polk State Office Bldg., Rm. 600, 505 Deaderick Street, Nashville, TN 37219
Texas	Dept. of Highways & Public Transportation, La Costa Annex, 11th & Brazos Streets, Austin, TX 78701–2483
Utah	Dept. of Public Safety, Calvin A. Rampton Bldg., 4501 South 2700 West, Salt Lake City, UT 84119
Vermont	Agency of Transportation, State Administration Bldg., 133 State Street, Montpelier, VT 05601
Virginia	Dept. of Motor Vehicles, 2300 West Broad Street, Rm. 730, PO Box 27412, Richmond, VA 23269
Washington	Traffic Safety Commission, 1000 South Cherry Street, Mail Stop PD-11, Olympia, WA 98504
West Virginia	Criminal Justice & Highway Safety Office, 5790A MacCorkle Avenue S.E., Charleston, WV 25304
Wisconsin	Dept. of Transportation, Hill Farms State Transportation Bldg., 4802 Sheboygan Avenue, PO Box 7910, Madison, WI 53707
Wyoming	Highway Dept., Highway Dept. Bldg., 5300 Bishop Blvd., PO Box 1708, Cheyenne, WY 82002–9019
Puerto Rico	Dept. of Transportation & Public Works, PO Box 41269, Minillas Station, Santurce, PR 00940
Virgin Islands	Office of Highway Safety, Commercial Bldg., Rm. 218, Lagoon Complex, Frederiksted, St Croix, VI 00840

State Offices for Motor Vehicles

State	Agency and Location
Alabama	Dept. of Revenue, 2721 Gunter Park Drive West, Montgomery, AL 36109
Alaska	Dept. of Public Safety, 5700 East Tudor Road, Anchorage, AK 99507
Arizona	Dept. of Transportation, Motor Vehicle Div. Bldg., Rm. 264, 1801 West Jefferson Street, Phoenix, AZ 85007
Arkansas	Dept. of Finance & Administration, Joel Y. Ledbetter Bldg., Rm. 130, 7th & Wolfe Streets, PO Box 1272, Little Rock, AR 72203
California	Dept. of Motor Vehicles, 2415 1st Avenue, PO Box 932328, Sacramento, CA 94232–3280
Colorado	Motor Vehicle Division, Dept. of Revenue, 140 West 6th Avenue, Denver, CO 80204
Connecticut	Dept. of Motor Vehicles, 60 State Street, Wethersfield, CT 06109
Delaware	Dept. of Public Safety, Highway Administration Bldg., Route 113, PO Box 698, Dover, DE 19903
Dist. of Columbia	Transportation System Administration, Dept. of Public Works, Municipal Center, Rm. 1018, 301 C Street N.W., Washington, DC 20001
Florida	Dept. of Highway Safety & Motor Vehicles, Neil Kirkman Bldg., 2900 Apalachee Pkwy., Tallahassee, FL 32399–0600
Georgia	Dept. of Revenue, Trinity-Washington Bldg., Rm. 104, 270 Washington Street S.W., Atlanta, GA 30334
Idaho	Motor Vehicle Bureau, Transportation Dept., 3311 West State Street, PO Box 7129, Boise, ID 83707–1129
Illinois	Vehicles Services Dept., Centennial Bldg., Rm. 312, Springfield, IL 62756
Indiana	Bureau of Motor Vehicles, State Office Bldg., Rm. 401, 100 North Senate Avenue, Indianapolis, IN 46204
Iowa	Dept. of Transportation, 5268 N.W. 2nd Avenue, Des Moines, IA 50313
Kansas	Division of Vehicles, Dept. of Revenue, Docking State Office Bldg., 3rd Floor, 915 Harrison Street, Topeka, KS 66626
Kentucky	Dept. of Vehicle Regulation, State Office Bldg., Rm. 1001, Clinton & High Streets, Frankfort, KY 40622
Louisiana	Office of Motor Vehicles, Dept. of Public Safety, 1771 North Lobdell, PO Box 64886, Baton Rouge, LA 70896
Maine	Dept. of State, Transportation Bldg., Child Street, State House, Station 29, Augusta, ME 04333
Maryland	Motor Vehicle Administration, Dept. of Transportation, 6601 Ritchie Hwy. N.E., Glen Burnie, MD 21062
Massachusetts	Registry of Motor Vehicles, Dept. of Public Safety, 100 Nashua Street, Boston, MA 02114
Michigan	Bureau of Automotive Regulation, Dept. of State, 208 North Capitol Avenue, Lansing, MI 48918
Minnesota	Dept. of Public Safety, Transportation Bldg., Rm. 161, John Ireland Blvd., St. Paul, MN 55155
Mississippi	Motor Vehicles Commission Board, 4273 I-55 North, Suite 201, PO Box 16873, Jackson, MS 39205
Missouri	Dept. of Revenue, Harry S. Truman State Office Bldg., 301 West High Street, PO Box 629, Jefferson City, MO 65102

Montana	Motor Vehicle Division, Dept. of Justice, Scott Hart Bldg., 5th Floor, 303 North Roberts Street, Helena, MT 59620-1419
Nebraska	Dept. of Motor Vehicles, State Office Bldg., 301 Centennial Mall South, PO Box 94789, Lincoln, NE 68509
Nevada	Dept. of Motor Vehicles & Public Safety, 555 Wright Way, Carson City, NV 89711-0900
New Hampshire	Dept. of Safety, James H. Hayes Bldg., Hazen Drive, Concord, NH 03305
New Jersey	Division of Motor Vehicles, Dept. of Law & Public Safety, 25 South Montgomery Street, Trenton, NJ 08666
New Mexico	Dept. of Taxation & Revenue, Joseph M. Montoya Bldg., 1100 St. Francis Drive, PO Box 1028, Santa Fe, NM 87504
New York	Dept. of Motor Vehicles, Swan Street Bldg., Rm. 510, Empire State Plaza, Albany, NY 12228
North Carolina	Dept. of Transportation, Motor Vehicles Bldg., Rm. 220, 1100 New Bern Avenue, Raleigh, NC 27697
North Dakota	Motor Vehicle Dept., Capitol Grounds, Bismarck, ND 58505
Ohio	Dept. of Highway Safety, 4300 Kimberly Pkwy., PO Box 16520, Columbus, OH 43266-0020
Oklahoma	Motor Vehicle Commission, 4400 Will Rogers Pkwy., Suite 215, Oklahoma City, OK 73108
Oregon	Motor Vehicle Division, Dept. of Transportation, 1905 Lana Avenue N.E., Salem, OR 97314
Pennsylvania	Dept. of Transportation, Transportation & Safety Bldg., Rm. 316, Commonwealth & Forster Street, Harrisburg, PA 17123
Rhode Island	Dept. of Transportation, State Office Bldg., Rm. 110, 101 Smith Street, Providence, RI 02903
South Carolina	Dept. of Highway & Public Transportation, Silas N. Pearman Bldg., 955 Park Street, PO Box 1498, Columbia, SC 29216
South Dakota	Dept. of Revenue, Public Safety Bldg., 1st Floor, 118 West Capitol Avenue, Pierre SD 57501-2080
Tennessee	Dept. of Revenue, Andrew Jackson State Office Bldg., Rm. 604, 500 Deaderick Street, Nashville, TN 37242
Texas	Dept. of Highway & Public Transportation, 40th & Jackson Avenue, Austin, TX 78779
Utah	Motor Vehicle Division, State Tax Commission, State Fairgrounds, 1095 Motor Avenue, Salt Lake City, UT 84116
Vermont	Dept. of Motor Vechicles, State Office Bldg., 120 State Street, Montpelier, VT 05603-0001
Virginia	Dept. of Motor Vehicles, 2300 West Broad Street, PO Box 27412, Richmond, VA 23269-0001
Washington	Dept. of Licensing, Highways-Licenses Bldg., 12th Avenue & Franklin Street, Mail Stop PB-01, Olympia, WA 98504
West Virginia	Dept. of Motor Vehicles, State Office Bldg. 3, Rm. 126, 1800 Washington Street East, Charleston, WV, 25317
Wisconsin	Dept. of Transportation, Hill Farms State Transportation Bldg., Rm. 255, 4802 Sheboygan Avenue, Madison, WI 53702
Wyoming	Dept. of Revenue & Taxation, Herschler Bldg., 122 West 25th Street, Cheyenne, WY 82002-0110
Puerto Rico	Dept. of Transportation & Public Works, PO Box 41243, Minillas Station, Santurce, PR 00940-1243
Virgin Islands	Motor Vehicle Bureau, Police Dept., Sub Base, Charlotte Amalie, St. Thomas, VI 00801

State Offices for Occupational Safety & Health

State	Agency and Location
Alabama	Dept. of Industrial Relations, 1816 8th Avenue North, PO Box 10444, Birmingham, AL 35202
Alaska	Dept. of Labor, Dept. of Labor Bldg., Rm. 304, 1111 West 8th Street, PO Box 21149, Juneau, AK 99802–2149
Arizona	Div. of Occupational Safety & Health, 800 West Washington Street, Rm. 202, PO Box 19070, Phoenix, AZ 85005
Arkansas	Safety Division, Dept. of Labor, 1022 High Street, Little Rock, AR 72202
California	Dept. of Industrial Relations, 1006 4th Street, Sacramento, CA 95814
Connecticut	Division of Occupational Safety and Health, Dept. of Labor, 200 Folly Brook Blvd., Wethersfield, CT 06109
Delaware	Dept. of Labor, Elbert N. Carvel State Office Bldg., 820 North French Street, Wilmington, DE 19801
Dist. of Columbia	Dept. of Employment Services, 950 Upshur Street N.W., Washington, DC 20011
Florida	Dept. of Labor & Employment Security, Lafayette Bldg., Rm. 204, 2551 Executive Center Circle West, Tallahassee, FL 32399–0663
Georgia	Dept. of Human Resources, 878 Peachtree Street N.E., Suite 100, Atlanta, GA 30309
Hawaii	Dept. of Labor & Industrial Relations, Keelikolani Bldg., 830 Punchbowl Street, Rm. 423, Honolulu, HI 96813
Idaho	Dept. of Labor & Industrial Services, 277 North 6th Street, Statehouse Mail, Boise, ID 83720
Indiana	State Board of Health, Health Bldg., 1330 West Michigan Street, Indianapolis, IN 46206
Iowa	Employment Appeal Board, Lucas State Office Bldg., 2nd Floor, East 12th & Walnut Streets, Des Moines, IA 50319
Kansas	Industrial Safety and Health Section, Dept. of Human Resources, 512 West 6th Street, Topeka, KS 66603–3150
Kentucky	Occupational Safety & Health Review Commission, 4 Millcreek Road, Frankfort, KY 40601–9427
Louisiana	Dept. of Labor, Employment Security Bldg., 1001 North 23rd Street, PO Box 94094, Baton Rouge, LA 70804
Maine	Dept. of Labor, 102 Sewall Street, State House, Station 82, Augusta, ME 04333
Maryland	Division of Labor & Industry, Dept. of Licensing & Regulation, 501 St. Paul Place, Baltimore, MD 21202
Massachusetts	Dept. of Labor & Industries, Leverett Saltonstall State Office Bldg., Rm. 1107, 100 Cambridge Street, Boston, MA 02202
Michigan	Dept. of Public Health, Baker-Olin North Bldg., 3500 North Logan Street, PO Box 30035, Lansing, MI 48909
Minnesota	Dept. of Labor & Industry, Space Center Bldg., 444 Lafayette Road, St. Paul, MN 55101
Mississippi	Dept. of Health, 305 West Lorenz Blvd., Jackson, MS 39206
Missouri	Dept. of Labor & Industrial Relations, 621 East McCarty Street, PO Box 449, Jefferson City, MO 65102
Montana	Dept. of Health & Environmental Sciences, Cogswell Bldg., Rm A113, Lockey Street, Helena, MT 59620

Nebraska	Dept. of Labor, State Office Bldg., 301 Centennial Mall South, PO Box 95024, Lincoln, NE 68509
Nevada	Dept. of Industrial Relations, 1370 South Curry Street, Carson City, NV 89710
New Hampshire	Inspection Division, Dept. of Labor, 19 Pillsbury Street, Concord, NH 03301
New Jersey	Workplace Standards Division, Dept. of Labor, Station Plaza 4, 3rd Floor, CN 386, Trenton, NJ 08625
New Mexico	Health & Environment Dept., Harold Runnels Bldg., 1190 St. Francis Drive, PO Box 968, Santa Fe, NM 87504-0968
New York	Dept. of Labor, State Office Bldg., Rm. 512, Albany, NY 12240
North Carolina	Dept. of Labor, Shore Bldg., 214 West Jones Street, Raleigh, NC 27603
North Dakota	Dept. of Health, Missouri Office Bldg., 1200 Missouri Avenue, PO Box 5520, Bismarck, ND 58502-5520
Ohio	Industrial Commission, 246 North High Street, 4th Floor, Columbus, OH 43215
Oklahoma	Dept. of Health, 1000 N.E. 10th Street, PO Box 53551, Oklahoma City, OK 73152
Oregon	Dept. of Insurance & Finance, Labor & Industries Bldg., Salem, OR 97310
Pennsylvania	Dept. of Labor & Industry, Labor & Industry Bldg., Rm. 1528, 7th & Forster Streets, Harrisburg, PA 17120
Rhode Island	Dept. of Health, Cannon Bldg., Rm. 206, 75 Davis Street, Providence, RI 02908
South Carolina	Labor Dept., Landmark Center, 3600 Forest Drive, PO Box 11329, Columbia, SC 29211
Tennessee	Dept. of Labor, 501 Union Bldg., 6th Floor, Nashville, TN 37219-5385
Texas	Occupational Safety and Health Division, Dept. of Health, 1100 West 49th Street, Austin, TX 78756
Utah	Div. of Occupational Safety & Health, Heber M. Wells Bldg., 160 East 300 South, PO Box 45580, Salt Lake City, UT 84145
Vermont	Dept. of Health, Administration Bldg., 10 Baldwin Street, Montpelier, VT 05602
Virginia	Dept. of Labor & Industry, 205 North Fourth Street, PO Box 12064, Richmond, VA 23241
Washington	Dept. of Labor & Industries, Office Bldg. 6, 805 Plum Street S.E., PO Box 207, Olympia, WA 98504
West Virginia	Industrial Hygiene Division, Dept. of Health, 151 11th Avenue, South Charleston, WV 25303
Wisconsin	Dept. of Industry, Labor, & Human Relations, General Executive Facility I, 201 East Washington, PO Box 7969, Madison, WI 53707
Wyoming	Occupational Health & Safety Commission, 604 East 25th Street, Cheyenne, WY 82002
Puerto Rico	Dept. of Labor & Human Resources, Prudencio Rivera Martinez Bldg., 505 Munoz Rivera Avenue, Hato Rey, PR 00918
Virgin Islands	Dept. of Labor, Government Complex, Bldg. 2, Rm. 207, Lagoon Street, Frederiksted, St. Croix, VI 00840

State Offices for Public Utilities

State	Agency and Location
Alabama	Public Service Commission, State Office Bldg., Rm. 712, 501 Dexter Avenue, PO Box 991, Montgomery, AL 36102
Alaska	Public Utilities Commission, Dept. of Commerce & Economic Development, 420 L Street, Suite 100, Anchorage, AK 99501
Arizona	Arizona Corporation Commission, 1200 West Washington Street, Phoenix, AZ 85007
Arkansas	Public Service Commission, 1000 Center Street, PO Box C-400, Little Rock, AR 72203
California	Public Utilities Commission, 505 Van Ness Avenue, San Francisco, CA 94102
Colorado	Dept. of Regulatory Agencies, Logan Tower, Office Level 2, 1580 Logan Street, Denver, CO 80203
Connecticut	Dept. of Public Utility Control, One Central Park Plaza, New Britain, CT 06051
Delaware	Public Service Commission, Dept. of Administrative Services, 1560 South DuPont Hwy., Dover, DE 19901
Dist. of Columbia	Public Service Commission, 450 5th Street, N.W., Washington, DC 20001
Florida	Public Service Commission, Fletcher Bldg., Rm. 116, 101 East Gaines Street, Tallahassee, FL 32399–0850
Georgia	Public Service Commission, State Office Bldg. Annex, Rm. 162, 244 Washington Street S.W., Atlanta, GA 30334
Hawaii	Public Utility Commission, Kekuanaoa Bldg., 465 South King Street, Honolulu, HI 96813
Idaho	Public Utilities Commission, 472 West Washington Street, Boise, ID 83720
Illinois	Commerce Commission, State of Illinois Center, Suite 9-100, 100 West Randolph Street, Chicago, IL 60601
Indiana	Utility Regulatory Commission, State Office Bldg., Rm. 913, 100 North Senate Avenue, Indianapolis, IN 46204–2284
Iowa	Dept. of Commerce, Lucas State Office Bldg., 5th Floor, East 12th & Walnut Streets, Des Moines, IA 50319
Kansas	Utilities Div., Corporation Commission, State Office Bldg., 4th Floor, 915 Harrison Street, Topeka, KS 66612
Kentucky	Public Service Commission, 730 Schenkel Lane, PO Box 615, Frankfort, KY 40602
Maine	Public Utilities Commission, 242 State Street, State House, Station 18, Augusta, ME 04333
Maryland	Public Service Commission, American Bldg., 231 East Baltimore Street, Baltimore, MD 21202
Massachusetts	Dept. of Public Utilities, Leverett Saltonstall State Office Bldg., 12th Floor, 100 Cambridge St, Boston, MA 02202
Michigan	Public Service Commission, Dept. of Commerce, 6545 Mercantile Way, PO Box 30221, Lansing, MI 48909
Minnesota	Dept. of Public Service, American Center Bldg., Rm. 790, 150 East Kellog Blvd., St. Paul, MN 55101
Mississippi	Public Service Commission, Walter Sillers State Office Bldg., Rm. 1729, 550 High Street, PO Box 1174, Jackson, MS 39215–1174
Missouri	Dept. of Economic Development, Harry S. Truman State Office Bldg., Rm. 530, 301 West High Street, PO Box 360, Jefferson City, MO 65102

Montana	Dept. of Public Service Regulation, Dept. of Highways Bldg., 2701 Prospect Avenue, Helena, MT 59620–2601
Nebraska	Public Service Commission, The Atrium, Rm. 300, 1200 North Street, PO Box 94927, Lincoln, NE 68509
Nevada	Public Service Commission, 727 Fairview Drive, Capitol Complex, Carson City, NV 89710
New Hampshire	Public Utilities Commission, 8 Old Suncook Road, Concord, NH 03301
New Jersey	Board of Public Utilities, Dept. of Energy, 2 Gateway Center, Newark, NJ 07104
New Mexico	Public Service Commission, Marian Hall, 224 East Palace Avenue, Santa Fe, NM 87503
New York	Public Service Commission, Dept. of Public Service, Agency Bldg. 3, Empire State Plaza, Albany, NY 12223
North Carolina	Utilities Commission, Dept. of Commerce, Dobbs Bldg., 430 North Salisbury Street, PO Box 29510, Raleigh, NC 27626–0510
North Dakota	Public Service Commission, State Capitol, Bismarck, ND 58505
Ohio	Public Utilities Commission, Borden Bldg., 15th Floor, 180 East Broad Street, Columbus, OH 43266–0573
Oklahoma	Corporation Commission, Jim Thorpe Office Bldg., Rm. 500, 2101 North Lincoln Blvd., Oklahoma City, OK 73105
Oregon	Public Utilities Commission, Labor & Industries Bldg., Rm. 300, Salem, OR 97310
Pennsylvania	Public Utility Commission, North Office Bldg., Rm. G-18, Commonwealth Ave. & North Street, PO Box 3265, Harrisburg, PA 17120
Rhode Island	Public Utilities Commission, 100 Orange Street, Providence, RI 02903
South Carolina	Public Service Commission, 111 Doctors Circle, PO Drawer 11649, Columbia, SC 29211
South Dakota	Public Utilities Commission, Dept. of Commerce & Regulation, State Capitol, 500 East Capitol Avenue, Pierre, SD 57501
Tennessee	Public Service Commission, 460 James Robertson Pkwy., Nashville, TN 37219
Texas	Railroad Commission, William B. Travis Bldg., 1701 North Congress Avenue, PO Box 42967, Capitol Station, Austin, TX 78711
Utah	Public Service Commission, Heber M. Wells Bldg., 4th Floor, 160 East 300 South, PO Box 45585, Salt Lake City, UT 84145–0801
Vermont	Dept. of Public Service, State Office Bldg., 120 State Street, Montpelier, VT 05602
Virginia	State Corporation Commission, Jefferson Bldg., 1220 Bank Street, PO Box 1197, Richmond, VA 23209
Washington	Utilities & Transportation Commission, Chandler Plaza Bldg., 1300 South Evergreen Park Drive, Olympia, WA 98504
West Virginia	Public Service Commission, 201 Brooks Street, PO Box 812, Charleston, WV 25323
Wisconsin	Public Service Commission, Hill Farms State Transportation Bldg., Rm. 475, 4802 Sheboygan Avenue, PO Box 7854, Madison, WI 53707
Wyoming	Public Service Commission, Herschler Bldg., 1st Floor, 122 West 25th Street, Cheyenne, WY 82002
Puerto Rico	Public Service Commission, La Electronica Bldg., Road No. 1 Km 14.7, Rio Piedras, PO Box C P, Hato Rey, PR 00919–3806
Virgin Islands	Public Services Commission, PO Box 40, St. Thomas, VI 00801

State Offices for Transportation

State	Agency and Location
Alabama	Public Service Commission, State Office Bldg., Rm. 782, 501 Dexter Avenue, Montgomery, AL 36130
Alaska	Dept. of Transportation Public Facilities, 3132 Channel Drive, PO Box Z, Juneau, AK 99811
Arizona	Dept. of Transportation, 206 South 17th Avenue, Phoenix, AZ 85007
Arkansas	Transportation Safety Agency, Justice Bldg., Rm. 100, Little Rock, AR 72201
California	Dept. of Transportation, Transportation Bldg., 1120 North Street, Sacramento, CA 95814
Colorado	Dept. of Highways, 4201 East Arkansas Avenue, Denver, CO 80222
Connecticut	Dept. of Transportation, 24 Wolcott Hill Road, Wethersfield, CT 06109
Delaware	Dept. of Transportation, Highway Administration Bldg., Rm. 202, Rt. 113, PO Box 778, Dover, DE 19903
Dist. of Columbia	Dept. of Public Works, Municipal Center, Rm. 1018, 301 C Street N.W., Washington, DC 20001
Florida	Dept. of Transportation, Haydon-Burns Bldg., 605 Suwannee Street, Mail Station 57, Tallahassee, FL 32399–0450
Georgia	Dept. of Transportation, 2 Capitol Square S.W., Atlanta, GA 30334
Hawaii	Dept. of Transportation, Aliiaimoku Hale, 869 Punchbowl Street, Honolulu, HI 96813
Idaho	Transportation Dept., 3311 West State Street, Rm. 201, PO Box 7129, Boise, ID 83707
Illinois	Dept. of Transportation, IDOT Administration Bldg., 2300 South Dirksen Pkwy., Springfield, IL 62764
Indiana	Dept. of Transportation, Harrison Bldg., Rm. 300, 143 West Market Street, Indianapolis, IN 46204
Iowa	Dept. of Transportation, 800 Lincoln Way, Ames, IA 50010
Kansas	Dept. of Transportation, Docking State Office Bldg., 7th Floor, 915 Harrison Street, Topeka, KS 66612
Kentucky	Dept. of Highways, Transportation Cabinet, State Office Bldg., 10th Floor, Clinton & High Streets, Frankfort, KY 40622
Louisiana	Dept. of Transportation & Development, 1201 Capitol Access Road, PO Box 94245, Baton Rouge, LA 70804–9245
Maine	Dept. of Transportation, Transportation Bldg., Child Street, State House, Station 16, Augusta, ME 04333
Maryland	Dept. of Transportation, PO Box 8755, BWI Airport, MD 21240–0755
Massachusetts	Executive Office of Transportation & Construction, Transportation Bldg., Rm. 3510, 10 Park Plaza, Boston, MA 02116–3969
Michigan	Dept. of Transportation, Transportation Bldg., 425 West Ottawa Street, P.O. Box 30050, Lansing, MI 48909
Minnesota	Dept. of Transportation, Transportation Bldg., Rm. 411, John Ireland Blvd., St. Paul, MN 55155
Mississippi	Dept. of Energy & Transportation, Dickson Bldg., 510 George Street, Jackson, MS 39202

Missouri	Highway & Transportation Dept., Highway & Transportation Bldg., PO Box 270, Jefferson City, MO 65102
Montana	Transportation Division, Public Service Commission, Dept. of Highways Bldg., 2701 Prospect Avenue, Helena, MT 59620–2601
Nebraska	Dept. of Roads, S Jct. US77 & N-2, PO Box 94759, Lincoln, NE 68509–4759
Nevada	Dept. of Transportation, 1263 South Stewart Street, Carson City, NV 89712
New Hampshire	Dept. of Transportation, John O'Morton Bldg., Rm. 102, Hazen Drive, PO Box 483, Concord, NH 03301
New Jersey	Dept. of Transportation, 1035 Parkway Avenue, CN 600, Trenton, NJ 08625
New Mexico	Taxation & Revenue Dept., 1200 South St. Francis Drive, PO Box 630, Santa Fe, NM 87509
New York	Dept. of Transportation, State Campus Bldg. 5, Albany, NY 12232
North Carolina	Dept. of Transportation, Highway Bldg., Rm. 154, One South Wilmington Street, PO Box 25201, Raleigh, NC 27611
North Dakota	Highway Dept., Highway Bldg., 600 East Boulevard Avenue, Bismarck, ND 58505–0700
Ohio	Dept. of Transportation, 25 South Front Street, PO Box 0415, Columbus, OH 43266
Oklahoma	Dept. of Transportation, Dept. of Transportation Bldg., 200 N.E. 21st Street, Oklahoma City, OK 73105
Oregon	Dept. of Transportation, Transportation Bldg., Rm. 135, Capitol Mall, Salem, OR 97310
Pennsylvania	Dept. of Transportation, Transportation & Safety Bldg., Rm. 1200, Commonwealth Avenue & Forster Street, Harrisburg, PA 17120
Rhode Island	Dept. of Transportation, State Office Bldg. Rm. 210, 101 Smith Street, Providence, RI 02903
South Carolina	Dept. of Highways & Public Transportation, Silas N. Pearman Bldg., 955 Park Street, PO Box 191, Columbia, SC 20202
South Dakota	Dept. of Transportation, Transportation Bldg., 700 East Broadway, Pierre, SD 57501–2586
Tennessee	Dept. of Transportation, James K. Polk State Office Bldg., Rm. 700, 505 Deaderick Street, Nashville, TN 37219
Texas	Dept. of Highways & Public Transportation, Dewitt C. Greer Bldg., 11th & Brazos Streets, Austin, TX 78701–2483
Utah	Dept. of Transportation , DOT/Public Safety Bldg., 4501 South 2700 West, Salt Lake City, UT 84119
Vermont	Agency of Transportation, State Administration Bldg., 133 State Street, Montpelier, VT 05602
Virginia	Dept. of Transportation, 1401 East Broad Street, Rm. 311, Richmond, VA 23219
Washington	Dept. of Transportation, Transportation Bldg., Maple Park, Mail Stop KF-01, Olympia, WA 98504–9990
West Virginia	Transportation Division, Dept. of Finance & Administration, State Parking Bldg., 212 California Avenue, Charleston, WV 25305
Wisconsin	Dept. of Transportation, Hill Farms State Transportation Bldg., Rm. 120B, 4802 Sheboygan Ave., PO Box 7910, Madison, WI 53707–7910
Wyoming	Highway Dept., Highway Department Bldg., 5300 Bishop Blvd., PO Box 1708, Cheyenne, WY 82002–9019
Puerto Rico	Dept. of Transportation & Public Works, Minillas Governmental Center, South Bldg., De Diego Ave, PO Box 41269, Minillas Station, Santurce, PR 00940
Virgin Islands	Dept. of Property & Procurement, Sub Base, Bldg. One, Charlotte Amalie, PO Box 1437, St. Thomas, VI 00801

State Offices for Weights and Measures

State	Agency and Location
Alabama	Dept. of Agriculture & Industries, Richard Beard Bldg., Rm. 126, 1445 Federal Drive, PO Box 3336, Montgomery, AL 36193
Alaska	Dept. of Commerce & Economic Development, 12050 Industry Way, PO Box 111686, Anchorage, AK 99511
Arizona	Dept. of Weights & Measures, 1951 West North Lane, Phoenix, AZ 85021
Arkansas	Bureau of Standards, 4608 West 61st Street, Little Rock, AR 72209
California	Division of Measurement Standards, Dept. of Food & Agriculture, 8500 Fruitridge Road, Sacramento, CA 95826
Colorado	Inspection and Consumer Services Division, Dept. of Agriculture, 3125 Wyandot Street, Denver, CO 80211
Connecticut	Dept. of Consumer Protection, State Office Bldg., Rm G-17, 165 Capitol Avenue, Hartford, CT 06106
Delaware	Division of Standards and Inspection, Dept. of Agriculture, Agriculture Bldg., 2320 Dupont Hwy., Dover, DE 19901
Dist. of Columbia	Dept. of Consumer & Regulatory Affairs, 1110 U Street S.E., Washington, DC 20020
Florida	Dept. of Agriculture & Consumer Services, Laboratory Complex, Administration Bldg., 3125 Conner Blvd., Tallahassee, FL 32301
Georgia	Weights & Measures Division, Dept. of Agriculture, Agriculture Bldg., Rm. 327, Capitol Square, Atlanta, GA 30334
Hawaii	Dept. of Agriculture, Princess Nahienaena Bldg., 725 Ilalo Street, Honolulu, HI 96813
Idaho	Bureau of Weights & Measures, Dept. of Agriculture, 2216 Kellog Lane, Boise, ID 83712
Illinois	Dept. of Agriculture, State Fairgrounds, PO Box 19281, Springfield, IL 62794-9281
Indiana	State Board of Health, Health Bldg., Rm. 136, 1330 West Michigan Street, Indianapolis, IN 46206
Iowa	Dept. of Agriculture & Land Stewardship, Wallace State Office Bldg., East 9th & Grand Avenue, Des Moines, IA 50319
Kansas	Weights & Measures, Board of Agriculture, 2016 West 37th Street, Topeka, KS 66611-2570
Kentucky	Division of Weights & Measures, Dept. of Agriculture, 106 West 2nd Street, Frankfort, KY 40601-2882
Louisiana	Dept. of Agriculture, 9181 Interline Drive, Suite 204, PO Box 44456, Baton Rouge, LA 70804
Maine	Dept. of Agriculture, Deering Bldg., AMHI Complex, State House, Station 28, Augusta, ME 04333
Maryland	Weights & Measures Section, Dept. of Agriculture, 50 Harry S. Truman Pkwy., Annapolis, MD 21401
Massachusetts	Executive Office of Consumer Affairs, John W. McCormack State Office Bldg., Rm. 1115, One Ashburton Place, Boston, MA 02108
Michigan	Dept. of Agriculture, Ottawa Bldg. North, 611 West Ottawa Street, PO Box 30017, Lansing, MI 48909
Minnesota	Division of Weights & Measures, Dept. of Public Service, 2277 Highway 36, Roseville, MN 55113
Mississippi	Dept. of Agriculture & Commerce, Walter Sillers State Office Bldg., Rm. 1603, 550 High Street, PO Box 1609, Jackson, MS 39215-1609

Missouri	Dept. of Agriculture, Dept. of Agriculture Bldg., 1616 Missouri Blvd., PO Box 630, Jefferson City, MO 65102
Montana	Bureau of Weights & Measures, Dept. of Commerce, 1424 9th Avenue, Helena, MT 59620-0423
Nebraska	Dept. of Agriculture, State Office Bldg., 301 Centennial Mall South, PO Box 94757, Lincoln, NE 68509
Nevada	Bureau of Weights & Measures, Dept. of Agriculture, PO Box 11100, Reno, NV 89510
New Hampshire	Dept. of Agriculture, Concord Center, 10 Ferry Street, Caller Box 2042, Concord, NH 03302-2042
New Jersey	Office of Weights & Measures, Dept. of Law & Public Safety, 187 West Hanover Street, Trenton, NJ 08625
New Mexico	Dept. of Agriculture, Box 30005, Dept. 3170, Las Cruces, NM 88003-0005
New York	Bureau of Weights & Measures, Dept. of Agriculture & Markets, Capitol Plaza, One Winner Circle, Albany, NY 12235
North Carolina	Dept. of Agriculture, Agriculture Bldg., Rm. 310, One West Edenton Street, PO Box 27647, Raleigh, NC 27611-7647
North Dakota	Weights & Measures Division, Public Service Commission, State Capitol, Bismarck, ND 58505
Ohio	Division of Weights & Measures, Dept. of Agriculture, 8995 East Main Street, Reynoldsburg, OH 43068
Oklahoma	Bureau of Standards, Dept. of Agriculture, 2800 North Lincoln Blvd., Oklahoma City, OK 73105
Oregon	Measurement Standards Division, Dept. of Agriculture, Agriculture Bldg., 635 Capitol Street N.E., Salem, OR 97310
Pennsylvania	Weights & Measures Bureau, Dept. of Agriculture, Agriculture Bldg., Rm. 206, 2301 North Cameron Street, Harrisburg, PA 17110
Rhode Island	Dept. of Labor, 220 Elmwood Avenue, Providence, RI 02907
South Carolina	Dept. of Agriculture, 1101 Williams Street, PO Box 11280, Columbia, SC 29211-1280
South Dakota	Dept. of Commerce & Regulation, Public Safety Bldg., 118 West Capitol Avenue, Pierre, SD 57501
Tennessee	Dept. of Agriculture, Ellington Agricultural Center, Hogan Road, PO Box 40627, Melrose Station, Nashville, TN 37204
Texas	Dept. of Agriculture, Stephen F. Austin State Office Bldg., 1700 N. Congress Avenue, PO Box 12847, Austin, TX 78711
Utah	Weights & Measures Section, Dept. of Agriculture, William Spry Bldg., 350 North Redwood Road, Salt Lake City, UT 84116
Vermont	Division of Weights & Measures, Dept. of Agriculture, Agriculture Bldg., 116 State Street, Montpelier, VT 05602
Virginia	Dept. of Agriculture & Consumer Services, Washington Bldg.-Capitol Square, Rm. 402, 1100 Bank Street, PO Box 1163, Richmond, VA 23209
Washington	Dept. of Agriculture, General Administration Bldg., Rm. 406, 11th Avenue & Columbia Street, Mail Stop GG-11, Olympia, WA 98504
West Virginia	Weights & Measures Division, Dept. of Labor, 1800 Washington Street East, Charleston, WV 25305
Wisconsin	Dept. of Agriculture, Trade & Consumer Protection, 801 West Badger Road, PO Box 8911, Madison, WI 53708
Wyoming	Weights & Measures Division, Dept. of Agriculture, 2219 Carey Avenue, Cheyenne, WY 82002
Puerto Rico	Dept. of Consumer Affairs, Minillas Governmental Center, North Bldg., De Diego & Baldorioty de Castro Ave, PO Box 41059, Santurce, PR 00940
Virgin Islands	Dept. of Licensing & Consumer Affairs, Golden Rock Shopping Center, Christiansted, St. Croix, VI 00820

STATE FIRE MARSHALS

Alabama

State Fire Marshal
135 South Union Street, Room 140
Montgomery, AL 36130-3401

Alaska

State Fire Marshall
Division of Fire Prevention
Department of Public Safety
5700 East Tudor Road
Anchorage, AK 99507-1225

Arizona

State Fire Marshall
Department of Building and Fire Safety
701 East Jefferson Street (Suite 200)
Phoenix, AZ 85034

Arkansas

Commander, Fire Marshal Section
Department of State Police
3 Natural Resources Drive
P.O. Box 5901
Little Rock, AR 72215

California

State Fire Marshall
State of California
7171 Bowling Drive (Suite 600)
Sacramento, CA 95823

Colorado

Director
Division of Fire Safety
Department of Public Safety
700 Kipling Street (Suite 3000)
Denver, CO 80215

Connecticut

State Fire Marshal
Bureau of State Fire Marshal and
Safety Services
Division State Police
Department of Public Safety
294 Colony Street
Meriden, CT 06450

Delaware

State Fire Marshall
Delaware Fire Prevention Commission
R.D. 2, Box 166A
Dover, DE 19901

District of Columbia

Fire Marshal
Fire Prevention Division
District of Columbia Fire Department
613 G Street, N.W. (Room 810)
Washington, DC 20001

Florida

Director
Division of State Fire Marshal
Larson Building (Room 541)
Tallahassee, FL 32399-0300

Georgia

State Fire Marshall
Office of the Comptroller General
Fire Marshal's Office
No. 2 Martin Luther King Jr. Drive
Floyd Building (620 West Tower)
Atlanta, GA 30334

Hawaii

NO STATE FIRE MARSHAL

Office of the State Fire Council
City and County of Honolulu Fire Department
1455 South Beretania Street (Room 305)
Honolulu, HI 96814

Idaho

State Fire Marshal
Department of Insurance
500 South Tenth Street
Boise, ID 83720

Illinois

State Fire Marshal
Office of the State Fire Marshal
Division of Fire Prevention
3150 Executive Park Drive
Springfield, IL 62703-4599

Indiana

State Fire Marshal
Department of Fire Prevention and Building Safety
1099 North Meridian Street (Suite 900)
Indianapolis, IN 46204

Iowa

State Fire Marshal
Division of State Fire Marshal
Department of Public Safety
Wallace State Office Building
Des Moines, IA 50319

Kansas

State Fire Marshal
State of Kansas
700 S.W. Jackson (Suite 600)
Topeka, KS 66603-3714

Kentucky

State Fire Marshall
Department of Housing, Buildings, and Construction
127 Building
Highway 127-South
Frankfort, KY 40601

Louisiana

State Fire Marshal
Office of the State Fire Marshal
Department of Public Safety and Corrections
P.O. Box 66614
Baton Rouge, LA 70806

Maine

State Fire Marshal
Office of the State Fire Marshal
99 Western Avenue
State House Station #52
317 State Street
Augusta, ME 04333

Maryland

State Fire Marshal
Department of Public Safety and Correctional Services
6776 Reisterstown Road (Suite 314)
Baltimore, MD 21215-2339

Massachusetts

State Fire Marshal
Division of Fire Prevention
Department of Public Safety
1010 Commonwealth Avenue
Boston, MA 02215

Michigan

Commanding Officer, Fire Marshall Division
Department of State Police
General Office Building
7150 Harris Drive
Lansing, MI 48913

Minnesota

State Fire Marshal
State Fire Marshal Division
Department of Public Safety
Market House
289 East 5th Street
St. Paul, MN 55101

Mississippi

State Fire Marshal
Department of Insurance
1804 Walter Sillers Building
P.O. Box 79
Jackson, MS 39205-0079

Missouri

State Fire Marshal
Department of Public Safety
Truman State Office Building
301 West High Street (8th Floor)
P.O. Box 844
Jefferson City, MO 65102

Montana

State Fire Marshal
Fire Marshal Bureau
Department of Justice
Scott Hart Building (Room 371)
303 North Roberts
Helena, MT 59602-1417

Nebraska

State Fire Marshal
State of Nebraska
246 South 14th Street
Lincoln, NE 68508

Nevada

State Fire Marshal
State Fire Marshal Division
Capitol Complex
1937 North Carson Street (Suite 244)
Carson City, NV 89710

New Hampshire

State Fire Marshal
Division of Safety Services
Department of Safety
James H. Hayes Building
Hazen Drive
Concord, NH 03305

New Jersey

Director, Fire Safety
Division of Housing & Community Development
Department of Community Affairs, CN 809
1313 Princeton Pike
Trenton, NJ 08625-0809

New Mexico

State Fire Marshal
Office of the State Fire Marshal
Department of Insurance
P.O. Drawer 1269
Santa Fe, NM 87504-1269

New York

State Fire Administrator
Office of Fire Prevention and Control
New York State Department of State
162 Washington Avenue
Albany, NY 12231

North Carolina

Administrative Officer
Fire-Rescue Services/Communications
Department of Insurance
Box 26387
Raleigh, NC 26711

North Dakota

State Fire Marshal
Fire Marshal's Office/Attorney General
State of North Dakota
State Capitol
Bismarck, ND 58505

Ohio

State Fire Marshal
Division of State Fire Marshal
Department of Commerce
8895 East Main Street
P.O. Box 525
Reynoldsburg, OH 43068

Oklahoma

State Fire Marshal
State of Oklahoma
4030 North Lincoln Boulevard
Building 3 (Suite 100)
Oklahoma City, OK 73105-5285

Oregon

State Fire Marshal
Fire Marshal Division
State of Oregon
3000 Market Street N.E. (Suite 534)
Salem, OR 97310

Pennsylvania

Director, Fire Marshal Division
Pennsylvania State Police
Department Headquarters
1800 Elmerton Avenue
Harrisburg, PA 17110

Rhode Island

State Fire Marshal
State of Rhode Island and Providence Plantations
Division of Fire Safety
1270 Mineral Springs Avenue
North Providence, RI 02904

South Carolina

State Fire Marshal
Division of State Fire Marshal
AT&T Building
1201 Main Street (Suite 810)
Columbia, SC 29201

South Dakota

State Fire Marshal
Department of Public Safety
118 West Capitol Avenue
Pierre, SD 57501

Tennessee

Commissioner for Fire Prevention
Department of Commerce and Insurance
500 James Robertson Parkway (5th Floor)
Nashville, TN 37219-5319

Texas

State Fire Marshal
State Board of Insurance
1110 San Jacinto Boulevard
Austin, TX 78701-1998

Utah

State Fire Marshal
Department of Public Safety
4501 South—2700 West
Salt Lake City, UT 84119

Vermont

Director of Fire Prevention
Fire Prevention Division
Department of Labor and Industry
120 State Street
Montpelier, VT 05602

Virginia

Chief Fire Marshal
Office of the State Fire Marshal
Department of Housing and Community Development
205 North Fourth Street
Richmond, VA 23219

Washington

Director of Fire Prevention
Department of Community Development
9th and Columbia Building
MS/GH-51 (Room 300)
Olympia, WA 98503-4151

West Virginia

State Fire Marshal
West Virginia State Fire Commission
Capitol Complex
2100 Washington Street, East
Charleston, WV 25305

Wisconsin

Chief of Fire Protection
Department of Industry, Labor and Human Relations
201 East Washington Avenue
P.O. Box 7969
Madison, WI 53707-7969

Wyoming

State Fire Marshal
Department of Fire Prevention and Electrical Safety
Herschler Building (2nd Floor)
Cheyenne, WY 82002

U.S. FOOD AND DRUG ADMINISTRATION
5600 Fishers Lane, Rockville, MD 20857

District Offices

Atlanta
60 8th Street, N.E.
Atlanta, GA 30309

Baltimore
900 Madison Avenue
Baltimore, MD 21201

Boston
One Montvale Avenue
Stoneham, MA 02180

Buffalo
599 Delaware Avenue
Buffalo, NY 14202

Chicago
433 West Van Buren, Rm. 1222
Chicago, IL 60607

Cincinnati
1141 Central Parkway
Cincinnati, OH 45202-1097

Dallas
3032 Bryan Street
Dallas, TX 75204

Denver
P.O. Box 25087
Denver, CO 80225-0087

Detroit
1560 East Jefferson Avenue
Detroit, MI 48207

Kansas City
1009 Cherry Street
Kansas City, MO 64106

Los Angeles
1521 West Pico Boulevard
Los Angeles, CA 90015-2486

Minneapolis
240 Hennepin Avenue
Minneapolis, MN 55401

Nashville
297 Plus Park Boulevard
Nashville, TN 37217

New Orleans
4298 Elysian Fields Avenue
New Orleans, LA 70122

New York
850 Third Avenue
Brooklyn, NY 11232-1593

Newark
61 Main Street
West Orange, NJ 07052

Orlando
7200 Lake Ellenor Drive
Suite 120
Orlando, FL 32809

Philadelphia
U.S. Customhouse
2nd & Chestnut Streets, Room 900
Philadelphia, PA 19106

San Francisco
500 U.N. Plaza
Federal Office Building, Rm. 526
San Francisco, CA 94102

San Juan
P.O. Box 5719
Puerta de Tierra Station
San Juan, PR 00906-5719

Seattle
22201 23rd Drive S.E.
Bothell, WA 98021-4421

National Fire Protection Association Codes Pertaining to Compressed Gases

Designation Number	Title
NFPA 43C	Code for the Storage of Gaseous Oxidizing Materials
NFPA 45	Standard on Fire Protection for Laboratories Using Chemicals
NFPA 49	Standard for Hazardous Chemicals Data
NFPA 50	Standard for Bulk Oxygen at Consumer Sites
NFPA 50A	Standard for Gaseous Hydrogen Systems at Consumer Sites
NFPA 50B	Standard for Liquefied Hydrogen Systems at Consumer Sites
NFPA 51	Standard for the Design and Installation of Oxygen-Fule Gas Systems for Welding, Cutting and Allied Processes
NFPA 51A	Standard for Acetylene Cylinder Charging Plants
NFPA 51B	Standard for Fire Prevention in Use of Cutting and Welding Processes
NFPA 52	Standard for Compressed Natural Gas (CNG) Vehicular Fuel Gas Systems
NFPA 54	National Fuel Gas Code
NFPA 53M	Standard on Fire Hazards in Oxygen-Enriched Atmospheres
NFPA 58	Standard for the Storage and Handling of Liquefied Petroleum Gases
NFPA 59	Standard for the Storage and Handling of Liquefied Petroleum Gases at Gas Utility Plants
NFPA 59A	Standard for the Production, Storage, and Handling of Liquefied Natural Gas (LNG)
NFPA 99	Standard for Health Care Facilities (supersedes NFPA 3M, 56A, 56B, 56C, 56D, 56E, 56G, 56HM, 56K, 76A, 76B, and 76C

Contact NFPA, Batterymarch Park, Quincy, MA 02269 (Telephone: 1-800-344-3555).

RELEVANT CODES OF OTHER ORGANIZATIONS

ASME Boiler and Pressure Vessel Code (Section VIII, Unfired Pressure Vessels), American Society of Mechanical Engineers, 345 East 47th Street, New York, NY 10017-2392

Uniform Fire Code, Western Fire Chiefs Association, 5360 South Workman Mill Road, Whittier, CA 90601

National Board Inspection Code—A Manual for Boiler and Pressure Vessel Inspectors, National Board of Boiler and Pressure Vessel Inspectors, 1055 Crupper Avenue, Columbus, OH 43229

ANSI/ASME B31.3, *Chemical Plant and Petroleum Refinery Piping,* American Society of Mechanical Engineers, 345 East 47th Street New York, NY 10017.

ASTM G88, *Guide for Designing Systems for Oxygen Service,* ASTM, 1916 Race Street, Philadelphia, PA 19103-1187

API 620, *Recommended Rules for Design and Construction of Large, Welded Low Pressure Storage Tanks,* American Petroleum Institute, 1220 L Street, N.W., Washington, DC 20005.

NFPA 70, *National Electrical Code,* National Fire Protection Association, Batterymarch Park, Quincy MA 02269.

APPENDIX 4

Publications of the Compressed Gas Association

CYLINDER SERIES

C-1 Methods for Hydrostatic Testing of Compressed Gas Cylinders. U.S. Department of Transportation (DOT) and Transport Canada (TC) regulations require certain cylinders to be periodically retested to requalify them for continued service. This publication gives details on the requirements for test equipment, method of operation, and test records. (20 pages)

C-2 Recommendations for the Disposition of Unserviceable Compressed Gas Cylinders with Known Contents. Outlines procedures for the safe disposition of unserviceable cylinders, including discharge of known cylinder content and methods of handling when content cannot be discharged. (8 pages)

C-3 Standards for Welding on Thin Walled Cylinders. Standards applicable to DOT welded or brazed compressed gas cylinders designed for 500 psig service pressure or less and having 1000 lb water capacity maximum; also minimum wall thickness under ⅜ inch. Covers procedure and operator qualification, radiographic inspection, and container repair. (24 pages)

C-4 American National Standard Method of Marking Portable Compressed Gas Containers to Identify the Material Contained. Sponsored by CGA, this standard covers containers having a water capacity of up to 1000 lb. Describes the standard method of marking containers. (4 pages)

C-5 Cylinder Service Life—Seamless, Steel, High Pressure Cylinders. Contains detailed methods of estimating wall thickness that can be applied with accuracy and simplicity to the retesting of compressed gas cylinders to determine their suitability for continued service. Includes information regarding elastic expansion limits and the method of determining and checking "k" factors. Tabulations of commercially used "k" factors and elastic expansion limits are included. (12 pages)

C-6 Standards for Visual Inspection of Steel Compressed Gas Cylinders. A guide for establishing cylinder inspection procedures and standards in order to meet cylinder inspection requirements of the U.S. Department of Transportation. Covers all types of steel compressed gas cylinders including those which may be visually inspected in lieu of hydrostatic retest. (16 pages)

C-6.1 Standards for Visual Inspection of High Pressure Aluminum Compressed Gas Cylinders. A guide for the visual inspection of aluminum cylinders having a service pressure of 1800 psig (12 411 kPa) and over. It is general in nature and will not cover all circumstances for each individual cylinder type or lading. Each inspection agency may modify these guidelines to fit their conditions of service, which may be more severe than encountered in transportation. (8 pages)

C-6.2 Guidelines for Visual Inspection and Requalification of Fiber Reinforced High Pressure Cylinders. This guideline provides data and details about fiber-reinforced cylinders to help the cylinder retester examine and hydrostatically retest composite cylinders. Contains tables listing allowable defects and photos of defects. (20 pages)

C-7 Guide to the Preparation of Precautionary Labeling and Marking of Compressed Gas Containers. Covers use of precautionary labels to warn of principal hazards. Includes general principles and illustrative labels for several types of gases. (32 pages)

C-8 Standard for Requalification of DOT-3HT Seamless Steel Cylinders. A guide to cylinder users for establishing their own cylinder inspection procedures and standards. Intended to cover visual inspection requirements and other service life limitations prescribed by U.S. DOT regulations. (6 pages)

C-9 Standard Color Marking of Compressed Gas Containers Intended for Medical Use. Describes the uniform method of color marking of compressed medical gas cylinders which will facilitate the recognition of cylinders intended for medical use. (8 pages)

C-10 Recommended Procedures for Changes of Gas Service for Compressed Gas Cylinders. Provides guidelines for the procedures used for changing cylinders from one gas service to another, including inspection and contaminant removal. (20 pages)

C-11 Recommended Practices for Inspection of Compressed Gas Cylinders at Time of Manufacture. Outlines U.S. DOT inspection requirements for cylinders as interpreted and practiced by manufacturers and inspectors. Qualification of certifying inspectors is covered as well as inspection requirements for seamless, welded, and brazed and nonrefillable cylinders. (20 pages)

C-12 Qualification Procedure for Acetylene Cylinder Design. Describes qualification tests for use by manufacturers of acetylene cylinders as required when a new cylinder design or significant design change occurs. Tests include a proof of the mechanical strength of the filler, a flashback test, an impact stability test, and a fire test. (16 pages)

C-13 Guidelines for Periodic Visual Inspection and Requalification of Acetylene Cylinders. Provides a guide for a reinspection procedure for requalification of acetylene cylinders and fillers for use by charging companies and reinspection agencies. (20 pages)

C-14 Procedures for Fire Test of DOT Cylinder Pressure Relief Device Systems. Describes a new set of test procedures and apparatus for fire testing of compressed gas cylinder pressure relief devices as required by U.S. DOT regulations, 49 CFR, Section 173.34(d). The procedures are applicable for cylinders which are less than 500 lb internal water volume and are designed to provide a means of testing to U.S. DOT requirements, with reliable test data and repeatable test results. (12 pages)

C-15 Procedures for Cylinder Design Proof and Service Performance Tests. Provides a list of tests that the cylinder user or manufacturer may choose from to identify the capability of a container for a specific gas service or industry. The publication explains the purpose of the test and when the test should be performed; it defines and describes the test method and cites the results that may be expected. (16 pages)

REGULATORS AND HOSE LINE EQUIPMENT SERIES

E-1 Standard Connections for Regulator Outlets, Torches and Fitted Hose for Welding and Cutting Equipment. Describes the standard connections for regulator outlets, torches, and fitted hose for welding and cutting equipment. (16 pages)

E-2 Hose Line Check Valve Standards for Welding and Cutting. Describes check valves for use at pressures of 200 psig or less, designed to fit the oxygen-fuel gas torch inlets or the outlets of both the oxygen and fuel gas regulators. (4 pages)

E-3 Pipeline Regulator Inlet Connection Standards. Describes the inlet connections to be used on removable pipeline regulators used in the welding, cutting, and related process industries and where the pipeline pressure does not exceed 200 psig (1379 kPa). (4 pages)

E-4 Standard for Gas Regulators for Welding and Cutting. Describes design and performance requirements for regulators intended for use in welding and cutting and allied applications to reduce supply pressure from a gas storage system, pipeline, or other source to the pressure required for the application. (4 pages)

E-5 Torch Standard for Welding and Cutting. Describes design, marking, performance, and stability testing of manual oxygen-fuel gas torches (and cutting attachments) intended for cutting, welding, scarfing, heating, and other allied processes. (6 pages)

E-6 Standard for Hydraulic Type Pipeline Protective Devices. Provides information on hydraulic type (NFPA designated P_f) pipeline protective devices, including materials and construction, design requirements, marking, and maintenance. (4 pages)

E-7 Standard for Medical Gas Regulators and Flowmeters. Describes requirements for materials of construction, cleanliness, performance, etc., for flowmeters, pressure-reducing regulators, and regulator/flowmeter and regulator/flowgauge combinations for the administration of medical gases. (6 pages)

GASES SERIES

G-1 Acetylene. Contains information on the properties, manufacture, transportation, storage, handling, and use of acetylene. (16 pages)

G-1.1 Commodity Specification for Acetylene. Covers specification requirements for acetylene. Presents data on quality verification, sampling, analytical procedures, and containers with supplemental tables. (8 pages)

G-1.2 Recommendations for Chemical Acetylene Metering. Presents data on types of meters for acetylene service, meter accessories, materials of construction for meters and meter accessories, meter installation, and meter maintenance. (16 pages)

G-1.3 Acetylene Transmission for Chemical Synthesis. Comprehensively presents recommended minimum safe practices governing design of acetylene piping systems, gas holder storage, and booster systems. (20 pages)

G-1.5 Carbide Lime—Its Value and Its Uses. Describes the use of calcium hydroxide (a by-product from acetylene generation) in agriculture, building construction, industrial and chemical processes, and miscellaneous uses. (16 pages)

G-1.6 Recommended Practices for Mobile Acetylene Trailer Systems. Provides information on design and construction, charging, transporting and discharging, markings, and fire safety practices. Recommendations on equipment at fill stations and discharge sites are also provided. (8 pages)

G-2 Anhydrous Ammonia. Presents information on the properties, manufacture, transportation, storage, and use of anhydrous ammonia. (48 pages)

G-2.1 American National Standard Safety Requirements for the Storage and Handling of Anhydrous Ammonia, ANSI K61.1. Sponsored by CGA, this publication includes standards for the location, design, construction, and operation of anhydrous ammonia systems. Sections on refrigerated storage systems, systems mounted on farm vehicles, tank motor vehicles, and tank rail cars for transportation purposes are also included. The standard does not apply to ammonia-manufacturing plants, refrigerating, or air-conditioning systems. (48 pages)

G-2.2 Guideline Method for Determining Minimum of 0.2% Water in Anhydrous Ammonia. Describes a suitable method for this purpose. (12 pages)

G-3 Sulfur Dioxide. Presents information on the properties, manufacture, physiological effects, transportation, storage, handling, and use of sulfur. Also covers emergency actions and first aid. (44 pages)

G-4 Oxygen. Presents information on the properties, manufacture, transportation, storage, handling, and use of oxygen. Section on liquid oxygen is included. (16 pages)

G-4.1 Cleaning Equipment for Oxygen Service. Describes cleaning processes, agents, methods, inspections, tests, and safety. An important reference for anyone assembling an oxygen system. (16 pages)

G-4.3 Commodity Specification for Oxygen. Describes specification requirements for all types and grades of commercially available gaseous and liquid oxygen. Presents data on quality verification, sampling, analytical procedures, and containers, with supplemental tables and charts. (12 pages)

G-4.4 Industrial Practices for Gaseous Oxygen Transmission and Distribution Piping Systems. Presents information on the current practices used in gaseous oxygen transmission and distribution piping systems such as encountered at various chemical facilities, steel mills, petroleum product refineries, etc. (16 pages)

G-4.5 Commodity Specification for Oxygen Produced by Chemical Reaction. Describes specification requirements for gaseous respiratory oxygen produced by chemical reaction. (4 pages)

G-5 Hydrogen. Presents information on the properties, manufacture, transportation, storage, handling, and use of hydrogen and includes a section on liquefied hydrogen. (16 pages)

G-5.3 Commodity Specification for Hydrogen. Lists specification requirements for all types and grades of commercially available gaseous and liquid hydrogen. Describes quality verification systems, sampling, and analytical procedures, with supplemental specification data. (12 pages)

G-6 Carbon Dioxide. Presents information on the properties, manufacture, transportation, storage, handling, and use of carbon dioxide. (24 pages)

G-6.1 Standard for Low Pressure Carbon Dioxide Systems at Consumer Sites. Concerns minimum requirements and practices for design, construction, installation, operation, and maintenance of low pressure CO_2 supply systems at consumer sites. (8 pages)

G-6.2 Commodity Specification for Carbon Dioxide. Describes specification requirements for gaseous, liquid, and solid CO_2. Presents data on quality verification, sampling, analytical procedures, and containers, with supplemental tables and charts. (16 pages)

G-6.3 Carbon Dioxide Cylinder Filling and Handling Procedures for Beverage Plants, NSDA TD01. This publication, co-sponsored with the National Soft Drink Association, presents detailed procedures for inspection, filling, care, proper handling, recommended shipping and storage practices, and DOT marking and labeling requirements for carbon dioxide cylinders as used in the beverage industry. (20 pages)

G-7 Compressed Air for Human Respiration. Covers special recommendations when air is intended for human use. To be used in conjunction with CGA G-7.1. (12 pages)

G-7.1 Commodity Specification for Air. Describes specification requirements for all types and grades of gaseous air known to be commercially available. Provides data concerning quality verification systems, sampling, analytical procedures, and containers. Supplemental specification tables and charts are included. (12 pages)

G-8.1 Standard for Nitrous Oxide Systems at Consumer Sites. Covers the general design principles recommended for the installation of nitrous oxide central supply systems—either cylinders connected to a common manifold or systems fed by bulk liquid containers—at medical or industrial consumer sites, with special reference to protection of the system, patients, personnel, and property from fire originating from sources apart from the system itself. (12 pages)

G-8.2 Commodity Specification for Nitrous Oxide. Describes the specification requirements for nitrous oxide manufactured by the thermal decomposition of nitrous oxide grade ammonium nitrate and filled into standard pressurized containers. Data regarding quality verification systems, sampling, analytical procedures, and containers are included. (12 pages)

G-9.1 Commodity Specification for Helium. Covers specification requirements for commercially produced grades of helium. Data regarding quality verification systems, sampling, analytical procedures, supplemental specifications, tables, and charts are included. (12 pages)

G-10.1 Commodity Specification for Nitrogen. Designates specification requirements for all types and grades of commercially available gaseous and liquid nitrogen. Material includes quality verification systems, analytical procedures, containers, and supplemental data. (12 pages)

G-11.1 Commodity Specification for Argon. Lists specification requirements for all types and grades of commercially available gaseous and liquefied argon. Describes quality verification systems, sampling, analytical procedures and supplemental specification data. (12 pages)

G-12 Hydrogen Sulfide. Presents information on the properties, manufacture, transportation, storage, handling, and use of hydrogen sulfide. (20 pages)

PROTECTION AND HANDLING SERIES

P-1 Safe Handling of Compressed Gases in Containers. Primarily for the guidance of users of compressed gases in cylinders, although some general precautions are included for tank car handling. Presents basic rules for safe handling and regulations applying to compressed gases. (20 pages)

P-2 Characteristics and Safe Handling of Medical Gases. Intended for the guidance of personnel engaged in and responsible for the handling and use of compressed medical gases. Contains information relating to the properties of medical gases as well as information on containers and regulations governing these commodities. Recommended safe practices are included. (28 pages)

P-2.1 Recommendations for Medical-Surgical Vacuum Systems in Health Care Facilities. Applies to piped vacuum systems in hospitals but not to systems used for vacuum cleaning or as a vacuum condensate return. Covers requirements for central vacuum supply systems with control equipment, piping system to points in the hospital where suction may be required, and suitable outlet valves at point of use. (8 pages)

P-2.5 Transfilling of High Pressure Gaseous Oxygen to Be Used for Respiration. This guide is intended for persons and organizations who transfill oxygen gas to be used for respiration and describes the primary hazards involved, as well as the proper equipment and procedures which must be followed to ensure compliance with government regulations. Includes sections covering personnel qualifications, hazards of transfilling, applicable government regulations, transfilling systems and equipment, materials compatibility, storage and maintenance of equipment, and references to applicable standards and codes. (12 pages)

P-2.6 Transfilling of Liquid Oxygen to Be

Used for Respiration. Describes the primary hazards involved in transfilling pressurized liquid oxygen and minimum performance requirements for transfilling equipment and advises that it is essential to be familiar with the hazardous properties of oxygen, required safety precautions, and applicable government regulations. (12 pages)

P-5 Suggestions for the Care of High Pressure Air Cylinders for Underwater Breathing. A guide to users of underwater breathing air cylinders and to those responsible for the care, maintenance, and recharging of such cylinders. (8 pages)

P-6 Standard Density Data, Atmospheric Gases and Hydrogen. Density data recommended in this publication were compiled by the Compressed Gas Association to provide uniform values of liquid and gas density for atmospheric gases and hydrogen for the benefit of suppliers and users of these commodities. Tables present standard density data and volumetric conversion factors. (4 pages)

P-7 Standard for Requalification of Cargo Tank Hose Used in the Transfer of Compressed Gases. A guide for cargo tank hose users in establishing hose requalification procedures and standards in compressed gas transfer service applications. (8 pages)

P-8 Safe Practices Guide for Air Separation Plants. Applies to safety in the design, construction, location, installation, and operation of cryogenic air separation plants regardless of whether the product is withdrawn as gas, liquid, or both. (48 pages)

P-9 The Inert Gases Argon, Nitrogen and Helium. Presents general information regarding the characteristics and handling of the inert gases argon, nitrogen, and helium and is intended primarily for the users of these gases. It describes the properties, manufacture, and commercial uses of these gases and the containers authorized, and it includes a section on storage and handling of liquefied inert gases. (16 pages)

P-10 Standard for Vinyl Chloride Monomer Tank Car Manway Cover and Protective Housing Arrangement and Emergency Safety Kit. Describes a standard manway cover and protective housing arrangement for new VCM tank cars and for retrofitting certain existing VCM cars. This standard arrangement will accommodate a safety kit, also described, for emergency use in capping leaking fittings. Includes 9 pages of detailed engineering drawings. (16 pages)

P-11 Metric Practice Guide for the Compressed Gas Industry. This general guide sets forth guidelines concerning units that are to be used to express (1) the volume and/or mass of gas in compressed gas containers, (2) pressure, and (3) container water capacities. It includes volume, pressure, mass, and other conversion factors pertinent to the compressed gas industry. (6 pages)

P-12 Safe Handling of Cryogenic Liquids. A general guide to the safe handling of cryogenic liquids commonly used in industry, including their properties, safety standards, general safety practices and first aid procedures, fire prevention and fire-fighting procedures, and recommendations for safe handling of these liquids in containers and storage systems. It is intended for use by consumers, shippers, carriers, distributors, and others desiring an introductory knowledge of cryogenic liquids. (28 pages)

P-13 Safe Handling of Liquid Carbon Monoxide. A guideline for safe handling of liquid carbon monoxide which points out special requirements for storage and handling, emergency procedures, and properties of the liquid. (12 pages)

P-14 Accident Prevention in Oxygen-Rich and Oxygen-Deficient Atmospheres. Identifies the hazards inherent in oxygen-rich and oxygen-deficient atmospheres and provides recommendations for safe working practices and ways of preventing accidents. (12 pages)

P-15 Filling of Industrial and Medical Nonflammable Compressed Gas Cylinders. Covers prefill inspection procedures and describes the steps necessary to ensure that cylinders of air, argon, carbon dioxide, helium, nitrous oxide, and oxygen are filled in a safe manner. Also provides filling pressure tables from 0 to 130°F and 1800 to 2640 psig for each gas. (16 pages)

PRESSURE RELIEF DEVICE SERIES

S-1.1 Pressure Relief Device Standards—Part 1—Cylinders for Compressed Gases. Minimum recommended requirements for pressure relief devices for use on TC/DOT cylinders having water capacities of 1000 lb or less. Describes the various types of pressure relief devices, their design, construction, and application for various cylinders. Also covers relief device test and maintenance requirements. (36 pages)

S-1.2 Pressure Relief Device Standards—Part 2—Cargo and Portable Tanks for Compressed Gases. Minimum recommended requirements for pressure relief devices for use on cargo tanks (tank trucks) and portable tanks (skid tanks) designed to TC/DOT specifications. These requirements are recommended for application to cargo and portable tanks that do not come within U.S. Department of Transportation and Transport Canada jurisdiction. (16 pages)

S-1.3 Pressure Relief Device Standards—Part 3—Compressed Gas Storage Containers. Minimum recommended requirements for pressure relief devices for storage containers constructed in accordance with the relevant ASME or API-ASME codes. (16 pages)

S-7 Method for Selecting Pressure Relief Devices for Compressed Gas Mixtures in Cylinders. Describes a method and procedures for selecting the appropriate pressure relief device for mixtures of two or more gases in a cylinder. Includes numerous examples and a sample worksheet. (32 pages)

CGA SAFETY BULLETINS SERIES

SB-1 Hazards of Refilling Compressed Refrigerant (Halogenated Hydrocarbons) Gas Cylinders. Presents recommendations for eliminating injury to persons and property when refrigerant gas is transferred from large to small cylinders in the field—a common but potentially hazardous practice of distributors and service organizations in the refrigerating and air-conditioning fields. (2 pages)

SB-2 Oxygen-Deficient Atmospheres. Depletion of the oxygen content in air by combustion or displacement with inert gas is a potential hazard to personnel throughout industry. This bulletin deals with recommended practices for the protection of personnel working in a potentially oxygen-deficient atmosphere. Supplemental to CGA P-14. (2 pages)

SB-3 Evidence of Ownership of Compressed Gas Cylinders. Lists methods that should be used to indicate cylinder ownership and authorization to charge cylinders. Also available in a poster version as SB-3.1. (2 pages)

SB-4 Handling Acetylene Cylinders in Fire Situations. Provides guidance to welders, other users, and those individuals concerned with safety and fire protection on how to handle acetylene cylinders in fire situations. (4 pages)

SB-5 Hazards of Reusing Disposable Refrigerant (Halogenated Hydrocarbon) Gas Cylinders. Provides warnings of the dangers of reusing disposable refrigerant cylinders and cites applicable DOT regulations. (2 pages)

SB-6 Nitrous Oxide Security and Control. A safety bulletin for manufacturers, repackagers, distributors, and those using nitrous oxide for medical and commercial-industrial purposes. It warns of the misuse and abuse of nitrous oxide so that effective steps can be taken to prevent theft or improper use. (2 pages)

SB-7 Rupture of Oxygen Cylinders in the Diving Industry. A safety bulletin developed in response to incidents wherein oxygen cylinders have failed violently following their use in offshore diving operations due to users having allowed sea water to flow back into the cylinders after the oxygen has been depleted. (2 pages)

SB-8 Use of Oxy-Fuel Gas Welding and Cutting Apparatus. Failure to use Oxy-fuel welding and cutting equipment according to the manufacturer's operating instructions can result in serious personal injury. This bulletin lists "dos" and "don'ts" to follow when using this type of equipment. (2 pages)

SB-9 Recommended Practice for the Outfitting and Operation of Vehicles Used in the Transportation and Transfilling of Liquid Oxygen to Be Used for Respiration. Special precautions need to be taken regarding the reactive properties exhibited by pure oxygen. This bulletin describes recommended practice for the outfitting and operation of vehicles used in transporting and transfilling liquid oxygen. (2 pages)

SB-10 Correct Labeling and Proper Fittings on Cylinders/Containers. Highlights five considerations that must be observed to ensure that compressed gases and cryogenic liquids are safely used. (2 pages)

SB-11 Use of Rubber Welding Hose. Provides a list of "dos" and "don'ts" concerning proper handling, use, and maintenance of rubber welding hose. Also lists additional reference sources. In 11 × 17 inch format suitable for posting. (2 pages)

CGA TECHNICAL BULLETINS SERIES

TB-2 Guidelines for Inspection and Repair of MC-330 and MC-331 Cargo Tanks. To assist those having the responsibility for the inspection and repair of these tanks. (8 pages)

TB-3 Hose Line Flashback Arrestors. Describes intent and use of hose line flashback arrestors. (2 pages)

TB-4 Certification for Exchange Product or Customer Pickup of Bulk Medical Liquids. Gives guidelines for uniform bulk product loading identification requirements for transporting compressed gas products in satisfaction of the Food and Drug Administration's Good Manufacturing Practice regulations. (2 pages)

VALVE CONNECTIONS SERIES

V-1 American National, Canadian, and Compressed Gas Association Standard for Compressed Gas Cylinder Valve Outlet and Inlet Connections. Detailed dimensioned drawings of 65 valve outlet connections for almost 200 different products. The standard covers threaded connections, yoke outlets, and the Pin Index Safety System for flush outlet valves of the yoke type used for medical gases. (92 pages)

V-5 Diameter Index Safety System. Describes a standard to provide noninterchangeable connections where removable exposed threaded connections are employed in conjunction with individual gas lines of medical gas administering equipment at pressures of 200 psig or less, such as outlets from medical gas regulators and connectors for anesthesia, resuscitation, and therapy apparatus. Detailed dimensioned drawings are included. In addition to the medical gases covered by the Pin Index Safety System, this standard includes air and suction. (20 pages)

V-6 Standard Cryogenic Liquid Transfer Connections. Gives cryogenic liquid connections which provide mechanical protection for product integrity. Engineering specifications and cutaway drawings are provided. (16 pages)

V-6.1 Standard Carbon Dioxide Transfer Connections. Specifies the transfer connections for liquid and vapor carbon dioxide for use with bulk liquid transport equipment. Engineering specifications and cutaway drawings are provided. (12 pages)

V-7 Standard Method of Determining Cylinder Valve Outlet Connections for Industrial Gas Mixtures. Provides standard procedure for selection of cylinder valve outlet connections for mixtures of two or more industrial gases. This standard supplements CGA V-1 but does not cover medical gases. (20 pages)

INSULATED CARGO TANKS

CGA-341 Standard for Insulated Cargo Tank Specification for Cryogenic Liquids. Contains minimum recommended requirements for insulated cargo tanks for the transportation of cold liquefied gases. (12 pages)

SAFETY POSTER SERIES

SP-1-12 Safety Posters. A series of 12 safety posters useful in reminding plant and shop personnel of important safety rules for handling, storing, labeling, transporting, and using compressed gases such as oxygen, acetylene, and argon. These two-color posters are sold in sets and are available in 8½ × 11 and 17 × 23-inch sizes. Larger format fits standard frames for "Size B" National Safety Council safety posters.

AUDIOVISUALS OF THE COMPRESSED GAS ASSOCIATION

AV-1 Safe Handling and Storage of Compressed Gases. Subjects covered include procedures for labeling and storage, precautionary safety warnings, personnel safety equipment, and handling procedures. In formats of videotape or 105 audio-sequenced slides. (21:50 minutes)

AV-2 Pre-Trip Inspection of Compressed Gas Tank Cars. Subjects covered include pre-loading inspections, safety equipment, standard markings, key inspection points, typical trouble spots, and warning signs. Format is 78 audio-sequenced slides. (19:50 minutes)

AV-3 Filling of Industrial & Medical Nonflammable Compressed Gas Cylinders. Covers numerous aspects and considerations with respect to prefill inspections and filling procedures. Includes Leader's Guide and study materials. (30:32 minutes)

AV-4 Characteristics and Safe Handling of Medical Gases. Subjects covered include general rules for handling, procedures for withdrawing gases, hazards of transfilling, regulations, and characteristics of medical gases. In formats of videotape or 78 audio-sequenced slides. (18:01 minutes)

AV-5 Safe Handling of Liquefied Nitrogen and Argon. Coverage includes properties of cryogenic liquids, containers, safety precautions, standard connections, materials compatibility, first aid, and hazards of oxygen-deficient atmospheres. In formats of videotape or 80 audio-sequenced slides. (23:05 minutes)

AV-6 Hazardous Material Shipping Papers and Placards for Cylinder Truck Operations. Covers procedures for preparation and use of hazardous materials shipping papers, applicable regulations, and proper posting of placards. Includes Leader's Guide and study materials. Available in several videotape formats. (34:46 minutes)

AV-7 Characteristics and Safe Handling of Carbon Dioxide. Subjects covered include chemical and physical properties, uses, physiological effects, safe practices, containers, compatible materials, and storage facilities. Available in several videotape formats. (27:29 minutes)

AV-8 Characteristics and Safe Handling of Cryogenic Liquid and Gaseous Oxygen. Subjects covered include characteristics, uses, containers, transfer systems and procedures, safe handling, compatible materials, equipment cleaning for oxygen service, hazards of oxygen-enriched atmospheres, and first aid and emergency response procedures. Available in several videotape formats. (23:13 minutes)

AV-9 Handling Acetylene Cylinders in Fire Situations. Provides guidance and practical information on procedures to use when acetylene cylinders are involved in a fire. Includes description of acetylene cylinder construction and potential sources of ignitable leaks from fusible metal pressure relief devices, valve components, etc. (approx. 20 minutes)

Index

In Case of a Transportation Emergency Involving a Compressed Gas

In the United States, ask for advice through CHEMTREC, the Chemical Transportation Emergency Center at the Chemical Manufacturers Association in Washington, DC

48 contiguous states, Puerto Rico, Virgin Islands, Alaska, Hawaii, and if transporting Canadian products in the United States (toll free)	(800) 424-9300
District of Columbia and foreign locations (exclusive of Canada)	(202) 483-7616
For non-emergency information only, call The Chemical Referral Center	(800) 262-8200
(If in District of Columbia, or (collect) if foreign location other than Canada)	(202) 887-1315

In Canada, ask for advice through CANUTEC, Transport of Dangerous Goods Branch, Transport Canada, Ottawa, Ontario.

In an emergency, from all points within Canada, call collect 24 hours a day	(613) 996-6666
For non-emergency information only, call	(613) 992-4624